PROGRESS IN BRAIN RESEARCH

VOLUME 128

NEURAL PLASTICITY AND REGENERATION

Other volumes in PROGRESS IN BRAIN RESEARCH

Volume 100: Neuroscience: From the Molecular to the Cognitive, by F.E. Bloom (Ed.) – 1994, ISBN 0-444-81678-X.
Volume 101: Biological Function of Gangliosides, by L. Svennerholm et al. (Eds.) – 1994, ISBN 0-444-81658-5.
Volume 102: The Self-Organizing Brain: From Growth Cones to Functional Networks, by J. van Pelt, M.A. Corner, H.B.M. Uylings and F.H. Lopes da Silva (Eds.) – 1994, ISBN 0-444-81819-7.
Volume 103: Neural Regeneration, by F.J. Seil (Ed.) – 1994, ISBN 0-444-81727-1.
Volume 104: Neuropeptides in the Spinal Cord, by F. Nyberg, H.S. Sharma and Z. Wiesenfeld-Hallin (Eds.) – 1995, ISBN 0-444-81719-0.
Volume 105: Gene Expression in the Central Nervous System, by A.C.H. Yu et al. (Eds.) – 1995, ISBN 0-444-81852-9.
Volume 106: Current Neurochemical and Pharmacological Aspects of Biogenic Amines, by P.M. Yu, K.F. Tipton and A.A. Boulton (Eds.) – 1995, ISBN 0-444-81938-X.
Volume 107: The Emotional Motor System, by G. Holstege, R. Bandler and C.B. Saper (Eds.) – 1996, ISBN 0-444-81962-2.
Volume 108: Neural Development and Plasticity, by R.R. Mize and R.S. Erzurumlu (Eds.) – 1996, ISBN 0-444-82433-2.
Volume 109: Cholinergic Mechanisms: From Molecular Biology to Clinical Significance, by J. Klein and K. Löffelholz (Eds.) – 1996, ISBN 0-444-82166-X.
Volume 110: Towards the Neurobiology of Chronic Pain, by G. Carli and M. Zimmermann (Eds.) – 1996, ISBN 0-444-82149-X.
Volume 111: Hypothalamic Integration of Circadian Rhythms, by R.M. Buijs, A. Kalsbeek, H.J. Romijn, C.M.A. Pennartz and M. Mirmiran (Eds.) – 1996, ISBN 0-444-82443-X.
Volume 112: Extrageniculostriate Mechanisms Underlying Visually-Guided Orientation Behavior, by M. Norita, T. Bando and B.E. Stein (Eds.) – 1996, ISBN 0-444-82347-6.
Volume 113: The Polymodal Receptor: A Gateway to Pathological Pain, by T. Kumazawa, L. Kruger and K. Mizumura (Eds.) – 1996, ISBN 0-444-82473-1.
Volume 114: The Cerebellum: From Structure to Control, by C.I. de Zeeuw, P. Strata and J. Voogd (Eds.) – 1997, ISBN 0-444-82313-1.
Volume 115: Brain Function in Hot Environment, by H.S. Sharma and J. Westman (Eds.) – 1998, ISBN 0-444-82377-8.
Volume 116: The Glutamate Synapse as a Therapeutical Target: Molecular Organization and Pathology of the Glutamate Synapse, by O.P. Ottersen, I.A. Langmoen and L. Gjerstad (Eds.) – 1998, ISBN 0-444-82754-4.
Volume 117: Neuronal Degeneration and Regeneration: From Basic Mechanisms to Prospects for Therapy, by F.W. van Leeuwen, A. Salehi, R.J. Giger, A.J.G.D. Holtmaat and J. Verhaagen (Eds.) – 1998, ISBN 0-444-82817-6.
Volume 118: Nitric Oxide in Brain Development, Plasticity and Disease, by R.R. Mize, T.M. Dawson, V.L. Dawson and M.J. Friedlander (Eds.) – 1998, ISBN 0-444-82885-0.
Volume 119: Advances in Brain Vasopressin, by I.J.A. Urban, J.P.H. Burbach and D. De Wied (Eds.) – 1999, ISBN 0-444-50080-4.
Volume 120: Nucleotides and their Receptors in the Nervous System, by P. Illes and H. Zimmermann (Eds.) – 1999, ISBN 0-444-50082-0.
Volume 121: Disorders of Brain, Behavior and Cognition: The Neurocomputational Perspective, by J.A. Reggia, E. Ruppin and D. Glanzman (Eds.) – 1999, ISBN 0-444-50175-4.
Volume 122: The Biological Basis for Mind Body Interactions, by E.A. Mayer and C.B. Saper (Eds.) – 1999, ISBN 0-444-50049-9.
Volume 123: Peripheral and Spinal Mechanisms in the Neural Control of Movement, by M.D. Binder (Ed.) – 1999, ISBN 0-444-50288-2.
Volume 124: Cerebellar Modules: Molecules, Morphology and Function, by N.M. Gerrits, T.J.H. Ruigrok and C.E. De Zeeuw (Eds.) – 2000, ISBN 0-444-50108-8.
Volume 125: Volume Transmission Revisited, by L.F. Agnati, K. Fuxe, C. Nicholson and E. Syková (Eds.) – 2000, ISBN 0-444-50314-5.
Volume 126: Cognition, Emotion and Autonomic Responses: the Integrative Role of the Prefrontal Cortex and Limbic Structures, by H.B.M. Uylings, C.G. Van Eden, J.P.C. De Bruin, M.G.P. Feenstra and C.M.A. Pennartz (Eds.) – 2000, ISBN 0-444-50332-3.
Volume 127: Neural Transplantation II. Novel Cell Therapies for CNS Disorders, by S.B. Dunnett and A. Björklund (Eds.) – 2000, ISBN 0-444-50109-6.

PROGRESS IN BRAIN RESEARCH

VOLUME 128

NEURAL PLASTICITY AND REGENERATION

EDITED BY

FREDRICK J. SEIL

Office of Regeneration Research Programs, VA Medical Center, and Departments of Neurology and Cell and Developmental Biology, Oregon Health Sciences University, Portland, OR 97201, USA

ELSEVIER
AMSTERDAM – LAUSANNE – NEW YORK – OXFORD – SHANNON – SINGAPORE – TOKYO
2000

ELSEVIER SCIENCE B.V.
Sara Burgerhartstraat 25
P.O. Box 211, 1000 AE Amsterdam, The Netherlands

© 2000 Elsevier Science B.V. All rights reserved.

This work is protected under copyright by Elsevier Science, and the following terms and conditions apply to its use:

Photocopying
Single photocopies of single chapters may be made for personal use as allowed by national copyright laws. Permission of the Publisher and payment of a fee is required for all other photocopying, including multiple or systematic copying, copying for advertising or promotional purposes, resale, and all forms of document delivery. Special rates are available for educational institutions that wish to make photocopies for non-profit educational classroom use.

Permissions may be sought directly from Elsevier Science Global Rights Department, PO Box 800, Oxford OX5 1DX, UK; phone: (+44) 1865 843830, fax: (+44) 1865 853333, e-mail: permissions@elsevier.co.uk. You may also contact Global Rights directly through Elsevier's home page (http://www.elsevier.nl), by selecting 'Obtaining Permissions'.

In the USA, users may clear permissions and make payments through the Copyright Clearance Center, Inc., 222 Rosewood Drive, Danvers, MA 01923, USA; phone: (978) 7508400, fax: (978) 7504744, and in the UK through the Copyright Licensing Agency Rapid Clearance Service (CLARCS), 90 Tottenham Court Road, London W1P 0LP, UK; phone: (+44) 171 631 5555, fax: (+44) 171 631 5500. Other countries may have a local reprographic rights agency for payments.

Derivative Works
Tables of contents may be reproduced for internal circulation, but permission of Elsevier Science is required for resale or distribution of such material.
Permission of the Publisher is required for all other derivative works, including compilations and translations.

Electronic Storage or Usage
Permission of the Publisher is required to store or use electronically any material contained in this work, including any chapter or part of a chapter.

Except as outlined above, no part of this work may be reproduced, stored in a retrieval system or transmitted in any form or by any means, electronic, mechanical, photocopying, recording or otherwise, without prior written permission of the Publisher.
Address permissions requests to: Elsevier Science Global Rights Department, at the mail, fax and e-mail addresses noted above.

Notice
No responsibility is assumed by the Publisher for any injury and/or damage to persons or property as a matter of products liability, negligence or otherwise, or from any use or operation of any methods, products, instructions or ideas contained in the material herein. Because of rapid advances in the medical sciences, in particular, independent verification of diagnoses and drugs dosages should be made.

First edition 2000

Library of Congress Cataloging in Publication Data
A catalog record of the Library of Congress has been applied for.

ISBN: 0-444-50209-2 (volume)
ISBN: 0-444-80104-9 (series)

∞ The paper used in this publication meets the requirements of ANSI/NISO Z39.48-1992 (Permanence of Paper).
Printed in The Netherlands.

List of Contributors

S.A. Azizi, Center for Gene Therapy, MCP Hahnemann University, Philadelphia, PA 19102, USA

M.S. Beattie, Department of Neuroscience, Ohio State University, 333 West 10th Avenue, Columbus, OH 43210, USA

A. Belhaj-Saïf, Department of Molecular and Integrative Physiology, University of Kansas Medical Center, Kansas City, KS 66160-7336, USA

J.R. Bethea, The Miami Project to Cure Paralysis, University of Miami School of Medicine, 1600 NW 10th Avenue, R-48, Miami, FL 33136, USA

M.M. Bolton, Department of Neurobiology, Box 3209, Duke University Medical Center, 101 Research Drive, Durham, NC 27710, USA

J.C. Bresnahan, Department of Neuroscience, Ohio State University, Columbus, OH 43210, USA

J.G. Broton, The Miami Project to Cure Paralysis and Department of Neurological Surgery, University of Miami School of Medicine, 1600 NW 10th Avenue, R-48, Miami, FL 33136, USA

B. Calancie, The Miami Project to Cure Paralysis and Department of Neurological Surgery, University of Miami School of Medicine, 1600 NW 10th Avenue, R-48, Miami, FL 33136, USA

J.K. Chapin, Department of Physiology and Pharmacology, SUNY Downstate Health Science Center, 450 Clarkson Avenue, Brooklyn, NY 11203, USA

P.D. Cheney, Mental Retardation and Human Development Research Center, University of Kansas Medical Center, 3901 Rainbow Blvd., Kansas City, KS 66160-7336, USA

S. De Lacalle, Department of Biology and Microbiology, California State University, Los Angeles, CA 90032, USA

M. Del Rosario Molano, The Miami Project to Cure Paralysis and Department of Neurological Surgery, University of Miami School of Medicine, 1600 NW 10th Avenue, R-48, Miami, FL 33136, USA

B.H. Dobkin, Department of Neurology and Reed Neurologic Research Center, University of California, Los Angeles, 710 Westwood Plaza, Los Angeles, CA 90095, USA

R. Drake-Baumann, Neurology Research, VA Medical Center, Portland, OR 97201, USA

M.E. Emborg, Department of Neurological Sciences and Research Center for Brain Repair, Rush University, Chicago, IL 60612, USA

I. Fischer, Department of Neurobiology and Anatomy, MCP Hahnemann University, 3200 Henry Avenue, Philadelphia, PA 19129, USA

H.M. Geller, Department of Pharmacology, UMDNJ-Robert Wood Johnson Medical School, 675 Hoes Lane, Piscataway, NJ 08854, USA

W. Gottschalk, Unit on Synapse Development and Plasticity, Laboratory of Developmental Neurobiology, NICHD, NIH, Bethesda, MD 20892-4480, USA

R.S. Hartley, The Center for Neurodegenerative Disease Research, Department of Pathology and Laboratory of Medicine, University of Pennsylvania School of Medicine, Philadelphia, PA 19104, and Layton Bioscience, Inc., Atherton, CA 94025, USA

J. Hill-Karrer, Department of Molecular and Integrative Physiology, University of Kansas Medical Center, Kansas City, KS 66160-7336, USA

J.A. Hoffer, School of Kinesiology, Faculty of Applied Sciences, Simon Fraser University, 8888 University Drive, Burnaby, BC V5A 1S6, Canada, and NeuroStream Technologies, Inc., Burnaby, BC, Canada

L.B. Jakeman, Department of Physiology and Cell Biology, Ohio State University, Columbus, OH 43210, USA

J.H. Kaas, Department of Psychology, Vanderbilt University, 301 Wilson Hall, 111 21st Avenue South, Nashville, TN 37240, USA

K.W. Kafitz, Institute for Physiology, Technical University of Munich, Biedersteinerstrasse 29, D-80802 Munich, Germany

K. Kallesøe, School of Kinesiology, Simon Fraser University, Burnaby, BC V5A 1S6, Canada

A. Konnerth, Institute for Physiology, Technical University of Munich, Biedersteinerstrasse 29, D-80802 Munich, Germany

G.C. Kopen, Center for Gene Therapy, MCP Hahnemann University, Philadelphia, PA 19102, USA

J.H. Kordower, Department of Neurological Sciences and Research Center for Brain Repair, Rush University, 2242 West Harrison Street, Chicago, IL 60612, USA

D.J. Krupa, Department of Neurobiology, Box 3209, Duke University Medical Center, Bryan Research Building, Room 333, 101 Research Drive, Durham, NC 27710, USA

V.M.-Y. Lee, Center for Neurodegenerative Disease Research, Department of Pathology and Laboratory Medicine, University of Pennsylvania Hospital, 3rd Floor Maloney Building, 3600 Spruce Street, Philadelphia, PA 19104-4283, USA

Q. Li, Department of Neuroscience, Ohio State University, Columbus, OH 43210, USA

Y. Liu, Department of Neurobiology and Anatomy, MCP Hahnemann University, Philadelphia, PA 19129, USA

D.C. Lo, Department of Neurobiology, Duke University Medical Center, Durham, NC 27710, USA

F.M. Longo, Department of Neurology, VA Medical Center and University of California San Francisco, V-127, 4150 Clement Street, San Francisco, CA 94121, USA

B. Lu, Unit on Synapse Development and Plasticity, Laboratory of Developmental Neurobiology, NICHD, NIH, Bldg 49, Rm 5A38, 49 Convent Drive, MSC 4480, Bethesda, MD 20892-4480, USA

S. Marty, INSERM U-106, Hôpital de la Salpêtrière, Bâtiment de Pédiatrie, 47 Boulevard de l'Hôpital, 73651 Paris Cedex 13, France

M. Mayer-Proschel, Department of Oncological Sciences (HCI), University of Utah Medical School, Salt Lake City, UT 84132, USA

B.J. McKiernan, Department of Molecular and Integrative Physiology, University of Kansas Medical Center, Kansas City, KS 66160-7336, USA

D.M. McTigue, Department of Physiology and Cell Biology, Ohio State University, Columbus, OH 43210, USA

S. Meiners, Department of Pharmacology, UMDNJ-Robert Wood Johnson Medical School, Piscataway, NJ 08854, USA

M.L.T. Mercado, Department of Pharmacology, UMDNJ-Robert Wood Johnson Medical

School, Piscataway, NJ 08854, USA
S. Müller, Department of Physiology, University of Bonn, D-53111 Bonn, and Klinikum Karlsbad-Langensteinbach, Karlsbad-Langensteinbach, Germany
M. Murray, Department of Neurobiology and Anatomy, MCP Hahnemann University, 3200 Henry Avenue, Philadelphia, PA 19129, USA
A. Nanassy, Department of Physiology, University of Bonn, D-53111 Bonn, and Klinikum Karlsbad-Langensteinbach, Karlsbad-Langensteinbach, Germany
M.A.L. Nicolelis, Department of Neurobiology, Duke University Medical Center, Durham, NC 27710, USA
M.C. Park, Department of Molecular and Integrative Physiology, University of Kansas Medical Center, Kansas City, KS 66160-7336, USA
K.G. Pearson, Department of Physiology, University of Alberta, Edmonton, AB T6G 2H7, Canada
D.G. Phinney, Center for Gene Therapy, MCP Hahnemann University, Philadelphia, PA 19102, USA
P.G. Popovich, Department of Molecular Virology, Immunology and Medical Genetics, College of Medicine and Public Health, Ohio State University, Columbus, OH 43210, USA
D.J. Prockop, SL99, Tulane University Medical Center, 1430 Tulane Avenue, New Orleans, LA 70112, USA
A. Ramón-Cueto, Neural Regeneration Group, Institute of Biomedicine, Spanish Council for Scientific Research (CSIC), Jaime Roig 11, 46010 Valencia, Spain
M.S. Rao, Department of Neurobiology and Anatomy, University of Utah Medical School, 50 North Medical Drive, Salt Lake City, UT 84132, USA
C.R. Rose, Institute for Physiology, Technical University of Munich, Biedersteinerstrasse 29, D-80802 Munich, Germany
R.A. Saavedra, 10301 Grosvenor Place, Apt. 712, Rockville, MD 20852, USA
M. Schwartz, Department of Neurobiology, Weizmann Institute of Science, Rehovot 76100, Israel
E.J. Schwarz, Center for Gene Therapy, MCP Hahnemann University, Philadelphia, PA 19102, USA
F.J. Seil, Office of Regeneration Research Programs (P3-R&D-35), VA Medical Center and Departments of Neurology and Cell and Developmental Biology, Oregon Health Sciences University, 3710 SW US Veterans Hospital Road, Portland, OR 97201, USA
N.T. Sherwood, Division of Biology, 216-76, California Institute of Technology, Pasadena, CA 91125, USA
B.T. Stokes, Department of Physiology and Cell Biology, Ohio State University, 228 Meiling Hall, 370 W. 9th Avenue, Columbus, OH 43210-1238, USA
A. Tessler, Department of Neurobiology and Anatomy, MCP Hahnemann University, and VA Medical Center, Philadelphia, PA 19129, USA
H. Thoenen, Max-Planck-Institute of Neurobiology, Department of Neurobiochemistry, Am Klopferspitz 18a, D-82152 Martinsried, Germany
P.A. Tresco, W.M. Keck Center for Tissue Engineering, University of Utah, 20 South 2030 East, Biopolymers Research Building, Rm 108D, Salt Lake City, UT 84112, USA
J.Q. Trojanowski, Center for Neurodegenerative Disease Research, Department of Pathology and Laboratory Medicine, University of Pennsylvania School of Medicine, Philadelphia, PA 19104, USA

A. Wernig, Department of Physiology, University of Bonn, Wilhelmstrasse 31, D-53111 Bonn, and Klinikum Karlsbad-Langensteinbach, Karlsbad-Langensteinbach, Germany

Y. Xie, Department of Neurology, VA Medical Center and University of California San Francisco, San Francisco, CA 94121, USA

Preface

The proceedings of the Eighth International Symposium on Neural Regeneration are presented in this volume. The meeting, which was held at the Asilomar Conference Center in Pacific Grove, California from 8 to 12 December, 1999, was cosponsored by the US Department of Veterans Affairs (Medical Research Service), the Paralyzed Veterans of America (Spinal Cord Research Foundation), the National Institutes of Health (National Institute of Neurological Disorders and Stroke), the Christopher Reeve Paralysis Foundation and the Eastern Paralyzed Veterans Association. The Program Planning Committee for the symposium included Drs. Susan V. Bryant (University of California, Irvine), Mary Bartlett Bunge (University of Miami School of Medicine), Edward D. Hall (Parke-Davis Pharmaceutical Research, Ann Arbor), Marston Manthorpe (Vical, Inc., San Diego), Ken Muneoka (Tulane University), Marion Murray (MCP Hahnemann University), John H. Peacock (VA Medical Center and University of Nevada, Reno), Fredrick J. Seil (VA Office of Regeneration Research Programs, VA Medical Center, Portland and Oregon Health Sciences University) and Suzanne Szollar (VA Medical Center and University of California, San Diego). Guest participants during the planning process were Drs. Vivian Beyda (Eastern Paralyzed Veterans Association) and Paul M. Hoffman (Medical Research Service, Department of Veterans Affairs) and Ms. Lisa A. Hudgins (Paralyzed Veterans of America).

The volume is organized into 6 topic sections, including: (I) Strategies for Spinal Cord Injury Repair, (II) Plasticity of the Injured Spinal Cord: Retraining Neural Circuits to Promote Motor Recovery, (III) Impact of Neuroprosthetic Applications on Functional Recovery, (IV) Neurotrophins and Activity-Dependent Plasticity, (V) Candidate Cells for Transplantation into the Injured CNS, and (VI) New Directions in Regeneration Research. Both clinical and experimental animal studies are presented in the first three sections, while predominantly basic research is the focus of the second half of the book. Not included in this volume are the keynote address by Dr. Martin Raff (University College, London) and featured presentations by Drs. Glenn I. Hatton (University of California, Riverside) and John R. Sladek, Jr. (The Chicago Medical School), plus the poster presentations, all of which contributed greatly to the excitement of the symposium, but are beyond the scope of a proceedings publication.

While a cure for spinal cord injury remains elusive, the contents of the volume convey a sense of progress toward this goal. More has been learned about the primary and secondary consequences of spinal cord injury and more is being understood about recovery mechanisms that are intrinsic to the nervous system and that might be further encouraged. Expanding the control capacity of uninjured portions of the nervous system may be one approach to improving the functional capabilities of those afflicted with this disorder. New therapies in the form of transplantable cells that can encourage growth or myelination or prevent secondary damage or that can substituted for injured cells appear promising for future applications. Genetic and tissue engineering studies give us further hope, and under continuous development are novel drugs with greater specificity and fewer detrimental effects and improved delivery methods for such drugs. Much more progress can be expected in the two year interval before the next of these symposia.

Fredrick J. Seil
Director, VA Office of
Regeneration Research Programs
Portland, OR, USA

Contents

List of Contributors v

Preface ix

Section I. Strategies for spinal cord injury repair

1. Strategies for spinal cord injury repair
   D.M. McTigue, P.G. Popovich, L.B. Jakeman and B.T. Stokes
   (Columbus, OH, USA) 3

2. Cell death and plasticity after experimental spinal cord injury
   M.S. Beattie, Q. Li and J.C. Bresnahan (Columbus, OH, USA) 9

3. The multi-domain structure of extracellular matrix molecules: implications for nervous system regeneration
   S. Meiners, M.L.T. Mercado and H.M. Geller (Piscataway, NJ, USA) 23

4. Spinal cord injury-induced inflammation: a dual-edged sword
   J.R. Bethea (Miami, FL, USA) 33

5. Immunological regulation of neuronal degeneration and regeneration in the injured spinal cord
   P.G. Popovich (Columbus, OH, USA) 43

Section II. Plasticity of the injured spinal cord: retraining neural circuits to promote motor recovery

6. Plasticity of neuronal networks in the spinal cord: modifications in response to altered sensory input
   K.G. Pearson (Edmonton, AB, Canada) 61

7. Neural plasticity as revealed by the natural progression of movement expression — both voluntary and involuntary — in humans after spinal cord injury
   B. Calancie, M. Del Rosario Molano and J.G. Broton (Miami, FL, USA) 71

8. Laufband (LB) therapy in spinal cord lesioned persons
   A. Wernig, A. Nanassy and S. Müller (Bonn and Karlsbad-Langensteinbach, Germany) 89

9. Spinal and supraspinal plasticity after incomplete spinal cord injury: correlations between functional magnetic resonance imaging and engaged locomotor networks
   B.H. Dobkin (Los Angeles, CA, USA) 99

Section III. Impact of neuroprosthetic applications on functional recovery

10. Impact of neuroprosthetic applications on functional recovery
    J.K. Chapin (Brooklyn, NY, USA) 115

11. Nerve cuffs for nerve repair and regeneration
    J.A. Hoffer and K. Kallesøe (Burnaby, BC, Canada) 121

12. Cortical motor areas and their properties: implications for neuroprosthetics
    P.D. Cheney, J. Hill-Karrer, A. Belhaj-Saïf, B.J. McKiernan, M.C. Park and J.K. Marcario (Kansas City, KS, USA) 135

13. Network level properties of short-term plasticity in the somatosensory system
    D.J. Krupa and M.A.L. Nicolelis (Durham, NC, USA) 161

14. The reorganization of somatosensory and motor cortex after peripheral nerve or spinal cord injury in primates
    J.H. Kaas (Nashville, TN, USA) 173

Section IV. Neurotrophins and activity-dependent plasticity

15. Neurotrophins and activity-dependent plasticity
    H. Thoenen (Martinsried, Germany) 183

16. Differences in the regulation of neuropeptide Y, somatostatin and parvalbumin levels in hippocampal interneurons by neuronal activity and BDNF
    S. Marty (Créteil, France) 193

17. Long-term regulation of excitatory and inhibitory synaptic transmission in hippocampal cultures by brain-derived neurotrophic factor
    M.M. Bolton, D.C. Lo and N.T. Sherwood (Durham, NC, USA) 203

18. Neurotrophins and activity-dependent inhibitory synaptogenesis
    F.J. Seil and R. Drake-Baumann (Portland, OR, USA) 219

19. Modulation of hippocampal synaptic transmission and plasticity by neurotrophins
    B. Lu and W. Gottschalk (Bethesda, MD, USA) 231

20. Neurotrophin-evoked rapid excitation of central neurons
    K.W. Kafitz, C.R. Rose and A. Konnerth (Munich, Germany) 243

Section V. Candidate cells for transplantation into the injured CNS

21. Candidate cells for transplantation into the injured CNS
    I. Fischer (Philadelphia, PA, USA) 253

22. Autoimmune involvement in CNS trauma is beneficial if well controlled
    M. Schwartz (Rehovot, Israel) 259

23. Olfactory ensheathing glia transplantation into the injured spinal cord
    A. Ramón-Cueto (Valencia, Spain) 265

24. Precursor cells for transplantation
    M.S. Rao and M. Mayer-Proschel (Salt Lake City, UT, USA) 273

25. Potential use of marrow stromal cells as therapeutic vectors for diseases of the central nervous system
    D.J. Prockop, S.A. Azizi, D.G. Phinney, G.C. Kopen and E.J. Schwarz (New Orleans, LA and Philadelphia, PA, USA) 293

26. Neurobiology of human neurons (NT2N) grafted into mouse spinal cord: implications for improving therapy of spinal cord injury
    V.M.-Y. Lee, R.S. Hartley and J.Q. Trojanowski (Philadelphia, PA and Sunnyvale, CA, USA) 299

27. Grafting of genetically modified fibroblasts into the injured spinal cord
    Y. Liu, M. Murray, A. Tessler and I. Fischer (Philadelphia, PA, USA) 309

Section VI. New directions in regeneration research

28. Delivery of therapeutic molecules into the CNS
    M.E. Emborg and J.H. Kordower (Chicago, IL, USA) 323

29. Neurotrophin small-molecule mimetics
    Y. Xie and F.M. Longo (San Francisco, CA, USA).................. 333

30. Tissue engineering strategies for nervous system repair
    P.A. Tresco (Salt Lake City, UT, USA) 349

31. In vivo neuroprotection of injured CNS neurons by a single injection of a DNA plasmid encoding the *Bcl-2* gene
    R.A. Saavedra, M. Murray, S. De Lacalle and A. Tessler (Philadelphia, PA and Los Angeles, CA, USA) 365

Subject Index 373

SECTION I

Strategies for spinal cord injury repair

CHAPTER 1

Strategies for spinal cord injury repair

Dana M. McTigue [1], Phillip G. Popovich [2], Lyn B. Jakeman [1] and Bradford T. Stokes [1,*]

[1] *Department of Physiology and Cell Biology, The Ohio State University, 304 Hamilton Hall, 1645 Neil Avenue, Columbus, OH 43210, USA*
[2] *Department of Molecular Virology, Immunology and Medical Genetics, The Ohio State University, 228 Meiling Hall, 370 W 9th Avenue, Columbus, OH 43210, USA*

Introduction

Traumatic injury to the spinal cord results in varying degrees of motor and sensory loss below the site of injury. Although the deficits associated with a spinal cord injury (SCI) may diminish somewhat over time due to incipient reparative events, the majority of the neurological problems remain with patients for the rest of their lives. Recent progress in experimental strategies to repair the injured spinal cord have shed new light on possible mechanisms for improving the outcome after SCI. Indeed, the potential use of stem cells, antagonists of inhibitory myelin proteins, polyethylene glycol, and various forms of transplantation (e.g., Schwann cells, olfactory ensheathing cells, peripheral nerves, fetal spinal tissue and cells genetically engineered to produce neurotrophins) have recently led to exciting results in the field of spinal cord trauma (Bregman et al., 1995; Cheng et al., 1996; Giovanini et al., 1997; Grill et al., 1997; Menei et al., 1998; Ramón-Cueto et al., 1998; McDonald et al., 1999; Shi et al., 1999). As an overview of the section on strategies for spinal cord injury repair, we will review additional approaches that may also prove beneficial in repairing the injured spinal cord.

Rat contusion injury model

Preclinical studies in reliable animal models of SCI are a necessary component of all clinical trial strategies. One such commonly used and well characterized model is contusion injury of the rat spinal cord (Noble and Wrathall, 1985; Bresnahan et al., 1991). This injury paradigm results in a typical pathology consisting of marked cavitation at the lesion site, i.e., epicenter, and a surrounding rim of surviving axons, many of which may be dysfunctional. The lesion typically extends into the rostral and caudal spinal gray and white matter as a function of time after injury, particularly in the dorsal funiculus. The three-dimensional morphology of the lesion in rodents (Bresnahan et al., 1991) also mimics that seen in a large proportion of human spinal injuries (Kakulas, 1984; Bunge et al., 1993, 1997). The animals exhibit an initial paralysis followed by a predicable improvement in hindlimb function that is proportional to the severity of the lesion (Basso et al., 1996). This injury model has been used to evaluate and describe deleterious and reparative events that occur endogenously after SCI (Beattie et al., 1997; Crowe et al., 1997; Wrathall et al., 1998; Popovich et al., 1999). For instance, in the acutely injured spinal cord, several mechanisms of destruction are initiated including hemorrhage and edema due to disrupted

---

* Corresponding author: Dr. Bradford T. Stokes, Department of Physiology and Cell Biology, The Ohio State University, 228 Meiling Hall, 370 W 9th Avenue, Columbus, OH 43210-1238, USA. Fax: +1-614-292-0577; E-mail: bstokes@smtp.med.ohio-state.edu

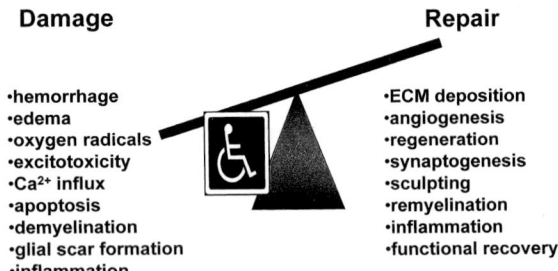

Fig. 1. After trauma to the spinal cord, several events occur that may lead to further tissue destruction or, in contrast, may result in tissue remodeling and repair. Damaging events include acute phenomena such as hemorrhage and Ca^{2+} influx as well as more protracted events, e.g., apoptosis and demyelination. Reparative processes such as angiogenesis and axon growth also typically occur in the injured spinal cord. Depending on the extent of these processes, functional recovery may be limited to nearly complete. Interestingly, inflammation in the injured spinal cord precipitates both deleterious and beneficial events in time and space after SCI.

blood vessels at the epicenter, lipid peroxidation and calcium influx/excitotoxic cell death (Fig. 1). Neurons and certain glial cells are highly susceptible to excitotoxicity in the central nervous system (CNS), which contributes to the massive cell death observed in the traumatized spinal cord.

Oligodendrocyte death, demyelination and remyelination

In addition to the necrotic cell death known to be prevalent after SCI, recent evidence from Beattie and colleagues has revealed that cells are lost through apoptosis both acutely within the lesion epicenter and in the rostro-caudal lesion extensions (see Chapter 2). Interestingly, this distal apoptosis can occur for up to 6–8 weeks after injury and typically occurs in oligodendrocytes and microglia found in regions of progressive Wallerian axonal degeneration (Crowe et al., 1997; Shuman et al., 1997). The apoptotic loss of oligodendrocytes may contribute to the demyelination that is consistently found in acutely injured spinal cords from a variety of species (Gledhill et al., 1973; Bresnahan et al., 1976; Balentine, 1978; Banik et al., 1980; Blight, 1985).

Following traumatic SCI, the demyelinated axons do, for the most part, become remyelinated by both oligodendrocytes and Schwann cells (Gledhill et al., 1973; Gledhill and McDonald, 1977; Harrison and McDonald, 1977; Griffiths and McCulloch, 1983). Oligodendrocyte remyelination likely occurs as a result of proliferation and differentiation of oligodendrocyte progenitors normally found throughout the CNS (Gensert and Goldman, 1997; Carroll et al., 1998; Keirstead et al., 1998; McTigue et al., 1999), although some would argue that mature differentiated oligodendrocytes can play a role in remyelination (Wood and Bunge, 1991; Wood and Mora, 1993). However, the endogenous processes of remyelination are insufficient to produce adequate functional recovery. This may be due in part to the failure of normal cues for axon remyelination. For example, the remyelinated internodes do not achieve the same thickness or length as in the normal CNS (Gledhill and McDonald, 1977; Harrison and McDonald, 1977). Recent trials with the drug 4-aminopyridine (4-AP), which increases axonal conduction by blocking K^+ channels, reveal some functional enhancements suggesting that endogenous remyelination processes can be improved upon (Blight et al., 1991; Hansebout et al., 1993; Hayes et al., 1994). Thus, if oligodendrocyte progenitors are important for remyelination, agents that can further augment their proliferation and differentiation could potentially be important treatments for SCI. For example, recent preliminary data from our laboratory revealed that the neurotrophins, brain-derived neurotrophic factor and neurotrophin-3 may augment oligodendrocyte progenitor proliferation after SCI (McTigue et al., 1999).

Extracellular matrix in the injured spinal cord

In addition to changes in the cellular components at the injury site, another facet of remodeling is the formation of a glial scar and the deposition of extracellular matrix (ECM) molecules. The total number and molecular composition of ECM molecules that accumulate is unknown, is likely to be injury-specific, and may depend on the animal model of SCI. For instance, recent experiments using contusion injury in mice reveal a very different lesion evolution than that seen in rats. Rather than cavitation, the lesioned area typically undergoes significant deposition of ECM molecules leading to marked scar formation (Kuhn and Wrathall, 1998; Ma et al., 1999; Steward et al., 1999; Jakeman et al., 2000).

This mouse contusion injury model may be useful in modeling certain types of human SCI, particularly those in which the meninges are compromised and a subsequent solid tissue scar forms at the lesion site (Kakulas, 1984; Bunge et al., 1997).

The ECM molecule, laminin, is consistently found in either type of injury site and is typically thought to be permissive for growing axons. However, Geller et al. (see Chapter 3), using an in vitro neurite growth system, have identified specific subregions of laminin that appear to be growth inhibitory. It is likely that other ECM molecules (e.g., tenascin-C) thought to be principally growth inhibitory (or permissive) will be shown to have varying effects on axon growth depending on the molecular makeup of the in vivo microenvironment after SCI. Such variability might be precipitated by the differential appearance of a variety of proteases known to be present after SCI. Thus, ECM deposition may be considered to either contribute to the problems after SCI if the majority of molecules are growth inhibitory, or to be part of the endogenous reparative processes if the ECM mostly consists of growth permissive molecules. Evidence suggests that it is likely to be both, particularly over time as the lesion evolves. The particular types of ECM proteins deposited in the lesioned tissue are also likely to be influenced by the distribution and local concentration of surface and secreted molecules, such as growth factors and cytokines, which are found within the injury site. Therefore, with the proper molecular tools, it may be possible to manipulate the ECM composition such that a supportive environment for sprouting and regenerating axons is created.

Inflammatory involvement after SCI

Finally, inflammation represents an important process in CNS trauma that likely plays a role in all the events mentioned above. The initiation of an inflammatory response by injury to the spinal cord has been known for some time, but only recently has it received widespread attention. The issue is complicated by the fact that certain portions of the inflammatory response likely promote tissue sparing and functional repair while other events may contribute to tissue destruction (see Chapter 2). Part of the complexity of this response stems from the plethora of chemokines and cytokines (both pro- and anti-inflammatory) that are present locally in the spinal parenchyma and in the circulation after spinal cord trauma (Semple-Rowland et al., 1995; McTigue et al., 1998, 2000; Streit et al., 1998). In a recent study by Bethea et al. (1999) (also see Chapter 4), a single dose of the anti-inflammatory cytokine interleukin-10 (IL-10), administered 30 min after injury, significantly improved anatomical and behavioral outcome in the rat. This would suggest that inflammatory events early after SCI are deleterious. However, the issue is not so straightforward. When an additional dose of IL-10 was given at 3 days after injury, the protective effects were completely lost. The explanation of these results is not yet understood. Thus, the immune response to SCI is clearly complex and further work to elucidate sites of action of immunomodulators is of obvious importance.

Because of the complex interactions of the immune cells in response to SCI, it has been important to clearly identify the time course of microglial and macrophage activation, astroglial response and leukocyte infiltration in the contused spinal cord (Popovich et al., 1997). Some of these results are summarized in Fig. 2. Early after SCI, neutrophils invade and then quickly exit the injured tissue. T-lymphocytes also enter after SCI, peaking at approximately 1 week after injury. This cellular response corresponds closely to the time course of chemokine and cytokine expression in the same injury model (Semple-Rowland et al., 1995; McTigue et al., 1998, 2000; Streit et al., 1998), suggesting that these cells may be responding to chemoattractive molecules produced in the injured spinal cord.

As can be seen in Fig. 2, microglia are activated early after SCI while monocyte infiltration is delayed over a period of days. Unfortunately, once these cells become phagocytic, it is impossible using routine immunohistochemical techniques and light microscopy to distinguish between microglia and blood-derived monocytes in the spinal cord. However, Popovich has hypothesized that while these cells look similar morphologically, they play very different roles in post-injury events (see Chapter 5). To better identify the role of microglia in the injury site and minimize the influence of infiltrating monocytes, he has used a technique in rats that depletes circulating monocytes acutely after spinal contusion

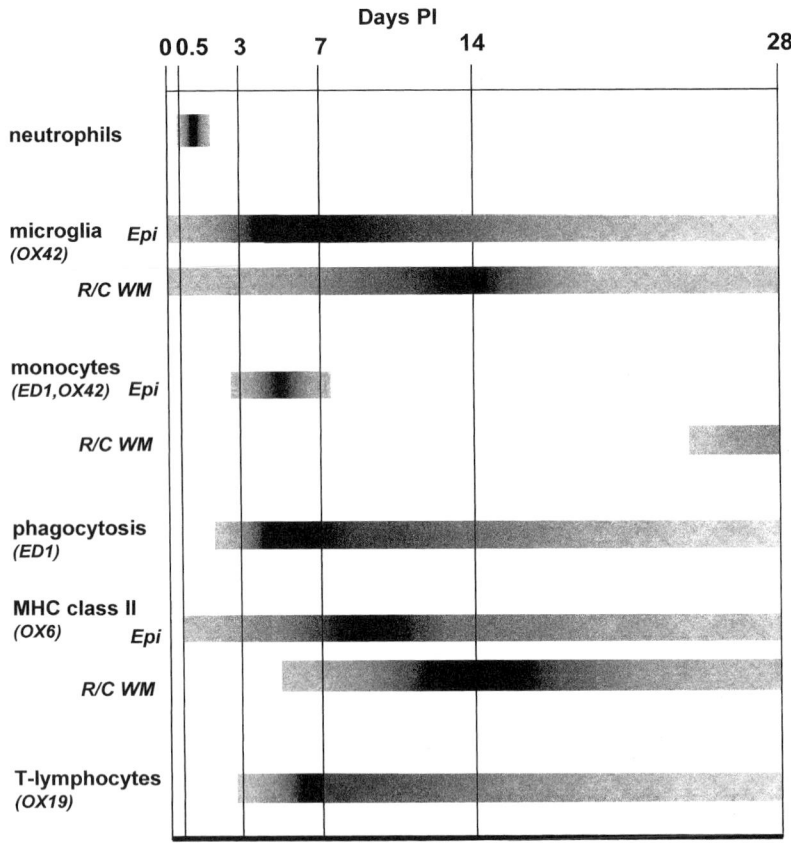

Fig. 2. Illustration of temporal appearance and relative amounts of activated resident and infiltrating inflammatory cells in the epicenter (*Epi*) and rostral and caudal white matter (*R/C WM*) of rats receiving spinal contusion injuries. The intensity of staining (and presumably cellular activation) is indicated by the density of lines for each cell type. Post-injury time (days PI) is indicated at the top of the figure. The immunohistochemical stain used for identification is indicated below each category. Early after injury, neutrophils invade and then quickly exit the injury site. Activation of microglia and initiation of phagocytosis begins early and continues for protracted times after injury. Circulating monocytes and T-lymphocytes enter the injured spinal cord during the first week after injury, while upregulation of MHC class II is observed during the first month after contusion injury. Collectively, these data illustrate that activated inflammatory cells rapidly enter the injured spinal cord, spread throughout the rostral/caudal axis, and persist for extended periods of time after injury.

injury without affecting resident microglia (Popovich et al., 1999; see also Chapter 5). The results demonstrate that reducing the contribution of peripherally derived macrophages produces an increase in tissue sparing and promotes axonal growth at the epicenter with significant improvements in locomotor recovery. This targeted approach suggests an apparent dichotomy in the actions of inflammatory cells in the injured spinal cord and underlines the importance of understanding the timing of entry, activation, and interactions of immune cells and mediators (e.g., cytokines, chemokines and growth factors) in the injured spinal cord.

Thus, it is clear that while great strides have been made in understanding processes involved in destruction and repair of the injured spinal cord (Schwab and Bartholdi, 1996; Olson, 1997), there is much to be learned about basic etiologic factors. More fundamental studies are clearly needed, in both animals and available human tissue, to reveal the events occurring endogenously in the injured CNS. Armed with this knowledge, we may then begin to make more educated attempts at enhancing the apparently limited reparative processes while concomitantly downregulating those events that act either passively or actively to restrict repair.

Acknowledgements

This work was supported by the Christopher Reeve Paralysis Foundation (MAR2-9803-1; DMM and BTS), NS33696 (PGP and BTS), NS37846 (PGP), and the Paralyzed Veterans of America (SCRF1892; LBJ and BTS).

References

Balentine, J.D. (1978) Pathology of experimental spinal cord trauma, II. Ultrastructure of axons and myelin. *Lab. Invest.*, 39: 254–266.

Banik, N.L., Powers, J.M. and Hogan, E.L. (1980) The effects of spinal cord trauma on myelin. *J. Neuropathol. Exp. Neurol.*, 39: 232–244.

Basso, D.M., Beattie, M.S. and Bresnahan, J.C. (1996) Graded histological and locomotor outcomes after spinal cord contusion using the NYU weight-drop device versus transection. *Exp. Neurol.*, 139: 244–256.

Beattie, M.S., Bresnahan, J.C., Komon, J., Tovar, C.A., Van Meter, M., Anderson, D.K., Faden, A.I., Hsu, C.Y., Noble, L.J., Salzman, S.K. and Young, W. (1997) Endogenous repair after spinal cord contusion injuries in the rat. *Exp. Neurol.*, 148: 453–463.

Bethea, J.R., Nagashima, H., Acosta, M.C., Briceno, C., Gomez, F., Marcillo, A.E., Loor, K., Green, J. and Dietrich, W.D. (1999) Systemically administered interleukin-10 reduces tumor necrosis factor-alpha production and significantly improves functional recovery following traumatic spinal cord injury in rats. *J. Neurotrauma*, 16: 851–863.

Blight, A.R. (1985) Delayed demyelination and macrophage invasion: a candidate for secondary cell damage in spinal cord injury. *CNS Trauma*, 2: 299–315.

Blight, A.R., Toombs, J.P., Bauer, M.S. and Widmer, W.R. (1991) The effects of 4-aminopyridine on neurological deficits in chronic cases of traumatic spinal cord injury in dogs: a phase I clinical trial. *J. Neurotrauma*, 8: 103–119.

Bregman, B.S., Kunkel-Bagden, E., Schnell, L., Dai, H.N., Gao, D. and Schwab, M.E. (1995) Recovery from spinal cord injury mediated by antibodies to neurite growth inhibitors. *Nature*, 378: 498–501.

Bresnahan, J.C., King, J.S., Martin, G.F. and Yashon, D. (1976) A neuroanatomical analysis of spinal cord injury in the Rhesus monkey (*Macaca mulatta*). *J. Neurol. Sci.*, 28: 521–542.

Bresnahan, J.C., Beattie, M.S., Stokes, B.T. and Conway, K.M. (1991) Three-dimensional computer-assisted analysis of graded contusion lesions in the spinal cord of the rat. *J. Neurotrauma*, 8: 91–101.

Bunge, R.P., Puckett, W.R., Becerra, J.L., Marcillo, A. and Quencer, R.M. (1993) Observations on the pathology of human spinal cord injury: a review and classification of 22 new cases with details from a case of chronic cord compression with extensive focal demyelination. In: Seil, F.J. (Ed.), *Neural Injury and Regeneration. Advances in Neurology, Vol. 59*. Raven Press, New York, pp. 75–89.

Bunge, R.P., Puckett and W.R., Hiester, E.D. (1997). Observations on the pathology of several types of human spinal cord injury, with emphasis on the astrocyte response to penetrating injuries. In: Seil, F.J. (Ed.), *Neuronal Regeneration, Reorganization, and Repair. Advances in Neurology, Vol. 72*. Lippincott-Raven, Philadelphia, PA, pp. 305–315.

Carroll, W.M., Jennings, A.R. and Ironside, L.J. (1998) Identification of the adult resting progenitor cell by autoradiographic tracking of oligodendrocyte precursors in experimental CNS demyelination. *Brain*, 121: 293–302.

Cheng, H., Cao, Y.H. and Olson, L. (1996) Spinal cord repair in adult paraplegic rats: partial restoration of hind limb function. *Science*, 273: 510–513.

Crowe, M.J., Bresnahan, J.C., Shuman, S.L., Masters, J.N. and Beattie, M.S. (1997) Apoptosis and delayed degeneration after spinal cord injury in rats and monkeys. *Nat. Med.*, 3: 73–76.

Gensert, J.M. and Goldman, J.E. (1997) Endogenous progenitors remyelinate demyelinated axons in adult CNS. *Neuron*, 19: 197–203.

Giovanini, M.A., Reier, P.J., Eskin, T.A., Wirth, E. and Anderson, D.K. (1997) Characteristics of human fetal spinal cord grafts in the adult rat spinal cord: influences of lesion and grafting conditions. *Exp. Neurol.*, 148: 523–543.

Gledhill, R.F. and McDonald, W.I. (1977) Morphological characteristics of central demyelination and remyelination: a single-fiber study. *Ann. Neurol.*, 1: 552–560.

Gledhill, R.F., Harrison, B.M. and McDonald, W.I. (1973) Demyelination and remyelination after acute spinal cord compression. *Exp. Neurol.*, 38: 472–487.

Griffiths, I.R. and McCulloch, M.C. (1983) Nerve fibres in spinal cord impact injuries. *J. Neurol. Sci.*, 58: 335–349.

Grill, R., Murai, K., Blesch, A., Gage, F.H. and Tuszynski, M.H. (1997) Cellular delivery of neurotrophin-3 promotes corticospinal axonal growth and partial functional recovery after spinal cord injury. *J. Neurosci.*, 17: 5560–5572.

Hansebout, R.R., Blight, A.R., Fawcett, S. and Reddy, K. (1993) 4-Aminopyridine in chronic spinal cord injury: a controlled, double-blind, crossover study in eight patients. *J. Neurotrauma*, 10: 1–18.

Harrison, B.M. and McDonald, W.I. (1977) Remyelination after transient experimental compression of the spinal cord. *Ann. Neurol.*, 1: 542–551.

Hayes, K.C., Potter, P.J., Wolfe, D.L., Hsieh, J.T.C., Delaney, G.A. and Blight, A.R. (1994) 4-Aminopyridine-sensitive neurologic deficits in patients with spinal cord injury. *J. Neurotrauma*, 11: 433–446.

Jakeman, L.B., Guan, Z., Wei, P., Ponnappan, R., Dzwonczyk, R., Popovich, P.G. and Stokes, B.T. (2000) Traumatic spinal cord injury produced by controlled contusion in mouse. *J. Neurotrauma*, 17: 303–323.

Kakulas, B.A. (1984) Pathology of spinal injuries. *CNS Trauma*, 1: 117–129.

Keirstead, H.S., Levine, J.M. and Blakemore, W.F. (1998) Response of the oligodendrocyte progenitor cell population (defined by NG2 labeling) to demyelination of the adult spinal cord. *Glia*, 22: 161–170.

Kuhn, P.L. and Wrathall, J.R. (1998) A mouse model of graded contusive spinal cord injury. *J. Neurotrauma*, 15: 125–140.

Ma, M., Walters, P., Basso, D.M., Stokes, B.T. and Jakeman, L.B. (1999) Recovery of function and histological outcome following controlled contusion injury in mice. *Soc. Neurosci. Abstr.*, 25: 311.

McDonald, J.W., Liu, X.Z., Qu, Y., Liu, S., Mickey, S.K., Turetsky, D., Gottlieb, D.I. and Choi, D.W. (1999) Transplanted embryonic stem cells survive, differentiate and promote recovery in injured rat spinal cord. *Nat. Med.*, 5: 1410–1412.

McTigue, D.M., Tani, M., Krivacic, K., Chernosky, A., Kelner, G.S., Maciejewski, D., Maki, R., Ransohoff, R.M. and Stokes, B.T. (1998) Selective chemokine mRNA accumulation in the rat spinal cord after contusion injury. *J. Neurosci. Res.*, 53: 368–376.

McTigue, D.M., Jakeman, L.B., Horner, P.J., Wei, P., Shah, A., Gage, F.H. and Stokes, B.T. (1999) Neurotrophin-producing transplants stimulate oligodendrocyte progenitor proliferation and Schwann cell myelination in the contused rat spinal cord. *Soc. Neurosci. Abstr.*, 25: 512.

McTigue, D.M., Popovich, P.G., Morgan, T.E. and Stokes, B.T. (2000) Localization of transforming growth factor-β1 and receptor mRNA after experimental spinal cord injury. *Exp. Neurol.*, 163: 220–230.

Menei, P., Montero-Menei, C.N., Whittemore, S.R., Bunge, R.P. and Bunge, M.B. (1998) Schwann cells genetically modified to secrete human BDNF promote enhanced axonal regrowth across transected adult rat spinal cord. *Eur. J. Neurosci.*, 10: 607–621.

Noble, L.J. and Wrathall, J.R. (1985) Spinal cord contusion in the rat: morphometric analyses of alterations in the spinal cord. *Exp. Neurol.*, 88: 135–149.

Olson, L. (1997) Regeneration in the adult central nervous system: experimental repair strategies. *Nat. Med.*, 3: 1329–1335.

Popovich, P.G., Wei, P. and Stokes, B.T. (1997) The cellular inflammatory response after spinal cord injury in Sprague–Dawley and Lewis rats. *J. Comp. Neurol.*, 377: 443–464.

Popovich, P.G., Guan, Z., Wei, P., Huitinga, I., van, R.N. and Stokes, B.T. (1999) Depletion of hematogenous macrophages promotes partial hindlimb recovery and neuroanatomical repair after experimental spinal cord injury. *Exp. Neurol.*, 158: 351–365.

Ramón-Cueto, A., Plant, G.W., Avila, J. and Bunge, M.B. (1998) Long-distance axonal regeneration in the transected adult rat spinal cord is promoted by olfactory ensheathing glia transplants. *J. Neurosci.*, 18: 3803–3815.

Schwab, M.E. and Bartholdi, D. (1996) Degeneration and regeneration of axons in the lesioned spinal cord. *Physiol. Rev.*, 76: 319–370.

Semple-Rowland, S.L., Mahatme, A., Popovich, P.G., Green, D.A., Hassler, G., Stokes, B.T. and Streit, W.J. (1995) Analysis of *TGF-β1* gene expression in contused rat spinal cord using quantitative RT-PCR. *J. Neurotrauma*, 12: 1003–1014.

Shi, R., Borgens, R.B. and Blight, A.R. (1999) Functional reconnection of severed mammalian spinal cord axons with polyethylene glycol. *J. Neurotrauma*, 16: 727–738.

Shuman, S.L., Bresnahan, J.C. and Beattie, M.S. (1997) Apoptosis of microglia and oligodendrocytes after spinal cord contusion in rats. *J. Neurosci. Res.*, 50: 798–808.

Steward, O., Schauwecker, P.E., Guth, L., Zhang, Z., Fujiki, M., Inman, D., Wrathall, J., Kempermann, G., Gage, F.H., Saatman, K.E., Raghupathi, R. and McIntosh, T. (1999) Genetic approaches to neurotrauma research: opportunities and potential pitfalls of murine models. *Exp. Neurol.*, 157: 19–42.

Streit, W.J., Semple-Rowland, S.L., Hurley, S.D., Miller, R.C., Popovich, P.G. and Stokes, B.T. (1998) Cytokine mRNA profiles in contused spinal cord and axotomized facial nucleus suggest a beneficial role for inflammation and gliosis. *Exp. Neurol.*, 152: 74–87.

Wood, P.M. and Bunge, R.P. (1991) The origin of remyelinating cells in the adult central nervous system: the role of the mature oligodendrocyte. *Glia*, 4: 225–232.

Wood, P.M. and Mora, J. (1993) Source of remyelinating oligodendrocytes. In: Seil, F.J. (Ed.), *Neural Injury and Regeneration. Advances in Neurology, Vol. 59*, Raven Press, New York, pp. 113–123.

Wrathall, J.R., Li, W. and Hudson, L.D. (1998) Myelin gene expression after experimental contusive spinal cord injury. *J. Neurosci.*, 18: 8780–8793.

CHAPTER 2

Cell death and plasticity after experimental spinal cord injury

Michael S. Beattie *, Qun Li and Jacqueline C. Bresnahan

Department of Neuroscience, The Ohio State University, 333 West 10th Avenue, Columbus, OH 43210, USA

Introduction

Models of spinal cord injury (SCI) have been developed over the years in order to study different aspects of recovery of function, the biology of injury, regeneration, and therapeutics. Our laboratory has been engaged in work relating to the bases of recovery of function after SCI for many years. Studies have ranged from analyses of changes in spinal cord anatomical circuits after spinal cord hemisections in cats and monkeys to more recent experiments that attempt to address the cell biology of the evolving lesion after spinal cord contusion injuries in rats. One theme has been the distributed nature of both the lesion itself and its remote effects on the rest of the CNS and the periphery. For example, a contusion lesion or spinal transection denervates rostral and caudal neural centers, providing an opportunity for collateral sprouting in both the brain and in segmental circuits that control posture, locomotion, and sensory processing. Recent work with antibodies that block the inhibition of axonal growth have reemphasized how important such remote remodeling may be for recovery (Thallmair et al., 1998). Another example of remote effects is the programmed cell death of oligodendrocytes associated with Wallerian degeneration in the long tracts of the cord after injury (reviewed in Beattie et al., 1998). There are also remote effects on the periphery, most clearly on muscle (Roy et al., 1999), but also on sensory systems, for example, the remodeling of bladder afferent systems after SCI (Steers et al., 1990). In this chapter, we will review work, mainly from our laboratory, that touches on the progression of secondary injury and endogenous repair at the lesion site as well as examples of remote remodeling that may play a role in positive and negative aspects of recovery of function. The final, functional outcome after SCI, in both models and man, must be determined by the complex interplay of all these dynamic events.

A framework for studying spinal cord injury

Fig. 1 presents a schematized nervous system, illustrating the complexity of the results of a spinal cord contusion injury, and emphasizing the importance of both local injury and remote effects of damage. The spinal cord gray matter is contained within the long descending and ascending spinal tracts, so that injuries confined to the central region may damage local gray matter circuits and yet leave long tract communication intact. The mechanical characteristics of the stiff white matter surrounding the pliable gray contribute to the production of a central focus of injury even when force is applied to the cord dorsum or ventrum. This forms the central degeneration seen so commonly in spinal contusion injuries. Secondary injury can spread centripetally, involving the white

---

* Corresponding author: Dr. Michael S. Beattie, Department of Neuroscience, The Ohio State University, 333 West 10th Avenue, Columbus, OH 43210, USA; Fax: +1-614-292-9202; E-mail: beattie.2@osu.edu

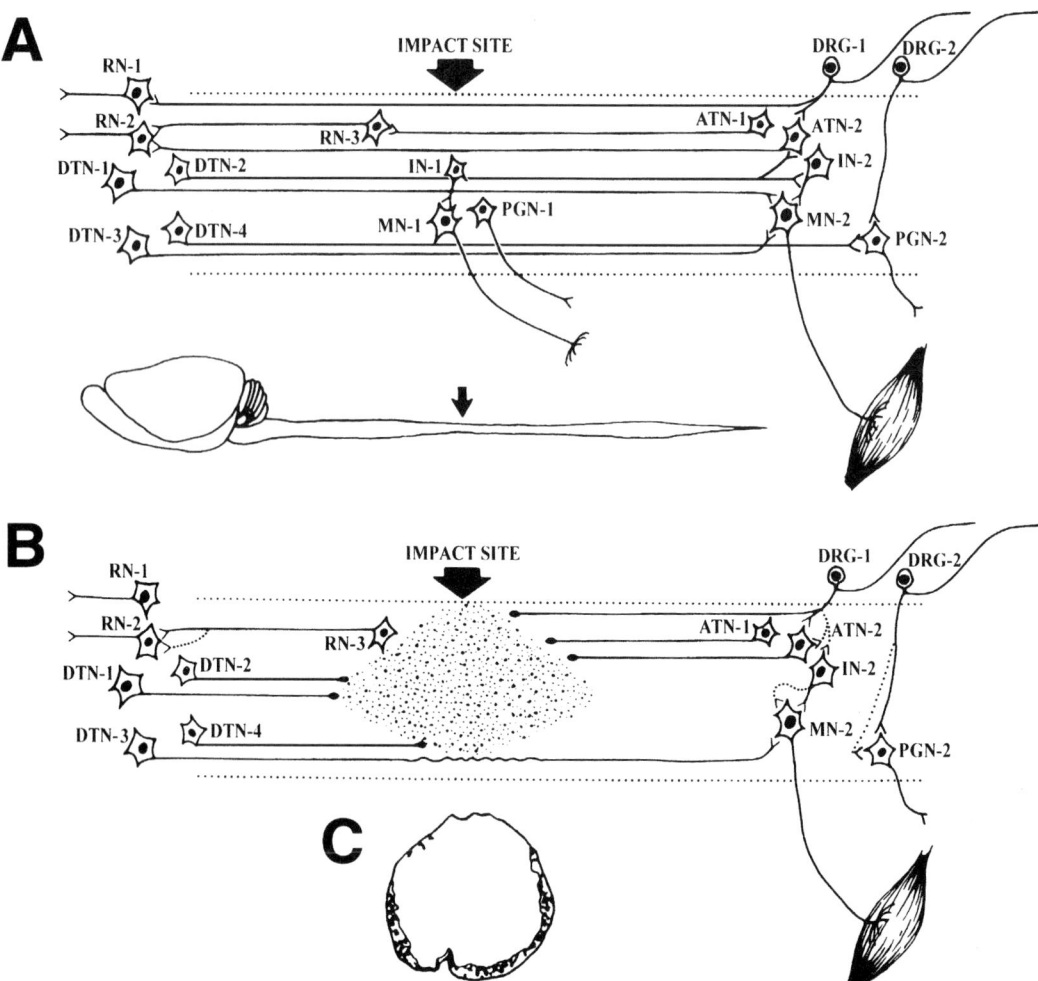

Fig. 1. Schematic diagram of a lesion in the thoracic spinal cord of a rat. (A) The normal spinal cord with several neuronal types and examples of some circuits are shown. At the right, sensory inputs from dorsal root ganglion cells (*DRG-1* and *DRG-2*) are shown projecting to spinal cord preganglionic neurons (*PGN-2*) and ascending tract neurons (*ATN-2*), as well as through the dorsal columns to the brain stem relay neuron (*RN-1*). Other ascending tract neurons (*ATN-1*), spinal cord interneurons (*IN-1* and *IN-2*), motor neurons (*MN-1* and *MN-2*), and preganglionic neurons (*PGN-1*) are shown. Additional relay neurons in the rostral spinal cord (*RN-3*) and brainstem (*RN-1 & 2*) are indicated as well. Descending tract neurons are shown projecting to ascending tract cells and interneurons (*DTN-2*) and motor (*DTN-1* and *DTN-3*) and preganglionic (*PGN-2*) neurons. (B) After a spinal cord injury in the thoracic region at the site indicated by the arrows, frank necrosis, axotomy, denervation, and compensatory processes ensue. The final status of some of these processes is indicated in (B) and the residual tissue at the lesion center is shown in (C). The ascending tract neurons (*AT-1* and *AT-2*) are axotomized, and the rostrally projecting collaterals of dorsal root ganglion cells (*DRG-1*) are pruned. At the lesion site, the interneurons (*IN-1*), motor neurons (*MN-1*), and preganglionic neurons (*PGN-1*) are destroyed, as are most axons passing through the lesion. Some axons at the periphery of the lesion are spared (*DTN-3* and (C)) but may be demyelinated due to the loss of oligodendrocytes at the lesion center, and therefore exhibit altered conduction properties. The degeneration of the distal segments of the interrupted axons produces denervation of neurons both rostrally (*RN-1* and *RN-3*) and caudally (*IN-2*, *MN-2* and *PGN-2*). Vacated postsynaptic sites may induce compensatory collateral sprouting of intact neurons (*IN-2*, *DRG-2*, dashed lines) or of neurons that have had their axonal arbors pruned by the lesion (*DRG-1*). Other functional alterations produced by the lesion include denervation supersensitivity and unmasking of previously ineffective synapses. (Adapted from Beattie et al., 1988.)

matter later. As the diagram indicates, the destruction of axons in the white matter leads to the loss of projections to and from the brain and caudal spinal cord. The loss of these projections may lead to reorganization of the affected circuits by synaptic sprouting or other changes in synaptic strength. Recent studies by Merzenich, Kaas, and other groups (e.g., Jain et al., 1997; Kaas, this volume) have emphasized the reorganization of neocortical and thalamic regions after chronic injury. There is also good evidence for reorganization at the spinal level (Goldberger et al., 1993), and evidence for the ability of descending systems to effect changes in altered spinal circuitry (e.g., Chen et al., 1999).

The evolving contusion lesion

The concept of secondary injury has been important in strategies for studying spinal cord injury in the laboratory. Original observations by Allen (1911) and McVeigh (1923) showed that immediately after a contusion injury to the cord, there was only minor obvious damage. More modern studies report the presence of petechial hemorrhage in the gray matter at the lesion site at minutes to hours after injury, with a rapid centripetal and rostro-caudal development of frank necrotic damage (see Fig. 2). This leads eventually to the formation of cystic cavities (Noble and Wrathall, 1985; Bresnahan et al., 1991; Guizar-Sahagun et al., 1994). The secondary expansion of the lesion cavity is associated with the invasion of peripheral immune cells, as well as the activation of resident microglia (Blight, 1992; Popovich et al., 1997, and this volume; Shuman et al., 1997). The rather lengthy time course of lesion expansion over days post-injury suggests that there is more to this secondary injury than simply the spread of necrotic damage. Recent studies of programmed cell death in models of stroke and head injury (Li et al., 1995a; Rink et al., 1995; Yakovlev et al., 1997) provided the background for a series of studies in spinal cord injury that have shown a clear pattern of apoptotic or programmed cell death after various kinds of SCI (Li et al., 1996a; Crowe et al., 1997; Liu et al., 1997; Shuman et al., 1997; Springer et al., 1999). Neurons die early at the lesion center by both necrosis and apoptosis, and at the lesion margins by apoptosis. White matter tracts exhibit many apoptotic cells at long times after injury (Crowe et al., 1997; Shuman et al., 1997; Emery et al., 1998). These dying cells are associated with the dying axons that have been injured by the lesion and with activated microglia.

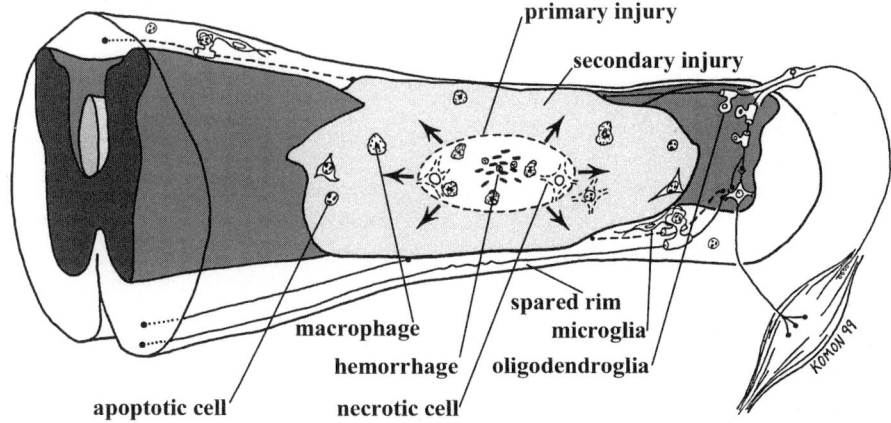

Fig. 2. Schematic drawing of a contusion lesion in the spinal cord showing some of the reactions of the cellular elements. Contusion lesions are characterized by a central hemorrhagic region that expands over time (arrows) due to activation of secondary injury processes including necrosis and apoptosis. Besides the blood elements that escape from damaged vessels, recruitment of immune cells occurs early after injury. Within a few days the injury site is filled with a large number of macrophages and activated microglia extend along the edges of the lesion as well as rostrally and caudally within fiber tracts undergoing Wallerian degeneration. The microglia may participate in the apoptotic death of oligodendroglia, which also occurs along the degenerating fiber tracts (dashed lines), extending into regions far remote from the actual injury site. Intact segmental circuitry caudal to the injury and circuitry rostral to the injury are denervated from loss of descending and ascending axons, respectively, initiating reorganization in those systems. (From Beattie and Bresnahan, 2000.)

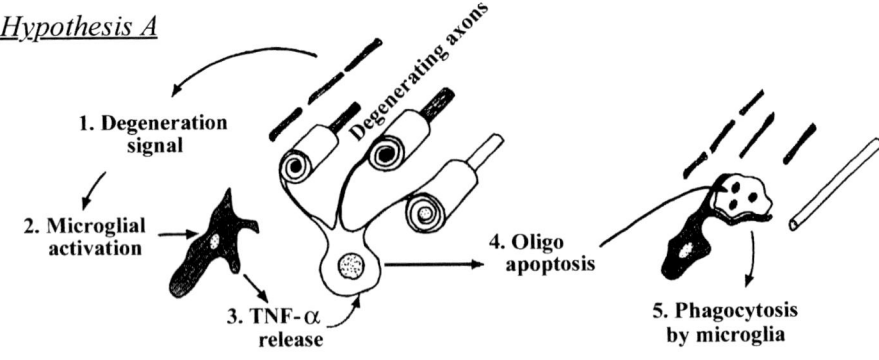

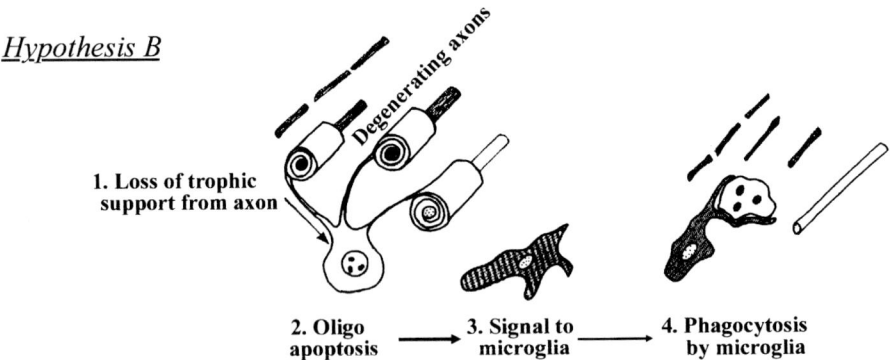

Fig. 3. Two proposed ways for microglia to induce apoptosis of oligodendrocytes in the white matter tracts undergoing degeneration after an injury. (A) Activated microglia may secrete cytokines (such as TNF-α) in response to degenerating axons and induce apoptosis in oligodendrocytes. (B) Alternatively, the oligodendrocyte may respond to loss of trophic support from the axons it myelinates as they undergo degeneration. In either case, the apoptotic oligodendrocytes may be engulfed by phagocytic cells which could lead to focal demyelination of surviving axons. (From Shuman et al., 1997.)

Double labeling studies show that the cells undergoing apoptosis in the white matter tracts in association with Wallerian degeneration are oligodendrocytes and microglia (see Shuman et al., 1997). This long-term example of secondary injury may be important for several reasons. First, the cell death after injury may be related to demyelination of long tracts. Demyelination has long been thought to be potentially important in the dysfunction seen after partial cord injuries like those produced in contusion models (Blight and Decrescito, 1986). Second, apoptosis is under the control of cell-signaling events and intracellular pathways that may offer new and unique targets for drug therapy. Third, the association with microglial activation suggests an analogy with demyelinating diseases like multiple sclerosis. However, the role of oligodendrocyte cell death and demyelination in the dysfunction seen following contusion injuries in rats or humans has yet to be definitively proven. The basis for the suggestion, however, can be seen in Fig. 3, where the death of oligodendrocytes that myelinate multiple axons has resulted in the denuding of several axons that survived the initial injury and axonal destruction. Cytokines like tumor necrosis factor TNF-α may play an important role in the initiation of neuronal cell death as well. A recent study has shown quite pronounced neuroprotection and behavioral sparing when interleukin (IL)-10, a potent anti-inflammatory cytokine, is given after SCI (Bethea et al., 1999; Bethea, this volume).

The suggestion that oligodendrocyte cell death is due to cytokine action from activated microglia has been put forth by a number of groups (Liu et al., 1997; Shuman et al., 1997). However, an alternative hypothesis is based on the work of Barres and Raff (Barres et al., 1993; Raff et al., 1993), which seems to show a dependence of oligodendrocytes on their ensheathed axons, at least during development. Barres et al. (1993) saw reduced numbers of oligodendrocytes in optic nerves in which the retinal ganglion cell axons had been severed. They attributed this loss to active cell death (apoptosis) related to the loss of trophic factors provided by the axons.

These alternative hypotheses are intriguing, since they both rely on intracellular pathways leading to the same fate (e.g., caspase activation and internucleosomal cleavage) that are initiated by very different mechanisms. In the first case, apoptosis is receptor-mediated, perhaps via the intracellular death domains associated with TNF-α receptor (see review by Pettmann and Henderson, 1998). In the other case, lack of ligand is the initiating process, analogous to the death of sympathetic neurons deprived of nerve growth factor (NGF) (see Deshmukh and Johnson, 1997). There is considerable current interest in this question, and it will be difficult to rule out one or the other mechanism, especially in the complex context of the contusion models in rats and mice. However, the use of transgenic and knockout animals and selective agonists and antagonists to cytokines and growth factors may soon yield insights into how oligodendrocytes (and neurons) respond to the lesion environment in adult animals. Recently, the notion that axons are necessary for oligodendrocyte maintenance has been challenged (Ueda et al., 1999). On the other hand, the role of TNF-α in the induction of cell death after injury is brought into question by studies of TNF-α receptor knockout mice (Bruce et al., 1996).

Interestingly, the role of growth and neurotrophic factors like NGF may be quite complicated. While NGF has been shown to be neuroprotective and may encourage axonal regeneration in the spinal cord (e.g., Grill et al., 1997a), there is also evidence that NGF can act via the low affinity neurotrophic factor receptor, p75, as a death signaling molecule (see Casaccia-Bonnefil et al., 1996; Chao et al., 1998; Yoon et al., 1998) when it is expressed in the absence of the 'high affinity' NGF receptor tyrosine kinase, TrkA. p75 is not normally expressed in oligodendrocytes in the adult rat spinal cord. However, we have recently reported that p75 is induced in oligodendrocytes in the dorsal columns following dorsal hemisections that cut dorsal column axons (Beattie et al., 1999). Some of these oligodendrocytes had apoptotic morphologies. Since NGF may be expressed by astrocytes and microglia after spinal injury (Krenz and Weaver, 2000), this provides yet another alternative for the induction of cell death. Like the cytokines (see the chapters by Popovich and Bethea in this volume), trophic factors might have dual roles in degeneration and repair after SCI.

Endogenous repair

In this chapter we will concentrate mostly on repair and recovery in segmental circuits remote to the lesion. However, we and others have seen substantial evidence for repair at the lesion site after contusion injuries. In amphibians, it has long been known that the spinal cord can repair itself under some conditions. In *Xenopus laevis*, the African clawed frog, effective repair and axonal regeneration occur only during active metamorphosis in tadpoles; the ependymal cells lining the central canal appear to be the source of the repairing cells (Michel and Reier, 1979; Beattie et al., 1990). In rats and rabbits, contusion injury also induces proliferation of cells surrounding the central canal (Matthews et al., 1979; Vaquero et al., 1981; Beattie et al., 1997, 1998). These proliferating cells appear to add to tissue bridges (trabeculae) that begin to fill in the contusion lesion cavity. Schwann cells and invading sensory axons can follow these cellular tracks. At very long times after injury, these strings of cells within the cavity can even support the growth of a few axons from within the central nervous system, i.e., the corticospinal tract and fibers from the brainstem reticular formation (Hill et al., 1998). The role of this kind of repair in recovery after contusion lesions is not yet known. When fibers are induced by neurotrophins to grow around dorsal hemisections in rats, there is some evidence for enhanced function (Grill et al., 1997b).

Some examples of anatomical plasticity after SCI

The loss of descending axons after spinal cord injury provides the possibility of reorganization of caudal segmental circuitry. Axonal and synaptic sprouting has long been thought to play a role in changes in spinal reflexes after injury, and may promote recovery and/or produce dysfunctional outcomes such as spasticity (see reviews by Goldberger and Murray, 1988 and Goldberger et al., 1993). We have studied changes in synaptic inputs to pudendal motoneurons in the lumbosacral spinal cord of rats and cats that control the pelvic musculature that mediates sexual and eliminative functions. Sacral spinal reflexes are disrupted after SCI in man, and bladder-sphincter dyssynergia and problems with defecation can

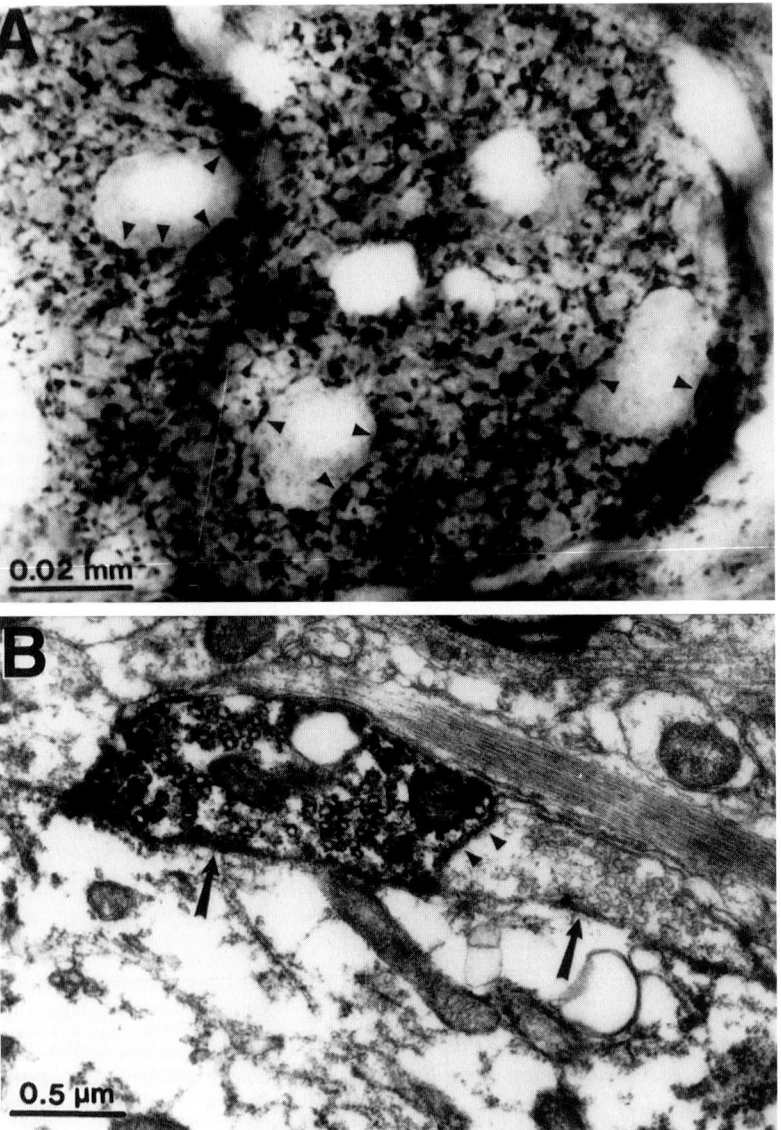

Fig. 4. GABAergic inputs to pudendal motoneurons in the cat sacral spinal cord. (A) Light micrograph of a GAD immunostained 60-μm-thick plastic embedded section showing immunostained terminal elements densely aggregated around the motoneuron cell bodies (arrowheads) and dendrites in Onuf's nucleus. (B) At the electron microscopic level, immunostained terminals were observed to be in synaptic contact with the motoneuron cell bodies (long arrow) as well as with adjacent presynaptic elements (arrowheads).

be severe complications that require constant attention. These reflexes therefore serve both as a model for synaptic reorganization and as a target for intervention (e.g., De Groat et al., 1998; Holmes et al., 1998a,b; Yoshiyama et al., 1999). In our model of SCI in the rat, for example, the external anal sphincter muscle becomes spastic after spinal cord transection and exhibits no recovery (Holmes et al., 1998a). After a contusion injury, however, the spasticity is reduced over time, concurrently with partial recoveries of other reflexes and locomotion (Holmes et al., 1998b). If we knew the synaptic organization of sacral reflexes, and their alteration after SCI, it might be possible to use this information to predict useful pharmacological treatment strategies.

Some of the most dramatic evidence for synaptic plasticity of the inputs to pudendal motoneurons comes from studies of hormonal regulation of sexual reflexes in rats. Leedy et al. (1987) and Matsumoto et al. (1988) showed that the number of synapses on motoneurons innervating the bulbocavernosus (BC) muscle was reduced by testosterone deprivation (castration), but that these synapses reappeared within two days after testosterone replacement. These changes were correlated with the loss and return of penile reflex behaviors mediated by the BC muscle. Electron microscopic (EM) examination of the motoneurons suggested that glial elements were involved in the removal of synapses, as had been shown for hypothalamic neurons whose inputs were changed by dehydration (Hatton, 1997) and for phrenic neurons after spinal cord lesions (Goshgarian et al., 1989). Interestingly, castration does not seem to radically alter the activation of microglia surrounding pudendal motoneurons, although section of the pudendal nerve does produce a mild activation (Brown et al., 1995). For a recent review of the potential role of microglia in synaptic plasticity, see Aldskogius et al. (1999).

We also showed that there appeared to be alterations in synaptic inputs to sacral motoneurons in the cat after spinal cord transection, and these changes could be observed by four days after injury (Beattie et al., 1993). Qun Li (Li et al., 1995b) followed these observations with EM-immunocytochemical studies of synaptic inputs to the sacral motoneurons in Onuf's nucleus and parasympathetic preganglionic neurons after spinal cord hemisection in

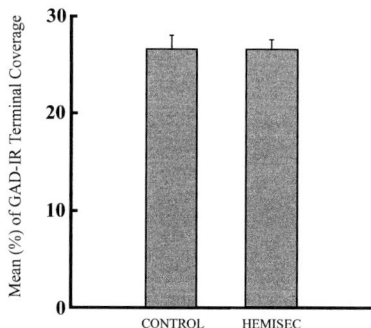

A. Two Days Post - SCI (n=4; p>0.05)

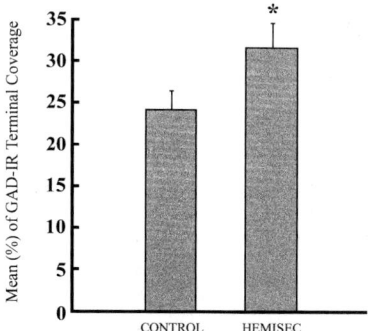

B. Six Weeks Post - SCI (n=4; p<0.05)

Fig. 5. Graphs indicating an increase in the proportion of the Onuf's nucleus motoneuron surface covered by GAD immunoreactive elements from two days (A) to six weeks (B) after a spinal cord hemisection in the cat. Measurements from motoneurons on the side contralateral to the hemisection are indicated on the graphs as the control. Data from four cases (25 cells per side per case) are shown. P values are for paired t-tests.

the cat. Antibodies to γ-aminobutyric acid (GABA) or glutamic acid decarboxylase (GAD) were used to label a subset of synaptic terminals apposed to Onuf's nucleus motoneurons. There is a substantial GABAergic input to these motoneurons that is almost twice as dense as the inputs onto other, more 'classic' motoneurons in the lumbosacral cord (Li et al., 1995b, and in preparation). Fig. 4A shows a field of motoneurons within Onuf's nucleus, surrounded by dark, GABA-positive boutons. These are shown to be synaptic terminals in contact with the motoneurons (Fig. 4B). Measurements of the two-

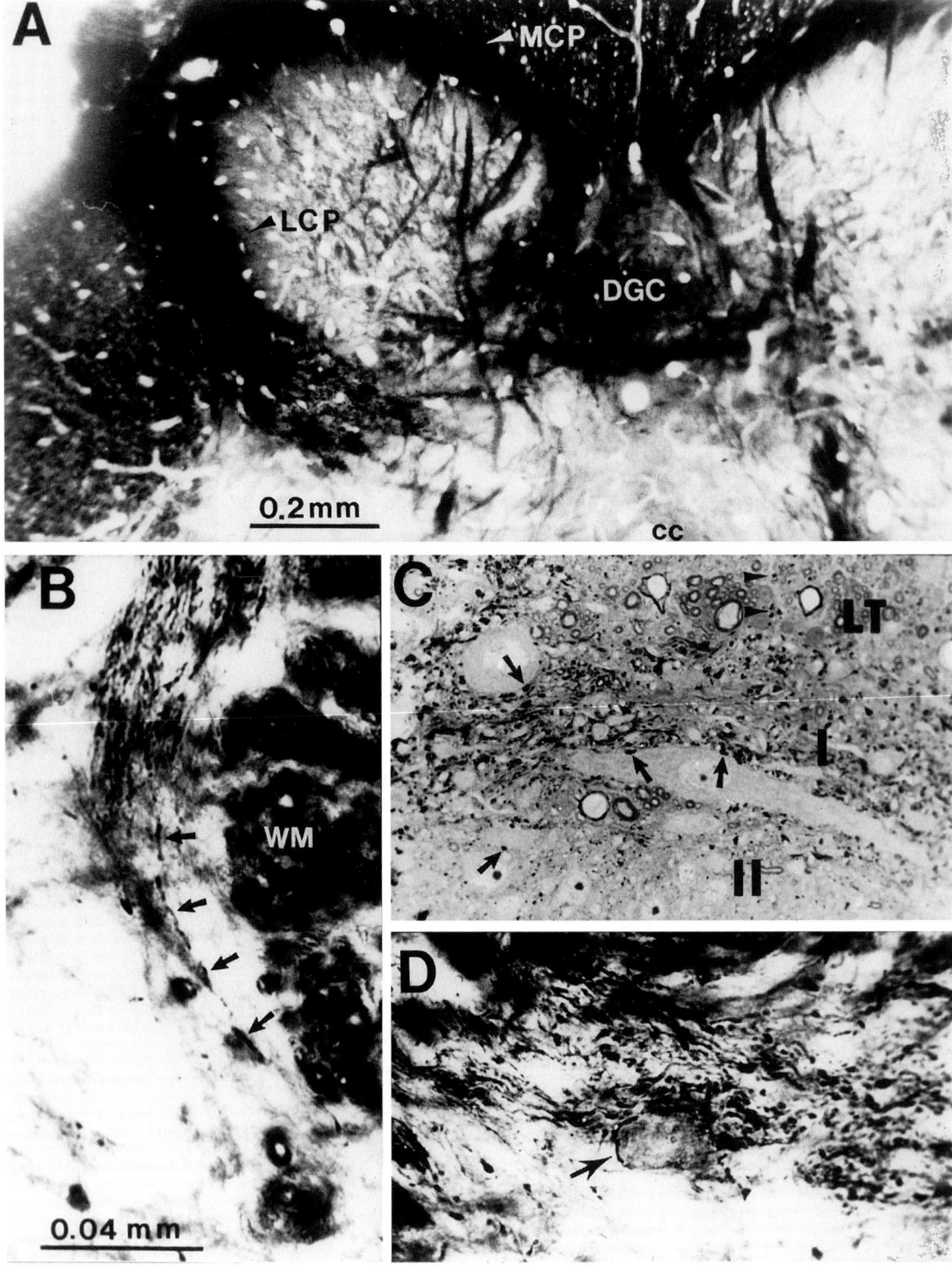

dimensional coverage of motoneuron profiles were made in plastic sections and also in EM montages in cats that had received thoracic spinal cord hemisections two days or six weeks prior to sacrifice. Fig. 5 shows the mean GABAergic coverage of Onuf's motoneurons on the intact and hemisected sides of the sacral cord in acute and chronic cases. There is an increase in the % terminal coverage by GABA terminals on the hemisected side, suggesting that this class of inputs to the motoneurons has reorganized as a consequence of removing descending fibers. This is similar to the reorganization of GABA inputs seen in the phrenic nucleus in the rostral spinal cord after injury (Tai and Goshgarian, 1996).

In order to examine the potential for sprouting of dorsal root afferents involved in sacral reflexes, we used antibodies to calcitonin gene-related peptide (CGRP; Li et al., 1996b). CGRP is contained in a subset of small- and medium-diameter primary afferent fibers (Traub et al., 1989). A 'lateral collateral pathway' (Morgan et al., 1981; Mawe et al., 1986) enters the region of the sacral parasympathetic nucleus (SPN) in both the cat and rat. Many of these fibers contain CGRP. CGRP is also contained in dorsal root afferents that innervate the sympathetic neurons within the intermediolateral cell column (Krenz and Weaver, 1998). Since CGRP fibers innervate the visceral and somatic periphery involved in sacral reflexes controlling bladder, bowel and sexual function, changes in the numbers or central targets of these fibers and their terminals could be responsible in part for the reflex changes seen after SCI.

Fig. 6A shows the lateral collateral pathway containing CGRP in the cat. Fig. 6B and D show the termination of a CGRP fiber onto a neuron located within the SPN. We examined the numbers of contacts between CGRP-positive fibers and neurons within the SPN in cats after acute (two days) or chronic (six weeks) spinal cord hemisections, com-

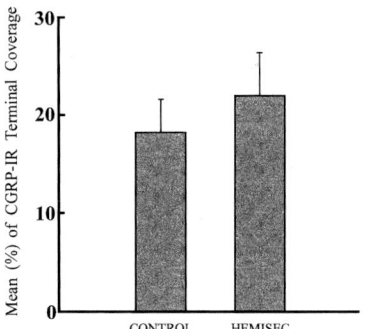

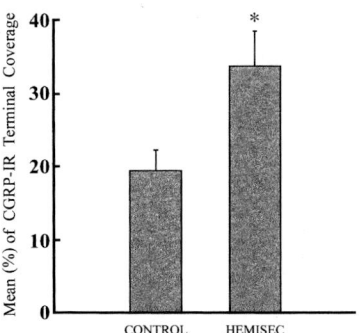

Fig. 7. Graphs indicating an increase in the proportion of the sacral parasympathetic neuron surface covered by CGRP immunoreactive elements from two days (A) to six weeks (B) after a spinal cord hemisection in the cat. Measurements from neurons on the side contralateral to the hemisection are indicated on the graphs as the control. Data shown are from four cases (25 cells per side per case). P values are for paired t-tests.

paring the hemisected to the intact side. As shown in Fig. 7, there was a significant increase in number of contacts on the side of spinal hemisection, most prominently at six weeks after injury. This suggests

Fig. 6. CGRP-immunostaining in the sacral spinal cord of the cat. (A) Dense input to the region of the sacral parasympathetic nucleus (SPN) at the base of the dorsal horn is provided by the lateral collateral pathway (*LCP*). Also shown is the medial collateral pathway (*MCP*) distributing to the region of the dorsal commissural gray (*DGC*). 60-μm-thick plastic section immunostained for CGRP. (B,D) Higher-power photomicrographs of the CGRP inputs to SPN neurons from similar sections. Arrows indicate CGRP immunoreactive axoms. (C) 1-μm-thick plastic section immunostained for CGRP and counterstained with toluidine blue from the area of the LCP indicated in (A). Arrows indicate CGRP immunoreactive axons. *I* and *II* indicate the position of Rexed's laminae I and II. *LT* indicates the position of Lissauer's tract.

that there may be substantial sprouting of CGRP afferents into this region of the cord. While it is difficult to relate these purely anatomical results to possible functional consequences, the presence of apparent sprouting does provide for the generation of more specific hypotheses about the role of different neurotransmitters in reflex alterations. Strategies can then be based on pharmacologic or other means directed to specific circuit alterations. For example, it has recently been shown that CGRP fibers appear to sprout within the rat intermediolateral cell column (Krenz and Weaver, 1998) and that this may be related to the development of autonomic dysreflexia in cardiovascular reflexes. The sprouting may be related to NGF; indeed, blockade of NGF action with intrathecal application of antibodies to NGF has been reported to reduce the sprouting and concomitantly reduce the autonomic dysreflexia (Krenz et al., 1999).

Changes in CGRP fibers also indicate that sensory systems undergo alterations after SCI and contribute to the functional outcomes. Such changes can be observed in alterations of the afferent limb of reflex systems as well as in altered inputs to systems regulating the conscious perception of pain, temperature, touch, and proprioception. This factor is especially important in incomplete lesions as modeled by the contusion lesions which spare some ascending fibers. Thus denervation and altered balance of inputs can affect not only the motor systems by producing spasticity but can affect sensation by producing allodynia and anomalous pain after SCI (Christensen et al., 1996; Yezierski, 1996; Lindsey et al., 1998).

These findings support the now rather old contention (Liu and Chambers, 1958) that collateral axonal sprouting is involved in the reorganization of spinal reflexes after injury. With the tools of modern molecular biology available, this work can be extended to understanding more about the plasticity within synapses themselves, and receptor systems associated with changes in cord function. Studies of reflex changes and concomitant changes in spinal circuitry after injury are likely to provide useful information that will translate into effective strategies for treating the consequences of SCI in man.

Conclusion

Spinal cord contusion injury initiates a complex series of interrelated biological responses that lead to secondary expansion of the lesion at its center, loss of supporting cells along injured white matter tracts, and loss of synaptic inputs to remote regions of the CNS. These degenerative events are accompanied by (and may initiate) reparative responses that include cellular proliferation at the lesion site and the reorganization of synaptic inputs in both spinal and brain circuits. The result of this complex interplay of death and repair can be seen by following the course of recovery from partial contusion lesions. Locomotor ability (Basso et al., 1995, 1996) as well as sacral reflex functions (Holmes et al., 1998b) slowly return toward normal, with the degree of recovery dependent upon the severity of injury. The models that have been developed in many laboratories over the past years to study this recovery can be used to study the parallel roles of cellular mechanisms, synaptic reorganization and plasticity, and learning and rehabilitation in the recovery process.

References

Aldskogius, H., Liu, L. and Svensson, M. (1999) Glial responses to synaptic damage and plasticity. *J. Neurosci. Res.*, 58: 33–41.

Allen, A.R. (1911) Surgery of experimental lesion of spinal cord equivalent to crush injury or fracture-dislocation of spinal column: a preliminary report. *J.A.M.A.*, 57: 870–880.

Barres, B.A., Jacobson, M.D., Schmid, R., Sendtner, M. and Raff, M.C. (1993) Does oligodendrocyte survival depend on axons?. *Curr. Biol.*, 3: 489–497.

Basso, D.M., Beattie, M.S. and Bresnahan, J.C. (1995) A new sensitive locomotor rating scale for locomotor recovery after spinal cord contusion injuries in rats. *J. Neurotrauma*, 12: 1–21.

Basso, D.M., Beattie, M.S. and Bresnahan, J.C. (1996) Histological and locomotor studies of graded spinal cord contusion using the NYU weight-drop device versus transection. *Exp. Neurol.*, 139: 224–256.

Beattie, M.S. and Bresnahan, J.C. (2000) Cell death, repair and recovery of function after spinal cord injury in rats. In: Kalb, R.G. and Strittmatter, S.M. (Eds.), *Neurobiology of Spinal Cord Injury*. Humana Press, Totowa, NJ, pp. 1–21.

Beattie, M.S., Stokes, B.T. and Bresnahan, J.C. (1988) Experimental spinal cord injury: strategies for acute and chronic intervention based on anatomic, physiologic, and behavioral studies. In: Stein, D. and Sabel, B. (Eds.), *Pharmacological*

Approaches to the Treatment of Brain and Spinal Cord Injury. Plenum Press, New York, pp. 43–74.

Beattie, M.S., Lopate, G. and Bresnahan, J.C. (1990) Metamorphosis alters the response to spinal cord transection in *Xenopus laevis* frogs. *J. Neurobiol.*, 21: 1108–1122.

Beattie, M.S., Leedy, M.G. and Bresnahan, J.C. (1993) Evidence for alterations of synaptic inputs to sacral spinal reflex circuits after spinal cord transection in the cat. *Exp. Neurol.*, 123: 35–50.

Beattie, M.S., Bresnahan, J.C., Koman, J., Tovar, C.A., Van Meter, M., Anderson, D.K., Faden, A.I., Hsu, C.Y., Noble, L.J., Salzman, S. and Young, W. (1997) Endogenous repair after spinal cord contusion injuries in the rat. *Exp. Neurol.*, 148: 453–463.

Beattie, M.S., Shuman, S.L. and Bresnahan, J.C. (1998) Apoptosis and spinal cord injury. *Neuroscientist*, 4: 163–171.

Beattie, M.S., Boyce, S.L., Yoon, S.O., Longo, F.M., Yeo, T.T. and Bresnahan, J.C. (1999) Expression of p75 NTFR in neurons and oligodendrocytes after spinal cord injury in rats: neurotrophin-induced apoptosis. *Soc. Neurosci. Abstr.*, 25: 762.

Bethea, J.R., Nagashima, H., Acosta, M.C., Briceno, C., Gomez, F., Marcillo, A.E., Loor, K., Green, J. and Dietrich, W.D. (1999) Systemically administered interleukin-10 reduces tumor necrosis factor-alpha production and significantly improves functional recovery following traumatic spinal cord injury in rats. *J. Neurotrauma*, 16: 851–863.

Blight, A.R. (1992) Macrophages and inflammatory damage in spinal cord injury. *J. Neurotrauma*, 9(suppl.1): S83–S92.

Blight, A.R. and Decrescito, V. (1986) Morphometric analysis of experimental spinal cord injury in the cat: the relation of injury intensity to survival of myelinated axons. *Neuroscience*, 19: 321–341.

Bresnahan, J.C., Beattie, M.S., Stokes, B.T. and Conway, K.M. (1991) Three-dimensional computer-assisted analysis studies of graded contusion lesions in spinal cord of the rat. *J. Neurotrauma*, 8: 91–101.

Brown, T.J., Bresnahan, J.C. and Beattie, M.S. (1995) Changes in the expression of rat microglial immunomolecules after unilateral pudendal nerve section or castration. *Soc. Neurosci. Abstr.*, 21: 1201.

Bruce, A.J., Boling, W., Kindy, M.S., Peschon, J., Kraemer, P.J., Carpenter, M.K., Holtsberg, F.W. and Mattson, M.P. (1996) Altered neuronal and microglial responses to excitotoxic and ischemic brain injury in mice lacking TNF receptors. *Nat. Med.*, 2: 788–794.

Casaccia-Bonnefil, P., Carter, B.D., Dobrowsky, R.T. and Chao, M.V. (1996) Death of oligodendrocytes mediated by the interaction of nerve growth factor with its receptor p75. *Nature*, 383: 716–719.

Chao, M., Casaccia-Bonnefil, P., Carter, B., Chittka, A., Kong, H.Y. and Yoon, S.O. (1998) Neurotrophin receptors: mediators of life and death. *Brain Res. Rev.*, 26: 295–301.

Chen, X.Y., Wolpaw, J.R., Jakeman, L.B. and Stokes, B.T. (1999) Operant conditioning of H-reflex increase in spinal cord-injured rats. *J. Neurotrauma*, 16: 175–186.

Christensen, M., Everhart, A., Pickelman, J. and Hulsebosch, C. (1996) Mechanical and thermal allodynia in chronic central pain following spinal cord injury. *Pain*, 68: 97–107.

Crowe, M.J., Bresnahan, J.C., Shuman, S.L., Masters, J.N. and Beattie, M.S. (1997) Apoptosis and delayed degeneration after spinal cord injury in rats and monkeys. *Nat. Med.*, 3: 73–76.

De Groat, W.C., Araki, I., Vizzard, M.A., Yoshiyama, M., Yoshimura, N., Sugaya, K., Tai, C.F. and Roppolo, J.R. (1998) Developmental and injury induced plasticity in the micturition reflex pathway. *Behav. Brain Res.*, 92: 127–140.

Deshmukh, M. and Johnson, E.M. (1997) Programmed cell death in neurons: focus on the pathway of nerve growth factor deprivation-induced death of sympathetic neurons. *Mol. Pharmacol.*, 51: 897–906.

Emery, E., Aldana, P., Bunge, M.B., Puckett, W., Srinivasan, A., Kean, R.W., Bethea, J. and Levi, A. (1998) Apoptosis after traumatic human spinal cord injury. *J. Neurosurg.*, 89: 911–920.

Goldberger, M.E. and Murray, M. (1988) Patterns of sprouting and implication for recovery of function. In: Waxman, S.G. (Ed.), *Advances in Neurology, Vol. 46*. Raven Press, New York, pp. 361–385.

Goldberger, M.E., Murray, M., Tessler, A. (1993) Sprouting and regeneration in the spinal cord: their roles in recovery of function after spinal injury. In: Gorio, A. (Ed.), Neuroregeneration. Raven Press, New York, pp. 241–264.

Goshgarian, H.G., Yu, X.-J. and Rafols, J. (1989) Neuronal and glial changes in the rat phrenic nucleus occurring within hours after spinal cord injury. *J. Comp. Neurol.*, 284: 902–921.

Grill, R., Blesch, A. and Tuszynski, M.H. (1997a) Robust growth of chronically injured spinal cord axons induced by grafts of genetically modified NGF secreting cells. *Exp. Neurol.*, 148: 444–452.

Grill, R., Murai, K., Blesch, A., Gage, F.H. and Tuszynski, M.H. (1997b) Cellular delivery of neurotrophin-3 promotes corticospinal axonal growth and partial functional recovery after spinal cord injury. *J. Neurosci.*, 17: 5560–5572.

Guizar-Sahagun, G., Grijalva, O., Madrazo, E., Oliva, E. and Zepeda, A. (1994) Development of post-traumatic cysts in the spinal cord of rats subjected to severe spinal cord contusion. *Surg. Neurol.*, 41: 241–249.

Hatton, G.I. (1997) Function-related plasticity in hypothalamus. *Annu. Rev. Neurosci.*, 20: 375–397.

Hill, C.E., Hermann, G.E., Beattie, M.S. and Bresnahan, J.C. (1998) Cortical and brainstem axons grow into lesion cavities after spinal cord contusion injury in the adult rat. *Soc. Neurosci. Abstr.*, 24: 1729.

Holmes, G.M., Rogers, R.C., Bresnahan, J.C. and Beattie, M.S. (1998a) External anal sphincter hyper-reflexia following spinal transection in the rat. *J. Neurotrauma*, 15: 451–457.

Holmes, G.M., Bresnahan, J.C., Stephens, R.L., Rogers, R.C. and Beattie, M.S. (1998b) Comparison of locomotor and pudendal reflex recovery in chronic spinally contused rats. *J. Neurotrauma*, 15: 874.

Jain, N., Catania, K.C. and Kaas, J.H. (1997) Deactivation and reactivation of somatosensory cortex after dorsal spinal cord injury. *Nature*, 386: 495–498.

Krenz, N.R. and Weaver, L.C. (1998) Sprouting of primary afferent fibers after spinal cord transection in the rat. *Neuroscience*, 85: 443–458.

Krenz, N.R. and Weaver, L.C. (2000) Nerve growth factor in glia and inflammatory cells of the injured rat spinal cord. *J. Neurochem.*, 74: 730–739.

Krenz, N.R., Meakin, S.O., Krassioukov, A.V. and Weaver, L.C. (1999) Neutralizing intraspinal nerve growth factor blocks autonomic dysreflexia caused by spinal cord injury. *J. Neurosci.*, 19: 7405–7414.

Leedy, M.G., Beattie, M.S. and Bresnahan, J.C. (1987) Testosterone-induced plasticity of synaptic inputs to adult mammalian motoneurons. *Brain Res.*, 424: 386–390.

Li, Y., Sharov, V.G., Jiang, N., Zaloga, C., Sabbah, H.N. and Chopp, M. (1995a) Ultrastructural and light microscopic evidence of apoptosis after middle cerebral artery occlusion in the rat. *Am. J. Pathol.*, 146: 1045–1051.

Li, Q., Beattie, M.S. and Bresnahan, J.C. (1995b) Onuf's nucleus (ON) motoneurons (MNs) have more GABA-ergic synapses than other somatic MNs: a quantitative immunocytochemical study in the cat. *Soc. Neurosci. Abstr.*, 21: 1201.

Li, G.L., Brodin, G., Farooque, M., Funa, K., Holtz, A., Wang, W.L. and Olsson, Y. (1996a) Apoptosis and expression of Bcl-2 after compression trauma to rat spinal cord. *J. Neuropathol. Exp. Neurol.*, 55: 280–289.

Li, Q., Beattie, M.S. and Bresnahan, J.C. (1996b) Ultrastructure of CGRP terminals in cat sacral spinal cord with evidence for sprouting in the sacral parasympathetic nucleus (SPN) after spinal hemisection. *Soc. Neurosci. Abstr.*, 22: 1843.

Lindsey, A.E., LoVerso, R.L., Tovar, C.A., Beattie, M.S. and Bresnahan, J.C. (1998) Mechanical allodynia and thermal hyperalgesia in rats with contusion spinal cord injury. *J. Neurotrauma*, 15: 880.

Liu, C.N. and Chambers, W.W. (1958) Intraspinal sprouting of dorsal root axons. *Arch. Neurol. Psychiatry*, 79: 46–61.

Liu, X.Z., Xu, X.M., Hu, R., Du, C., Zhang, S.X., McDonald, J.W., Dong, H.X., Wu, Y.J., Fau, G.S., Jacquin, M.F., Hsu, C.Y. and Choi, D.W. (1997) Neuronal and glial apoptosis after traumatic spinal cord injury. *J. Neurosci.*, 17: 5395–5406.

Matsumoto, A., Micevych, P. and Arnold, A. (1988) Androgen regulates synaptic input to motoneurons of adult rat spinal cord. *J. Neurosci.*, 8: 4168–4176.

Matthews, M.A., St. Onge, M.F. and Faciane, C.L. (1979) An electron microscopic analysis of abnormal ependymal cell proliferation and envelopment of sprouting axons following spinal cord transection in the rat. *Acta Neuropathol.*, 45: 27–36.

Mawe, G.M., Bresnahan, J.C. and Beattie, M.S. (1986) A light and electron microscopic analysis of the sacral parasympathetic nucleus after labelling primary afferent and efferent elements with HRP. *J. Comp. Neurol.*, 250: 33–57.

McVeigh, J.F. (1923) Experimental cord crushes — with especial reference to the mechanical factors involved and subsequent changes in the areas of the cord affected. *Arch. Surg. (Chic.)*, 7: 573–600.

Michel, M.E. and Reier, P.J. (1979) Axonal–ependymal associations during early regeneration of the transected spinal cord in *Xenopus laevis* tadpoles. *J. Neurocytol.*, 8: 529–548.

Morgan, C., Nadelhaft, I. and De Groat, W.C. (1981) The distribution of visceral primary afferents from the pelvic nerve to Lissauer's tract and the spinal gray matter and its relationship to the sacral parasympathetic nucleus. *J. Comp. Neurol.*, 201: 415–440.

Noble, L. and Wrathall, J. (1985) Spinal cord contusion in the rat: morphometric analyses of alterations in the spinal cord. *Exp. Neurol.*, 88: 135–149.

Pettmann, B. and Henderson, C.E. (1998) Neuronal cell death. *Neuron*, 20: 633–647.

Popovich, P.G., Wei, P. and Stokes, B.T. (1997) Cellular inflammatory response after spinal cord injury in Sprague–Dawley and Lewis rats. *J. Comp. Neurol.*, 377: 443–464.

Raff, M.C., Barres, B.A., Burne, J.F., Coles, H.S., Ishizaki, Y. and Jacobson, M.D. (1993) Programmed cell death and the control of cell survival: lessons from the nervous system. *Science*, 262: 695–700.

Rink, A., Fung, K.-M., Trojanowski, J.Q., Lee, V.M.-Y., Neugebauer, E. and McIntosh, T.K. (1995) Evidence of apoptotic cell death after experimental traumatic brain injury in the rat. *Am. J. Pathol.*, 147: 1575–1583.

Roy, R.R., Talmadge, R.J., Hodgson, J.A., Oishi, Y., Baldwin, K.M. and Edgerton, V.R. (1999) Differential response of fast hindlimb extensor and flexor muscles to exercise in adult spinalized cats. *Muscle Nerve*, 22: 230–241.

Shuman, S.L., Bresnahan, J.C. and Beattie, M.S. (1997) Apoptosis of microglia and oligodendrocytes after spinal cord injury in rats. *J. Neurosci. Res.*, 50: 798–808.

Springer, J.E., Azbill, R.D. and Knapp, P.E. (1999) Activation of the caspase-3 apoptotic cascade in traumatic spinal cord injury. *Nat. Med.*, 5: 943–946.

Steers, W.D., Ciambotti, J., Erdman, S. and De Groat, W.C. (1990) Morphological plasticity in efferent pathways to the urinary bladder of the rat following urethral obstruction. *J. Neurosci.*, 10: 1943–1951.

Tai, Q. and Goshgarian, H.G. (1996) Ultrastructural quantitative analysis of glutamatergic and GABAergic synaptic terminals in the phrenic nucleus after spinal cord injury. *J. Comp. Neurol.*, 372: 343–355.

Thallmair, M., Metz, G.A.S., Z'Graggen, W.J., Raineteau, O., Kartje, G.L. and Schwab, M.E. (1998) Neurite growth inhibitors restrict plasticity and functional recovery following corticospinal tract lesions. *Nat. Neurosci.*, 1: 124–131.

Traub, R.J., Solodkin, A. and Ruda, M.A. (1989) Calcitonin gene-related peptide immunoreactivity in the cat lumbosacral spinal cord and the effects of multiple dorsal rhizotomies. *J. Comp. Neurol.*, 287: 225–237.

Ueda, H., Levine, J.M., Miller, R.H. and Trapp, B.D. (1999) Rat optic nerve oligodendrocytes develop in the absence of viable retinal ganglion cell axons. *J. Cell Biol.*, 146: 1365–1374.

Vaquero, J., Ramiro, M., Oya, S. and Cavezudo, J.M. (1981) Ependymal reaction after experimental spinal cord injury. *Acta Neurochir.*, 55: 295–302.

Yakovlev, A.G., Knoblach, S.M., Fan, L., Fox, G.B., Goodnight, R. and Faden, A.I. (1997) Activation of CPP32-like caspases contributes to neuronal apoptosis and neurological dysfunction after traumatic brain injury. *J. Neurosci.*, 17: 7415–7424.

Yezierski, R. (1996) Pain following spinal cord injury: the clinical problem and experimental studies. *Pain*, 68: 185–194.

Yoon, S.O., Casaccia-Bonnefil, P., Carter, B. and Chao, M. (1998) Competitive signaling between TrkA and p75 nerve growth factor receptors determines cell survival. *J. Neurosci.*, 18: 3273–3281.

Yoshiyama, M., Nezu, F.M., Yokoyama, O., Chancellor, M.B. and De Groat, W.C. (1999) Influence of glutamate receptor antagonists on micturition in rats with spinal cord injury. *Exp. Neurol.*, 159: 250–257.

CHAPTER 3

The multi-domain structure of extracellular matrix molecules: implications for nervous system regeneration

Sally Meiners, Mary Lynn T. Mercado and Herbert M. Geller *

Department of Pharmacology, UMDNJ–Robert Wood Johnson Medical School, 675 Hoes Lane, Piscataway, NJ 08854, USA

Introduction

From the early investigations of Ramón y Cajal (1928), most recently confirmed by Davies et al. (1999), it has become clear that axons can grow for some distance in the uninjured mammalian central nervous system, but axonal growth after a lesion is ultimately thwarted by the glial scar (Reier and Houle, 1988). While the mechanical disruption of cut fiber tracts combined with the presence of hypertrophied astrocytes were thought to be the primary impediments to regeneration, more recent work has revealed that orderly regrowth may be influenced primarily by molecules within the glial scar, even when mechanical disruption is minimal (Davies et al., 1996). Thus, it is likely that molecules in the region of the injury, rather than a physical scar, signal neurons to stop growing (an effect on axonal outgrowth) or to alter their direction of growth (an effect on axonal guidance).

The astrocyte-derived molecules most strongly implicated in directing axonal growth after injury are found in the extracellular matrix of the glial scar (Stichel and Muller, 1998). The two that are invariably upregulated in astrocytes following central nervous system injury are tenascin-C (Brodkey et al., 1995) and several different chondroitin sulfate proteoglycans (CSPGs) (McKeon et al., 1995); these molecules are closely associated with laminins to form an extracellular matrix (McKeon et al., 1995; McKerracher et al., 1996). The localization of these extracellular matrix molecules within areas of brain or spinal cord injury implies that they are involved in the regulation of axonal regrowth following injury. While laminins are considered to be promotional for neuronal growth (Lein et al., 1992), and CSPGs are considered inhibitory (Pindzola et al., 1993; Fernaud-Espinosa et al., 1998; Zuo et al., 1998), the function of tenascin-C after injury is more controversial, with some work indicating that it is inhibitory and other work suggesting that it is promotional (Gates et al., 1996; Zhang et al., 1997).

Extracellular matrix molecules are characterized by a multi-domain structure. As such, we attribute their specific, and in the case of tenascin-C, sometimes contradictory roles to the biological actions of individual domains, as discussed below. Tenascin-C, for example, is comprised of six identical arms; each arm consists of 14 epidermal growth factor (EGF) domains, 8–15 fibronectin type III (FN-III) domains depending on species and alternative RNA splicing, and a single fibrinogen domain (Fig. 1). Laminins are heterotrimeric glycoproteins composed of three chains: an α chain, a β chain and a γ chain arranged in a cross-structure. The structure of laminin-1, comprised of an α1, β1 and γ1 chain, is shown in Fig. 2. Similarly, many other extracellular matrix molecules such as thrombospondin, F-spondin,

---

* Corresponding author: Dr. Herbert M. Geller, Department of Pharmacology, UMDNJ–Robert Wood Johnson Medical School, 675 Hoes Lane, Piscataway, NJ 08854, USA; Fax: +1-732-235-4073; E-mail: geller@umdnj.edu

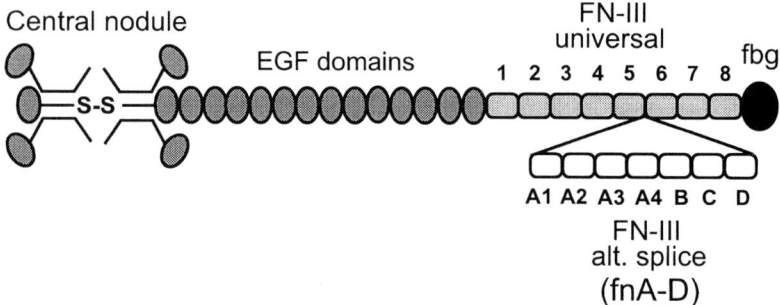

Fig. 1. Multi-domain structure of human tenascin-C. The amino termini of three arms are joined to form a trimer, and two trimers are connected to form a hexamer. The universal fibronectin type-III (FN-III) domains 1–5 and 6–8 (fn1–5 and fn6–8) are present in all tenascin-C splice variants. The large splice variant contains seven alternatively spliced FN-III domains (designated *A1, A2, A3, A4, B, C,* and *D*, or fnA–D) which are missing in the smallest splice variant. (From Meiners et al., 1996.)

CSPGs, fibronectin, etc., are multi-domain proteins composed of modular mosaics comprised of several smaller domains (Engel, 1996). The active domains of these molecules interact in concert with receptors or other matrix molecules to cause the net biological action of the parent molecule.

Domains or regions of extracellular matrix molecules can retain particular interactions or may assume altered conformations that result in biological activities that differ from the parent molecule. Given that the expression of certain proteases is increased in the central nervous system following injury (Turgeon and Houenou, 1997; Yuguchi et al., 1997), we surmise that fragments of extracellular matrix proteins may be generated which then impact on neuronal regeneration along with the parent molecule, as has been suggested for laminin-1 (Chen and Strickland, 1997). For this reason, we have initiated studies on the regulation of neuronal growth by laminin-1, tenascin-C, and their smaller fragments and domains.

Laminin-1 and neuronal growth

Laminins are a family of molecules that mediate many different cellular processes such as cell adhesion, migration, proliferation, differentiation, and cell survival (Malinda and Kleinman, 1996). Laminins are one of the major components of the basement membrane, and they are expressed in discrete areas of the nervous system during development and after injury (Schittny and Yurchenco, 1989). While several different laminins composed of various α, β, and γ chains have been identified, laminin-1 (α1–β1–γ1) is the major species in the central nervous system (Luckenbill-Edds, 1997). There, laminin-1 has been shown to promote neurite formation, augment neurite outgrowth, mediate neuronal adhesion, promote neuronal survival, and guide pathfinding axons or migrating cell bodies (Ernsberger et al., 1989). Recent immunocytochemical

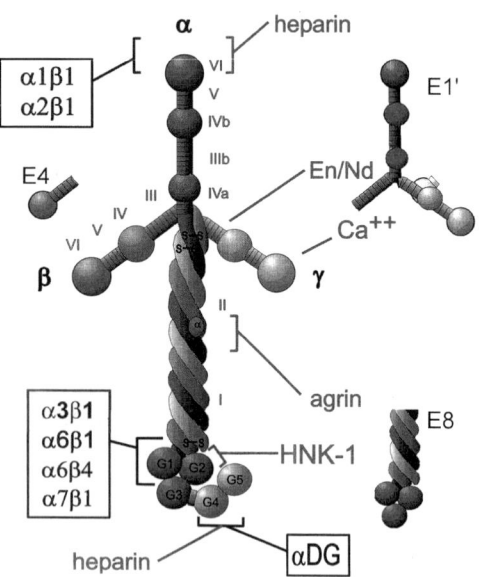

Fig. 2. Diagram of laminin-1 structure and binding sites. The structure of laminin-1 is depicted, along with its enolase-digested fragments. The E1′ fragment contains the α1β1 and α2β1 integrin binding sites, while the E8 fragment contains the α3β1, α6β1, α6β4 and α7β1 binding sites. The E4 fragment has no defined integrin binding sites. (From Colognato-Pyke et al., 1995.)

experiments have suggested that laminin-2 (merosin) ($\alpha 2$–$\beta 1$–$\gamma 1$) may function in the central nervous system as well (Hagg et al., 1997). However, its predominant expression is in the peripheral nervous system, where it has an influence on nerve growth and Schwann cell migration (Anton et al., 1994). Therefore, this manuscript will focus on the known properties of laminin-1 that may apply to studies of central nervous system regeneration.

Laminin regions and neuronal growth

The ability of intact laminin-1 to influence various aspects of neuronal growth may be ascribed to diverse functions of different laminin-1 regions, as demonstrated by studies using enolase-digested fragments of laminin-1 and function-blocking antibodies directed against specific regions of laminin-1. For instance, when olfactory epithelial explants were plated on native laminin-1 or the proteolytic fragments E1' (the short arms of the $\alpha 1$, $\beta 1$, and $\gamma 1$ chains), E4 (domains V and VI of the $\beta 1$ chain), E3 (the globular domains LG4 and LG5 of the $\alpha 1$ chain), or E8 (the coiled-coil domain of laminin-1) (see Fig. 2), only native laminin-1 and E8 elicited neuronal migration (Calof et al., 1994). These results indicate that E8 is responsible for neuronal migration on laminin-1. The E8 fragment also mimicked laminin-1's induction of aggregation, proliferation, and neuritic sprouting by murine neuroepithelial cells (Drago et al., 1991). Powell et al. (1998) provided further evidence that the neurite outgrowth-promoting activity maps to E8 by demonstrating that cerebellar granule cell outgrowth on E8 is equal to that on laminin-1. Blocking antibodies against the E1' domain enhanced olfactory epithelial neuronal migration and neurite outgrowth on laminin-1. Therefore, migration and outgrowth-suppressing activities were mapped to the E1' fragment. A further study by Ivins et al. (1998) showed that adding a blocking antibody against the N-terminus of the $\alpha 1$ chain induced neuronal attachment and neurite outgrowth of late embryonic retinal neurons on a laminin substrate. This indicated that laminin-1 could be 'activated' by obstructing inhibitory regions in the $\alpha 1$ chain.

Studies in our laboratory have demonstrated that the E1' fragment of laminin-1 behaves differently than laminin-1 in the regulation of neuronal growth. For example, neurite outgrowth is impaired on the E1' fragment but enhanced on laminin-1. Furthermore, E1' and laminin-1 differ in their ability to alter the trajectory of growing neurites in culture. The experimental paradigm (described in Meiners et al., 1999a,b) is to evaluate the behavior of neurites on either side of an interface between poly-l-lysine (PLL) and laminin-1 or its E1', E4, or E8 fragments. Initial time lapse studies demonstrate that neurites on the PLL side of a PLL/laminin-1 or a PLL/E8 interface have no particular preference for either substrate (data not shown). However, neurites on the laminin-1 (Fig. 3) or E8 side of the interface turn to avoid PLL, choosing to remain on the more permissive laminin-1 substrate. This indicates that

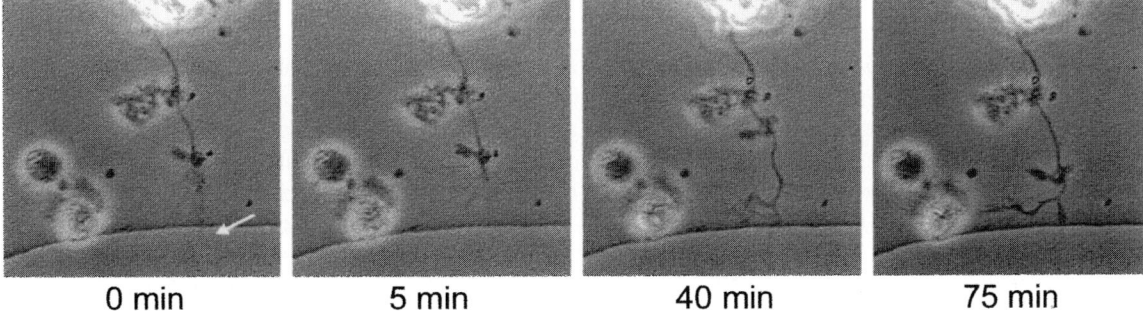

Fig. 3. Dynamics of growth cones at a boundary. Time lapse images of a rat cerebellar granule neuronal growth cone as it grows from laminin-1 towards PLL. 0 min: at the beginning of the recording, the growth cone is fan-shaped with long filopodia, several of which sample PLL (arrow). 5 min: the growth cone has become more club-shaped and withdraws from the interface. 40 min: after touching PLL, the growth cone becomes more elongated and turns away from the interface. 75 min: the growth cone completes its turn away from the interface. (Excerpted from a movie located at http://www2.umdnj.edu/~geller/lab/boundmov.htm.)

neurite behavior on one side of an interface cannot necessarily predict neurite behavior on the other side. Neurites at an interface between E4 and PLL exhibit random movement, suggesting an absence of guidance cues from E4. On the other hand, neurites on the PLL side of a PLL/E1′ interface avoid E1′, regarding it as a non-permissive substrate, whereas neurites on the E1′ side of the interface have no preference for either substrate. Hence, laminin-1 and its E1′ fragment provide widely divergent guidance cues to growing neurites. These data indicate that native laminin-1 serves to contain neurites and that this effect is attributed to E8, whereas E1′ acts as an inhibitory boundary to advancing growth cones. We suggest that proteolytic fragments of laminin-1 may similarly provide divergent guidance cues to regrowing neurites in an area of injury to the mammalian central nervous system.

Receptors and binding proteins for laminin-1

The actions of laminin-1 regions on neuronal growth are primarily mediated by interactions with specific integrin receptors. Discrete integrin binding sites along the laminin-1 molecule have been mapped using enolase-digested proteolytic fragments and affinity chromatography. Binding sites for the α3β1, α6β1, α6β4, and α7β1 integrins have been mapped to the globular domains of the E8 fragment (Calof et al., 1994; Chao et al., 1996; Yao et al., 1996; Ivins et al., 1998). The migration of olfactory epithelial neurons on E8 has been shown to be mediated by α6β1 integrin (Calof et al., 1994). The α6β1 integrin also mediates attachment and spreading of embryonic retinal neurons on the E8 region of laminin-1, whereas the α3β1 integrin mediates neurite outgrowth (Ivins et al., 1998). Involvement of a β1 integrin in neurite outgrowth on E8 is further supported by work showing that addition of a blocking antibody to β1 reduces neurite outgrowth from retinal explants (Bates and Meyer, 1997). The E1′ fragment has been shown to contain binding sites for the integrin receptors α1β1 and α2β1, and therefore its migration and outgrowth-suppressing activities may be mediated by these integrins. Our data has also demonstrated that a blocking antibody against the β1 integrin chain increased the percentage of neurites crossing from PLL to E1′, suggesting that a β1 integrin mediates the boundary forming properties of this fragment.

As part of its function in the extracellular matrix, laminin-1 binds to other matrix molecules and cell-adhesion molecules. These interactions may then alter the biological actions of laminin-1 by blocking active sites for the regulation of neuronal growth. Amongst the matrix molecules that laminin-1 binds to are a variety of proteoglycans including the heparan sulfate proteoglycans, perlecan (Battaglia et al., 1992; Yurchenco et al., 1993; Couchman et al., 1996; Ettner et al., 1998), syndecan (Hoffman et al., 1998), and agrin (Denzer et al., 1997), and the chondroitin sulfate proteoglycan, bamacan (Couchman et al., 1996). Grumet et al. (1993) discovered that laminin binds to Ng-CAM, a cell-adhesion molecule expressed by astrocytes and neurons. Laminin-1 has also been shown to interact with type IV collagen via both E1′ and E8 (Tomaselli et al., 1990). Finally, laminin-1 has been found to complex with entactin/nidogen via E1′ (Gerl et al., 1991). Thus, binding to collagen or entactin/nidogen might, for example, reverse the promotion of neurite outgrowth by E8 or the suppression of neurite outgrowth by E1′.

Tenascin-C and neuronal growth

Tenascin-C is another extracellular matrix molecule that is invariably upregulated following injury to the adult central nervous system (McKeon et al., 1991; Laywell et al., 1996). However, as mentioned previously, its role in the injury response remains an open question. Recent work (Gates et al., 1996) strongly suggests that tenascin-C inhibits neurite regeneration following injury. For example, when living striatal scars from adult mice were removed before or after upregulation of tenascin-C and used as in vitro substrates for embryonic neurons, neurite outgrowth decreased as tenascin-C was upregulated. Furthermore, blocking tenascin-C with antibodies increased neurite growth. On the other hand, regenerating spinal cord (Zhang et al., 1997) and entorhinal cortex (Deller et al., 1997) neurons were found to be closely associated with tenascin-C immunoreactivity, thus suggesting that tenascin-C may promote regrowth.

Our work has centered upon understanding how tenascin-C regulates neuronal growth. The contradictory actions of this molecule are likely explained by a combination of several properties: (1) tenascin-C has a complicated, multi-domain structure (Fig. 1), some domains of which may inhibit and others may promote neuronal process outgrowth (Meiners and Geller, 1997); (2) tenascin-C is not a single molecule, but is a family of alternatively spliced variants containing different combinations of fibronectin type III (FN-III) domains (Chung et al., 1996; Gotz et al., 1996; Meiners and Geller, 1997); (3) the actions of tenascin-C may differ depending upon whether it is part of an extracellular matrix or as an isolated molecule (Meiners et al., 1999b); (4) its actions may differ depending upon whether it is present as a soluble or substrate-bound molecule (Lochter et al., 1991; Meiners and Geller, 1997); and (5) proteolytic fragments of tenascin-C may be generated following injury having functions that differ considerably from those of the parent molecule.

Domains of the tenascin-C molecule

Similarly to laminin-1, regulation of neuronal growth by intact tenascin-C has been ascribed to several different domains of the molecule (Phillips et al., 1995; Dorries et al., 1996; Gotz et al., 1996). Anti-adhesive and growth cone repelling properties have been attributed to the EGF domains and FN-III domains 4–5 (fn4–5) (Dorries et al., 1996; Gotz et al., 1996). Adhesion of peripheral neurons is mediated by fn3, whereas adhesion of both peripheral and central neurons is mediated by fn6 (Phillips et al., 1995). Neurite outgrowth promotion has been mapped to the fn6–8 region (Varnum-Finney et al., 1995; Dorries et al., 1996; Meiners and Geller, 1997) as well as to the alternatively spliced region, fnA–D (Gotz et al., 1996, 1997; Meiners et al., 1999b). Hence the concerted action of several domains, as well as the presence of different receptors for the domains on different classes of neurons, is probably responsible at least in part for the diverse and sometimes conflicting neuronal responses to tenascin-C.

To add to this complexity, work using proteolytic fragments of tenascin-C or portions of tenascin-C expressed as recombinant proteins has shown that certain tenascin-C regions, when presented alone, function differently than the intact molecule. For example, a proteolytic fragment of tenascin-C containing the FN-III domains was more potent than intact tenascin-C in terms of neuronal adhesion (Friedlander et al., 1988). Similarly, forebrain and dorsal root ganglion neurons bound more avidly to a recombinant protein corresponding to fn6 than to tenascin-C (Phillips et al., 1995), and cerebellar granule neurons (Xiao et al., 1996) and neuroblastoma cell lines (Prieto et al., 1992) bound more avidly to the fibrinogen domain. The EGF domains (Gotz et al., 1996) and fn3 (Phillips et al., 1995), expressed as recombinant proteins, promoted neurite outgrowth from hippocampal and dorsal root ganglion neurons, respectively, but they had no effect on neurite outgrowth in the context of intact tenascin-C (Gotz et al., 1997). Furthermore, our recent work demonstrated that in addition to promoting neurite growth, an fnA–D recombinant protein supplied permissive guidance (separate from growth-promoting) cues to cerebellar granule neurites (Meiners et al., 1999a). Neurites demonstrated a strong preference for fnA–D when given a choice at an interface between fnA–D and PLL (Meiners et al., 1999a), but purified fnA–D-containing tenascin-C splice variants had the opposite effect and formed barriers to advancing growth cones (Dorries et al., 1996; Gotz et al., 1996; Meiners et al., 1999a). It seems highly plausible, then, that the impact of tenascin-C on neuronal regrowth following injury is due to both the upregulation of the parent molecule and the generation of smaller fragments with independent activities.

Binding sites for tenascin-C regions

Tenascin-C regions exert their actions on neurons by interacting with specific neuronal integrin receptors and other effector molecules. For example, fn6 apparently interacts with a β1 integrin on forebrain and dorsal root ganglion neurons to mediate cellular adhesion, whereas fn3 interacts with a different integrin on dorsal root ganglion neurons to mediate adhesion. This conclusion was supported by data showing that neuronal attachment to fn6 was inhibited in the presence of a function-blocking antibody against β1 integrin, whereas attachment to fn3 was inhibited by RGD-containing peptides but not by the β1 integrin antibody (Phillips et al., 1995). Integrin receptors are also implicated in the promotion of neurite outgrowth by tenascin-C regions. The fn6–8 region of tenascin-C facilitates neuronal process extension by interacting with the α8β1 integrin receptor on sensory and motor neurons (Varnum-Finney et al., 1995), and our unpublished data has indicated that the fnA–D region facilitates process extension (but not neurite guidance) by interacting with a β1 integrin receptor on cerebellar granule neurons.

Actions of tenascin-C regions can also be mediated through non-integrin receptors. FnA–D binds to annexin II to facilitate migration of endothelial cells and loss of focal adhesions (Chung et al., 1996), but annexin II did not appear to serve as a receptor on cerebellar granule or cerebral cortical neurons in our studies (S. Meiners, unpublished data). The smallest tenascin-C splice variant (Fig. 1) binds to the immunoglobulin superfamily neuronal adhesion molecule, contactin/F11, via fn5–6 (Weber et al., 1996); interruption of fn5 and fn6 by the alternatively spliced region eliminates binding of larger variants. Contactin/F11 and the smallest tenascin-C splice variants are upregulated in overlapping regions during retinal development (D'Alessandri et al., 1995), which is consistent with a putative role for contactin/F11–tenascin-C interactions in the establishment of synaptic layers in the retina. Tenascin-C's close cousin, tenascin-R, also binds to contactin/F11 via the EGF domains (Xiao et al., 1996); the anti-adhesive and growth cone repelling properties of the EGF domains in tenascin-R are apparently mediated by this receptor (Xiao et al., 1997), and the same may be true for the EGF domains in tenascin-C.

Like laminin-1, specific tenascin-C regions have also been shown to bind to extracellular matrix proteins and proteoglycans, which may then modify the functional properties of the regions by blocking active sites for the regulation of neuronal outgrowth. Fn4–5, which is repulsive for neuronal attachment and demonstrates growth cone repelling properties (Gotz et al., 1996), binds to the proteoglycan neurocan (Rauch et al., 1997). Fn3–5 binds to fibronectin (Chung et al., 1995) and the heparan sulfate proteoglycan, perlecan (Chung and Erickson, 1997); binding is mediated by both fn3, which comprises an attachment site for dorsal root ganglion neurons (Phillips et al., 1995), and fn4–5. Fn5 by itself also binds to heparin and heparan sulfate, inhibiting the interaction of fn5–6 with contactin (Weber et al., 1996). The cell attachment-promoting fibrinogen domain (Prieto et al., 1992; Faissner, 1993) interacts with heparin and heparan sulfate proteoglycans (Aukhil et al., 1993) as well as neurocan and phosphacan (Milev et al., 1997). We ourselves have shown that fnA–D binds to an as yet unidentified matrix protein on cerebral cortical astrocytes (Meiners et al., 1999b), blocking a neurite outgrowth promoting site in fnA1–A4. Thus the net actions of tenascin-C and its regions on neuronal growth and regeneration must be considered in terms of the cellular and molecular microenvironment, and the types of other extracellular matrix molecules incorporated with tenascin-C.

Conclusion

Extracellular matrix molecules that are upregulated in the vicinity of an injury to the central nervous system, such as laminin-1 and tenascin-C, are highly likely to be involved in the regulation of axonal regrowth. Concomitant upregulation of proteases following central nervous system injury is suggested to generate extracellular matrix protein fragments. These protein fragments may interact with specific neuronal receptors or other extracellular matrix or cell adhesion molecules to elicit actions distinct from those of the parent molecule. Thus, increased expression of the parent molecule, as well as potential generation of smaller fragments, likely influences the extent of neuronal regeneration. Moreover, small regions derived from matrix molecules that, by themselves, influence neuronal growth, such as the E8 region of laminin-1 and the fnA–D region of tenascin-C, may find utility as reagents to improve regeneration after injury.

References

Anton, E.S., Sandrock Jr., A.W. and Matthew, W.D. (1994) Merosin promotes neurite growth and Schwann cell migration in vitro and nerve regeneration in vivo: evidence using an antibody to merosin, ARM-1. *Dev. Biol.*, 164: 133–146.

Aukhil, I., Joshi, P., Yan, Y. and Erickson, H.P. (1993) Cell- and heparin-binding domains of the hexabrachion arm identified by tenascin expression proteins. *J. Biol. Chem.*, 268: 2542–2553.

Bates, C.A. and Meyer, R.L. (1997) The neurite-promoting effect of laminin is mediated by different mechanisms in embryonic and adult regenerating mouse optic axons in vitro. *Dev. Biol.*, 181: 91–101.

Battaglia, C., Mayer, U., Aumailley, M. and Timpl, R. (1992) Basement-membrane heparan sulfate proteoglycan binds to laminin by its heparan sulfate chains and to nidogen by sites in the protein core. *Eur. J. Biochem.*, 208: 359–366.

Brodkey, J.A., Laywell, E.D., O'Brien, T.F., Faissner, A., Stefansson, K., Dorries, H.U., Schachner, M. and Steindler, D.A. (1995) Focal brain injury and upregulation of a developmentally regulated extracellular matrix protein. *J. Neurosurg.*, 82: 106–112.

Calof, A.L., Campanero, M.R., O'Rear, J.J., Yurchenco, P.D. and Lander, A.D. (1994) Domain-specific activation of neuronal migration and neurite outgrowth-promoting activities of laminin. *Neuron*, 13: 117–130.

Chao, C., Lotz, M.M., Clarke, A.C. and Mercurio, A.M. (1996) A function for the integrin alpha6beta4 in the invasive properties of colorectal carcinoma cells. *Cancer Res.*, 56: 4811–4819.

Chen, Z.L. and Strickland, S. (1997) Neuronal death in the hippocampus is promoted by plasmin-catalyzed degradation of laminin. *Cell*, 91: 917–925.

Chung, C.Y. and Erickson, H.P. (1997) Glycosaminoglycans modulate fibronectin matrix assembly and are essential for matrix incorporation of tenascin-C. *J. Cell Sci.*, 110: 1413–1419.

Chung, C.-Y., Zardi, L. and Erickson, H.P. (1995) Binding of tenascin-C to soluble fibronectin and matrix fibrils. *J. Biol. Chem.*, 270: 29012–29017.

Chung, C.-Y., Murphy-Ullrich, J.E. and Erickson, H.P. (1996) Mitogenesis, cell migration, and loss of focal adhesions induced by tenascin-C interacting with its cell surface receptor, annexin II. *Mol. Biol. Cell*, 7: 883–892.

Colognato-Pyke, H., O'Rear, J.J., Yamada, Y., Carbonetto, S., Cheng, Y.S. and Yurchenco, P.D. (1995) Mapping of network-forming, heparin-binding, and alpha 1 beta 1 integrin-recognition sites within the alpha-chain short arm of laminin-1. *J. Biol. Chem.*, 270: 9398–9406.

Couchman, J.R., Kapoor, R., Sthanam, M. and Wu, R.R. (1996) Perlecan and basement membrane-chondroitin sulfate proteoglycan (bamacan) are two basement membrane chondroitin/dermatan sulfate proteoglycans in the Engelbreth–Holm–Swarm tumor matrix. *J. Biol. Chem.*, 271: 9595–9602.

D'Alessandri, L., Ransche, B., Winterhalter, K.H. and Vaughan, L. (1995) Contactin/F11 and tenascin-C co-expression in the chick retina correlates with formation of the synaptic plexiform layers. *Curr. Eye Res.*, 14: 911–926.

Davies, S.J.A., Field, P.M. and Raisman, G. (1996) Regeneration of cut adult axons fails even in the presence of continuous aligned glial pathways. *Exp. Neurol.*, 142: 203–216.

Davies, S.J., Goucher, D.R., Doller, C. and Silver, J. (1999) Robust regeneration of adult sensory axons in degenerating white matter of the adult rat spinal cord. *J. Neurosci*, 19: 5810–5822.

Deller, T., Haas, C.A., Naumann, T., Joester, A., Faissner, A. and Frotscher, M. (1997) Up-regulation of astrocyte-derived tenascin-C correlates with neurite outgrowth in the rat dentate gyrus after unilateral entorhinal cortex lesion. *Neuroscience*, 81: 829–846.

Denzer, A.J., Brandenberger, R., Gesemann, M., Chiquet, M. and Ruegg, M.A. (1997) Agrin binds to the nerve-muscle basal lamina via laminin. *J. Cell Biol.*, 137: 671–683.

Dorries, U., Taylor, J., Xioa, Z., Lochter, A., Montag, D. and Schachner, M. (1996) Distinct effects of recombinant tenascin domains on neuronal cell adhesion, growth cone guidance, and neuronal polarity. *J. Neurosci. Res.*, 43: 420–438.

Drago, J., Nurcombe, V. and Bartlett, P.F. (1991) Laminin through its long arm E8 fragment promotes the proliferation and differentiation of murine neuroepithelial cells in vitro. *Exp. Cell Res.*, 192: 256–265.

Engel, J. (1996) Domain organizations of modular extracellular matrix proteins and their evolution. *Matrix Biol.*, 15: 295–299.

Ernsberger, U., Edgar, D. and Rohrer, H. (1989) The survival of early chick sympathetic neurons in vitro is dependent on a suitable substrate but independent of NGF. *Dev. Biol.*, 135: 250–262.

Ettner, N., Gohring, W., Sasaki, T., Mann, K. and Timpl, R. (1998) The N-terminal globular domain of the laminin alpha1 chain binds to alpha1beta1 and alpha2beta1 integrins and to the heparan sulfate-containing domains of perlecan. *FEBS Lett.*, 430: 217–221.

Faissner, A. (1993) Tenascin glycoproteins in neural pattern formation: facets of a complex picture. *Perspect. Dev. Neurobiol.*, 1: 155–164.

Fernaud-Espinosa, I., Nieto-Sampedro, M. and Bovolenta, P. (1998) A neurite outgrowth-inhibitory proteoglycan expressed during development is similar to that isolated from adult brain after isomorphic injury. *J. Neurobiol.*, 36: 16–29.

Friedlander, D.R., Hoffman, S. and Edelman, G.M. (1988) Functional mapping of cytotactin: proteolytic fragments active in cell-substrate adhesion. *J. Cell Biol.*, 107: 2329–2340.

Gates, M.A., Fillmore, H. and Steindler, D.A. (1996) Chondroitin sulfate proteoglycan and tenascin in the wounded adult mouse neostriatum in vitro: dopamine neuron attachment and process outgrowth. *J. Neurosci.*, 16: 8005–8018.

Gerl, M., Mann, K., Aumailley, M. and Timpl, R. (1991) Localization of a major nidogen-binding site to domain III of laminin B2 chain. *Eur. J. Biochem.*, 202: 167–174.

Gotz, B., Scholze, A., Clement, A., Joester, A., Schutte, K., Wigger, F., Frank, R., Spiess, E., Ekblom, P. and Faissner, A. (1996) Tenascin-C contains distinct adhesive, anti-adhesive, and neurite outgrowth promoting sites for neurons. *J. Cell Biol.*, 132: 681–699.

Gotz, M., Bolz, J., Joester, A. and Faissner, A. (1997) Tenascin-C synthesis and influence on axon growth during rat cortical development. *Eur. J. Neurosci.*, 9: 496–506.

Grumet, M., Friedlander, D.R. and Edelman, G.M. (1993) Evidence for the binding of Ng-CAM to laminin. *Cell Adhes. Commun.*, 1: 177–190.

Hagg, T., Portera-Cailliau, C., Jucker, M. and Engvall, E. (1997) Laminins of the adult mammalian CNS; laminin-alpha2 (merosin M-) chain immunoreactivity is associated with neuronal processes. *Brain Res.*, 764: 17–27.

Hoffman, M.P., Nomizu, M., Roque, E., Lee, S., Jung, D.W., Yamada, Y. and Kleinman, H.K. (1998) Laminin-1 and laminin-2 G-domain synthetic peptides bind syndecan-1 and are involved in acinar formation of a human submandibular gland cell line [published erratum appears in J. Biol. Chem., 1999, Apr 30; 274(18), 12950]. *J. Biol. Chem.*, 273: 28633–28641.

Ivins, J.K., Colognato, H., Kreidberg, J.A., Yurchenco, P.D. and Lander, A.D. (1998) Neuronal receptors mediating responses to antibody activated laminin-1. *J. Neurosci.*, 18: 9703–9715.

Laywell, E.D., Friedman, P., Harrington, K., Robertson, J.T. and Steindler, D.A. (1996) Cell attachment to frozen sections of injured adult mouse brain — effects of tenascin antibody and lectin perturbation of wound-related extracellular matrix molecules. *J. Neurosci. Methods*, 66: 99–108.

Lein, P.J., Banker, G.A. and Higgins, D. (1992) Laminin selectively enhances axonal growth and accelerates the development of polarity by hippocampal neurons in culture. *Dev. Brain Res.*, 69: 191–197.

Lochter, A., Vaughan, L., Kaplony, A., Prochiantz, A., Schachner, M. and Faissner, A. (1991) J1/tenascin in substrate-bound and soluble form displays contrary effects on neurite outgrowth. *J. Cell Biol.*, 113: 1159–1171.

Luckenbill-Edds, L. (1997) Laminin and the mechanism of neuronal outgrowth. *Brain Res. Rev.*, 23: 1–27.

Malinda, K.M. and Kleinman, H.K. (1996) The laminins. *Int. J. Biochem. Cell Biol.*, 28: 957–959.

McKeon, R.J., Schreiber, R.C., Rudge, J.S. and Silver, J. (1991) Reduction of neurite outgrowth in a model of glial scarring following CNS injury is correlated with the expression of inhibitory molecules on reactive astrocytes. *J. Neurosci.*, 11: 3398–3411.

McKeon, R.J., Höke, A. and Silver, J. (1995) Injury-induced proteoglycans inhibit the potential for laminin-mediated axon growth on astrocytic scars. *Exp. Neurol.*, 136: 32–43.

McKerracher, L., Chamoux, M. and Arregui, C.O. (1996) Role of laminin and integrin interactions in growth cone guidance. *Mol. Neurobiol.*, 12: 95–116.

Meiners, S. and Geller, H.M. (1997) Long and short splice variants of human tenascin differentially regulate neurite outgrowth. *Mol. Cell. Neurosci.*, 10: 100–116.

Meiners, S., Mercado, M.L., Kamal, M.S. and Geller, H.M. (1999a) Tenascin-C contains domains that independently regulate neurite outgrowth and neurite guidance. *J. Neurosci.*, 19: 8443–8453.

Meiners, S., Powell, E.M., Geller, H.M., 1999b. Neurite outgrowth promotion by the alternatively spliced region of tenascin-C is influenced by cell type specific binding. Matrix Biology, in press.

Milev, P., Fischer, D., Haring, M., Schulthess, T., Margolis, R.K., Chiquet-Ehrismann, R. and Margolis, R.U. (1997) The fibrinogen-like globe of tenascin-C mediates its interactions with neurocan and phosphacan/protein-tyrosine phosphatase-zeta/beta. *J. Biol. Chem.*, 272: 15501–15509.

Phillips, G.R., Edelman, G.M. and Crossin, K.L. (1995) Separate cell binding sites within cytotactin/tenascin differentially promote neurite outgrowth. *Cell Adhes. Commun.*, 3: 257–271.

Pindzola, R.R., Doller, C. and Silver, J. (1993) Putative inhibitory extracellular matrix molecules at the dorsal root entry zone of the spinal cord during development and after root and sciatic nerve lesions. *Dev. Biol.*, 156: 34–48.

Powell, S.K., Williams, C.C., Nomizu, M., Yamada, Y. and Kleinman, H.K. (1998) Laminin-like proteins are differentially regulated during cerebellar development and stimulate granule cell neurite outgrowth in vitro. *J. Neurosci. Res.*, 54: 233–247.

Prieto, A.L., Andersson-Fisone, C. and Crossin, K.L. (1992) Characterization of the multiple adhesive and counteradhesive domains in the extracellular matrix protein cytotactin. *J. Cell Biol.*, 119: 663–678.

Ramón y Cajal, S. (1928) *Degeneration and Regeneration of the Nervous System.* Oxford University Press, London.

Rauch, U., Clement, A., Retzler, C., Frolich, L., Fassler, R., Gohring, W. and Faissner, A. (1997) Mapping of a defined neurocan binding site to distinct domains of tenascin-C. *J. Biol. Chem.*, 272: 26905–26912.

Reier, P.J. and Houle, J.D. (1988) The glial scar: its bearing on axonal elongation and transplantation approaches to CNS repair. In: Waxman, S.G. (Ed.), *Functional Recovery in Neurological Disease.* Raven, New York, pp. 87–138.

Schittny, J.C. and Yurchenco, P.D. (1989) Basement membranes: molecular organization and function in development and disease. *Curr. Opin. Cell Biol.*, 1: 983–988.

Stichel, C.C. and Muller, H.W. (1998) The CNS lesion scar: new vistas on an old regeneration barrier. *Cell Tissue Res.*, 294: 1–9.

Tomaselli, K.J., Hall, D.E., Flier, L.A., Gehlsen, K.R., Turner, D.C., Carbonetto, S. and Reichardt, L.F. (1990) A neuronal cell line (PC12) expresses two beta 1-class integrins — alpha 1 beta 1 and alpha 3 beta 1 — that recognize different neurite outgrowth-promoting domains in laminin. *Neuron*, 5: 651–662.

Turgeon, V.L. and Houenou, L.J. (1997) The role of thrombin-like (serine) proteases in the development, plasticity and pathology of the nervous system. *Brain Res. Rev.*, 25: 85–95.

Varnum-Finney, B., Venstrom, K., Muller, U., Kypta, R., Backus, C., Chiquet, M. and Reichardt, L.F. (1995) The integrin receptor alpha 8 beta 1 mediates interactions of embryonic chick motor and sensory neurons with tenascin-C. *Neuron*, 14: 1213–1222.

Weber, P., Ferber, P., Fischer, R., Winterhalter, K.H. and Vaughan, L. (1996) Binding of contactin/F11 to the fibronectin type III domains 5 and 6 of tenascin is inhibited by heparin. *FEBS Lett.*, 389: 304–308.

Xiao, Z.C., Taylor, J., Montag, D., Rougon, G. and Schachner, M. (1996) Distinct effects of recombinant tenascin-R domains in neuronal cell functions and identification of the domain interacting with the neuronal recognition molecule F3/11. *Eur. J. Neurosci.*, 8: 766–782.

Xiao, Z.C., Hillenbrand, R., Schachner, M., Thermes, S., Rougon, G. and Gomez, S. (1997) Signaling events following the interaction of the neuronal adhesion molecule F3 with the N-terminal domain of tenascin-R. *J. Neurosci. Res.*, 49: 698–709.

Yao, C.C., Ziober, B.L., Squillace, R.M. and Kramer, R.H. (1996) Alpha7 integrin mediates cell adhesion and migration on specific laminin isoforms. *J. Biol. Chem.*, 271: 25598–25603.

Yuguchi, T., Kohmura, E., Yamada, K., Otsuki, H., Sakaki, T., Yamashita, T., Nonaka, M., Sakaguchi, T., Wanaka, A. and Hayakawa, T. (1997) Expression of tPA mRNA in the facial nucleus following facial nerve transection in the rat. *NeuroReport*, 8: 419–422.

Yurchenco, P.D., Sung, U., Ward, M.D., Yamada, Y. and O'Rear, J.J. (1993) Recombinant laminin G domain mediates myoblast adhesion and heparin binding. *J. Biol. Chem.*, 268: 8356–8365.

Zhang, Y., Winterbottom, J.K., Schachner, M., Lieberman, A.R. and Anderson, P.N. (1997) Tenascin-C expression and axonal sprouting following injury to the spinal dorsal columns in the adult rat. *J. Neurosci. Res.*, 49: 433–450.

Zuo, J., Neubauer, D., Dyess, K., Ferguson, T.A. and Muir, D. (1998) Degradation of chondroitin sulfate proteoglycan enhances the neurite-promoting potential of spinal cord tissue. *Exp. Neurol.*, 154: 654–662.

CHAPTER 4

Spinal cord injury-induced inflammation: a dual-edged sword

John R. Bethea *

The Miami Project to Cure Paralysis, University of Miami School of Medicine, 1600 NW 10th Avenue, R-48, Miami, FL 33136, USA

Introduction

In this chapter on inflammation following spinal cord injury (SCI), I will discuss the disparate effects inflammation has on tissue destruction, loss of function, and functional recovery following SCI. Inflammation is a very complex biological process with effects ranging from, but not limited to, host defense to wound healing to autoimmune disease processes. An important regulator of inflammatory responses is a family of proteins called cytokines. Cytokines play a central role in the initiation, perpetuation, regulation and attenuation of immune and inflammatory responses. Pro-inflammatory cytokines such as tumor necrosis factor-alpha (TNF-α), interleukin-1 (IL-1) and interleukin-6 (IL-6) are generally produced during the effector phase of an immune response. However, anti-inflammatory cytokines such as interleukin-4 (IL-4), interleukin-10 (IL-10) and interleukin-13 (IL-13), that are essential for shutting off an immune response, are synthesized during the attenuation phase. Because cytokines have such diverse biological functions, their expression is tightly regulated and not constitutive, as is the case for many growth factors. While several cytokines that appear to be regulators of immune-mediated injury and recovery following SCI will be discussed, specific attention will be devoted to the pro-inflammatory cytokine, TNF-α, and the anti-inflammatory cytokine, IL-10. Another important point that will be discussed is how timing of an inflammatory response may help determine if destructive or reparative events will be initiated.

Inflammation following SCI

Injury-induced inflammation is thought to be a contributing factor to the resulting neuropathology and secondary necrosis that occurs after SCI (Blight, 1985; Dusart and Schwab, 1994; Barna et al., 1994; Bartholdi and Schwab, 1997; Popovich et al., 1997; Zhang et al., 1997; Bethea et al., 1998; Taoka et al., 1998). One of the hallmarks of SCI is the progressive secondary neuronal necrosis that develops after the initial injury (Zhang et al., 1997). It has been proposed that inflammation is an important effector of secondary neuronal necrosis (Zhang et al., 1997). Central nervous system (CNS) inflammatory responses that occur after SCI are initiated by peripherally derived immune cells (macrophages, neutrophils and T-cells) and activated glial cells (astrocytes and microglia) that migrate into the lesion site following injury. T-cells are essential for activating macrophages and mounting a cellular immune response. Macrophages and neutrophils have been proposed to participate in tissue destruction and enlargement of the lesion. Macrophages and microglia

---

* Corresponding author. Dr. John R. Bethea, Miami Project to Cure Paralysis, University of Miami School of Medicine, 1600 NW 10th Ave., R-48, Miami, FL 33136, USA. Fax: +1-305-243-4427;
E-mail: jbethea@miamiproj.med.miami.edu

contribute to the secondary pathological and inflammatory response through the release of cytokines, for example, TNF-α, IL-1, IL-6 and IL-10 (Giulian, 1990; Blight, 1992, 1994; Giulian et al., 1993, 1996; Popovich et al., 1994, 1998; Yakovlev and Faden, 1994; Zhang et al., 1997; Bethea et al., 1999). Cytokines facilitate CNS inflammatory responses by inducing the expression of additional cytokines, chemokines, nitric oxide, and reactive oxygen and nitrogen species (Benveniste, 1992; Kaltschmidt et al., 1993; Rothwell and Relton, 1993; Wu, 1993; Hsu et al., 1994; Salzman and Faden, 1994; Berkman et al., 1995; Hopkins and Rothwell, 1995; Cross et al., 1996; Minc-Glomb et al., 1996; Ransohoff and Benveniste, 1996; Sato et al., 1996; Stalder et al., 1996; C.X. Wang et al., 1996; O'Neill and Kaltschmidt, 1997). Activated leukocytes also secrete growth factors and proteolytic enzymes that are important for wound healing. The timing of leukocyte infiltration into the injured cord is very well described. The infiltration of activated leukocytes into the injured cord is strictly coordinated by the expression of a family of cytokines called chemokines.

Chemokines are another important class of proinflammatory molecules that are synthesized in response to CNS trauma or disease (Veenam et al., 1992; Berkman et al., 1995; Ransohoff and Benveniste, 1996; McTigue et al., 1998). Chemokines recruit white blood cells or leukocytes from the bloodstream to sites of inflammation or injury. Neutrophils arrive within hours and peak 1 day after injury; monocytes and T-lymphocytes begin to appear in the spinal cord as early as 2–3 days after injury and persist for several weeks (Popovich et al., 1997). Since activated leukocytes probably participate in both the neurodestructive and wound-healing events following SCI, a better understanding of the chemical signals that regulate their recruitment is essential.

Another important mediator of inflammation, tissue destruction and wound healing, peripherally and possibly in the spinal cord, are the matrix metalloproteinases (MMPs). MMPs are a family of zinc-containing endo-proteinases that are essential for normal immune function and tissue remodeling (Clark, 1996; Goetzl et al., 1996). MMPs are thought to play an essential role in the initiation of inflammation, angiogenesis, deposition of new matrix and tissue repair, all essential components of the wound-healing process (Clark, 1996). For example, MMPs are essential for TNF-α processing and activation (Goetzl et al., 1996). With respect to disease of or injury to the CNS, MMPs are thought to participate in the pathogenesis of multiple sclerosis, Alzheimer's disease, malignant gliomas, and the pathology and subsequent inflammatory response following trauma to the brain or spinal cord (Rosenberg, 1995; Goetzl et al., 1996; Yong et al., 1998). Following injury to the CNS, the blood–brain barrier (BBB) is disrupted and inflammatory cells and circulating inflammogens enter the CNS. MMPs are thought to play a critical role in BBB breakdown, tissue destruction and activation of inflammatory responses within the brain and spinal cord.

Wound healing following SCI

Following traumatic injury to the spinal cord, as in other organs and tissues, wound repair begins. This is not typically a regenerative process but a fibroproliferative event that develops into a fibrotic scar that 'patches' rather than restores the damaged tissue (Clark, 1996). In the CNS this can potentially have devastating functional consequences. Wound healing is an integration of a well coordinated sequence of events involving the release of soluble mediators, extracellular matrix molecules and cellular interactions. The wound-healing process follows a very precise, temporally restricted, sequence of events and can be categorized into three main groups in their order of initiation: inflammation, tissue formation and tissue remodeling (Clark, 1996). These three phases of the repair process are not independent of one another but are overlapping in time and cellular makeup (Clark, 1996). There are two phases of the inflammatory response, 'early' and 'late'. The 'early' phase begins within minutes to hours of injury and peaks approximately 1 day later. This response can be characterized by the very early production of cytokines, chemokines and the infiltration of neutrophils into the site of injury. In support of this we have recently demonstrated that nuclear factor-κB (NF-κB)p65 activation occurs as early as 30 min after SCI (Bethea et al., 1998). The NF-κB family of transcription factors are central regulators of inflammatory responses peripherally and in the CNS. Furthermore, we have determined that TNF-α and monocyte chemoattractant protein (MCP)-1 levels are significantly elevated in the spinal cord as early as 1 h after injury (Bethea et al., 1999; and unpublished observations). The early production of MCP-1 may be important for attracting microglial cells to the injury site. The infiltration of macrophages, which peaks between 2 and 3 days, characterizes the 'late' inflammatory response. In the spinal cord, macrophages may persist for several weeks following the initial injury (Popovich et al., 1997). Macrophages secrete growth factors, proteolytic enzymes and cytokines which are essential for tissue formation and tissue remodeling to occur. Because of the unique neuronal environment within the CNS, we propose that the 'early' inflammatory response may be neurotoxic in nature. Furthermore, the 'late' inflammatory response may be essential for repair and recovery of function. In the final two phases of wound repair (tissue formation and tissue modeling), events occur that may prevent functional regeneration of CNS axons. For example, glial 'scar' tissue is deposited and may form a functional barrier to axonal regeneration.

TNF-α and TNF receptors

TNF-α is a 17,000 Da peptide produced primarily by activated macrophages, T-lymphocytes, astrocytes and microglia. TNF-α is an active participant in many inflammatory responses. Specifically, TNF-α can enhance the permeability of endothelial cells and promotes binding of neutrophils, T-lymphocytes and macrophages to endothelial cells, thereby facilitating the transendothelial migration of activated leukocytes into the injured spinal cord. TNF-α can also induce the expression of other cytokines and other soluble pro-inflammogens, either alone or in combination with other cytokines such as interferon-γ.

TNF-α's biological responses are mediated by two distinct receptors: TNF-R1 (55 kDa) and TNF-R2 (75 kDa). The cytoplasmic domains of these receptors and the adaptor proteins that interact with their respective cytoplasmic domains are not identical, suggesting that different signaling pathways are activated and distinct biological effects are initiated. TNF-α signaling through the TNF-R1 initiates apoptosis, gene expression and cytokine production. Signaling through the TNF-R2 is not well understood outside of T-lymphocyte development.

TNF-α expression following SCI

Several studies have demonstrated that TNF-α mRNA and protein are elevated in the cord following SCI (Yakovlev and Faden, 1994; C.-Y. Wang et al., 1996; Streit et al., 1998; Taoka et al., 1998; Bethea et al., 1999). We have recently demonstrated that following SCI, TNF-α protein is expressed within the cord and by activated monocytes/macrophages (Bethea et al., 1999). This early synthesis of TNF-α may be an important signal for initiating a progressive inflammatory response within the spinal cord. We detected TNF-α protein in the cord as early as 1 h after injury with levels persisting for up to 7 days after injury. There are several possible sources for TNF-α in the spinal cord at these early times. Both astrocytes and microglia have been shown to secrete TNF-α in response to CNS injury or infection (Benveniste, 1992; Ghirnikar et al., 1996). Neutrophils enter the spinal cord within 6 h of injury (Cunha et al., 1992) and are another potential source of TNF-α at these early time points. While monocytes are not actively recruited to the cord until much later, we detected monocytes/macrophages within 30 min of injury (Bethea et al., 1998). The early presence of monocytes could be due to circulating monocytes that entered the cord as a consequence of blood–spinal barrier disruption. The production of TNF-α between 1 and 3 days after injury could be due to infiltrating neutrophils and monocytes as well as endogenous glial cells. TNF-α is also spontaneously secreted by SCI-activated monocytes 3 and 7 days after injury. While we were unable to detect the spontaneous release of TNF-α 1 day after injury, we did not examine earlier time points. Peripherally synthesized TNF-α can also enter the spinal cord following SCI through saturable TNF-α transporters (Pan et al., 1999). In this study, these authors demonstrated that there is a very early (within minutes) influx of TNF-α and a later wave peaking between 1 and 5 days after SCI. They did not look beyond 5 days. These studies suggest that there is a peripheral source of TNF-α that contributes to the overall concentration in the spinal cord following SCI. Our monocyte studies demonstrate that this cell type can spontaneously secrete TNF-α following SCI and therefore may be a potential peripheral source of this cytokine (Bethea et al., 1999).

Role of TNF-α in CNS neuropathology

The role that TNF-α plays in mediating CNS neuropathology is controversial. Several studies demonstrated that TNF-α may be neuroprotective in vitro. For example, TNF-α prevents glutamate-induced cell death and the cytotoxic effects of amyloid β-peptide and iron by regulating intracellular calcium levels and suppressing the accumulation of reactive oxygen species in pure hippocampal cultures (Cheng et al., 1994; Barger et al., 1995). The neuroprotective effects of TNF-α may be dependent on the ratio of TNF-R1 to TNF-R2 levels. One possibility is that higher levels of TNF-R2 may facilitate a cytoprotective response and increased levels of TNF-R1 may induce a cytotoxic response. In support of this, recent in vivo genetic studies using mice that are nullizygous for each of the TNF receptors and TNF-α suggest that signaling through TNF-R1 may elicit some of the cytotoxic properties of TNF-α, while signaling through TNF-R2 may be neuroprotective (Eugster et al., 1999). Studies are underway to investigate the cellular distribution of TNF receptors and what effect injury and cytokines have on their expression.

To further complicate matters, when macrophages or microglia are present in neuronal cultures, TNF-α is not neuroprotective (Toulmond et al., 1996; M. Mattson, pers. commun.). In support of this, microglial cells stimulated with either lipopolysaccharide, prion protein or β-amyloid peptides secrete neurotoxic compounds (Combs et al., 1999). Importantly, lipopolysaccharide, prion protein or β-amyloid peptides are not cytotoxic when cultured with neurons alone. Downen et al. (1999) recently demonstrated that IL-1 and IFN-γ stimulate mixed glial cells to secrete neurotoxic substances and that the lethality of this response is reduced in the presence of a neutralizing antibody to TNF-α. There is also a considerable body of evidence suggesting that TNF-α is very deleterious in stroke, brain and spinal cord injury (Liu et al., 1994; Dawson et al., 1996; Barone et al., 1997; Nawashiro et al., 1997; Bethea et al., 1998; Lavine et al., 1998). Both in vitro and in vivo studies with TNF-α demonstrate that this cytokine is a potent mediator of microgliosis, astrogliosis and cell death. For example, Lavine et al. (1998) demonstrated that by blocking TNF-α systemically with a neutralizing antibody, neurological outcome improved following ischemic reperfusion injury. Following SCI, inhibition of TNF-α either by activated protein C or IL-10 significantly improved functional recovery (Taoka et al., 1998; Bethea et al., 1999). While the mechanisms of TNF-α-mediated cell death in neurons and glial cells have not been elucidated, it has been reported that TNF-α induces nuclear and DNA fragmentation, a characteristic feature of apoptosis (D'Souza et al., 1995). TNF-α can also have profound effects on CNS metabolism (Tureen, 1995). Finally, TNF-α injected into the cisterna magna of rabbits resulted in decreased oxygen uptake, reduced cerebral blood flow and an increase in intracranial pressure (Tureen, 1995). Therefore, following CNS injury, when activated monocytes, microglia and astrocytes are in close proximity to neurons and oligodendrocytes, the release of TNF-α and other pro-inflammatory molecules may be cytotoxic to these cell types.

Interleukin-10

Interleukin-10 is a potent anti-inflammatory cytokine that has been used in vivo to successfully reduce inflammation and improve functional outcome in humans and animal models of human inflammatory diseases (Rott et al., 1994; Berg et al., 1995; Crisi et al., 1995; Rogy et al., 1995; Balasingam and Yong, 1996; Berman et al., 1996; Cua et al., 1996; Issazazeh et al., 1996; Koedel et al., 1996; Rudick et al., 1996; Bethea et al., 1998, 1999). IL-10 is synthesized by numerous cell types including T helper lymphocytes (Th$_2$), monocytes/macrophages, astrocytes, and microglia (Liang et al., 1991; Geng et al., 1994). In the peripheral immune system, IL-10 suppresses the majority of monocyte/macrophage inflammatory responses. IL-10 blocks the production of multiple cytokines, MMPs and chemokines (Frei et al., 1994; Geng et al., 1994; Mertz et al., 1994; Berkman et al., 1995; Issazazeh et al., 1996; Bethea et al., 1998). One of the ways in which IL-10 reduces peripheral inflammation is by reducing the phosphorylation, translocation and subsequent DNA binding of NF-κB family members, while having little or no effect on other transcription factors (Wang et al., 1995; Romano et al., 1996). Additionally, IL-10 can abrogate the activation of monocytes by blocking tyrosine kinase activity and the Ras signaling pathway (Geng et al., 1994). With respect to the CNS, IL-10 reduces TNF-α production by astrocytes and antigen presentation by both astrocytes and microglia, and prevents experimental allergic encephalomyelitis in Lewis rats (Rott et al., 1994; Crisi et al., 1995; Issazaseh et al., 1996). In in vivo studies of cerebral malaria, it was shown that a single systemic (but not intrathecal) dose of IL-10 is sufficient to block cytokine production and infiltration of inflammatory cells in a rat model of CNS inflammation (Koedel et al., 1996).

IL-10 exerts its biological responses by binding to its receptor, a member of the class II cytokine receptor family (Bazan, 1990a,b; Thoreau et al., 1991; Findbloom and Winestock, 1995; Kotenko et al., 1997). It was recently demonstrated that, like other members of this family, the IL-10 receptor has two chains, designated alpha and beta (Kotenko et al., 1997). Furthermore, through the use of chimeric receptor constructs it was determined that the beta chain is essential for IL-10 signal transduction to occur (Kotenko et al., 1997). Upon interacting with its receptor, IL-10 induces phosphorylation and the subsequent activation of Jak 1 and Tyk 2 and STAT 1, STAT 3 and, in certain cells, STAT 5 (Ho and Moore, 1994; Findbloom and Winestock, 1995; Kotenko et al., 1997; O'Farrel et al., 1998). Jak1 and the STATs interact with the alpha chain and Tyk 2 binds to the beta chain (Ho and Moore, 1994; Findbloom and Winestock, 1995; Kotenko et al., 1997; O'Farrel et al., 1998). One hypothesis is that the beta chain is required for juxtapositioning Tyk 2 with the alpha chain so signal transduction can be activated (Kotenko et al., 1997). Site-directed mutagenesis of the membrane-distal tyrosine residues in the alpha chain along with STAT 3 dominant negative mutations demonstrated that these tyrosines are essential for the anti-inflammatory properties of IL-10 and that STAT 3 is not required (O'Farrel et al., 1998). While the anti-inflammatory potential of IL-10 is clear, the mechanism(s) through which it exerts these effects have not been totally elucidated and appear to be cell type- and target gene-dependent.

Il-10 and neuroprotection

Following SCI, IL-10 has recently been demonstrated to reduce an 'early' inflammatory response and improve functional recovery (Bethea et al., 1999). In these studies we showed that a single dose of IL-10 given 30 min after SCI attenuates injury-induced TNF-α synthesis in the spinal cord by activated macrophages and significantly improves hindlimb motor function 2 months after injury. In contrast to these data, while multiple doses of IL-10 attenuated more chronic systemic inflammatory responses, there was no effect on functional recovery or tissue sparing. These data suggest that attenuation of an early (e.g., 3 days or earlier) inflammatory response may be neuroprotective and that inhibition of a more chronic (e.g., one occurring at 4 days or later) inflammatory response, that may be essential for tissue recovery and wound healing, may not improve the outcome. In addition to reducing the synthesis of potentially cytotoxic substances, a single injection of IL-10 may initiate the synthesis of growth factors that participate in the wound-healing process and ultimately in functional recovery. Furthermore, one could hypothesize that a second dose of IL-10 administered at more chronic times could alter or turn off 'wound healing' initiated by the initial dose of IL-10.

The inability of two doses of IL-10 to promote functional recovery was surprising. Several other studies have shown that multiple doses of IL-10 were beneficial in treating CNS injury (Knoblach and Faden, 1998; Serpa et al., 1998). The administration of IL-10 in these studies either preceded injury and was accompanied by a second dose within several hours of injury. Therefore, the timing of IL-10 delivery and the subsequent reduction of an inflammatory response may have been within a critical window where the inflammatory response was neurodestructive. Finally, in support of our data, we have determined that repeated doses of IL-10 significantly worsens injury outcome in an excitotoxic model of SCI (R.P. Yezierski and J.R. Bethea, unpubl. observations).

IL-10's chronic neuroprotective properties were recently demonstrated in several independent studies. First, using an excitotoxic model of SCI, we demonstrated that a single injection of IL-10 significantly reduced spontaneous pain behaviors and neuronal loss weeks after injury (R.P. Yezierski and J.R. Bethea, unpubl. observations). More importantly, we demonstrated that if IL-10 was given at the time of pain onset (~2 weeks after injury), we reduced the severity of this spontaneous pain and significantly reduced tissue loss within the cord. In studies with a model of transient global ischemia (Dietrich et al., 1999), rats were subjected to 12.5 min of transient global ischemia and received either no treatment (normothermia), mild hypothermia (32–34°C) for 4 h, IL-10 (5.0 µg, ip) or a combination of IL-10 and hypothermia. Rats were allowed to survive for 2 months before histopathological analysis was performed. Chronic neuroprotection was only demonstrated in the group of rats receiving both IL-10 and mild hypothermia. The inability of IL-10 alone to induce hippocampal neuroprotection may be due to the reduced inflammatory response that occurs following brain injury compared to spinal cord injury.

In addition to being a potent anti-inflammatory cytokine and potentially exerting its neuroprotective effects by attenuating inflammation, recent studies suggest that IL-10 promotes some of its neuroprotective effects by acting directly on neurons (Barbieri et al., 1998). Using neurons isolated from the brains of mice deficient in IL-10, it was determined that these cells were more susceptible to N-methyl-D-aspartate (NMDA) toxicity and to deprivation of oxygen- and glucose-induced cell death. Furthermore, using cerebellar granule cells isolated from wild-type mice, IL-10, in a dose-dependent manner, protected these cells from excitotoxic cell death. Similarly, IL-10 was shown to prevent apoptosis, as determined by TUNEL staining, in cultures of cerebellar granule cells (Sanna et al., 1998). In fact, IL-10 elicited greater neuroprotection than brain-derived neurotrophic factor (BDNF) or fibroblast growth factor 2 (FGF2) when high concentrations (300 mM) of glutamate were used (Sanna et al., 1998). In each of these studies, the neuroprotective effects of IL-10 were observed only when the cells were pretreated with IL-10 for 12–24 h. This suggests that IL-10 is 'altering' the cellular environment and making neurons more resistant to glutamate-mediated neuronal toxicity. Two possible ways in which IL-10 could be improving neuronal survival would be by the downregulation of glutamate receptors and/or the upregulation of Bcl-2, Bcl-xl or other anti-apoptotic proteins. In support of this, IL-10 has been shown to prevent apoptosis by upregulating Bcl-2 expression in other systems (Levy and Brouet, 1994; Taga et al., 1994).

The precise mechanisms by which IL-10 is inducing improved motor function have not been elucidated. There are, however, several additional explanations that could account for the neuroprotective effects of IL-10. IL-10 could increase the local synthesis of nerve growth factors, reduce astrogliosis and attenuate secondary injury mechanisms following SCI. For example, it has recently been demonstrated that IL-10 induced nerve growth factor production in astrocytes and reduced reactive astrogliosis following brain injury in rats (Balasingam and Yong, 1996; Brodie, 1996). Additionally, IL-10 has also been shown to protect neurons from excitotoxic cell death in vitro (Barbieri et al., 1998; Sanna et al., 1998). In vivo studies indicated that neurons from mice nullizygous for IL-10 were more susceptible to NMDA toxicity and to deprivation of oxygen- and glucose-induced cell death (Barbieri et al., 1998). Therefore, it appears that IL-10 may improve functional recovery by attenuating inflammation, promoting tissue sparing, and reducing glial reactivity and neuronal cell death.

Beneficial effects of inflammation following SCI

Recent studies suggest that inflammation following SCI may participate in reparative processes (Dusart and Schwab, 1994; Hirschberg et al., 1994; Guth et al., 1995; Berman et al., 1996; Ghirnikar et al., 1996; Klusman and Schwab, 1997; Streit et al., 1998). Whether or not inflammation participates in neurodestructive or neuroconstructive processes may be dependent on several important variables; for example, timing of the inflammatory response, cellular components and what soluble or cell surface factors are contributing to this immune reaction.

Studies by Klusman and Schwab (1997) suggest that following SCI, the induction of a well timed, cytokine-specific inflammatory response in the spinal cord may be neuroprotective. They demonstrated that when a cytokine cocktail (TNF-α, IL-1 and IL-6) is injected into the spinal cord 4 days after injury, there was a reduction in the recruitment of peripheral macrophages, microglial activation and tissue loss. However, the injury-induced inflammatory response was enhanced and there was greater tissue loss when the cytokines were delivered 1 day after injury. In support of this, our studies (Bethea et al., 1999; Brewer et al., 1999; Lee et al., 1999) suggest that the timing of an inflammatory response may be a critical variable in determining if an inflammatory response is destructive or constructive. For example, we have shown that an early (e.g., 3 days or earlier) inflammatory response initiated in the spinal cord and by SCI-activated macrophages may participate in injury-induced neuropathology and loss of motor function. The 'early' synthesis of TNF-α and other cytokines by resident glial cells and activated macrophages may further facilitate inflammatory responses and cell death within the CNS by inducing the expression of cell adhesion molecules on astrocytes (Shrikant et al., 1995), prostaglandin E_2 (PGE_2) synthesis by microglia (Minghetti et al., 1998) and by activating MMPs to further disrupt the blood–spinal barrier (Rosenberg, 1995). We are not suggesting that inhibition of TNF-α is the primary means by which IL-10 is neuroprotective. Rather, we suggest that TNF-α serves as an indicator of spinal cord inflammation and macrophage activation. A recent study using mice nullizygous for TNF-α supports these findings (Scherbel et al., 1999). Following traumatic brain injury, wild-type mice had a poorer neurological outcome than TNF-α-deficient mice 2 days after injury. However, wild-type mice recovered to almost normal levels between 2 and 4 weeks after injury, while TNF-α-deficient mice did not recover beyond the levels detected at 2 days. Furthermore, TNF-α-deficient mice had a more severe histopathological outcome than wild-type mice at the end of the study. This study suggests that the production of TNF-α between 2 and 5 days may be neurodestructive and that the synthesis of TNF-α at later time points may be essential for reparative processes to occur. Our results support these data. Delivery of IL-10 30 min after injury significantly reduces TNF-α protein in the spinal cord between 6 h and 1 day after injury and the production of TNF-α by activated macrophages 3 days after SCI, resulting in partial functional recovery. While a second dose of IL-10 3 days after SCI attenuated more chronic levels of TNF-α, it had no effect on functional recovery.

These data suggest that the timing and components of an inflammatory response may greatly determine the outcome.

Future directions

The studies described in this chapter demonstrated that we know very little about the consequences of an inflammatory reaction following CNS injury. It is clear that inflammation has both beneficial and detrimental consequences after injury. Developing a more thorough understanding of those different processes will be essential for developing effective therapeutic and reparative strategies for treating SCI.

References

Balasingam, V. and Yong, V.W. (1996) Attenuation of astroglial reactivity by interleukin-10. *J. Neurosci.*, 16: 2945–2955.

Barbieri, I., Brusa, R., Basudev, H., Casati, C. and Grilli, M. (1998) Neuroprotective action of interleukin-10 in mouse primary neuronal cultures. *Soc. Neurosci. Abstr.*, 24: 1446.

Barger, S.W., Horster, D., Furukawa, K., Goodman, Y., Krieglstein, J. and Mattson, M.P. (1995) Tumor necrosis factor α and β protect neurons against amyloid β-peptide toxicity: evidence for involvement of a MB-binding factor and attenuation of peroxide and Ca^{2+} accumulation. *Proc. Natl. Acad. Sci. USA*, 92: 9328–9332.

Barna, B.P., Pettay, J., Barnett, G.H., Zhou, P., Iwasaki, K. and Estes, M.L. (1994) Regulation of monocyte chemoattractant protein-1 expression in adult human non-neoplastic astrocytes is sensitive to tumor necrosis factor (TNF) or antibody to the 55-kDa TNF receptor. *J. Neuroimmunol.*, 50: 101–107.

Barone, F.C., Arvin, B., White, R.F., Miller, A., Webb, C.L., Willete, R.N., Lysko, P.G. and Feuerstein, G.Z. (1997) Tumor necrosis factor α: a mediator of focal ischemic brain injury. *Stroke*, 28: 1233–1244.

Bartholdi, D. and Schwab, M.E. (1997) Expression of pro-inflammatory cytokine and chemokine mRNA upon experimental spinal cord injury in mouse: an in situ hybridization study. *Eur. J. Neurosci.*, 9: 1422–1438.

Bazan, J.F. (1990a) Structural design and molecular evolution of a cytokine receptor superfamily. *Proc. Natl. Acad. Sci. USA*, 87: 6934–6938.

Bazan, J.F. (1990b) Shared architecture of hormone binding domains in type I and II interferon receptors. *Cell*, 61: 753–754.

Benveniste, E.N. (1992) Inflammatory cytokines within the central nervous system: sources, function, and mechanism of action. *Am. J. Physiol.*, 263: C1–C16.

Berg, D.J., Kuhn, R., Rajewsky, K., Muller, W., Menon, S., Davidson, N., Grunig, G. and Rennick, D. (1995) Interleukin-10 is a central regulator of the response to LPS in murine models of endotoxic shock and the Schwartzman reaction but not endotoxin tolerance. *J. Clin. Invest.*, 96: 2339–2347.

Berkman, N., John, M., Roesems, G., Jose, P.J., Barnes, P.J. and Chung, K.F. (1995) Inhibition of macrophage inflammatory protein-1 a expression by IL-10. *J. Immunol.*, 155: 4412–4418.

Berman, R.M., Suzuki, T., Tahara, H., Robbins, P.D., Narula, S.K. and Lotze, M.T. (1996) Systemic administration of cellular IL-10 induces an effective specific, and long-lived immune response against established tumors in mice. *J. Immunol.*, 157: 231–238.

Bethea, J.R., Castro, M., Keane, R.W., Lee, T.T., Dietrich, W.D. and Yezierski, R.P. (1998) Traumatic spinal cord injury induces nuclear factor-κ-B activation. *J. Neurosci.*, 18: 3251–3260.

Bethea, J.R., Nagashima, H., Acosta, M.C., Briceno, C., Gomez, F., Marcillo, A.E., Loor, K., Green, J. and Dietrich, W.D. (1999) Systemically administered interleukin-10 reduces tumor necrosis factor-alpha production and significantly improves functional recovery following traumatic spinal cord injury in rats. *J. Neurotrauma*, 16: 851–863.

Blight, A.R. (1985) Delayed demyelination and macrophage invasion: a candidate for secondary cell damage in spinal cord injury. *CNS Trauma*, 2: 299–315.

Blight, A.R. (1992) Macrophages and inflammatory damage in spinal cord injury. *J. Neurotrauma*, 9: S83–91.

Blight, A.R. (1994) Effects of silica on the outcome from experimental spinal cord injury: implication of macrophages in secondary tissue damage. *Neuroscience*, 60: 263–273.

Brewer, K., Yezierski, R.P. and Bethea, J.R. (1999) Neuroprotective effects of interleukin-10 following excitotoxic spinal cord injury. *Exp. Neurol.*, 159: 484–493.

Brodie, C. (1996) Differential effects of Th$_1$ and Th$_2$ derived cytokines on NGF synthesis by mouse astrocytes. *FEBS Lett.*, 394: 117–120.

Cheng, B., Christakos, S. and Mattson, M.P. (1994) Tumor necrosis factors protect against metabolic-excitotoxic insults and promote maintenance of calcium homeostasis. *Neuron*, 12: 139–153.

Clark, R.A.F. (Ed.) (1996) *The Molecular and Cellular Biology of Wound Repair*, 2nd ed. Plenum Press, New York.

Combs, C.K., Johnson, D.E., Cannady, S.B., Lehman, T.M. and Landreth, G.E. (1999) Identification of microglial signal transduction pathways mediating a neurotoxic response to amyloidogenic fragments of β-amyloid and prion proteins. *J. Neurosci.*, 19: 928–939.

Crisi, G.M., Santambrogio, L., Hochwald, G.M., Smith, S.R., Carlino, J.A. and Thorbecke, G.J. (1995) Staphylococcal enterotoxin B and tumor-necrosis factor-α-induced relapses of experimental allergic encephalomyelitis: protection by transforming growth factor-β and interleukin-10. *Eur. J. Immunol.*, 25: 3035–3040.

Cross, A.H., Keeling, R.M., Goorha, S., San, M., Rodi, C., Wyatt, P.S., Manning, P.T. and Misko, T.P. (1996) Inducible nitric oxide synthase gene expression and enzyme activity correlate with disease activity in murine experimental autoimmune encephalomyelitis. *J. Neuroimmunol.*, 71: 145–153.

Cua, D.J., Coffman, R.L. and Stohlman, S.A. (1996) Exposure to T helper 2 cytokines in vivo before encounter with antigen selects for T helper subsets via alterations in antigen-presenting cell function. *J. Immunol.*, 157: 2830–2836.

Cunha, F.G., Moncada, S. and Liew, F.Y. (1992) Interleukin-10 inhibits the induction of nitric oxide synthase by interferon-γ in murine macrophages. *Biochem. Biophys. Res. Commun.*, 183: 1155–1159.

Dawson, D.A., Martin, D. and Hallenbeck, J.M. (1996) Inhibition of tumor necrosis factor-alpha reduces focal cerebral ischemic injury in the spontaneously hypertensive rat. *Neurosci. Lett.*, 218: 41–44.

Dietrich, W.D., Busto, R. and Bethea, J.R. (1999) Postischemic hypothermia and IL-10 treatment provide long-lasting neuroprotection of CA1 hippocampus following transient global ischemia in rats. *Exp. Neurol.*, 158: 444–450.

Downen, M., Amaral, T.D., Hua, L.L., Zhao, M. and Lee, S.C. (1999) Neuronal death in cytokine-activated primary human brain cell culture: role of tumor necrosis factor-α. *Glia*, 28: 114–127.

D'Souza, S., Alinauskas, K., McCrea, E., Goodyer, C. and Antel, J.P. (1995) Differential susceptibility of human CNS-derived cell populations to TNF-dependent and independent immune-mediated injury. *J. Neurosci.*, 15: 7293–7300.

Dusart, I. and Schwab, M.E. (1994) Secondary cell death and the inflammatory reaction after dorsal hemisection of the rat spinal cord. *Eur. J. Neurosci.*, 7: 12–24.

Eugster, H., Frei, K., Bachmann, R., Bluethman, H., Lassmann, H. and Fontana, A. (1999) Severity of symptoms and demyelination in MOG-induced EAE depends on TNF-R1. *Eur. J. Immunol.*, 29: 626–632.

Findbloom, D.S. and Winestock, K.D. (1995) IL-10 induces the tyrosine phosphorylation of Tyk 2 and Jak 1 and the differential assembly of STAT1a and STST3 complexes in human T cells and monocytes. *J. Immunol.*, 155: 1079–1090.

Frei, K., Lins, H., Schwerdel, C. and Fontana, A. (1994) Antigen presentation in the central nervous system: the inhibitory effect of IL-10 on MHC class II expression and production of cytokines depends on the inducing signals and the type of cell analyzed. *J. Immunol.*, 152: 2720–2728.

Geng, Y., Gulbins, E., Altman, A. and Lotz, M. (1994) Monocyte deactivation by interleukin 10 via inhibition of tyrosine kinase activity and the Ras signaling pathway. *Proc. Natl. Acad. Sci. USA*, 91: 8602–8606.

Ghirnikar, R.S., Lee, Y.L., He, T.R. and Eng, L.F. (1996) Chemokine expression in rat stab wound brain injury. *J. Neurosci. Res.*, 46: 727–733.

Giulian, D. (1990) Microglia, cytokines, and cytotoxins: modulators of cellular responses after injury to the central nervous system. *J. Immunol. Immunopharmacol.*, 10: 15–21.

Giulian, D., Vaca, K. and Corpuz, M. (1993) Brain glia release factors with opposing actions upon neuronal survival. *J. Neurosci.*, 13: 29–37.

Giulian, D., Yu, J., Li, X., Tom, D., Li, J., Wendt, E., Lin, S., Schwarcz, R. and Noonan, C. (1996) Study of receptor-mediated neurotoxins released by HIV-1-infected mononuclear phagocytes found in human brain. *J. Neurosci.*, 16: 3139–3153.

Goetzl, E.J., Banda, M.J. and Leppert, D. (1996) Matrix metalloproteinases in immunity. *J. Immunol.*, 156: 1–4.

Guth, L., Zhang, Z. and Roberts, E. (1995) Key role for pregnenolone in combination therapy that promotes recovery after spinal cord injury. *Proc. Natl. Acad. Sci. USA*, 91: 12308–12312.

Hirschberg, D.L., Yoles, E., Belkin, M. and Schwartz, M. (1994) Inflammation after azonal injury has conflicting consequences for recovery of function: rescue of spared azons is impaired but regeneration is supported. *J. Neuroimmunol.*, 50: 9–16.

Ho, A.S.Y. and Moore, K.W. (1994) Interleukin-10 and its receptor. *Ther. Immunol.*, 1: 173–185.

Hopkins, S.J. and Rothwell, N.J. (1995) Cytokines and the nervous system, I. Expression and recognition. *Trends Neurosci.*, 18: 83–88.

Hsu, C.Y., Lin, T.-N., Xu, J., Chao and J., Hogan, E.L. (1994) Kinins and related inflammatory mediators in central nervous system injury. In: Salzman, S.K. and Faden, A.I. (Eds.), *The Neurobiology of Central Nervous System Trauma*. Oxford University Press, New York, pp. 145–154.

Issazaseh, S., Lorentzen, J.C., Mustafa, M.I., Höjeberg, B., Müssener, Å. and Olsson, T. (1996) Cytokines in relapsing experimental autoimmune encephalomyelitis in DA rats: persistent mRNA expression of proinflammatory cytokines and absent expression of interleukin-10 and transforming growth factor-β. *J. Neuroimmunol.*, 69: 103–115.

Kaltschmidt, B., Baeuerle, P.A. and Kaltschmidt, C. (1993) Potential involvement of the transcription factor NF-κB in neurological disorders. *Molec. Aspects Med.*, 14: 171–190.

Klusman, I. and Schwab, M.E. (1997) Effects of pro-inflammatory cytokines in experimental spinal cord injury. *Brain Res.*, 762: 173–184.

Knoblach, S.M. and Faden, A.I. (1998) Interleukin-10 improves outcome and alters proinflammatory cytokine expression after experimental traumatic brain injury. *Exp. Neurol.*, 153: 143–151.

Koedel, U., Bernatowicz, A., Frei, K., Fontana, A. and Pfister, H.-W. (1996) Systemically (but not intrathecally) administered IL-10 attenuates pathophysiologic alterations in experimental pneumococcal meningitis. *J. Immunol.*, 157: 5185–5191.

Kotenko, K.V., Krause, C.D., Izotova, L.S., Pollack, B.P., Wu, W. and Pestka, S. (1997) Identification and functional characterization of a second chain of the interleukin-10 receptor complex. *EMBO J.*, 16: 5894–5903.

Lavine, S.D., Hofman, F.M. and Zlokovic, B.V. (1998) Circulating antibody against tumor necrosis factor alpha protects rat brain from reperfusion injury. *J. Cereb. Blood Flow Metab.*, 18: 52–58.

Lee, T.T., Green, B.A., Dietrich, W.D. and Yezierski, R.P. (1999) Neuroprotective effects of cytokines and neurotrophic factors following spinal cord contusion injury in the rat. *J. Neurotrauma*, 16: 347–356.

Levy, Y. and Brouet, J.C. (1994) Interleukin-10 prevents spontaneous death of germinal center B cells by induction of the BCL-2 protein. *J. Clin. Invest.*, 93: 424–428.

Liang, F.Y., Moret, V., Wiesendanger, M. and Rouiller, E.M. (1991) Corticomotoneuronal connections in the rat: evidence from double-labeling of motor neurons and corticospinal axon arborizations. *J. Comp. Neurol.*, 311: 356–366.

Liu, T., Clark, R.K., McDonnell, P.C., Young, P.R., White, R.F., Barone, F.C. and Feuerstein, G.Z. (1994) Tumor necrosis factor α expression in ischemic neurons. *Stroke*, 25: 1481–1488.

McTigue, D.M., Tani, M., Kricac, K., Chernosky, A., Kelner, G.S., Maciejewski, D., Maki, R., Ransohoff, R.M. and Stokes, B.T. (1998) Selective chemokine mRNA accumulation in the rat spinal cord after contusion injury. *J. Neurosci. Res.*, 53: 368–376.

Mertz, P.M., DeWitt, D.L., Stetler-Stevenson, W.G. and Wah, L.M. (1994) Interleukin 10 suppression of monocyte prostaglandin H synthase-2. *Biochemistry*, 269: 21322–21329.

Minc-Glomb, D., Yadid, G., Tsarfaty, I., Reseau, J.H. and Schwartz, J.P. (1996) In vivo expression of inducible nitric oxide synthase in cerebellar neurons. *J. Neurochem.*, 66: 1504–1509.

Minghetti, L., Polizzi, E., Nicolini, A. and Levi, G. (1998) Opposite regulation of prostaglandin E_2 synthesis by transforming growth factor-1 and interleukin 10 in activated microglial cultures. *J. Neuroimmunol.*, 82: 31–39.

Nawashiro, H., Martin, D. and Hallenbeck, J.M. (1997) Inhibition of tumor necrosis factor and amelioration of brain infarction in mice. *J. Cereb. Blood Flow Metab.*, 17: 229–232.

O'Farrel, A.M., Liu, Y., Moore, K.W. and Mui, A.L.F. (1998) IL-10 inhibits macrophage activation and proliferation by distinct signaling mechanisms: evidence for Stat3-dependent and independent pathways. *EMBO J.*, 17: 1006–1018.

O'Neill, L.A.J. and Kaltschmidt, C. (1997) NF-κB: a crucial transcription factor for glial and neuronal cell function. *Trends Neurosci.*, 20: 252–258.

Pan, W., Kastin, A., Bell, R. and Olson, R. (1999) Upregulation of tumor necrosis factor-α transport across the blood barrier after acute compressive spinal cord injury. *J. Neurosci.*, 19: 3649–3655.

Popovic, P.G., Reinhard, J., Flanagan, E.M. and Stokes, B.T. (1994) Elevation of the neurotoxin quinolinic acid occurs following spinal cord trauma. *Brain Res.*, 633: 348–352.

Popovich, P.G., Wei, P. and Stokes, B.T. (1997) Cellular inflammatory response after spinal cord injury in Sprague–Dawley and Lewis rats. *J. Comp. Neurol.*, 377: 433–464.

Popovich, P.G., Whitacre, C.C. and Stokes, B.T. (1998) Is spinal cord injury an autoimmune disorder?. *Neuroscientist*, 4: 71–76.

Ransohoff, R.M. and Benveniste, E.N. (1996) *Cytokines and the CNS*. CRC Press, Boca Raton, New York, 339 pp.

Rogy, M.A., Auffenberg, T., Espat, N.J., Philip, R., Remick, D. and Wollenberg, G.K. (1995) Human tumor necrosis factor receptor (p55) and interleukin 10 gene transfer in the mouse reduces mortality to lethal endotoxemia and also attenuates local inflammatory responses. *J. Exp. Med.*, 181: 2289–2293.

Romano, M.F., Lamberti, A., Petrella, A., Bisogni, R., Tassone, P.F., Formisano, S., Venuta, S. and Turco, M.C. (1996) IL-10 inhibits nuclear factor-κB/Rel nuclear activity in CD2-stimulated human peripheral T lymphocytes. *J. Immunol.*, 156: 2119–2123.

Rosenberg, G.A. (1995) Matrix metalloproteinases in brain injury. *J. Neurotrauma*, 12: 833–842.

Rothwell, N.J. and Relton, J.K. (1993) Involvement of interleukin-1 and lipocortin-1 in ischaemic brain damage. *Cerebrovasc. Brain Metab. Rev.*, 5: 178–198.

Rott, O., Fleisher, B. and Cash, E. (1994) Interleukin-10 prevents experimental allergic encephalomyelitis in rats. *Eur. J. Immunol.*, 24: 1434–1440.

Rudick, R.A., Ransohoff, R.M., Peppler, R., Medendorp, S.V., Lehmann, P. and Alam, J. (1996) Interferon beta induces interleukin-10 expression: relevance to multiple sclerosis. *Ann. Neurol.*, 40: 618–627.

Salzman, S.K. and Faden, A.I. (1994) *The Neurobiology of Central Nervous System Trauma*. Oxford University Press, New York, 347 pp.

Sanna, A., Pflug, B., Colangelo, A.M. and Mocchetti, I. (1998) Interleukin-10 prevents glutamate-induced apoptosis in cerebellar granule cells. *Soc. Neurosci. Abstr.*, 24: 1781.

Sato, I., Himi, T. and Murota, S. (1996) Lipopolysaccharide-induced nitric oxide synthase activity in cultured cerebellar granule neurons. *Neurosci. Lett.*, 205: 45–48.

Scherbel, U., Raghupathi, R., Nakamura, M., Saatman, K.E., Trojanowski, J.Q., Neugebauer, E., Mario, M.W. and McIntosh, T.K. (1999) Differential acute and chronic responses of tumor necrosis factor-deficient mice to experimental brain injury. *Proc. Natl. Acad. Sci. USA*, 96: 8721–8726.

Serpa, P.A., Ellison, J.A., Feuerstein, G.Z. and Barone, F.C. (1998) IL-10 reduces rat brain injury following focal stroke. *Neurosci. Lett.*, 251: 189–192.

Shrikant, P., Weber, E., Jilling, T. and Benveniste, E.N. (1995) Intercellular adhesion molecule-1 gene expression by glial cells. *J. Immunol.*, 155: 1489–1501.

Stalder, A.K., Pagenstecher, A. and Campbell, I.L. (1996) Lymphocytic meningoencephalomyelitis induced by transgenic expression of TNF-α in the CNS. *Soc. Neurosci. Abstr.*, 22: 1455.

Streit, W.J., Semple-Rowland, S., Hurley, S.D., Miller, R.C., Popovich, P.G. and Stokes, B.T. (1998) Cytokine mRNA profiles in contused spinal cord and axotomized facial nucleus suggest a beneficial role for inflammation and gliosis. *Exp. Neurol.*, 152: 74–87.

Taga, K., Chretien, J. and Cherney, B. (1994) Interleukin-10 inhibits apoptotic cell death in infectious mononucleosis T cells. *J. Clin. Invest.*, 94: 251–260.

Taoka, Y., Okajima, K., Uchiba, M., Harada, N., Johno, M. and Naruo, M. (1998) Activated protein C reduces the severity of compression injury in rats by inhibiting activation of leukocytes. *J. Neurosci.*, 18: 1393–1398.

Thoreau, E., Petridou, B., Kelly, P.A. and Mornon, J.P. (1991) Structural symmetry of the extracellular domain of the cytokine/growth hormone/prolactin receptor family and interferon receptors revealed by hydrophobic cluster analysis. *FEBS Lett.*, 282: 26–31.

Toulmond, S., Parnet, P. and Linthorst, A.C.E. (1996) When cytokines get on your nerves: cytokine networks and CNS pathologies. *Trends Neurosci.*, 19: 409–410.

Tureen, J. (1995) Effect of recombinant human tumor necrosis factor-alpha on cerebral oxygen uptake, cerebrospinal fluid lactate, and cerebral blood flow in the rabbit: role of nitric oxide. *J. Clin. Invest.*, 95: 1086–1091.

Veenam, C.L., Reiner, A. and Honig, M.G. (1992) Biotinylated dextran amine as an anterograde tracer for single- and double-labeling studies. *J. Neurosci. Methods*, 41: 239–254.

Wang, P., Wu, P., Siegel, M.I., Egan, R.W. and Billah, M.M. (1995) Interleukin (IL)-10 inhibits nuclear factor κB (NF-κB) activation in human monocytes. *J. Biol. Chem.*, 270: 9558–9563.

Wang, C.X., Nuttin, B., Heremans, H., Dom, R. and Gybels, J. (1996) Production of tumor necrosis factor in spinal cord following traumatic injury in rats. *J. Neuroimmunol.*, 69: 151–156.

Wang, C.-Y., Mayo, M.W. and Baldwin, A.S. (1996) TNF- and cancer therapy-induced apoptosis: potentiation by inhibition of NF-κB. *Science*, 274: 784–787.

Wu, W. (1993) Expression of nitric oxide synthase (NOS) in injured CNS neurons as shown by NADPh diaphorase histochemistry. *Exp. Neurol.*, 120: 153–159.

Yakovlev, A.G. and Faden, A.I. (1994) Sequential expression of c-fos protooncogene, TNF-alpha, and dynorphin genes in spinal cord following experimental traumatic injury. *Mol. Chem. Neuropathol.*, 23: 179–190.

Yong, V.W., Krekoski, C.A., Forsyth, P.A., Bell, R. and Edwards, D.R. (1998) Matrix metalloproteinases and diseases of the CNS. *Trends Neurosci.*, 21: 75–80.

Zhang, Z., Krebs, C.J. and Guth, L. (1997) Experimental analysis of progressive necrosis after spinal cord trauma in the rat: etiological role of the inflammatory response. *Exp. Neurol.*, 143: 141–152.

CHAPTER 5

Immunological regulation of neuronal degeneration and regeneration in the injured spinal cord

Phillip G. Popovich [*]

Department of Molecular Virology, Immunology and Medical Genetics, College of Medicine and Public Health, The Ohio State University, Columbus, OH 43210, USA

Introduction

Inflammation is classically defined as a host defense mechanism triggered by injury or infection whereby leukocytes and humoral factors attempt to restore homeostasis to the affected site. In its acute form, inflammation is characterized by pain, swelling, heat and loss of function. These latter features of inflammation are caused by the collective and parallel processes of vessel dilation, hyperemia, plasma extravasation and leukocyte infiltration, particularly neutrophils. If the acute response is ineffective in restoring tissue homeostasis, chronic inflammation ensues, provoking monocyte infiltration and concomitant activation of adaptive immunity governed by lymphocyte activation. At this point the boundaries between physiological and pathological inflammation/immunity can become ambiguous. In the injured spinal cord, the cells and mediators of the immune system can influence the molecular, biochemical and anatomical cues that are essential for successful regeneration. Thus, as we continue to explore the biological impact of leukocyte infiltration into the traumatically injured spinal cord, it will be important to be able to distinguish between inflammation, physiological immunity, and pathological immunity. However, to accomplish this, one must appreciate basic immunological principles as well as the various systemic factors and microenvironmental cues within the injury site that can influence immune function. In this chapter, an overview of the interrelationship between immune and inflammatory processes will be discussed as it pertains to macrophage and lymphocyte-mediated reparative and degenerative influences on the injured spinal cord. When relevant, the roles played by cytokines, chemokines, and/or cell adhesion molecules will be presented. Although important in the acute pathophysiology of cerebrovascular ischemia and perhaps in acute compressive spinal cord injury (SCI), a discussion of neutrophils will not be undertaken as excellent reviews already exist on this topic (Taoka et al., 1997, 1998) In the latter part of the chapter, a discussion of inherent physiological variables present in clinical and experimental models of SCI will be discussed from the perspective of immune-mediated modulation of neural regeneration. For example, SCI-induced alterations in the neuroendocrine system, intrinsic differences in the location and severity of injury, genetic factors, immunological competence at time of injury, age and gender may influence the nature, magnitude and ultimately the regenerative potential of post-traumatic inflammation and immunity. This type of theoretical analysis is necessary to facilitate the development and interpretation of

[*] Corresponding author: Dr. Phillip G. Popovich, Department of Molecular Virology, Immunology and Medical Genetics, College of Medicine and Public Health, The Ohio State University, Columbus, OH 43210, USA. Fax: +1-614-292-9805; E-mail: popovich.2@osu.edu

immune-based therapies designed to minimize neuropathology and promote recovery of function after SCI.

Innate/natural immunity in the injured spinal cord

The onset and perpetuation of inflammation following traumatic central nervous system (CNS) injury involves the coordinated efforts of phagocytic cells (neutrophils, monocytes/macrophages), lymphocytes and soluble mediators/proteins (e.g., cytokines, complement) that serve to amplify the cellular response. Collectively, these cells and their secretory products represent the natural and acquired immune systems (Table 1). Within the injured CNS, resident microglia, extravasated complement proteins, and infiltrating neutrophils and blood monocytes constitute the natural immune system. Representing the first line of defense against injury and infection, these cells and proteins interact to eliminate pathogenic organisms in the affected site (if present) while simultaneously priming the site for repair. Unfortunately, the unrefined nature in which these cells operate can irreversibly damage neurons and glia. For example, spillover of nitrogen and oxygen free radicals, cytokines, proteolytic enzymes, and neuronal excitotoxins is a byproduct of natural immunity and may cause collateral damage (also referred to as 'secondary' injury) to cells spared by the primary traumatic insult. While this type of bystander damage is easily repaired in peripheral tissues, CNS neurons are less likely to regenerate. Macrophages also appear to play a role in the formation of an extracellular environment that is inhibitory to axon growth (Fitch and Silver, 1997). Therefore, since SCI elicits a robust macrophage response, it is conceivable that these cells mediate post-traumatic secondary injury and/or contribute to the failure of CNS axonal regeneration. Indeed, attenuation of the macrophage response is effective in improving neurologic function and in decreasing axon/myelin damage after experimental SCI (Giulian and Robertson, 1990; Blight, 1994; Popovich et al., 1999).

Microglia in the traumatically injured spinal cord

Previously, we demonstrated spatial correlations between microglial/macrophage activation and discrete regions of anatomically and physiologically compromised spinal cord (Popovich et al., 1993, 1996a, 1997). However, we could not determine if these cellular reactions signified tissue damage or repair or whether a specific macrophage subtype was involved. Making these distinctions is difficult due to the variability of the macrophage response to injury, i.e., the phenotype, morphology, and density of responding cells change as a function of time after injury, as does their distribution within or adjacent to the site of impact.

Within the CNS, microglia are the resident tissue macrophages and are derived from bone-marrow precursors during development. They represent as much as 20% of the total glial population and are widely distributed throughout the CNS, making them ideal sentinel cells (Lawson et al., 1990). The scope of microglial surveillance is exemplified by activation of spinal cord microglia following brain injury (Schmitt et al., 1998). Even the onset of systemic pathology (e.g., heart disease) can trigger microglial activation

TABLE 1

Natural and acquired immunity in the injured spinal cord

Natural immunity. Innate mechanisms of defense against infection and injury. Both vertebrates and invertebrates possess some form of natural immunity.
    Physical barriers: vertebral bodies (spinal column), skull, meninges, blood–brain barrier, cerebrospinal fluid, 'immune privilege factor'
    Circulating proteins: complement
    Cells: microglia, neutrophils, NK (natural killer) cells, monocytes/macrophages, dendritic cells
    Humoral mediators: cytokines derived from microglia, macrophages, astrocytes, endothelia, neurons

Acquired immunity. All vertebrates, but not invertebrates, possess some form of acquired immunity.
    Circulating proteins: immunoglobulins (antibodies)
    Cells: lymphocytes (B- and T-cells)
    Humoral mediators: lymphocyte-derived cytokines ('lymphokines')

(Streit and Sparks, 1997). Microglial-derived cytokines, neurotransmitters, proteolytic enzymes and neurotrophins can influence the differentiation and survival of neurons, astrocytes and oligodendrocytes (Banati and Graeber, 1994). Recent data also suggest that microglia are part of a resident neurotrophin network. Indeed, microglia express p75, trkA, trkB and trkC receptors and respond to neurotrophin-3 (NT-3) and brain-derived neurotrophic factor (BDNF) (Elkabes et al., 1996; Nakajima et al., 1998). Because microglial activation precedes that of any other cell type in the CNS in response to injury or infection, it is conceivable that the rate, magnitude and the ultimate effect (supporting regeneration or degeneration) of subsequent inflammatory processes at the injury site are controlled by microglia.

When evaluating microglial influences on posttraumatic inflammation, one must not discount the importance of the cellular microenvironment in which these cells become activated. For example, due to differences in cytoarchitecture or microvascular density, gray and white matter microglia express distinct morphologies and different levels of cell surface antigens (Perry et al., 1993a). These phenotypic differences exist before and after SCI and likely correspond with unique cell functions (Streit et al., 1988; Popovich et al., 1997). Thus, anterograde axonal degeneration may trigger unique functions in white matter microglia compared to the effects of ischemia or necrotizing injury on gray matter microglia. In addition, expression of chemokine and cytokine mRNA and evidence of cytokine signaling (e.g., NFκB expression) are more pronounced in gray matter after SCI (Bartholdi and Schwab, 1997; Bethea et al., 1998). Perhaps because of these differences, anti-inflammatory agents exert unique effects at the injury site (mostly gray matter injury) compared with regions undergoing Wallerian degeneration (primarily white matter injury) (Zhang et al., 1996). Taken together, these data support the idea that metabolic changes in microglial function are differentially affected at the injury site and in degenerating white matter tracts removed from the site of trauma. The extent of microglial activation also may be dependent on the extent of blood–brain barrier (BBB) injury, which varies between the gray and white matter (Popovich et al., 1996a). Given differences in gray-to-white matter ratios in cervical, thoracic, lumbar and sacral spinal cord, the level at which a SCI occurs could affect the molecular profile of the CNS inflammatory response.

Blood–brain barrier injury and inflammation

Injury to the BBB is an important variable to consider when discussing neuroinflammation because of the introduction of cells and proteins into the previously 'sequestered' spinal tissue. Such proteins and cellular byproducts can influence the nature of the inflammatory response. Spinal trauma causes extravasation of red blood cells and accumulation of iron-containing hemoglobin at the injury site. In the presence of iron, neutrophils and macrophages generate inordinate quantities of toxic hydroxyl radicals ($\cdot OH$). Unlike superoxide ($\cdot O_2-$), which is a weak oxidant, $\cdot OH$ radicals can cause considerable DNA, protein, and lipid damage resulting in neuronal and glial cell death (Halliwell and Gutteridge, 1985; Liu et al., 1994).

Blood–brain barrier injury also facilitates the extravasation of immunoglobulins and complement proteins into the neural parenchyma. Accumulation of complement fragments (specifically C3a and C5a fragments) in the CNS could be deleterious by augmenting cell-mediated inflammatory bystander damage (e.g., influencing macrophage cytokine profiles), directly mediating lysis of neurons and glia via formation of the membrane attack complex (MAC) and/or causing cell damage through disruption of the lipid bilayer. Serum complement has been implicated as a contributing factor in the onset of CNS autoimmune demyelination (Brosnan et al., 1983; Linington et al., 1989). A similar role for complement in the acute neuropathology of spinal trauma must not be overlooked as elevated levels of complement in the serum and cerebrospinal fluid (CSF) of SCI patients has been described (Rebhun et al., 1991) and discrete zones of oligodendrocyte loss colocalize to regions of elevated complement receptor expression on microglia and macrophages (Koshinaga and Whittemore, 1995; Crowe et al., 1997; Popovich et al., 1997; Carlson et al., 1998). However, a physiological role for complement must also be considered, given the recent findings of constitutive expression of C3a and C5a receptors on spinal cord neurons (Nataf et al., 1999).

Complement proteins also can influence cytokine production by microglia and infiltrating leukocytes. This is significant as recent evidence suggests that the cytokine environment in the CNS determines the phenotype of the inflammatory infiltrate. For example, intraspinal injections of tumor necrosis factor (TNF)-α induced monocyte infiltration while interleukin (IL)-1β recruited mostly neutrophils (Schnell et al., 1999).

Thus, BBB injury precipitates the onset of various pathophysiological processes that operate upstream of chronic inflammation and may ultimately influence the nature of the inflammatory response. If BBB injury and tissue trauma are minimized, inflammation is reduced and CNS axon regeneration occurs, even within the normally inhibitory environment of the white matter, reinforcing the notion that BBB disruption plays a role in the extent of tissue damage and repair occurring after SCI (Davies et al., 1997). Since plasma protein extravasation decreases rostral and caudal to the injury site, each of the aforementioned effects are likely to be graded as a function of distance from the lesion center (Noble and Wrathall, 1989).

CNS macrophage heterogeneity: microglia vs. hematogenous macrophages

Even under controlled conditions (e.g., in vitro), microglia and macrophages can be triggered to produce different levels and types of cytokines and neurotoxic molecules (e.g., quinolinate, superoxide) (Giulian et al., 1989, 1995; Banati et al., 1993; Alberati-Giani et al., 1996; Mosley and Cuzner, 1996). Coupled to the release of these compounds are differences in phagocytic potential, both to particulate antigens and to myelin (Mosley and Cuzner, 1996; DeJong and Smith, 1997). At the ultrastructural level, microglia and hematogenous (blood-derived) macrophages harvested from sites of CNS trauma or infection are morphologically distinct and respond differently to modulatory signals from astrocytes (Giulian et al., 1995; DeWitt et al., 1998). Previously, we described marked heterogeneity in the morphology and expression of surface antigens on microglia/macrophages after SCI (Popovich et al., 1997). Heterogeneous neurotrophin expression in vivo and differential responses to neurotrophins by microglia in vitro suggest that these cells may elicit unique functional properties in the pathological CNS even if exposed to similar factors at the injury site (Elkabes et al., 1996). Future studies are needed to clarify the potential for macrophage/microglial functional heterogeneity and to determine if a disproportionate activation of one cell type relative to the other influences indices of neuronal plasticity and regeneration.

Manipulating the microglia and macrophage response after spinal cord injury

In the periphery, macrophages are essential for wound repair and regeneration of injured tissues. After SCI, intrinsic repair programs are initiated that result in the formation of a matrix of endothelial sprouts, ependymal cords emanating from the central canal, reactive glia (astrocytes and microglia), and infiltrating Schwann cells (Guth et al., 1985; Beattie et al., 1997). These repair programs precede hematogenous macrophage infiltration but occur subsequent to or in parallel with the activation of resident microglia. Therefore, interventions that alter the kinetics or nature of the acute inflammatory response to trauma might indirectly affect later processes or repair and regeneration. Since the cellular matrix that forms after injury is not sufficient to maintain axonal growth, it is possible that infiltrating macrophages antagonize the efforts of resident cells that are attempting to repair the injury site. Previously, we and others have suggested that the inefficient progression of endogenous repair and the formation of an axon restrictive growth environment is mediated by BBB damage, acute infiltration of blood monocytes, and accumulation of inhibitory extracellular matrix molecules (Fitch and Silver, 1997; Popovich et al., 1999). For example, acute depletion or inhibition of monocytically derived macrophages (i.e., hematogenous cells) or serum complement may limit chondroitin sulfate proteoglycan deposition and reduce the phagocytosis-coupled release of antibacterial agents (e.g., superoxide, hydrogen peroxide, hypochlorous acid), quinolinic acid, or proteolytic enzymes. These latter compounds, although innocuous in the regenerative tissues of the periphery, could cause inefficient repair, progressive necrosis/apoptosis, and destruction

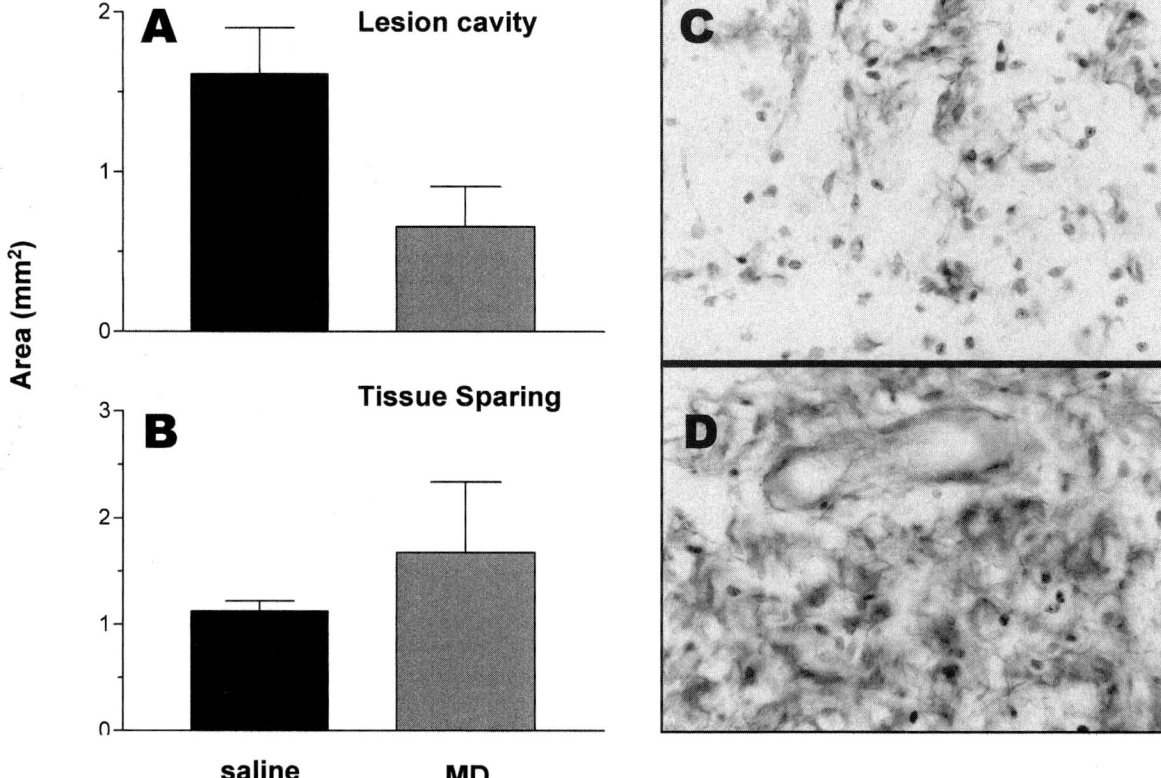

Fig. 1. Depletion of blood-derived macrophages reduces secondary injury and promotes axon regeneration in the lesion center following SCI. Selective depletion of hematogenous macrophages using intravenous injections of liposome-encapsulated chlodronate significantly reduces macrophage infiltration into the injured spinal cord (Popovich et al., 1999). When compared to animals receiving saline injections (A–C), there is marked attenuation of necrotic cavitation (defined by outlining regions of necrosis with image analysis); (A) greater preservation of white matter (B) and robust axon growth (D) in macrophage-depleted animals (A,B,D) (see Popovich et al., 1999 for a more comprehensive analysis using this treatment).

of healthy tissues and neural/glial progenitors within the CNS. In a model of spinal contusion injury, observations of robust axon growth at the lesion center and reduced necrotic cavitation caudal to the site of impact following selective depletion of monocytically derived macrophages supports this hypothesis (Popovich et al., 1999) (Fig. 1).

Recent studies in which successful anatomical repair and functional recovery were noted following transplantation of macrophages or microglia demonstrate that the CNS regenerative failure may be partially overcome if the quality of the macrophage response is altered (Rabchevsky and Streit, 1997; Rapalino et al., 1998). Indeed, the microenvironment in which microglia and/or macrophages are activated influences their neurotrophic or neurotoxic effector potential (Zhang and Fedoroff, 1996; Zietlow et al., 1999). A recent report by Rapalino et al. (1998) provides an excellent example of how 'environment' dictates macrophage effector function. In their studies, blood monocytes were incubated with degenerating peripheral nerve segments and then transplanted into a transected spinal cord. Axonal regeneration was supported and partial recovery of locomotor function was achieved. Furthermore, the regenerative potential of the macrophages was realized only after activation by peripheral nerve but not optic nerve segments (Zeev-Brann et al., 1998), presumably because peripheral nerves lack the undefined inhibitory factors which suppress macrophage function in the CNS (Hirschberg and Schwartz, 1995; Lazarov-Spiegler et al., 1998). These studies illustrate the need for more investigations that seek to define the signals controlling macrophage activation within the injury site.

Adaptive/acquired immunity and spinal cord injury

Unlike natural immunity, adaptive immune responses can discriminate between multiple pathogens and 'non-self' proteins (i.e., antigens) (Table 1). T- and B-lymphocytes are the principal cells involved in adaptive or acquired immune responses. For lymphocytes to participate in CNS inflammation they must be activated; otherwise, they cannot cross the BBB or be maintained within the brain or spinal cord (Wekerle et al., 1986; Hickey et al., 1991). T-cell activation is dependent on ligation of the T-cell receptor (TCR) by peptide and major histocompatibility complex (MHC) molecules found on antigen-presenting cells (APCs). Within the pathological CNS, the most likely APCs are microglia, macrophages and dendritic cells (Kreutzberg, 1996). In addition to the interaction between the TCR and APC, successful T-cell activation depends on co-stimulatory molecules present on the T-cell (e.g., CD28) and the APC (e.g., B7-1, B7-2), cytokines (proinflammatory Th1 vs. immunosuppressive Th2), and the microenvironment in which antigen recognition occurs. Generally, once the TCR is engaged by the MHC/antigen complex, the T-cell undergoes a series of intracellular events leading to cell proliferation. Subsequent release of Th1 cytokines (IL-2, interferon [IFN]γ) by the activated T-cells results in activation of nearby APCs and amplification of the immune response. However, lack of costimulation or the presence of transforming growth factor (TGF)-β or IL-4 (Th2 cytokines) may cause T-cells to become functionally inactive (anergized) or deleted via apoptosis. If antigen-specific triggering of the TCR occurs in the presence of Th2 cytokines or high glucocorticoid levels, an immunosuppressive T-cell population may develop which can persist as long as the antigenic signal is present. Because circulating glucocorticoid levels are elevated following trauma, the latter consideration may play a significant role in shaping the repertoire of the T-cell response following SCI (Cruse et al., 1992, 1996).

Traumatic spinal cord injury and lymphocyte reactions

If the peptide recognized by lymphocytes is of CNS origin (e.g., a myelin peptide fragment), an autoimmune response may develop. The hypothesis that immune system recognition of neural antigens after SCI might contribute to secondary neuropathology and/or the inability of the CNS to regenerate is not without precedent. Based on the animal model for multiple sclerosis (MS), i.e., experimental autoimmune encephalomyelitis (EAE), exposure of T-lymphocytes to CNS myelin results in an acute paralytic disease paralleled by marked intraparenchymal inflammation, demyelination, neuronal cell death and axonal transection (Pender et al., 1990; Trapp et al., 1998). Despite the negative connotation that accompanies 'autoimmunity', these reactions need not imply an insidious pathology. Indeed, pathological autoimmune disease develops in only 3% of the population despite the presence of 'self-reactive' lymphocytes in all individuals (Pette et al., 1990; Theofilopoulos, 1993). However, following SCI, lymphocyte exposure to previously sequestered proteins of the CNS may be sufficient to facilitate the expansion of multiple autoreactive T-cell populations and the subsequent onset of autoimmune disease.

Trauma-induced autoimmunity: a mechanism of injury or repair?

Previously, we demonstrated that myelin basic protein (MBP) reactive T-lymphocytes isolated from lymph nodes of acute spinal injured rats induced a paralytic disease reminiscent of EAE when injected into naive rats (Popovich et al., 1997). Still, the disease-inducing capability of these cells was restricted such that transfer of disease was only possible using T-cells isolated from animals injured 7 days earlier. The inability to transfer disease at later post-injury intervals suggests the development of active regulatory mechanisms which reduce or eliminate the encephalitogenicity of these T-cells. A similar time-dependent regulation of autoimmunity has been described after peripheral nerve injury in animals and humans (Olsson et al., 1992, 1993). Although autoimmunity against neural antigens has been implicated in the acute pathophysiology of SCI (Popovich et al., 1996b), the absence of chronic progressive encephalomyelitis after SCI implies that the myelin-reactive cells become tolerized or actively suppressed by mechanisms presently undefined in models of CNS injury and neurodegeneration (Popovich et al., 1997, 1998).

Given the role played by activated T-cells in modulating macrophage function, microvascular endothelial integrity (Naparstek et al., 1984), axonal conduction (Yarom et al., 1983; Koller et al., 1997), and antibody production by B-lymphocytes (see below), it is important to also consider a role for lymphocytes in repair of the pathological CNS. In EAE it is clear that regulatory T-cells and their cytokines modulate the severity and duration of adverse immunological reactions (Sun et al., 1999). In fact, the majority of infiltrating lymphocytes are not specific for CNS proteins (Cross et al., 1990). These 'irrelevant' lymphocytes may represent a surveillance or active immunoregulatory network. Intraperitoneal injection of MBP-reactive T-cells following injury to the optic nerve resulted in significant neuroanatomical and electrophysiological preservation of retinal ganglion cells (Moalem et al., 1999a). Using this same approach, we and others have demonstrated behavioral improvement and neuroprotection following experimental SCI (Hauben et al., 1999; P.G. Popovich and D.M. Basso, unpubl. observations). Preliminary data from this laboratory indicate a sustained increase in hindlimb function in paraplegic rats using standardized scales for locomotor recovery and tests to evaluate the integrity of descending motor pathways (inclined plane testing of rubrospinal projections). How MBP-reactive T-cells afford neuroprotection in these models of CNS injury remains undefined but may be related to local production of neurotrophins. Indeed, recent data suggest that in addition to releasing pro-inflammatory cytokines, myelin reactive T-cells can produce BDNF (Kerschensteiner et al., 1999).

Although the majority of the examples given above relate to T-cell reactions targeting MBP, neuroantigen-specific T-cell reactions after SCI may not be limited to this protein. As in EAE, a dynamic autoreactive T-cell repertoire could develop as a function of time after injury. For example, as whole myelin is degraded at the injury site, the self-reactive T-cell repertoire may expand to include multiple myelin proteins (e.g., MBP, proteolipid protein [PLP], myelin oligodendrocyte glycoprotein [MOG], and/or peptide fragments of these proteins). This concept of intramolecular and intermolecular spreading results in the formation of various T-cell clones capable of causing autoimmune pathology (Lehmann et al., 1993). However, consistent with the theme of this chapter, T-cell recognition of 'cryptic' or previously sequestered myelin fragments can also be associated with neuroprotection. Moalem et al. (1999a,b) demonstrated similar neuroprotection in the optic nerve model using MBP-reactive T-cells and T-cells that recognized a cryptic epitope of the MBP molecule.

The degree to which these different T-cell clones become pathological will depend on the cytokine environment in which they become activated or the extent to which circulating glucocorticoids accumulate at the injury site, since, as stated above, regulatory T-cell clones can be induced in the presence of increased glucocorticoids (Ramirez et al., 1996; also reviewed in Popovich et al., 1996b). Although glucocorticoid synthesis has not been measured in the periphery in models of experimental SCI, elevated cortisol and plasma adrenocorticotrophic hormone (ACTH) in human SCI correlates with reduced immune function (Cruse et al., 1992).

Antigen binding to immunoglobulins on B-lymphocytes triggers antibody production. However, not all vertebrates develop the full composite of acquired immunity. For example, only higher vertebrates develop antibody diversity (multiple immunoglobulin classes) while lower vertebrates (e.g., common fish) produce only IgM antibodies. Whether evolutionary divergence in the development of acquired immunity came at the expense of CNS regeneration in adult mammals is not known. Like T-cell responses, if the antigen that binds is derived from a 'self' protein, autoantibody production may occur. This axis of autoimmune activation may participate in secondary injury of the traumatized spinal cord through interactions with extravasated serum complement. After SCI, anti-myelin antibodies are present on myelinated axons, providing an ideal environment for complement activation (Mizrachi et al., 1983; Palladini et al., 1987). Experimental evidence supports the pathological potential of this reaction; co-injections of serum complement and antibodies to the myelin sphingolipid, galactocerebroside, have been used successfully to produce focal areas of demyelination (Keirstead and Blakemore, 1997).

Repair of the injured spinal cord using myelin autoantibodies may also be possible. This hypothesis is supported by studies showing that exogenous administration or augmented production of myelin autoantibodies promote CNS remyelination in a model of viral encephalomyelitis (Miller and Rodriguez, 1995) and regeneration of corticospinal axons after spinal hemisection, respectively (Huang et al., 1999). In the viral encephalomyelitis model, there appears to be regional susceptibility to viral mediated injury that can be overcome by unique arms of the immune system. For example, in regions of the brain containing abundant white matter (e.g., brainstem, cerebellum) MHC class-I cell-mediated immunity was more protective than antibody, while other regions containing greater densities of neurons and less myelin (e.g., striatum) were protected to a greater extent by antibody treatment (Drescher et al., 1999). Whether similar regional specificity of immune-mediated protection or injury occurs after traumatic injury to the CNS has not been determined but may partially account for the apparent dichotomy of post-traumatic inflammation.

Inflammation and spinal cord injury: beyond the injury site

The natural migration of inflammatory leukocytes to the injured CNS offers unique possibilities for treating SCI and traumatic brain injury (TBI). Because these cells become colocalized with damaged neurons and axons, they could serve as natural vehicles for delivering trophic or mitogenic signals. Some success in this area has been achieved in EAE where myelin-reactive lymphocytes, genetically modified to secrete anti-inflammatory cytokines (e.g., IL-4, TGF-β, IL-10) or oligodendroglial mitogens (e.g., platelet-derived growth factor), have alleviated the histopathological and neurological consequences of disease (Chen et al., 1998; Tuohy and Mathisen, 1998). For this type of strategy to be useful in SCI and TBI patients, it would be valuable to first consider how inflammatory cell recruitment and function can be affected by physiological and pathological factors prevalent in the clinical and experimental literature.

Neuroendocrine dysfunction, CNS injury and the immune response

The delayed degeneration and repair of the injured spinal cord is clearly influenced by inflammation. However, for this local inflammation to occur, leukocytes must first migrate into the injury site. While emigration of leukocytes across the BBB is affected locally by the expression of vascular cell adhesion molecules, leukocyte migration to regions of tissue damage (i.e., trafficking) is influenced by activation of the hypothalamic–pituitary adrenal (HPA) and sympathetic–adrenal medullary (SAM) axes and subsequent elevations of circulating glucocorticoids (GCs) and catecholamines (epinephrine and norepinephrine; EPI/NEPI) (Fig. 2). Following a stressful event (e.g., SCI or TBI), hypothalamic relays to the pituitary and limbic forebrain (e.g., hippocampus, amygdala) stimulate the release of GC. Since the HPA axis is tempered by hippocampal and other limbic innervation (Herman and Cullinan, 1997), neuronal degeneration in any of these regions following TBI would significantly impact immune responses. The hypothalamus also activates the sympathetic nervous system via connections with the locus coeruleus and limbic forebrain (Sternberg et al., 1992). These higher brain centers project to sympathetic neurons in the spinal cord that innervate the adrenal medulla, thymus, spleen and lymph nodes (Bellinger et al., 1997). In this context, the consequences of injury location (spinal level) on CNS inflammation are more global, i.e., the higher the level of SCI, the more damage to supraspinal pathways that influence autonomic projections at thoracic/lumbar levels. Indeed, cervical SCI disrupts immune function more than thoracic SCI, causing more notable impairments in wound healing and HPA/SAM function (Kliesch et al., 1996). Also, peripheral inflammation (e.g., bone/muscle trauma) and associated increases of circulating cytokines (e.g., IL-10, TNF, IL-6) are usually coupled with TBI and SCI. How these reactions and parallel changes in systemic pathophysiology (e.g., fever, muscle metabolism, wound healing, host defense) affect inflammatory processes within the CNS is not known (Plata-Salaman, 1998) (Fig. 2). Clearly, neuroimmunological studies of SCI and brain trauma transcend cellular reactions at the injury site.

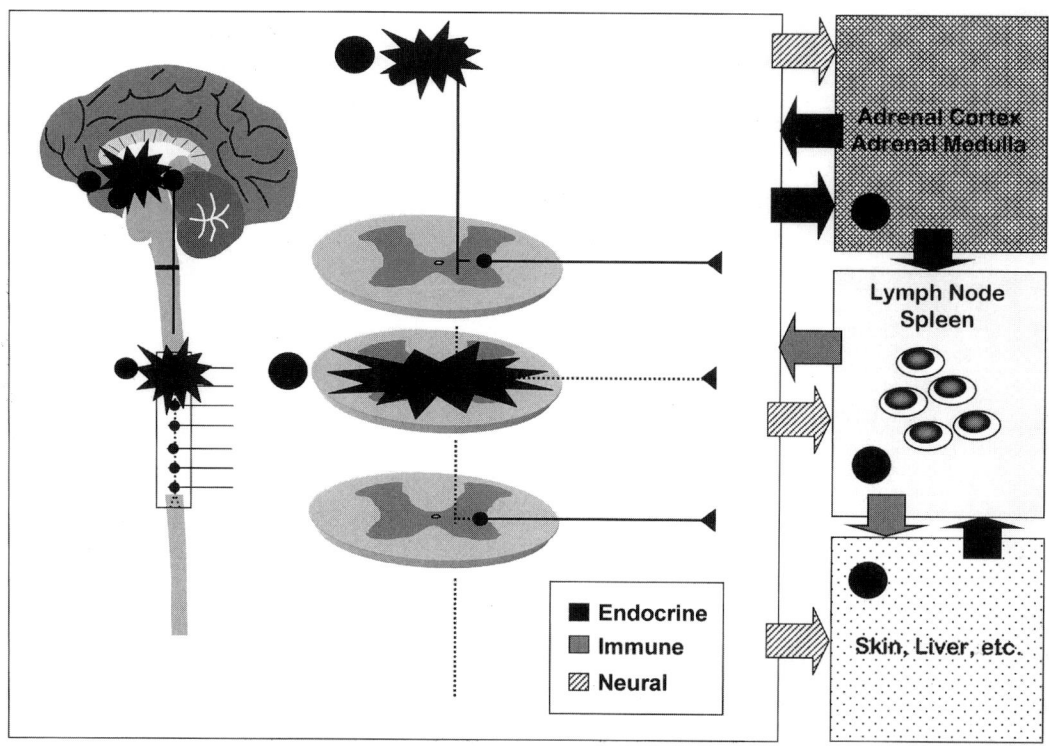

Fig. 2. SCI and TBI affect central and peripheral inflammation through disruption of the HPA and SAM axes. Brain injury causes neuronal cell loss in the hypothalamus and limbic forebrain (e.g., hippocampus) disrupting descending projections to autonomic neurons in the spinal cord. SCI also disrupts sympathetic projections to the adrenal gland and peripheral lymphoid tissues. Higher levels of SCI will produce more comprehensive damage to neuroendocrine pathways. Subsequent changes in leukocyte trafficking and activation will alter the 'quality' of CNS inflammation and wound healing responses in the periphery (hatched arrows). Stress-induced activation of endocrine pathways also occurs following TBI and SCI (black arrows). The degree to which cortisol (from adrenal cortex) and systemic cytokines (induced by acute phase response from liver) affect inflammation after CNS injury has not been studied in detail. Arrows represent sympathetic connections from the CNS (hatched), endocrine hormones (e.g., cortisol, norepinephrine, epinephrine from the adrenal gland and IL-6, IL-1, TNF from the liver; black), and inflammatory mediators (cytokines, neuropeptides, etc.; gray).

Inflammatory reactions in the injured brain and spinal cord are distinct

Despite our tendency to generalize characteristics of inflammation in the injured brain to all models of CNS trauma, neuroinflammation in the injured spinal cord appears to be unique. Studies by Schnell et al. (1999) reveal a clear distinction between inflammation in the brain and spinal cord following identical traumatic insults or injections of pro-inflammatory cytokines (Fig. 3). Extravasation of leukocytes and BBB injury were extensive in the spinal cord but were delayed in onset and of lesser magnitude in the brain. Interestingly, intracerebral injections of monocyte chemoattractant protein (MCP)-1, macrophage inflammatory protein (MIP)-2, and IL-8 can overcome deficits in leukocyte recruitment to the brain, suggesting that the contrast between brain and spinal cord inflammation may be explained by differences in chemokine production (Bell et al., 1996).

The degree of mechanical trauma imparted to the brain or spinal cord also will vary due to fundamental differences in their structure (Blight, 1996; Povlishock, 1996). For instance, while the white matter lies directly beneath the meninges of the spinal cord, gray matter occupies this position in the brain. Gray and white matter differences in metabolic activity, vascularization, cell density and composition could influence inflammatory processes (see above). Unique lymphatic drainage pathways may explain

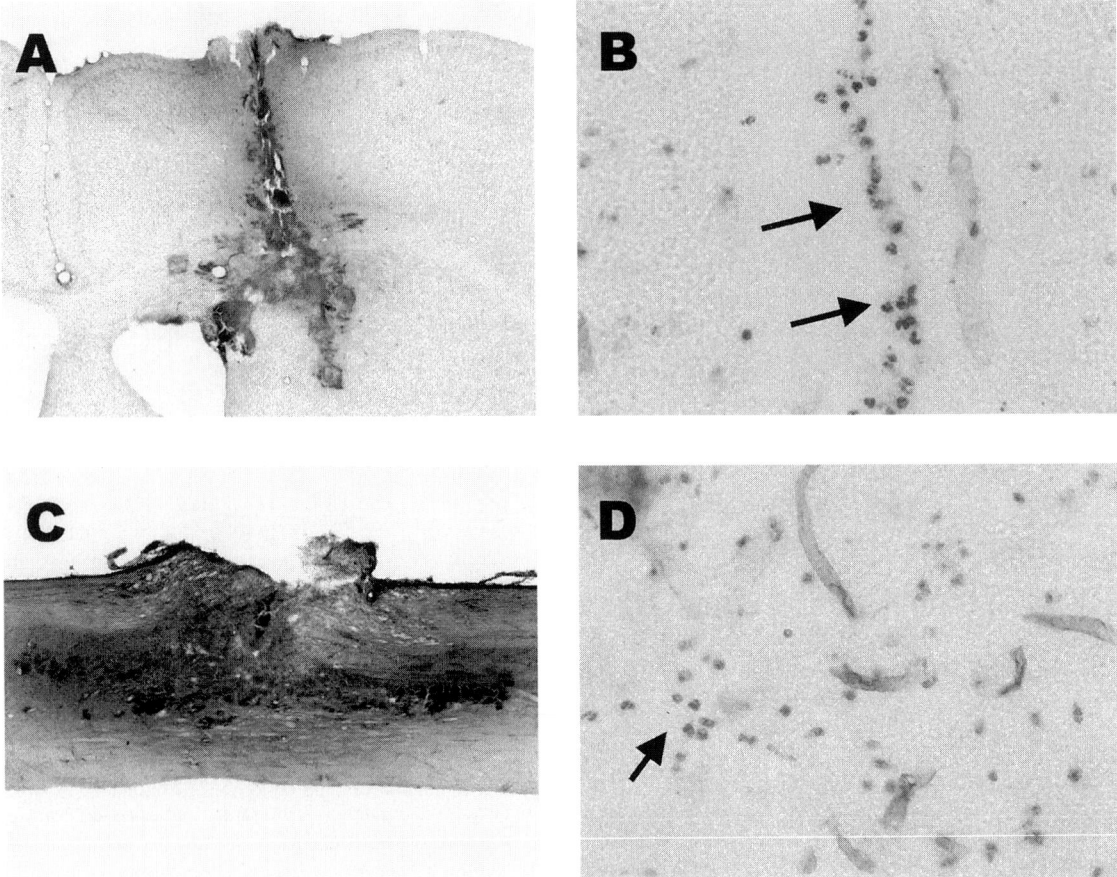

Fig. 3. Inflammation in the injured spinal cord is more robust than in the injured brain. Representative sections from the traumatically injured rat brain (A,B) and spinal cord (C,D) 1 day after injury. Note the distinct patterns of blood–brain barrier permeability (HRP; A,C) and neutrophil extravasation (cresyl violet counterstain; B,D). In the injured brain, HRP extravasation is restricted to the injury site (A) but is found throughout the gray and white matter of the injured spinal cord (C). Neutrophil recruitment to the injured brain is not accompanied by migration of these tissues into the brain parenchyma (B). Instead, neutrophils (arrow) remain marginated within the cerebral vessels. In the injured spinal cord (D), neutrophils are scattered throughout the injured tissue (arrow). Original magnification 40× (A,C) and 800× (B,D). (Photomicrographs were kindly provided by Dr. Lisa Schnell.)

differential trafficking of immune cells to brain and spinal cord. From the brain, large molecular weight proteins and, presumably, cells flow from the interstitium into the CSF, then into venous blood or lymph along extensions of the cranial nerves (Cserr and Knopf, 1992). Similar pathways are suspected from the spinal cord but to different regional lymph nodes. Thus, the consequences of CNS inflammation will be affected by how cells and proteins get into and out of the CNS (Wekerle et al., 1986; Harling-Berg et al., 1991).

Species variations and genetic determinants that influence inflammation and CNS regeneration

Unlike injured peripheral nerves or the CNS of lower vertebrates (e.g., fish and frogs), significant axon regeneration is not observed in the injured adult mammalian CNS. The reasons for regenerative failure are likely to include the presence of growth inhibitory molecules and a cellular/molecular substrate that is hostile to regrowing axons (Aubert et al., 1995; Fitch and Silver, 1999). Absence of these factors in the CNS of lower vertebrates may be explained by the presence of a unique immune–CNS axis. For instance, in the normal white matter of fish, microglia and macrophages outnumber oligodendrocytes 3 : 1 but are seldom found in gray matter (Dowding and Scholes, 1993). In contrast, microglia are more prevalent in the gray matter of mammalian brain and spinal cord (Lawson et al., 1990). Lymphocytes are also abundant in normal fish spinal cord, reaching cell densities 5000-fold greater than in rats (Dowding and Scholes, 1993).

In addition to differences in regenerative capacity between the CNS of higher and lower vertebrates, inter-species differences exist in mammals that might be explained by inflammatory mechanisms. Spinal trauma elicits a robust inflammatory reaction in both rats and mice; however, the morphology and cellular composition of the evolving lesions are distinct (reviewed in Steward et al., 1999). Rat SCI results in the formation of a necrotic/cavitating lesion whereas the injured mouse spinal cord initiates a 'wound-healing' response, i.e., the lesion decreases in size and fills in with inflammatory cells and connective tissue (Zhang et al., 1996; Zhang and Guth, 1997; Jakeman et al., 2000). Species differences in the production of inflammatory neurotoxins and cytokines also may underlie the distinct pathophysiology and associated patterns of motor recovery following SCI in rats and guinea pigs (Popovich et al., 1994; Blight et al., 1997). Classical examples of inter-species differences in the neuroinflammatory response exist in EAE where CNS pathology varies in onset, duration, and magnitude between rats and mice. Moreover, different animal strains (e.g., Lewis, Sprague–Dawley and Fischer rats; C57BL/6 vs. DBA/2J mice) exhibit various degrees of susceptibility to inflammatory-mediated neuropathology; this can be partially explained by strain-dependent variations in HPA axis activation (Sternberg et al., 1989; Mason, 1991; Popovich et al., 1997). Recent evidence also demonstrates strain variations in susceptibility to neuronal cell death and lesion evolution following traumatic SCI, TBI and excitotoxic brain injury (Steward et al., 1999). How species and strain variations influence the immune–CNS regeneration/degeneration axis awaits further study. Genetic approaches may prove useful in identifying differences in the 'quality' of inflammation in the injured CNS of mice and rats. Whether specific immune effector mechanisms influence the pathophysiology of SCI and TBI may become clearer by studying SCI and TBI in mice with targeted deletions of cytokines or *MHC* genes. Also, transgenic approaches could be used in which specific transgenes (e.g., cytokines) are placed under the control of regulatable promoters in cells resident (e.g., astrocytes) or recruited (e.g., lymphocytes) to the injured CNS.

Aging, inflammation and CNS regeneration

The pathophysiology of SCI is typically modeled in young adult animals to reflect the predominant cohort of human SCI. However, as the average life expectancy of SCI patients increases, it is essential that we understand how aging impacts the biological variables associated with CNS regeneration. It is well established that cellular immune function declines with increasing age. Aging also affects the morphology and presumably the function of resident microglia. Upregulated expression of MHC molecules, acquisition of an activated morphology, increased phagocytosis, and increased cytokine production are all characteristics of aged microglia (Perry et al., 1993b; Ye and Johnson, 1999). How well and to what extent leukocytes infiltrate the pathologic CNS also may be age-dependent. Studies by Anthony et al. (1997) suggest an age-related 'window of susceptibility' in rats for cytokine-induced alterations in BBB permeability. In aged rats, impaired macrophage recruitment is accompanied by incomplete remyelination of damaged CNS axons (Gilson and Blakemore, 1993). Age-related changes in HPA and SAM axes, due to progressive atrophy and dysfunctional CNS neurons, could influence the efficacy of inflammatory processes in the injured CNS and peripheral tissues. Chronic elevations of hypothalamic corticotrophin releasing factor (CRF) and decreased noradrenergic innervation of lymphoid tissues in aged animals support this hypothesis and may explain the impaired inflammatory response and decreased healing of decubiti and urinary tract infections in chronic spinal patients.

Systemic immunosuppression and spinal cord injury: can it do more harm than good?

The numerous physiological and injury-related influences on immune cell function described above illustrate the number of variables that must be taken into consideration when interpreting experimental data related to post-traumatic CNS inflammation. However, these same variables represent potential obstacles that may need to be overcome in developing immune-based therapies for CNS injury. Despite the widespread use of the glucocorticoid, methylprednisolone, to treat acute SCI, the consequences of systemic immunosuppression on neural regeneration and/or immune-mediated secondary injury have received little consideration. Because glucocorticoids and immunosuppressive cytokines (e.g., IL-10, TGF-β) globally suppress cytokine production, adhesion molecule expression, cellular proliferation, blood vessel growth, and DNA synthesis, indices of wound healing and tissue repair are likely to be markedly suppressed in animals or patients receiving chronic systemic doses of these compounds. Similarly, despite the acute neurological improvements observed following systemic cytokine administration in TBI and SCI (e.g., IL-10) (Knoblach and Faden, 1998; Bethea et al., 1999), it will be important to define how bolus or chronic infusions of these molecules affect the repair properties of CNS inflammation that will be inhibited in parallel with the undesirable aspects of inflammation. Unlike hormones, cytokines are predominantly autocrine and paracrine mediators that are used by inflammatory cells to signal localized cellular responses. Systemic infusion of these proteins represents an assault on the coordinated activities of inflammation that may ultimately do more harm than good in the chronically injured CNS. Future studies should determine whether antagonizing and/or promoting the activity of specific pathways in the inflammatory cascade would result in more notable improvements in preservation of neural tissue and subsequent recovery of function. Such targeted or precision immunotherapies may be a useful alternative or adjunct strategy for treating various traumatic injuries or neurodegenerative conditions of the CNS.

Acknowledgements

The work described in this chapter was supported in part by the NIH (NS37846, NS-33696, NS-10165), Sandoz Pharmaceutical Corporation and Sandoz Research Institute, and the Medical Directors Support Fund of the Ohio State University. The author would also like to acknowledge the continued collaboration and support of Drs. Dana McTigue, Bradford Stokes, Caroline Whitacre, Michele Basso and Sandra Kostyk. The photomicrographs in Fig. 3 were kindly provided by Dr. Lisa Schnell at the Brain Research Institute, University of Zurich and Swiss Federal Institute of Technology, Zurich, Switzerland.

References

Alberati-Giani, D., Ricciardi-Castagnoli, P., Kohler, C. and Cesura, A.M. (1996) Regulation of the kynurenine metabolic pathway by interferon-gamma in murine cloned macrophages and microglial cells. *J. Neurochem.*, 66: 996–1004.

Anthony, D.C., Bolton, S.J., Fearn, S. and Perry, V.H. (1997) Age-related effects of interleukin-1β on polymorphonuclear neutrophil-dependent increases in blood–brain barrier permeability in rats. *Brain*, 120: 435–444.

Aubert, I., Ridet, J.L. and Gage, F.H. (1995) Regeneration in the adult mammalian CNS: guided by development. *Curr. Opin. Neurobiol.*, 5: 625–635.

Banati, R.B. and Graeber, M.B. (1994) Surveillance, intervention and cytotoxicity: is there a protective role of microglia?. *Dev. Neurosci.*, 16: 114–127.

Banati, R.B., Gehrmann, J., Schubert, P. and Kreutzberg, G.W. (1993) Cytotoxicity of microglia. *Glia*, 7: 111–118.

Bartholdi, D. and Schwab, M.E. (1997) Expression of pro-inflammatory cytokine and chemokine mRNA upon experimental spinal cord injury in mouse: an in situ hybridization study. *Eur. J. Neurosci.*, 9: 1422–1438.

Beattie, M.S., Bresnahan, J.C., Komon, J., Tovar, C.A., Van Meter, M., Anderson, D.K., Faden, A.I., Hsu, C.Y., Noble, L.J., Salzman, S.K. and Young, W. (1997) Endogenous repair after spinal cord contusion injuries in the rat. *Exp. Neurol.*, 148: 453–463.

Bell, M.D., Taub, D.D. and Perry, V.H. (1996) Overriding the brain's intrinsic resistance to leukocyte recruitment with intraparenchymal injections of recombinant chemokines. *Neuroscience*, 74: 283–292.

Bellinger, D.L., Felten, S.Y., Lorton, D. and Felten, D.L. (1997) Innervation of lymphoid organs and neurotransmitter–lymphocyte interactions. In: Keane, R.W. and Hickey, W.F. (Eds.), *Immunology of the Nervous System*. Oxford University Press, New York, pp. 226–329.

Bethea, J.R., Castro, M., Keane, R.W., Lee, T.T., Dietrich, W.D. and Yezierski, R.P. (1998) Traumatic spinal cord injury induces nuclear factor-κB activation. *J. Neurosci.*, 18: 3251–3260.

Bethea, J.R., Nagashima, H., Acosta, M.C., Briceno, C., Gomez, F., Marcillo, A.E., Loor, K., Green, J. and Dietrich, W.D. (1999) Systemically administered interleukin-10 reduces tumor necrosis factor-alpha production and significantly improves functional recovery following traumatic spinal cord injury in rats. *J. Neurotrauma*, 16: 851–863.

Blight, A.R. (1994) Effects of silica on the outcome from experimental spinal cord injury: implication of macrophages in secondary tissue damage. *Neuroscience*, 60: 263–273.

Blight, A.R. (1996) An overview of spinal cord injury models. In: Narayan, R.K., Wilberger, J.E. and Povlishock, J.T. (Eds.), *Neurotrauma*. McGraw-Hill, New York, pp. 1367–1379.

Blight, A.R., Leroy, E.C. and Heyes, M.P. (1997) Quinolinic acid accumulation in injured spinal cord: time course, distribution, and species differences between rat and guinea pig. *J. Neurotrauma*, 14: 89–98.

Brosnan, C.F., Traugott, U. and Raine, C.S. (1983) Analysis of humoral and cellular events and the role of lipid haptens during CNS demyelination. *Acta Neuropathol.*, 9: 59–70.

Carlson, S.L., Parrish, M.E., Springer, J.E., Doty, K. and Dossett, L. (1998) Acute inflammatory response in spinal cord following impact injury. *Exp. Neurol.*, 151: 77–88.

Chen, L.Z., Hochwald, G.M., Huang, C., Dakin, G., Tao, H.C.C., Simmons, W.J., Dranoff, G. and Thorbecke, G.J. (1998) Gene therapy in allergic encephalomyelitis using myelin basic protein-specific T-cells engineered to express latent transforming growth factor-β1. *Proc. Natl. Acad. Sci. USA*, 95: 12516–12521.

Cross, A.H., Cannella, B., Brosnan, C.F. and Raine, C.S. (1990) Homing to central nervous system vasculature by antigen specific lymphocytes, I. Localization of C14-labeled cells during acute, chronic, and relapsing experimental allergic encephalomyelitis. *Lab. Invest.*, 63: 162–170.

Crowe, M.J., Bresnahan, J.C., Shuman, S.L., Masters, J.N. and Beattie, M.S. (1997) Apoptosis and delayed degeneration after spinal cord injury in rats and monkeys. *Nat. Med.*, 3: 73–76.

Cruse, J.M., Lewis, R.E., Bishop, G.R., Kliesch, W.F. and Gaitan, E. (1992) Neuroendocrine–immune interactions associated with loss and restoration of immune system function in spinal cord injury and stroke patients. *Immunol. Res.*, 11: 104–116.

Cruse, J.M., Keith, J.C., Bryant Jr., M.L. and Lewis Jr., R.E. (1996) Immune system–neuroendocrine dysregulation in spinal cord injury. *Immunol. Res.*, 15: 306–314.

Cserr, H.F. and Knopf, P.M. (1992) Cervical lymphatics, the blood–brain barrier and the immunoreactivity of the brain: a new view. *Immunol. Today*, 13: 507–512.

Davies, S.J., Fitch, M.T., Memberg, S.P., Hall, A.K., Raisman, G. and Silver, J. (1997) Regeneration of adult axons in white matter tracts of the central nervous system. *Nature*, 390: 680–683.

DeJong, B.A. and Smith, M.E. (1997) A role for complement in phagocytosis of myelin. *Neurochem. Res.*, 22: 491–498.

DeWitt, D.A., Perry, G., Cohen, M., Doller, C. and Silver, J. (1998) Astrocytes regulate microglial phagocytosis of senile plaque cores of Alzheimer's disease. *Exp. Neurol.*, 149: 329–340.

Dowding, A.J. and Scholes, J. (1993) Lymphocytes and macrophages outnumber oligodendrocytes in normal fish spinal cord. *Proc. Natl. Acad. Sci. USA*, 90: 10183–10187.

Drescher, K.M., Murray, P.D., David, C.S., Pease, L.R. and Rodriguez, M. (1999) CNS cell populations are protected from virus-induced pathology by distinct arms of the immune system. *Brain Pathol.*, 9: 21–31.

Elkabes, S., DiCicco-Bloom, E.M. and Black, I.B. (1996) Brain microglia/macrophages express neurotrophins that selectively regulate microglial proliferation and function. *J. Neurosci.*, 16: 2508–2521.

Fitch, M.T. and Silver, J. (1997) Activated macrophages and the blood–brain barrier: inflammation after CNS injury leads to increases in putative inhibitory molecules. *Exp. Neurol.*, 148: 587–603.

Fitch, M.T. and Silver, J. (1999) Beyond the glial scar: cellular and molecular mechanisms by which glial cells contribute to CNS regenerative failure. In: Tuszynski, M.H. and Kordower, J.H. (Eds.), *CNS Regeneration: Basic Science and Clinical Advances*. Academic Press, New York, pp. 55–88.

Gilson, J. and Blakemore, W.F. (1993) Failure of remyelination in areas of demyelination produced in the spinal cord of old rats. *Neuropathol. Appl. Neurobiol.*, 19: 173–181.

Giulian, D. and Robertson, C. (1990) Inhibition of mononuclear phagocytes reduces ischemic injury in the spinal cord. *Ann. Neurol.*, 27: 33–42.

Giulian, D., Chen, J., Ingeman, J.E., George, J.K. and Noponen, M. (1989) The role of mononuclear phagocytes in wound healing after traumatic injury to adult mammalian brain. *J. Neurosci.*, 9: 4416–4429.

Giulian, D., Li, J., Bartel, S., Broker, J., Li, X. and Kirkpatrick, J.B. (1995) Cell surface morphology identifies microglia as a distinct class of mononuclear phagocyte. *J. Neurosci.*, 15: 7712–7726.

Guth, L., Barrett, C.P., Donati, E.J., Anderson, F.D., Smith, M.V. and Lifson, M. (1985) Essentiality of a specific cellular terrain for growth of axons into a spinal cord lesion. *Exp. Neurol.*, 88: 1–12.

Halliwell, B. and Gutteridge, J.M.C. (1985) Oxygen radicals and the nervous system. *Trends Neurosci.*, 8: 22–26.

Harling-Berg, C.J., Knopf, P.M. and Cserr, H.F. (1991) Myelin basic protein infused into cerebrospinal fluid suppresses experimental autoimmune encephalomyelitis. *J. Neuroimmunol.*, 35: 45–51.

Hauben, E., Yoles, E., Nevo, U., Moalem, G., Agranov, G., Neeman, M., Axelrod, S., Mor, F., Cohen, I. and Schwartz, M. (1999) Autoimmune T-cells are neuroprotective in spinal cord injury. *J. Neurotrauma*, 16: 980.

Herman, J.P. and Cullinan, W.E. (1997) Neurocircuitry of stress: central control of the hypothalamo–pituitary–adrenocortical axis. *Trends Neurosci.*, 20: 78–84.

Hickey, W.F., Hsu, B.L. and Kimura, H. (1991) T-lymphocyte entry into the central nervous system. *J. Neurosci. Res.*, 28: 254–260.

Hirschberg, D.L. and Schwartz, M. (1995) Macrophage recruitment to acutely injured central nervous system is inhibited by a resident factor: basis for an immune–brain barrier. *J. Neuroimmunol.*, 61: 89–96.

Huang, D.W., McKerracher, L., Braun, P.E. and David, S. (1999) A therapeutic vaccine approach to stimulate axon regeneration in the adult mammalian spinal cord. *Neuron*, 24: 639–647.

Jakeman, L.B., Guan, Z., Wei, P., Ponnappan, R., Dzwonczyk, R., Popovich, P.G. and Stokes, B.T. (2000) Traumatic spinal cord injury produced by controlled contusion in mouse. *J. Neurotrauma*, 17: 303–323.

Keirstead, H.S. and Blakemore, W.F. (1997) Identification of post-mitotic oligodendrocytes incapable of remyelination within the demyelinated adult spinal cord. *J. Neuropathol. Exp. Neurol.*, 56: 1191–1201.

Kerschensteiner, M., Gallmeier, E., Behrens, L., Leal, V.V., Misgeld, T., Klinkert, W.E.F., Kolbeck, R., Hoppe, E., Oropeza-Wekerle, R.-L., Bartke, I., Stadelmann, C., Lassman, H., Wekerle, H. and Hohlfeld, R. (1999) Activated human T cells, B cells, and monocytes produce brain-derived neurotrophic factor in vitro and in inflammatory brain lesions: a neuroprotective role of inflammation?. *J. Exp. Med.*, 189: 865–870.

Kliesch, W.F., Cruse, J.M., Lewis, R.E., Bishop, G.R., Brackin, B. and Lampton, J.A. (1996) Restoration of depressed immune function in spinal cord injury patients receiving rehabilitation therapy. *Paraplegia*, 34: 82–90.

Knoblach, S.M. and Faden, A.I. (1998) Interleukin-10 improves outcome and alters proinflammatory cytokine expression after experimental traumatic brain injury. *Exp. Neurol.*, 153: 143–151.

Koller, H., Siebler, M. and Hartung, H.-P. (1997) Immunologically induced electrophysiological dysfunction: implications for inflammatory diseases of the CNS and PNS. *Prog. Neurobiol.*, 52: 1–26.

Koshinaga, M. and Whittemore, S.R. (1995) The temporal and spatial activation of microglia in fiber tracts undergoing anterograde and retrograde degeneration following spinal cord lesion. *J. Neurotrauma*, 12: 209–222.

Kreutzberg, G.W. (1996) Microglia: a sensor for pathological events in the CNS. *Trends Neurosci.*, 19: 312–318.

Lawson, L.J., Perry, V.H., Dri, P. and Gordon, S. (1990) Heterogeneity in the distribution and morphology of microglia in the normal adult mouse brain. *Neuroscience*, 39: 151–170.

Lazarov-Spiegler, O., Rapalino, O., Agranov, G. and Schwartz, M. (1998) Restricted inflammatory response in the CNS: a key impediment to axonal regeneration?. *Mol. Med. Today*, 4: 337–342.

Lehmann, P.V., Sercarz, E.E., Forsthuber, T., Dayan, C.M. and Gammon, G. (1993) Determinant spreading and the dynamics of the autoimmune T-cell repertoire. *Immunol. Today*, 14: 203–208.

Linington, C., Morgan, B.P., Scolding, N.J., Wilkins, P., Piddlesden, S. and Compston, D.A. (1989) The role of complement in the pathogenesis of experimental allergic encephalomyelitis. *Brain*, 112: 895–911.

Liu, D., Yang, R., Yan, X. and McAdoo, D.J. (1994) Hydroxyl radicals generated in vivo kill neurons in the rat spinal cord: electrophysiological, histological and neurochemical results. *J. Neurochem.*, 62: 37–44.

Mason, D. (1991) Genetic variation in the stress response: susceptibility to experimental allergic encephalomyelitis and implications for human inflammatory disease. *Immunol. Today*, 12: 57–60.

Miller, D.J. and Rodriguez, M. (1995) A monoclonal autoantibody that promotes central nervous system remyelination in a model of multiple sclerosis is a natural autoantibody encoded by germline immunoglobulin genes. *J. Immunol.*, 154: 2460–2469.

Mizrachi, Y., Ohry, A., Aviel, A., Rozin, R., Brooks, M.E. and Schwartz, M. (1983) Systemic humoral factors participating in the course of spinal cord injury. *Paraplegia*, 21: 287–293.

Moalem, G., Leibowitz-Amit, R., Yoles, E., Mor, F., Cohen, I.R. and Schwartz, M. (1999a) Autoimmune T-cells protect neurons from secondary degeneration after central nervous system axotomy. *Nat. Med.*, 5: 49–55.

Moalem, G., Monsonego, A., Shani, Y., Cohen, I.R. and Schwartz, M. (1999b) Differential T-cell response in central and peripheral nerve injury: connection with immune privilege. *FASEB J.*, 13: 1207–1217.

Mosley, K. and Cuzner, M.L. (1996) Receptor-mediated phagocytosis of myelin by macrophages and microglia: effect of opsonization and receptor blocking agents. *Neurochem. Res.*, 21: 481–487.

Nakajima, K., Kikuchi, Y., Ikoma, E., Honda, S., Ishikawa, M., Liu, Y. and Kohsaka, S. (1998) Neurotrophins regulate the function of cultured microglia. *Glia*, 24: 272–289.

Naparstek, Y., Cohen, I.R., Fuks, Z. and Vlodavsky, I. (1984) Activated T lymphocytes produce a matrix-degrading heparan sulphate endoglycosidase. *Nature*, 310: 241–244.

Nataf, S., Stahel, P.F., Davoust, N. and Barnum, S.R. (1999) Complement anaphylatoxin receptors on neurons: new tricks for old receptors?. *Trends Neurosci.*, 22: 397–402.

Noble, L.J. and Wrathall, J.R. (1989) Distribution and time course of protein extravasation in the rat spinal cord after contusive injury. *Brain Res.*, 482: 57–66.

Olsson, T., Diener, P., Ljungdahl, A., Hojeberg, B., Van der Meide, P.H. and Kristensson, K. (1992) Facial nerve transection causes expansion of myelin autoreactive T cells in regional lymph nodes and T cell homing to the facial nucleus. *Autoimmunity*, 13: 117–126.

Olsson, T., Sun, J.B., Solders, G., Xiao, B.G., Hojeberg, B., Ekre, H.P. and Link, H. (1993) Autoreactive T and B cell responses to myelin antigens after diagnostic sural nerve biopsy. *J. Neurol. Sci.*, 117: 130–139.

Palladini, G., Grossi, M., Maleci, A., Lauro, G.M. and Guidetti, B. (1987) Immunocomplexes in rat and rabbit spinal cord after injury. *Exp. Neurol.*, 95: 639–651.

Pender, M.P., Stanley, G.P., Yoong, G. and Nguyen, K.B. (1990) The neuropathology of chronic relapsing experimental allergic encephalomyelitis induced in the Lewis rat by inoculation with whole spinal cord and treatment with cyclosporin A. *Acta Neuropathol.*, 80: 172–183.

Perry, V.H., Andersson, P.B. and Gordon, S. (1993a) Macrophages and inflammation in the central nervous system. *Trends Neurosci.*, 16: 268–273.

Perry, V.H., Matyszak, M.K. and Fearn, S. (1993b) Altered antigen expression of microglia in the aged rodent CNS. *Glia*, 7: 60–67.

Pette, M., Fujita, K., Kitze, B., Whitaker, J.N., Albert, E., Kappos, L. and Wekerle, H. (1990) Myelin basic protein-specific T lymphocyte lines from MS patients and healthy individuals. *Neurology*, 40: 1770–1776.

Plata-Salaman, C.R. (1998) Brain injury and immunosuppression. *Nat. Med.*, 4: 768–769.

Popovich, P.G., Streit, W.J. and Stokes, B.T. (1993) Differential expression of MHC Class II antigen in the contused rat spinal cord. *J. Neurotrauma*, 10: 37–46.

Popovich, P.G., Reinhard Jr., J.F., Flanagan, E.M. and Stokes, B.T. (1994) Elevation of the neurotoxin quinolinic acid occurs following spinal contusion injury. *Brain Res.*, 633: 348–352.

Popovich, P.G., Horner, P.J., Mullin, B.B. and Stokes, B.T. (1996a) A quantitative spatial analysis of the blood–spinal cord barrier, I. Permeability changes after experimental spinal contusion injury. *Exp. Neurol.*, 142: 258–275.

Popovich, P.G., Stokes, B.T. and Whitacre, C.C. (1996b) Concept of autoimmunity following spinal cord injury: possible roles for T lymphocytes in the traumatized central nervous system. *J. Neurosci. Res.*, 45: 349–363.

Popovich, P.G., Wei, P. and Stokes, B.T. (1997) The cellular inflammatory response after spinal cord injury in Sprague–Dawley and Lewis rats. *J. Comp. Neurol.*, 377: 443–464.

Popovich, P.G., Whitacre, C.C. and Stokes, B.T. (1998) Is spinal cord injury an autoimmune disorder?. *Neuroscientist*, 4: 71–76.

Popovich, P.G., Guan, Z., Wei, P., Huitinga, I., van Rooijen, N. and Stokes, B.T. (1999) Depletion of hematogenous macrophages promotes partial hindlimb recovery and neuroanatomical repair after experimental spinal cord injury. *Exp. Neurol.*, 158: 351–365.

Povlishock, J.T. (1996) An overview of brain injury models. In: Narayan, R.K., Wilberger and J.E., Povlishock, J.T. (Eds.), *Neurotrauma*. McGraw Hill, New York, pp. 1325–1336.

Rabchevsky, A.G. and Streit, W.J. (1997) Grafting of cultured microglial cells into the lesioned spinal cord of adult rats enhances neurite outgrowth. *J. Neurosci. Res.*, 47: 34–48.

Ramirez, F., Fowell, D.J., Puklavec, M., Simmonds, S. and Mason, D. (1996) Glucocorticoids promote a TH2 cytokine response by $CD4^+$ T cells in vitro. *J. Immunol.*, 156: 2406–2412.

Rapalino, O., Lazarov-Spiegler, O., Agranov, E., Velan, G.J., Yoles, E., Fraidakis, M., Solomon, A., Gepstein, R., Katz, A., Belkin, A., Hadani, M. and Schwartz, M. (1998) Implantation of stimulated homologous macrophages results in partial recovery of paraplegic rats. *Nat. Med.*, 4: 814–821.

Rebhun, J., Madorsky, J.G. and Glovsky, M.M. (1991) Proteins of the complement system and acute phase reactants in sera of patients with spinal cord injury. *Ann. Allergy*, 66: 335–338.

Schmitt, A.B., Brook, G.A., Buss, A., Nacimiento, W., Noth, J. and Kreutzberg, G.W. (1998) Dynamics of microglial activation in the spinal cord after cerebral infarction are revealed by expression of MHC class II antigen. *Neuropathol. Appl. Neurobiol.*, 24: 167–176.

Schnell, L., Fearn, S., Schwab, M., Perry, V.H. and Anthony, D.C. (1999) Cytokine-induced acute inflammation in the brain and spinal cord. *J. Neuropathol. Exp. Neurol.*, 58: 245–254.

Sternberg, E.M., Young, W.S., Bernardini, R., Calogero, A.E., Chrousos, G.P., Gold, P.W. and Wilder, R.L. (1989) A central nervous system defect in biosynthesis of corticotropin-releasing hormone is associated with susceptibility to streptococcal cell wall-induced arthritis in Lewis rats. *Proc. Natl. Acad. Sci. USA*, 86: 4771–4775.

Sternberg, E.M., Chrousos, G.P., Wilder, R.L. and Gold, P.W. (1992) The stress response and the regulation of inflammatory disease. *Ann. Int. Med.*, 117: 854–866.

Steward, O., Schauwecker, P.E., Guth, L., Zhang, Z., Fujiki, M., Inman, D., Wrathall, J., Kempermann, G., Gage, F.H., Saatman, K.E., Raghupathi, R. and McIntosh, T.K. (1999) Genetic approaches to neurotrauma research: opportunities and potential pitfalls of murine models. *Exp. Neurol.*, 157: 19–42.

Streit, W.J. and Sparks, D.L. (1997) Activation of microglia in the brains of humans with heart disease and hypercholesterolemic rabbits. *J. Mol. Med.*, 75: 130–138.

Streit, W.J., Graeber, M.B. and Kreutzberg, G.W. (1988) Functional plasticity of microglia: a review. *Glia*, 1: 301–307.

Sun, D., Whitaker, J.N. and Wilson, D.B. (1999) Regulatory T cells in experimental allergic encephalomyelitis, II. T cells functionally antagonistic to encephalitogenic MBP-specific T cells show persistent expression of FasL. *J. Neurosci. Res.*, 58: 357–366.

Taoka, Y., Okajima, K., Uchiba, M., Murakami, K., Kushimoto, S., Johno, M., Naruo, M., Okabe, H. and Takatsuki, K. (1997) Role of neutrophils in spinal cord injury in the rat. *Neuroscience*, 79: 1177–1182.

Taoka, Y., Okajima, K., Uchiba, M., Murakami, K., Harada, N., Johno, M. and Naruo, M. (1998) Activated protein C reduces the severity of compression-induced spinal cord injury in rats by inhibiting activation of leukocytes. *J. Neurosci.*, 18: 1393–1398.

Theofilopoulos, A.N. (1993) Molecular pathology of autoimmunity. In: Bona, C.A., Siminovitch, K.A., Zanetti, M. and Theofilopoulos, A.N. (Eds.), *The Molecular Pathology of Autoimmune Diseases*. Harwood Academic Publishers, Chur, pp. 1–12.

Trapp, B.D., Peterson, J., Ransohoff, R.M., Rudick, R., Mork, S. and Bo, L. (1998) Axonal transection in the lesions of multiple sclerosis. *N. Engl. J. Med.*, 338: 278–285.

Tuohy, V.K. and Mathisen, P.M. (1998) T-cell design: optimizing the therapeutic potential of autoreactive T cells by genetic modification. *Res. Immunol.*, 149: 834–842.

Wekerle, H., Linington, C., Lassmann, H. and Meyermann, R. (1986) Cellular immune reactivity within the CNS. *Trends Neurosci.*, 9: 271–277.

Yarom, Y., Naparstek, Y., Lev-Ram, V., Holoshitz, J., Ben-Nun, A. and Cohen, I.R. (1983) Immunospecific inhibition of nerve conduction by T lymphocytes. *Nature*, 303: 246–247.

Ye, S.-M. and Johnson, R.W. (1999) Increased interleukin-6 expression by microglia from brain of aged mice. *J. Neuroimmunol.*, 93: 139–148.

Zeev-Brann, A.B., Lazarov-Spiegler, O., Brenner, T. and Schwartz, M. (1998) Differential effects of central and peripheral nerves on macrophages and microglia. *Glia*, 23: 181–190.

Zhang, S.C. and Fedoroff, S. (1996) Neuron–microglia interactions in vitro. *Acta Neuropathol.*, 91: 385–395.

Zhang, Z. and Guth, L. (1997) Experimental spinal cord injury:

Wallerian degeneration in the dorsal column is followed by revascularization, glial proliferation, and nerve regeneration. *Exp. Neurol.*, 147: 159–171.

Zhang, Z., Fujiki, M., Guth, L. and Steward, O. (1996) Genetic influences on cellular reactions to spinal cord injury: a wound-healing response present in normal mice is impaired in mice carrying a mutation (WldS) that causes delayed Wallerian degeneration. *J. Comp. Neurol.*, 371: 485–495.

Zietlow, R., Dunnett, S.B. and Fawcett, J.W. (1999) The effect of microglia on embryonic dopaminergic neuronal survival in vitro: diffusible signals from neurons and glia change microglia from neurotoxic to neuroprotective. *Eur. J. Neurosci.*, 11: 1657–1667.

SECTION II

Plasticity of the injured spinal cord: retraining neural circuits to promote motor recovery

CHAPTER 6

Plasticity of neuronal networks in the spinal cord: modifications in response to altered sensory input

Keir G. Pearson *

Department of Physiology, University of Alberta, Edmonton, AB T6G 2H7, Canada

Introduction

A fundamental problem in motor control is to establish how the magnitude of activity in a set of muscles is scaled for a particular movement. The underlying mechanisms must account for the facts that (1) numerous muscles contract simultaneously so that the action of any one muscle is dependent on the actions of other active muscles, and (2) there is often no unique pattern of activity in a group of muscles for a specific movement. There is also the need to explain how the motor output is matched to complex changes in the mechanics of the system as a movement evolves. This enormous complexity has led to the view that learning is partially involved in establishing the appropriate motor patterns for coordinated movements (Jordan, 1996; Mussa-Ivaldi and Bizzi, 1997). Evidence for this position has come from recent studies on the development of motor patterns for reaching in infants (Konczak et al., 1997) and modifications of motor patterns for reaching in adults in response to altered external conditions (Thoroughman and Shadmehr, 1999). It is also consistent with modern views on the development of sensory systems (Knudsen, 1994). In the latter, the high level of plasticity in young animals is considered necessary to refine neuronal networks according to use, and to match the functioning of these networks to changes in the anatomy of the sensory apparatus. Thus in both motor and sensory systems, mechanisms are required to adapt the motor patterns to changes in body mechanics and/or anatomy. Currently, our understanding of the underlying adaptive mechanisms is rudimentary. Most progress has been made on determining the mechanisms of plasticity in the oculomotor system of vertebrates (Du Lac et al., 1995; Raymond, 1998) and the sound-localization system of the barn owl (Feldman and Knudsen, 1997; Brainard and Knudsen, 1998; Knudsen, 1998).

An important issue is the extent to which the acquisition and maintenance of the appropriate motor patterns for coordinated limb movements depends on plasticity of neuronal networks in the spinal cord (Georgopoulos, 1996; Wolpaw, 1997). This must be likely because there is now considerable evidence that injury and use can modify neuronal systems in the spinal cord. One of the strongest examples of spinal plasticity is the ability of cats to recover the capacity to step with their hind legs following transection of the spinal cord (see review by Barbeau and Rossignol, 1994). In young animals this ability returns spontaneously (Forssberg et al., 1980), while regular training is required in adult cats (De Leon et al., 1998). Partial transections of the spinal cord also lead to compensatory changes in stepping which are considered to depend to a large extent on alterations in the activation and functioning of spinal

---

* Corresponding author: Dr. Keir G. Pearson, Department of Physiology, University of Alberta, Edmonton, AB T6G 2H7, Canada. Fax: +1-780-492-8915;
E-mail: keir.pearson@ualberta.ca

neuronal networks (Jiang and Drew, 1996; Brustein and Rossignol, 1998). Similar adaptive mechanisms may exist in the spinal cord of humans because locomotor performance in spinal cord-injured patients can be improved by regular training (Barbeau and Rossignol, 1994; Wernig et al., 1995, and this volume; Dietz et al., 1998). Additional evidence for plasticity in spinal networks has come from studies on alterations in locomotor patterns and modifications of spinal reflex pathways regulating stepping following damage to peripheral nerves in cats (Goldberger, 1988a,b; Carrier et al., 1997; Pearson et al., 1999), on classical and instrumental conditioning of spinal reflexes in rats, cats and monkeys (Durkovic and Damianopoulos, 1986; Wolpaw, 1997), and on the development of nociceptive withdrawal reflexes in rats (Holmberg et al., 1997).

The purpose of this article is to review recent investigations on adaptive plasticity in the spinal cord of the cat that utilized procedures for modifying sensory input from leg receptors. This topic has also been discussed to some extent in a number of recent articles (Pearson et al., 1998; Bouyer and Rossignol, 2000; Pearson, 2000).

Adaptive modification of stepping following peripheral nerve injury

The first series of detailed studies on the long-term effects of peripheral nerve lesions on the walking system of the cat was carried out by Goldberger (Goldberger, 1988a,b; Goldberger and Murray, 1988). These studies, together with a number of isolated studies by other investigators (Wetzel et al., 1976; Rasmussen et al., 1986), clearly demonstrated a remarkable adaptation of stepping following removal of sensory input from a hind leg by transection of dorsal roots. These adaptations compensated to a large extent the deficit produced by the nerve transections and, provided that at least one dorsal root of a hind leg was left intact (spared-root preparation), stepping movements often returned toward normal within a few weeks. Although effective stepping movements also returned following complete deafferentation, these differed in form compared to normal movements. This is an example of behavioral substitution, and it is distinctly different from the adaptive modification of the neuronal system producing normal stepping movements that occurs in spared-root preparations (Goldberger, 1988a,b).

An important issue that arose from these studies was whether the improvement in stepping depended at all on modification of neuronal networks in the spinal cord. The strongest indication for adaptive changes in spinal networks came from studies on spared-root preparations. If an ipsilateral hemisection of the thoracic cord was carried out following recovery of stepping after partial deafferentation, the hind leg continued to step with movements similar to those prior to the hemisection. This finding strongly indicated that neuronal networks in the ipsilateral lumbar cord had been modified during the period of recovery. The mechanisms responsible for this recovery have not been established, but they may depend to some extent on sprouting of sensory afferents in the spared root (Goldberger and Murray, 1988). Unfortunately, no electrophysiological studies were done on these animals so no information was obtained on the physiological mechanisms that might have been associated with functional recovery.

Plasticity of neuronal networks in the spinal cord has also been revealed by adaptive responses induced by transection of peripheral nerves innervating small groups of muscles (Whelan et al., 1995; Carrier et al., 1997; Whelan and Pearson, 1997; Bouyer and Rossignol, 2000; Pearson et al., 1999). A particularly interesting set of phenomena has been observed following denervation of the ankle flexor muscles in intact and chronic spinal cats (Carrier et al., 1997). Immediately following the nerve transections there is little flexion movement at the ankle, but within a few days stepping movements of the partially denervated hind leg return close to normal due to compensatory movements around the knee and hip. When these animals are spinalized and trained to step with their hind legs on a treadmill, the stepping movements of the denervated leg are disorganized and very different from the movements in trained spinal animals with normal innervation of the ankle flexor muscles. The disorganized stepping movements persist for at least 1 month. However, if the ankle flexors are denervated after spinalization and step training, the disorganized movements do not occur. These observations indicate that the adaptive processes occurring prior to spinalization involve a modification of spinal circuitry but not in a manner that allows this circuitry by itself to generate a relatively normal stepping pattern in the spinal animal. The maladaptive alterations in the spinal circuitry appear to be compensated by corrective supraspinal signals to yield a more-or-less normal motor pattern for stepping.

Plasticity in afferent pathways regulating stance duration

Although we have little information on the mechanisms associated with adaptive plasticity in the walking system of the cat, a number of recent studies have suggested that modification of afferent pathways regulating the stepping motor pattern may be a significant factor. Direct evidence for plasticity in afferent pathways was first obtained by examining alterations in the influence of these pathways on stance duration following denervation of ankle extensor muscles of a hind leg (Whelan et al., 1995; Whelan and Pearson, 1997). Increasing the loading of the medial gastrocnemius (MG) muscle by denervating synergist muscles (lateral gastrocnemius, LG, and soleus, SOL) resulted in an increase in the effectiveness of MG group I afferents in regulating the duration of extensor bursts. Normally, stimulation of these afferents during walking in decerebrate animals has only a weak effect on extensor burst duration (Fig. 1A).

However, in animals in which the MG has been subjected to increased loading for 3 to 7 days, stimulation of the MG group I afferents has a powerful effect on extensor burst duration, often maintaining extensor activity for the duration of long stimulus trains (Fig. 1B). This increase in effectiveness is appropriate for compensating for the loss of afferent feedback from the denervated synergists. The opposite effects were seen for the severed afferents from the LG and SOL muscles. In normal animals, stimu-

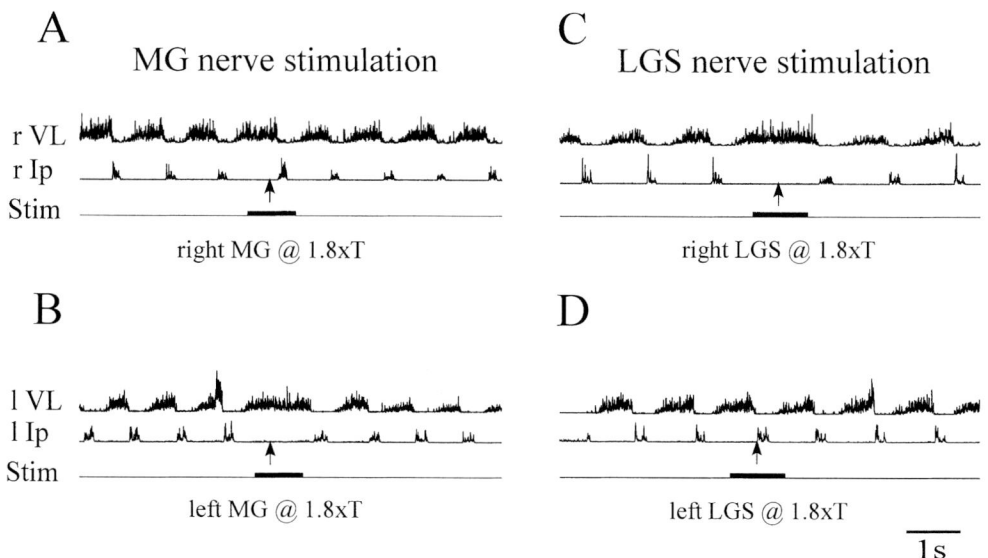

Fig. 1. Modification of strength of reflexes regulating the duration of extensor bursts in a decerebrate walking cat. Recordings made 21 days after cutting the nerve to the LG and SOL muscles (LGS nerve) of the left hind leg. (A,B) Records showing an increase of the effectiveness of MG group I afferents in influencing the duration of extensor bursts. In the control leg (A) stimulation of the MG group I afferents has only a weak effect of extensor burst duration, whereas in the experimental leg (B) stimulation of the MG group I afferents prolongs extensor burst duration for the duration of the stimulus train. (C,D) Records showing a decrease in the effectiveness of LGS group I afferents in influencing the duration of extensor bursts. In the control leg (C) stimulation of the LGS group I afferent prolongs extensor burst duration for the duration of the stimulus train, whereas in the experimental leg (D) stimulation of the LGS group I afferents has no effect on extensor burst duration. In each set of records the top two traces are rectified and filtered electromyograms (EMGs) from the knee extensor vastus lateralis (*VL*) and hip flexor iliopsoas (*Ip*) muscles and the bottom trace is a marker for the 200 Hz stimulus train. Each arrow indicates the time an Ip burst would have occurred in the absence of the stimulus train (for details see Whelan et al., 1995).

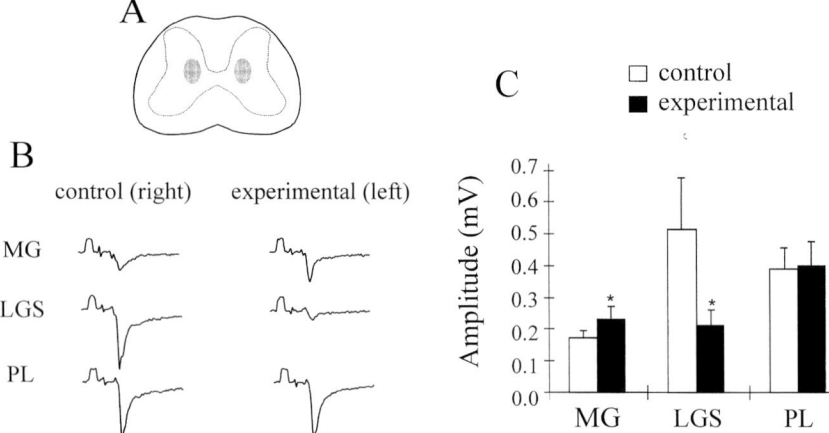

Fig. 2. Modification of group I field potentials in the intermediate nucleus (shaded region in (A) of lumbar segment 6 five days after cutting the LGS nerve of the left hind leg. (B) Field potentials recorded in the intermediate nuclei of right (control) and left (experimental) sides in response to stimulation of group I afferents in the ipsilateral MG, LGS and PL nerves. The amplitude of the initial calibration pulse is 0.1 mV. Note the increase in the MG field potential in the left (experimental) side relative to right (control) side. (C) Histograms showing averages of the maximum field potential amplitudes recorded from multiple tracks through the intermediate nuclei in one animal. The maximum amplitudes of the MG and LGS group I field potentials were significantly increased and decreased, respectively, on the experimental (left) side. The field potentials from plantaris (PL) group I afferents were similar on both sides. (Modified from Fouad and Pearson, 1997.)

lation of the group I afferents from LG and SOL has a powerful effect on the duration of extensor bursts (Fig. 1C), but beginning a few days after axotomy these afferents have a progressively weaker effect on extensor burst duration, eventually becoming almost completely ineffective (Fig. 1D). This decline in effectiveness of LG/SOL group I afferents is most likely due to degenerative events in the spinal cord associated with axotomy (Fouad and Pearson, 1997; Whelan and Pearson, 1997).

The enhanced effectiveness of the MG group I afferents persisted in some animals after transection of the spinal cord and the induction of stepping with L-DOPA. Thus at least one site for this plasticity is in the spinal cord, but exactly what changes occur in the spinal cord remains to be established. Enhanced transmission from group I afferents to spinal interneurons is one possibility because monosynaptic field potentials evoked by stimulation of MG group I afferents in the intermediate regions of the spinal cord are increased in amplitude following chronic loading of the MG muscle (Fig. 2). However, this increase is relatively weak and it has not been proven that this increase is involved in enhancing the effectiveness of MG group I afferents on extensor burst generation. Another possible mechanism is that the excitability of interneurons in the afferent pathway from the MG group I afferents to the system generating the extensor bursts is increased.

The alterations in the effectiveness of the group I afferents from the MG muscle and from the LG and SOL muscles on the spinal network generating the locomotor rhythm is shown schematically in Fig. 3. In this figure it is assumed that the basis for the locomotor rhythm is mutual inhibition between an extensor (E) and flexor (F) half-center, and the group I regulation of extensor burst duration is assumed to be due to excitatory influences from the group I afferents onto the extensor half-center.

Plasticity in pathways regulating extensor burst amplitude

Immediately after denervating the LG, SOL and plantaris (PL) muscles in a cat hind leg there is a large increase in yield (flexion) at the ankle joint during the initial part of the stance phase. This exaggerated yield progressively decreases over a period of 1 to 2 weeks and ankle movements return close to normal. Associated with this recovery is an increase

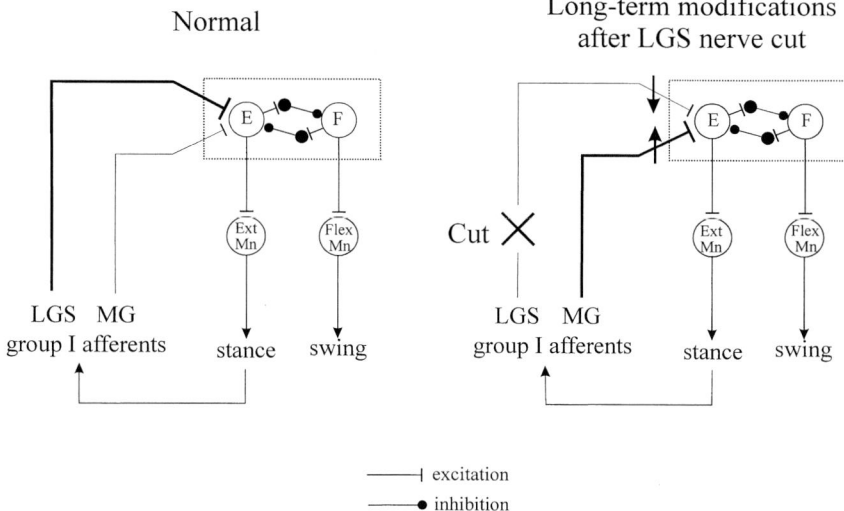

Fig. 3. Schematic diagram summarizing the modification of the strength of group I afferent pathways on the central pattern generator after cutting the LGS nerve. The central pattern generator (dotted rectangle) is assumed to consist of mutual inhibition between extensor (E) and flexor (F) half-centers, and the influence of group I afferents is assumed to be mediated via excitatory effects on the extensor half-center. In normal animals (left) the influence of LGS group I afferents is strong (thick line) and the influence of MG group I afferents is weak (thin line). The long-term modifications produced by cutting the LGS nerve (right) are an increase in the influence of MG group I afferents (thick line) and a decrease in the influence of the cut LGS afferents (thin line).

in the magnitude of the only remaining functional ankle extensor muscle, MG (Fig. 4A). The interesting feature of this adaptive change in MG burst activity is a difference in the rate of increase in the initial and late components of the MG bursts (Fig. 4B,C). The increase in the late component occurs relatively quickly and is most likely due to an increase in the gain of afferent pathways contributing to the generation of the MG bursts. The increase in the initial component of the MG bursts occurs more slowly. The initial component begins before ground contact and establishes the stiffness at the ankle joint during the early part of stance. The increase in this component is primarily responsible for the decrease in the rate of ankle yield during early stance (Fig. 5).

The mechanisms underlying the increase in the initial component of the MG bursts are unknown. One attractive possibility is that an error signal generated by increased feedback from the abnormally stretched MG muscle recalibrates the magnitude of a *feedforward* motor command generating the initial component. Consistent with this possibility is that the adaptive changes in the MG burst do not occur when normal sensory feedback is eliminated by immobilization of the leg. This finding also demonstrates that the adaptive changes are not simply due to regenerative events triggered by axotomy. This conclusion is supported by the observation of similar adaptive changes in the MG bursts following the weakening of the LG, SOL and PL muscles by the injection of botulinum toxin (Misiaszek and Pearson, 1999).

Because the adaptive changes in the magnitude of burst activity in the MG muscle have only been observed in conscious walking animals, an obvious question is whether these changes are due to modification in spinal neuronal circuits. Two facts suggest that this is a likely possibility. The first is that afferent feedback contributes substantially to the generation of the late component of the MG bursts via spinal reflex pathways (Pearson, 1995; Pearson et al., 1999). Preliminary observations suggest that the increase in the late component of the MG bursts is associated with an increase gain in these reflex pathways. One of these observations is that the strength of stretch reflexes from the MG muscle of the adapted leg is often greater than from the MG muscle of the contralateral leg (see initial data in Pearson et al., 1998). Another is that the slope of the relationship between ankle yield and the magnitude

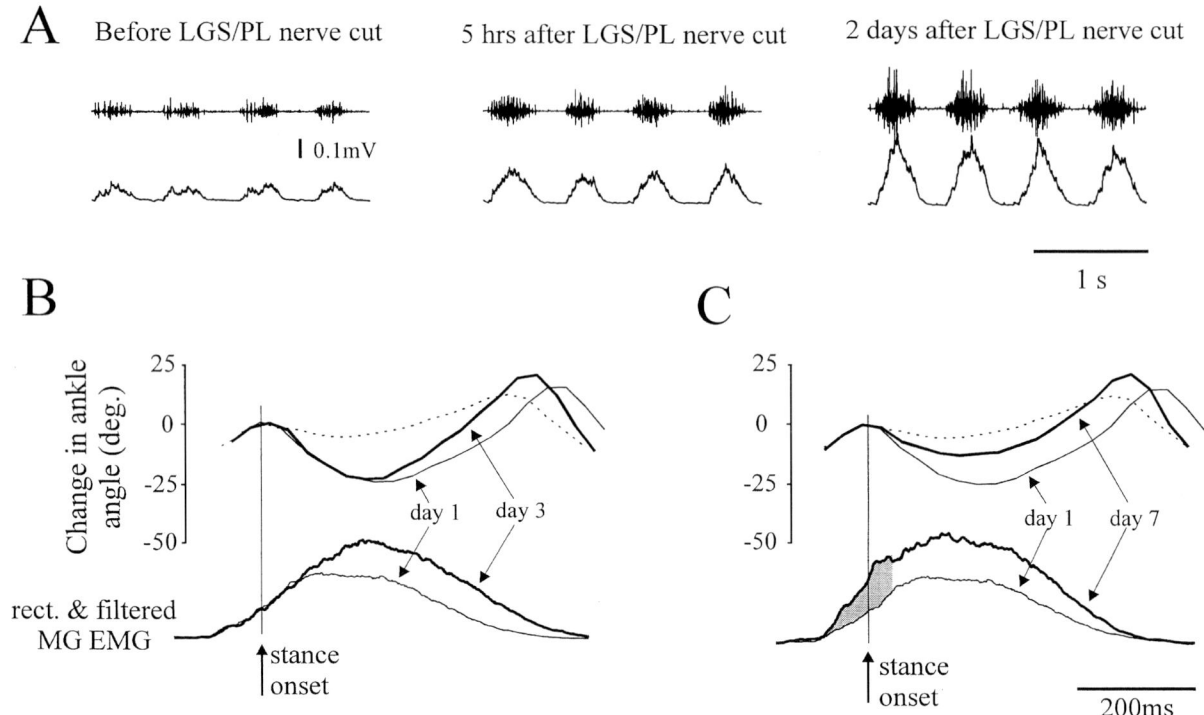

Fig. 4. Increase in the magnitude in MG bursts after cutting nerves to synergist muscles. (A) Raw EMGs (top) and rectified and filtered EMGs (bottom) from the MG muscle in a conscious walking cat before cutting the LGS and PL nerves (left), 5 h after cutting the nerves (middle), and 2 days later (right). (B) Superimposed averages of the rectified and filtered MG EMG bursts (bottom) and of the changes in ankle angle (top) on day 1 (thin traces) and day 3 (thick traces) after cutting the LGS and PL nerves. Note the increase in the magnitude of the EMG on day 3 relative to day 1 begins after stance onset and that the initial flexion of the ankle was similar on both days. (C) Superimposed averages of the rectified and filtered MG EMG bursts (bottom) and of the changes in ankle angle (top) on day 1 (thin traces) and day 7 (thick traces) after cutting the LGS and PL nerves. Note the increase in the magnitude of the initial component of the EMG on day 7 (shaded) and a corresponding decrease in the magnitude of ankle flexion during early stance. The dotted lines in (B) and (C) show the change in ankle angle before the nerves were cut. (Modified from Pearson et al., 1999.)

of the late component increases in the days following transection of the nerves to the LG and SOL muscles (K. Pearson and J. Misiaszek, unpubl.). This indicates that the length increase of the MG muscle during early stance contributes more to the generation of the MG bursts a few days after the nerve transections than the same length increase on the day of the nerve transections. Another fact suggesting the spinal origin for the adaptive increase in MG activity is that a neuronal network in the spinal cord generates the initial component of the MG bursts (Forssberg et al., 1980; Hiebert et al., 1994). Thus it is reasonable to expect that the increase in the initial component of the MG bursts is due to an increased drive from this network. One obvious approach to establish the spinal origin of the adaptive changes in the MG bursts is to determine whether similar events occur in chronic spinal cats that have been trained to step with their hind legs. In a preliminary study (Rossignol et al., 1997) increases in the initial and late components of the MG bursts were observed following transection of the LG/SOL nerve in chronic spinal cats. However, it remains to be established whether these changes occur consistently from animal to animal, and whether the time-courses of changes are similar to those in intact animals.

Fig. 6 summarizes the events in the spinal cord that may underlie the increase in MG burst magnitude after cutting the LG, SOL and PL nerves. In the first few days after the nerve cuts, the increase in MG activity occurs mainly after ground contact. This increase in the late component is likely due

increased. This is likely due to enhanced input from the extensor half-center (E) to the MG motoneurons (indicated by the up arrow in the pathway from E to MG in the right diagram) as a result of a persistent error in ankle yield signaled by increased activity in the MG group I afferents.

Mechanisms for spinal plasticity

Currently very little is known about the cellular mechanisms in the spinal cord that are associated with adaptive changes in the motor pattern for stepping. Recent work has demonstrated that step-training in spinal cats reduces the effectiveness of a glycinergic inhibitory system (De Leon et al., 1999). The general notion is that step training results in a progressive removal of tonic inhibition from the spinal pattern generating network. How this is achieved mechanistically is unknown. The fact that training is necessary for the recovery of stepping suggests that the diminution of inhibition is dependent on the pattern of sensory feedback from the moving legs. Indeed, recovery of stepping does not occur when animals are trained only to stand. In addition to reducing the tonic inhibition on the spinal pattern generating network, patterned sensory feedback is necessary for establishing some specific features of the trained motor program. The best example of this is the need for feedback from cutaneous receptors in the paw to enable correct plantar placement of the foot (Bouyer and Rossignol, 1998). Again, however, we know nothing about how cutaneous signals are integrated into spinal networks to produce the appropriate motor pattern for foot placement. The modifications in the motor pattern of the MG muscle following transection of the LG, SOL and PL nerves are also dependent on use because they do not occur if the leg is immobilized (Pearson et al., 1999). Numerous mechanisms could underlie this use-dependent plasticity. For example, the increase in the strength of transmission in reflex pathways that contribute to the generation of the late component of the MG bursts could be due to a removal of presynaptic inhibition from primary afferents, changes in postsynaptic receptor sensitivity or number, or modification of the cellular and synaptic properties of interneurons in the reflex pathways. It is unlikely that modifications of the MG motoneu-

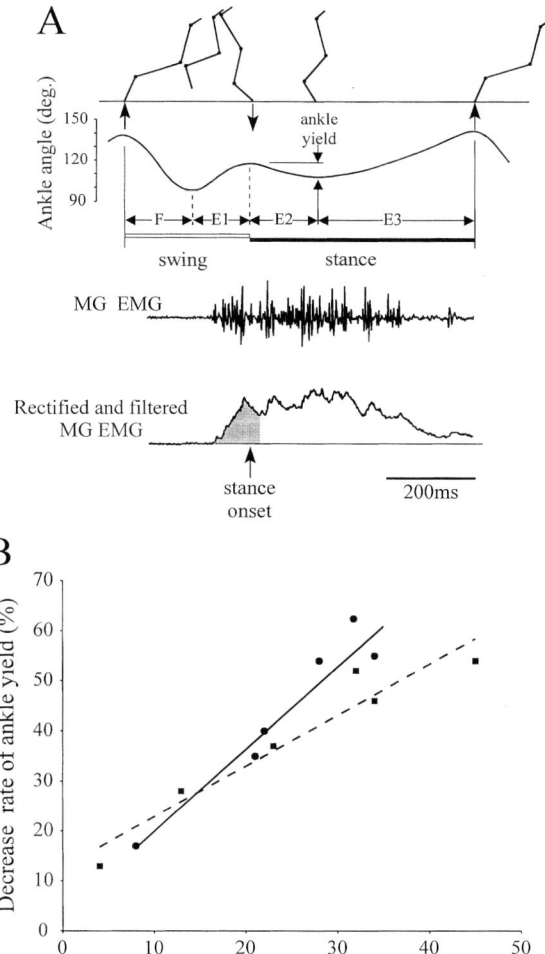

Fig. 5. The decrease in the rate of ankle yield (flexion) during the week following transection of the LGS and PL nerves is linearly related to the increase in the initial component of the MG EMG. (A) Diagram showing the relationship between ankle kinematics and the MG EMG. Note the EMG burst begins about 80 ms before stance onset. The initial component of the MG EMG is shaded in the rectified and filtered record. (B) Plots showing the linear relationship between the decrease rate of ankle yield during early stance and the increase in the amplitude of the initial component of the MG EMG for two animals. Each data point represents average values obtained on a different day during the first week after the nerve transections.

to an increase in the gain of transmission in MG group I afferent pathways that function to reinforce the generation of activity in the MG motoneurons (indicated by the up arrow in the left diagram). By 1 week the initial component of the MG bursts is also

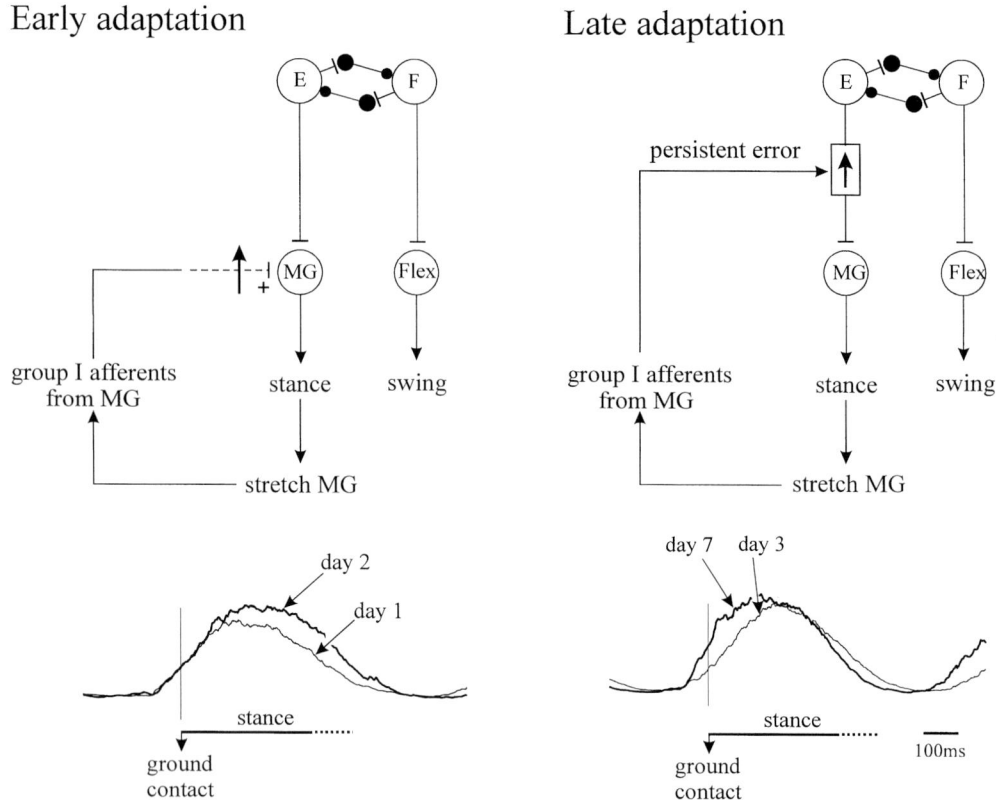

Fig. 6. Schematic diagram summarizing the possible mechanisms underlying the increase in the magnitude of MG bursts following cutting of the LGS and PL nerves. In the first few days after the nerve cuts, the increase in magnitude occurs mainly in the component of the MG bursts that begins after ground contact (bottom left). An increase in the gain in the group I pathway contributing to MG burst generation is postulated to be responsible for the increase in magnitude of the late component (top left). The initial component of the MG bursts, beginning before ground contact, increases more slowly than the late component and is significantly enhanced by 7 days (bottom right). One proposal for explaining this late adaptive response is that persistent errors in ankle flexion signaled by additional activity in group I afferents results in an increase in the central drive from the extensor half-center to the MG motoneurons (top right).

rons themselves is an important factor because the magnitude of the late component of the MG bursts often increases without significant changes in the initial component of these bursts.

The difficulty with identifying the mechanisms underlying plasticity in the spinal cord is well illustrated by results from investigation on the mechanisms underlying instrumental conditioning of the H-reflexes and stretch reflexes in rats, monkeys and humans (Wolpaw, 1997). Not only does up- and down-conditioning of the H-reflex depend on different mechanisms, but many of the changes are not directly related to changes leading to the modification in H-reflex amplitude. Wolpaw (1997) has classified the modifications in neuronal circuits into three categories: primary, compensatory and reactive. The primary changes are responsible for the rewarded behavior, the compensatory changes maintain the previous behavioral repertoire, and the reactive changes are induced as a consequence of the primary and compensatory changes. The reactive changes may even be maladaptive. With examples of these different types of plasticity in the systems regulating the strength of the monosynaptic reflex pathways, we can anticipate just as elaborate modifications in the neuronal systems regulating adaptive plasticity in more complex systems. It is likely that plasticity is expressed at many different sites in the spinal cord, and that multiple mechanisms are involved.

Conclusions

Many studies have now demonstrated adaptive plasticity in neuronal networks in the spinal cord. Adaptive plasticity in spinal networks can be expressed during normal development, during functional recovery from injury to the central and peripheral nervous system, and during instrumental and classical conditioning of spinal reflexes. In all these situations the adaptive modification of behavior depends either on use or on temporal contingencies of multiple sensory signals. Thus plasticity is induced and regulated by the pattern of sensory feedback from peripheral receptors. The mechanisms underlying spinal cord plasticity are, in general, poorly understood. Nevertheless, it seems very likely that multiple mechanisms are involved, and that changes occur at many sites in the spinal cord. There is nothing to suggest that these mechanisms are unique to the spinal cord. The adaptive modification of behavior in normal animals almost certainly involves modification of supraspinal as well as spinal networks. The relative importance of plastic changes at spinal versus supraspinal sites remains unclear. One attractive possibility is that spinal plasticity is largely involved in local adaptation of activity in one or a few muscles, while supraspinal plasticity acts more globally to coordinate activity in muscles throughout the body.

Acknowledgements

I thank Tania Lamb and John Misiaszek for their helpful comments on a draft of the manuscript. Supported by a grant from the Medical Research Council of Canada.

References

Barbeau, H. and Rossignol, S. (1994) Enhancement of locomotor recovery following spinal cord injury. *Curr. Opin. Neurol.*, 7: 517–524.

Bouyer, L. and Rossignol, S. (1998) The contribution of cutaneous inputs to locomotion in the intact and the spinal cat. *Ann. N.Y. Acad. Sci.*, 860: 508–512.

Bouyer, L. and Rossignol, S. (2000) Spinal cord plasticity associated with locomotor compensation to peripheral nerve lesions in the cat. In: Patterson, M.M. (Ed.), *Spinal Cord Plasticity: Alterations in Reflex Function*. Kluwer Academic Publishers, New York, in press.

Brainard, M.S. and Knudsen, E.I. (1998) Sensitive periods for visual calibration of the auditory space map in the barn owl optic tectum. *J. Neurosci.*, 15: 3929–3942.

Brustein, E. and Rossignol, S. (1998) Recovery of locomotion after ventral and ventrolateral spinal lesion in the cat, I. Deficits and adaptive mechanisms. *J. Neurophysiol.*, 80: 1245–1267.

Carrier, L., Brustein, E. and Rossignol, S. (1997) Locomotion of the hindlimbs after neurectomy of ankle flexors in intact and spinal cats: model for the study of locomotor plasticity. *J. Neurophysiol.*, 77: 1979–1993.

De Leon, R.D., Hodgson, J.A., Roy, R.R. and Edgerton, V.R. (1998) Locomotor capacity attributable to step training versus spontaneous recovery after spinalization in adult cats. *J. Neurophysiol.*, 79: 1329–1340.

De Leon, R.D., Tamaki, H., Hodgson, J.A., Roy, R.R. and Edgerton, V.R. (1999) Hindlimb locomotor and postural training modulates glycinergic inhibition in the spinal cord of the adult spinal cat. *J. Neurophysiol.*, 82: 359–369.

Dietz, V., Wirz, M., Colombo, G. and Curt, A. (1998) Locomotor capacity and recovery of spinal cord function in paraplegic patients: a clinical and electrophysiological evaluation. *Electroenceph. Clin. Neurophysiol.*, 109: 140–153.

Du Lac, S., Raymond, J.L., Sejnowski, T.J. and Lisberger, S.G. (1995) Learning and memory in the vestibuo-ocular reflex. *Annu. Rev. Neurosci.*, 18: 409–442.

Durkovic, R.G. and Damianopoulos, E.N. (1986) Forward and backward classical conditioning of the flexion reflex in the spinal cat. *J. Neurosci.*, 6: 2921–2925.

Feldman, D.E. and Knudsen, E.I. (1997) An anatomical basis for visual calibration of the auditory space map in the barn owl's midbrain. *J. Neurosci.*, 17: 6820–6837.

Forssberg, H., Grillner, S. and Halbertsma, J. (1980) The locomotion of the low spinal cat, I. Coordination within a hindlimb. *Acta Physiol. Scand.*, 108: 269–281.

Fouad, K. and Pearson, K.G. (1997) Modification of group I field potentials in the intermediate nucleus of the cat spinal cord after chronic axotomy of an extensor nerve. *Neurosci. Lett.*, 236: 9–12.

Georgopoulos, A.P. (1996) On the translation of directional motor cortical commands to activation of muscles via spinal interneuronal systems. *Cognit. Brain Res.*, 3: 151–155.

Goldberger, M.E. (1988a) Spared root deafferentation of a cat's hindlimb: hierarchical regulation of pathways mediating recovery of motor behavior. *Exp. Brain Res.*, 73: 329–342.

Goldberger, M.E. (1988b) Partial and complete deafferentation of cat hindlimb: the contribution of behavioral substitution to recovery of motor function. *Exp. Brain Res.*, 73: 343–353.

Goldberger, M.E. and Murray, M. (1988) Patterns of sprouting and implications for recovery of function. In: Waxman, S.G. (Ed.), *Functional Recovery in Neurological Disease*. Raven Press, New York, pp. 361–385.

Hiebert, G.W., Gorassini, M.A., Jiang, W., Prochazka, A. and Pearson, K.G. (1994) Corrective responses to loss of ground support during walking, II. Comparison of intact and chronic spinal cats. *J. Neurophysiol.*, 71: 611–622.

Holmberg, H., Schouenborg, J., Yu, Y.B. and Weng, H.R. (1997)

Developmental adaptation of rat nociceptive withdrawal reflexes after neonatal tendon transfer. *J. Neurosci.*, 17: 2071–2078.

Jiang, W. and Drew, T. (1996) Effects of bilateral lesions of the dorsolateral funiculi and dorsal columns at the level of the low thoracic spinal cord on the control of locomotion in the adult cat, I. Treadmill walking. *J. Neurophysiol.*, 76: 849–866.

Jordan, M.I. (1996) Computational aspects of motor control and motor learning. In: Heuer, H. and Keele, S.W. (Eds.), *Handbook of Perception and Action. Motor Skills, Vol. 2*, Academic Press, London, pp. 71–120.

Knudsen, E.I. (1994) Supervised learning in the brain. *J. Neurosci.*, 14: 3985–3997.

Knudsen, E.I. (1998) Capacity for plasticity in the adult owl auditory system expanded by juvenile experience. *Science*, 279: 1531–1534.

Konczak, J., Borutta, M. and Dichgans, J. (1997) The development of goal-directed reaching in infants, II. Learning to produce task-adequate patterns of joint torque. *Exp. Brain Res.*, 113: 465–474.

Misiaszek, J.E. and Pearson, K.G. (1999) Injecting botulinum toxin into ankle extensors mimics the effects of axotomy. *Soc. Neurosci. Abstr.*, 25: 122.

Mussa-Ivaldi, F.A. and Bizzi, E. (1997) Learning newtonian mechanics. In: Morasso, P. and Sanguineti, V. (Eds.), *Self-Organization, Computational Maps, and Motor Control*. Elsevier, New York, pp. 191–237.

Pearson, K.G. (1995) Proprioceptive regulation of locomotion. *Curr. Opin. Neurobiol.*, 5: 786–791.

Pearson, K.G. (2000) Neural adaptation in the generation of rhythmic behavior. *Annu. Rev. Physiol.*, 62: 723–753.

Pearson, K.G., Misiaszek, J.E. and Fouad, K. (1998) Enhancement and resetting of locomotor activity by muscle afferents. *Ann. N.Y. Acad. Sci.*, 860: 203–215.

Pearson, K.G., Fouad, K. and Misiaszek, J.E. (1999) Adaptive changes in motor activity associated with functional recovery following muscle denervation in walking cats. *J. Neurophysiol.*, 82: 370–381.

Rasmussen, S.A., Goslow, G.E. and Hannon, P. (1986) Kinematics of locomotion in cats with partially deafferented spinal cords: the spared-root preparation. *Neurosci. Lett.*, 65: 183–188.

Raymond, J.L. (1998) Learning in the oculomotor system: from molecules to behavior. *Curr. Opin. Neurobiol.*, 8: 770–776.

Rossignol, S., Bouyer, L.J.G., Whelan, P.J. and Pearson, K.G. (1997) Chronic spinal cats can recover locomotor function following transection of an extensor nerve. *Soc. Neurosci. Abstr.*, 23: 761.

Thoroughman, K.A. and Shadmehr, R. (1999) Electromyographic correlates of learning an internal model of reaching movements. *J. Neurosci.*, 19: 8573–8588.

Wernig, A., Muller, S., Nanassy, A. and Cagol, E. (1995) Laufband therapy based on 'rules of spinal locomotion' is effective in spinal cord injured persons. *Eur. J. Neurosci.*, 7: 823–829.

Wetzel, M.C., Atwater, A.E., Wait, J.V. and Stuart, D.G. (1976) Kinematics of locomotion by cats with a single hindlimb deafferented. *J. Neurophysiol.*, 39: 667–678.

Whelan, P.J. and Pearson, K.G. (1997) Plasticity in reflex pathways controlling stepping in the cat. *J. Neurophysiol.*, 78: 1643–1650.

Whelan, P.J., Hiebert, G.W. and Pearson, K.G. (1995) Plasticity of the extensor group I pathway controlling the stance to swing transition in the cat. *J. Neurophysiol.*, 74: 2782–2787.

Wolpaw, J.R. (1997) The complex structure of a simple memory. *Trends Neurosci.*, 20: 588–594.

CHAPTER 7

Neural plasticity as revealed by the natural progression of movement expression — both voluntary and involuntary — in humans after spinal cord injury

Blair Calancie *, Maria Del Rosario Molano and James G. Broton

The Miami Project to Cure Paralysis and Department of Neurological Surgery, University of Miami School of Medicine, 1600 NW 10th Avenue, R-48, Miami, FL 33136, USA

Introduction

Research into effective treatments for spinal cord injury (SCI) has been focused on two primary target populations: those with acute injury, and those with non-acute injury. For the former approach, common sense dictates that one might benefit from interventions which minimize loss of central nervous system (CNS) tissue that might have survived the initial physical insult, but is now further threatened by metabolic consequences of that insult. This approach is commonly referred to as 'neuroprotective'. Alternatively, at some time-point following the initial trauma, the CNS tissue loss will have stabilized. Beyond this time, deliberate therapeutic interventions designed to replace lost CNS tissue or promote new growth in surviving CNS tissue (and the functions that such tissue subserves) can be referred to as 'neurorestorative'.

Clinical trials in this country and abroad have utilized both neuroprotective and neurorestorative interventions in efforts to minimize deficits following

either the initial trauma (Bracken et al., 1990; Geisler et al., 1991; George et al., 1995) or long-term complications of such trauma (Falci et al., 1997). In neuroprotective studies, the outcome measures employed rely upon subjective evaluations of strength, sensation, and function, which may be either difficult to interpret, or may lack sensitivity to small differences in spinal cord functional properties across and below the lesion (Daverat et al., 1988; Davis et al., 1993; Waters et al., 1993; Mizukami et al., 1995; Wells and Nicosia, 1995; Donovan et al., 1997).

Interpretation of trials initiated during the acute (hours to days) or subacute period following SCI (i.e., within the first 6 months) is further hampered for two reasons. First, spontaneous recovery of neurologic function (improved sensation and/or strength) may continue for long periods of time following injury (Piepmeier and Jenkins, 1988). Second, substitution of muscle action by non-paralyzed limb segments is a well-known strategy used by persons with SCI to achieve functional goals (Kendall and McCreary, 1983; Nixon, 1985; Somers, 1992), yet it can be difficult to recognize, and should not be confused with a restoration of neurologic function below the spinal lesion.

To address questions of muscle substitution and to lend a more objective aspect to motor and sensory testing, neurophysiologic approaches offer certain advantages to the study of recovery following spinal

---

* Corresponding author: Dr. Blair Calancie, The Miami Project to Cure Paralysis, University of Miami School of Medicine, 1600 NW 10th Avenue, R-48, Miami, FL 33136, USA. Fax: +1-305-545-8347; E-mail: bcalancie@miami.edu

cord injury beyond those which rely exclusively on clinical examination. There are many published studies of spinal cord neurophysiologic properties following human spinal cord injury, dating back 20 years and more (Kuhn, 1950; Lance et al., 1966; Diamantopoulos and Van der Olsen, 1967; Herman, 1969; Eidelberg, 1981). For the most part, these studies typically involve one-time measures from their sample population, with limited follow-up. This approach is useful for documenting alterations in spinal cord properties at a given point in time, but does not allow one to follow the evolution of those properties over time; to do this, detailed, repeated neurophysiologic examinations of spinal cord properties are necessary in a large population of persons following SCI.

For several years, we have been conducting clinical and neurophysiologic studies on persons with acute SCI who are admitted to our hospital complex (University of Miami/Jackson Memorial Hospital, or UM/JMH). This chapter details our protocol, and presents some findings representative of the population examined to date. Emphasis will be placed on emergence and/or demonstration of voluntary contractions and involuntary movement and the time-points when such movements become evident. Details of central motor conduction recovery as tested with transcranial magnetic stimulation are beyond the scope of this chapter, and will be presented elsewhere.

Measurement techniques

The UM/JMH complex includes a Level I trauma center and a dedicated SCI rehabilitation unit. For these reasons, the great majority of persons with SCI and/or suspected spine fracture in Greater Miami/Dade County (population ≈ 3 million) are admitted to UM/JMH. Additional subjects include those patients with acute spine trauma sustained in Caribbean and Latin American countries who arrange for transport to Miami and admission to UM/JMH.

Details of instrumentation are described elsewhere (Calancie et al., 1996, 1999) and will be only briefly summarized here. All neurophysiologic data were based on muscle electromyogram (EMG) recorded with pairs of self-adhesive surface electrodes (S'offset; Medi-Trace). Data were amplified, filtered and stored on digital tape for off-line analysis. During recording, data were presented in real time both visually (12-channel computer display) and audibly (12-channel audio mixer with speaker output). Tibial nerve electrical stimulation (both single-pulse and high-frequency) was delivered via a Grass S88 stimulator (used for preliminary studies) or a Digitimer D185 constant-voltage stimulator (used for most subjects). Deep tendon reflexes were tested with a hammer fit with a microswitch to provide a trigger pulse for synchronization.

For persons with cervical injury or fracture, and barring casts or dressings, EMG was recorded from the following left-side muscles: elbow flexors (including biceps brachii — Biceps), elbow extensors (triceps brachii — Triceps), wrist extensors (including extensor carpi radialis — ECR), wrist flexors (including flexor carpi radialis — FCR), thenar eminence (including abductor pollicis brevis — APB), hypothenar eminence (including abductor digiti minimi — ADM), hip flexors (including psoas major — Psoas), knee extensors (quadriceps — QUADS), knee flexors (hamstring — HAMS), ankle dorsi-flexors (including tibialis anterior — TA), ankle plantar flexors (including soleus — Soleus), and foot intrinsic great-toe plantar-flexor (abductor hallucis — AbH). After completing the protocol described below, electrodes were switched to comparable right-side muscles, and the protocol repeated. For persons with thoracic or thoracolumbar injury, EMG was recorded bilaterally from the six lower-limb muscles mentioned above, without switching sides of electrodes.

Subjects were usually tested either fully supine or partially reclining; the former position was used almost exclusively for the initial evaluation. After electrode placement, subjects were asked to attempt an isolated yet maximal contraction in each of the muscle groups, and were given auditory feedback of just that target muscle's activity. Care was taken to have the subject both contract and relax on command, as volitional relaxation is a useful means to differentiate a true, volitional contraction from a triggered spasm (see also Waters et al., 1993). Following voluntary contractions, deep tendon reflexes were tested at the knees (the contralateral knee was tested to look for cross-over responses), the ankle, and the wrist. For persons with cervical injury, H-reflexes and M-waves in the soleus to single-pulse tibial nerve stimulation at the popliteal fossa were elicited. Using a stimulus intensity adequate to evoke a large M-wave, a series of three-pulse trains (2 ms interpulse interval) was applied to the tibial nerve via surface electrodes, using an interval between trains of 5–10 s. This was repeated for the right tibial nerve. These inputs are highly effective at eliciting interlimb reflexes (Calancie, 1991; Calancie et al., 1996) in persons in whom such reflexes have emerged following injury.

Clinical examinations of sensory perception (light-touch and pinprick) and muscle strength were carried out using routine methods (Ditunno et al., 1994). Muscle testing of the foot intrinsic muscles (AbH) for plantar flexion of the great toe was also done.

We attempt to carry out follow-up evaluations at 1 week, 2 weeks and 4 weeks following the initial evaluation. Beyond this point the frequency of continued follow-up depends upon the person's neurologic status. If they have recovered or retained volitional contraction in at least one lower-limb muscle — a condition which we term 'motor-incomplete' — then follow-up is more frequent than if they remain 'motor-complete' (i.e., American Spinal Injury Association impairment scale [ASIA] 'A' or 'B'). At most, we attempt an additional five follow-up evaluations on persons who are motor-incomplete between 1 and 12 months after the injury, compared to three follow-ups over the next 11 months on persons who are motor-complete at 1 month after the injury. Beyond 1 year after the injury, follow-ups are attempted every 6 (motor-incomplete) or 12 months (motor-complete).

During off-line analysis, EMG voluntary recruitment records were assigned values of 0 (no recruitment) through 5 (normal interference pattern) as detailed elsewhere (Alexeeva et al., 1997). Reflex amplitudes to tendon taps (peak-to-peak) were measured with cursors from the single largest response (rather than from the average response). Latency was measured at the take-off of the compound EMG waveform from the background activity. Note that the same person (B.C.) delivered the tendon taps in all but four trials (i.e., >99% of the measures).

Cumulative data

A total of 174 subjects with acute spine and/or spinal cord injury consented for neurophysiologic and clinical testing at the time of this writing. Reasons for excluding persons with SCI included traumatic brain injury ($n = 59$), declined consent ($n = 49$), and early discharge or transfer ($n = 34$). A common reason subjects refused consent was that they considered themselves to have normal neurologic function, already knew of their impending hospital discharge, and chose not to undergo the inconvenience of coming back to the laboratory for follow-up testing. The majority of subjects were tested within 7 days of their injury (0–2 days $= 65$; 3–7 days $= 81$). Of the remaining subjects, all but nine were initially tested during the second week after the injury. In cases when subjects had surgery to stabilize and/or decompress the spinal canal, we postponed testing (either the initial testing or follow-up) until the second day following the surgery. Fourteen subjects who were neurologically intact (referred to as ablebodied [A-B]) were included for examination of deep tendon response properties for comparison to persons with SCI.

The SCI sample population was made up of 133 males and 41 females. Ages ranged between 7 and 83 years (mean $= 39.6 \pm 17.5$). The incidence within the sample population of injury mechanism, ASIA category, and initial injury level (or radiologic level if the subject was neurologically intact) is shown in Table 1. Combining all initial and follow-up evaluations, a total of 544 evaluations of spinal cord neurophysiologic properties were carried out at UM/JMH by our group at this writing, the majority (458) within the past 2 years.

Voluntary contractions

The neurologic deficits of persons who initially present as ASIA 'A' (no sensory or motor sparing caudal to the injury) are often thought of as being representative of human spinal cord injury. However, in our sample population this group made up only 36% of the total population at the time of initial evaluation. Examples of muscle contraction waveforms at initial and most recent follow-up evaluations in a subject categorized as ASIA 'A' are shown in Fig. 1.

TABLE 1

SCI population analyzed

Mechanism	Incidence	ASIA category	Incidence	Injury level	Incidence
MVA	74	A	62	C1–C2	25
Fall	33	B	8	C3–C4	17
Gunshot	20	C	35	C5–C6	64
Diving	11	D	61	C7–T1	9
Ped hit by car	10	E	8	T2–T5	8
Sport	6			T6–T9	9
Water – other	4			T10–T12	15
Motorcycle	4			L1–L3	20
Bicycle	1			L4–S2	2
Other	11			Unknown	5

Number of subjects for each grouping, including: injury cause (Mechanism; left), injury severity (ASIA category; center), and injury level (Injury level; right).

These data were obtained from a 38-year-old male with C2 spinal cord injury first examined 6 days after the injury; he was injured in a motor vehicle accident. At this time he was ventilator-dependent,

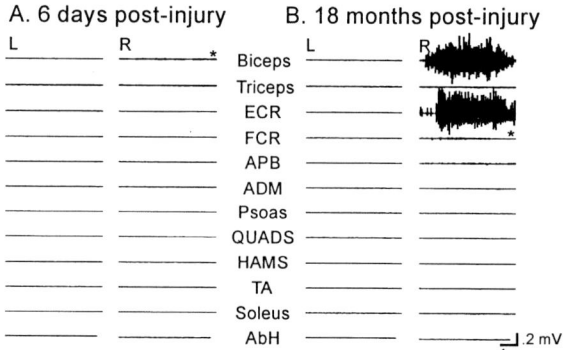

Fig. 1. EMG from 12 left- (L) and right-side (R) muscles measured at 6 days (A) and 18 months (B) following the injury. Subject sustained severe cervical spinal cord injury leading to ASIA category of 'A' at the initial evaluation (A), and remained classified as ASIA 'A' at the 18-month evaluation (B). Muscle abbreviations in this and subsequent figures are: ECR = wrist extensors; FCR = wrist flexors; APB = thenar eminence of the hand, including abductor pollicis brevis; ADM = hypothenar eminence of the hand, including abductor digiti minimi; $QUADS$ = quadriceps; $HAMS$ = hamstring; TA = tibialis anterior; and AbH = abductor hallucis of the foot. Subject was asked to contract and relax each muscle in isolation. Individual traces marked with an asterisk (*) denote trials in which a trace amount of EMG was recorded, but the resolution of the traces shown makes visualization of this activity difficult. Records were not obtained simultaneously, but have been aligned vertically using a 'cut-and-paste' approach in this and two subsequent figures.

and only the right biceps showed trace signs of volitional recruitment during this evaluation (indicated by asterisk). During the most recent examination (approximately 18 months later), the subject showed good volitional recruitment in both his right biceps and wrist extensors; wrist flexors showed trace EMG activity upon voluntary contraction attempts. There was still no sign of volitional recruitment in any of his left-side muscles examined. Fig. 2 depicts the EMG values during voluntary contraction attempts across each of the evaluations conducted in this subject. At 34 days after the injury, the right biceps was still the only muscle to show activity. Two months later, trace EMG was seen in the right wrist extensors and flexors, and the right biceps EMG interference pattern had increased to a score of '2'. The subject was still ventilator-dependent at this time, and was weaned from ventilator support at approximately 6 months following injury. Since then, the subject showed increased strength in these muscles, but no additional muscles were being recruited at the most recent examination (Fig. 2, 18-month record).

In our experience, it is not uncommon for persons to demonstrate considerable improvement in neurologic motor function following SCI. Fig. 3 shows EMG records from left-side muscles of a 42-year-old man with C5 SCI first examined 8 days after the injury; he was struck by a car while riding a bicycle. He was initially classified as ASIA 'A'. At this time, he could recruit only his biceps on the left side. Note that we were unable to test some muscles at this acute stage due to a cast on his right forearm

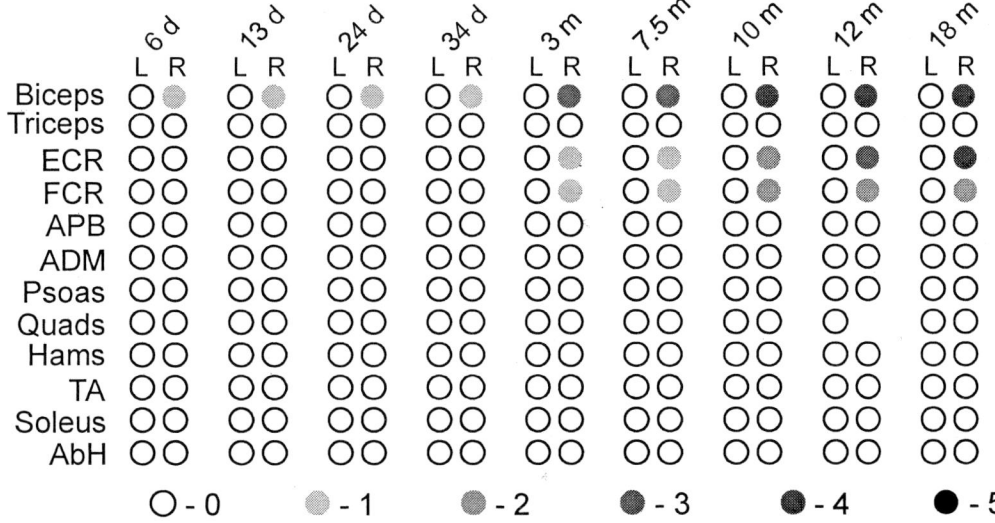

Fig. 2. Circle-diagram EMG summary of recruitment pattern seen in each of the muscles tested for the same subject whose data are shown in Fig. 1. Time of testing relative to injury date is indicated above each column of circles (d = days; m = months). Darker fill indicates more pronounced EMG recruitment; legend of fill patterns shown along the bottom. By way of example, this subject had an assigned EMG grade of '3' in the right biceps and '1' in the right ECR and FCR muscles at the 3-month and 7.5-month tests.

and hand and multiple dressings on his legs and hips. By 5.5 months after the injury, this person now showed well-defined recruitment in the biceps, ECR and FCR muscle groups on the left side, along with trace recruitment in the APB and (in the lower limb) the AbH muscles. By 21 months after the injury, this person was able to recruit all muscles examined on his left side.

Fig. 4 summarizes the time-course of muscle recruitment recovery as seen over the ten examinations conducted on this subject since the time of his injury and whose data are shown in Fig. 3. Several aspects of this figure are noteworthy. First, many of the muscles in which strong recruitment was ultimately seen did not show any signs of recruitment until almost 1 year (10.5 months) after the injury. Second, there was asymmetry between left- and right-hand sides in the pattern of recovery and ultimate signals obtained. Third, the first muscles of the lower limbs in which volitional recruitment was seen were the AbH muscles of the feet. Comparison of records between 12 and 21 months suggests a reduction in strength at the more recent follow-up. While this may be due to neurologic deterioration secondary to spinal cord tethering or cyst formation, the more likely explanation is that the subject was simply not contracting as hard in these muscles as he was able. Such a finding serves as a reminder to have subjects contract as hard as they can in all trials and for all muscles when carrying out this type of evaluation.

Persons with acute SCI who fall into ASIA categories 'C' through 'E' have sparing of sensation at their most inferior neurologic segment (S4–S5), as well as varying degrees of motor function in muscles below the injury at the time of testing. Fig. 5 shows EMG records from such a person (ASIA 'C'; 42 years old; injured at C5 in a water-related accident) at the earliest (2 days) and most recent (5.5 months) times examined after the injury. As for earlier figures, records in which EMG is present but is not easily seen with this level of graphic reduction are indicated with an asterisk. As summarized in Fig. 6, much of the improvement in strength and distribution of EMG contractions was seen between 1.5 and 3.75 months after the injury in this subject. Even so, 3 of the 24 muscles examined were still not being recruited during volitional contraction attempts at 3.75 months, but were successfully recruited by 5.5 months after the injury.

We have seen 16 examples to date of persons with acute injury categories of ASIA 'A' or 'B' (injury at or above T10) who have gone on to recover motor

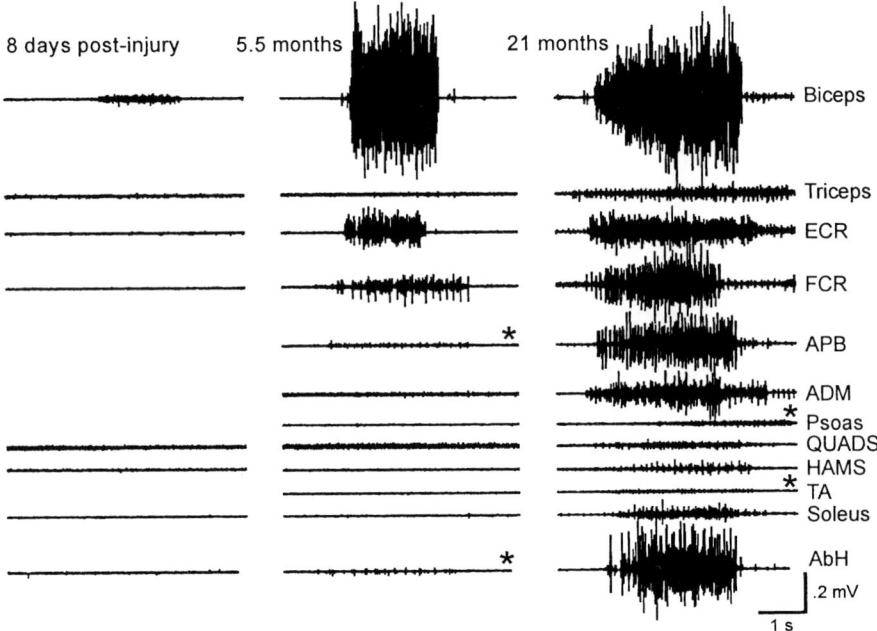

Fig. 3. EMG from 12 left-side muscles measured at 3 different times after the injury: 8 days (initial), 5.5 months, and 21 months. Multiple records are missing from the 8-day series of traces due to a cast on the subject's left hand, and multiple dressings on his legs and hips. Note that AbH was the first of this subject's lower-limb muscles to show voluntary recruitment. As for Fig. 1, trace EMG recruitment is denoted with an asterisk.

function in one or more lower-limb muscles since our initial evaluation. In many of these cases, and as seen in the subject whose data are depicted in Figs. 3 and 4, the first muscle to recover volitional recruitment is the AbH, an intrinsic muscle of the foot. Fig. 7 shows that of the sixteen instances in which we have confirmed recovery of volitional recruitment in at least one lower-limb muscle following a period of complete lower-limb paralysis, AbH was the first recruited in twelve of these trials (i.e., 75%). Moreover, in the four cases in which AbH was not the first recruited, it was being recruited volitionally at the time of the next follow-up examination (not shown).

As an extreme example of this pattern, Fig. 8 illustrates EMG records from a 79-year-old woman with a C1 fracture sustained in a motor vehicle accident 2.5 months previously. Since the time of injury, she had been fully ventilator-dependent, as she was at the time of the examination depicted in Fig. 8. Attempts to contract all left-side muscles showed no activity at any time (not shown). Right-side contraction attempts (site of attempted contraction indicated above the traces) also showed no recruitment in any of the target muscles until she attempted to plantar-flex her right ankle (i.e., contract the soleus). At this time, EMG activity was seen in the AbH muscle (which can be considered a synergist for ankle plantar-flexion). Three successive attempts to contract the AbH muscle revealed consistent EMG signal, which could be started and stopped upon voluntary command. Expanding the time-scale and amplitude of the EMG waveforms in AbH shown in Figs. 8 and 9 confirms that this EMG activity was due to the discharge of one single motor unit, without signs of any additional motor recruitment. The discharge rate during this voluntary contraction was approximately 6 Hz.

The subject's ability to stop the discharge on command was crucial in our conclusion that this activity was due to volitional signals from supraspinal motor centers, and did not instead reflect some form of triggered spasm. Thus despite this subject's total ventilator dependence and absence of volitional contraction in any upper-limb muscle — including an inability to shrug her shoulders, an action mediated by the spinal accessory (i.e., cranial) nerve —

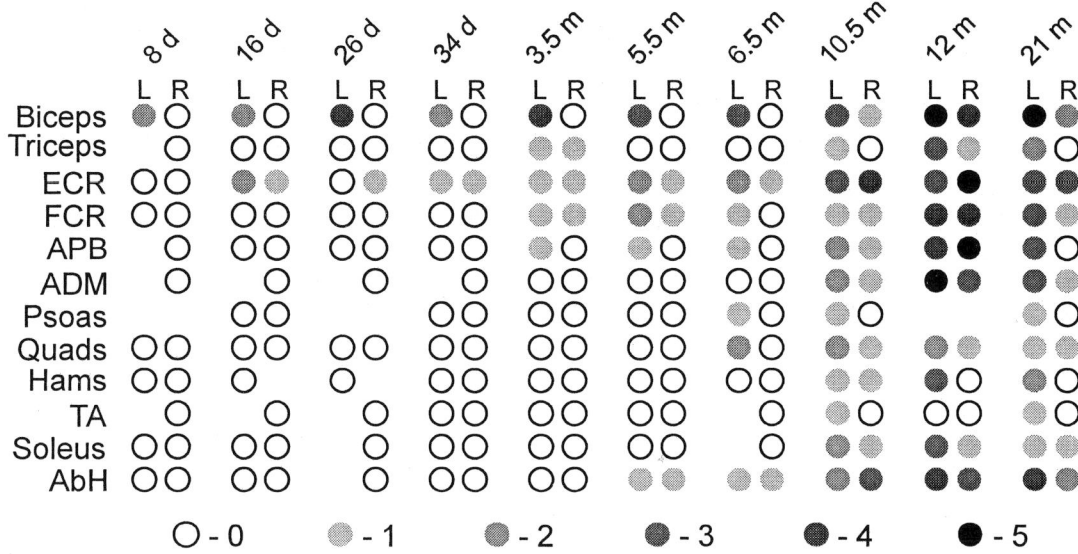

Fig. 4. Circle-diagram EMG summary of recruitment for the same subject whose data are shown in Fig. 3. Multiple muscles could not be monitored in the initial evaluation due to injuries sustained. AbH muscles were first of all lower-extremity muscles to be recruited, at 5.5 months after the injury.

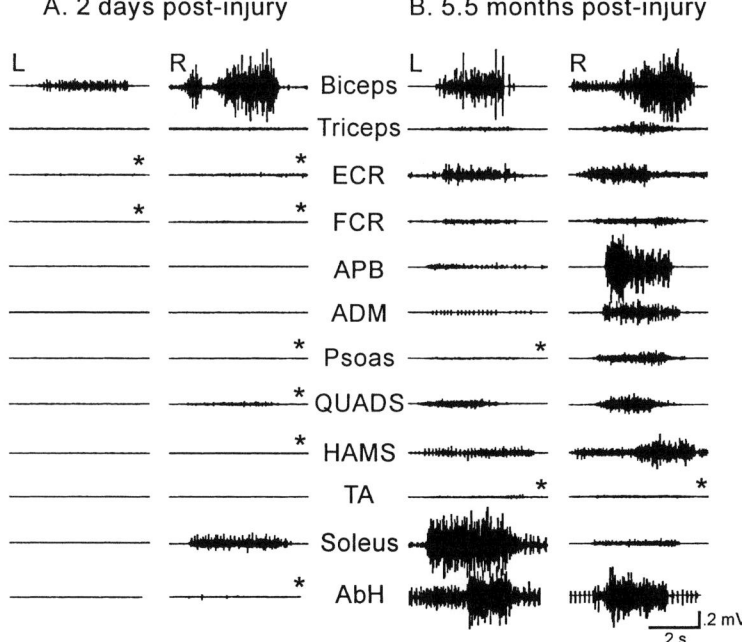

Fig. 5. Left- and right-side EMG at two time periods in a subject with ASIA 'C' injury upon initial evaluation (2 days after the injury; A), who recovered adequate leg function to be reclassified as ASIA 'D' at 5.5 months. This subject had considerable sparing of right-side muscle function in his arm and leg at the earliest time studied, although the magnitude of interference pattern was minimal (indicated by asterisks). He was able to recruit all 24 muscles studied at the 5.5-month time-point, although left-side recruitment was still markedly diminished compared to right-side contractions.

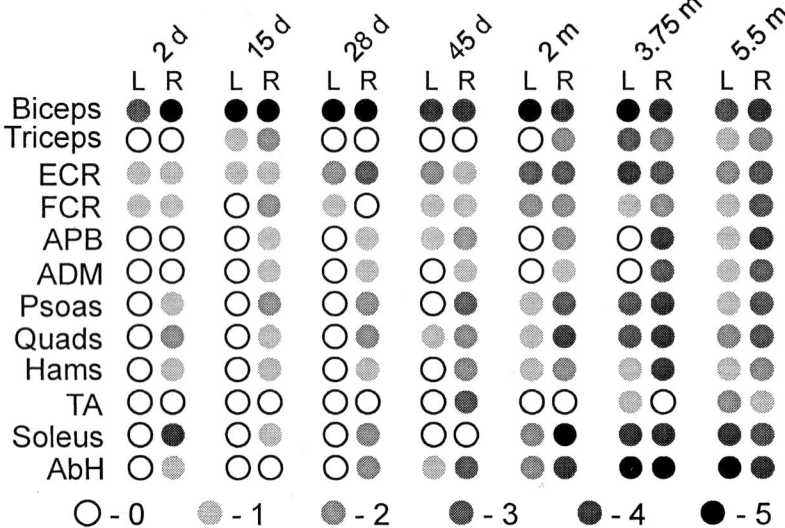

Fig. 6. Circle-diagram EMG summary of recruitment for the same subject whose data are shown in Fig. 5. Right-side recovery was evident at the initial examination period, and continued in this manner until the most recent evaluation. Note that for the first three evaluations, no EMG in the left leg was evident; once contraction became evident in this leg (45 days after the injury), it was seen in both the quadriceps and abductor hallucis. Note also that the left-side ADM was slow to recover (no recruitment at 3.75 months after the injury) compared to most upper-limb muscles, and was rated as 'trace' contraction at the 5.5-month study.

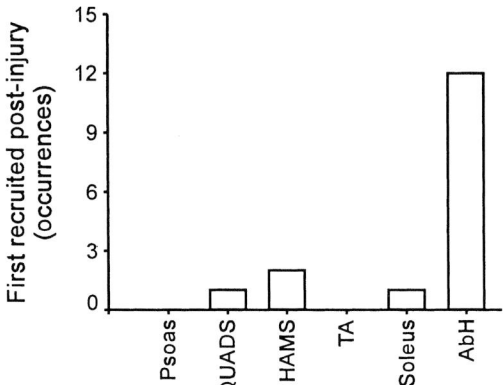

Fig. 7. Summary of those cases in which complete paralysis of a lower extremity was noted on at least one post-injury examination, followed by restoration of volitional contraction in only one muscle of that leg. In 75% of these cases, the first muscle in which EMG activity was noted was AbH. Hamstring was noted to contract first in 2 of the 16 observations (12.5%), and one each observation was made for quadriceps and soleus. In each of the four cases in which AbH was noted to not be first recruited, it was being recruited (along with the original muscle seen) by the next evaluation.

there was preservation of function in at least one descending motor axon innervating the right AbH motoneuron pool in this subject's lumbar enlargement.

Subsequent evaluations over the following 5 months showed no change in this recruitment pattern; the subject has since been lost to further follow-up.

Involuntary (or reflex) contractions

Reflex responses to patellar and Achilles tendon taps at each examination time-point are shown in Fig. 10 for the three subjects whose voluntary contractions were described above. These values represent the mean of the responses collected from both sides (except those few cases for which EMG from a target muscle [quadriceps or soleus] was not recorded). Data are expressed in millivolts; that is, there is no normalization to a maximal evoked waveform, such as an M-wave. Note also that in order to demonstrate better the responses to tendon taps at the earlier post-injury time-points, data along the *x*-axis (time after the injury) are plotted on a logarithmic scale.

At the acute stage, there was no response to tendon taps in either the quadriceps or soleus muscles bilaterally for both subjects depicted with ASIA 'A' category injuries during the initial evaluation (Fig. 10A,B). As early as 1 week later, Achilles tendon tap responses were evoked in the person whose

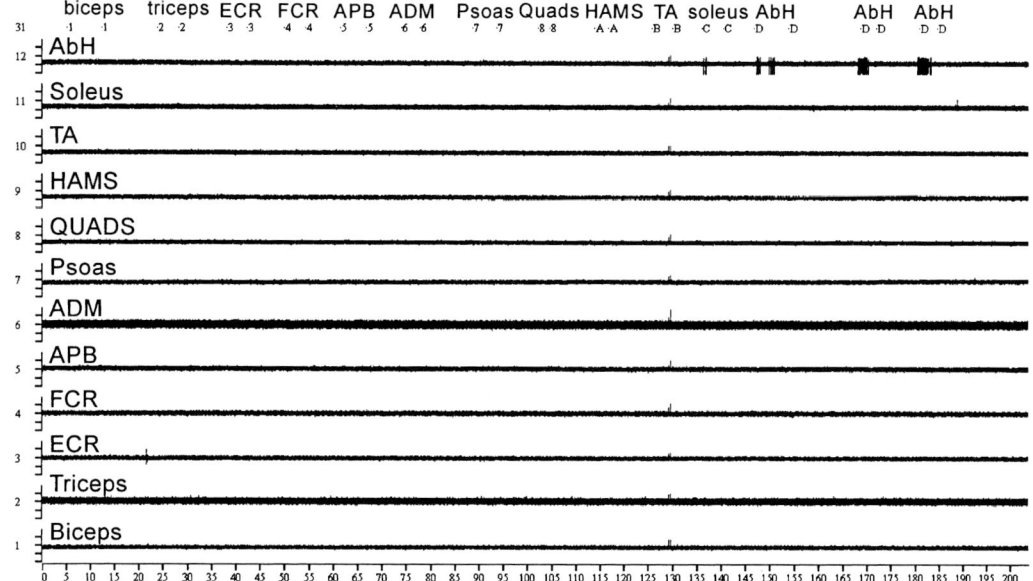

Fig. 8. Continuous record of EMG captured during a series of attempted voluntary contractions in right-side muscles in a subject with C1 spinal cord injury who was fully ventilator-dependent, examined 2.5 months following a motor vehicle accident. She had been examined three times previously, without evidence of voluntary contraction. The muscles in which contractions were being attempted are noted above the top (i.e., AbH) trace. Thus biceps was attempted first, followed by triceps, wrist extensors (ECR), etc. Three contraction attempts of AbH ended this record. Only the AbH muscle showed reproducible activity which was started and stopped on command. At the time of examination, this subject was unable to even shrug her shoulders. The repeated numbers/letters under the muscle names above the AbH records correspond to the time on the tape record when the verbal command to contract (first occurrence) and relax (second occurrence) was given to the subject. This accounts for the apparent delay between contraction/relaxation commands for the AbH muscle (denoted by D/D) and the actual EMG activity seen in AbH at these times.

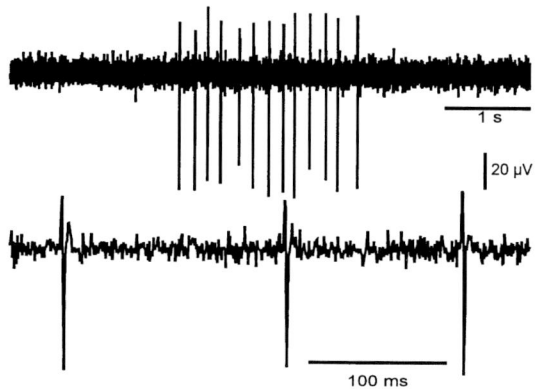

Fig. 9. Higher magnification of EMG record from right AbH during one of the three voluntary contraction attempts of this muscle depicted in Fig. 8. Upper record: total contraction attempt; lower record: expanded time scale part way through the upper record. The lower record shows that only one single motor unit is discharging during this contraction. Vertical calibration bar applies to both records; horizontal calibration bars differ for each record as indicated.

injury status had remained ASIA 'A' (Fig. 10A; triangles); patellar tendon responsiveness was slower to emerge in this individual, only becoming evident during the examination approximately 32 weeks after the injury (Fig. 10A; circles). Subsequent examinations in this individual revealed relative stability in response amplitude of patellar and Achilles tendon taps. In the person with ASIA 'A' injury that improved to ASIA 'C' status by 5.5 months after the injury, tendon responses (Fig. 10B) were roughly comparable at the acute stage to those of the person whose results are depicted in Fig. 10A. In contrast, tendon tap responses could be reliably elicited from the subject with ASIA 'C' spinal cord injury as early as 2 days after the injury (Fig. 10C). The response amplitudes were consistently 0.5 mV or greater, values not evident in most of the measures of the two persons with ASIA 'A' injury initially (Fig. 10A,B).

Fig. 11 shows the absolute response amplitude to taps of the patellar (A) and Achilles (B) tendons,

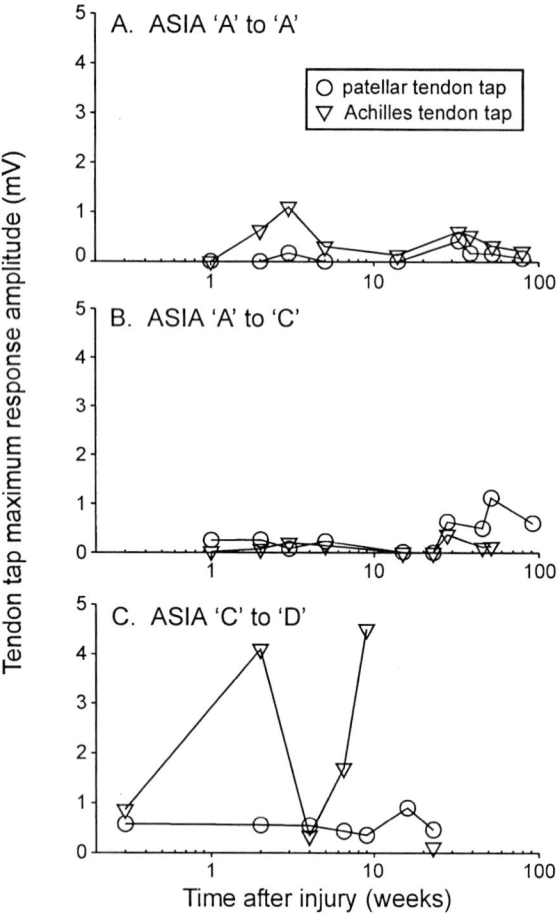

Fig. 10. Average of left- and right-side response amplitudes to patellar (circles) and Achilles (triangles) tendon taps as measured in each evaluation for the same three subjects whose EMG records are illustrated in Figs. 1–6. Response amplitudes are the peak-to-peak maxima for each series of trials. Responses seen in the subject with ASIA 'C' injury (C) were considerably higher in amplitude at the initial evaluation than for the two subjects with 'motor-complete' injury (A,B). The Achilles tap amplitudes in (C) show wide fluctuations; otherwise, other measures show relatively constant response amplitudes across measures.

averaged across subjects grouped into the different ASIA categories at the time of initial (i.e., acute) evaluation. The numbers above the standard error bars represent the number of observations included in each group. For comparison, data from neurologically intact subjects, or able-bodied (A-B) subjects are shown. Calculations for this figure excluded data from persons with injury below T10, and from those persons whose injury was greater than 1 week prior to the date of initial evaluation. For each of the 'motor-incomplete' categories (ASIA 'C'–'E'), tendon responses were large in amplitude, reaching values roughly comparable to those of able-bodied subjects for taps to the patellar tendon (Fig. 11A). While Achilles tendon tap response amplitudes were, on average, somewhat diminished from those of A-B subjects (for which the mean amplitude was 3.6 mV), responses were usually well defined and 'brisk'. In marked contrast, tendon tap amplitudes in persons with ASIA 'A' spinal cord injury at this initial evaluation rarely exceeded 0.2 mV in any one person for patellar taps, or 0.4 mV for Achilles taps. The very small average response amplitudes and standard errors reflect these findings.

The findings just presented for tendon tap responses focused on only that muscle group from which responses would be anticipated: ipsilateral quadriceps for taps to the patellar tendon, and ipsilateral soleus for taps to the Achilles tendon. In some subjects with incomplete SCI at the acute stage, however, we observed short-latency EMG activity in multiple muscles beyond the 'target' (i.e., homonymous) muscles following different tendon taps. Fig. 12 shows examples of this tendon tap 'spread', taken from an 81-year-old male subject with acute SCI who was classified as ASIA 'D' at the initial evaluation (5 days after the injury). Taps to his left patellar tendon led to well-defined responses in both his quadriceps and hamstring muscle groups (Fig. 12A; single trial). In addition to the expected response in soleus, both the quadriceps and hamstring also showed evoked responses to taps at the Achilles tendon (Fig. 12B; single trial). Finally, taps to the right patellar tendon led to short-latency EMG in the contralateral (i.e., left-side) hip flexors, quadriceps and hamstring group, as shown in Fig. 12C. This record illustrates the superimposition of five successive trials, demonstrating the consistency of responses from tap to tap (taps delivered approximately 3–5 s apart).

The incidence of short-latency quadriceps responses to taps of the contralateral patellar tendon in the population studied is shown in Fig. 13. Three such responses were seen out of 28 examinations in A-B subjects (left-side tap and right-side response in 3 of 14 individuals studied). Minimum latencies

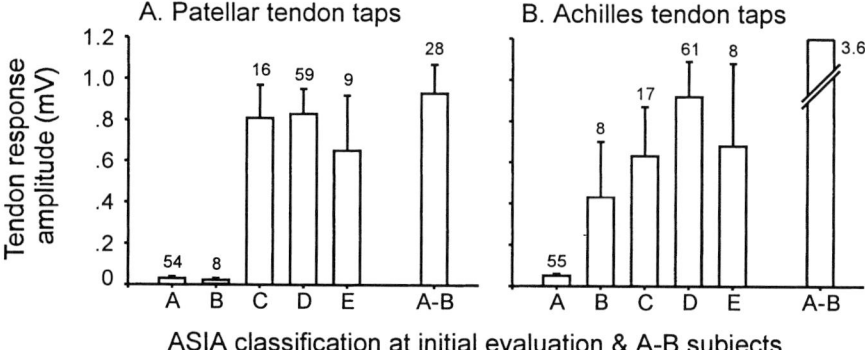

Fig. 11. Average response amplitude to patellar (A) and Achilles tendon taps (B), grouped according to ASIA category at the time of initial evaluation. The great majority of observations were taken from ASIA 'A' and ASIA 'D' subjects. Subjects whose injury was more than 7 days prior to the date of initial evaluation were excluded, as were those with injury at or below T11. Those bars designated 'A-B' represent responses obtained from persons without neurologic injury (i.e., able-bodied, or control subjects). The average Achilles tendon tap response amplitude measured in A-B subjects was 3.6 mV (B). Error bars represent standard errors of the means. Numbers above the error bars show the number of observations made for each group.

were somewhat later (typically 28–30 ms) than the ipsilateral (i.e., homonymous) response latency routinely seen in A-B subjects (~20 ms; not shown). These 'cross-over' responses were almost never seen in persons with ASIA 'A' injury. Of the two responses seen in ASIA 'A' subjects, both were from a person who improved to ASIA 'C' status 6 months after the injury. No cross-over was seen at the acute stage in any ASIA 'A' person who remained in the 'A' category beyond 1 month after the injury. In contrast, short-latency responses to taps of the contralateral patellar tendon were frequently seen in persons with motor-incomplete injury (ASIA 'C' through 'E'), being present in 35% to 55% of the examinations made at the acute stage following SCI. As for Fig. 11, data for Fig. 13 were restricted to persons with SCI at or above T10, and in whom the injury had occurred within 1 week of examination.

In the subject with ASIA 'A' injury whose data are depicted in Figs. 1, 2 and 10, electrical stimulation with single shocks of his left-side tibial nerve elicited the expected M-wave and H-reflexes from the ipsilateral soleus and AbH muscles (not shown). Delivery of a brief, high-frequency stimulus train to this nerve (three pulses at 500 Hz; 1 ms pulse duration, 150 V pulse intensity) also evoked time-locked EMG activity in this person's APB and ADM muscles of his right hand, as shown in Fig. 14. This figure shows the superimposition of 11 successive stimulus trains of identical parameters, separated from each other in time by approximately 3–4 s. Two motor units were recruited by this input, a spontaneously discharging (i.e., tonic) unit in the right APB, and a non-tonic (or 'phasic') unit in the right ADM. The arrows in Fig. 14 indicate the time of discharge of these respective units relative to the delivery of the electrical stimulus to the back of the contralateral knee. The APB motor unit tended to discharge at a latency of 55–60 ms relative to stimulus delivery, whereas the ADM unit discharged at a latency of approximately 51 ms. The difference in latencies of these two waveforms in the different muscles proves that they did not arise from a common source (e.g., a large muscle unit from an interosseus muscle midway between the APB and ADM surface recording electrodes). At the earlier, 12 months after the injury evaluation, such interlimb reflexes were not evident in this subject, indicating their emergence at some point between 12 and 18 months after the injury, at a time long after this subject had developed spasticity (as indicated by the emergence of tendon reflexes; Fig. 10A).

The 'interlimb reflex' activity depicted in Fig. 14 was familiar to us, and is almost always observed in persons with chronic, motor-complete injury to the cervical spinal cord (roughly 98% of the people examined who fit this clinical picture). We have now observed similar responses to lower-limb stimulation in a small number of persons with cervical spinal cord injury in which some recovery of lower-limb voluntary contraction is seen (i.e., subjects are mo-

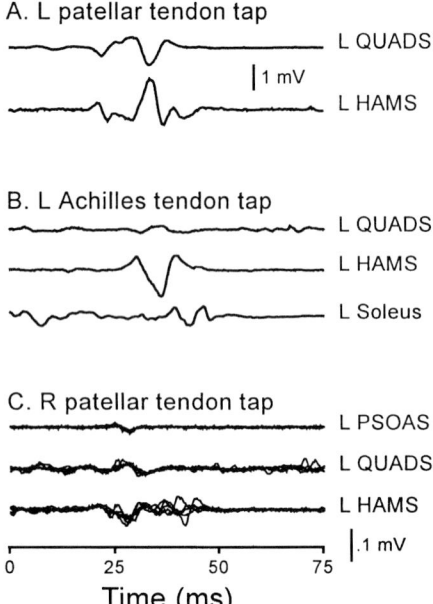

Fig. 12. Individual responses to tendon taps (delivered at time 0) at sites indicated from an 81-year-old male with C5 fracture and ASIA 'D' classification at the initial evaluation. Taps to the ipsilateral patellar tendon cause short-latency responses in the quadriceps and hamstring groups (A). Taps to the Achilles tendon cause short-latency responses in each of the soleus, quadriceps and hamstring groups (B). Taps to the right-side patellar tendon cause short-latency responses in the hip flexors (psoas), quadriceps and hamstring muscle groups of the contralateral (i.e., left) side (C). Note that responses to five successive taps of the contralateral tendon were superimposed, to show reproducibility of the responses from trial to trial. Note also the different vertical calibration for records in (C) (bar = 0.1 mV) compared to those of (A) and (B) (bar = 1 mV).

tor-incomplete). Taken from the ASIA 'C' subject (initially) whose data are depicted in Figs. 5, 6 and 10C, Fig. 15 illustrates short-latency (~75 ms) recruitment of a single motor unit in this subject's left-side ADM muscle in response to stimulation of his right tibial nerve with a 3-pulse train (superimposition of 15 trials). Following this initial response, the unit frequently discharged a brief train of pulses, as seen by the later activity in the L ADM (110–160 ms; Fig. 15B). Comparable stimulation of the left-side tibial nerve (Fig. 15A; superimposition of 11 trials) failed to elicit consistent motor unit discharge; the four occurrences of this motor unit's waveform between approximately 90 and 120 ms following the stimulation indicate that this motor unit was dis-

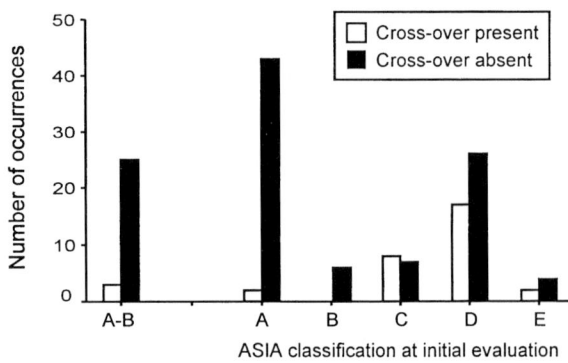

Fig. 13. Summary of subjects from whom a quadriceps contraction in response to taps of the contralateral patellar tendon was observed. Data obtained from able-bodied subjects (A-B) and subjects grouped according to ASIA category as shown. Open bars denote the number of observations in which a cross-over (i.e., contralateral) response to at least one tendon tap was seen in a given subject; closed bars denote those cases in which no cross-over response was evident for any of the taps delivered.

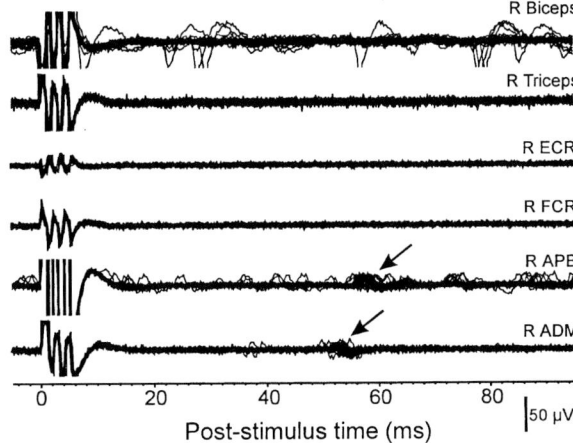

Fig. 14. Responses in six arm muscles to electrical stimulation of the tibial nerve at the contralateral popliteal fossa, from a subject with ASIA 'A' category spinal cord injury at the time (18 months after the injury) when these records were obtained. Multiple trials are superimposed to allow assessment of reproducibility. The arrows above the APB and ADM records point to periods when the probability of discharge of a motor unit in each of these two muscles was particularly high, indicating recruitment of each unit by the stimulus pattern. The right APB motor unit was discharging in a slow, rhythmic manner (i.e., it was 'tonic') throughout this series of stimuli, whereas there was no tonic activity in the ADM muscle during this time. The calibration bar applies to each of the six records shown. A brief train of three stimuli was delivered in each trial, consisting of 1 ms square-wave pulses, separated in time by 2 ms (corresponding to an instantaneous frequency of 500 Hz).

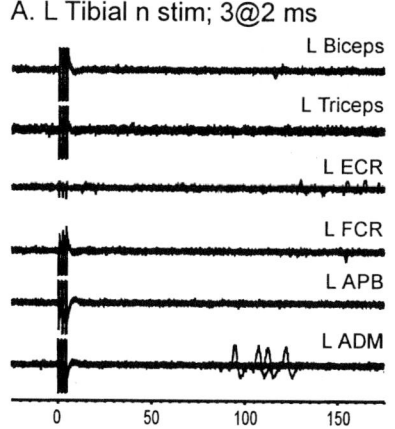

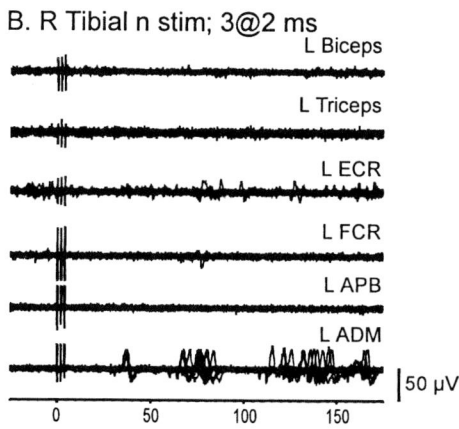

Fig. 15. EMG records in left-side upper-extremity muscles following train (3 pulses at 2 ms intervals) electrical stimulation to the left- (A) and right-side (B) tibial nerves in a subject with cervical SCI who corresponded to an ASIA 'D' category at the time of testing (5.5 months after the injury). Multiple trials are superimposed to allow assessment of reproducibility. Vertical calibration bar (50 μV) applies to all the records. Left-side stimulation did not lead to consistent responses in any of the left-side upper-limb muscles examined. In contrast, right-side stimulation of the tibial nerve caused recruitment of a motor unit in this subject's left ADM muscle at a latency of approximately 75 ms, followed by a later period of discharge beginning at approximately 125 ms after the stimulus. There were two trials when this motor unit discharged at an earlier latency (approximately 40 ms), which may also be associated with the stimulation. Follow-up examinations will look for additional activity at this very early latency.

charging slowly at the time of stimulation (i.e., it was tonic). No other upper-limb muscle examined showed any evidence of time-locked responsiveness to electrical stimulation of left- or right-side tibial

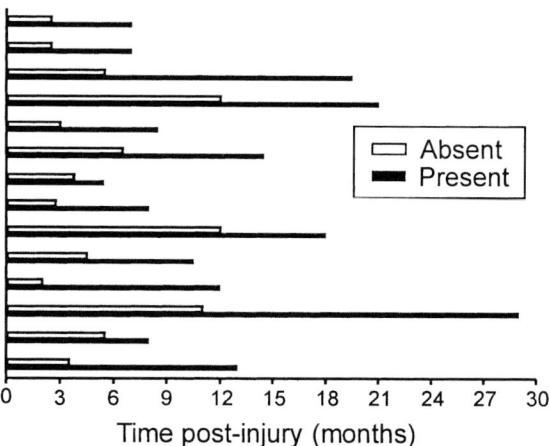

Fig. 16. Summary of the time after the injury at which interlimb reflexes are first seen following lower-limb sensory stimulation (typically electrical stimulation). Closed bars denote the time at which interlimb reflexes were first seen following injury. Open bars denote the time at which such reflexes were looked for and not seen. Thus the expression of interlimb reflexes emerged at some period representing the difference between closed and open bars for each of the 14 subjects in which such activity has been seen.

nerves at the knee in this subject, examined at 5.5 months after the injury; nor was there any sign of this activity at the prior examination (at 3.75 months after the injury).

Finally, we have consistently seen that such interlimb reflexes are not evident immediately after SCI, but take some time to develop. Based on 14 subjects studied since their injury, Fig. 16 shows the time after the injury at which interlimb reflexes were not seen (open bars) and the time at which such responses were first observed (closed bars) in each of these subjects. In no cases were interlimb reflexes observed at any time earlier than 5.5 months after the injury. Typically, such responses were seen 8–12 months after the injury, long after subjects had developed spasticity.

Discussion and conclusions

We are currently studying neurologic recovery in persons with acute SCI who are admitted to the UM/JMH complex, using both clinical (ASIA) and neurophysiologic criteria. There are numerous reports documenting findings from repeated clinical evaluations of persons with acute SCI, whereas repeated examinations using electrophysiology in this population has not been done, to our knowledge. In anticipation of initiating clinical trials for restoring neurologic function after injury, studies described herein and the methods used may serve multiple purposes: (1) the ability to differentiate between spontaneous recovery of function and that due to the intervention(s) used should be improved; (2) this information should aid in selection of the more appropriate subjects for a particular intervention; (3) findings may contribute toward a better understanding of the mechanism(s) underlying spontaneous recovery of function, information which might help direct the most appropriate animal experimentation for achieving further improvement in function; and (4) in the absence of intervention(s), this information may lead to more accurate prediction of long-term outcome, affecting rehabilitation planning and allocation of resources.

One finding from these observations of humans after acute SCI is that in the absence of post-injury complications (such as respiratory arrest or severe gastrointestinal bleeding), there is virtually always some improvement in neurologic function after the injury. This appears to contradict results from many animal studies which describe a protracted series of histopathological alterations extending for days and weeks following traumatic SCI (reviewed in Schwab and Bartholdi, 1996). However, if one postulates that human spinal cord axons may be particularly sensitive to even mild trauma or insult, such events might influence short-term conduction properties without necessitating permanent axonal disruption. Examples of just such axonal sensitivity to modest changes in spinal cord blood pressure can be found in descriptions from intraoperative monitoring studies designed to protect the human CNS from inadvertent injury during elective or emergency surgery (Calancie et al., 1998). That is to say, secondary deterioration of CNS tissue might not be evident in the context of human SCI if the tissue that is degenerating has not been functional since the time of initial trauma. Under these circumstances, subjects may not show signs of clinical deterioration, yet processes of cellular degeneration may be ongoing.

The improvement in voluntary contraction strength seen after spinal cord injury often spans many months (Piepmeier and Jenkins, 1988). On average, and in the absence of medical complications, persons with motor-incomplete SCI at the acute stage can expect to recover more strength from a wider range of muscles innervated from levels below the injury than persons with motor-complete injuries; this is a given. It is also clear that this process continues for a longer period in motor-incomplete persons than for persons with motor-complete injury. Although summary data were not presented due to the limited amount of long-term follow-up data we have accumulated to date, we find such improvement extends for periods beyond 12 months after the injury in most, if not all of the subjects who fit this criterion.

This recovery may be due to a combination of: (a) restoration or improvement of conduction in axons which are present and traverse the injury site; (b) strengthening of synaptic connections between axons which are functional and traverse the injury site and their target neurons (see also Little et al., 1999); (c) sprouting or regeneration of fibers near the target motoneuron populations; or (d) peripheral changes in neuromuscular properties, such as sprouting and reinnervation of denervated muscle fibers by a non-paralyzed motoneuron and/or hypertrophy of non-paralyzed muscle fibers (Thomas et al., 1997). Given this protracted time-course, regeneration of descending motor actions mediating contraction attempts cannot be ruled out. Finally, many of these same processes may contribute to the emergence of involuntary movements, some of which can, through appropriate application, aid in the accomplishment of certain functional tasks, such as transfers (Somers, 1992).

We recently reported on the pattern of motor preservation of lower-limb muscles in a large sample of persons with chronic SCI who were motor-incomplete (Calancie et al., 1999). In that report, both voluntary contractions and responses to transcranial magnetic stimulation were used to assess voluntary movements. We found that of all lower-limb muscles examined, the abductor hallucis (AbH) had a higher probability of responding to voluntary contraction efforts or transcranial magnetic stimulation compared to the other lower-limb muscles examined. What we were unable to address in that study, though, was whether or not AbH was the first muscle to show recovery in those persons whose initial EMG study indicated an absence of voluntary recruitment in any of the lower-limb muscles examined. Although our sample size is relatively small ($n = 16$), our initial findings are that AbH is indeed the first to be recruited in the majority of trials which meet these conditions. In those few cases when AbH was not the first to show recovery, it was being recruited by the next evaluation. These findings, and those of others (Stauffer, 1975), do not agree with other reports that lower-extremity motor recovery following acute SCI proceeds in a proximal-to-distal direction (Graziani et al., 1996; Little et al., 1999). While we are not stating that the reverse is true, our findings show a marked tendency toward AbH being the first to show volitional contraction following SCI in which there is a prolonged period of complete paralysis in a subject's lower extremity.

The combination of the absolute reflex amplitudes and the spread of reflex excitability was very effective for separating the SCI subject population into motor-complete and motor-incomplete at the time of initial evaluation. Subjects with injury more than 7 days prior to the time of initial evaluation were excluded out of recognition that spinal cord reflex excitability, even in persons with ASIA 'A' type injury, can increase rapidly in the 2–3-week period following trauma. We further restricted our comparisons to include only those persons with injury at or above T10. Injuries below this level can, and usually do, involve nerve roots which make up the cauda equina. Spine fracture(s) in this region may leave the spinal cord relatively intact, yet may result in significant root pathology, rendering interpretation of reflexes problematic. While the presence of a large, well-defined reflex response to tap of a tendon in the leg signifies that the afferent (sensory) and efferent (motor) pathways mediating that response are intact, the absence of a response does not differentiate between spinal cord hyporeflexia and nerve root involvement (afferent, efferent, or both).

Enhanced tendon reflexes have long been associated with spinal cord pathology (Guttmann, 1976; Byrne and Waxman, 1990). Reports of large-amplitude tendon reflexes following acute SCI are inconsistent with the classic notion of 'spinal shock', a profound absence of reflexes below the injury (Guttmann, 1976). Based on findings presented herein, we would argue that the term 'spinal shock' is inappropriate for most persons with motor-incomplete SCI, and that use of this term should be restricted to apply only to those subjects with motor-complete SCI at the most acute stage following injury. In our sample population, this accounts for only 36% of the population studied. An earlier study from this laboratory showed that presynaptic inhibition of segmental reflexes was markedly enhanced in the most acute period following SCI, for both persons with motor-complete and for motor-incomplete subjects (Calancie et al., 1993). This argues that the hyporeflexia seen in this study for motor-complete subjects does not simply reflect the withdrawal of a tonic, excitatory influence as has often been postulated.

There is very little information related to short-latency responses from muscles of the lower limb to contralateral sensory perturbations. Recent reports describe such interactions as primarily inhibitory in nature in humans (Harrison et al., 1986) and arising from proprioceptive afferents (Evans and Harrison, 1999) or of mixed excitatory and inhibitory nature in animal studies (Harrison et al., 1986). Emergence of such responses in motor-incomplete SCI subjects as seen in the present study, and the limited observation of such activity in persons without injury (i.e., able-bodied), is an area of research that needs greater attention.

In our experience, all persons with SCI at the cervical level who remain motor-complete develop short-latency contractions in one or more muscles of the distal upper limb in response to a wide variety of sensory stimulation (cutaneous or proprioceptive) applied to the lower limb or limbs, provided the distal upper-limb muscles are not fully denervated as a result of the injury. Often the responses are seen in the contralateral limb, although ipsilateral recruitment is not unusual (Calancie, 1991; Calancie et al., 1996). We previously reported that these interlimb reflexes were not seen in persons with cervical cord injury who recovered volitional contraction in at least one muscle of either leg (that is, motor-incomplete subjects). Based on a much larger sample population, coupled with the routine use of a stimulus pattern optimal for eliciting such reflexes (three rapid stimuli delivered to the tibial nerve), we are now in a position to revise our earlier reported incidence to include persons with motor-incomplete SCI.

We have observed interlimb reflexes similar to those depicted in Figs. 14 and 15 in four subjects with motor-incomplete SCI at the cervical level. In all cases, however, the upper-limb muscle(s) from which EMG could be evoked by lower-limb stimulation had failed to recover volitional contraction beyond an EMG score of '2'. Even when some recovery was seen, it was relatively late in time after the injury. For the subject depicted in Fig. 15, whose L ADM muscle was demonstrating this interlimb reflex activity, reference to Fig. 6 shows that this muscle (L ADM) was very late to recover any sign of voluntary contraction, and was graded at '1' at the most recent examination. These findings suggest that if interlimb reflexes are indeed due to new growth of ascending afferent fibers originating from the lower limb (or their second-order target cells), as we have proposed previously, there need not be a target population completely devoid of direct, supraspinal motor innervation in order for such sprouting and reinnervation to occur.

Fig. 16 shows that there was a protracted period of time prior to emergence of interlimb reflexes in persons with SCI. The most common interval was from 8 to 12 months. We have not calculated an average time of emergence, since the intervals from follow-up to follow-up were not consistent within or between subjects.

In order for such novel and functional connections to be established, it is first necessary for the original synaptic inputs (in this case thought to be corticospinal tract axons) to degenerate. Work from Richard Bunge and others (Miklossy et al., 1991; Quencer et al., 1992; Bunge et al., 1993; Quencer and Bunge, 1996) showed that the rate of axon/myelin breakdown in persons with SCI or other CNS trauma was prolonged by a considerable margin compared to the rate often reported in animal studies. Post-injury periods of 7 months and longer, in which cellular debris was still evident in the human cord below the lesion epicenter, were not uncommon. Additional studies along these lines showed that there was preservation of normal morphology within the population of sympathetic preganglionic neurons in spinal segments below a T3 transection of 23 years duration (Krassioukov et al., 1999). This is also suggestive that axons remain near the injury subsequent to establishment of novel synaptic connections.

Finally, we would expect that the same sensory inputs adequate to evoke interlimb reflexes as described above and elsewhere (Calancie, 1991; Calancie et al., 1996) would also be capable of exciting sympathetic preganglionic neurons of the thoracic spinal cord, as this population of neurons also undergoes substantial denervation from supraspinal inputs following severe cervical spinal cord injury. Indeed, we have recent evidence (not shown) of such effects based on transient increases in blood pressure (as measured with an arterial line) following brief stimulus trains to the tibial nerve at the knee in persons with proven interlimb reflexes. Thus we propose that autonomic dysreflexia, a vexing complication of complete, cervical spinal cord injury that can be life threatening, is caused by the same spontaneous sprouting of lower-limb afferents onto these autonomic neurons as that which leads to short-latency recruitment of motoneurons to the distal muscles of the upper limb following adequate lower-limb stimulation. A similar conclusion based on more direct evidence from an animal model was recently reported by Weaver, Krassioukov and colleagues (Weaver et al., 1997; Krenz et al., 1999). Nevertheless, and despite the multiple arguments in favor of our explanation for interlimb reflexes, direct evidence for this proposed regenerative sprouting following cervical SCI in humans is lacking.

The findings presented above reflect our experiences in documenting recovery following SCI in a typical large urban hospital in this country. Data collection has been going on for two years, and will continue for another three. This chapter was written to describe an overview of the study, provide some basis for why the study is being done, and present preliminary data related to acute presentation and emergence of both voluntary and involuntary movements. Data were restricted to voluntary contractions seen at the most acute stage, and emergence of involuntary movement alterations. Findings of central motor conduction changes (as assessed with transcranial magnetic stimulation) were beyond the scope of this report. Moreover, we have not attempted to express time courses of motor recovery for different populations of subjects, because this too is beyond the scope of the present report. For the data presented, some sections have relatively low sample sizes, which must be taken into consideration, as this affects the strength of conclusions we can draw. As we continue to accrue subjects, we hope to have addressed these limited sample sizes by the end of the study.

Acknowledgements

This work was supported by grants from NIH (RO1 HD31240 and RO1 NS 36542) and by The Miami Project to Cure Paralysis.

References

Alexeeva, N., Broton, J.G., Suys, S. and Calancie, B. (1997) Central cord syndrome of cervical spinal cord injury: widespread changes in muscle recruitment studied by voluntary contractions and transcranial magnetic stimulation. *Exp. Neurol.*, 148: 399–406.

Bracken, M.B., Shepard, M.J., Collins, W.F., Holford, T.R., Young, W., Baskin, D.S., Eisenberg, H.M., Flamm, E., Leo-Summers, L., Maroon, J., Marshall, L.F., Perot, P.L., Piepmeier, J., Sonntag, V.K.H., Wagner, F.C., Wilberger, J.E. and Winn, H.R. (1990) A randomized, controlled trial of methylprednisolone or naloxone in the treatment of acute spinal-cord injury. *N. Engl. J. Med.*, 322: 1405–1411.

Bunge, R.P., Puckett, W.R., Becerra, J.L., Marcillo, A. and Quencer, R.M. (1993) Observations on the pathology of human spinal cord injury. A review and classification of 22 new cases with details from a case of chronic cord depression with extensive focal demyelination. In: Seil, F.J. (Ed.), *Advances in Neurology, Vol. 59*. Raven Press, New York, pp. 75–89.

Byrne, T.N. and Waxman, S.G. (1990) *Spinal Cord Compression: Diagnosis and Principles of Management*. F.A. Davis, Philadelphia, PA.

Calancie, B. (1991) Interlimb reflexes following cervical spinal cord injury in man. *Exp. Brain Res.*, 85: 458–469.

Calancie, B., Broton, J.G., Klose, K.J., Traad, M., Difini, J. and Ayyar, D.R. (1993) Alterations in presynaptic inhibition contribute to segmental hypo- and hyperexcitability after spinal cord injury in man. *Electroenceph. Clin. Neurophysiol.*, 89: 177–186.

Calancie, B., Lutton, S. and Broton, J.G. (1996) Central nervous system plasticity after spinal cord injury in man: interlimb reflexes and the influence of cutaneous stimulation. *Electroenceph. Clin. Neurophysiol.*, 101: 304–315.

Calancie, B., Harris, W., Broton, J.G., Alexeeva, N. and Green, B.A. (1998) 'Threshold-level' multipulse transcranial electrical stimulation of motor cortex for intraoperative monitoring of spinal motor tracts: description of method and comparison to somatosensory evoked potential monitoring. *J. Neurosurg.*, 88: 457–470.

Calancie, B., Alexeeva, N., Broton, J.G., Suys, S., Hall, A. and Klose, K.J. (1999) Distribution and latency of muscle responses to transcranial magnetic stimulation of motor cortex after spinal cord injury in humans. *J. Neurotrauma*, 16: 49–67.

Daverat, P., Sibrax, M.C., Dartigues, J.F., Mazaux, J.M., Marit, E., Debelleix, X. and Barat, M. (1988) Early prognostic factors for walking in spinal cord injuries. *Paraplegia*, 26: 255–261.

Davis, L.A., Warren, S.A., Reid, D.C., Oberle, K., Saboe, L.A. and Grace, M.G.A. (1993) Incomplete neural deficits in thoracolumbar and lumbar spine fractures. Reliability of Frankel and Sunnybrook scales. *Spine*, 18: 257–263.

Diamantopoulos, E. and Van der Olsen, P.Z. (1967) Excitability of motor neurones in spinal shock in man. *J. Neurol. Neurosurg. Psychiatry*, 30: 427–431.

Ditunno, J.F., Young, W., Donovan, W.H. and Creasey, G. (1994) The International Standards Booklet for Neurologic and Functional Classification of Spinal Cord Injury. *Paraplegia*, 32: 70–80.

Donovan, W.H., Brown, D.J., Ditunno, J.F., Dollfus, P. and Frankel, H.L. (1997) Neurological issues. *Spinal Cord*, 35: 275–281.

Eidelberg, E. (1981) Consequences of spinal cord lesions upon motor function with special reference to locomotor activity. *Prog. Neurobiol.*, 17: 185–202.

Evans, P. and Harrison, P.J. (1999) Crossed group I reflexes in lower limb pathways in humans. *Soc. Neurosci. Abstr.*, 25: 121.

Falci, S., Holtz, A., Akesson, E., Azizi, M., Ertzgaard, P., Hultling, C., Kjaeldgaard, A., Levi, R., Ringden, O., Westgren, M., Lammertse, D. and Seiger, A. (1997) Obliteration of a posttraumatic spinal cord cyst with solid human embryonic spinal cord grafts: first clinical attempts. *J. Neurotrauma*, 14: 875884.

Geisler, F.H., Dorsey, F.C. and Coleman, W.P. (1991) Recovery of motor function after spinal-cord injury — a randomized, placebo-controlled trial with GM-1 ganglioside. *N. Engl. J. Med.*, 324: 1829–1838.

George, E.R., Scholten, D.J., Buechler, M., Jordan-Tibbs, J., Mattice, C. and Albrecht, R.M. (1995) Failure of methylprednisolone to improve the outcome of spinal cord injuries. *Am. Surg.*, 61: 659–664.

Graziani, V., Crozier, K.C. and Selby-Silverstein, L. (1996) Lower extremity function following spinal cord injury. *Top. Spinal Cord Inj. Rehab.*, 1: 46–55.

Guttmann, L. (1976) *Spinal Cord Injuries: Comprehensive Management and Research*, 2nd ed. Blackwell, London.

Harrison, P.J., Jankowska, E. and Zytnicki, D. (1986) Lamina VIII interneurones interposed in crossed reflex pathways in the cat. *J. Physiol. (Lond.)*, 371: 147–166.

Herman, R. (1969) Relationship between the H reflex and the tendon jerk response. *Electromyography*, 9: 359–370.

Kendall, F.P. and McCreary, E.K. (1983) *Muscles. Testing and Function*, 3rd ed. Williams and Wilkins, Baltimore, MD.

Krassioukov, A.V., Bunge, R.P., Puckett, W.R. and Bygrave, M.A. (1999) The changes in human spinal sympathetic preganglionic neurons after spinal cord injury. *Spinal Cord*, 37: 6–13.

Krenz, N.R., Meakin, S.O., Krassioukov, A.V. and Weaver, L.C.

(1999) Neutralizing intraspinal nerve growth factor blocks autonomic dysreflexia caused by spinal cord injury. *J. Neurosci.*, 19: 7405–7414.

Kuhn, R.A. (1950) Functional capacity of the isolated human spinal cord. *Brain*, 73: 1–51.

Lance, J.W., De Gail, P. and Neilson, P.D. (1966) Tonic and phasic spinal cord mechanisms in man. *J. Neurol. Neurosurg. Psychiatry*, 29: 535–544.

Little, J.W., Ditunno, J.F., Stiens, S.A. and Harris, R.M. (1999) Incomplete spinal cord injury: neuronal mechanisms of motor recovery and hyperreflexia. *Arch. Phys. Med. Rehabil.*, 80: 587–599.

Miklossy, J., Clarke, S. and Van der Loos, H. (1991) The long distance effects of brain lesions: visualization of axonal pathways and their terminations in the human brain by the Nauta method. *J. Neuropathol. Exp. Neurol.*, 50: 595–614.

Mizukami, M., Kawai, N., Iwasaki, Y., Yamamoto, Y., Yoshida, Y., Koyama, N., Sekiguchi, S., Kimura, T. and Nihei, R. (1995) Relationship between functional levels and movement in tetraplagic patients. A retrospective study. *Paraplegia*, 33: 189–194.

Nixon, V. (1985) *Spinal Cord Injury: A Guide to Functional Outcomes in Physical Therapy Management. Rehabilitation Institute of Chicago Procedure Manual.* Aspen Publishers, Rockville, MD.

Piepmeier, J.M. and Jenkins, N.R. (1988) Late neurological changes following traumatic spinal cord injury. *J. Neurosurg.*, 69: 399–402.

Quencer, R.M. and Bunge, R.P. (1996) The injured spinal cord: imaging, histopathologic, clinical correlates, and basic science approaches to enhancing neural function after spinal cord injury. *Spine*, 21: 2064–2066.

Quencer, R.M., Bunge, R.P., Egnor, M., Green, B.A., Puckett, W., Naidich, T.P., Post, M.J.D. and Norenberg, M. (1992) Acute traumatic central cord syndrome: MRI–pathological correlations. *Neuroradiology*, 34: 85–94.

Schwab, M.E. and Bartholdi, D. (1996) Degeneration and regeneration of axons in the lesioned spinal cord. *Physiol. Rev.*, 76: 319–370.

Somers, M.F. (1992) *Spinal Cord Injury: Functional Rehabilitation.* Appleton and Lange, Norwalk, CT.

Stauffer, E.S. (1975) Diagnosis and prognosis of acute cervical spinal cord injury. *Clin. Orthop.*, 112: 9–15.

Thomas, C.K., Broton, J.G. and Calancie, B. (1997) Motor unit forces and recruitment patterns after cervical spinal cord injury. *Muscle Nerve*, 20: 212–220.

Waters, R.L., Adkins, R.H., Yakura, J.S. and Sie, I. (1993) Motor and sensory recovery following complete tetraplegia. *Arch. Phys. Med. Rehab.*, 74: 242–247.

Weaver, L.C., Cassarn, A.K., Krassioukov, A.V. and Llewellyn Smith, I.J. (1997) Changes in immunoreactivity for growth associated protein-43 suggest reorganization of synapses on spinal sympathetic neurons after cord transection. *Neuroscience*, 81: 535–551.

Wells, J.D. and Nicosia, S. (1995) Scoring acute spinal cord injury: a study of the utility and limitations of five different grading systems. *J. Spinal Cord Med.*, 18: 33–41.

CHAPTER 8

Laufband (LB) therapy in spinal cord lesioned persons

Anton Wernig [1,*], Andreas Nanassy [2] and Sabina Müller [2]

[1] *Department of Physiology, University of Bonn, Wilhelmstrasse 31, D-53111 Bonn, Germany*
[2] *Klinikum Karlsbad-Langensteinbach, Karlsbad-Langensteinbach, Germany*

Introduction

Nonhuman primates relearn treadmill-stepping after most severe, though incomplete, spinal cord transection (Eidelberg et al., 1981). In lower vertebrates like cat, such motor learning occurs even after complete transection of the low thoracic spinal cord (Lovely et al., 1986; Barbeau and Rossignol, 1987; Edgerton et al., 1991), indicating the existence of spinal locomotor programs capable, together with afferent information, of generating step-like movements in hindlimbs. In man after complete spinal transection, only elements of stepping short of full-step cycles, may be evocable on the treadmill (Wernig and Müller, 1996) while even with little voluntary activity remaining, full stepping may be entrained (Wernig and Müller, 1991, 1992; Wernig et al., 1995). Improvement of stepping in paralyzed cats and monkeys was achieved by intensive treadmill-stepping, while daily standing was not successful or even negative (Edgerton et al., 1991). Translated into strategies for motor rehabilitation of spinal cord-injured patients, relearning of walking can be done only by training of upright walking.

Two independent groups of researchers have recently adopted aided treadmill locomotion (which in connection with humans will be called Laufband therapy) for spinal cord-injured persons (Visintin and Barbeau, 1989; Barbeau and Blunt, 1991; Wernig and Müller, 1991, 1992). In a recent study, Laufband (LB) therapy was applied to 89 incompletely paralyzed chronic and acute para- and tetraplegics; from the comparison with 64 patients treated conventionally and from consecutively applying both therapies in individual patients, a considerable superiority of LB therapy for restoration or improvement of walking over ground became obvious (Wernig et al., 1995).

In the present review, effects of LB therapy in 87 chronic and acute patients are described (see Wernig et al., 1995, 1998). Of these patients, results of follow-up evaluations are reported, which were performed 0.5 year up to 6.5 years after discharge from the hospital where these patients had undergone LB therapy, either in the course of renewed rehabilitation therapy (initially chronic patients) or in the course of post-acute rehabilitation (initially acute patients). Follow-up investigations were important, since improvements in locomotion achieved by LB therapy had occurred under controlled clinical conditions, usually within a few weeks only, and in patients who often had little voluntary muscle activity remaining. It was particularly important to know, therefore, whether initially wheelchair-bound patients who had gained some independent walking could maintain their new capabilities in domestic surroundings.

---

* Corresponding author: Dr. Anton Wernig, Department of Physiology, University of Bonn, Wilhelmstrasse 31, D-53111 Bonn, Germany. Fax: +49-228-287-2208; E-mail: wernig@physio.uni-bonn.de

Techniques, assessments, patients

Techniques and equipment for LB therapy

The principles of LB therapy have been published previously (Wernig and Müller, 1991, 1992; Wernig et al., 1995). In brief, exercise of upright walking is performed on a motor-driven treadmill (No. 5 in Fig. 1) with a range of speed at the onset of therapy between 0.2 and some 1.2 km/h, depending on the patient's condition. Patients are supported by a harness suspended from the ceiling and initially by one or two therapists helping with limb movements and placement, if necessary. During walking, 'rules of spinal locomotion' are enforced by therapists regardless of the amount of voluntary muscle activity the patient is capable of contributing; these rules, derived from animal experiments (for review see: Grillner, 1981; Lovely et al., 1986; Barbeau and Rossignol, 1987; Edgerton et al., 1991; Gossard and Hultborn, 1991), imply extension of hip and knee joints with optimal limb loading during stance (No. 4 in Fig. 1), unloading, and shift of body weight onto the contralateral limb. Additional afferent stimulation, like skin irritation by pinching or pressing, may be effective in inducing or facilitating flexion movements; often special shoes with thin leather soles (Bonmed) are used for this purpose. Patients are allowed to hold onto the lateral frame (No. 3 in Fig. 1) using their arms for balance, but are strongly discouraged to use them for body weight support. Body weight support (BWS) is set when the patient is in an upright position with knees fully extended. BWS supplied via the harness is adjusted to the needs of the patient. Chronic patients already ambulating or accustomed to standing upright often perform better with little or even without body weight support (even without BWS the harness is still important to secure the patient). In any case, BWS up to about 40%, used especially in the initial phase of therapy in acute and chronic patients, is gradually reduced in the course of weeks according to the capabilities of the patient. The harness (No. 2 in Fig. 1) is critical in that it must not cause pain for the patient while BWS is supplied and at the same time must allow maximal freedom for limb movements. The harness that we have designed for this purpose is described in detail in www.meb.uni-bonn.de/wernig. The suspension system (No. 1 in Fig. 1) may consist of a pair of simple manually operated pulleys and a spring balance. The latter provides sufficient yield, due to some vertical displacement under load together with some movement of the body within the harness, to allow a small amount of body displacement in the vertical axis as occurs during physiological gait. In other words, this suspension is a limited but sufficiently dynamic system necessary for vertical body swing during walking. Other dynamic suspension systems employ pneumatic devices of variable complexity; these often allow vertical displacements that are too large and are thus of limited use, even with additional security chains, for training of severely paralyzed patients whose legs cannot carry body weight. From a practical and economical point of view, it is advisable for larger institutions to have two simple set-ups running in parallel rather than one complex one; this way a single therapist can often operate and supervise two advanced or less paralyzed patients at the same time.

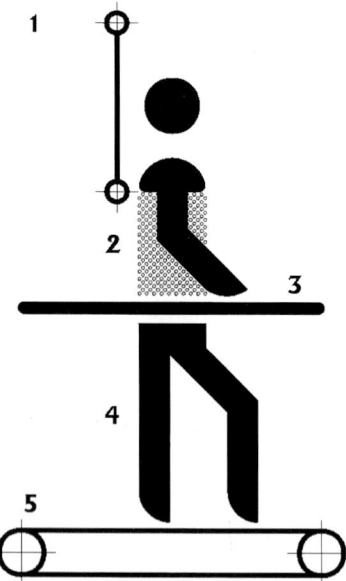

Fig. 1. Schematic representation of equipment and principles of Laufband (LB) therapy. *1* = suspension system; *2* = harness; *3* = frame allowing use of arms for balance but not body weight support; *4* = knee is extended to allow full loading during stance phase; *5* = moving belt of treadmill. Speed in the beginning of therapy is between 0.2 and 1.2 km/h, depending on the patient's capability.

Protocol of LB therapy

LB therapy was performed usually once, less often twice daily for periods of 30 min during 5 days per week. At least once a week, walking over ground is attempted, even if massive help by two therapists for balance and weight support is necessary. As soon as a few steps can be made with moderate help only, walking over ground is regularly attempted immediately before or after a training session on the treadmill. With further improvement, walking over ground increasingly replaces walking on the treadmill which, however, is often maintained to train for endurance and speed. Stair case climbing may be attempted surprisingly early after gain of stepping capability over ground and can be achieved even in severely paralyzed patients (Wernig et al., 1995). When walking over ground, the same 'rules of spinal locomotion' are applied; for practical reasons, therefore, walking is initially only allowed during the LB therapy session and under the guidance of therapists specially trained in LB therapy. It is an important goal to teach patients to generally maintain these rules during all walking activities. Apart from LB therapy, all patients participated in the regular conventional rehabilitation program for indoor patients, which includes training of functions for every-day living, sports and other activities aimed to enhance muscle strength and mobility. In general, patients who took part in the program for LB therapy obtained the same total amount of individual and other therapy as all other indoor patients.

Patients

Our current criteria for selecting chronic spastic paretic patients to enter LB therapy are presence of some voluntary muscle activity in the lower limbs, particularly the quadriceps femoris, mobility of joints, no severe muscle shortenings, and no skin ulcerations or other severe diseases. Missing voluntary hip flexion can be tolerated initially, especially when it can be elicited by facilitating measures in the initial testing on the treadmill (see above). With all patients we thoroughly discuss the possible therapeutic goals in order not to raise expectations that are too high. Thus for tetraplegic patients with low amounts of voluntary activity in their legs and with arm and/or rump paralyses hindering the use of crutches or rollators, gain of independent walking is an unlikely outcome. In severely paralyzed paraplegics, even with the use of arms, the possible entraining of stepping may allow limited walking over short distances only. However, even aided walking with the help of another person, including or not including stair case climbing, or independent walking for even a few steps only, would be of advantage in daily life, and are thus acceptable therapeutic goals. The leading principle may thus be to enable each patient to reach his/her highest level of individual walking capability by intensive and aided training of upright walking.

Criteria for selecting acute patients are basically similar to those described for chronic patients, taking into account spontaneous recovery continuing for several weeks after spinal cord damage. LB therapy was started as soon as some voluntary movements in lower limbs appeared, rather than waiting for spontaneous recovery of motor functions to plateau. In acute patients who have suffered trauma of the spinal column, the safety of the procedure has to be assured by the orthopedic surgeon. With surgical stabilization of the vertebral column (Harms, 1992), the start of walking exercise and other physical therapy was usually allowed within a few weeks after trauma (for details see Wernig et al., 1995). Also with acutely spinal cord-lesioned patients, LB therapy was usually performed for 5 days a week from the very beginning, which was well tolerated. The cause of spinal cord injury most frequently was trauma followed by non-progressive myelitis, tumors, vascular disorders and other causes. Patients with components of flaccid paralysis are not included in this report, but have been reported to profit from LB therapy under certain conditions (Wernig et al., 1995, 1999).

Assessments

Walking capabilities before and after LB therapy, as well as in the follow-up investigations, were classified into six easily assessable classes (0–5) solely based on the patient's ability to walk over ground (Table 1). Accordingly, wheelchair-bound patients not capable of standing up and walking without help by others are separated from those not

TABLE 1

Classification of walking capabilities

Dependent
0 = lower limbs cannot support body weight for walking even with moderate help by two therapists
1 = capable of walking only with the help of two therapists
2 = walking at the railing with help of one therapist

Not dependent
3 = walking with the rollator/walker or reciprocal frame
4 = walking with two regular canes or four-point canes
5 = walking without devices (free walking) for more than five steps

wheelchair-bound and capable of standing up from the wheelchair by themselves. Assessments were done by two independent evaluators from the video film records; in the case of discrepancies, the higher class was assumed before start of therapy and the lower class later on. The amount of voluntary muscle activity was evaluated from the force and range of single-joint movements evocable upon verbal command in defined resting positions (horizontal and sitting), avoiding readily evocable spastic extension or flexion patterns (Kendall et al., 1971). Under these rating conditions, values were defined as follows: 0 = no muscle contraction visible or palpable; 1 = muscle contraction visible and/or palpable, no movement of limbs; 2 = some joint angle movement with passive support by the therapist balancing gravity; 3 = full range of joint angle movement against gravity; 4 = full movement plus maintenance of position against moderate applied resistance; 5 = like 4, against maximal applied resistance. Values in between were allowed and valued as half points.

As an overall measure for muscle function in a limb, values of the following eight major flexor and extensor muscles/muscle groups assessed as single-joint movements were summed: gluteus maximus, gluteus medius and minimus, iliopsoas, sartorius, quadriceps, hamstrings, tibialis anterior, and triceps surae (values shown in Table 2). Follow-up investigations were performed the same day patients arrived at the clinic for routine ambulatory check-ups or within the first days in the clinic in case of renewed indoor treatment. In addition to assessment of voluntary activity in defined resting positions, patients were asked to walk over ground at the optimal speed with their preferred devices used in their domestic surrounding. Video recordings were made from all performances.

Results

Gain and maintenance of walking capability in initially chronic patients

Thirty-five para- and tetraplegic patients with spastic paresis and no obvious signs of flaccid paralysis are reported who performed LB therapy usually in the course of a repeated period of rehabilitation (5.5 months to 15 years, median 1.5 years, after spinal cord damage). Of 25 wheelchair-bound patients (classes 0–2, upper histogram in Fig. 2), 20 became independent after LB therapy (classes 3–5, middle histogram). Patients already independent before LB therapy (classes 3–5 in the upper histogram in Fig. 2) usually remained within their original functional class but improved in speed and endurance of walking (data not shown). Median duration of LB therapy was 12 weeks (range 8–20 weeks) for wheelchair-bound patients (depicted in Table 2) and 10 weeks (range 4–12 weeks) for the others. Antispastic medication applied in 27 of the 35 patients was maintained in eight, but could be reduced in the course of LB therapy in nineteen.

The follow-up investigations performed 0.5 to 6.5 years later (median 20 months) show that the improvements achieved under clinical conditions were generally maintained (Fig. 2, lower histogram; Table 2). Only one patient fell back from class 3 to 2 while one patient improved from 2 to 3 and two patients improved from 3 and 4 to 5.

Voluntary muscle activity

In the course of LB therapy, marked increases in voluntary muscle activity surprisingly occurred only in a few patients (marked 0/A, 0/D left limb, 0/F, 2/B and 2/D in Table 2). Noteworthy, in four of these five patients, antispastic medication was markedly reduced. Strikingly, on the other hand, limbs often remained almost completely paralyzed, i.e., had cumulative values of 10 and lower for the 8 major flexor and extensor muscles in this limb, with no single muscle exceeding 2.5 (patients coded 0/C,

TABLE 2

Motor capabilities before and following Laufband therapy and at follow-up investigations of 25 chronic, initially non-independent walkers

Code	Walking capabilities[a]			Stair case walking[b]			Cumulative voluntary muscle activity[c]					
							right limb			left limb		
	before LB	after LB	follow up	before LB	after LB	follow up	before LB	after LB	follow up	before LB	after LB	follow up
0/A	0	4	5	no	yes	yes	13	26	29.5	19.5	30	32.5
0/B	0	3	3	no	yes	no	4.5	10.5	–	5	8.5	–
0/C	0	3	3	no	yes	yes	3	4	–	2.5	3.5	–
0/D	0	3	3	no	no	no	1	4	4.5	5	18.5	18.5
0/E	0	3	3	no	no	no	24	29.5	29.5	0	0	0
0/F	0	5	5	no	yes	yes	15	23.5	23.5	18.5	26.5	27.5
0/G	0	2	2	no	no	no	8.5	12.5	15	5	7	7.5
0/H	0	2	2	no	no	no	14	17	–	16	17.5	–
0/I[d]	0	2	3	no	no	no	7	10	20.5	5.5	7.5	18.5
0/J[d]	0	1	1	no	yes	yes	8	13.5	–	10	15	–
1/A	1	4	4	no	yes	yes	4	5.5	9	15	15.5	17.5
1/B	1	4	4	no	yes	yes	8.5	10.5	13	24	25.5	27
1/C	1	3	3	no	yes	yes	13.5	14	16.5	12.5	13	14.5
1/D	1	3	3	no	no	no	5.5	7	–	11.5	14.5	–
1/E	1	3	3	no	no	no	13.5	16.5	19.5	13	17.5	16.5
1/F	1	3	2	no	no	no	7.5	8	–	11.5	15	–
1/G	1	3	3	no	no	no	15	20.5	–	8.5	9	–
1/H[d]	1	2	2	no	no	no	11	11	9.5	4.5	5.5	6.5
2/A	2	5	5	no	yes	yes	19	21.5	–	20	22.5	–
2/B	2	5	5	no	yes	yes	8	17.5	19	5	16	17
2/C	2	5	5	no	yes	yes	24	27.5	29.5	7.5	10	19.5
2/D	2	4	4	no	yes	–	12.5	23.5	–	13.5	24	–
2/E	2	4	4	no	yes	yes	27.5	31	28.5	7	14.5	13
2/F	2	4	4	no	yes	yes	6	10	13	29	30	30
2/G	2	3	3	no	yes	yes	6.5	6	7	8	12.5	15

[a] Walking capability: 6 main classes defined in the text (see Table 1).
[b] Staircase walking with or without supporting persons.
[c] Cumulative force of 8 major flexors and extensors determined in the resting position.
[d] Severe paralysis of arms.

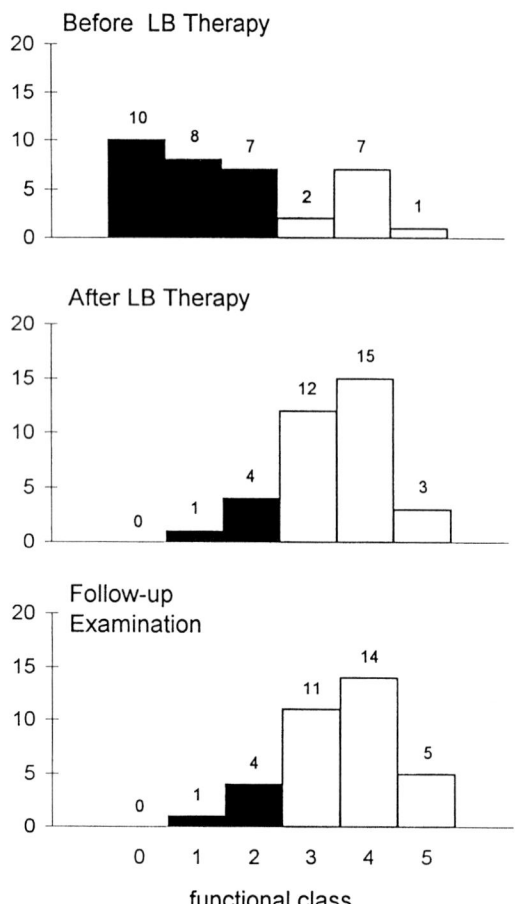

Fig. 2. Walking capabilities in 35 chronic para- and tetraplegic patients before (upper histogram) and after LB therapy performed in the clinic (middle histogram) as well as 0.5 to 6.5 years (median 20 months) after discharge from the hospital (lower histogram, follow-up investigation). Classes 0–2 (black bars): non-independent walkers. Classes 3–5 (white bars): independent walkers (Table 1). Number of patients on top of each column. (From Wernig et al., 1998.)

0/D, 0/E, 0/H, 0/I, 1/A, 1/F, 1/G, 1/H and 2/G in Table 2). This, obviously, is too low to account for the new walking capabilities several of these patients had achieved (Table 2). This phenomenon was most striking in the patient coded 0/C, who learned to walk for distances of 20–40 m with the help of a rollator, though his voluntary activity in the lower limbs did not exceed 3.5 and 4 (Table 2; see also patient Z in Wernig and Müller, 1991, 1992). Obviously, motor programs causing complex limb movements which contribute to the gait cycle may be activated or become facilitated during stepping when applying 'rules of spinal locomotion' (see Discussion).

LB therapy in acute patients

Forty-one para- and tetraplegic patients without prominent peripheral nerve lesions started LB therapy 3–16 weeks (median 8) after spinal cord damage. They performed LB therapy as part of their post-acute rehabilitation for periods of 3 to 22 weeks (median 10). Fig. 3 (upper histogram) shows their walking capability at the onset, and the middle histogram at the end of LB therapy. On the basis of comparison with a matched group of conventionally treated patients, or based on the considerably greater improvement achieved by LB therapy closely following conventional therapy, it appeared that LB therapy was superior to conventional therapy in acute and post-acute patients (Wernig et al., 1995). At the follow-up investigation 0.5 to 6 years after discharge from the clinic (median 17 months), only six of eight patients had remained wheelchair-bound (classes 0–2), fewer patients needed a walker (class 3), and clearly more patients could walk for short distances without any devices (class 5). It is obvious that these patients, more than the initially chronic patients, not only had maintained but further improved their walking capability in domestic surroundings.

Discussion

This report summarizes findings showing that both acute and chronic spinal cord-lesioned incompletely paralyzed patients may improve their walking capability considerably by intensive training of upright walking following 'rules of spinal locomotion'. Previous comparisons with conventionally treated control groups had shown a superiority of LB therapy (Wernig et al., 1995). It is suggested, therefore, that LB therapy be performed as soon as possible following spinal cord lesion. The fact that after one or several periods of conventional physiotherapy chronic patients also significantly improved further by LB therapy shows that therapeutic goals can now be redefined. As a principle, every patient, acute or chronic, may be brought to his/her individual limit to establish some amount of walking by in-

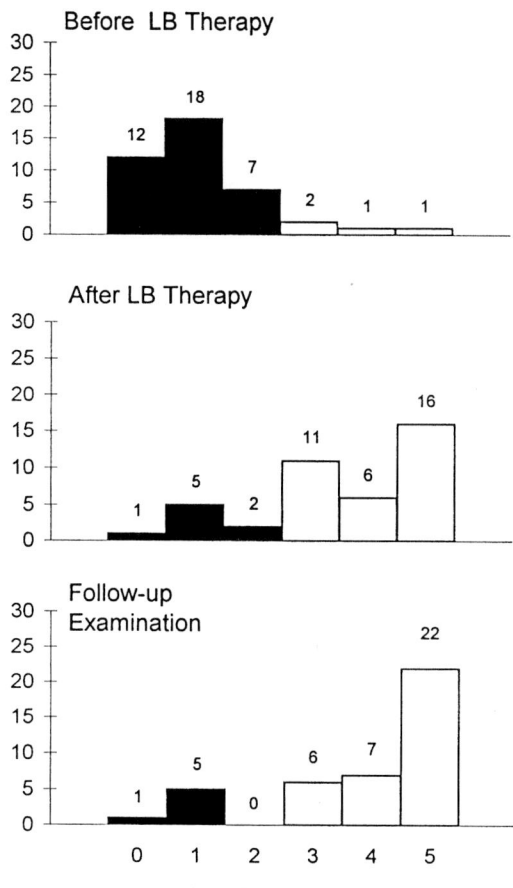

Fig. 3. Walking capabilities in 41 acute para- and tetraplegic patients. Upper histogram: at the beginning of LB therapy during post-acute rehabilitation 3 to 16 weeks (median = 8) after spinal cord damage; middle histogram: the same patients at discharge from the clinic; lower histogram: follow-up investigations 0.5 to 6 years (median 17 months) after discharge from the clinic. Number of patients on top of each column. (From Wernig et al., 1998.)

tensive locomotor therapy; once established, walking is the best further training to maintain and further improve walking. This implies that any spinal cord-injured person who has remaining motor functions in the lower limbs, particularly active knee extension, needs to be tested and may benefit from properly performed LB therapy.

Two features obviously connected to the improvements in walking have been identified: one is an increase in the voluntary activity of individual muscles; the other is the involvement of reflex-like components contributing to limb movements in the gait cycle (motor programs). Increase in voluntary muscle activity, as tested in defined resting positions rather than during walking, may be brought about by an increase in muscle mass or improved recruitment of motor units. Some increase in voluntary muscle activity occurred in all patients, but large increases were rare. On the other hand, a striking enhancement was found in two patients (marked 0/I and 2/C in Table 2) long after the LB therapy period in domestic surroundings (where patients had continued walking); this clearly speaks for enhanced motor unit recruitment due to activity-related synaptic learning (in descending supraspinal connections). Similarly, related to activity-dependent motor learning, progress in locomotion in all patients almost certainly involves adjustment in motor strategies to optimize remaining motor functions in limb and rumps for locomotion. In several patients an increase in voluntary muscle activity was moderate. It appears, therefore, that 'reflex' components or motor programs and facilitating maneuvers are involved in locomotion of these patients. Interestingly, in this respect, attempted single-joint movements often lead to complex multi-joint movements instead, indicating that motor patterns like multi-joint limb flexion can become recruited rather than single muscles (A. Wernig, S. Müller, unpubl. observations; see also below). In fact, it was previously observed (Dimitrijevic et al., 1984) that some patients may evoke complete flexion patterns by employing Jendrassik maneuvers even in the absence of functional voluntary muscle activity (see also patient Z in Wernig and Müller, 1991, 1992). 'Rules of spinal locomotion' define a sequence of events in the gait cycle which has been shown in spinal animals to maintain locomotor movements (Grillner, 1981; Lovely et al., 1986; Barbeau and Rossignol, 1987; Gossard and Hultborn, 1991). It appears that proprioceptive inputs are driving spinal motor programs and modulate spinal rhythm generators. In the presence of remaining voluntary muscle activity — as was the case for all but one of the patients reported here — the correct proprioceptive inputs, i.e., correct limb positions, might in addition gate and facilitate remaining descending connections and thus account for the discrepancy in muscle activity evocable in the resting position vs. locomotion. In the nearly complete absence of supraspinal connections, the 'rules of spinal locomotion' might become even more prominent in governing locomotor movements. This phenomenon was most striking in almost completely paralyzed patients. The patient coded 0/C (Table 2) learned to walk over ground for distances of 20–40 m with the help of a rollator, although his voluntary activity in the lower limbs did not exceed 4 and 3.5. Similar stepping was observed previously in a patient with zero voluntary activity in one limb (also assessed electromyographically), and a total value of 2 in the other (Wernig and Müller, 1991, 1992). Obviously, training stabilizes the influence of proprioceptive information on the gait cycle. Applying these rules, elements of stepping with hip flexion and knee extension short of complete step cycles could be evoked in a 13-year-old girl completely paralyzed below spinal level T6 (Wernig and Müller, 1996). Similarly, elements of stepping after complete spinal cord transection have repeatedly been reported, including alternating phasic extensor and flexor activities observed from electromyographic recordings during aided stepping on the treadmill (Wernig and Müller, 1991, 1992; Dietz et al., 1995; Dobkin et al., 1995; Wernig et al., 1995; Harkema et al., 1997). The question then arises why such persons are not capable of performing complete stepping sequences like a cat does. Do we have to increase the amount of training? The 13-year-old patient was trained on the treadmill 2 times a day, 5 days per week for 4 months without ever becoming able to walk on the treadmill completely independently. It appears that more supraspinal inputs are needed in primates than in lower vertebrates to bring spinal motor programs above threshold. Thus any means of enhancing such inputs, be it enhanced axonal sprouting, with or without implantation of new cells, or transmitting such signals via electronic devices, could promote further progress (keeping in mind, however, that activity-related learning will have to make such signals functionally meaningful).

In the light of these findings, strategies of rehabilitation for incomplete para- and tetraplegic persons need to include intensive training of upright walking and emphasizing the 'rules of spinal locomotion' even during aided walking on the treadmill. This, of course, can also, though involving considerably more effort, be achieved by other means than training on a treadmill with body weight supported by a harness. There are, however, obvious obstacles involved in other strategies. Walking in water, most helpful initially with severe paralysis, precludes proper limb loading and entrains 'wrong' motor programs. When already capable of some walking, patients training on parallel bars or with canes or a rollator often have the tendency to maintain upright body position using the arms for weight support, precluding sufficient loading of the lower limb. Such insufficient loading might also occur during motor-supported bicycling.

Results of the follow-up investigations on chronic and acute patients indicate that walking capability generally can be maintained under domestic surroundings. The precise amount of carry-over effects is not known since not all initially trained patients returned for the follow-up investigations (see Wernig et al., 1998). Still, the present results support the notion that upright walking, once it is achieved, is the best training to maintain and improve walking capability. However, use of the treadmill with the stabilizing harness allows more intensive and safer training than any other means. Therefore, especially severely paralyzed patients, including those with arm and shoulder paralyses, might significantly further improve their motor capabilities by continuing training on the treadmill after discharge from the hospital. Following the successful application of LB therapy in spinal cord-injured persons, it has also been employed for patients with stroke (Richards et al., 1993; Hesse et al., 1994; Visintin et al., 1998), traumatic and hypoxic brain damage, and multiple sclerosis (A. Wernig, unpubl.). It is quite obvious that a harness supporting the patient to maintain an upright position, the moving surface of the treadmill, and the help with limb setting by therapists is a method suitable for training of patients with motor deficits of different causes, including convalescent and elderly people without specific diseases.

Acknowledgements

This work was supported in part by a fund from the Deutsche Stiftung Querschnittlähmung. Drs. Harms and Stoltze are thanked for their continuous interest in and support of our work.

References

Barbeau, H. and Blunt, R. (1991) A novel interactive locomotor approach using body weight support to retrain gate in spastic paretic subjects. In: Wernig, A. (Ed.), *Plasticity of Motoneuronal Connections. Restorative Neurology, Vol. 5*. Elsevier, Amsterdam, pp. 461–474.

Barbeau, H. and Rossignol, S. (1987) Recovery of locomotion after chronic spinalization in the adult cat. *Brain Res.*, 412: 844–895.

Dietz, V., Colombo, G., Jensen, L. and Baumgartner, L. (1995) Locomotor capacity of spinal cord in paraplegic patients. *Ann. Neurol.*, 37: 574–586.

Dimitrijevic, M.R., Dimitrijevic, M.M., Faganel, J. and Sherwood, A.M. (1984) Suprasegmentally induced motor unit activity in paralyzed muscles of patients with established spinal cord injury. *Ann. Neurol.*, 16: 216–221.

Dobkin, B.H., Harkema, S.J., Requejo, P.S. and Edgerton, V.R. (1995) Modulation of locomotor-like EMG activity in subjects with complete and incomplete spinal cord injury. *J. Neurorehabil.*, 9: 183–190.

Edgerton, V.R., Roy, R.R., Hodgson, J.A., Gregor, R.J. and De Guzman, C.P. (1991) Recovery of full weight-supporting locomotion of the hindlimbs after complete thoracic spinalization of adult and neonatal cats. In: Wernig, A. (Ed.), *Plasticity of Motoneuronal Connections. Restorative Neurology, Vol. 5*. Elsevier, Amsterdam, pp. 405–418.

Eidelberg, E., Walden, J.G. and Nguyen, L.H. (1981) Locomotor control in macaque monkeys. *Brain*, 104: 647–663.

Gossard, J.P. and Hultborn, H. (1991) On the organisation of spinal rhythm generation in locomotion. In: Wernig, A. (Ed.), *Plasticity of Motoneuronal Connections. Restorative Neurology, Vol. 5*. Elsevier, Amsterdam, pp. 385–404.

Grillner, S. (1981) Control of locomotion in bipeds, tetrapods, and fish. In: Brookhart, J.E., Mountcastle, V.B., Brooks, V.B. and Geiger, S.R. (Eds.), *Handbook of Physiology*, Section 1, Vol. 2, Part 2. American Physiological Society, Bethesda, MD, pp. 1127–1236.

Harkema, S.J., Requejo, P.S., Hurley, S.L., Patel, U.K., Dobkin, B.H. and Edgerton, V.R. (1997) Human lumbosacral spinal cord interprets loading during stepping. *J. Neurophysiol.*, 77: 797–811.

Harms, J. (1992) Screw-threaded rod system in spinal fusion surgery. *Spine: State Art Rev.*, 6: 541–575.

Hesse, S.T., Bertelt, C., Schaffrin, A., Malezik, M. and Mauritz, K.H. (1994) Restoration of gait in nonambulatory hemiparetic patients by treadmill training with partial body weight support. *Arch. Phys. Med. Rehabil.*, 75: 1087–1093.

Kendall, H.O., Kendall, F.P. and Wadsworth, G.E. (1971) *Muscles — Testing and Function*. Williams and Wilkins, Amsterdam.

Lovely, R.G., Gregor, R.J., Roy, R.R. and Edgerton, V.R. (1986) Effects of training on the recovery of full weight-bearing stepping in the spinal adult cat. *Exp. Neurol.*, 92: 421–435.

Richards, C.L., Malouin, F., Wood-Dauphinee, S., Bouchard, J.P. and Brunet, D. (1993) Task specific physical therapy for optimization of gait recovery in acute stroke patients. *Arch. Phys. Med. Rehabil.*, 74: 612–620.

Visintin, M. and Barbeau, H. (1989) The effects of body weight support on the locomotor pattern of spastic paretic patients. *Can. J. Neurol. Sci.*, 16: 315–325.

Visintin, M., Barbeau, H., Korner-Bitensky, N. and Mayo, N.E. (1998) A new approach to retrain gait in stroke patients through body weight support and treadmill stimulation. *Stroke*, 29: 1122–1128.

Wernig, A. and Müller, S. (1991) Improvement of walking in spinal cord injured persons after treadmill training. In: Wernig, A. (Ed.), *Plasticity of Motoneuronal Connections. Restorative Neurology, Vol. 5*. Elsevier, Amsterdam, pp. 475–486.

Wernig, A. and Müller, S. (1992) Laufband locomotion with body weight support improved walking in persons with spinal cord injuries. *Paraplegia*, 30: 229–238.

Wernig, A. and Müller, S. (1996) 'Laufband' therapy based on the 'rules of spinal locomotion' is effective in spinal cord injured persons. *Eur. J. Neurosci. Suppl.* 9: 57 (Abstr.).

Wernig, A., Müller, S., Nanassy, A. and Cagol, E. (1995) Laufband therapy based on 'rules of spinal locomotion' is effective in spinal cord injured persons. *Eur. J. Neurosci.*, 7: 823–829.

Wernig, A., Nanassy, A. and Müller, S. (1998) Maintenance of locomotor abilities following Laufband (treadmill) therapy in para- and tetraplegic persons: follow-up studies. *Spinal Cord*, 36: 744–749.

Wernig, A., Nanassy, A. and Müller, S. (1999) Laufband (treadmill) therapy in incomplete paraplegia and tetraplegia. *J. Neurotrauma*, 16: 719–726.

CHAPTER 9

Spinal and supraspinal plasticity after incomplete spinal cord injury: correlations between functional magnetic resonance imaging and engaged locomotor networks

Bruce H. Dobkin *

Department of Neurology and Reed Neurologic Research Center, University of California, Los Angeles, 710 Westwood Plaza, Los Angeles, CA 90095, USA

Introduction

Fueled by a remarkable decade of neuroscientific insights into the biological signals, cascades, and cells that might repair an injured spinal cord (Gimenez y Ribotta and Privat, 1998), clinicians in neurorehabilitation must consider how to employ and test the coming armamentarium of interventions in patients with physical impairments and serious disabilities. Most animal models of spinal cord injury (SCI) use repair strategies designed to improve locomotor capabilities of the hindlimbs or bowel and bladder function. For example, interventions such as transplants of embryonic tissue or neurotrophin-secreting fibroblasts within or near a lesion in the thoracic spinal cord would, ideally, provide descending input to the truncal muscles for postural stability and to the lumbar circuits, sometimes referred to as pattern generators, that drive the flexor and extensor motor pools to the lower extremities (Bregman et al., 1995, 1997; Tuszynski et al., 1996; Grill et al., 1997; Li et al., 1997).

Research with biological interventions after experimental SCI that target locomotion are motivated by the fact that the recovery of the ability to walk is among the most important goals of people with SCI. By a month after an acute SCI, no one with a complete sensorimotor loss below the SCI (graded 'A' by the American Spinal Injury Association [ASIA] scale) and fewer than 10% with some pin or touch sensation (ASIA 'B') will recover an energy-efficient, reciprocal stepping gait (Consortium for Spinal Cord Medicine, 1999). These groups account for 55% or more of all people with SCI. Of those graded ASIA 'C', with movement in some lower-extremity muscles against gravity, from 40 to 60% recover to walk at least 150 ft. Of those graded ASIA 'D', with more useful leg strength, from 75 to 90% recover. Walking speeds and endurance, however, are often well below normal, and the need for braces and assistive devices is common. Thus, conventional therapies for walking are successful only in patients with good selective motor control and strength, meaning the ability to at least flex the hips and flex and extend the knees against gravity (Waters et al., 1989, 1994).

This chapter provides a general review of the most prominent spinal and supraspinal nodes of the locomotor networks (Fig. 1). The lower-thoracic and upper-lumbar motor neurons are among the most important integral parts of the networks. They are likely to play a pivotal role in the recovery of walking after

* Corresponding author: Dr. Bruce H. Dobkin, Department of Neurology, University of California, Los Angeles, Reed Neurologic Research Center, 710 Westwood Plaza, Los Angeles, CA 90095, USA. Fax: +1-310-794-9486; E-mail: bdobkin@ucla.edu

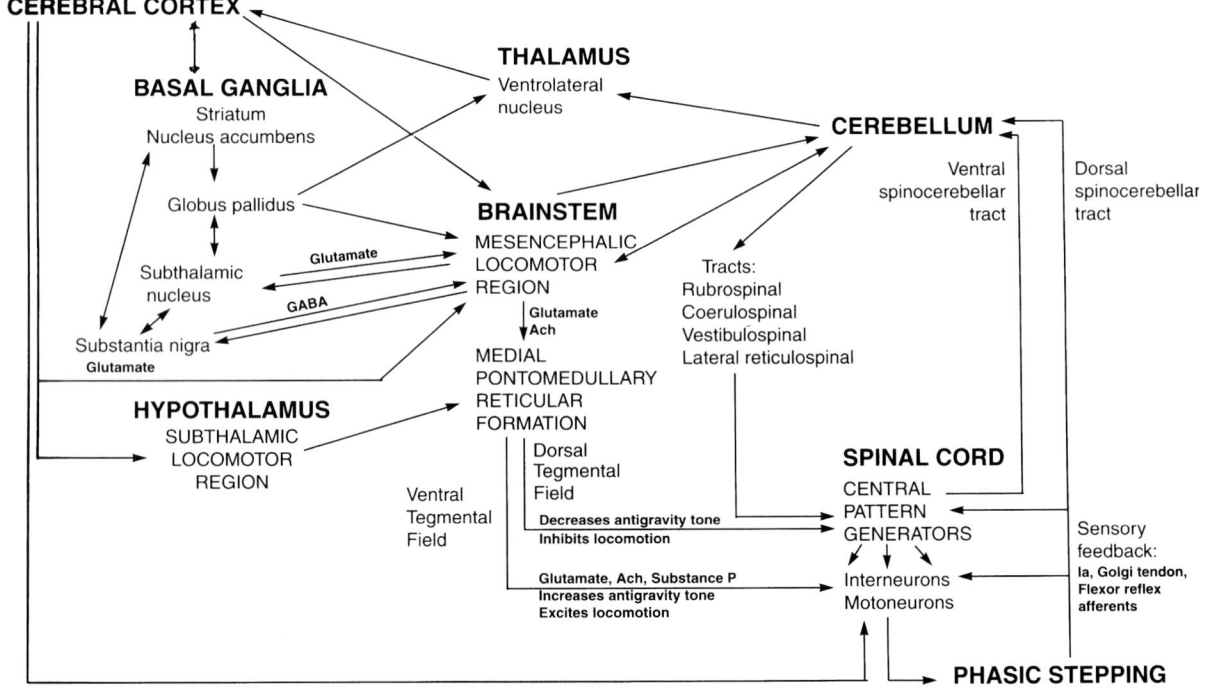

Fig. 1. Simplified diagram of locomotor networks emphasizing sensory inputs to the spinal locomotor circuits and cerebellum and supraspinal modulators of walking that can be imaged during fMRI activation studies. (Adapted from Dobkin, 1996a.)

SCI and will be a key target for biological interventions. Particular segmental sensory inputs during locomotor retraining can optimize their activity and drive the reorganization of the networks after stroke or SCI.

Functional neuroimaging can be employed to visualize the nodes in the locomotor networks, including the cortical neuronal assemblies that represent the leg and thoracolumbar trunk. Passive and voluntary movements of the distal lower extremity during functional magnetic resonance imaging (fMRI) may allow clinicians to monitor the evolution of sensory inputs to supraspinal structures and the cerebral motor contributions to the recovery of walking. Functional neuroimaging may also reveal the efficacy of employing a pharmacologic intervention or a physical therapy, such as treadmill training with partial body weight support, that provides the best of possible sensory input and practice conditions to drive reorganization and functional gains. Imaging could also monitor the success of biologic interventions that target the thoracolumbar motor pools when such strategies are combined with physical therapies that engage and, in a sense, reeducate the locomotor networks to function under new circumstances.

Segmental and ascending spinal cord locomotor signals

In all mammals that have been studied, including a nonhuman primate (Fedirchuk et al., 1998), a rhythm-generating network has been localized to the lumbar spinal cord. Rhythm generators for simple, rhythmically alternating flexion and extension may work at each lower-extremity joint and interact across joints during fictive locomotion and stepping (Orlovsky et al., 1999). They supply the lumbar motor pools with alternating inputs for excitation and inhibition and help control interlimb coordination for stepping. The precise distribution of these spinal networks is the subject of many studies. One recent experiment in neonatal rats suggested that the origin of patterned motor output extended over the entire lumbar region and into caudal thoracic segments (Kjaerulff and Kiehn, 1996). This distribution, extending into the thoracic spine, might offer a more cephalad target for biological interventions aimed at restoring walking.

Sensory inputs from the hips, knees, and the dorsum and soles of the feet interact with the rhythm generators of the cord (Pearson, 1995; Nielsen et al., 1997; Van Wezel et al., 1997; Pearson et al., 1998). These stretch, load, positional, and other inputs contribute to the timing of leg movements during the stance and swing phases of the gait cycle. Small movements at the hip, for example, influence the rhythm-generating network, especially the stance to swing transition when the hip is extended at the end of stance. During locomotion at ordinary speeds, the mechanism for swinging the leg forward is not triggered until a particular degree of posterior positioning of the limb is reached. Also, the magnitude of activity in knee and ankle extensors and the duration of extensor muscle bursts during stance help generate a normal motor pattern for walking. Flexor activity does not occur for swing in one hindlimb until the extensors are unloaded in the stance limb. Even stretch reflexes are modulated by the phase of the step cycle. The same sensory input from the foot will increase hip and knee flexion if applied during the swing phase, but increase activation of the extensor muscles if applied during the stance phase. A locomotor retraining technique should provide these sensory inputs.

After a low-thoracic spinal cord transection, these segmental sensory inputs have been used to train cats (Lovely et al., 1986; Barbeau and Rossignol, 1987; Hodgson et al., 1994; DeLeon et al., 1998) and rats (Giszter et al., 1998a,b) to step independently on a moving treadmill belt over a range of speeds. In people with a complete SCI, sensory inputs such as levels of loading on the stance leg and the degree of hip extension prior to the swing phase lead to alterations in step-like electromyographic activity (Dobkin et al., 1992a,b, 1995; Dietz et al., 1995; Harkema et al., 1995, 1997) and can produce minimally assisted stepping (Harkema et al., 1998). This approach has been operationalized using body weight-supported treadmill training (BWSTT) in people with incomplete SCI (Dobkin et al., 1995; Barbeau et al., 1998; Dietz et al., 1998; Wernig et al., 1998; Dobkin, 1999). Carried out properly, BWSTT appears to allow subjects to make optimal use of segmental sensory inputs and residual supraspinal motor control. Often, subjects who have a high SCI improve in their thoracolumbar postural control along with stepping ability for at least the swing phase of gait, suggesting a link in how the trunk and hip motor pools respond to upright practice. A randomized clinical trial in North America aims to determine whether or not BWSTT is superior to conventional physical therapy in those with recent incomplete SCI (Dobkin et al., 1999). If so, this therapeutic approach would be expected to be accompanied by an evolution of changes in the locomotor networks of the brain that could be mapped by serial functional neuroimaging studies. What is the nature of the nodes of the supraspinal network that might be imaged?

Feedback from alpha and gamma motor neurons and Ia interneurons for reciprocal inhibition, as well as from segmental afferents, is appreciated by the efference copy system of the cerebellum (Jueptner and Weiller, 1998) and higher motor centers (Muir and Steeves, 1997), including the motor cortices and the locomotor regions of the dorsolateral midbrain and reticulospinal neurons of the pons (Fig. 1). The neocerebellum is activated when monitoring the outcome of movements via afferent sensory information and optimizes movements from proprioceptive feedback. With passive ankle movements and during locomotion, the neurons of the dorsal spinocerebellar tract in the dorsolateral funiculus of the lumbar cord fire in relation to both Ia and Ib activity (Orlovsky et al., 1999). This activity provides the cerebellum, then, with information about the performance of leg movements. The ventral spinocerebellar neurons, projecting to the cerebellar cortex in the contralateral lateral funiculus, burst during locomotion, reflecting activity in the spinal rhythm generators. Spinoreticulocerebellar pathways also carry bilateral information predominantly from the spinal circuits for stepping.

The cerebellum appears to have a powerful gating action on the motor cortex, affecting locomotor cycle-related signals that arise from both spinal locomotor and segmental afferent inputs. The Purkinje cells of the cerebellum are rhythmically active throughout the step cycle (Orlovsky et al., 1999). Neurons of the fastigial and interpositus nuclei burst primarily during the flexor phase of stepping and project to the ventrolateral nucleus of the thalamus. Activation of these cerebellar regions can be appreciated by functional neuroimaging.

Descending locomotor network signals

The reticulospinal and propriospinal projections to the lumbosacral motor pools are necessary for the initiation of locomotion (Jordan, 1991). The dorsolaterally located mesencephalic locomotor region (MLR) activates glutamatergic reticulospinal neurons in the pons. In animal experiments, electrical and pharmacologic stimulation studies of these two regions, as well as of the cerebellar fastigial nucleus, subthalamic nucleus, and other neuronal projections to the reticulospinal neurons have produced locomotor hindlimb activity. The lateral reticulospinal tract also contains fibers that descend from the locus coeruleus. The MLR and locomotor region of the pons are also influenced by inputs from corticoreticulospinal and basal ganglia projections. Presumably, they play a functional role in walking. The regions that participate in the initiation of stepping also participate in the control of body orientation, equilibrium, and postural tone. The vestibulospinal and rubrospinal neurons are rhythmically modulated by cerebellar inputs, primarily for extension and flexion, respectively. In addition, chains of polysynaptically interacting propriospinal neurons have been identified in the lateral tegmentum of the pons and medulla, reaching the upper cervical cord, as well as within the dorsal portion of the lateral funiculus in the spinal cord. Of interest in relation to SCI, the reticulospinal and propriospinal fibers are intermingled widely on the periphery of the ventral and lateral spinal tracts, where reticulospinal paths may come to be replaced by propriospinal ones (Nathan et al., 1996). The shortest propriospinal fibers are closest to the gray matter. In studies of cats, these fibers connect motor neurons to axial, girdle, and thigh muscles. In man, the short propriospinal fibers descend for one or two segments. In a sense, the axial and proximal leg motor pools are hard wired to interact together. They are externally linked by propriospinal pathways, among others, and internally linked within the gray matter as parts of the rhythm-generating apparatus.

Routal and Pal (1999) showed that continuous columns are found medially in the ventral gray horns from cervical (C)1 to lumbar (L)3 (column 1), and more laterally from C8 to sacral (S)3 (column 2), L1 to S2, and L4 to S3. Of interest for locomotor control, column 2 appears to innervate the erector spinae and hip muscles. As noted above, the caudal thoracic and the lumbar motor pools are linked within the circuitry of the spinal rhythm generators. Short propriospinal connections across these segments are also likely to link and coordinate multimuscle group and multijoint movements. This organization probably allows for considerable computational flexibility and is a source of plasticity when descending activity is diminished after SCI and when segmental afferent activity becomes a more dominant input. These columns could be important targets for biological interventions that reinstate some supraspinal input. Task-oriented practice for walking, such as BWSTT, might then drive the new inputs to organize the continuous thoracic and lumbosacral columns and their rhythmic circuits for successful locomotion. The columnar distribution of the motor neurons within the spinal cord and orgenization around muscle synergies (Loeb et al, 2000) probably also contribute to the ability of cortical inputs to control movements.

Supraspinal centers are activated during locomotion especially to provide equilibrium and visuomotor control and to coordinate leg responses, for example, on uneven surfaces and when confronted with obstacles. An increase in cortical activity in moving from rather stereotyped to more skilled lower-extremity movements would become necessary as a hemiplegic or paraplegic person relearned to walk with a reciprocal gait. Some pyramidal neurons of the primary motor cortex (M1) in Brodmann's area 4, which is activated during functional neuroimaging movement paradigms, have been shown to be rhythmically active during stepping from inputs via the ventrolateral nucleus of the thalamus, among others. Cortical neurons fire especially during a visually induced perturbation from steady walking, during either the stance or swing phase, as needed. For the lower extremity, these neurons are especially important for flexor control. For example, transcranial magnetic stimulation (TMS) studies in man show greater activation of corticospinal input to the tibialis anterior, compared to the gastrocnemius (Brouwer and Ashby, 1992), but all of the thoracic paraspinal and lower-extremity muscles can be activated (Houlden et al., 1999). Indeed, a single corticospinal axon can branch across the contralateral lumbar motor pools to multiple muscles. This contributes to the ability of pyramidal neurons to be organized in distinct groups that participate in the computation of directional movements, as well as for joint kinematics and torques, rather than only for the activation of specific muscles (Kakei et al., 1999). This branching, however, tends not to include projections from a pyramidal cell to both cervical and lumbar cord motor pools. Strick and colleagues (He et al., 1993) found only 0.2% of primary motor cortex cells that were double labeled retrogradely in macaques from both lower cervical and lower lumbar segments, compared to 4% that were double labeled from the upper and lower cervical segments.

M1 in humans receives inputs from the adjacent Brodmann's area 3a, which receives primarily deep receptor inputs. M1 may receive projections from the primary somatosensory area 3b, which receives cutaneous inputs from the skin. Area 3b does project to the anterior and ventral parietal cortical fields that in turn project topographically to area 4, as well as to the supplementary motor area (SMA), area 6 anterior to area 4, and to the putamen. The cortical sensory representations for surfaces of the hand, and presumably for the foot, have been shown to be substantially remodeled during the acquisition of a motor skill (Xerri et al., 1999). Repetitive practice that leads to the acquisition of a motor skill alters the cortical receptive fields or representations for the skin surfaces that are engaged by the task, as well as the cortical representations for movements that have been learned so as to produce success in solving the motor task. Functional imaging studies of the brain during the acquisition of a skill involving the upper or lower extremity might be expected to reveal remodeling changes in some of these sensorimotor areas.

How much descending input and which tracts are necessary to carry out overground locomotion with reciprocal stance and swing movements and an upright posture? Chronic paralysis in a mid-thoracic contusion model in the cat was associated with severe loss of axons, selective loss of large fibers, and demyelination, but some animals recovered effective locomotion when about 10% of axons survived, most of them at a depth of less than 600 μm from the surface of the cord (Blight, 1983). In a neuropathological study of people with complete and incomplete SCI and controls, the minimum number of corticospinal fibers needed to allow volitional movement with leg strength graded '2' (muscle twitch) to '3' (movement against gravity but not resistance) on the British Medical Council Scale was 3.5 to 10% of the total number of axons at the thoracic (T)4 level (Kaelan et al., 1989). In a series of 20 subjects with traumatic cervical SCI, a general correlation was made between the degree of sparing of long motor and sensory tracts and the clinical examination of residual function, but four cases had clinically complete with neuropathologically incomplete lesions (Hayes and Kakulas, 1997). Kakulas (1999) has attempted to relate fiber counts in specific tracts to sensorimotor function. Six chronic clinically incomplete and four complete SCI subjects were examined after death. At the T4 level, controls had about 41,000 nerve fibers in the lateral corticospinal tract. Sparing of about 3000 fibers on one side was associated with voluntary foot motion. Subjects with complete motor loss had about 2000 residual fibers. When touch and vibration were intact, about 117,000 of 452,000 fibers in the controls were spared. Lesions of the spinal cord that completely spare the lateral or ventral funiculi in nonhuman primates permit walking, and as little as about 25% of white matter tracts allow this (Vilensky et al., 1992). Spared descending inputs that cross a spinal lesion are likely to gain greater synaptic control over the thoracolumbosacral motor pools, especially if locomotor practice reproduces the segmental sensory inputs that typically arise during normal ambulation.

Propriospinal and reticulospinal pathways that persist on the periphery of the spinal cord after trauma might play a critical role in the recovery of walking after a severe SCI. Activity in the paraspinal muscles above a severe cord lesion during stepping training could mechanically stretch muscles and joints below the lesion, driving propriospinal fibers and the thoracolumbar rhythm generators below the lesion. In a similar vein, reconstitution of any descending input to the motor neuron column for the paraspinal and trunk muscles by a biological intervention across the lesion might drive the rhythmic locomotor circuits when combined with optimal step retraining. Giszter and colleagues (Giszter et al., 1998a) reported the adoption of a tight in-phase knee–hip coupling in rats who received a bridging fetal spinal cord transplant after a spinal cord transection as neonates and recovered the ability to bear weight and step on their hindlimbs. Using intracranial microstimulation to map the motor cortex, they found no cortical representation of the hindlimbs or tail. Rats that had good weight support did, however, have axial motor responses that extended below the T8–T10 transection (Giszter et al., 1998b). Poor weight bearers, whether or not they received a transplant, did not have these mid- to low-axial muscle motor representations. Thus, they found a strong association between cortical control of axial motor neurons and improved stepping. Corticospinal or corticoreticulospinal connections may have either crossed the transplant or engaged propriospinal fibers. Or, afferent feedback from segmental sensory inputs and stretch of the trunk muscles now reached the primary sensorimotor cortex and changed the responses to stimulation over the cortical representational map for the axial muscles. In this experiment, the effects of the implant on the locomotor network of the cortex could also have been imaged by a new technique, micro-positron emission tomography (microPET) with fluorodeoxyglucose (FDG) during hindlimb stepping.

Mechanisms of cortical representational plasticity

Therapies aimed at improving arm and hand sensorimotor function after a stroke have, to date, received far more attention than studies of recovery of the lower extremity for walking after SCI (Dobkin, 1996a). Motor learning and changes in the cortical maps that represent movements during the training of skilled hand movements have been observed by functional neuroimaging (Karni et al., 1995, 1998). One of the synaptic activity-dependent mechanisms involved in the acquisition of a hand movement skill is long-term potentiation (LTP). This fundamental cellular mechanism for learning in the hippocampus has also been demonstrated in normal somatosensory, motor, and visual cortices (Malenka and Nicoll, 1999). Mechanisms for spinal cord learning are also likely to include repeated sensory inputs during practice that produce LTP and long-term depression (Randic et al., 1993; Grillner, 1997). Use-dependent sensorimotor learning mechanisms, including LTP, have been shown to contribute to cortical representational plasticity during skills learning, even after a supraspinal injury (Dobkin, 1998). The strength of connections across the primary motor cortex representation for the hand or leg may be achieved by the long-term potentiation of excitatory synaptic transmission in both the horizontal cortico–cortical connections in layers II/III and ascending cortical pathways (Hess et al., 1996; Donoghue, 1997). Motor learning, compared to motor activity, leads to a significant increase in the number of synapses per neuron in layers II/III (Kleim et al., 1996). Activity was also associated with a transient increase in the percentage of neurons that expressed the immediate early gene, c-*fos*, which regulates late genes that are involved in modifying cell structures and functions.

Asanuma and Pavlides (1997) showed that the stimulus parameters in sensory cortex for producing LTP in the motor cortex were within the range of discharges of sensory cortical neurons when they respond to ordinary peripheral stimulation. They also showed that repetitive activity of the pyramidal neurons can produce LTP in spinal interneurons. While tetanic stimulation of the ventrolateral nucleus of the thalamus (VL) alone did not induce LTP, they produced associative LTP in the VL when they combined VL and sensory cortex stimulation. They proposed that repeated practice of a particular movement increased the excitability of a selected group of VL terminals by associative LTP, so that the VL's untrained, diffuse input became able, with training, to excite selected cortical efferent zones without further input from the sensory cortex. They hypothesized that circuits are initially diffuse, leading to excessive muscle contractions during a new movement, but become more specific as LTP is induced by practice. Increases in cortical synaptic efficacy and dendritic arborization among the neuronal assemblies that represent the movement are likely to accompany this reorganization (Dobkin, 1998). Thus, it is not surprising that clinically significant gains in movement with practice have run in parallel with reorganized brain activations using a variety of functional imaging methods (Cramer et al., 1997; Shadmehr and Holcomb, 1997; Classen et al., 1998).

Neurotrophins also play a role in cortical representation of plasticity. Neural activity associated with physical activity in rodents has increased the expression of neurotrophins such as brain-derived neurotrophic factor and nerve growth factor (Neeper et al., 1996). The combination of a motor activity in a learning situation, such as swimming in a water maze, led to greater activity-induced expression of the neurotrophins during the learning phase. This expression likely affects the molecular and cellular events that influence cortical plasticity, motor learning, and memory, including greater synaptic efficacy and dendritic and axonal sprouting.

A growing number of studies point to the augmented effectiveness of residual corticospinal and afferent activity after stroke and SCI on the spinal and cortical regulation of facilitation and recruitment (Jain et al., 1997; Davey et al., 1999). These alterations could expand the cortical and subcortical movement representations for locomotion, much as upper-extremity training and recovery are accompanied by changes in the size and extent of sensorimotor networks after a focal cerebral injury caused by a stroke (Weiller et al., 1993; Nudo et al., 1996). The allocation of cortical space in M1 and somatosensory areas depends on the synaptic efficacy among neuronal assemblies that represent a movement or skin surface. Temporally coincident inputs to the assemblies of the sensorimotor cortex during practice produce synergistic, multijoint movements (Buonomano and Merzenich, 1998). This appears to hold as true after a SCI as a stroke. Cortical representations for the hand and trunk have shown considerable plasticity in people with complete SCI (Cohen et al., 1991; Bruehlmeier et al., 1998). PET scans performed in people with complete SCI revealed an expansion of the hand's topographic map toward the leg area during hand movements and greater bilateral activation of the thalamus and cerebellum (Bruehlmeier et al., 1998), perhaps from altered input by spinothalamic and spinocerebellar inputs to primary motor cortex. Changes in representational maps were also found by intracortical microstimulation of the leg area after amputation of a hindlimb in monkeys (Wu and Kaas, 1999). Cortical representations in M1 for the hindlimb region that had been deafferented and deefferented evoked hip stump, trunk, and tail movements.

Functional imaging of the locomotor network

Might functional neuroimaging open a real-time window on the changes in cortical and subcortical activity induced by a drug, a biological or a retraining regimen for walking? PET can measure resting cerebral glucose or oxygen utilization and blood flow, as well as the increase in synaptic and glial activity in regions where metabolic demand rises. Several neurotransmitter receptors have been imaged, such as those for dopamine and the benzodiazepines. In the near future, the technique may allow the imaging of proteins and gene products. However, PET exposes subjects to radiation, making serial imaging studies risky, and the equipment is highly specialized and not generally available. TMS does not allow precise mapping of the primary sensorimotor cortex of the leg representation, because this region is deep within the interhemispheric fissure. TMS has been very useful in stimulation paradigms of the hand region, language areas, and visual cortex on the more superficial lateral surface of the hemispheres. The technique has excellent temporal resolution, but can map only one small region at a time. Functional MRI, on the other hand, can be repeated as often as desired since no radiation is involved and the technique offers a view of the entire brain.

Functional MRI using the BOLD or blood oxygenation level-dependent contrast technique measures the small-percentage changes in blood flow that are coupled to neuronal activation. Unlike PET scanning, this technique cannot measure actual blood flow or metabolism. Spatial resolution in-plane is from 1 to 3 mm. The patient is usually in the MRI scanner for less than 1 h. Anatomical and reference scanning takes about 15 min. Each trial of 30 s of rest followed by 30 s of movement is usually repeated four times and averaged across trials. Each pixel of activation is analyzed by contrasting MRI signal values obtained at rest with values obtained during movement, using a method such as Student's t-test. Activated pixels are defined as those where the increase in MRI signal during task performance is significant at $P < 0.001$. Within each region, the volume of activated brain is determined by counting the number of significantly activated pixels. For each region and each task, a distribution is made of the control values. Each subject is then compared against the control distribution, with the 95% upper limit in controls used to define increased activation in a subject.

In studies of normal subjects and those recovering from stroke that assessed finger movements, equivalent regions have been activated by PET and fMRI (Cramer et al., 1997; Dettmers et al., 1997; Cao et al., 1998). A variety of interconnected sensorimotor regions become activated under different conditions, such as whether or not a task is a skilled or unskilled movement, highly practiced or being learned, and rapid or forceful, among others. These include, on the motor side, M1, area 6, the supplementary motor area (SMA), area 24 of the cingulate cortex, ventral and dorsal premotor cortex, dorsolateral frontal cortex, and distinct regions of insular cortex. On the sensory side, these include areas 3b, 3a, S1, and superior and inferior parietal cortex.

Far fewer studies of walking or of foot movements have been reported. A PET study of postural standing revealed significant activation of the vermis (Ouchi et al., 1999). We injected FDG while a subject walked for 30 to 40 min before placing the subject in the scanner. Activations were greatest in the paramedian primary sensorimotor cortices for the legs, the cerebellar vermis and hemispheres, the occipital cortex for vision, and the temporo–parieto–occipital junction for integrating sensory input (Dobkin, 1996b). This finding is consistent with the ascending and descending pathways and nodes of the locomotor network. A larger FDG–PET study of twelve normal subjects compared treadmill walking to resting supine and found a similar distribution of activations (Ishii et al., 1995). An fMRI study of the lower extremity, of course, cannot be accomplished during walking. Other methods could be adapted to at least test the representational plasticity of the lower extremity. For example, self-paced dorsiflexion of each foot when the subject was supine in the PET scan gantry produced significant increases in regional cerebral blood flow in the primary sensorimotor cortex of the leg region, in SMA, the cingulate gyrus, and the cerebellar vermis (Honda et al., 1995). The act of imagining locomotor movements of the legs while being scanned with ^{15}O-labeled water also activated the right visual association area. In some subjects these areas were more highly activated when walking was imagined than during actual foot movements.

Voluntary repetitive movements of the toes, much like the fingers, have produced activations during fMRI studies. Zigzag movements of the right index finger and large toe have been compared to the activations induced by writing a subject's signature with the finger or toe (Rijntjes et al., 1999). Somatotopic segregation of the toe and finger was found in the contralateral sensorimotor cortex, adjacent superior parietal lobe (area 5), SMA, anterior cingulate gyrus, thalamus, basal ganglia, cerebellar hemisphere, vermis, the dorsal lateral premotor cortex, and the frontal operculum and secondary sensory cortex.

Passive movements of the fingers and wrist have also produced primary sensorimotor cortical activations. In patients with SCI who are plegic or exhibit only extensor or flexor synergistic movements that often cause head motion artifacts during fMRI scanning, a passive movement testing strategy could serve as a paradigm for the study of all patients. As noted earlier, the dorsal spinocerebellar tract carries passive ankle movement, so the cerebellum and cortex should become activated. Fig. 2 shows the significant activations during an fMRI study in which passive, alternating dorsiflexion and plantar flexion movements of the large toe and forefoot for 20° with less than 10° of ankle motion were compared to the resting state in seven normal subjects. The contralateral primary sensorimotor and anterior cingulate (area 24) cortices, bilateral SMA and posterior insular cortices and, in some cases, portions of the vermis and ipsilateral cerebellum were activated. The medial activations were best appreciated on sagittal scans. The studies did not cover most of the cerebellum for technical reasons. The anterior speech area was activated in subjects who noted the up- versus down-position of the foot in silent speech. When normal and SCI subjects imagined walking while supine in the scanner gantry and did not move, our fMRI studies to date have revealed activations that can include the bilateral M1, SMA, inferior parietal, and visual cortices. With PET and fMRI procedures, the mental state, vision, and subtle sensorimotor cues

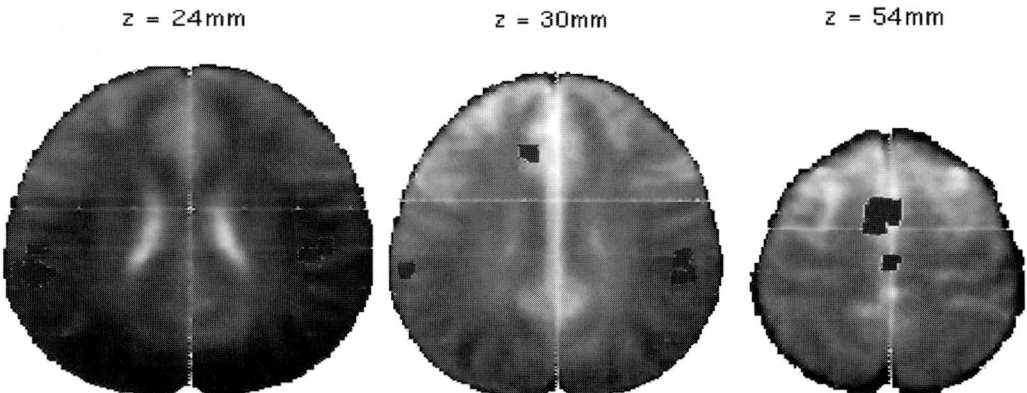

Fig. 2. Group average activations during an fMRI sequence of rest versus passive alternating dorsiflexion and plantar flexion of the right ankle and forefoot. Axial views of the brain are shown. Significant activations were seen in the contralateral primary sensorimotor cortex in the medial foot region (54 mm above the anterior–posterior commissure line), the supplementary motor region just anterior to this, and anteromedial cingulate motor cortex below and anterior to the SMA. The intensity of activation was highest in the contralateral M1. Bilateral secondary sensorimotor regions showed modest activations at 24 and 30 mm above the anterior–posterior commissure line.

must be controlled and taken into account in order to make legitimate comparisons during rest, imagining, and active or passive movement paradigms.

Fig. 3 compares passive and voluntary movements of the forefoot in a SCI subject with an incomplete lesion at T6 who had been trained with BWSTT to walk over ground. Passive movement engaged a much larger representation than found in the control subjects that included the foot, leg, and low-trunk movement representations of the primary sensorimotor cortex. Voluntary compared to passive movement activated an expanded, simultaneous representation

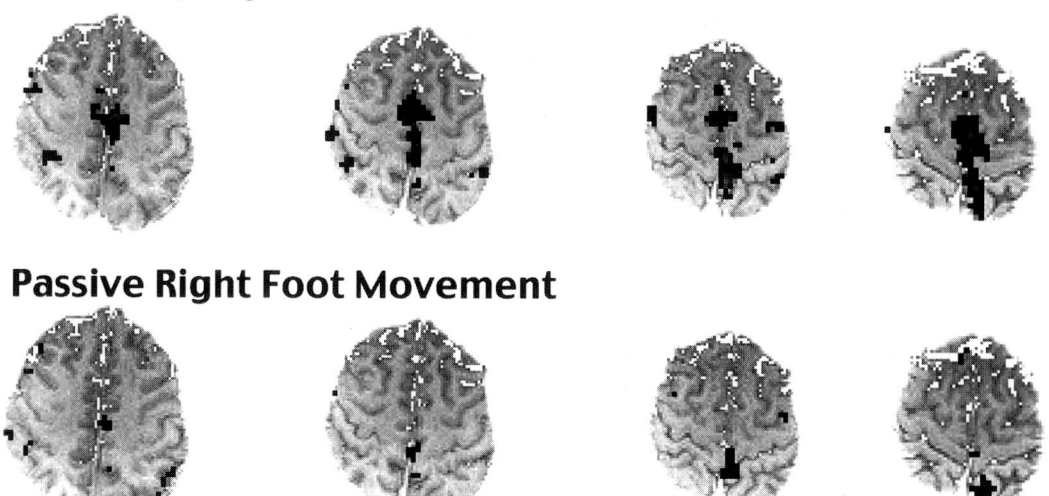

Fig. 3. fMRI activations during voluntary (top row) and passive (bottom row) 20° flexor and extensor movements of the right forefoot and toes in an incomplete T6 SCI subject who walks with a spastic paretic gait. Compared to control subjects in Fig. 2, the activations reveal a larger representation that includes the entire leg and lower trunk with passive movement, as well as much larger-than-normal regions of activation with voluntary forefoot motion. The whiter areas within the darker areas represent relatively greater levels of activation.

for the right foot, leg and trunk primary sensorimotor map, along with larger-than-normal activations in the distributed motor system that included the SMA, insula, and premotor cortex. He had better motor control of the left foot. Of interest, the voluntary foot task on the left produced a smaller area of representational activation than that elicited by the right foot, consistent with upper-extremity studies that sometimes reveal a shrinking representation as a new or recovering motor skill improves.

We also used this passive movement paradigm in a SCI subject 25 years after the injury and who could not walk. He had no voluntary motion or ability to tell sharp from dull or deep pressure below C7. He described a sense of appreciation of large lower-extremity movements, but could not tell the direction of proprioceptive movements at the hips and below. We found a large expansion of the contralateral primary sensorimotor cortical representation for the foot, extending 2.5 cm to include the representations of the entire leg and trunk, along with activation of the SMA and cingulate cortex. Profound, but perhaps not complete sensory deprivation for many years apparently altered synaptic dynamics from both horizontally and vertically projecting inputs to the cortical representations from the lower extremity (Finnerty et al., 1999). The columnar organization of thoracolumbar motor neurons and interconnecting propriospinal fibers may also have altered the efficacy of synaptic activity within the cord and feedback to supraspinal centers via residual projections. If a biological intervention were available to restore supraspinal motor input to the thoracolumbar locomotor circuits, a subject who had demonstrable cortical activation with passive foot movement would seem to be a better candidate than one who had no activation and, thus, had less chance to form and use an ascending–descending pathway during rehabilitation.

Serial studies during the period of step training might reveal an evolution in the foot map for active or passive movement that expands into the leg and trunk map or an expansion of the trunk representation into the foot map. As noted above, the foot and ankle dorsiflexors have a large pyramidal neuron representation which interacts cortically and within the lumbar motor pools with other locomotor movement representations. Also, segmental sensory inputs from the foot have considerable influence on the spinal rhythm generators, the cerebellum, brainstem locomotor regions, and cortical motor networks. The evolution, distribution, magnitude, and size of activations over time might reveal information that could be used by the therapist to assess the efficacy of retraining, as well as by the clinician when looking for evidence that a biological intervention was leading to in vivo repair.

Conclusion

What are the best interventions that therapists can use to train people to walk after a SCI? Should we expect training and biological approaches to induce adaptations within the spinal locomotor pools and in supraspinal representations for locomotor movements? By monitoring the response to treatment with functional imaging activation paradigms, rehabilitative strategies for SCI, as well as stroke, may come to be designed based in part upon whether or not they engage the distributed network needed for successful walking. Functional MRI can also reveal residual ascending and descending synaptic activity across a SCI lesion. In the near future, efforts in neural repair after SCI using cell implants, bridges, genetic manipulations, and other biologic and pharmacologic interventions will require training paradigms such as BWSTT to enhance the synaptic efficacy of new inputs on neuronal targets. Given the connectivity and plasticity of the thoracolumbar locomotor pools and the distributed sensorimotor regions involved in locomotion, even a slight increase in ascending or descending activity might substantially reduce disability. Functional neuroimaging could inform clinical scientists about the in vivo evolution of efficacy of these interventions.

References

Asanuma, H. and Pavlides, C. (1997) Neurobiological basis of motor learning in mammals. *NeuroReport*, 8(43): I–VI.
Barbeau, H. and Rossignol, S. (1987) Recovery of locomotion after chronic spinalization in the adult cat. *Brain Res.*, 412: 84–95.
Barbeau, H., Pepin, A., Norman, K.E., Ladoucer, M. and Leroux, A. (1998) Walking after spinal cord injury: control and recovery. *Neuroscientist*, 4: 14–24.
Blight, A. (1983) Cellular morphology of chronic spinal cord in-

jury in the cat: analysis of myelinated axons by line-sampling. *Neuroscience*, 10: 521–543.

Bregman, B.S., Kunkel-Bagden, E., Schnell, L., Dai, H.N., Gao, D. and Schwab, M.E. (1995) Recovery from spinal cord injury mediated by antibodies to neurite growth inhibitors. *Nature*, 378: 498–501.

Bregman, B., Diener, P.S., McAtee, M., Dai, H.N., James, C., 1997. Intervention strategies to enhance anatomical plasticity and recovery of function after spinal cord injury. In: Seil, F.J. (Ed.), Neural Regeneration, Reorganization, and Repair. Advances in Neurology, vol. 72, Lippencott-Raven, Philadelphia, PA, pp. 257–275.

Brouwer, B. and Ashby, P. (1992) Corticospinal projections to lower limb motoneurons in man. *Exp. Brain Res.*, 89: 649–654.

Bruehlmeier, M., Dietz, V. and Leenders, K. (1998) How does the human brain deal with a spinal cord injury?. *Eur. J. Neurosci.*, 10: 3918–3922.

Buonomano, D. and Merzenich, M. (1998) Cortical plasticity: from synapses to maps. *Annu. Rev. Neurosci.*, 21: 149–186.

Cao, Y., D'Olhaberriague, L., Vikingstad, B., Levine, S. and Welch, K.M. (1998) Pilot study of functional MRI to assess cerebral activation of motor function after poststroke hemiparesis. *Stroke*, 29: 112–122.

Classen, J., Liepert, J., Wise, S., Hallett, M. and Cohen, L.G. (1998) Rapid plasticity of human cortical movement representation induced by practice. *J. Neurophysiol.*, 79: 1117–1123.

Cohen, L., Topka, H., Cole, R. and Hallet, M. (1991) Paresthesias induced by magnetic brain stimulation in patients with thoracic spinal cord injury. *Neurology*, 41: 1283–1288.

Consortium for Spinal Cord Medicine (1999) *Outcomes Following Traumatic Spinal Cord Injury: Clinical Practices Guidelines for Health-care Professionals.* Paralyzed Veterans of America, Washington, DC.

Cramer, S., Nelles, G. and Benson, R. (1997) A functional MRI study of subjects recovered from hemiparetic stroke. *Stroke* 28: 2518–2527.

Davey, N., Smith, H., Savic, G., Maskill, D., Ellaway, P. and Frankel, H. (1999) Comparison of input–output patterns in the corticospinal system of normal subjects and incomplete spinal cord injured patients. *Exp. Brain Res.*, 127: 382–390.

DeLeon, R., Hodgson, J., Roy, R. and Edgerton, V.R. (1998) Locomotor capacity attributable to step training versus spontaneous recovery after spinalization in adult cats. *J. Neurophysiol.*, 79: 1329–1340.

Dettmers, C., Stephan, K., Lemon, R. and Frackowiak, R. (1997) Reorganization of the executive motor system after stroke. *Cerebrovasc. Dis.*, 7: 187–200.

Dietz, V., Colombo, D., Jensen, L. and Baumgartner, L. (1995) Locomotor capacity of spinal cord paraplegic patients. *Ann. Neurol.*, 37: 574–582.

Dietz, V., Wirz, M., Curt, A. and Colombo, G. (1998) Locomotor pattern in paraplegic patients: training effects and recovery of spinal cord function. *Spinal Cord*, 36: 380–390.

Dobkin, B.H., 1996a. Neurologic Rehabilitation. F.A. Davis, Oxford University Press, Philadelphia, PA.

Dobkin, B.H. (1996b) Recovery of locomotor control. *Neurologist*, 2: 239–249.

Dobkin, B.H. (1998) Activity-dependent learning contributes to motor recovery. *Ann. Neurol.*, 44: 158–160.

Dobkin, B.H. (1999) Overview of treadmill locomotor training with partial body weight support: a neurophysiologically sound approach whose time has come for randomized clinical trials. *Neurorehabil. Neural Repair*, 13: 157–165.

Dobkin, B.H., Edgerton, V.R. and Fowler, E. (1992a) Sensory input during treadmill training alters rhythmic locomotor EMG output in subjects with complete spinal cord injury. *Soc. Neurosci. Abstr.*, 18: 1043.

Dobkin, B., Edgerton, V., Fowler, E. and Hodgson, J. (1992b) Training induces rhythmic locomotor EMG patterns in subjects with complete SCI. *Neurology*, 42(Suppl. 3): 207–208.

Dobkin, B., Harkema, S., Requejo, P. and Edgerton, V.R. (1995) Modulation of locomotor-like EMG activity in subjects with complete and incomplete chronic spinal cord injury. *J. Neurol. Rehabil.*, 9: 183–190.

Dobkin, B., Apple, D., Barbeau, H., Saulino, M., Fugate, L. and Scott, M. (1999) Randomized trial of body weight-supported treadmill training after acute spinal cord injury. *Neurorehabil. Neural Repair*, 13: 50.

Donoghue, J. (1997) Limits of reorganization in cortical circuits. *Cereb. Cortex*, 7: 97–99.

Fedirchuk, B., Nielson, J., Petersen, N. and Hultborn, H. (1998) Pharmacologically evoked fictive motor patterns in the acutely spinalized marmoset monkey. *Exp. Brain Res.*, 122: 351–361.

Finnerty, G., Roberts, L. and Connors, B. (1999) Sensory experience modifies the short-term dynamics of neocortical synapses. *Nature*, 400: 367–371.

Gimenez y Ribotta, M. and Privat, A. (1998) Biological interventions for spinal cord injury. *Curr. Opin. Neurol.*, 11: 647–654.

Giszter, S., Graziani, V., Kargo, W., Hockensmith, G. and Davies, M. (1998a) Primitives, pattern generators, and cortical maps in spinal injured rats. *J. Spinal Cord Med.*, 21: 378.

Giszter, S., Kargo, W., Davies, M. and Shibayama, M. (1998b) Fetal transplants rescue axial muscle representations in M1 cortex of neonatally transected rats that develop weight support. *J. Neurophysiol.*, 80: 3021–3030.

Grill, R., Murai, K., Blesch, A., Gage, F.H. and Tuszynski, M.H. (1997) Cellular delivery of neurotrophin-3 promotes corticospinal axonal growth and partial functional recovery after spinal cord injury. *J. Neurosci.*, 17: 5560–5572.

Grillner, S. (1997) Ion channels and locomotion. *Science*, 278: 1087–1088.

Harkema, S., Requejo, P., Dobkin, B. and Edgerton, V. (1995) Load and phase dependent modulation of motor pool output by the human lumbar spinal cord during manually assisted stepping. *Proc. Int. Symp. Neurons, Networks, and Motor Behavior.* The University of Arizona, Tucson.

Harkema, S., Hurley, S., Dobkin, B. and Edgerton, V.R. (1997) Human lumbosacral spinal cord interprets loading during stepping. *J. Neurophysiol.*, 77: 797–811.

Harkema, S.J., Dobkin, B.H. and Edgerton, V.R. (1998) Activity

dependent plasticity after complete human spinal cord injury. *Exp. Neurol.*, 151: 156.
Hayes, K. and Kakulas, B. (1997) Neuropathology of human spinal cord injury sustained in sports-related activities. *J. Neurotrauma*, 14: 235–248.
He, S.-Q., Dum, R. and Strick, P. (1993) Topographic organization of corticospinal projections from the frontal lobe: motor areas on the lateral surface of the hemisphere. *J. Neurosci.*, 13: 952–980.
Hess, G., Aizenman, C. and Donoghue, J. (1996) Conditions for the induction of long-term potentiation in layer II/III horizontal connections of the rat cortex. *J. Neurophysiol.*, 75: 1765–1778.
Hodgson, J., Roy, R., Dobkin, B. and Edgerton, V.R. (1994) Can the mammalian spinal cord learn a motor task?. *Med. Sci. Sports Exerc.*, 26: 1491–1497.
Honda, M., Freund, H.-J. and Shibasaki, H. (1995) Cerebral activation by locomotive movement with the imagination of natural walking. *Hum. Brain Mapp.*, 2(Suppl. 1): 318.
Houlden, D.A., Schwartz, M.L., Tator, C.H., Ashby, P. and MacKay, W.A. (1999) Spinal cord-evoked potentials and muscle responses evoked by transcranial magnetic stimulation in 10 awake human subjects. *J. Neurosci.*, 19: 1855–1862.
Ishii, K., Senda, M. and Toyama, H. (1995) Brain function in bipedal gait: a PET study. *Hum. Brain Mapp.*, 2(Suppl 1): 321.
Jain, N., Catania, K. and Kaas, J.H. (1997) Deactivation and reactivation of somatosensory cortex after dorsal spinal cord injury. *Nature*, 386: 495–498.
Jordan, L. (1991) Brainstem and spinal cord mechanisms for the initiation of locomotion. In: Shimamura, M., Grillner, S. and Edgerton, V. (Eds.), *Neurobiological Basis of Human Locomotion*. Japan Scientific Societies Press, Tokyo, pp. 3–20.
Jueptner, M. and Weiller, C. (1998) A review of differences between basal ganglia and cerebellar control of movements as revealed by functional imaging studies. *Brain*, 121: 1437–1449.
Kaelan, C., Jacobsen, P., Morling, P. and Kakulas, B. (1989) A quantitative study of motoneurons and corticospinal fibres related to function in human spinal cord injury. *Paraplegia*, 27: 148–149.
Kakei, S., Hoffman, D. and Strick, P. (1999) Muscle and movement representations in the primary motor cortex. *Science*, 285: 2136–2139.
Kakulas, B. (1999) A review of the neuropathology of human spinal cord injury with emphasis on special features. *J. Spinal Cord Med.*, 22: 119–124.
Karni, A., Meyer, G., Jezzard, P. and Ungleider, S. (1995) Functional MRI evidence for adult motor cortex plasticity during motor skill learning. *Nature*, 377: 155–158.
Karni, A., Meyer, G., Hipolito, C.R., Jezzard, P. and Adams, M.M. (1998) The acquisition of skilled motor performance: fast and slow experience-driven changes in primary motor cortex. *Proc. Natl. Acad. Sci. USA*, 95: 861–868.
Kjaerulff, O. and Kiehn, O. (1996) Distribution of networks generating and coordinating locomotor activity in the neonatal rat spinal cord in vitro: a lesion study. *J. Neurosci.*, 16: 5777–5794.
Kleim, J.A., Lussnig, E., Schwarz, E.R., Comery, T.A. and Greenough, W.T. (1996) Synaptogenesis and FOS expression in the motor cortex of the adult rat after motor skill learning. *J. Neurosci.*, 16: 4529–4535.
Li, Y., Field, P. and Raisman, G. (1997) Repair of adult rat corticospinal tract by transplants of olfactory ensheathing cells. *Science*, 277: 2000–2002.
Loeb, E.P., Giszter, S., Saltiel, P., Bizzi, E. and Mussa-Ivaldi, F. (2000) Output units of motor behavior: An experimental and modeling study. *J. Cogn. Neurosci.*, 12: 78–97.
Lovely, R., Gregor, R., Roy, R. and Edgerton, V.R. (1986) Effects of training on the recovery of full-weight-bearing stepping in the adult spinal cat. *Exp. Neurol.*, 92: 421–435.
Malenka, R. and Nicoll, R. (1999) Long-term potentiation — a decade of progress?. *Science*, 285: 1870–1874.
Muir, G. and Steeves, J. (1997) Sensorimotor stimulation to improve locomotor recovery after spinal cord injury. *Trends Neurosci.*, 20: 72–77.
Nathan, P., Smith, M. and Deacon, P. (1996) Vestibulospinal, reticulospinal and descending propriospinal nerve fibres in man. *Brain*, 119: 1809–1833.
Neeper, S., Gomez-Pinella, F., Choi, J. and Cotman, C. (1996) Physical activity increases mRNA for brain-derived neurotrophic factor and nerve growth factor in rat brain. *Brain Res.*, 726: 49–56.
Nielsen, J., Petersen, N. and Fedirchuk, B. (1997) Evidence suggesting a transcortical pathway from cutaneous foot afferents to tibialis anterior motoneurones in man. *J. Physiol. (Lond.)*, 501: 473–484.
Nudo, R., Wise, B., SiFuentes, F. and Milliken, G. (1996) Neural substrates for the effects of rehabilitative training on motor recovery after ischemic infarct. *Science*, 272: 1791–1794.
Orlovsky, G., Deliagina, T. and Grillner, S. (1999) *Neuronal Control of Locomotion: From Mollusc to Man*. Oxford University Press, Oxford.
Ouchi, Y., Okada, Y., Yoshikawa, E., Nobezawa, S. and Futatsubashi, M. (1999) Brain activation during maintenance of standing postures in humans. *Brain*, 122: 329–338.
Pearson, K. (1995) Proprioceptive regulation of locomotion. *Curr. Opin. Neurobiol.*, 5: 786–791.
Pearson, K., Misiaszek, J. and Fouad, K. (1998) Enhancement and resetting of locomotor activity by muscle afferents. *Ann. N.Y. Acad. Sci.*, 860: 203–215.
Randic, M., Jiang, C. and Cerne, R. (1993) Long-term potentiation and long-term depression of primary afferent neurotransmission in the rat spinal cord. *J. Neurosci.*, 13: 5228–5241.
Rijntjes, M., Dettmers, C. and Buchel, C. (1999) A blueprint for movement: functional and anatomical representations in the human motor system. *J. Neurosci.*, 19: 8043–8048.
Routal, R. and Pal, G. (1999) A study of motoneuron groups and motor columns of the human spinal cord. *J. Anat.*, 195: 211–224.
Shadmehr, R. and Holcomb, H. (1997) Neural correlates of motor memory consolidation. *Science*, 277: 821–825.
Tuszynski, M., Gabriel, K., Gage, F., Suhr, S., Meyer, S. and Rosetti, A. (1996) Nerve growth factor delivery by gene trans-

fer induces differential outgrowth of sensory, motor, and noradrenergic neurites after adult spinal cord injury. *Exp. Neurol.*, 137: 157–173.

Van Wezel, B., Ottenhoff, F. and Duysens, J. (1997) Dynamic control of location-specific information in tactile cutaneous reflexes from the foot during human walking. *J. Neurosci.*, 17: 3804–3814.

Vilensky, J., Moore, A., Eidelberg, E. and Walden, J. (1992) Recovery of locomotion in monkeys with spinal cord lesions. *J. Motor Behav.*, 24: 288–296.

Waters, R., Yakura, J., Adkins, R. and Barnes, G. (1989) Determinants of gait performance following spinal cord injury. *Arch. Phys. Med. Rehabil.*, 70: 811–818.

Waters, R., Adkins, R., Yakura, J. and Sie, I. (1994) Motor and sensory recovery following incomplete paraplegia. *Arch. Phys. Med. Rehabil.*, 75: 67–72.

Weiller, C., Ramsay, S., Wise, R. and Frackowiak, R. (1993) Individual patterns of functional reorganization in the human cerebral cortex after capsular infarction. *Ann. Neurol.*, 33: 181–189.

Wernig, A., Nanassy, A. and Müller, S. (1998) Maintenance of locomotor abilities following laufband (treadmill) therapy in para- and tetraplegic persons: follow-up studies. *Spinal Cord*, 36: 744–749.

Wu, C. and Kaas, J. (1999) Reorganization in primary motor cortex of primates with long-standing therapeutic amputations. *J. Neurosci.*, 19: 7679–7697.

Xerri, C., Merzenich, M., Jenkins, W. and Santucci, S. (1999) Representational plasticity in cortical area 3b paralleling tactual-motor skill acquisition in adult monkeys. *Cereb. Cortex*, 9: 264–276.

SECTION III

Impact of neuroprosthetic applications on functional recovery

CHAPTER 10

Impact of neuroprosthetic applications on functional recovery

John K. Chapin *

Department of Physiology and Pharmacology, SUNY Downstate Health Science Center, 450 Clarkson Avenue, Brooklyn, NY 11203, USA

Introduction

Neuroprostheses are machines designed to artificially restore lost neurological functions. Following upon the success of heart-stimulating devices (pacemakers), much recent interest has focused upon the implantation of therapeutic stimulating devices in the nervous system. In recent years, neuroprosthetic devices for functional electrical stimulation of muscles have been successfully moved from the laboratory to the clinic, allowing some restoration of motor functions such as hand grasping in paralyzed patients (Bhadra and Peckham, 1997). Such devices are typically controlled through voluntary movement of some unparalyzed part of the body, such as the shoulder or eyes. As such, only a limited repertoire of movements can be controlled, and even they may be difficult to learn and fatiguing to carry out.

More recent interest has focused on development of neuroprosthetic devices that utilize information electrically recorded directly from the central nervous system (CNS), either through macroelectrodes placed on the scalp surface (Wolpaw et al., 1998) or microelectrodes implanted within the actual motor control centers of the brain (Schmidt, 1980). An obvious application for such devices would be to build electronic 'bridges' capable of carrying neural information across neurological lesions, such as spinal cord injuries. For example, it has long been known that the motor cortices of the brain contain representations of intended limb movements (Evarts, 1968; Georgopoulos et al., 1986). Therefore, devices capable of electronically extracting such motor commands from the brains of paralyzed individuals could be used to physically manifest the intended movements by actuating equivalent movement of a robot arm. In this paper I will review our recent demonstration of the feasibility of such 'neurorobotic' control in experimental rats. I will also address the various issues surrounding the possible use of neuroprostheses for restoration of movement after paralysis, such as that caused by spinal cord injury.

'Neurorobotic' control in experimental rats

In our recently published paper (Chapin et al., 1999) we describe use of 'motor' signals derived from simultaneously recorded primary motor (M1) cortical neurons to allow rats to control real-time movement of a robot arm purely through brain activity. As schematized in Fig. 1, an experimental paradigm was developed wherein rats were trained to press down a lever to receive a water reward. The lever was spring loaded, and connected to a pivot containing an angular transducer whose output was used to radially move a horizontally oriented one-dimensional 'robot

---

* Corresponding author: Dr. John K. Chapin, Department of Physiology and Pharmacology, SUNY Downstate Health Science Center, 450 Clarkson Avenue, Brooklyn, NY 11203, USA.
E-mail: john_chapin@netmail.hscbklyn.edu

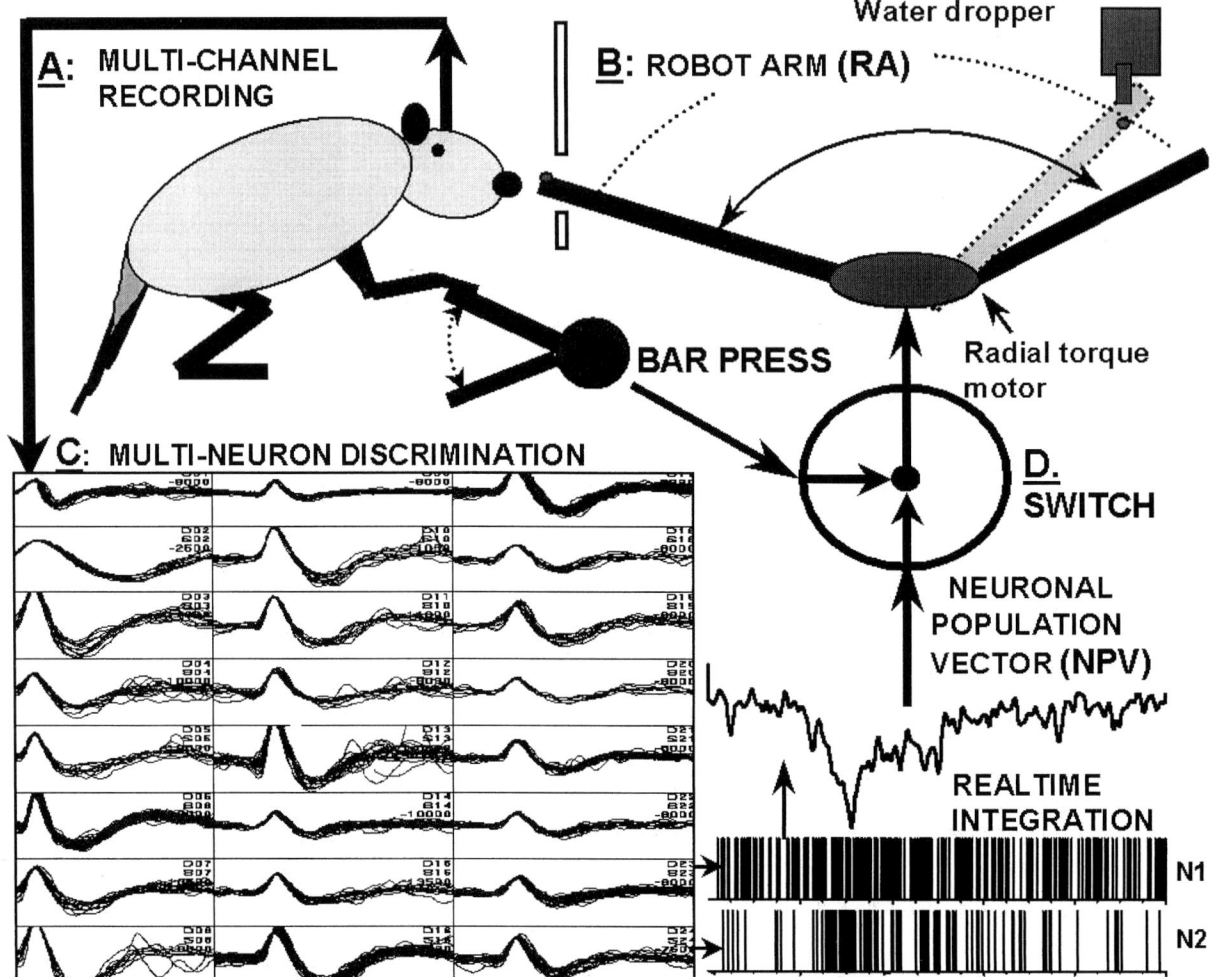

Fig. 1. Experimental paradigm. (A) Rats are chronically implanted with multi-electrode recording arrays (8 or 16 channels) in the M1 cortex and/or VL thalamus, M2 cortex or cerebellum for a total of 24 or 32 channels. (B) Initially the rats are trained to press down a spring-loaded lever (*BAR*) for a water reward brought from a water dropper to the mouth by a 'robot arm' (*RA*). The robot arm position is continuously variable and may be moved beyond the water dropper. Positioning is actuated by a radial torque motor driven by an analog voltage output from an angular transducer on the bar. (C) Neuronal spike waveforms are discriminated online, and transmitted to a neural population vector (*NPV*) integration circuit consisting of a 32-channel weighted summing amplifier and an integrator with a 20-ms time constant. (D) A switch is used to determine whether the robot arm position will be controlled from the BAR or the NPV integrator. In experiments, rats typically begin working in the BAR → RA mode, in which the rat obtains water by pressing the bar to position the robot arm. Without warning the paradigm was then switched to the NPV → RA mode, in which the RA is positioned by brain-derived signals.

arm'. The tip of this arm carried a small hollowed cup capable of carrying a drop of water. When the lever was depressed, this robot arm was proportionally moved in a horizontal plane to, or beyond, a water dropper. If the animal pressed the lever down to a level where the robot arm was positioned directly under the water dropper, a drop of water was loaded in the cup. Upon releasing the lever, the robot arm sprang back to its resting position with its tip just protruding through an opening in a Plexiglas barrier. The animal was then able to drink the water.

Once the animals were trained in this task, the aim was to determine whether a similar accuracy of robot arm positioning could be accomplished using signals derived from multiple neuron recordings in motor areas of the brain. In 6 trained rats, 23–46 single neurons were simultaneously recorded through microelectrode arrays (from NB Labs, Denison, TX) chronically implanted in the forelimb representations of the M1 cortex and ventrolateral (VL) thalamus. Most neurons recorded from these areas exhibited significant increases in spiking discharge over the lever movement task, exhibiting their strongest overall discharge rate during the 100 ms just before the animal began to press down the lever.

The next issue to address was whether these lever movement-related multi-neuronal activity patterns could be electronically combined to create a 'neuronal population vector' that could be used to directly move the robot arm. For this we developed a multi-channel weighted neural activity integrator. The inputs to this device consisted of 32 channels of 5 V transistor–transistor logic (TTL) pulse trains created by a 64-channel online multi-neuron discriminator (from Plexon Inc., Dallas, TX). Each pulse represented the occurrence of a single discriminated action potential recorded from one of the 32 different neurons selected for inclusion into the population vector. The gain (amplification) of each of these 32 input channels could be adjusted to any level (with positive or negative polarity) determined through experimentation with various mathematical and statistical schemes for population vector encoding (Chapin, 1998). After this weighted summation of simultaneously recorded neuronal spike trains, the signal was integrated (with a 25 ms time constant), low-pass filtered, and then amplified for scaling purposes. The output was a single time-varying signal that represented the linearly weighted and integrated activity of 32 simultaneously recorded neurons.

This signal was then routed to a switch which provided the input to the angular servo-motor controlling movement of the robot arm. At the beginning of experimentation each day, this switch was set in the 'lever-movement-controls-robot-arm' position. After a few minutes of this behavior, the switch was then suddenly changed to the 'brain-activity-controls-robot-arm' position. This allowed us to test the accuracy of the weight-integrated neuronal population vector signal in terms of its ability to mimic the movement of the robot arm.

Fortuitously, our investigations demonstrated that the peak amplitude of the integrated neuronal activity that just preceded the onset of lever movement was highly proportional to the displacement of the lever movement over the first 300–500 ms following lever movement onset. This was extremely useful for the success of the 'brain-activity-controls-robot-arm' experiment because it allowed the rat's fine-positioning of the lever to be directly reflected in the amplitude of the 'neurorobotic control signal' that was used to control the robot arm position. To obtain the neuronal weightings for this control signal, principal components analysis (PCA) was used in a manner previously described (Nicolelis et al., 1995; Chapin and Nicolelis, 1996, 1999) to optimally capture this covariant multi-neuron signal while rejecting irrelevant signals and noise. For this reason, four of our six animals were able to repeatedly use their brain activity alone to position the robot arm under the water dropper, and then return the water to their mouths. On experimental days, these animals obtained most of their daily water in this mode.

The success of this paradigm among the six animals was mainly dependent on the number of recorded neurons exhibiting the appropriate pre-movement activity pattern. All 32 neurons exhibited this pattern in the animal that retrieved the water reward with 100% accuracy. In contrast, fewer than 20 such neurons were recorded in both of the animals whose performance was so inaccurate that they were unable to maintain this brain-actuated water-fetching task. This underscores the importance of recording from neuronal populations, as opposed to single neurons. On the other hand, it suggests that reasonable success might be obtained with as few as 25 neurons for each dimension of robot arm movement. Thus, even though the native neuronal populations of these areas may measure in the millions, a much smaller number might be sampled to achieve reasonable robotic accuracy.

Another interesting result was the slow dissociation between brain activity and lever movement that was observed over successive days of training in this 'neurorobotic mode'. This apparently resulted from the fact that the neural population signal used to control the robot arm consisted mainly of a ~100 ms duration activity peak that just preceded onset of actual lever pressing. During the neurorobotic mode training, the robot arm tended to move to the water dropper and back to the mouth before or around the actual onset of lever pressing. Thus, the animals eventually learned to obtain their water reward without actually making the originally trained lever movement. It was interesting, however, that they still made the normal preparatory movements in this behavior, including reaching the forepaw to the lever. We interpret this finding to indicate that the M1 cortex maintains some ability to modify its normal association with limb movement, but that the overall context of movement is still important for expression of this activity. This importance of overall motor context is amplified by our (unpublished) observation that it is very difficult to train rats to perform such neurorobotic control movements in the absence of previous training in the lever movement paradigm.

Multi-neuron recordings

The results of the above studies underscore the importance of neuronal population recordings, rather than single-neuron recordings, to achieve accurate control of a neurorobotic device (see Humphrey et al., 1970; Dormont et al., 1982). Thus, the necessary antecedent of this feasibility demonstration was our development of multi-single-neuron recording technology beginning in the mid-1980s (Chapin and Patel, 1987; Shin and Chapin, 1990a,b; Nicolelis et al., 1993a,b). Though the possible importance of multi-neuron recording had been recognized by many, there were several technological hurdles. First, it was unclear whether sufficient recording stability could be obtained with traditional single-unit recording electrodes, which typically consist of metal needles completely insulated save for a 5–20 µm tip. While such electrodes are optimal for recording well-isolated single neurons in acute recording experiments, they tend to be susceptible to movement and other forms of artifact, and also tend to lose their recording ability when chronically implanted in the brain. This is presumed to be the result of biological encapsulation of the electrode tips with scar tissue, similar to the immunologically mediated processes that are triggered by contact of any foreign material in the body.

In order to get around this difficulty we have adopted the use of chronically implanted microwires, which have been widely used by several investigators, especially in hippocampus recordings (Kubie, 1984). This technique involves implantation of multiple microwires formed into bundles or arrays, allowing neuronal activity to be sampled from a small brain region, rather than a single spot. The microwire recording tips are easily prepared, usually by simply cutting them with scissors to yield a flat or partially oblique recording contact. Interestingly, these electrodes were observed to allow recording of good single-neuron action potentials for long periods of time, their discriminability even increasing over time.

Since large numbers of microwires can be implanted, this technique is amenable to large-scale multi-neuron population recordings, which are essential for reading out neural control signals for neurorobotic applications. Our investigations suggest that at least 25 task-dependent neurons must be simultaneously recorded to obtain movements in one dimension that are sufficiently well controlled to prevent extinction of operantly conditioned lever-pressing behavior after switching to neurorobotic mode. In fact, we anticipate that 250–500 neurons may be needed for control of multi-joint movements in three-dimensional space.

Control signals in human subjects

Which areas of the brain will ultimately be most appropriate for obtaining neurorobotic control signals? The motor cortex is the most commonly mentioned brain area for achieving such control. It has been widely reported to contain a representation of intended movements, and a certain amount of detail about the direction and force of such movements. Even the motor cortex, however, is differentiated into several subregions, including a rostral and a caudal subregion of primary motor cortex, plus several divisions of the premotor and supplementary motor cortices. Each of these is functionally distinct, mainly in relation to the kind of neural processing involved in converting sensory or internal signals into plans for motor output. Moreover, there is still much disagreement about the fundamental parameters of movement that are controlled by the motor cortex: direction or force (Schmidt et al., 1975; Scott and Kalaska, 1995; Kakei et al., 1999).

Of course, the brain contains a large number of other motor areas beyond the motor cortex. These are distributed throughout the cerebrum, brainstem and spinal cord. Most of these areas (especially the basal ganglia, cerebellum and spinal cord) are themselves the subject of major investigations into the different parameters of movement that they control. Thus, the choice of brain areas to record is dependent on the kind of neurorobotic task to be performed, and the type of robotic technology that is envisioned. If, for example, our goal is only to control reaching of a robot arm to a particular target in space, a task easily accomplished by existing robotic technology, then the only information that need come from the brain would be the spatial position of the desired reach. If so, the appropriate brain recordings might be obtained from the parietal cortex, which is known to contain representations of the positions of objects in extrapersonal space.

On the other hand, if we wish to use neuroprosthetic technology to control walking, a much more elaborate system would be required. First, there would be a decision over whether to develop a robotic exoskeleton to fit around the legs, or instead to develop stimulating devices to direct the contraction of the patients' own limb musculature. Moreover, even simple walking movements require delicately timed interactions between rhythmic motor output programs and sensory feedback. Thus, a robotic device that simulated walking would need to have access to sensory feedback from the limb and foot. It would also be necessary to sample neuronal activity in the locomotor coordinating regions of the brainstem, including the reticular and vestibular nuclei and the cerebellum. To conclude, if one wishes to utilize neurorobotic or neuroprosthetic technology to restore the full motor potential of paralyzed individuals, it will be necessary to more fully understand the 'whole brain' mechanisms of motor control. Though much research has been directed to understanding the functionality of individual brain areas, little effort has been made to understand motor control at the system-wide level. This may be alleviated through extensive use of the multi-site multi-electrode recordings such as those we have initiated for the neurorobotic studies described above. If so, the scientific understanding of brain mechanisms of motor control may be significantly advanced as a byproduct of applied research in neurorobotic or neuroprosthetic control.

Acknowledgements

This work was supported by NIH Contract NS62352, NIH Grant NS26722 and DARPA/ONR Grant N00014-98-1-0679.

References

Bhadra, N. and Peckham, P.H. (1997) Peripheral nerve stimulation for restoration of motor function. *J. Clin. Neurophysiol.*, 14: 378–393.

Chapin, J.K. (1998) Population-level analysis of multi-single neuron recording data: multivariate statistical methods. In: Nicolelis, M.A.L. (Ed.), *Neuronal Population Recording*. CRC Press, Boca Raton, FL, pp. 198–232.

Chapin, J.K. and Nicolelis, M.A.L. (1996) Neural network mechanisms of oscillatory brain states: characterization using simultaneous multi-single neuron recordings. *Electroenceph. Clin. Neurophysiol. Suppl.*, 45: 113–122.

Chapin, J.K. and Nicolelis, M.A.L. (1999) Population coding in simultaneously recorded neuronal ensembles in ventral posteromedial (VPM) thalamus: multidimensional sensory representations and population vectors. *J. Neurosci. Methods*, 94: 121–140.

Chapin, J.K. and Patel, I.M. (1987) Effects of ethanol on arrays of single cortical neurons recorded for several days. *Soc. Neurosci. Abstr.*, 13: 508.

Chapin, J.K., Markowitz, R.A., Moxon, K.A. and Nicolelis, M.A.L. (1999) Direct real-time control of a robot arm using signals derived from neuronal population recordings in motor cortex. *Nat. Neurosci.*, 2: 664–670.

Dormont, J.F., Schmied, A. and Condé, H. (1982) Motor command in the ventrolateral thalamic nucleus: neuronal variability can be overcome by ensemble average. *Exp. Brain Res.*, 48: 315–322.

Evarts, E.V. (1968) Relation of pyramidal tract activity to force exerted during voluntary movement. *J. Neurophysiol.*, 31: 14–27.

Georgopoulos, A.P., Kettner, R.E. and Schwartz, A.B. (1986) Neuronal population coding of movement direction. *Science*, 233: 1416–1419.

Humphrey, D.R., Schmidt, E.M. and Thompson, W.D. (1970) Predicting measures of motor performance from multiple cortical spike trains. *Science*, 170: 758–762.

Kakei, S., Hoffman, D.S. and Strick, P.L. (1999) Muscle and movement representations in the primary motor cortex. *Science*, 285: 2136–2169.

Kubie, J.L. (1984) A driveable bundle of microwires for collecting single-unit data from freely moving rats. *Physiol. Behav.*, 32: 115–118.

Nicolelis, M.A.L., Lin, C.-S., Woodward, D.J. and Chapin, J.K. (1993a) Peripheral block of ascending cutaneous information induces immediate spatio-temporal changes in thalamic networks. *Nature*, 361: 533–536.

Nicolelis, M.A.L., Lin, C.-S., Woodward, D.J. and Chapin, J.K. (1993b) Distributed processing of somatic information by networks of thalamic cells induces time-dependent shifts of their receptive fields. *Proc. Natl. Acad. Sci. USA*, 90: 2212–2216.

Nicolelis, M.A.L., Baccala, L.A., Lin, R.C.S. and Chapin, J.K. (1995) Synchronous neuronal ensemble activity at multiple levels of the rat somatosensory system anticipates onset and frequency of tactile exploratory movements. *Science*, 268: 1353–1358.

Schmidt, E.M. (1980) Single neuron recording from motor cortex as a possible source of signals for control of external devices. *Ann. Biomed. Eng.*, 8: 339–349.

Schmidt, E.M., Jost, R.G. and Davis, K.K. (1975) Reexamination of the force relationship of cortical cell discharge patterns with conditioned wrist movements. *Brain Res.*, 83: 213–223.

Scott, S.H. and Kalaska, J.F. (1995) Changes in motor cortex activity during reaching movements with similar hand paths but different arm postures. *J. Neurophysiol.*, 73: 2563–2567.

Shin, H.-C. and Chapin, J.K. (1990a) Modulation of afferent transmission to single neurons in the ventroposterior thalamus during movement in rats. *Neurosci. Lett.*, 108: 116–120.

Shin, H.-C. and Chapin, J.K. (1990b) Movement-induced modulation of afferent transmission to single neurons in the ventroposterior thalamus and somatosensory cortex in rat. *Exp. Brain Res.*, 81: 515–522.

Wolpaw, J.R., Ramoser, H., McFarland, D.J. and Pfurtscheller, G. (1998) EEG-based communication: improved accuracy by response verification. *IEEE Trans. Rehabil. Eng.*, 6: 326–333.

CHAPTER 11

Nerve cuffs for nerve repair and regeneration

Joaquín Andrés Hoffer [1,2,*] and Klaus Kallesøe [2]

[1] School of Kinesiology, Simon Fraser University, 8888 University Drive, Burnaby, BC V5A 1S6, Canada
[2] NeuroStream Technologies, Inc., Burnaby, BC, Canada

Introduction

In the nerve regeneration field, a 'nerve cuff' is commonly understood to mean a tubular conduit that is surgically installed around the proximal and distal stumps of a severed nerve. Such tubulization cuffs are commonly used to contain the regenerating axonal sprouts and orient and guide their elongation toward the distal stump (Fig. 1, 'T-cuff'). In addition to providing these mechanical functions, three other types of nerve cuffs are available. Nerve modulation cuffs provide chemical reservoirs or fluid delivery chambers used to modulate nerve functions with pharmacological agents (Fig. 1, 'M-cuff'). Nerve-stimulating cuffs contain electrodes that can selectively stimulate the enclosed axons (Fig. 1, 'S-cuff') and nerve-recording cuffs contain electrodes that can monitor the electrical signals that travel along the axons (Fig. 1, 'R-cuff').

Vigorous research is currently aimed at discovering and validating clinically effective methods to assist nerve repair, support axonal regeneration and minimize formation of painful neuroma. The four types of nerve cuffs mentioned above make up a family of powerful tools that may be used either singly or in various combinations in research and in clinical applications. We provide here an overview of properties, advantages, limitations and practical guidelines for using each type of nerve cuff, list examples of demonstrated uses, and suggest possible future uses in nerve regeneration studies and therapies.

Nerves are delicate structures that in the course of normal movements must bend and stretch considerably, and can be severely damaged by compressive forces or disruptions to their nutrient supply. Any surgical intervention, and in particular the implantation of a device in the proximity of a nerve, carries a risk of causing serious damage to the nerve. This risk, however, can be greatly minimized if the biomechanical properties of the nerve and surrounding tissues are well understood, the site for implant is correctly selected and the device to be implanted is carefully designed to match the requirements of the site and of the functions to be performed.

Electrical or chemical nerve cuff interfaces require three components: an insulating wall made of thin, flexible, biocompatible material, internal electrodes (or fluid ports) connected to insulated lead wires (or catheters), and a safe and reliable method for cuff installation. When the cuff, electrode, catheter and lead designs are correct for the anatomical site and appropriate surgical procedures are used, nerve cuffs can be the most reliable and stable of all available means for establishing permanent connections with peripheral nerves, cranial nerves or spinal roots (reviewed by: Hoffer and Loeb, 1980; Hoffer, 1990; Hoffer and Kallesøe, 2000).

---

* Corresponding author: Dr. J.A. Hoffer, School of Kinesiology, Faculty of Applied Sciences, Simon Fraser University, 8888 University Drive, Burnaby, BC V5A 1S6, Canada. Fax: +1 604 291-3040; E-mail: hoffer@sfu.ca

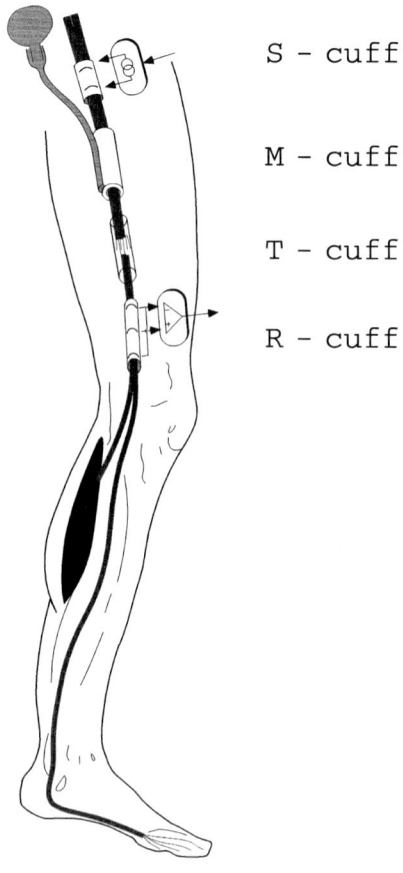

Fig. 1. Schematic representation of four types of nerve cuffs that can be used singly or in various combinations to assist nerve regeneration and to evaluate the return of function. *T-cuff* = tubulization cuff implanted around the nerve repair region is used to contain and orient regenerating axonal sprouts and to control neuroma formation. *M-cuff* = nerve modulation cuff is used as chemical reservoir to modulate nerve functions with pharmacological agents delivered through catheters attached to implanted pumps (shown above) or to external ports (not shown). The functions of T- and M-cuffs are often combined in a single cuff. *S-cuff* = nerve stimulation cuff contains electrodes used for electrical stimulation of regenerating axons for therapeutic or diagnostic purposes. *R-cuff* = nerve recording cuff contains electrodes appropriate to monitor action potentials traveling along the axons in order to track progress of the repair process.

Objectives for the treatment of injured nerves

The process and stages of peripheral nerve degeneration and regeneration after nerve injuries are becoming increasingly well understood and have been reviewed in detail (e.g., Bowe et al., 1989; Frykman, 1993; Dagum, 1998; Thanos et al., 1998; Diao and Vannuyen, 2000). Nerve axons, if transected, quickly degenerate distal to the injury. The proximal stump of transected axons will generally die back to the next node of Ranvier and, after a latent period of variable duration, will sprout growth terminals that regenerate vigorously and have potential for forming new connections. Ideally, the growth processes will find the distal nerve stump and proceed to grow down the path left by the degenerated distal stumps, all the way to appropriate endings that are then reinnervated. Two mechanisms are thought to guide this process: mechanical guidance and chemical guidance (Frykman, 1993). However, the axonal growth processes can easily lose their way and too often end up growing in inappropriate directions, never to find a suitable end-organ. In such cases, axons may send out many more processes until one makes a successful connection, at which time all the processes that did not make suitable connections are withdrawn.

Unfortunately, other processes also occur at the same time, such as fibrous proliferation that interferes with axonal regeneration. Lingering sprouts of unsuccessfully regenerated axons can form painful neuromas, a serious secondary complication after nerve injury. It has thus been long understood that successful nerve repair and regeneration often require surgical assistance and may benefit from administration of key growth-supporting substances (Frykman, 1993; Dagum, 1998).

The need for treatment can depend on the type, severity and location of the injury. Nerve injuries were classified by Sunderland (1978) into five degrees of severity. First-degree injuries cause local conduction block but spontaneous recovery can be expected because the axons remain intact, even though there may be segmental demyelination. In contrast, if a nerve is completely transected (fifth degree), surgical intervention is almost always necessary. Surgery may also be necessary after injuries of intermediate severity, e.g., after fourth-degree traction injuries if spontaneous axonal regeneration is impeded by ensuing fibrosis (Dagum, 1998).

Successful nerve regeneration requires four conditions: the central neuron has to survive, axons must sprout and grow into a supportive environment, the regenerated axons must make appropriate distal connections, and the central nervous system must appropriately integrate the new peripheral signals (Frykman, 1993). Considerable advances have been achieved in microsurgical nerve-repair methods that assist nerve regeneration and it is now generally agreed that further improvement must await progress in other fronts (Harris and Tindall, 1991; Dagum, 1998). We review in the following sections some recent research in which alternative methods for bridging large gaps in lacerated nerves were tested, substances that may enhance regeneration were evaluated, or electrical stimulation to direct, accelerate or enhance regeneration was used.

Each damaged axon must regenerate a new distal process to span the entire distance from the site of injury to a suitable distal end-organ. After a start-up period of variable duration (Al-Majed et al., 2000), axonal elongation typically proceeds at 1–3 mm/day. Thus, the time until functional connections are fully restored may be as long as 2–3 years after a thigh-level injury to a human sciatic nerve. It is important to find ways to speed up the regenerative process, because denervated muscles will not wait forever. Muscles undergo not only disuse atrophy, which is often reversible, but also progressive fibrosis, which is not. The extent of functional recovery will depend on how soon, how fast and how many regenerating axons find new targets. If an insufficient number of axons make suitable connections, fine motor control and sensory discrimination will not be restored (Sunderland, 1978).

Neuromas-in-continuity, often formed in incompletely transected nerves or as a consequence of partial nerve regeneration (Dagum, 1998), contain scar tissue and tangled axonal sprouts that could not establish connections with appropriate end-organs and can be exceedingly painful, to the point of being disabling. Progressive growth of a neuroma-in-continuity can cause compression neuropathy to the rest of the nerve, shown by declining compound action potential (CAP) amplitudes measured from repaired nerves over time (Smahel et al., 1993). Prevention of neuroma formation remains a focus of current research on effective methods to assist the nerve repair process.

The four types of nerve cuffs shown in Fig. 1 are available to assist in nerve repair and for monitoring the results of experimental therapies during the regenerative process after nerve lesions. The properties of nerve cuffs will be reviewed in the next section and then separately for each type of nerve cuff.

General properties of nerve cuffs

Nerve cuffs should generally be implanted away from joints, in locations where the nerve is not branching or receiving local blood supply and can be safely freed from its surrounding tissues. The shape, size, thickness, orientation and flexibility of a nerve cuff and its associated cables or catheters must be carefully matched to the anatomical site. It is important to stress that inappropriate decisions on site selection, surgical technique or cuff or lead design are almost guaranteed to cause some form of nerve damage (e.g., Larsen et al., 1998). On the other hand, if the cuff is correctly designed, dimensioned and installed, the long-term prognosis for the nerve can be excellent (e.g., Haugland and Sinkjær, 1995; Struijk et al., 1999).

Immediately following a surgical intervention, fibrous tissue quickly envelopes and penetrates cavities in any foreign objects. As a nerve cuff becomes ensheathed, the mechanical association between cuff and nerve is increasingly stabilized. The layer of fibrous tissue that is formed around the nerve replaces fluid inside the cuff and this causes increases in the electrical impedance inside the cuff and, desirably, in the recorded nerve signal amplitudes (Stein et al., 1978). On the other hand, excessive scar tissue formation outside the cuff is undesirable, as it could form adhesions to surrounding structures and reduce nerve mobilization. For this reason, the external surface of a cuff should be as smooth as possible.

To prevent compression neuropathy associated with post-surgical edema, the cuff lumen must accommodate some nerve swelling (Ducker and Hayes, 1968). Nerve compression, when it occurs, affects most severely the largest, most superficially located axons (Sunderland, 1978; Crouch, 1997; Crouch et al., 1997). Even mild compression can cause a noticeable reduction in axonal conduction velocities, while more severe compression can cause conduction block and even complete anterograde degeneration of affected axons.

Tubulization cuffs

Many experimenters have confirmed the classic findings of Ducker and Hayes (1968) and Lundborg (1988) who showed in animal experiments that a tube placed around a nerve juncture site helps orient the growing axonal processes and can improve the results of nerve repair. Tubulization can be beneficial when the ends of a severed nerve can be cut cleanly, approximated and sutured together without causing tension (Fig. 2A). Tubulization can also support regeneration and functional repair when the ends of a damaged nerve cannot be joined and the regenerating axons must cross a gap of up to several mm (Fig. 2B). Considerable research has been focused on (1) testing a great variety of implantable tube materials, both absorbable and non-absorbable, (2) on the regenerative ability of axons across a range of gap lengths, and (3) on the effects of tube wall thickness, cross-sectional area, length, shape and internal architecture (Ducker and Hayes, 1968; Stensaas et al., 1987, 1988; Terenghi, 1995; Danielsen, 1996; Hansson et al., 1997; Heath and Rutkowski, 1998; Stanec and Stanec, 1998; Hadlock et al., 1999).

When a nerve is injured, the body's normal tissue repair mechanisms, fibrocyte migration and accumulation of axonally transported materials all contribute to the considerable swelling and edema that typically ensue. When a tube is used to contain and orient the sprouting terminals, the consensus has been that the tube lumen must be larger than the nerve cross-sectional area by about a factor of 2.5, though not too much larger; otherwise nerve compression will occur (Ducker and Hayes, 1968). Smahel et al. (1993) found that initially 'mobile' cuffs that provided 'a snug fit' gave apparently good results, as assessed from the whole-nerve CAPs that were recorded with electrodes placed distal to the repair zone in acute experiments 6 weeks after the nerve repair procedure. However, the CAPs that Smahel et al. (1993) measured in other animal cohorts 10 weeks or 4 months after nerve repair had lower amplitudes, indicating that proliferation of axonal terminals and scar tissue formation inside the cuff had started to compress the axons that had managed to regenerate across the suture line and past the recording electrodes.

In addition to the characteristic swelling of cut nerves, some swelling should be expected to occur whenever any surgical procedure is done and, in particular, when uninjured portions of nerves need to be mobilized. Whenever a cuff is fitted around a freshly mobilized length of nerve, the post-surgical swelling can be accommodated with little or no compressive trauma to the nerve when longitudinally ridged, multi-chambered silicone rubber cuffs are used (Hoffer et al., 1998). In chambered cuffs, the inside wall of the cuff includes several longitudinal soft-edged ridges that project into the lumen and form parallel chambers between the nerve and the cuff wall (Fig. 3A). The chambers provide extra space for the nerve to swell into during the days that follow the surgery. Cross-sectional views of the sequential changes that occur after installing a chambered cuff around an intact nerve (e.g., cat sciatic nerve) are represented schematically in Fig. 3 for three time points: the day of implantation, 1 week later and 6–12 months later. After the nerve circumference is measured with a flexible ruler at the time of implantation, the cuff size is chosen to initially fit snugly around the nerve with the ridges just

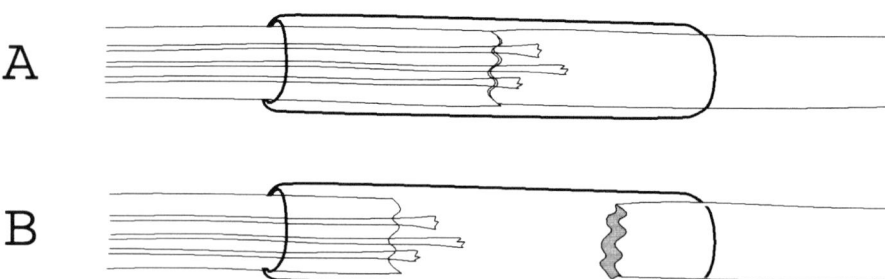

Fig. 2. (A) Tubulization cuff is often placed around the approximated ends of a severed nerve. (B) When the ends of a damaged nerve cannot be rejoined, a tubulization cuff can bridge the gap and help orient the growing axonal processes toward the distal nerve stump.

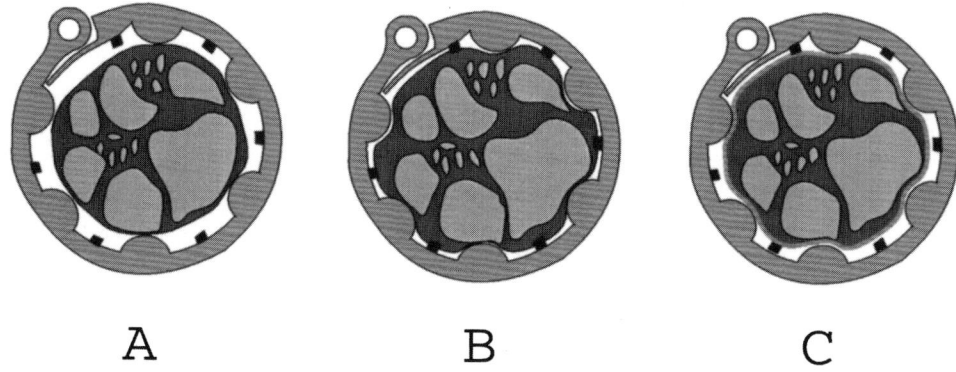

Fig. 3. Cross-sectional view of sequential changes that occur after installing a chambered cuff around a nerve. Chambered cuffs have longitudinal ridges that project into the lumen and form parallel chambers that provide extra space for the nerve to fill if edema causes inflammation as a result of surgery. (A) On the day of implantation, the cuff is sized to initially fit snugly. (B) One week later, edema has caused the nerve to swell into the chambers. (C) Months later, the nerve has returned to its original size and only a thin layer of connective tissue surrounds it.

touching the nerve (Fig. 3A). One week later, CAP and impedance measurements have suggested that the nerve has swollen and almost fills the chambered spaces between the ridges (Fig. 3B). Histological studies have shown that several months later the nerve has approximately regained its original size and a thin (about 100 μm; light gray) connective tissue sheath typically forms around the nerve inside the cuff. At the time of explantation, the chambered spaces are typically filled with clear fluid, not with connective tissue, and there are few or no adhesions found between the nerve and the cuff wall or electrodes that may be located inside the cuff (Fig. 3C).

Modulation cuffs

Numerous studies have investigated the possibility of enhancing nerve regeneration or otherwise improving the functional outcome of nerve repair through the application of specific neurotrophic or pharmacological substances to the nerve, e.g., ACTH and related peptides, gangliosides (Horowitz, 1989), nerve growth factor (Liuzzi and Tedeschi, 1991), arginine, S-adenosylmethionine, polyamines (Cestaro, 1994), leukemia inhibitory factor (Tham et al., 1997), thyroid hormones (Voinesco et al., 1998), and various neurotrophins (Yin et al., 1998), among others.

Implanted nerve cuffs can be used as reservoirs, sometimes called 'chambers', into which test substances are placed. The wall of the cuff can be 'doped' or the chamber space filled with a test substance at the time of implantation (e.g., Tham et al., 1997; Voinesco et al., 1998), or substances can be administered via a catheter terminating inside the cuff lumen. In the latter case, the other end of the catheter is either connected to an external access port, through which periodic injections can be made (e.g., Hoffer and Loeb, 1983; Loeb and Hoffer, 1985; Hoffer et al., 1990; Strange and Hoffer, 1999a), or to an implanted osmotic pump that provides continuous flow of a test substance (e.g., Betz et al., 1980; Barry and Ribchester, 1995; Hekimian et al., 1995). Nerve cuffs equipped with catheters can be placed either proximally to or encompassing the nerve repair region and used for focused delivery of pharmacological agents that may assist regeneration (Fig. 4A). Alternatively, cuffs can be used to infuse anesthetic substances, e.g., to selectively block the central transmission of action potentials along small-diameter pain fibers (Fig. 4B).

When an infusion cuff will be used during extended periods, care must be taken to maintain catheter patency. If the cuff catheter is connected to an external port, it is necessary to infuse once per day enough heparinized saline (1–5% heparin solution) to fill the catheter (Hoffer and Loeb, 1983; Hoffer et al., 1990). When the cuff is connected to an osmotic pump, the constant fluid flow is normally sufficient to prevent blockage.

Cuffs intended for fluid delivery provide a reser-

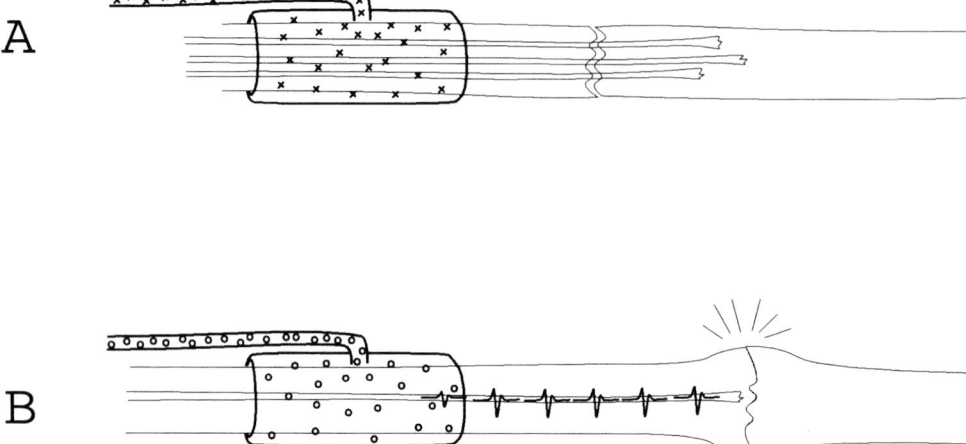

Fig. 4. Modulation cuffs are reservoirs equipped with catheters into which chemical substances can be delivered or from which fluid samples can be withdrawn. (A) Example of an application in which a nerve regeneration-promoting test substance is infused. (B) Example of an application in which an anesthetic substance is infused to selectively block the transmission of action potentials generated by pain fibers ending in a neuroma.

voir around the nerve for fresh solution to be contained. Tightly fitting rings at the ends of the cuff must be avoided, because they could cause nerve compression and because pressure could rise excessively during fluid infusion (Hoffer and Kallesøe, 2000). A multi-chambered wall structure (Fig. 3) provides additional reservoir volume while maintaining the alignment of the cuff around the nerve. Unless contact with neighboring nerves or other structures must be minimized, excess fluid may flow out the open ends of the cuff. Alternatively, additional catheters may be attached to the nerve cuff to drain excess fluid when fresh solution is administered (Hekimian et al., 1995). Drain catheters allow for transfer of outflow solution to a specific location, e.g., the abdominal cavity or an external sampling vial.

Stimulation cuffs

The application of electrical currents to enhance regeneration of damaged tissues remains an intriguing area for further research. In analogy to the well documented benefits from application of electrical currents to speed up healing of, e.g., skin or bone tissues, it is also possible to accelerate or enhance axonal growth with electrical treatment (e.g., Nix and Kopf, 1983). Progress in this area has been slow, partly because of practical difficulties in administering electrical currents to nerves safely and in well controlled ways but these hurdles are lessened by recent advances in the design, industrial fabrication and commercial availability of nerve S-cuffs (Hoffer and Kallesøe, 2000).

For some applications, the stimulation electrodes can be as simple as a pair of insulated wires with bared surfaces placed in the vicinity of the target tissues to be stimulated. However, S-cuffs offer several advantages that were reviewed in greater detail by Hoffer and Kallesøe (2000). First, when the stimulating electrodes are fixed inside a cuff, the placement of the electrodes with respect to the nerve is also stabilized and thus the electrodes cannot drift during movements or over time. Second, the cuff wall channels the flow of current into the target nerve, not through surrounding tissues, which can reduce the current requirement by an order of magnitude or more; this has desirable consequences on battery life. Third, unwanted stimulation of other nerves by current spillover is eliminated or markedly reduced. In addition, cuff stimulating electrodes provide more evenly distributed field gradients than, e.g., disk electrodes (Hurlbert and Tator, 1994).

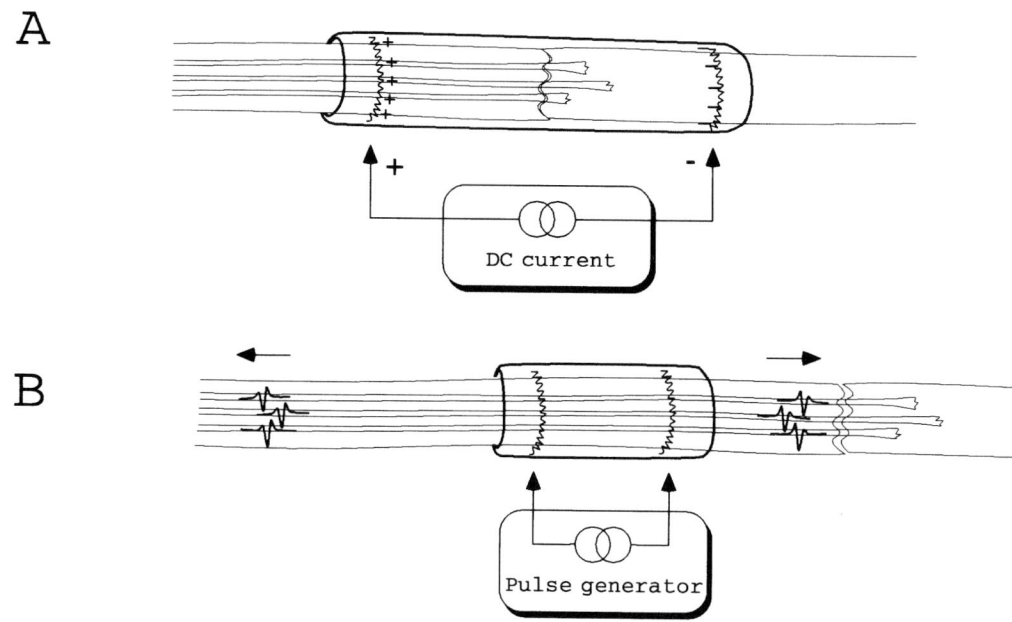

Fig. 5. (A) Stimulation cuffs can be used to apply weak direct current (DC) fields longitudinally along a nerve lesion. The S-cuff contains two electrodes. The anode (+) is placed proximal to the lesion site and the cathode (−) is placed distally so as to attract the regenerating axon sprouts. (B) Alternatively, S-cuffs can be used to apply trains of brief electrical stimuli that elicit nerve action potentials. Note that when an electrical stimulus is delivered, two action potentials travel in each stimulated axon, one in each direction.

Two different types of current delivery have been tested for effects on nerve regeneration: weak direct current (DC), and trains of brief current stimuli of sufficient amplitude to elicit action potentials. Beveridge and Politis (1988) showed that application of weak continuous electric current can assist the growth of axon projections across a neuroma-like lesion. Longitudinal 1.5 µA DC fields were applied by placing a nerve cuff electrode at each side of a crush site, with the anode cuff placed just proximal and the cathode cuff placed distal to the lesion site. The authors reported a fourfold increase in the number of myelinated fibers found distal to the neuroma after current application for up to 2 months. Zanakis (1990) also found significant increases in regenerating axon numbers after delivering DC currents as low as 1.4 µA to regenerating nerves for periods between 8 and 60 days, using a single nerve cuff that spanned the region of the lesion and contained both electrodes (Fig. 5A).

Patterned trains of electrical stimuli that elicit action potentials in motor axons can modify molecular and mechanical properties of motor neurons and their innervated motor units (Salmons and Vrbova, 1969). Gordon et al. (1999) used nerve cuff electrodes to stimulate regenerating motor axons in an investigation of roles of neuromuscular activity on the recruitment of motor units reinnervated after nerve crush. Al-Majed et al. (2000) found that 20 Hz trains of brief stimuli applied after rat femoral nerve section and repair had lasting effects on the speed with which axonal regeneration proceeded. Electrical stimulation of freshly cut nerves for periods as brief as 1 h dramatically reduced the period of axonal outgrowth that precedes axonal regeneration from 10 weeks to 3 weeks (Al-Majed et al., 2000).

In addition, electrical stimulation may have beneficial effects in maintaining the size and integrity of the proximal axon stumps and nerve cell bodies after nerve section. In nerves that were sectioned and ligated to prevent regeneration, it was shown that conduction velocity slows down and compound action potentials decline exponentially over the following months (Hoffer et al., 1979) and the proximal axon cross-sectional areas are progressively reduced (Gillespie and Stein, 1983). Significantly, the rate of decline is much more pronounced in sensory axons than in motor axons (Hoffer et al., 1979). This difference in rate of retrograde atrophy in sensory and motor axons may be related to the levels of residual activity present in the axons after nerve section, since the motor axons remain connected to their sources of nerve activity in the motoneuron cell bodies, whereas the severed sensory axons become disconnected from their sources of activity in the peripheral sensory receptors (Hoffer et al., 1979). Sensory receptors continue to remain viable and available for reinnervation for many years (Dagum, 1998), and severed sensory afferents ending in neuroma can survive and respond to electrical stimulation as long as 30 years after limb amputation (Stein et al., 1980). Therefore, electrical stimulation of the proximal stumps of transected sensory nerves may help reduce retrograde atrophy and enhance their regenerative capacity (Hoffer et al., 1979). To date this hypothesis remains untested.

Fig. 5B shows a method for application of trains of stimuli with a bipolar S-cuff placed proximally to a nerve lesion site. Note that when electrical stimuli are delivered to a nerve, action potentials will typically travel in both directions along stimulated axons. It remains to be determined whether the effects on axonal regeneration are produced by action potentials traveling distally, proximally or in both directions. Experimental tools for resolving this question are available; unidirectional action potentials can be elicited, for example, with asymmetric S-cuff electrodes (Sweeney and Mortimer, 1986).

Finally, electrical stimulation applied directly to denervated muscles may extend the window of time during which muscles can be successfully reinnervated by regenerating motor axons. Williams (1996a,b) showed that continuous electrical stimulation helps preserve muscle function in the absence of innervation.

Recording cuffs

Nerve-recording cuffs are designed to pick up very small-amplitude extracellular potentials generated by axons. Details are provided elsewhere (Hoffer, 1990; Hoffer and Kallesøe, 2000) and are only summarized here. R-cuffs are a particularly useful tool to monitor the time-dependent changes in axonal conductivity and reestablishment of peripheral connectivity after a nerve crush or nerve section and repair (Davis et al., 1978; Gordon et al., 1980).

R-cuffs typically contain at least three electrodes in a balanced tripolar configuration, with the recording electrode at the center of the cuff symmetrically placed between tied indifferent electrodes near the cuff ends to reduce pickup of unwanted signals produced by sources outside the cuff. An essential property is that the insulating wall be completely sealed along its length. If the wall is leaky, the small nerve-generated currents will be shunted out and much larger signals generated by muscles outside the cuff will enter and contaminate the recordings. This requirement poses practical difficulties, because R-cuffs must be simple to open, safe to install, and must remain securely and permanently sealed after installation around the nerve, more so than any other cuffs.

Recording cuffs were originally developed in the 1970s using longitudinally slit silicone tubing that was sealed by encircling the cuff with several sutures (Hoffer, 1990). The sutures could be used to pull the cuff open to load the nerve and then tied together to keep the cuff closed. However, sutures or knots often yielded and tongues of connective tissue grew into the cuff through the longitudinal opening, which degraded the recorded signals. The design also required a fairly thick, stiff tubing wall that would not collapse when sutures were tightened.

An improved cuff closure method uses a 'piano-hinge' system (Fig. 6A) that provides a simple, secure seal and is also quickly reversible (Kallesøe et al., 1996; Hoffer and Kallesøe, 2000). Interlocking sets of small tubular elements are used to draw the cuff edges together. This system reduces the structural strength requirements of the cuff wall and allows it to be made of thin, flexible silicone sheeting that is much better matched to the mechanical properties of nerves. To load it, the cuff is positioned under the nerve and opened by pulling apart on two temporary sutures passed through the hinge elements at either edge of the cuff. The nerve is allowed to gently drop into the cuff without any need for gripping, holding or pushing the nerve. Once the nerve is inside the cuff, the temporary sutures are removed and a flexible closure suture made of monofilament nylon or equivalent material is passed through the

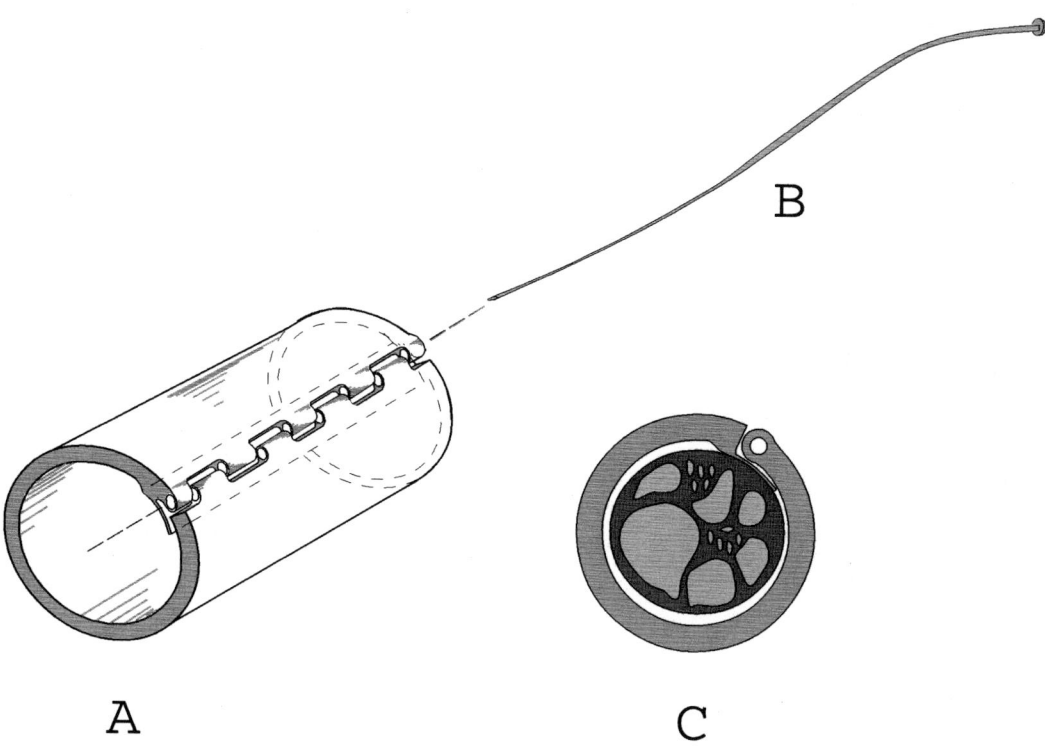

Fig. 6. (A) A simple, reversible nerve cuff-closing system that provides a patent seal consists of interlocking sets of small tubular elements (Kallesøe et al., 1996). (B) A flexible monofilament nylon suture is passed through the tubular elements, trimmed and sealed with focally applied heat. (C) An inside flap ensures that no connective tissue bridge can form across the longitudinal opening of the cuff. This feature provides the cuff wall with high insulative value which is especially required for R-cuffs.

interdigitated hinge elements (Fig. 6B). After insertion, the ends of the closure suture are trimmed and quickly and safely sealed with focally applied heat. This method ensures that no connective tissue can grow across the longitudinal slit (Fig. 6C). As well, the cuff has invariant lumen dimensions and therefore will not exert compressive forces on the nerve if correctly sized at the time of installation. The improved cuff wall and closing system recently developed for R-cuffs is also available for thin-walled T-cuffs, M-cuffs and S-cuffs.

The nerve signal amplitudes that can be recorded from intact peripheral nerves depend on many factors, most importantly axon sizes, numbers of active axons, cross-sectional area of the nerve and length of the R-cuff (Hoffer, 1990; Hoffer and Kallesøe, 2000). Electroneurographic (ENG) signal amplitudes recorded from mammalian limb nerves during normal movements typically range from about 5 µV (peak-to-peak), e.g., from the cat sciatic nerve using 4 mm I.D., 30-mm-long cuffs, to up to 90 µV (peak-to-peak) from fine nerves like the rabbit tenuissimus nerve using 0.3 mm I.D., 5-mm-long cuffs (Hoffer, 1990). With properly designed and implanted R-cuffs, such ENG signals can be recorded indefinitely (e.g., Gordon et al., 1980; Hoffer et al., 1981, 1990, 1996; Hoffer and Haugland, 1992; Sinkjær et al., 1993; Christensen, 1997; Strange and Hoffer, 1999b; Struijk et al., 1999).

Gordon et al. (1980) first used R-cuffs to follow the systematic changes in nerve activity that accompanied the regenerative process for several months after experimental nerve crush or nerve section and repair. Fig. 7 shows an example of a recording cuff configured for monitoring the return of sensory ENG activity that may be elicited by cutaneous stimulation and travel along regenerated axons. The R-cuff electrode and differential amplifier system

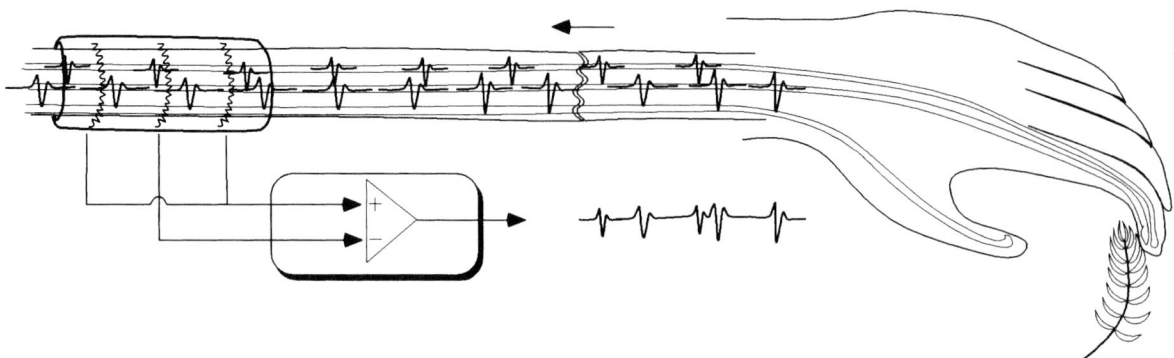

Fig. 7. Recording cuffs can be used to monitor the return of sensory nerve activity in response to cutaneous stimulation. The R-cuff electrode and differential amplifier system records multi-unit nerve activity from axons that have made suitable connections with peripheral receptors.

records an interference multi-unit ENG pattern that is the algebraic sum of the individual extracellular action potentials that travel in all axons that have made suitable connections with peripheral receptors in the area of skin that is being stimulated. The recorded ENG activity can be quantified on the basis of power spectrum analysis or other methods (Hoffer, 1990).

Additional information about any changes in the status of a nerve over time is also available from the CAPs that can be chronically recorded with implanted R-cuffs. A CAP is generated every time that the axons in the nerve are synchronously activated by a brief electrical stimulus delivered via electrodes in a second, S-cuff implanted some distance away on the same nerve (Fig. 8). This method was first used by Davis et al. (1978) and later by Krarup and Loeb (1988), Krarup et al. (1988, 1989) and Fugleholm et al. (1994) to follow in detail the time course of peripheral nerve degeneration and regeneration after crush lesions or after section and resuture in individual animals. Fig. 8A shows an implementation where both the S-cuff and R-cuff are placed proximally to the site of nerve lesion, as used by Davis et al. (1978). Fig. 8B shows an alternative implementation where the R-cuff is placed distally to the lesion. In either case it is possible to measure the peak-to-peak amplitude and latency to onset (or to first peak) of the CAPs recorded with the R-cuff each time the nerve is stimulated via the S-cuff and the systematic changes in these measured variables can be followed over weeks or months to monitor the regenerative process in a repaired nerve. Using pairs of implanted S-cuffs and R-cuffs, experiments such as done by Smahel et al. (1993), who tested the compressive effects of cuffs after 6 weeks, 10 weeks and 4 months on three groups of ten animals each, can be done using fewer animals, providing many more data points and with smaller dispersion (lower uncertainty), since each animal can serve as its own internal control during longitudinal studies.

Summary and discussion

Implanted nerve cuffs provide powerful methods for mechanical guidance, chemical delivery, electrical stimulation and continuous or periodic monitoring of the effects of therapeutic methods applied to injured nerves. In this review, T-cuffs, M-cuffs, S-cuffs and R-cuffs were described as separate devices (Fig. 1) and were generally treated as independent tools, but in practice their functions can be combined very effectively. Typical 'combined' cuffs include: T- and M-cuffs (e.g., Hekimian et al., 1995), T- and S-cuffs (e.g., Zanakis, 1990), R- and M-cuffs (Hoffer and Loeb, 1983; Loeb and Hoffer, 1985) and S- and M-cuffs (Strange and Hoffer, 1999a). R-cuffs can also be used to stimulate nerves. The only function that should usually not be incorporated into a single cuff is simultaneous stimulation and recording of CAPs because very large-amplitude stimulation artifacts would contaminate the recorded signal. It is preferable to implant separate S- and R-cuffs (as in examples in Fig. 8) and implant a separate grounding

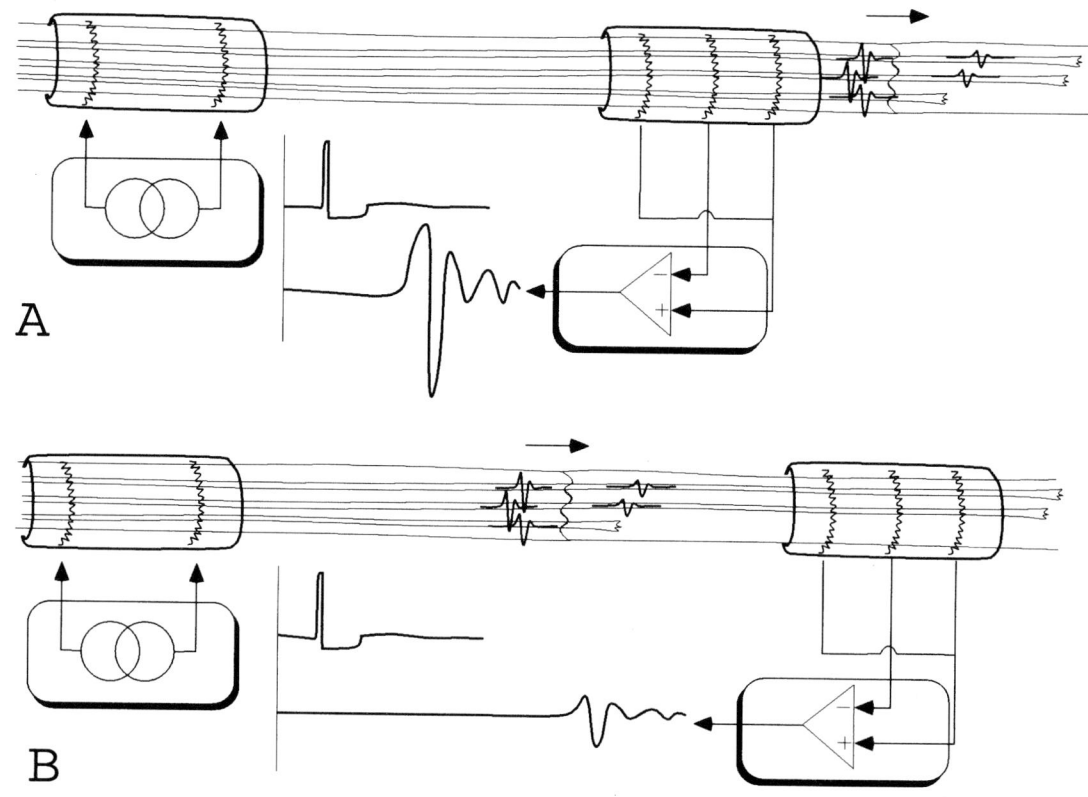

Fig. 8. (A) Implantation of two cuffs allows measuring CAP amplitudes and latencies in order to monitor progress in the nerve repair and regeneration process. Following each brief electrical stimulus (top trace) delivered to the nerve via the S-cuff, a CAP (bottom trace) is recorded by the R-cuff. The CAP sums all synchronously evoked action potentials in conducting nerve fibers. (B) Alternative implementation where the R-cuff is placed on the distal side of the lesion. In this location the recorded CAP can be expected to be of smaller amplitude and to occur later than when the configuration shown in (A) is used because all the nerve fibers may not have regenerated past the lesion site and the new distal processes may be smaller than their proximal parent axons.

electrode in between the two cuffs to further reduce stimulus artifact pickup (Hoffer, 1990).

A particularly powerful approach that uses nerve cuffs for studies of nerve regeneration involves implantation of a proximal S-cuff and a distal R-cuff in a first surgical procedure, followed by a second surgical procedure some days or weeks later when experimental crush or section of the nerve is performed at a location between the two cuffs, and a combined T- and M- or T- and S-cuff is implanted directly over the nerve repair site. This approach allows the experimenter to obtain control data (e.g., normal CAP amplitudes and latencies for the nerves studied) prior to the experimental procedure, followed by treatment of the nerve using test therapies that may include combinations of mechanical, chemical and/or electrical components. During the following days, weeks or months, the first implanted S-cuff and R-cuff can continue to be used to periodically monitor the extent and time course of nerve repair and return of function for each applied therapy.

The substantial body of nerve regeneration research data available to date from the use of the four types of nerve cuffs described here has been largely generated by pioneering researchers who fabricated the nerve cuff devices themselves. With the recent advent of commercial nerve cuffs, additional researchers are adopting their use, and future progress in the nerve regeneration field can be expected to reflect this increased use.

Acknowledgements

This study was funded by a National Institutes of Health research contract (NINDS-NO1-NS-6-2339, J.A. Hoffer, P.I.) and by NeuroStream Technologies, Inc. We thank Ms. W. Ng for valuable assistance with the preparation of this paper.

References

Al-Majed, A.A., Neumann, C.M., Brushart, T.M. and Gordon, T. (2000) Brief electrical stimulation promotes the speed and accuracy of motor axonal regeneration. *J. Neurosci.*, 20: 2602–2608.

Barry, J.A. and Ribchester, R.R. (1995) Persistent polyneuronal innervation in partially innervated rat muscle after reinnervation and recovery from prolonged nerve conduction block. *J. Neurosci.*, 15: 6327–6339.

Betz, W.J., Caldwell, J.H. and Ribchester, R.R. (1980) Sprouting of active nerve terminals in partially inactive muscles of the rat. *J. Physiol.*, 303: 281–297.

Beveridge, J.A. and Politis, M.J. (1988) Use of exogenous electric current in the treatment of delayed lesions in peripheral nerves. *Plast. Reconstr. Surg.*, 82: 573–579.

Bowe, C.M., Hildebrand, C., Kocsis, J.D. and Waxman, S.G. (1989) Morphological and physiological properties of neurons after long-term axonal regeneration: observations on chronic and delayed sequelae of peripheral nerve injury. *J. Neurol. Sci.*, 91: 259–292.

Cestaro, B. (1994) Effects of arginine, S-adenosylmethionine and polyamines on nerve regeneration. *Acta Neurol. Scand. Suppl.*, 154: 32–41.

Christensen, P.R. (1997) Sensory source identification from nerve recordings with multi-channel electrode arrays. M.A.Sc. Thesis, Simon Fraser University, Burnaby, BC.

Crouch, D. (1997) Morphometric analysis of neural tissue following the long-term implantation of nerve cuffs in the cat forelimb. M.Sc. Thesis, Simon Fraser University, ISBN 0612241130, Burnaby, BC.

Crouch, D., Strange, K.D. and Hoffer, J.A. (1997) Morphometric analysis of cat median nerves after long-term implantation of nerve cuff recording electrodes. Proc. International Functional Electrical Stimulation Society/V Neural Prostheses International Conference, Vancouver, BC, pp. 245–246.

Dagum, A.B. (1998) Peripheral nerve regeneration, repair, and grafting. *J. Hand Ther.*, 11: 111–117.

Danielsen, N. (1996) Nerve regeneration and repair. *Diabet. Med.*, 13: 677–678.

Davis, L.A., Gordon, T., Hoffer, J.A., Jhamandas, J. and Stein, R.B. (1978) Compound action potentials recorded from mammalian peripheral nerves following ligation or resuturing. *J. Physiol.*, 285: 543–559.

Diao, E. and Vannuyen, T. (2000) Techniques for primary nerve repair. *Hand Clin.*, 16: 53–66.

Ducker, T.B. and Hayes, G.J. (1968) Experimental improvements in the use of Silastic cuff for peripheral nerve repair. *J. Neurosurg.*, 28: 582–587.

Frykman, G.K. (1993) The quest for better recovery from peripheral nerve injury: current status of nerve regeneration research. *J. Hand Ther.*, 6: 83–88.

Fugleholm, K., Schmalbruch, H. and Krarup, C.J. (1994) Early peripheral nerve regeneration after crushing, sectioning, and freeze studied by implanted electrodes in the cat. *Neuroscience*, 14: 2659–2673.

Gillespie, M.J. and Stein, R.B. (1983) The relationship between axon diameter, myelin thickness and conduction velocity during atrophy of mammalian peripheral nerves. *Brain Res.*, 259: 41–56.

Gordon, T., Hoffer, J.A., Jhamandas, J. and Stein, R.B. (1980) Long-term effects of axotomy on neural activity during cat locomotion. *J. Physiol.*, 303: 243–263.

Gordon, T., Tyreman, N., Rafuse, V.F. and Munson, J.B. (1999) Limited plasticity of adult motor units conserves recruitment order and rate coding. *Prog. Brain Res.*, 123: 191–202.

Hadock, T.A. and Sundback, C. (1999) Multi-lumen polymeric guidance channel, method for promoting nerve regeneration and method of manufacturing a multi-lumen nerve guidance channel. United States Patent No. 5,925,053.[*]

Hansson, H.-A., Wells, M.R., Lynch, S.E., Antoniades, H.N., 1997. Device to promote drug-induced nerve regeneration. United States Patent No. 5,656,605.[*]

Harris, M.E. and Tindall, S.C. (1991) Techniques of peripheral nerve repair. *Neurosurg. Clin. N. Am.*, 2: 93–104.

Haugland, M.K. and Sinkjær, T. (1995) Cutaneous whole nerve recordings used for correction of footdrop in hemiplegic man. *IEEE Trans. Rehabil. Eng.*, 3: 307–317.

Heath, C.A. and Rutkowski, G.E. (1998) The development of bioartificial nerve grafts for peripheral-nerve regeneration. *Trends Biotechnol.*, 16: 163–168.

Hekimian, K.J., Seckel, B.R., Bryan, D.J., Wang, K.K., Chakalis, D.P. and Bailey, A. (1995) Continuous alteration of the internal milieu of a nerve-guide chamber using an osmotic pump and internal exhaust system. *J. Reconstr. Microsurg.*, 11: 93–98.

Hoffer, J.A. (1990) Techniques to record spinal cord, peripheral nerve and muscle activity in freely moving animals. In: Boulton, A.A., Baker, G.B. and Vanderwolf, C.H. (Eds.), *Neurophysiological Techniques: Applications to Neural Systems. Neuromethods, Vol. 15*. Humana Press, Clifton, NJ, pp. 55–145.

Hoffer, J.A. and Haugland, M.K. (1992) Signals from tactile sensors in glabrous skin: recording, processing, and applications for the restoration of motor functions in paralyzed humans. In: Stein, R.B., Peckham, P.H. and Popovic, D.P. (Eds.), *Neural Prostheses: Replacing Motor Function after Disease or Disability*. Oxford University Press, Oxford, pp. 99–125.

Hoffer, J.A. and Kallesøe, K. (2000) How to use nerve cuffs to stimulate, record or modulate neural activity. In: Chapin, J.K. and Moxon, C. (Eds.), *Neural Prostheses for Restoration of*

---

[*]United States patents can be accessed at http://patent.womplex.ibm.com

Sensory and Motor Function. CRC Press, Boca Raton, FL, pp. 139–175.

Hoffer, J.A. and Loeb, G.E. (1980) Implantable electrical and mechanical interfaces with nerve and muscle. *Ann. Biomed. Eng.*, 8: 351–360.

Hoffer, J.A. and Loeb, G.E. (1983) A technique for reversible fusimotor blockade during chronic recording from spindle afferents in walking cats. *Exp. Brain Res., Suppl.*, 7: 272–279.

Hoffer, J.A., Stein, R.B. and Gordon, T. (1979) Differential atrophy of sensory and motor fibers following section of cat peripheral nerves. *Brain Res.*, 178: 347–361.

Hoffer, J.A., Loeb, G.E. and Pratt, C.A. (1981) Single unit conduction velocities from averaged nerve cuff electrode records in freely moving cats. *J. Neurosci. Methods*, 4: 211–225.

Hoffer, J.A., Leonard, T.R., Cleland, C.L. and Sinkjær, T. (1990) Segmental reflex action in normal and decerebrate cats. *J. Neurophysiol.*, 64: 1611–1624.

Hoffer, J.A., Stein, R.B., Haugland, M.K., Sinkjær, T., Durfee, W.K., Schwartz, A.B., Loeb, G.E. and Kantor, C. (1996) Neural signals for command control and feedback in functional neuromuscular stimulation: a review. *J. Rehabil. Res. Dev.*, 33: 145–157.

Hoffer, J.A., Chen, Y., Strange, K. and Christensen, P.R. (1998) Nerve cuff having one or more isolated chambers. United States Patent No. 5,824,027.*

Horowitz, S.H. (1989) Therapeutic strategies in promoting peripheral nerve regeneration. *Muscle Nerve*, 12: 314–322.

Hurlbert, R.J. and Tator, C.H. (1994) Characterization of longitudinal field gradients from electrical stimulation in the normal and injured rodent spinal cord. *Neurosurgery*, 34: 471–483.

Kallesøe, K., Hoffer, J.A., Strange, K., Valenzuela, I., 1996. Implantable cuff having improved closure. United States Patent No. 5,487,756.*

Krarup, C. and Loeb, G.E. (1988) Conduction studies in peripheral cat nerve using implanted electrodes, I. Methods and findings in controls. *Muscle Nerve*, 11: 922–932.

Krarup, C., Loeb, G.E. and Pezeshkpour, G.H. (1988) Conduction studies in peripheral cat nerve using implanted electrodes, II. The effects of prolonged constriction on regeneration of crushed nerve fibers. *Muscle Nerve*, 11: 933–944.

Krarup, C., Loeb, G.E. and Pezeshkpour, G.H. (1989) Conduction studies in peripheral cat nerve using implanted electrodes, III. The effects of prolonged constriction on the distal nerve segment. *Muscle Nerve*, 12: 915–928.

Larsen, J.O., Thomsen, M., Haugland, M. and Sinkjær, T. (1998) Degeneration and regeneration in rabbit peripheral nerve with long-term nerve cuff electrode implant: a stereological study of myelinated and unmyelinated axons. *Acta Neuropathol.*, 96: 365–378.

Liuzzi, F.J. and Tedeschi, B. (1991) Peripheral nerve regeneration. *Neurosurg. Clin. N. Am.*, 2: 31–42.

Loeb, G.E. and Hoffer, J.A. (1985) Activity of spindle afferents from cat anterior thigh muscles, II. Effects of fusimotor blockade. *J. Neurophysiol.*, 54: 565–577.

Lundborg, G. (1988) *Nerve Injury and Repair.* Churchill Livingstone, New York.

Nix, W.A. and Kopf, H.C. (1983) Electrical stimulation of regenerating nerve and its effect on motor recovery. *Brain Res.*, 272: 21–25.

Salmons, S. and Vrbova, G. (1969) The influence of activity on some contractile characteristics of mammalian fast and slow muscles. *J. Physiol.*, 201: 535–549.

Sinkjær, T., Haugland, M.K. and Haase, J. (1993) Neural cuff electrode recordings as a replacement of lost sensory feedback in paraplegic patients. In: Bothe, H.-W., Samii, M. and Eckmiller, R. (Eds.), *Neurobionics.* Elsevier Science Publ., Amsterdam, pp. 267–277.

Smahel, J., Meyer, V.E. and Morgenthaler, W. (1993) Silicone cuffs for peripheral nerve repair: experimental findings. *J. Reconstr. Microsurg.*, 9: 293–297.

Stanec, S. and Stanec, Z. (1998) Reconstruction of upper-extremity peripheral-nerve injuries with ePTFE conduits. *J. Reconstr. Microsurg.*, 14: 227–232.

Stein, R.B., Charles, D., Gordon, T., Hoffer, J.A. and Jhamandas, J. (1978) Impedance properties of metal electrodes for chronic recording from mammalian nerves. *IEEE Trans. Biomed. Eng.*, 25: 532–537.

Stein, R.B., Charles, D., Hoffer, J.A., Arsenault, J., Davis, L.A., Moorman, S. and Moss, B. (1980) New approaches to controlling powered arm prostheses, particularly by high-level amputees. *Bull. Prosth. Res.*, 17: 51–62.

Stensaas, L.J., Todd, R.J. and Triolo, P.M. (1987) Prostheses and methods for promoting nerve regeneration. United States Patent No. 4,662,884.*

Stensaas, L.J., Todd, R.J. and Triolo, P.M. (1988) Prostheses and methods for promoting nerve regeneration and for inhibiting the formation of neuromas. United States Patent No. 4,778,467.*

Strange, K.D. and Hoffer, J.A. (1999a) Gait phase information provided by sensory nerve activity during walking: applicability as state controller feedback for FES. *IEEE Trans. Biomed. Eng.*, 46: 797–809.

Strange, K. and Hoffer, J.A. (1999b) Restoration of use of paralyzed limb muscles using sensory nerve signals for state control of FES-assisted walking. *IEEE Trans. Rehabil. Eng.*, 7: 289–300.

Struijk, J.J., Thomsen, M., Larsen, J.O. and Sinkjær, T. (1999) Cuff electrodes for long-term recording of natural sensory information *IEEE Eng. Med. Biol. Mag.*, 18: 91–98.

Sunderland, S. (1978) *Nerve and Nerve Injuries*, 2nd ed. Churchill Livingstone, New York.

Sweeney, J.D. and Mortimer, J.T. (1986) An asymmetric two electrode cuff for generation of unidirectionally propagated action potentials. *IEEE Trans. Biomed. Eng.*, 33: 541–549.

Terenghi, G. (1995) Peripheral nerve injury and regeneration. *Histol. Histopathol.*, 10: 709–718.

Tham, S., Dowsing, B., Finkelstein, D., Donato, R., Cheema, S.S., Bartlett, P.F. and Morrison, W.A. (1997) Leukemia inhibitory factor enhances the regeneration of transected rat sciatic nerve and the function of reinnervated muscle. *J. Neurosci. Res.*, 47: 208–215.

Thanos, P.K., Okajima, S. and Terzis, J.K. (1998) Ultrastructure and cellular biology of nerve regeneration. *J. Reconstr. Microsurg.*, 14: 423–436.

Voinesco, F., Glauser, L., Kraftsik, R. and Barakat-Walter, I. (1998) Local administration of thyroid hormones in silicone chamber increases regeneration of rat transected sciatic nerve. *Exp. Neurol.*, 150: 69–81.

Williams, H.B. (1996a) A clinical pilot study to assess functional return following continuous muscle stimulation after nerve injury and repair in the upper extremity using a completely implantable electrical system. *Microsurgery*, 17: 597–605.

Williams, H.B. (1996b) The value of continuous electrical muscle stimulation using a completely implantable system in the preservation of muscle function following motor nerve injury and repair: an experimental study. *Microsurgery*, 17: 589–596.

Yin, Q., Kemp, G.J. and Frostick, S.P. (1998) Neurotrophins, neurones and peripheral nerve regeneration. *J. Hand Surg.*, 23: 433–437.

Zanakis, M.F. (1990) Differential effects of various electrical parameters on peripheral and central nerve regeneration. *Acupunct. Electrother. Res.*, 15: 185–191.

CHAPTER 12

Cortical motor areas and their properties: implications for neuroprosthetics

Paul D. Cheney *, Jennifer Hill-Karrer, Abderraouf Belhaj-Saïf, Brian J. McKiernan, Michael C. Park and Joanne K. Marcario

Department of Molecular and Integrative Physiology and Mental Retardation and Human Development Research Center, University of Kansas Medical Center, 3901 Rainbow Boulevard, Kansas City, KS 66160-7336, USA

Introduction

Spinal cord injury is a major source of disability both in the United States and worldwide. While at present there is no treatment that will reverse the paralysis associated with spinal cord injury, as documented in this volume, several new approaches to treatment are showing considerable promise. One of these approaches is the use of signals recorded from populations of neurons to control neuroprosthetic devices. For the purposes of this paper, a neuroprosthetic device will be defined as the use of signals recorded from populations of neurons in the brain to control either a robotic limb or a biological limb through functional electrical stimulation (FES) of muscles. The basic goal is illustrated in Fig. 1, taken from Fetz (1999). As listed below, a number of neurobiological questions arise in relation to the development and implementation of such devices. The goal of this paper will be to review existing data relevant to these issues.

(1) What characteristics of movement-related neuronal signals might be best suited for neuroprosthetic control?

(2) What brain areas might provide the most optimal muscle representations and neuronal signals for controlling movements of a robotic arm or a paralyzed arm implanted with electrodes for FES?

(3) Are the neuronal signals and muscle maps in these areas fixed or adaptable? What conditions produce map plasticity?

(4) Do the properties of neurons change after spinal cord lesions and do they remain appropriate for controlling prosthetic devices?

(5) To what extent does the activity of neurons show adaptive learning that might be exploited for the purpose of obtaining optimal signals for prosthetic device control?

Not listed here are numerous technical challenges that will need to be overcome if such neuroprosthetic devices are to become a reality. These challenges include methods for long-term (decades) stable recording from populations of neurons, methods for telemetry of large numbers of channels to external devices for signal processing, technology for attaining complete subcutaneous implantation of multi-channel electrode arrays and processing electronics (e.g., automatic spike detection and separation, artificial neural networks), and implantable battery technology for powering the electronics and multiple channels of electrical stimuli for FES of many individual muscles. Still another problem that will need to be addressed in achieving optimal control of prosthetic limbs is the fact that the limbs

---

* Corresponding author: Dr. Paul D. Cheney, Mental Retardation and Human Development Research Center, University of Kansas Medical Center, 3901 Rainbow Boulevard, Kansas City, KS 66160, USA; Fax: +1-913-588-5677; E-mail: pcheney@kumc.edu

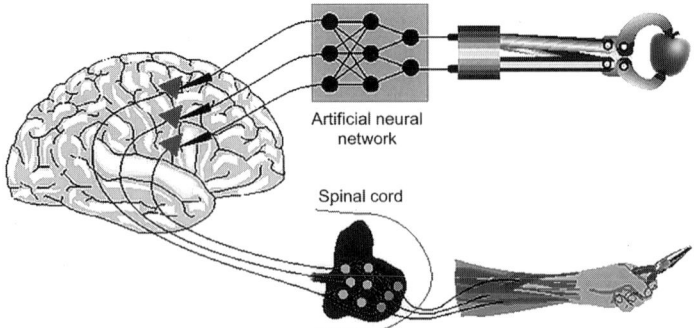

Fig. 1. Illustration of the concept of using the activity of neuronal populations recorded from motor cortex coupled with artificial neural networks to control robotic arms. (From Fetz, 1999.)

will encounter a wide range of unknown and varying loads. Although these are substantial hurdles, as technology continues to progress, it is reasonable to expect that solutions will be forthcoming.

Not many years ago this approach would have seemed like science fiction but now the reality of these devices seems within reach. For example, recent work by Chapin et al. (1999) has demonstrated that rats can learn to use multiple channels of neural data from motor cortex and motor thalamus to control movements of a single joint mechanical arm. While this was a relatively simple paradigm, the results, nevertheless, constitute a 'proof-of-concept' establishing the feasibility of this approach and its potential application to humans with spinal cord injury and other paralyzing conditions.

Characteristics of neuronal signals suited for prosthetic control

What properties would characterize the ideal neural structure for multi-channel recording and prosthetic control? This will depend, in part, on the complexity of the device to be controlled, which could range from a simple single joint robotic arm to control of a paralyzed arm and hand using FES of actual muscles. In the latter case, two components of forelimb motor control need to be distinguished. One component involves proximal muscles (shoulder and elbow muscles) which are used to position the hand at target locations in space. The second component of forelimb motor control is use of the digits once the hand has been positioned at the target object in space. These two components of forelimb motor control are dependent on each other and often overlap in their temporal patterns of activity. However, bringing the hand to a target and using the hand for manipulation are relatively independent operations. Consequently, neuronal signals that would be suitable for proximal control (bringing the hand to a target) cannot be easily used for distal control (grasp and manipulation). Therefore, one requirement of a forelimb neuroprosthetic control system would be separate neuronal population signals and corresponding artificial neuronal networks for control of the hand and the arm. This would apply to control of either a robotic arm and hand or control of a paralyzed arm and hand using FES of individual muscles. Given this requirement, the source of neuronal signals should be one where there is good separation of neurons representing distal and proximal muscles. The cortical area that provides the clearest separation of neurons supplying proximal and distal muscles is primary motor cortex.

Control of a paralyzed arm by FES of individual muscles is a more complicated problem than control of a robotic arm in the sense that more elements (individual muscles) require control, and the relationships between output signals, excitation of muscles and movement is more complex. However, the adaptability of the brain circuits and neuronal activity patterns provide encouragement that humans, in particular, might quickly learn relationships between neuronal activity and desired movement patterns, whether it be reaching with the arm or use of the digits. The nature of appropriate neuronal signals is another issue. Here, again, the logical assumption is that, regardless of the control system, neurons with the clearest and most reproducible relationship to movement will provide the most optimal signals for neuroprosthetic control.

What brain areas might be best suited as the source of signals for prosthetic control?

Given the presence of sufficient adaptive plasticity, it might be suggested that neurons in many areas of the brain, even areas not concerned with motor function, could potentially be trained or adapted for generating appropriate signals for controlling neuroprosthetic devices. This possibility is reinforced by some remarkable recent examples of adaptive plasticity in the brain. For example, Sadato et al. (1996) studied regional cerebral blood flow in a group of subjects blind since early infancy and a group of sighted volunteers. Not only did the blind subjects have a larger motor representation of the reading fingers, but Braille reading also activated primary and association visual areas. These results suggested that areas normally reserved for vision could be activated by other modalities, in this case, somatosensory input. The question remained as to whether activation of visual cortex in blind Braille readers was functionally important. This was tested by applying repetitive transcranial magnetic stimulation (TMS) to visual cortex during Braille reading. TMS applied to visual cortex was found to interfere with Braille reading (Sadato et al., 1998). Stimulation of other cortical areas did not interfere with Braille reading, suggesting the existence of a remarkable degree of developmental cross-modality plasticity.

Despite what appears to be a high level of ubiquitous adaptive plasticity throughout the brain (Hallet, 1999), neurons that are already part of motor system structures and used in the execution of movements should offer substantial advantages over non-movement-related neurons when presented with the task of adapting their activity for controlling prosthetic devices. Moreover, while deep brain motor system structures such as thalamus and basal ganglia might contain neurons with potentially useful control signals, the additional technical complications of placing large numbers of deep brain electrodes would probably be unwarranted. Therefore, cortical motor areas would seem to offer the optimal source of appropriate neuronal control signals.

The next question is which cortical motor areas might provide the best signals for control of prosthetic devices. Intuitively, primary motor cortex would seem to be the logical and obvious choice for obtaining signals to control prosthetic devices, but other possible targets are worthy of consideration. A new understanding of cortical motor areas has emerged in recent years. For example, the traditional view that primary motor cortex is the only significant source of corticospinal neurons for the control of movement has recently changed dramatically. Injection of retrograde tracers in the spinal cord of macaque monkeys has revealed that corticospinal neurons whose axons project to or near motoneuron pools in the spinal cord are not confined to primary motor cortex but, in fact, arise from multiple, separable regions within the frontal lobe of primates (Dum and Strick, 1991, 1996; He et al., 1993, 1995; Galea and Darian-Smith, 1994; Picard and Strick, 1996). Seven distinct forelimb motor representations within the frontal lobe have now been identified, including: primary M1 cortex; supplementary motor area (SMA); dorsal, ventral and rostral cingulate motor areas (CMAd, CMAv, and CMAr) on the medial aspect of the hemisphere; and dorsal and ventral premotor areas (PMd and PMv) on the lateral aspect of the hemisphere (Dum and Strick, 1991, 1996; He et al., 1993, 1995). Excluding primary motor cortex (M1) leaves six areas that are considered 'premotor' cortical areas because they project both to the spinal cord and to primary motor cortex (Dum and Strick, 1991). Support for these as distinct motor output representations of the forelimb is based not only on the distribution of corticospinal neurons, but also on cytoarchitectonics and patterns of projection to and from primary motor cortex, parietal cortex and thalamus (Gentilucci et al., 1988; Dum and Strick, 1991, 1996; Kurata, 1991; Luppino et al., 1991; He et al., 1993, 1995; Galea and Darian-Smith, 1994). Fig. 2 is taken from the work of Strick and colleagues (He et al., 1993; Dum and Strick, 1996) and summarizes their findings on the distribution of corticospinal neurons in the frontal lobe (about 70% of the total corticospinal projection). It is important to note that, unlike corticospinal neurons in primary somatosensory and parietal cortex, which terminate largely in the spinal cord dorsal horn (Fetz, 1969), frontal lobe corticospinal neurons make terminations in the ven-

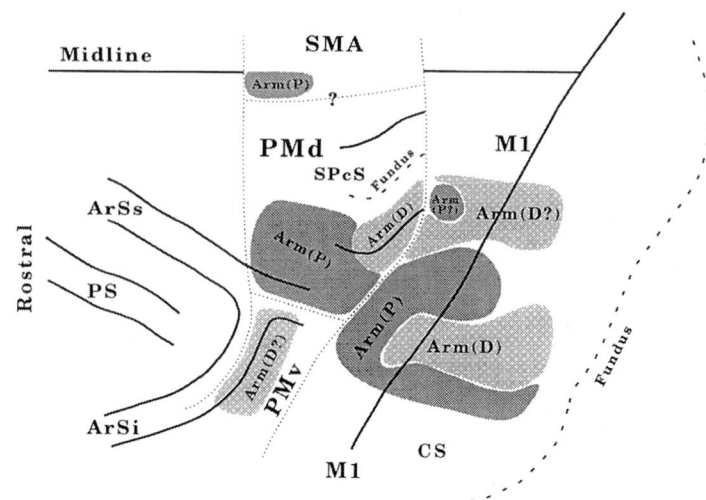

Fig. 2. (A) Origin of corticospinal projections from the motor areas on the medial wall of the hemisphere. This reconstruction of the frontal lobe of a macaque brain indicates the origin of corticospinal neurons (shaded regions) that project to the cervical segments of the spinal cord. In this view, the medial wall is unfolded and reflected upward to reveal the cingulate sulcus. The anterior bank of the central sulcus is also unfolded. A dashed line marks the fundus of each unfolded sulcus. The centers of the different cortical motor areas are designated by the circled letters. The boundaries between the motor areas and cytoarchitectonic areas (identified by numbers) are denoted with dotted lines. *ArGenu* (with arrow), level of the genu of the arcuate sulcus; *ArSi*, inferior limb of the arcuate sulcus; *ArSs*, superior limb of the arcuate sulcus; *CC*, corpus callosum; *CgG*, cingulate gyrus; *CgSd*, dorsal bank of the cingulate sulcus; *CgSv*, ventral bank of the cingulate sulcus; *CMAd*, cingulate motor area on the dorsal bank of the cingulate sulcus; *CMAr*, rostral cingulate motor area; *CMAv*, cingulate motor area on the ventral bank of the cingulate sulcus; *CS*, central sulcus; *M1*, primary motor cortex; *PMd*, dorsal premotor area; *PMv*, ventral premotor area; *PS*, principal sulcus; *SGm*, medial portion of the superior frontal gyrus; *SPcS*, superior precentral sulcus; *SMA*, supplementary motor area. (From Dum and Strick, 1996.) (B) Proximal and distal representation in the arm areas of the primary motor cortex and the premotor areas on the lateral surface of the hemisphere. This map is based on the peaks in the distribution of corticospinal neurons labeled following tracer injections into lower cervical and upper cervical segments. *M1*, primary motor cortex; *PMd*, dorsal premotor area; *PMv*, ventral premotor area; *SMA*, supplementary motor area; *ArSs*, superior arcuate sulcus; *ArSi*, inferior arcuate sulcus. (Modified from He et al., 1993; Dum and Strick, 1996.)

tral horn and intermediate zone of the spinal cord. In some cases the density of termination in the ventral horn from premotor areas is similar to that from primary motor cortex (Dum and Strick, 1996).

Finally, a rostral subdivision of SMA has also been described and termed pre-SMA or area F6 (Matsuzaka et al., 1992). Pre-SMA contains neurons that are predominately related to a sensory stimulus or the delay period between a stimulus and response, whereas neurons in SMA proper are more strongly related to aspects of movement execution. However, unlike the other premotor areas named above, pre-SMA lacks both a direct corticospinal projection and a projection to M1. Consequently, it does not meet the definition of a true premotor area. Evidence for these discrete corticospinal output zones is based in large part on exhaustive neuroanatomical labeling work (Dum and Strick, 1991, 1996; He et al., 1993, 1995; Galea and Darian-Smith, 1994), electrophysiological mapping studies (Preuss et al., 1996) and cytoarchitectonic studies (e.g., Matelli et al., 1991).

Output map of primary motor cortex

Note that the corticospinal map of M1 in the rhesus monkey (Fig. 2B) consists of a large core distal representation surrounded by a proximal muscle representation. In addition, the possibility of a large second representation of distal muscles is suggested. We have recently examined the output map of primary motor cortex in relation to 24 different forelimb muscles including shoulder, elbow, wrist, digit and intrinsic hand muscles. Stimulus-triggered averaging was used to compute averages of rectified electromyographic (EMG) activity while the monkey performed a reach and prehension task (Fig. 3). The advantage of stimulus-triggered averaging is that effects in individual muscles can be rigorously quantified and low rates of stimulation (20 Hz or less) avoid physiological spread of the stimulus by temporal summation. Both excitatory and inhibitory effects can be detected and quantified with this method. Maps derived from analysis of poststimulus effects show a core distal muscle representation located largely in the bank of the precentral gyrus and extending to the area 3a border (Fig. 4). The distal representation is surrounded by a 'horseshoe'-shaped proximal muscle representation matching the distribution of corticospinal neurons reported by He et al. (1993). A similar organization of distal and proximal muscle representations has been reported by Strick and Preston (1982), Nudo et al. (1992, 1996a,b) and Nudo and Milliken (1996) in the squirrel monkey. In the macaque monkey, Kwan et al. (1978) described a concentric organization consisting of a central core of digit representation surrounded by concentric circular zones of increasing diameter for the wrist, elbow and shoulder. Our results differ from this description in that we do not see clear separation of zones for distal and wrist muscles within the distal representation or zones for elbow and shoulder muscles within the proximal muscle representation. In addition, a substantial zone producing effects in both proximal and distal muscles is present in some parts of the forelimb M1 representation. We have found similar maps of proximal and distal muscle representation in four rhesus monkeys and conclude that these features represent a highly consistent aspect of intra-areal forelimb somatotopic organization. Maps of individual muscles within the distal core representation show extensive overlap and in some cases non-contiguous 'islands' of representation. In more recent studies using stimulus-triggered averaging of activity from 24 forelimb muscles in the macaque monkey (Park et al., 1999), we were unable to find evidence for a second, medially placed distal representation suggested by the anatomical maps of He et al. (1993).

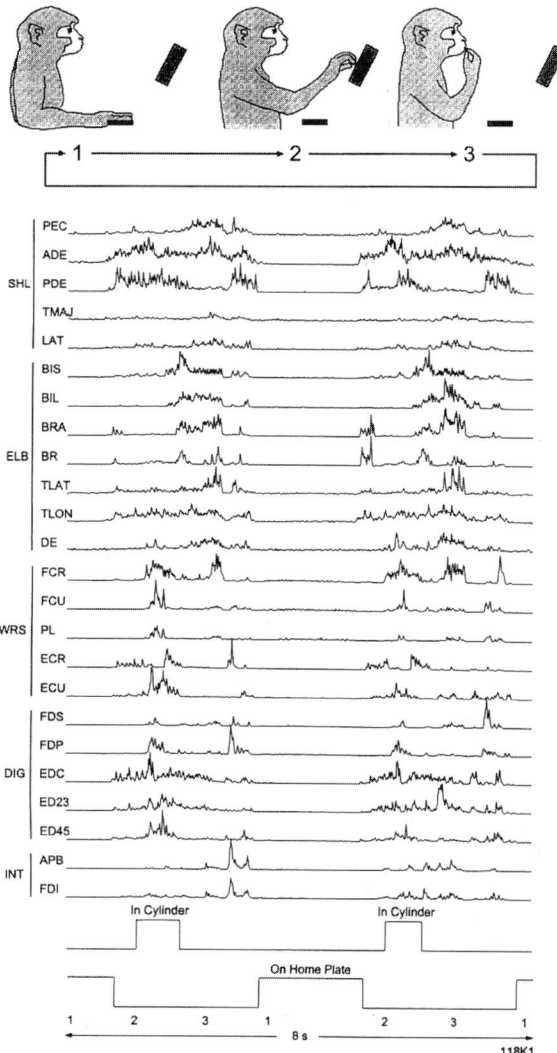

Fig. 3. Top: illustration of reach and prehension task. The monkey begins from a home plate at waist height, reaches for a food pellet delivered to a well, takes the food pellet to its mouth and then returns to home plate. Entry into the food well is detected by infrared beams. Contact with home plate is detected by a switch closure (lower records). Single trial records show changes in rectified and filtered EMG activity for all 24 forelimb muscles during performance of the reach and prehension task. Shoulder, elbow, wrist, digit and intrinsic hand muscles were recorded. Although the records for the intrinsic hand muscles look very similar, visual examination of the expanded records confirmed the lack of temporal synchronization in peaks. (From McKiernan et al., 1998.)

The properties of different premotor areas in comparison to primary motor cortex are summarized in Table 1. Galea and Darian-Smith (1994) measured

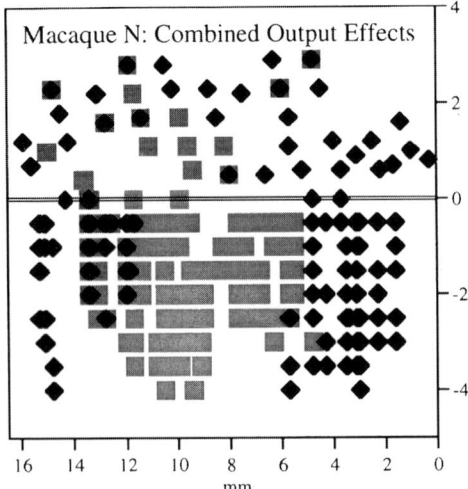

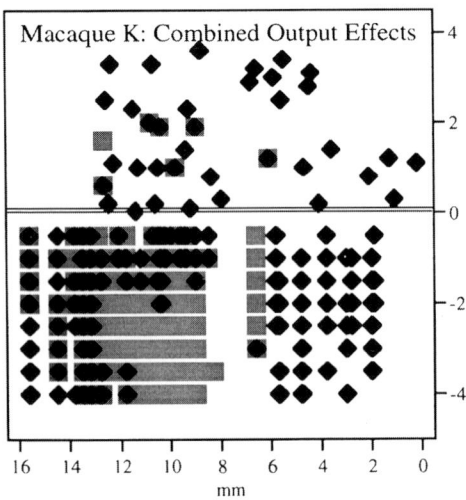

Fig. 4. Maps of primary motor cortex in two rhesus macaques based on effects in stimulus-triggered averages of EMG activity recorded simultaneously in 22–24 forelimb muscles. The precentral gyrus has been unfolded to represent the maps in two dimensions. Black squares represent effects in proximal muscles (shoulder and elbow muscles). Gray squares represent effects in distal muscles (wrist, digit and intrinsic hand muscles). Superimposed diamonds on squares represent sites where both distal and proximal effects were obtained. Note that in both macaques there is a heavy area of overlap between distal and proximal effects, particularly along the lateral segment of the proximal representation. Maps are based on both excitatory and inhibitory events. The horizontal line represents unfolding of M1 at the rostral convexity of the central sulcus. The area 3a/4 border corresponds to the bottom of each map. (From work of Hill-Karrer et al., 1995.)

the number of corticospinal neurons in different corticospinal output zones of the frontal lobe. They estimated that 70% of the total contralateral corticospinal projection originates from frontal and cingulate cortex; and of these neurons, 58% originate from primary motor cortex, 22% from SMA, 14% from cingulate motor areas and 4% from the arcuate premotor areas on the lateral surface of the hemisphere. These numbers are in general agreement with those of Strick and colleagues (Table 1). Dum and Strick (1991) reported that the total number of corticospinal neurons in the arm representation of the premotor areas (all frontal lobe areas except M1) equals or exceeds the total number in the arm representation of primary motor cortex. Also, the premotor cortical areas collectively constitute more than 60% of the frontal lobe that projects to the spinal cord. He et al. (1995) reported that the projections from CMAd and CMAv were as topographically organized as those from primary motor cortex. CMAr did not show a clear somatotopic organization, but this may have been due to the relative paucity of hindlimb corticospinal neurons in CMAr. The premotor areas also differed in the distribution of their projections to motoneurons of distal and proximal forelimb muscles (Table 1). Projections to distal muscles were identified based on injections of horseradish peroxidase (HRP) into upper cervical spinal cord segments (C2–C4). Proximal muscles were identified by injections of HRP into lower cervical segments (C7–T1). Hindlimb corticospinal neurons were identified by HRP injections into lumbar segments (L6–S1).

Mapping studies of premotor areas

To date, there have been relatively few mapping studies of premotor areas that have used electrical stimulation methods to test functional output. The original surface stimulation study by Woolsey et al. (1952) provided the first evidence for a supplementary motor area, but this study was limited by the resolution of the stimulation method. However, in more recent years, intracranial microstimulation (ICMS) has provided a higher-resolution method for functional output mapping and this approach has been widely applied in different forms to primary motor cortex (Asanuma and Rosen, 1972; Cheney, 1996). A limited number of studies have also used ICMS to map premotor ar-

TABLE 1

Properties of cortical premotor areas in relation to primary motor cortex (M1)

Cortical area	MI (F1)	SMA (F3, 6aβ)	CMAd (6c, 24d)	CMAv (23c, 24d)	CMAr (24c)	PMd (F2, 6aα)	PMv (F4–F5)
Total number of CS neurons:							
forelimb distal (LC[a])	15,918	5229	4629	2564	2186	6081	339
forelimb prox. (UC[a])	10,398	4970	1852	2310	2484	7214	2304
hindlimb (L6-S1)	23,870	5796	3740	2550	388	5164	6
% high density bins (cervical, upper 10%)	53	19	15	7	<1	5	<1
Cortical area occupied by forelimb CS neurons (mm^2)	84	44	22	14	24	45	18
% of total frontal lobe CS projection (frontal lobe = 70% of total)	46	15	9	7	4	17	2
Density of CS neurons (mean, cells/mm^2)	288	275	279	290	176	254	167
Electrical excitability (estimated average ICMS threshold, μA)	10	20	30	30	35	60	40
Functional activity:							
move execution	+++++	+++	−	++++	++	+++	+++
set related	++	+++	−	++	++++	++++	++
signal related	+	++	−	−	−	+	+++
Special functional role	move execution	♦ self-initiated ♦ selection ♦ sequence ♦ bilat. move	movement sequence from memory	−	reward based motor selection	visually guided reaching	visual grasp (F5)
Directional tuning	Y	Y	−	−	−	Y	−

Data based on Wiesendanger (1986), Mitz and Wise (1987), Gentilucci et al. (1988), Dum and Strick (1991), Luppino et al. (1991), He et al. (1993, 1995), Galea and Darian-Smith (1994), Gallese et al. (1996), Passingham (1996), Picard and Strick (1996), Preuss et al. (1996), Tanji and Shima (1996), Wise (1996a,b), Cadoret and Smith (1997), Murata et al. (1997), Wise et al. (1997), Shima and Tanji (1998), Kazenikov et al. (1999).
PMv = APA, PMd = SPcS, CMAv = CMAc.
− = not tested, CS = corticospinal.
[a] LC = lower cervical, UC = upper cervical.

eas. In all of these studies except one, mapping was based on the use of repetitive ICMS (brief trains of high-frequency stimuli) to evoke movements of different body parts. While this approach is useful for mapping broad features of somatotopic organization, it cannot be easily quantified and lacks specificity at the level of individual muscles provided by stimulus-triggered averaging of EMG activity (Cheney et al., 1991). Nevertheless, much has been learned by mapping the movements evoked from repetitive ICMS applied to the cortex. The following points attempt to summarize these findings.

(1) SMA has a complete motor representation with hindlimb movements located caudally, forelimb movements centrally and oro-facial movements rostrally (Hummelsheim et al., 1986; Mitz and Wise, 1987; Luppino et al., 1991).

(2) The majority of movements evoked from SMA appear to involve proximal joints (Hummelsheim et al., 1986; Mitz and Wise, 1987; Luppino et al., 1991). This finding is somewhat surprising in view of the relatively equal numbers of presumed distal and proximal corticospinal neurons (Table 1).

(3) To date, there have been few systematic mi-

crostimulation studies of the cingulate motor areas. Luppino et al. (1991) reported that movements could be evoked from two cingulate areas, one corresponding to the dorso-caudal portion of CMAr (24d) and the other corresponding to the ventro-rostral part of CMAr (24c). Movements of the hindlimb, forelimb and face could be evoked from area 24d. In contrast to conclusions from anatomical labeling studies (He et al., 1995), this representation appeared to be somatotopically organized. The somatotopic organization was less clear in 24c and thresholds for evoking movements were somewhat higher. Cadoret and Smith (1997) reported that the frequency of success in evoking movements from CMAv using ICMS was about the same as the success rate for SMA with an average threshold exceeding 15 µA in both areas.

(4) In the lateral (arcuate) premotor areas, movements of hindlimb, forelimb and face were evoked from PMd in a medio-lateral orientation resembling that of primary motor cortex (Preuss et al., 1996). The forelimb representation was much greater than the hindlimb representation. In contrast to PMd, PMv does not appear to contain a representation of the hindlimb. Only movements of the forelimb and face could be evoked from PMv (Gentilucci et al., 1988; Rizzolatti et al., 1988; Hepp-Reymond et al., 1994; Godschalk et al., 1995; Preuss et al., 1996). Movements could be evoked from PMd at thresholds as low as primary motor cortex.

(5) In general, compared to primary motor cortex, higher current levels were needed to evoke movements from premotor areas (Table 1).

Functional roles of different frontal lobe motor areas

While the location and distribution of corticospinal neurons in the frontal lobe have been described in great detail, delineation of the functional roles of these different areas has been limited. Probable synaptic connections with motoneurons have been demonstrated for corticospinal neurons in several of these discrete areas. For example, Dum and Strick (1996) showed labeling in the ventral horn (lamina IX) from anterograde tracer injections in SMA, CMAd and CMAv, although the majority of labeling was in the intermediate zone. These findings have prompted the notion that each area could potentially be directly involved in movement execution not only through projections to primary motor cortex but also through direct projections to spinal motoneurons.

Many studies have shown that cells in premotor areas are modulated in relation to relatively simple tasks such as key press, flexion–extension movements of the elbow and precision grip (Smith, 1979; Weinrich and Wise, 1982; Weinrich et al., 1984; Godschalk et al., 1985; Kurata et al., 1985; Kurata and Tanji, 1986; Kurata and Wise, 1988a,b; Rizzolatti et al., 1988; Kurata, 1989; Crutcher and Alexander, 1990; Shima et al., 1991; Tanji, 1994; Cadoret and Smith, 1995). These findings and the presence of corticospinal neurons suggest that premotor areas are not exclusively involved in the higher-order aspects of movement. Nevertheless, there is no question that premotor areas seem to be preferentially involved in more abstract aspects of motor control such as preparatory set and movement sequencing (Roland et al., 1980; Wise, 1985; Gentilucci et al., 1988; Kurata and Wise, 1988a,b; Kurata, 1994; Picard and Strick, 1997; Wise et al., 1997). For example, using the 2-deoxyglucose method (2DG), Picard and Strick (1997) reported that CMAd was preferentially activated when monkeys performed a sequence of arm movements from memory. Based on preliminary evidence, the authors suggest that 2DG uptake in CMAd is not as intense when monkeys, making similar movement sequences, are guided by visual cues. This is consistent with the report by Savaki et al. (1995) showing that 2DG labeling in CMAd is absent in monkeys performing simple movements guided by visual cues. Recently, Shima and Tanji (1998) reported an interesting reward-dependent property of neurons in CMAr using a task in which monkeys selected one of two possible movements based on internal preference, that is, no external cues were provided. During task performance, the reward rate for one movement was gradually reduced. Neurons in CMAr became activated during the intertrial interval (delivery of food reward to onset of new trial) for trials in which the reward was reduced and, presumably as a consequence, the monkey began selecting the alternate movement. Bilateral reversible inactivation of the forelimb part of CMAr by injection of muscimol, a γ-aminobutyric acid (GABA) agonist, resulted in failure to select the optimal movement, that is, the movement that would deliver a larger food reinforcement. Even after the reward was reduced substantially, the monkey continued to select the previously performed movement. These results strongly suggest that CMAr is critically involved in the process of reward-based selection of specific movements.

Another example of premotor area involvement in a specific type motor function is the role of PMv in grasp. PMv can be subdivided into two components, F4 and F5 (Rizzolatti et al., 1998). Neurons in F5 are active in relation to a variety of movements of the hand and face, whereas neurons in F4 are active in relation to arm, neck and oro-facial movements. The predominant projection to F5 is from the anterior intraparietal cortex (AIP). Reversible lesions of AIP abolish the ability to use visual input for pre-shaping the hand and fingers in a manner appropriate for grasp of a specific object (Gallese et al., 1994). However, with AIP blocked, tactile feedback upon contact with the target object could be used successfully to adjust the grasp. In addition to specificity for a specific grasp type, some neurons in F5 discharge when the monkey sits passively and observes a particular action carried out by another individual. These neurons discharge only for observing movements which engage the neuron when the movement is performed actively by the monkey (Gallese et al., 1996). Taken together, these findings suggest that F5 mediates central commands for grasp of specific objects under visual guidance.

Of course, SMA is the most extensively studied of the premotor areas (Wiesendanger, 1986; Porter and Lemon, 1993; Passingham, 1996; Wise, 1996a,b; Rizzolatti et al., 1998). These studies suggest that SMA might be considered most like primary motor cortex in that many neurons discharge in relation to simple movements without any apparent contextual dependence. However, many studies have demonstrated a variety of higher-order motor functions which appear to involve SMA, including bimanual movements, execution of learned movement sequences, execution of self-initiated movements, and movement selection (Roland et al., 1980; Brinkman, 1984; Wiesendanger, 1986; Sadato et al., 1997).

Generalizations about motor functions of premotor areas

In conclusion, some generalizations are possible. First, although activity in premotor areas may, under some conditions, closely resemble that in primary motor cortex, more exhaustive investigation generally reveals that the activity is context-dependent, that is, it only occurs for particular behavioral conditions; for example, self-initiated or self-paced versus guided by sensory cues, either visual, auditory or tactile. In other cases, activation may depend on performance of movement sequences but, again, with dependence on whether movements are performed from memory or guided by sensory cues. In contrast, the discharge of neurons in motor cortex is tightly muscle-dependent. We have noted that under a wide variety of conditions, the activity of output cells in motor cortex can be predicted based on the activity of the cells' target muscles, that is, the muscles which the cells facilitate or suppress in spike triggered averages of EMG activity (Cheney et al., 1985, 1988, 1991). This is also evident from Table 1, which shows that the fraction of cells with discharge related to sensory cues (signal-related) or preparatory set (delay period activity) compared to movement execution is greater in premotor areas than in primary motor cortex.

What types of signals might be best suited for controlling neuroprosthetic devices?

Control of proximal movements

Humphrey et al. (1970) first showed that correlations between cortical cell discharge rate and force or rate of change of force can be improved by considering a population of cells rather than any individual cell. Clearly, all movements involve thousands of cortical cells and, although the precision with which any given cell specifies a parameter of movement may show substantial variance, the sum total of the entire population of neurons involved in the movement specifies key parameters with little variance. As pointed out earlier, signals for bringing the hand to a target in space must be derived from a different population of neurons than signals for controlling hand grasp and manipulation. Positioning the hand in extrapersonal space depends on use of proximal muscles (shoulder and elbow) and might best be accomplished with signals that reliably represent movement direction in three-dimensional (3-D) space. Georgopoulos and colleagues have ex-

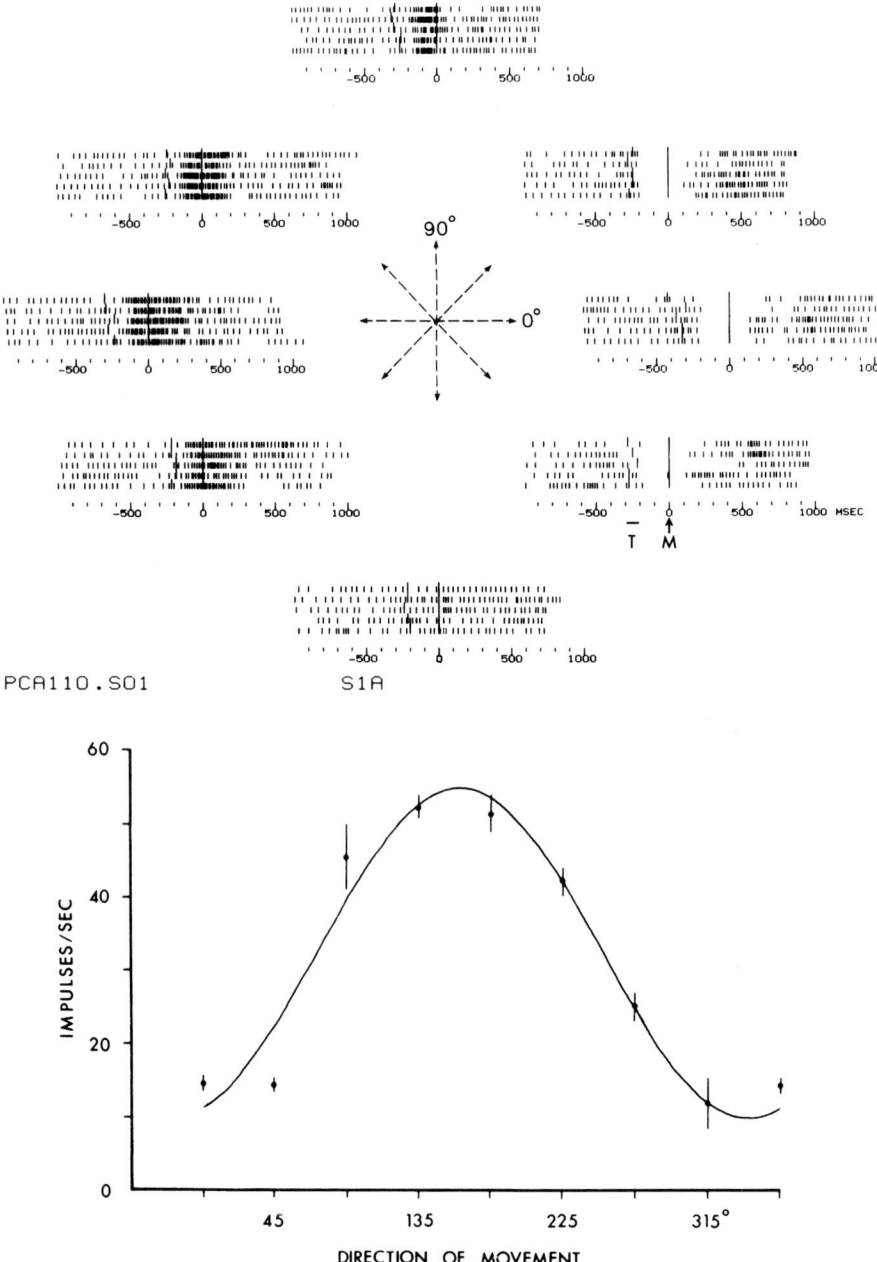

Fig. 5. Orderly variation in the frequency of discharge of a motor cortical cell with the direction of movement. Upper half: rasters are oriented to the movement onset, M, and show impulse activity during five repetitions of movements made in each of the eight directions indicated by the center diagram. Note the orderly variation in cell's activity during the RT (reaction time), MT (movement time) and TET (total experiment time). Lower half: directional tuning curve for the same cell. The discharge frequency is for TET. The data points are the mean ± SEM. The regression equation for the fitted sinusoidal curve is $D = 32.37 + 7.281 \sin\theta - 21.343 \cos\theta$, where D is the frequency of discharge and θ is the direction of movement or, equivalently, $D = 32.37 + 22.5 \cos(\theta - \theta_0)$, where θ_0 is the preferred direction ($\theta_0 = 161°$). (From Georgopoulos et al., 1982.)

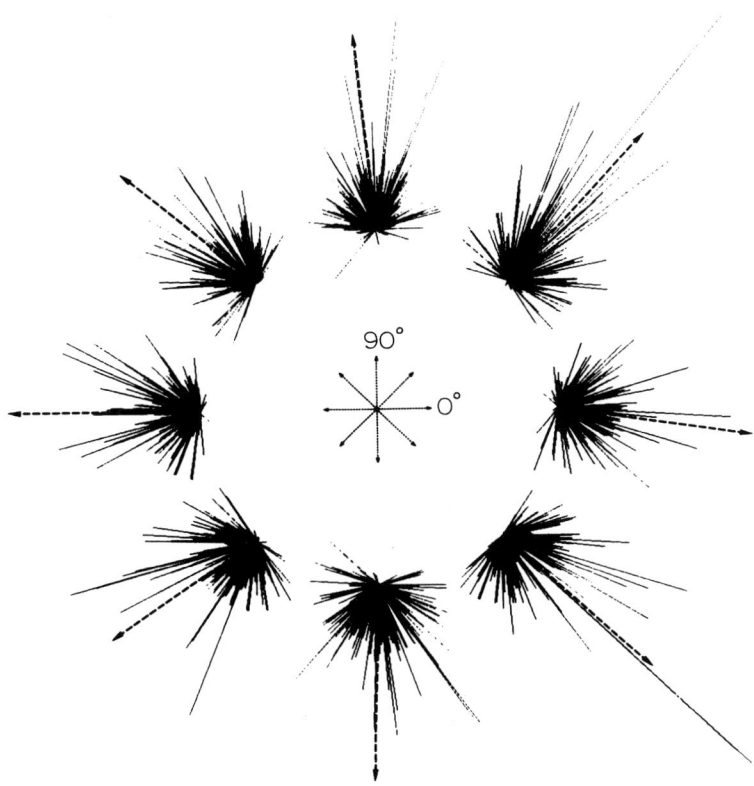

Fig. 6. Vector contributions of 241 directionally tuned motor cortical cells are shown for each of the eight movement directions tested. Note the spatial congruence between the direction of the vectorial sum (thick interrupted lines in each plot) and the direction of movement (small lines at center). (From Georgopoulos et al., 1983.)

amined the encoding of arm movement direction by populations of motor cortex neurons (Georgopoulos, 1986, 1995; Georgopoulos et al., 1981, 1982, 1983, 1986, 1988, 1999). These experiments are particularly relevant to the issue of whether signals from a group of neurons in motor cortex contain the appropriate signals for specifying arm movements that would bring the hand to specific target locations in space. Monkeys were trained to move a handle to each of eight targets in two-dimensional (2-D) space. The discharge of motor cortex neurons in relation to each direction was measured and plotted as shown in Fig. 5 for one neuron. This neuron, like many others in motor cortex, shows broad tuning of discharge with movement direction but with a best direction, for which discharge is greatest, in this case, for movements at 161 degrees. A cell's directional tuning curve can be described by a cosine function and represents its contribution to movement in different directions of 2-D space. The contribution of a cell to movement in a particular direction can be plotted as a vector whose length is the change in cell discharge rate for that movement direction, and whose direction is the cell's preferred movement direction. Fig. 6 shows the movement vectors for 241 motor cortex neurons recorded in relation to movement in eight different directions for the same task illustrated in Fig. 5. The vectorial sum of the whole population is illustrated by the dotted line. Note the high degree of congruence between the direction of the vectorial sum of the population and the actual movement direction indicated by the arrows in the center of the figure. These data demonstrate that movement vectors calculated from the discharges of a population of motor cortex neurons produce an excellent match to the actual movement direction.

This approach has also been extended to the analysis of movements in 3-D space (Georgopoulos et

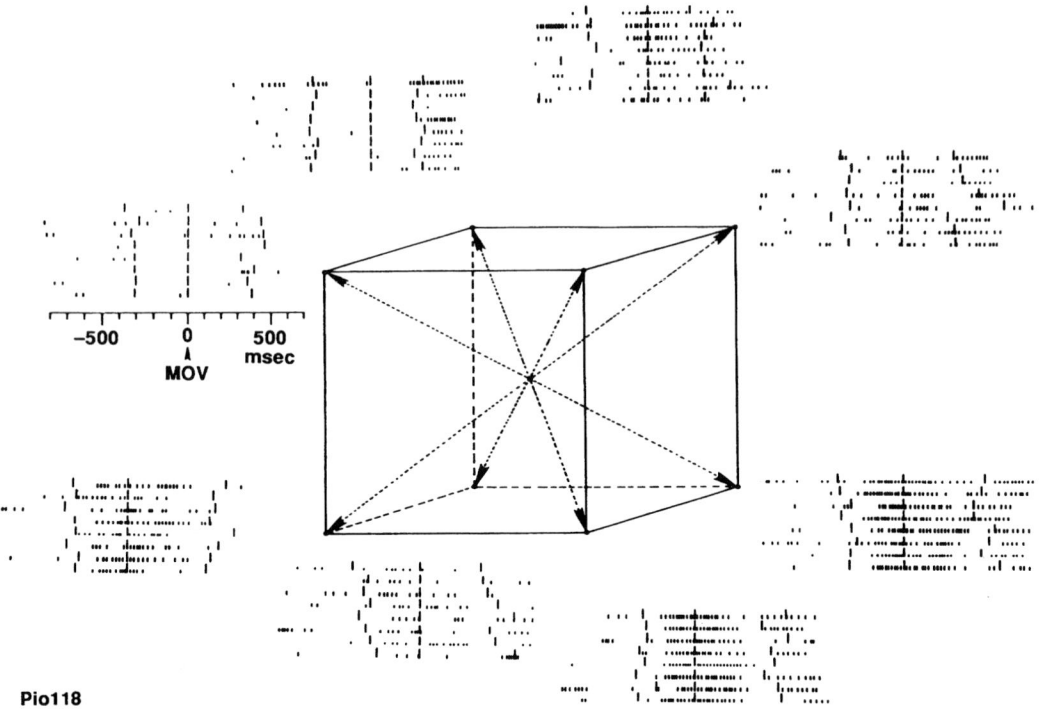

Fig. 7. Impulse activity of a single cell with movements in eight different directions indicated in the center drawing. Each line represents activity in one trial; eight trials for each movement direction are shown. Short bars indicate the occurrence of a spike; longer bars in each trial indicate, from left to right, the onset of target, the beginning of movement (MOV, aligned at zero), and the end of the movement. (The discharge after the end of the movement was not analyzed.) In the analysis of variance, the differences in activity among the eight movement directions were statistically significant ($P < 0.0001$, F-test). Each set of eight trials corresponds to the movement direction that points to it in the center diagram. (From Schwartz et al., 1988.)

al., 1986, 1988; Schwartz et al., 1988). Monkeys were trained to reach out and push red buttons that had been lit. A center button was located directly in front of each animal at shoulder level. Eight target buttons were placed at equal distances (12.5 cm) from the center button so that the direction of movements made from the center to targets sampled the 3-D space at approximately equal angular intervals (Fig. 7). The neuronal population vectors for each target in 3-D space were again found to closely match the movement vector. The average angle between the movement and population vectors for the eight directions of movement tested was 15.8 degrees with a range from 7.2 to 21.9 degrees. Using a smaller number of neurons substantially reduced the accuracy. This was particularly true for population sizes of less than 100 neurons. For example, using a random sample of 12 neurons from a population of 224, the average difference in angle between the population and movement vectors was 29.7 degrees. However, it is worth noting that these figures were derived from cells recorded sequentially over time and then added together as a population. Data from a population of simultaneously recorded neurons would most likely show better accuracy. Also of importance is the fact that the preferred directions of the 224 neurons recorded in relation to proximal arm movements evenly represented all directions of 3-D space. There was little or no tendency for preferential representation of any particular movement direction. These findings generalize to 3-D space, the previous findings with 2-D tasks and the power of neuronal population vectors as robust predictors of actual movement direction.

Schwartz has further extended this approach to more complex movements (Schwartz, 1994). Monkeys were trained to track a dot on a computer screen that was programmed to move in a spiral pattern of progressively increasing or decreasing radius. A population vector method was used to transform neuronal activity into the spatial dimension so the motor cortical representation of the hand through space could be visualized and compared with the actual movement. The preferred direction of each of 349 motor cortex cells was calculated from its activity on a standard center-out task (Fig. 5). The population vector computed from these cells for spiral movements closely matched the actual movement vectors. These findings demonstrate that data about directional preference derived from the standard center-out task can be used to accurately predict movement trajectories of the hand through space for other more complex tasks.

Based on this work we can conclude that a broad sample of motor cortical neurons concerned with control of shoulder and elbow muscles might provide optimal signals for bringing the hand (real or robotic) accurately to specific target locations in extrapersonal space given the use of an appropriate signal processing interface. Of course, this does not take into account the fact that variable loads may be encountered. However, the natural properties of neurons should help compensate for changes in load. Evarts (1968) originally showed that pyramidal tract neurons in motor cortex discharged at higher frequencies as the load opposing movement was increased. This finding has since been replicated in several other studies (Hepp-Reymond et al., 1978; Cheney and Fetz, 1980; Kalaska et al., 1989). Changes in discharge related to load changes would be exerted in the direction of the load and could be taken advantage of by the network to increase force in that direction. Implementation of this approach would be aided by the existence of visual feedback about the actual location of the limb in relation to the intended or 'commanded' position.

The feasibility of using activity from populations of motor cortex neurons to control a robotic arm has been confirmed in a modeling study by Lukasin et al. (1996). They showed that an artificial neural network could be trained to use experimentally recorded spike train data from a population of motor cortex cells to control a planar, two-joint, six-muscle simulated robotic arm. The actuator generated forces in close quantitative agreement with those exerted by trained monkeys. Moreover, output of the robotic arm could be controlled using as few as fifteen motor cortical cells. Lin et al. (1997) tested a self-organizing feature model of the neural representation of arm trajectories based on neuronal discharge rates. The self-organizing feature model selects the optimal weights determining each neuron's contribution to an overall movement representation and can extract not only direction-related information but also other information carried in the discharge of neurons that may be relevant to movement control. Unlike the population vector method, this method has the advantage that it does not assume any linear relationships between discharge rate and parameters such as movement speed and trajectory curvature. A recent study by Chapin et al. (1999, and this volume) has further confirmed the feasibility of using neuronal populations in motor cortex to control a robotic arm. They showed that signals from a population of neurons in motor cortex and thalamus of the rat could be used to control the movement of a single joint mechanical arm that provided a water reward. Interestingly, although the discharge of neurons was initially associated with movements of the forelimb, in some cases neural activity gradually became dissociated from limb movements. The rat was then able to produce changes in neuronal firing rate to control the robotic arm in the relative absence of limb movements.

Control of the hand and digits

The foregoing arguments present a relatively convincing case that directional activity in a population of motor cortex neurons could be used to bring the hand to a specific target location in 3-D space, but now we need to turn our attention to the problem of how the hand itself could be controlled. A population of neurons from the distal muscle representation of M1, separate from those for proximal control, would be best suited for control of wrist and digit movements. In the simplest case, this might involve hand grasp and release without precision grip (Porter and Lemon, 1993). But hand control is potentially a much more difficult problem because the anatomy of the hand does not lend itself nicely to a single population vector control model. Biological hands have five digits and a wrist, each of which can, in the simplest sense, flex and extend; but in reality

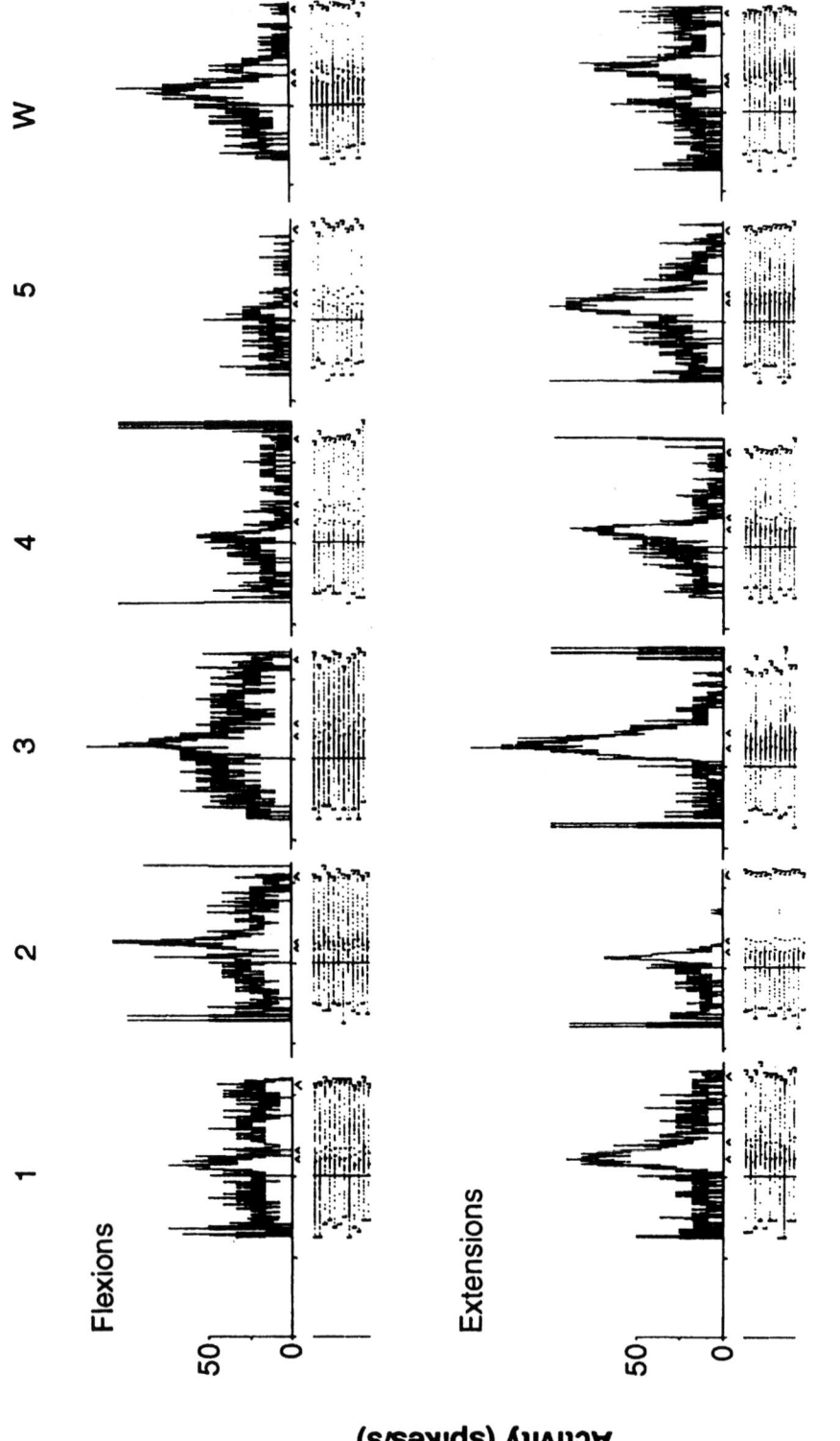

Fig. 8. Activity of an M1 neuron during individuated (relatively independent) flexion and extension movements of each of the digits (1–5) and the wrist (W). The traces show activity during instructed flexion (upper row) and during instructed extension (lower row) of each digit. In each frame, the dot raster below shows the neuron's discharge during 10 to 13 successful trials of the indicated instructed movement, aligned at the onset of the instruction signal (vertical line); the peri-event histogram above is formed from this rastered data (bin width, 10 ms). Tick marks in each raster line indicate (i) movement onset in the instructed digit, (ii) end of movement, and (iii) reward delivery; carat marks beneath each histogram indicate the average time of these events. (Reprinted with permission from Schieber and Hibbard (1993) Science 261: 489–492. Copyright 1993, American Association for the Advancement of Science.)

each digit and the wrist move in a 3-D space and in that sense each could be treated separately by a population vector model.

What do we know about the neural basis for control of the hand and digits? Unlike the orderly arrangement of digit representations in primary somatosensory cortex (Nelson et al., 1980), the representation of digits in the primary motor cortex does not appear orderly. Rather, the representation of individual digits is highly variable across monkeys and highly overlapping (Schieber and Hibbard, 1993). In this case, a measure of the representation of individual digits can be obtained by using ICMS to evoke movements (Nudo et al., 1992; Preuss et al., 1996) or by recording the activity of individual neurons in relation to individuated (relatively independent) finger movements (Schieber and Hibbard, 1993). Fig. 8 shows the activity of an M1 neuron in relation to individuated flexion and extension movements of the wrist and five digits. This neuron increased its discharge for almost all finger and wrist movements. This organization can be understood in relation to a number of motor system and corticospinal neuron properties. First, individual forearm digit muscles have branched tendons that supply multiple digits (Serlin and Schieber, 1993). Moreover, in one muscle where it was tested in the monkey (ED45), individual motor units generated tension across all the tendons, suggesting that the muscle is acting as a single multi-tendoned functional muscle rather than a compartmentalized muscle (Schieber et al., 1997). Consistent with these anatomical constraints is the finding that attempts to produce relatively independent finger movements involve the activity of a wide range of extrinsic and intrinsic hand muscles, some acting as agonists, others as antagonists, and still others as stabilizers (Schieber, 1995) (Fig. 9). Additionally, most individual corticomotoneuronal cells influence multiple muscles (Fetz and Cheney, 1980; Lemon et al., 1986; Porter and Lemon, 1993; McKiernan et al., 1998). Taking these findings into account, the distribution of neurons in the cortex representing particular digit movements will clearly be highly overlapping.

This leaves the question of whether a neuronal population vector model could be used to characterize movements of the wrist and digits in a manner similar to that which has been so effective for proximal joints. Georgopoulos et al. (1999) recently adapted the cosine tuning function model to the control of digit and wrist movements by adopting a geometric configuration of the hand in which the digits were modeled as lines with a common origin but separated by 45 degrees on a unit length radius semicircle (Fig. 10). Flexion and extension were represented as two different planes at equal, unit-length distances from a middle, neutral plane. The cosine tuning function approach was then applied to this model to determine the population movement vector for a group of 176 neurons recorded in relation to twelve individuated movements (five digits and the wrist for both flexion and extension). Seventy-five percent of the motor cortical neurons showed tuning in this movement space. Moreover, the population vector computed in this space predicted the instructed finger movement. Therefore, although single neurons may be related to several different finger movements and neurons of different types appear to be randomly distributed throughout the distal representation of motor cortex (Schieber and Hibbard, 1993), the activity of the neuronal population is capable of specifying particular wrist and finger movements.

Parameter encoding by motor cortex neurons

A continuing issue in motor control is whether movement direction is actually the parameter specified or encoded by motor cortex cells or if there is another more fundamental parameter. Unquestionably, most motor cortex cells show directional tuning as described above. However, it should be emphasized that individual muscles also show tuning curves with best directions not unlike those reported for motor cortex cells (Georgopoulos et al., 1984). It is also true that the activity of motor cortex cells is heavily influenced by the load against which the monkey moves (Cheney and Fetz, 1980; Cheney et al., 1988; Kalaska et al., 1989; Scott and Kalaska, 1995). Moreover, tuning curves can be generated for movements in isometric space so actual joint displacement and movement velocity are clearly not essential (Georgopoulos et al., 1992). It is also clear that the activity of motor cortex cells is closely tied to the activity of target muscles and that, over a broad range, the rate of discharge of wrist-related cortico-

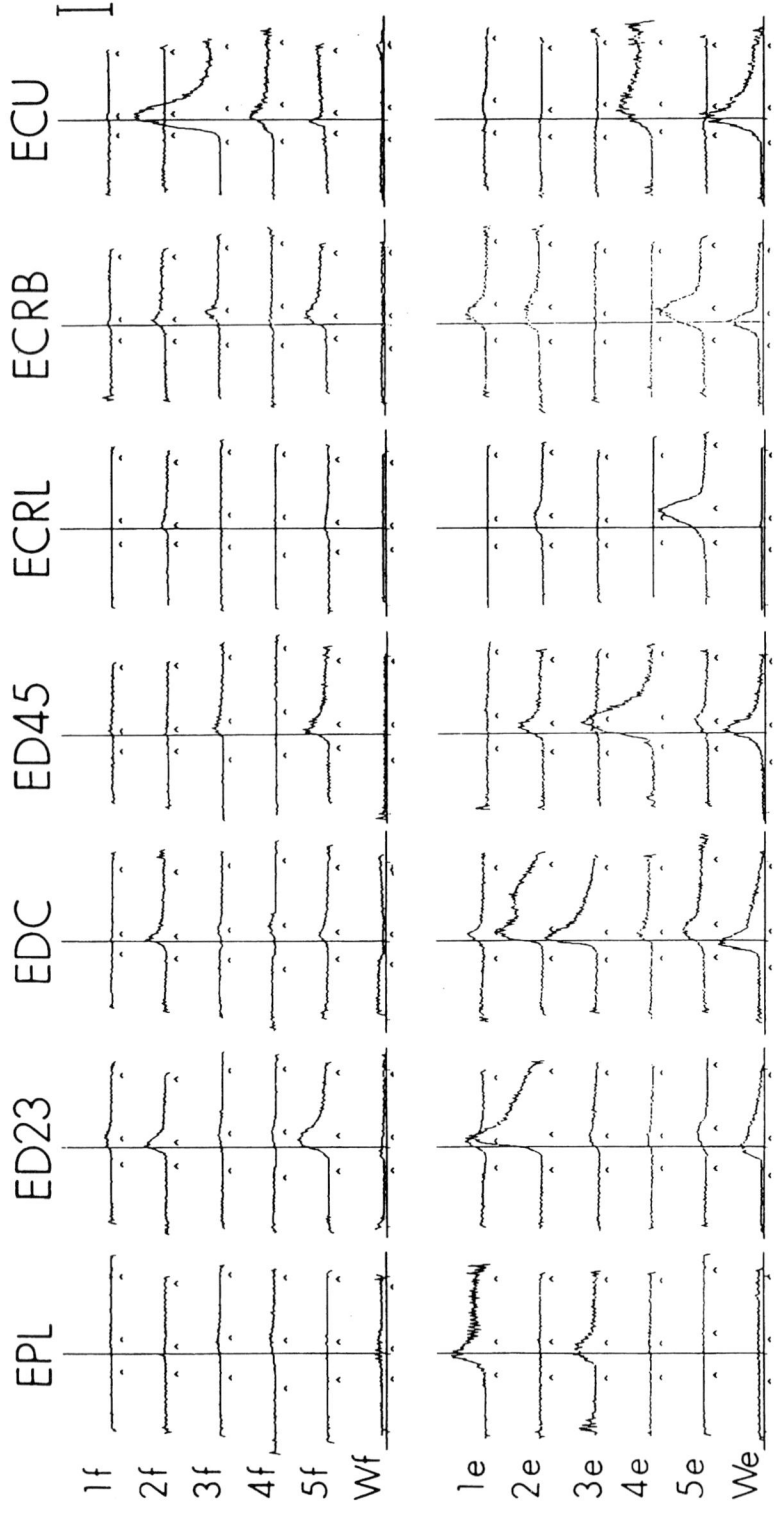

Fig. 9. Typical EMG recordings from each of seven extensor muscles studied in relation to a task that required individuated (relatively independent) flexion (*f*) and extension (*e*) movements of each of the five digits (*1–5*) and the wrist (*W*). Horizontal calibration, 1 sec; vertical calibration (arbitrary integrated EMG units): *EPL*, 250; *ED23*, 1000; *EDC*, 1000; *ED45*, 1000; *ECRL*, 1000; *ECRB*, 1000; *ECU*, 1000. Similar findings were obtained for seven flexor muscles (not shown). Muscle abbreviations: extensor carpi radialis longus (*ECRL*), extensor carpi radialis brevis (*ECRB*), extensor carpi ulnaris (*ECU*), extensor digitorum communis (*EDC*), extensor digitorum 2,3 (*ED23*), extensor digitorum 4,5 (*ED45*), extensor pollicis longus (*EPL*). (From Schieber, 1995.)

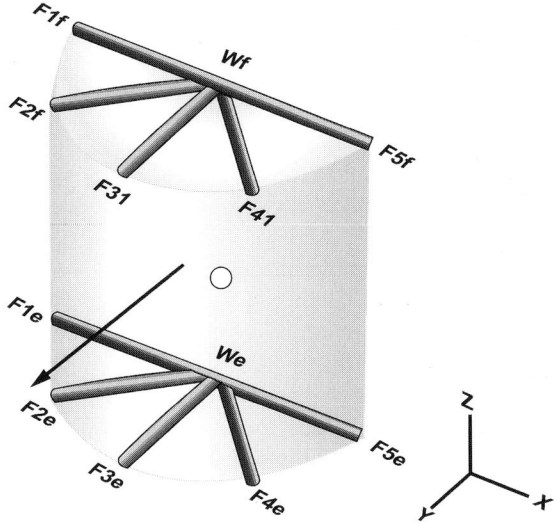

Fig. 10. Representation of instructed flexion and extension movements of the five fingers (F) and wrist (W) in a three-dimensional space that models the geometry of the hand. Each instructed movement is denoted by the number of the instructed digit (1 = thumb through 5 = little finger), and the first letter of the instructed movement direction (f = flexion, e = extension). The arrow exemplifies a vector from the origin of the axes to one finger movement ($F2e$). (From Georgopoulos et al., 1999.)

motoneuronal cells varies as a linear function of the torque produced (Cheney and Fetz, 1980; Cheney et al., 1985, 1991; Fetz and Cheney, 1980). Taken together, these findings support the notion that the most fundamental parameters encoded in the discharge of motor cortex cells, at least cells with a demonstrable synaptic linkage to motoneurons, are parameters in a muscle-based coordinate system (Miller and Houk, 1995). Nevertheless, the advantage and elegance of the population vector approach is that it accurately predicts the direction of hand trajectories in space and this is the most important requirement of any system designed to use neuronal population activity to control neuroprosthetic devices.

Are cortical motor output maps fixed or flexible?

While maps of motor cortex have revealed fine detail of motor output and, perhaps, convey an impression of stability, it is important to recognize that these maps are dynamic and show both rapid and long-term forms of adaptive plasticity under a variety of circumstances (Donoghue, 1995; Seitz and Freund, 1997; Hallet, 1999). Adaptive plasticity and altered representation of movements and muscles in motor cortex has been demonstrated in relation to: (1) use and practice (Karni et al., 1995; Nudo et al., 1996a; Classen et al., 1998; Rioult-Pedotti et al., 1998); (2) immobilization (Liepert et al., 1995); (3) stroke-induced and other types of injury to motor cortex (Seitz et al., 1995; Nudo and Milliken, 1996; Nudo et al., 1996b; Cicinelli et al., 1997; Rouiller et al., 1998); (4) spinal cord injury (Streletz et al., 1995; Bruehlmeier et al., 1998; Green et al., 1998); (5) limb amputation (Sanes et al., 1990a,b; Cohen et al., 1991; Fuhr et al., 1992; Kew et al., 1994; Ojemann and Silbergold, 1995; Schieber and Deuel, 1997); (6) altered somatosensory input (Keller et al., 1996; Rossini et al., 1996; Rossi et al., 1998; Ziemann et al., 1998a,b; Hamdy et al., 1999); (7) muscle denervation (Sanes et al., 1990a,b); (8) repetitive electrical stimulation of motor cortex (Nudo et al., 1990); (9) exercise (Zanette et al., 1995); (10) synchronized use of thumb and foot movements (Liepert et al., 1999); and (11) development of implicit knowledge of a motor sequence (Pascual-Leone et al., 1994).

Motor cortex map plasticity has been demonstrated in animals using ICMS and in humans using TMS and functional imaging (Hallet, 1999). For interpretation of results of ICMS experiments, it is important to remember that this technique is thought to activate corticospinal neurons predominantly indirectly through synaptic inputs rather than directly (Porter and Lemon, 1993). Therefore, changes in maps observed with this method may result from intracortical circuit reorganization or excitability changes rather than changes in corticospinal terminal organization at the spinal level. This interpretation is reinforced by the findings of Huntley (1997) showing that rapid changes in map boundaries related to peripheral nerve cuts in the rat correlated with the spatial extent of horizontally projecting intracortical axonal projections. For example, cutting the facial nerve supplying the vibrissal muscles in the rat produced a 1.2 mm expansion of the forelimb representation into the vibrissal representation. Horizontal neuronal projections from the adjacent forelimb region were found to extend 1.2 mm into the vibrissal area and presumably mediated the forelimb expansion after the facial nerve supplying the vibrissal muscles was cut. Stimulation outside the area containing intracortical horizontal projections from the forelimb representation failed to produce responses of either the vibrissae or forelimb in animals in which the facial nerve was cut. Similar shifts in map boundaries can be induced by injection of the GABA antagonist, bicuculline, suggesting that the excitability of these pathways is controlled by intracortical inhibitory mechanisms (Jacobs and Donoghue, 1991). Horizontal intracortical connections in layers II and III from rats trained on a skilled reaching paradigm were found to produce larger field potentials when stimulated than the untrained cortex (Rioult-Pedotti et al., 1998). This difference was thought to be related to the induction of long-term potentiation (LTP) and intracortical cholinergic mechanisms (Hess and Donoghue, 1999). Aroniadou and Keller (1995) showed that LTP in rat motor cortex is most effectively prevented when N-methyl-D-aspartate (NMDA) receptors are blocked.

Use-related plasticity parallels similar observations on plasticity in maps of primary somatosensory cortex associated with use. Nudo et al. (1996a,b) showed that training squirrel monkeys to perform a motor skill task produced an expansion in the map territories of muscles used in the task when compared to a task that required activation of the same muscles but without a requirement for skilled use of the digits. Similarly, immobilization has been shown to reduce the representation of the anterior tibial muscle compared to the unaffected leg (Liepert et al., 1995). Plasticity in motor maps has been reported both in relation to withdrawal and stimulation of somatosensory input. In humans, pharyngeal stimulation produced an expansion of the representation of the pharynx in motor cortex and a decrease in the esophageal representation lasting 30 minutes or more. These changes occurred in the absence of parallel changes in the excitability of brainstem-mediated reflexes, suggesting that the source of plasticity was cortical in origin. Anesthesia of the skin overlying a muscle reduces the cortical output representation of that muscle as demonstrated with TMS (Rossi et al., 1998). Ischemic deafferentation of the arm has been shown to produce a moderate increase in the size of the motor-evoked potential in the biceps muscle. This change was enhanced by conditioning low-frequency (0.1/s) TMS of the cortex on the side exhibiting plasticity and reduced by stimulation of the opposite cortex (Ziemann et al., 1998a,b). ICMS in the rat at a rate of 1/s for 1–3 h produced expansion of the representation of the cortical area stimulated (Nudo et al., 1990). Borders shifted by 210–670 μm and changes were progressive and reversible.

Motor cortical map changes associated with ischemic or traumatic injury of the cortex have been demonstrated both in humans and in animals (Hallet, 1999). Nudo and Milliken (1996) produced ischemic lesions of identified representations of the cortical motor output map in squirrel monkeys by coagulating small arterioles entering the surface of the cortex. Motor cortex was mapped with a track spacing of 250 μm both before and a few months after ischemic lesions of primary motor cortex (area 4). Subtotal ischemic lesions of the hand/digit representation resulted in a marked but transient deficit in use of the hand contralateral to the lesion, loss of digit representation in regions adjacent to the lesion and expansion of proximal muscle representations into zones that previously belonged to the digits. This loss of digit representation could be prevented by use of a 'rehabilitative' training program in which the monkey was required to use the affected hand in a skilled motor task (Nudo et al., 1996b).

Limb amputation and deafferentation also evoke plastic reorganization of motor cortex with expansion of adjacent representations into the territory of the amputated limb but without loss of the somatosensory representation. Ojemann and Silbergold (1995) reported a case of forelimb amputation below the shoulder in a patient 24 years prior to neurological surgery for removal of a tumor in which primary M1 and S1 cortex were mapped. They found that representations of the face and jaw had moved medially and the shoulder representation had moved laterally, partially filling in the former arm representation. However, a zone remained between the jaw and shoulder representations from which no movements could be elicited. Phantom sensations were evoked from stimulation of the upper limb region of S1 cortex. These findings are consistent with those of Kew et al. (1994) who studied cortical blood flow changes based on positron emission tomography (PET) in patients with upper limb amputation.

Shoulder movements on the amputated side produced blood flow increases over a wider area and greater magnitude in contralateral cortex than on the intact side. Fuhr et al. (1992) showed that muscles on the amputated side tended to show activation with TMS from a larger cortical region than the same muscles on the intact side. Motor-evoked potentials on the amputated side from contralateral cortex were also greater in magnitude and of shorter latency than the intact side. These changes could not be explained by changes in the excitability of the alpha motoneuron pools on the amputated side. In conclusion, cortical motor maps can no longer be viewed as static representations of output relationships. Rather, the size and strength of the cortical motor representation is dynamic and subject to influence from a variety of factors. Intracortical mechanisms account probably for most existing demonstrations of map plasticity; the extent to which corticospinal terminations with motoneurons may show plasticity is unknown.

Do corticospinal and rubrospinal neurons survive spinal cord injury?

The extent of corticospinal and rubrospinal cell death after spinal cord injury has been extensively studied in rodent models using retrograde tracers. Early studies emphasized the presence of cell death beginning 5–10 weeks after cord injury (Feringa et al., 1983, 1984; Goshgarian et al., 1983; Feringa and Vahlsing, 1985). For example, Feringa and Vahlsing (1985) reported that one year after T9 spinal cord transection, HRP inserted into the spinal cord at T3–T4 only labeled about 7% of the corticospinal neurons labeled in control rats. However, in more recent work using Fluorogold retrograde labeling, no evidence of either corticospinal or rubrospinal cell loss was found 10 or 20 weeks after T9 spinal cord transection in rats (Pruitt et al., 1988; McBride et al., 1990). Transection appears to alter the uptake and transport of some substances, including HRP, but the transport of Fluorogold does not seem to be affected. To definitively demonstrate the absence of corticospinal and rubrospinal cell loss in spinal cord injury, the somata of these cells were pre-labeled with Fluorogold 4 days before spinal cord transection to control for any changes in the ability of axons to transport substances (McBride et al., 1989). At 10 and 20 weeks after lesion, there was no cell loss of either corticospinal or rubrospinal neurons, although the size of these neurons was significantly reduced. Compression injury of the spinal cord also produces no significant loss of rubrospinal neurons (Theriault and Tator, 1994). The absence of progressive degeneration and loss of corticospinal neurons has also been demonstrated in human autopsy material (Fishman, 1987). Merline and Kalil (1990) demonstrated a developmental dependence of cell loss on spinal cord transection. They found death of corticospinal neurons if the spinal cord was transected before the neurons innervated their spinal targets. In conclusion, spinal cord injury, either by transection or compression, does not produce cell loss in the cortex or red nucleus unless injury occurs very early in development. Therefore, these neurons would represent potentially viable signal sources for control of neuroprosthetic devices in people with spinal cord injury.

Do the properties of neurons change after spinal cord lesions?

Work described in the previous section demonstrates that most corticospinal neurons survive interruption of their axons at the spinal cord level. Of course, even without corticospinal neurons there are many other cell types that presumably could provide appropriate signals for controlling neuroprosthetic devices. However, the fact that corticospinal cells survive severing of their axons in spinal cord injury does not mean that the cells are functionally normal or necessarily accessible to central control. Spinal cord injury may have either eliminated or grossly distorted sensory input to motor cortex and other cortical motor areas. In view of this, it is not clear that in spinal cord injury, cortical motor areas would show either adequate excitability or that their activity in relation to attempted movements would resemble the activity associated with normal movement of the intact limb. This issue has been inadequately studied to provide conclusive answers. Nevertheless, intuitively it is reasonable to assume that cortical motor areas affected by spinal cord injury would remain accessible to central control and would become active in relation to efforts to move a limb. While there is no direct confirmation of this assumption, some recent studies have provided limited relevant information. PET imaging in spinal cord injury has revealed an increased rather than decreased resting glucose metabolism in some motor system structures, including SMA, anterior cingulate and putamen, although cerebellar glucose metabolism was reduced (Roelche et al., 1997). A recent study by Green et al. (1998) using high-resolution electroencephalography (EEG) coregistered with magnetic resonance imaging (MRI) demonstrated normally located cortical movement potentials in quadraparetics during attempts to move the toes. These subjects were unable to produce any actual toe movement. This suggests that in a part of motor cortex in which normal motor output was severed at the spinal level, neuronal elements remain excitable and accessible to central motor commands. Recently, efforts have been made to image primary motor cortex with functional MRI in subjects with spinal cord injury and paralysis while subjects attempted to move the paralyzed limb. Preliminary results have demonstrated activation of cortical areas thought to be associated with the paralyzed limb (D.R. Humphrey, pers. commun.).

To what extent can the activity of neurons be adapted to new conditions?

Neurons develop a set of natural functional relationships that are the result of genetic and environmental influences on neural circuit specification throughout development. Through this natural process of development, neurons in motor cortex acquire strong and consistent activity relationships to movement. Of course, subjects are normally not aware of the activity of their neurons or even the activity of individual muscles — they are only aware of movements. But what if information about the activity of neurons were provided? Could subjects learn to control the activity of these neurons and does the activity remain linked to peripheral movements? This question is relevant to the development of neuroprosthetic devices in the case of paralysis or amputation because efforts to move will not be accompanied by any detectable movement. In these cases, it is of interest to know to what extent subjects can control the activity of neurons through operant conditioning techniques. As yet, this has not been attempted in subjects with spinal cord injury or amputation,

but it has been investigated in unimpaired monkeys. Operant conditioning of neural activity in the brain began with the work of Olds (1965) in the rat. More recently, in several different innovative studies, Fetz and colleagues applied this method to recordings of motor cortex cells in monkeys to examine relationships between neuronal and muscle activity (Fetz, 1969; Fetz and Finocchio, 1971, 1975; Fetz and Baker, 1973). These studies demonstrated that, given appropriate feedback about the firing rate of individual neurons, monkeys could learn to increase and decrease neuronal activity. Increases in neuronal activity were 50–500% over baseline and generally took the form of bursts. The bursts were often associated with specific movements reflecting the representation within which the cell was located. Of particular interest was the fact that the activity of neighboring cells in the cortex could be dissociated from one another. Moreover, the activity of muscles which were coactivated with the cortical neurons during voluntary movements or during conditioned bursts of neuronal activity could be dissociated from bursts of cortical cell activity by specifically rewarding increases in cell activity and decreases in muscle activity. Several different mechanisms might explain this dissociation. First, it is possible that these cells were not corticomotoneuronal cells and did not make synaptic connections with motoneurons. Second, the monkey may have been able to selectively activate such a small number of neurons that their activity was insufficient to raise the motoneurons to firing threshold. Finally, the spinal actions of these cells on motoneurons might have been actively inhibited, either postsynaptically at the motoneuron level, or presynaptically at the terminals of corticomotoneuronal axons. Conditioning the monkey to produce bursts of EMG activity without activation of cortical neurons proved to be more difficult, and complete suppression of cell activity was generally not possible. These results demonstrate a remarkable degree of voluntary control over cortical cell activity and suggest that humans presented with the same circumstances might establish a high degree of control and fractionation of activity within a population of recorded motor cortex neurons.

In addition to operant control of unit activity, other studies have demonstrated operant conditioning of EEG spindles in cats (Wyrwicka and Sterman,

1968) and the H-reflex in primates and rats (Carp and Wolpaw, 1994). The success of H-reflex conditioning is inversely related to the extent of spinal cord injury, emphasizing the role of descending systems as a factor underlying conditioning (Chen et al., 1999). Recently, Minor et al. (1998) demonstrated that human subjects could learn to switch between a mu rhythm in the EEG or a beta rhythm as a means of signaling yes or no answers to questions. Such applications of adaptive control over EEG signals should be applicable to people with conditions such as amyotrophic lateral sclerosis, who may have very limited motor abilities.

Conclusion

In this paper we have attempted to review the neurobiological properties of the primate cortical motor control system that are relevant to the development and implementation of neuroprosthetic control devices, including the use of robotic arms in people with limb amputations and control of paralyzed arms by FES in people with spinal cord injury. While the reality of such devices still seems a long way off, advances in our knowledge of the organization of cortical motor areas and the properties of neurons it contains provide strong support for the feasibility of this effort and provide reason for long-term optimism. This optimism is supported by recent successes both in animal models and with modeling approaches using spike trains from populations of real motor cortex neurons to control robotic or simulated limbs (Lukasin et al., 1996; Chapin et al., 1999). Further technical advances will be very important to continued progress in this field. Even if effective and safe technology were in place, the relatively invasive nature of the recording and muscle stimulation electrodes would probably be a deterrent for all but the most determined patients. Nevertheless, it is important to remind ourselves that progress in this field occurs in small steps across a variety of disciplines, and future technical advances may make this technology highly effective and much less invasive. For example, capsule-like injectable muscle stimulating electrodes with telemetry control have been tested and might become a reality for large numbers of muscles (Loeb et al., 1991). Similarly, alternative methods for recording the activity of individual or populations of neurons in real-time through the dura or even the skull might provide a much less invasive source of motor-related signals for neuroprosthetic control. These approaches hold great promise for the future, as long as we continue to take small steps forward.

Acknowledgements

This work was supported by NIH grant NS39023 and Paralyzed Veterans of America (Spinal Cord Research Foundation) grant PVA1657. The authors would like to thank Don Warn and Patrick Moonasar for their untiring help with the illustrations.

References

Aroniadou, V.A. and Keller, A. (1995) Mechanisms of LTP induction in rat motor cortex in vitro. *Cereb. Cortex*, 5: 353–362.

Asanuma, H. and Rosen, I. (1972) Topographical organization of cortical efferent zones projecting to distal forelimb muscles in the monkey. *Exp. Brain Res.*, 14: 243–256.

Brinkman, C. (1984) Supplementary motor area of the monkey's cerebral cortex: short- and long-term deficits after unilateral ablation and the effects of subsequent collosal section. *J. Neurosci.*, 4: 918–929.

Bruehlmeier, M., Dietz, V., Leenders, K.L., Roelcke, U., Missimer, J. and Curt, A. (1998) How does the human brain deal with spinal cord injury?. *Eur. J. Neurosci.*, 10: 3918–3922.

Cadoret, G. and Smith, A.M. (1995) Input–output properties of hand-related cells in the ventral cingulate cortex in the monkey. *J. Neurophysiol.*, 73: 2584–2590.

Cadoret, G. and Smith, A.M. (1997) Comparison of the neuronal activity of the SMA and the ventral cingulate cortex during prehension in the monkey. *J. Neurophysiol.*, 77: 153–166.

Carp, J.S. and Wolpaw, J.R. (1994) Motoneuron plasticity underlying operantly conditioned decrease in primate H-reflex. *J Neurophysiol.*, 72: 431–442.

Chapin, J.K., Moxon, K.A., Markowitz, R.S. and Nicolelis, M.A.L. (1999) Real-time control of a robot arm using simultaneously recorded neurons in the motor cortex. *Nat. Neurosci.*, 2: 664–670.

Chen, X.Y., Wolpaw, J.R., Jakeman, L.B. and Stokes, B.T. (1999) Operant conditioning of H-reflex increases in spinal cord-injured rats. *J. Neurotrauma*, 16: 175–186.

Cheney, P.D. (1996) Electrophysiological methods for mapping brain motor circuits. In: Toga, A.W. and Mazziotta, J.C. (Eds.), *Brain Mapping: The Methods*. Academic Press, New York, pp. 277–307.

Cheney, P.D. and Fetz, E.E. (1980) Functional classes of primate corticomotoneuronal cells and their relation to active force. *J. Neurophysiol.*, 44: 773–791.

Cheney, P.D., Kasser, R.J. and Fetz, E.E. (1985) Motor and

sensory properties of primate corticomotoneuronal cells. *Exp. Brain Res. Suppl.*, 10: 211–231.

Cheney, P.D., Mewes, K. and Fetz, E.E. (1988) Encoding of motor parameters by corticomotoneuronal (CM) and rubromotoneuronal (RM) cells producing postspike facilitation of forelimb muscles in the behaving monkey. *Behav. Brain Res.*, 28: 181–191.

Cheney, P.D., Fetz, E.E. and Mewes, K. (1991) Neural mechanisms underlying corticospinal and rubrospinal control of limb movements. *Prog. Brain Res.*, 87: 213–252.

Cicinelli, P., Traversa, R. and Rossini, P.M. (1997) Post-stroke reorganization of brain motor output to the hand: a 2–4 month follow-up with focal magnetic transcranial stimulation. *Electroenceph. Clin. Neurophysiol.*, 105: 438–450.

Classen, J., Liepert, J., Wise, S.P., Hallet, M. and Cohen, L.G. (1998) Rapid plasticity of human cortical movement representation induced by practice. *J. Neurophysiol.*, 79: 1117–1123.

Cohen, L.G., Bandinelli, S., Findley, T.W. and Hallet, M. (1991) Motor reorganization after upper limb amputations in man. *Brain*, 114: 615–627.

Crutcher, M.D. and Alexander, G.E. (1990) Movement-related neuronal activity selectively coding either direction or muscle pattern in three motor areas in the monkey. *J. Neurophysiol.*, 64: 151–163.

Donoghue, J.P. (1995) Plasticity of the adult sensorimotor representations. *Curr. Opin. Neurobiol.*, 5: 749–754.

Dum, R.P. and Strick, P.L. (1991) The origin of corticospinal projections from the premotor areas in the frontal lobe. *J. Neurosci.*, 11: 667–689.

Dum, R.P. and Strick, P.L. (1996) Spinal cord terminations of the medial wall motor areas in macaque monkeys. *J. Neurosci.*, 16: 6513–6525.

Evarts, E.V. (1968) Relation of pyramidal tract activity to force exerted during voluntary movement. *J. Neurophysiol.*, 31: 14–27.

Feringa, E.R. and Vahlsing, H.L. (1985) Labeled corticospinal neurons one year after spinal cord transection. *Neurosci. Lett.*, 58: 283–286.

Feringa, E.R., Gilbertie, W.J. and Vahlsing, H.L. (1983) Retrograde transport in corticospinal neurons after spinal cord transection. *Neurology*, 33: 478–482.

Feringa, E.R., Vahlsing, H.L. and Dauser, R.C. (1984) Histologic evidence for death of cortical neurons after spinal cord transection. *Neurology*, 34: 1002–1006.

Fetz, E.E. (1969) Operant conditioning of cortical unit activity. *Science*, 163: 955–958.

Fetz, E.E. (1999) Real-time control of a robotic arm by neuronal ensembles. *Nat. Neurosci.*, 2: 583–584.

Fetz, E.E. and Baker, M.A. (1973) Operantly conditioned patterns on precentral unit activity and correlated responses in adjacent cells and contralateral muscles. *J. Neurophysiol.*, 36: 179–204.

Fetz, E.E. and Cheney, P.D. (1980) Postspike facilitation of forelimb muscle activity by primate corticomotoneuronal cells. *J. Neurophysiol.*, 44: 773–791.

Fetz, E.E. and Finocchio, D.V. (1971) Operant conditioning of specific patterns of neural and muscular activity. *Science*, 174: 431–435.

Fetz, E.E. and Finocchio, D.V. (1975) Correlations between the activity of motor cortex cells and arm muscles during operantly conditioned response patterns. *Exp. Brain Res.*, 23: 217–240.

Fishman, P.S. (1987) Retrograde changes in the corticospinal tract of posttraumatic paraplegics. *Arch. Neurol.*, 10: 1082–1084.

Fuhr, P., Cohen, L.G., Dang, N., Findley, T.W., Haghighi, S., Oro, J. and Hallet, M. (1992) Physiological analysis of motor reorganization following lower limb amputation. *Electroenceph. Clin. Neurophysiol.*, 85: 53–60.

Galea, M.P. and Darian-Smith, I. (1994) Multiple corticospinal neuron populations in the macaque monkey are specified by their unique cortical origins, spinal terminations, and connections. *Cereb. Cortex*, 4: 166–194.

Gallese, V., Murata, A., Kaseda, M., Niki, N. and Sakata, H. (1994) Deficit in hand preshaping after muscimol injection in the monkey parietal cortex. *NeuroReport*, 5: 1525–1529.

Gallese, V., Fadiga, L., Fogassi, L. and Rizzolatti, G. (1996) Action recognition in the premotor cortex. *Brain*, 119: 593–609.

Gentilucci, M., Fogassi, G., Luppino, G., Matelli, M., Camarda, R. and Rizzzolatti, G. (1988) Functional organization of inferior area 6 in the macaque monkey, I. Somatotopy and control of proximal movements. *Exp. Brain Res.*, 71: 475–490.

Georgopoulos, A.P. (1986) On reaching. *Annu. Rev. Neurosci.*, 9: 147–170.

Georgopoulos, A.P. (1995) Current issues in directional motor control. *Trends Neurosci.*, 18: 506–510.

Georgopoulos, A.P., Kalaska, J.F. and Massey, J.T. (1981) Spatial trajectories and reaction times of aimed movements: effects of practice, uncertainty and change in target location. *J. Neurophysiol.*, 46: 725–743.

Georgopoulos, A.P., Kalaska, J.F., Caminiti, R. and Massey, J.T. (1982) On relations between the direction of two-dimensional arm movements and cell discharge in primate motor cortex. *J. Neurosci.*, 2: 1527–1537.

Georgopoulos, A.P., Caminiti, R., Kalaska, J.F. and Massey, J.T. (1983) Spatial coding of movement: a hypothesis concerning the coding of movement direction by motor cortical populations. *Exp. Brain Res. Suppl.*, 7: 327–336.

Georgopoulos, A.P., Kalaska, J.F., Crutcher, M.D., Caminiti, R. and Massey, J.T. (1984) The representation of movement direction in the motor cortex: single cell and population studies. In: Edelman, G.M., Gail, W.E. and Cowan, W.M. (Eds.), *Dynamic Aspects of Neocortical Function*. Wiley, New York, pp. 501–524.

Georgopoulos, A.P., Schwartz, A.B. and Kettner, R.E. (1986) Neuronal population coding of movement direction. *Science*, 233: 1416–1419.

Georgopoulos, A.P., Kettner, R.E. and Schwartz, A.B. (1988) Primate motor cortex and free arm movements to visual targets in three-dimensional space, II. Coding of movement by a neuronal population. *J. Neurosci.*, 8: 2928–2937.

Georgopoulos, A.P., Ashe, J., Smyrnis, N. and Taira, M. (1992)

The motor cortex and the coding of force. *Science*, 256: 1692–1695.

Georgopoulos, A.P., Pellizzer, G., Polliakov, A.V. and Schieber, M.H. (1999) Neural coding of finger and wrist movements. *J. Comput. Neurosci.*, 6: 279–288.

Godschalk, M., Lemon, R.N., Kuypers, H.G.J.M. and Van Der Steen, J. (1985) The involvement of monkey premotor cortex neurones in preparation for visually cued arm movements. *Behav. Brain Res.*, 18: 143–157.

Godschalk, M., Mitz, A.R., Van Duin, B. and Van der Burg, H. (1995) Somatotopy of monkey premotor cortex examined with microstimulation. *Neurosci. Res.*, 23: 269–279.

Goshgarian, H.G., Koistinen, J.M. and Schmidt, E.R. (1983) Cell death and changes in the retrograde transport of horseradish peroxidase in rubrospinal neurons following spinal cord hemisection in the adult rat. *J. Comp. Neurol.*, 214: 251–257.

Green, J.B., Sora, E., Bialy, Y., Ricamato, A. and Thatcher, R.W. (1998) Cortical sensorimotor reorganization after spinal cord injury: an electroencephalographic study. *Neurology*, 50: 1115–1121.

Jacobs, K.M. and Donoghue, J.P. (1991) Reshaping the cortical motor map by unmasking latent intracortical connections. *Science*, 251: 944–947.

Hallet, M. (1999) Plasticity in the human motor system. *Neuroscientist*, 5: 324–332.

Hamdy, S., Rothwell, J.C., Aziz, Q., Singh, K.D. and Thompson, D.G. (1999) . *Nat. Neurosci.*, 1: 64–68.

He, S.Q., Dum, R.P. and Strick, P.L. (1993) Topographic organization of corticospinal projections from the frontal lobe: motor areas on the lateral surface of the hemisphere. *J. Neurosci.*, 13: 952–980.

He, S.Q., Dum, R.P. and Strick, P.L. (1995) Topographic organization of corticospinal projections from the frontal lobe: motor areas on the medial surface of the hemisphere. *J. Neurosci.*, 15: 3284–3306.

Hepp-Reymond, M.-C., Wyss, U.R. and Anner, R. (1978) Neuronal coding of static force in primate motor cortex. *J. Physiol. (Lond.)*, 74: 287–291.

Hepp-Reymond, M.-C., Hosler, E.J., Maier, M.A. and Qi, H.X. (1994) Force related neuronal activity in two regions of the primate ventral premotor cortex. *Can. J. Physiol. Pharmacol.*, 72: 571–579.

Hess, G. and Donoghue, J.P. (1999) Facilitation of long-term potentiation in layer II/III horizontal connections of rat motor cortex following layer I stimulation: route of effect and cholinergic contributions. *Exp. Brain Res.*, 127: 279–290.

Hill-Karrer, J., McKiernan, B.J. and Cheney, P.D. (1995) Mapping motor cortex output zones with stimulus triggered averaging of EMG activity: distal and proximal forelimb muscle representations in the rhesus monkey. *Soc. Neurosci. Abstr.*, 21: 2074.

Hummelsheim, H., Wiesendanger, M., Bianchetti, M., Wiesendanger, R. and Macpherson, J. (1986) Further investigations of the efferent linkage of the supplementary motor area (SMA) with the spinal cord in the monkey. *Exp. Brain Res.*, 65: 75–82.

Humphrey, D.R., Schmidt, E.M. and Thompson, W.D. (1970) Predicting measures of motor performance from multiple cortical spike trains. *Science*, 179: 758–762.

Huntley, G.W. (1997) Correlation between patterns of horizontal connectivity and the extend of short-term representational plasticity in rat motor cortex. *Cereb. Cortex*, 7: 143–156.

Kalaska, J.F., Cohen, D.A., Hyde, M.L. and Prud'homme, M. (1989) A comparison of movement direction-related versus load direction-related activity in primate motor cortex using a two-dimensional reaching task. *J. Neurosci.*, 9: 2080–2102.

Karni, A., Meyer, G., Jezzard, P., Adams, M.M., Turner, R. and Ungerleider, L.G. (1995) Functional MRI evidence for adult motor cortex plasticity during motor skill learning. *Nature*, 377: 155–158.

Kazenikov, O., Hyland, B., Corboz, M., Babalian, A., Rouiller, E.M. and Wiesendanger, M. (1999) Neural activity of supplementary and primary motor areas in monkeys and its relation to bimanual and unimanual movement sequences. *Neuroscience*, 89: 661–674.

Keller, A., Weintraub, D. and Miyashita, E. (1996) Tactile experience determines the organization of movement representations in rat motor cortex. *NeuroReport*, 7: 2373–2378.

Kew, J.J., Ridding, M.C., Rothwell, J.C., Passingham, R.E., Leigh, P.N., Sooriakumaran, S., Frackowiak, R.S. and Brooks, D.J. (1994) Reorganization of cortical blood flow and transcranial magnetic stimulation maps in human subjects after upper limb amputation. *J. Neurophysiol.*, 72: 2517–2524.

Kurata, K. (1989) Distribution of neurons with set- and movement-related activity before hand and foot movements in the premotor cortex of rhesus monkeys. *Exp. Brain Res.*, 77: 245–256.

Kurata, K. (1991) Corticocortical inputs to the dorsal and ventral aspects of the premotor cortex of macaque monkeys. *Neurosci. Res.*, 12: 263–280.

Kurata, K. (1994) Information processing for motor control in primate premotor cortex. *Behav. Brain Res.*, 61: 135–142.

Kurata, K. and Tanji, J. (1986) Premotor cortex neurons in macaques: activity before distal and proximal forelimb movements. *J. Neurosci.*, 6: 403–411.

Kurata, K. and Wise, S.P. (1988a) Premotor cortex of rhesus monkeys: set-related activity during two conditional motor tasks. *Exp. Brain Res.*, 69: 327–343.

Kurata, K. and Wise, S.P. (1988b) Premotor and supplementary motor cortex in rhesus monkeys: neuronal activity during externally- and internally-instructed motor tasks. *Exp. Brain Res.*, 72: 237–248.

Kurata, K., Okano, K. and Tanji, J. (1985) Distribution of neurons related to hindlimb as opposed to forelimb movement in the monkey premotor cortex. *Exp. Brain Res.*, 60: 188–191.

Kwan, H.C., MacKay, W.A., Murphy, J.T. and Wong, Y.C. (1978) Spatial organization of precentral cortex in awake primates, II. Motor outputs. *J. Neurophysiol.*, 41: 1120–1131.

Lemon, R.N., Mantel, G.W.H. and Muir, R.B. (1986) Corticospinal facilitation of hand muscles during voluntary movement in the conscious monkey. *J. Physiol. (Lond.)*, 381: 497–527.

Liepert, J., Tegenthoff, M. and Marlin, J.P. (1995) Changes in cortical motor area size during immobilization. *Electroenceph. Clin. Neurophysiol.*, 97: 382–386.

Liepert, J., Terborg, C. and Weiller, C. (1999) Motor plasticity induced by synchronized thumb and foot movements. *Exp. Brain Res.*, 125: 435–439.

Lin, S., Si, J. and Schwartz, A.B. (1997) Self-organization of firing activities in monkey's motor cortex: trajectory computation from spike signals. *Neural Comput.*, 9: 607–621.

Loeb, G.E., Zamin, C.J., Schulman, J.H. and Troyk, P.R. (1991) Injectable microstimulator for functional electrical stimulation. *Med. Biol. Eng. Comput.*, 29: NS13–19.

Lukasin, A.V., Amirikian, B.R. and Georgopoulos, A.P. (1996) A simulated actuator driven by motor cortical signals. *NeuroReport*, 7: 1597–2601.

Luppino, G., Matelli, M., Camarda, R.M., Gallese, V. and Rizzolatti, G. (1991) Multiple representations of body movements in mesial area 6 and adjacent cingulate cortex: an intracortical microstimulation study in the macaque monkey. *J. Comp. Neurol.*, 311: 463–482.

Matelli, M., Luppino, G. and Rizzolatti, G. (1991) Architecture of superior and mesial area 6 and the adjacent cingulate cortex in the macaque monkey. *J. Comp. Neurol.*, 311: 445–462.

Matsuzaka, Y., Aizawa, H. and Tanji, J. (1992) A motor area rostral to the supplementary motor area (pre-supplementary motor area) in the monkey: neuronal activity during a learned motor task. *J. Neurophysiol.*, 68: 653–662.

McBride, R.L., Feringa, E.R., Garver, M.K. and Williams Jr., J.K. (1989) Prelabeled red nucleus and sensorimotor cortex neurons of the rat survive 10 and 20 weeks after spinal cord transection. *J. Neuropathol. Exp. Neurol.*, 48: 568–576.

McBride, R.L., Feringa, E.R., Garver, M.K. and Williams Jr., J.K. (1990) Retrograde transport of fluoro-gold in corticospinal and rubrospinal neurons 10 and 20 weeks after T-9 spinal cord transection. *Exp. Neurol.*, 108: 83–85.

McKiernan, B.J., Marcario, J.K., Hill-Karrer, J. and Cheney, P.D. (1998) Corticomotoneuronal (CM) postspike effects on shoulder, elbow, wrist, digit and intrinsic hand muscles during a reach and prehension task in the monkey. *J. Neurophysiol.*, 80: 1961–1980.

Merline, M. and Kalil, K. (1990) Cell death of corticospinal neurons is induced by axotomy before but not after innervation of spinal targets. *J. Comp. Neurol.*, 296: 506–516.

Miller, L.E. and Houk, J.C. (1995) Motor co-ordinates in primate red nucleus: preferential relation to muscle activation versus kinematic variables. *J. Physiol. (Lond.)*, 488: 533–548.

Minor, L.A., McFarland, D.J. and Wolpaw, J.R. (1998) Answering questions with an electroencephalogram-based brain–computer interface. *Arch. Phys. Med. Rehabil.*, 79: 1029–1033.

Mitz, A.R. and Wise, S.P. (1987) The somatotopic organization of the supplementary motor area: intracortical microstimulation mapping. *J. Neurosci.*, 7: 1010–1021.

Murata, A., Fadiga, L., Fogassi, L., Gallese, V., Raos, V. and Rizzolatti, G. (1997) Object representation in the ventral premotor cortex (area F5) of the monkey. *J. Neurophysiol.*, 78: 2226–2230.

Nelson, R.J., Sur, M., Felleman, D.J. and Kaas, J.H. (1980) Representations of the body surface in postcentral cortex of *Macaca fascicularis*. *J. Comp. Neurol.*, 192: 611–643.

Nudo, R.J. and Milliken, G.W. (1996) Reorganization of movement representations in primary motor cortex following focal ischemic infarcts in adult squirrel monkeys. *J. Neurophysiol.*, 75: 2144–2149.

Nudo, R.J., Jenkins, W.M. and Merzenich, M.M. (1990) Repetitive microstimulation alters the cortical representation of movements in adult rats. *Somatosens. Mot. Res.*, 7: 463–483.

Nudo, R.J., Jenkins, W.M., Merzenich, M.M., Prejean, T. and Grenda, R. (1992) Neurophysiological correlates of hand preference in primary motor cortex of adult squirrel monkeys. *J. Neurosci.*, 12: 2918–2947.

Nudo, R.J., Milliken, G.W., Jenkins, W.M. and Merzenich, M.M. (1996a) Use-dependent alterations of movement representations in primary motor cortex of adult squirrel monkeys. *J. Neurosci.*, 15: 785–807.

Nudo, R.J., Wise, B.M., SiFuentes, F. and Milliken, G.W. (1996b) Neural substrates for the effects of rehabilitative training on motor recovery after ischemic infarct. *Science*, 272: 1792–1794.

Ojemann, J.G. and Silbergold, D.L. (1995) Cortical stimulation mapping of phantom limb rolandic cortex. Case report. *J. Neurosurg.*, 82: 641–644.

Olds, J. (1965) Operant conditioning of single unit responses. *Excerpta Med. Int. Congr. Ser.*, 87: 372–380.

Park, M.C., Belhaj-Saïf, A. and Cheney, P.D. (1999) Absence of a second distal muscle representation in maps of primary motor cortex in rhesus macaques. *Soc. Neurosci. Abstr.*, 25: 1663.

Pascual-Leone, A., Grafman, J. and Hallett, M. (1994) Modulation of cortical motor output maps during development of implicit and explicit knowledge. *Science*, 263: 1287–1289.

Passingham, R.E. (1996) Functional specialization of the supplementary motor area in monkeys and humans. *Adv. Neurol.*, 70: 105–116.

Picard, N. and Strick, P.L. (1996) Motor areas of the medial wall: a review of their location and functional activation. *Cereb. Cortex*, 6: 342–353.

Picard, N. and Strick, P.L. (1997) Activation on the medial wall during remembered sequences of reaching movements in monkeys. *J. Neurophysiol.*, 77: 2197–2201.

Porter, R. and Lemon, R. (1993) *Corticospinal Function and Voluntary Movement*. Clarendon Press, Oxford.

Preuss, T.M., Stepniewska, I. and Kaas, J.H. (1996) Movement representation in the dorsal and ventral premotor areas of owl monkeys: a microstimulation study. *J. Comp. Neurol.*, 371: 649–676.

Pruitt, J.N., Feringa, E.R. and McBride, R.L. (1988) Corticospinal axons persist in cervical and high thoracic regions 10 weeks after a T-9 spinal cord transection. *Neurology*, 38: 946–950.

Rioult-Pedotti, M.S., Friedman, D., Hess, G. and Donoghue, J.P. (1998) Strengthening the horizontal cortical connections following sill learning. *Nat. Neurosci.*, 1: 230–234.

Rizzolatti, G., Camarda, R., Fogassi, L., Gentilucci, M., Luppino, G. and Matelli, M. (1988) Functional organization of inferior area 6 in the macaque monkey, II. Area F5 and the control of distal movements. *Exp. Brain Res.*, 71: 491–507.

Rizzolatti, G., Luppino, G. and Matelli, M. (1998) The organization of the cortical motor system: new concepts. *Electroenceph. Clin. Neurophysiol.*, 106: 283–296.

Roelche, U., Curt, A., Otte, A., Missimer, J., Maguire, R.P., Dietz, V. and Leenders, K.L. (1997) Influence of spinal cord injury on cerebral sensorimotor systems: a PET study. *J. Neurol. Neurosurg. Psychiatry*, 62: 61–65.

Roland, P.E., Larsen, B., Lassen, N.A. and Skinhoj, E. (1980) Supplementary motor area and other cortical areas in organization of voluntary movements in man. *J. Neurophysiol.*, 43: 118–136.

Rossi, S., Pasqualetti, P., Tecchio, F., Sabato, A. and Rossini, P.M. (1998) Modulation of corticospinal output to human hand muscles following deprivation of sensory feedback. *Neuroimage*, 87: 163–175.

Rossini, P.M., Rossi, S., Tecchio, F., Pasqualetti, P., Finazzi-Agro, A. and Sabato, A. (1996) Focal brain stimulation in healthy humans: motor map changes following partial hand sensory deprivation. *Neurosci. Lett.*, 214: 191–195.

Rouiller, E.M., Yu, X.H., Moret, V., Tempini, A., Wiesendanger, M. and Liang, F. (1998) Dexterity in adult monkeys following early lesion of the motor cortical hand area: the role of cortex adjacent to the lesion. *Eur. J. Neurosci.*, 10: 729–740.

Sadato, N., Pascual-Leone, A., Grafman, J., Ibanez, V., Deiber, M.P., Dold, G. and Hallett, M. (1996) Activation or primary visual cortex by Braille reading in blind subjects. *Nature*, 380: 526–528.

Sadato, N., Yonekura, Y., Waki, A., Yamada, H. and Ishii, Y. (1997) Role of the supplementary motor area and the right premotor cortex in the coordination of bimanual finger movements. *J. Neurosci.*, 17: 9667–9674.

Sadato, N., Pascual-Leone, A., Grafman, J., Deiber, M.P., Ibanez, V. and Hallett, M. (1998) Neural networks for Braille reading by the blind. *Brain*, 121: 1213–1229.

Sanes, J.N., Suner, S. and Donoghue, J.P. (1990a) Dynamic organization of primary motor cortex output to target muscles in adult rats, I. Long-term patterns of reorganization following motor or mixed peripheral nerve lesions. *Exp. Brain Res.*, 79: 479–491.

Sanes, J.N., Suner, S. and Donoghue, J.P. (1990b) Dynamic organization of primary motor cortex output to target muscles in adult rats, II. Rapid reorganization following motor nerve lesions. *Exp. Brain Res.*, 79: 492–503.

Savaki, H.E., Kennedy, C., Sokoloff, L. and Mishkin, M. (1995) Visually-guided reaching with the forelimb contralateral to a 'blind' hemisphere: a metabolic mapping study in monkeys. *J. Neurosci.*, 13: 2772–2789.

Schieber, M.H. (1995) Muscular production of individuated finger movements: the roles of extrinsic finger muscles. *J. Neurosci.*, 15: 284–297.

Schieber, M.H. and Deuel, R.K. (1997) Primary motor cortex reorganization in a long-term monkey amputee. *Somatosens. Mot. Res.*, 14: 157–167.

Schieber, M.H. and Hibbard, L.S. (1993) How somatotopic is the motor cortex hand area?. *Science*, 261: 489–492.

Schieber, M.H., Chua, M., Petit, J. and Hunt, C.C. (1997) Tension distribution of single motor units in multitendoned muscles: comparison of a homologous digit muscle in cats and monkeys. *J. Neurosci.*, 17: 1734–1747.

Schwartz, A.B. (1994) Direct cortical representation of drawing. *Science*, 265: 540–542.

Schwartz, A.B., Kettner, R.E. and Georgopoulos, A.P. (1988) Primate motor cortex and free arm movements to visual targets in three-dimensional space, I. Relations between single cell discharge and direction of movement. *J. Neurosci.*, 8: 2913–2927.

Scott, S.H. and Kalaska, J.F. (1995) Changes in motor cortex activity during reaching movements with similar hand paths but different arm postures. *J. Neurophysiol.*, 73: 2563–2567.

Seitz, R.J. and Freund, H.J. (1997) Plasticity of the human motor cortex. *Adv. Neurol.*, 73: 321–333.

Seitz, R.J., Huang, Y., Knorr, U., Tellmann, L., Herzog, H. and Freund, H.J. (1995) Large-scale plasticity of the human motor cortex. *NeuroReport*, 27: 742–744.

Serlin, D.M. and Schieber, M.H. (1993) Morphologic regions of the multitendoned extrinsic finger muscles in the monkey forearm. *Acta Anat.*, 146: 255–266.

Shima, K. and Tanji, J. (1998) Role of cingulate motor area cells in voluntary movement selection based on reward. *Science*, 282: 1335–1338.

Shima, K., Aya, K., Mushiake, H., Inase, M., Aizawa, H. and Tanji, J. (1991) Two movement-related foci in the primate cingulate cortex observed in signal-triggered and self-paced forelimb movements. *J. Neurophysiol.*, 65: 188–202.

Smith, A.M. (1979) The activity of supplementary motor area neurons during a maintained precision grip. *Brain Res.*, 172: 315–327.

Streletz, L.J., Belevich, J.K., Jones, S.M., Bhushan, A., Shah, S.H. and Herbison, G.J. (1995) Transcranial magnetic stimulation: cortical motor maps in acute spinal cord injury. *Brain Topogr.*, 7: 245–250.

Strick, P.L. and Preston, J.B. (1982) Two representations of the hand in area 4 of a primate, I. Motor output organization. *J. Neurophysiol.*, 48: 139–149.

Tanji, J. (1994) The supplementary motor area of the cerebral cortex. *Neurosci. Res.*, 19: 251–268.

Tanji, J. and Shima, K. (1996) Contrast of neuronal activity between the supplemental motor area and other cortical motor areas. *Adv. Neurol.*, 70: 95–103.

Theriault, E. and Tator, C.H. (1994) Persistence of rubrospinal projections following spinal cord injury in the rat. *J. Comp. Neurol.*, 342: 249–258.

Weinrich, M. and Wise, S.P. (1982) The premotor cortex of the monkey. *J. Neurosci.*, 2: 1329–1345.

Weinrich, M., Wise, S.P. and Mauritz, K.H. (1984) A neurophysiological study of the premotor cortex in the rhesus monkey. *Brain*, 107: 385–414.

Wiesendanger, M. (1986) Recent developments in studies of the supplementary motor area of primates. *Rev. Physiol. Biochem. Pharmacol.*, 103: 1–59.

Wise, S.P. (1985) The primate premotor cortex: past, present and preparatory. *Annu. Rev. Neurosci.*, 8: 1–19.

Wise, S.P. (1996a) Evolutionary and comparative neurobiology of the supplementary sensorimotor area. *Adv. Neurol.*, 70: 71–83.

Wise, S.P. (1996b) Corticospinal efferents of the supplementary sensorimotor area in relation to the primary motor area. *Adv. Neurol.*, 70: 57–69.

Wise, S.P., Boussaoud, D., Johnson, P.B. and Caminiti, R. (1997) Premotor and parietal cortex: corticocortical connectivity and combinatorial computations. *Annu. Rev. Neurosci.*, 20: 25–42.

Woolsey, C.N., Settlage, P.H., Meyer, D.R., Sencer, W., Pinto Hamuy, T.P. and Travis, A.M. (1952) Patterns of localization in precentral and 'supplementary' motor areas and their relation to the concept of a premotor area. *Res. Publ. Assoc. Ment. Dis.*, 30: 238–264.

Wyrwicka, W. and Sterman, M.B. (1968) Instrumental conditioning of sensorimotor cortex EEG spindles in the waking cat. *Physiol. Behav.*, 3: 703–707.

Zanette, G., Bonato, C., Polo, A., Tinazzi, M., Manganotti, P. and Fiaschi, A. (1995) Long-lasting depression of motor-evoked potentials to transcranial magnetic stimulation following exercise. *Brain Res.*, 107: 80–86.

Ziemann, U., Corwell, B. and Cohen, L.G. (1998a) Modulation of plasticity in human motor cortex after forearm ischemic nerve block. *J. Neurosci.*, 18: 1115–1123.

Ziemann, U., Hallet, M. and Cohen, L.G. (1998b) Mechanisms of deafferentation-induced plasticity in human motor cortex. *J. Neurosci.*, 18: 7000–7007.

CHAPTER 13

Network level properties of short-term plasticity in the somatosensory system

David J. Krupa* and Miguel A.L. Nicolelis

Department of Neurobiology, Box 3209, Duke University Medical Center, Bryan Research Building, Room 333, 101 Research Drive, Durham, NC 27710, USA

Introduction

Over the past three decades, numerous studies have demonstrated that sensory representations within the adult mammalian brain are capable of being dynamically shaped and altered by the loss or modification of normal ascending sensory input at the periphery, such as by a peripheral nerve cut or digit or limb amputation (for review see: Merzenich ct al., 1984; Wall, 1988; Kaas, 1991; Buonomano and Merzenich, 1998). These sensory deprivations typically result in a shift in the receptive field (RF) properties of sensory neurons such that, following the sensory loss, sensory neurons respond to stimuli that previously did not elicit significant neuronal responses. The time course of this plastic reorganization ranges from immediately following a sensory deprivation (Dostrovsky et al., 1976; Metzler and Marks, 1979; Merzenich et al., 1983b; Rasmusson and Turnbull, 1983; Calford and Tweedale, 1988; Nicolelis et al., 1993b; Faggin et al., 1997; Krupa et al., 1999) to weeks or months later (Merzenich et al., 1983a; Wall and Cusick, 1984; Wilson and Snow, 1987; Pons et al., 1991; Florence and Kaas, 1995). In addition, this sensory reorganization has been shown to occur in the somatosensory (Merzenich et al., 1983a, 1984; Calford and Tweedale, 1988; Pettit and Schwark, 1993), visual (Eysel, 1982; Kaas et al., 1990; Chino et al., 1992; Gilbert and Wiesel, 1992) and auditory systems (Robertson and Irvine, 1989; Rajan et al., 1993; Irvine and Rajan, 1997). Such types of plasticity have been demonstrated in a wide range of mammalian species, including humans (Elbert et al., 1994; Yang et al., 1994; Dostrovsky, 1999). In short, plastic reorganization of sensory representations following the loss of normal sensory input at the periphery appears to be a ubiquitous property of the mammalian central nervous system (CNS).

One of the primary goals of research in this field is to gain a detailed understanding of the underlying cellular and network level properties that might mediate this sensory plasticity. Understanding the mechanisms of sensory plasticity is important because such an understanding may lead to improvements in the rehabilitation and recovery of sensory and motor skills following peripheral nerve damage. Further, comprehension of the mechanisms mediating sensory plasticity might also contribute to the revelation of the underlying mechanisms that mediate the acquisition or modification of learned motor skills as well as contribute to the broad knowledge of cortically dependent memories.

To date, much of the study of the sensory plasticity that follows a sensory deprivation has focused primarily on cortical sensory areas (Merzenich et al.,

---

* Corresponding author: Dr. David J. Krupa, Department of Neurobiology, Box 3209, Duke University Medical Center, Bryan Research Building, Room 333, 101 Research Drive, Durham, NC 27710, USA. Fax: +1-919-684-5435; E-mail: Krupad@neuro.duke.edu

1983a, 1984; Calford and Tweedale, 1988; Robertson and Irvine, 1989; Kaas et al., 1990). However, sensory reorganization has also been shown to occur in subcortical structures (Wall and Egger, 1971; Devor and Wall, 1978; Wilson and Snow, 1987; Garraghty and Kaas, 1991a; Pettit and Schwark, 1993; Faggin et al., 1997; Parker et al., 1998). Thus, a fundamental question in our understanding of sensory plasticity is how plastic reorganization in one structure affects the reorganization of other structures within the sensory pathway. For instance, it might be that plasticity occurs only in cortical areas and that the reorganization of RFs seen in subcortical structures is simply a reflection of cortical plasticity that has been relayed to the subcortical structures via corticofugal feedback projections. Alternatively, the reorganization of RFs seen in cortical areas may simply be the result of plastic changes that have occurred in subcortical or spinal structures that have ascended to cortex. A third hypothesis is that both ascending and descending pathways contribute to this phenomenon. As described below, recent evidence indicates that the sensory plasticity that occurs immediately following a peripheral sensory deprivation appears to occur at multiple sites within a sensory system, and that corticofugal feedback projections, as well as ascending pathways, may play an important role in the plastic reorganization seen in subcortical structures (Nicolelis et al., 1993b; Faggin et al., 1997; Krupa et al., 1999; Parker and Dostrovsky, 1999).

A second issue related to sensory plasticity regards the time course of the plastic changes following a sensory deprivation. Many studies have examined plastic reorganization weeks or months after a sensory loss (Kalaska and Pomeranz, 1979; Merzenich et al., 1983a; Garraghty and Kaas, 1991b; Pons et al., 1991; Florence and Kaas, 1995). However, other studies have demonstrated that RF reorganization can occur essentially immediately after the peripheral sensory loss (Rasmusson and Turnbull, 1983; Calford and Tweedale, 1988; Nicolelis et al., 1993b; Pettit and Schwark, 1993; Faggin et al., 1997; Krupa et al., 1999). Thus, an important question regarding the nature of this plasticity is the time course of the plastic changes in different structures. For instance, is there a temporal cascade of plasticity, appearing first in one structure and then progressing later to a downstream structure? Since many of the characteristics of the reorganization seen immediately after a sensory deprivation are similar to those seen weeks or months later, it is possible that the sensory plasticity that immediately follows a peripheral sensory deprivation may provide the initial stimulus for triggering long-term modifications in the sensory pathway, such as changes in synaptic plasticity or even sprouting of terminals, that could further amplify the process of plastic reorganization over weeks to months after the sensory loss.

The nature of spatiotemporal receptive fields

As stated above, RF reorganization following a sensory deprivation has been shown to occur in the visual and auditory as well as the somatosensory system. Many of the basic properties of the RF reorganization seen within these different sensory systems are very similar. In this chapter, we will focus primarily on reorganization within the somatosensory system, in particular, within the trigeminal somatosensory system of the rat.

The basic measure typically used for describing and quantifying plasticity following a peripheral sensory deprivation is changes in the spatial and/or temporal characteristics of the RFs of individual sensory neurons. Typically, the RF of a somatosensory neuron is defined as the region of body surface that, when stimulated, induces a significant change in firing rate of the neuron. In the rat trigeminal somatosensory system (see Fig. 1, below), quantitative analysis of the RFs of neurons in the ventral posterior medial (VPM) nucleus of the thalamus and the primary sensory cortex, S1, have demonstrated that these neurons have large, multiwhisker RFs whose center is defined by the whisker that, when stimulated, elicits the strongest sensory response (Nicolelis et al., 1993a; Nicolelis and Chapin, 1994; Ghazanfar and Nicolelis, 1999). This whisker, known as the principal whisker of the RF, is used to identify the location of a given neuron in the topographic map of the whisker pad observed across the trigeminal system. However, the surround RF of typical VPM or S1 neurons encompasses a much larger region of the whisker pad than simply the principal whisker. On average, the RF surround of VPM neurons is defined by up to 13 whiskers whose

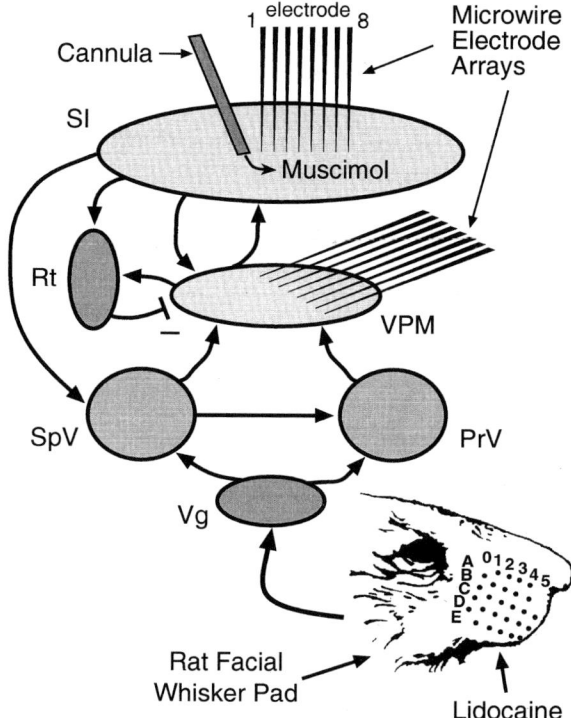

Fig. 1. Simplified schematic of the rat trigeminal somatosensory system. Both feedforward (from the periphery to cortex) and feedback (from cortex to subcortical nuclei) projections are illustrated to highlight the recurrent nature of this neural network. Also shown is the characteristic pattern of rows (A to E) and columns (0 to 5 shown) of the rat's facial whiskers. In different experiments, microwire electrode arrays were implanted in multiple structures within the trigeminal system, including the barrel region of the S1 cortex, VPM thalamus, and spinal trigeminal nucleus (not shown here). An infusion cannula was also implanted in S1 cortex to allow the infusion of the $GABA_A$ agonist, muscimol. Single unit activity was isolated on the individual electrodes. Then, up to 25 facial whiskers were individually stimulated multiple times and the resulting receptive field (RF) properties for each neuron were quantified. Then, depending upon the experiment, either the S1 cortex was inactivated via infusion of 150 ng of muscimol and/or a peripheral nerve block was induced by injection of 40 μl of 1% lidocaine. During these inactivations, the same set of whiskers was again stimulated and the resulting RFs quantified. Comparison of RF properties before and after a particular manipulation yielded statistically quantifiable changes in RF properties. Abbreviations: *PrV* = principal trigeminal nucleus; *Rt* = reticular thalamic nucleus; *S1* = primary somatosensory cortex; *SpV* = spinal trigeminal nucleus; *Vg* = trigeminal ganglion; *VPM* = ventroposterior medial thalamic nucleus. (Modified from Krupa et al., 1999.)

stimulation produces an excitatory response that is smaller in magnitude and longer in latency than the response produced by stimulation of the principal whisker (Nicolelis et al., 1993a; Nicolelis and Chapin, 1994). Similar results are obtained when recording from layer V S1 cortical neurons in the barrel cortex where the surround RF averages eight whiskers (Ghazanfar and Nicolelis, 1999). These large RFs suggest that individual VPM or S1 cortical neurons receive afferent input from a large region of the facial whisker pad. The existence of spatially diverse afferent input converging on VPM and S1 neurons has been confirmed by in vivo intracellular recordings which demonstrate that the subthreshold RFs of individual neurons in these structures are composed of many whiskers (Chiaia et al., 1991b; Moore and Nelson, 1998). In these experiments, excitatory postsynaptic potentials (EPSPs) of different magnitudes and latencies were elicited in single VPM or S1 neurons when many different individual whiskers were stimulated one at a time.

Recent evidence, using new methods to characterize the RF properties of single neurons, has demonstrated that, in addition to being larger than originally reported, the RFs of both cortical and subcortical sensory neurons are much more dynamic than first suspected (Nicolelis et al., 1993a; Ghazanfar and Nicolelis, 2000). For example, in a study in which this dynamic nature of RFs in VPM neurons was examined (Nicolelis and Chapin, 1994), the RF center of a large percentage of VPM neurons was found to shift over post-stimulus time. In these neurons, the principal whisker that was defined by the short-latency component of the response elicited by individually stimulating a large number of whiskers was different than the principal whisker defined by the long-latency component of the evoked response. In these VPM neurons, the principal whisker, defined by the responses recorded 5–15 ms following stimulus, was located in the caudal region of the whisker pad. However, in these same neurons, the principal whisker, defined by the responses recorded 20–50 ms following stimulus, shifted to the more rostral region of the whisker pad. In short, these neurons displayed a dynamic, post-stimulus time-dependent shift in their RF centers that migrated from a caudal to more rostral position within the overall whisker pad. A second class of neurons was recorded in the

VPM in which there was no long-latency shift in the RF center. In these neurons, the short-latency principal whisker was centered in the rostral region of the whisker pad. The principal whisker defined by the long-latency component of the response remained centered on the same rostral whisker as the short-latency principal whisker.

More recently, the RF dynamics of 197 cortical neurons located in layer V of the rat S1 cortex were studied with the same methods employed to characterize VPM RFs (Ghazanfar and Nicolelis, 1999). S1 layer V RFs were found to be quite large in lightly anesthetized animals (mean = 8.5 whiskers). Moreover, in 88% of layer V S1 neurons, the spatial domain of the RFs changed as a function of post-stimulus time. Interestingly, the patterns of these S1 cortical spatiotemporal RFs varied widely, unlike those reported for VPM, which shifted primarily in a caudal to rostral direction, or not at all. A total of four main directions of RF shifts were observed in this study: (1) rostral-to-caudal (17.7% of the sample); (2) caudal-to-rostral (16.0%); (3) dorsal-to-ventral (23.4%); and (4) ventral-to-dorsal (6.3%). Neurons whose RFs traversed an equal distance in two or more directions were defined as unclassifiable. Ghazanfar and Nicolelis (1999) also demonstrated that the direction of the spatiotemporal RF shift was dependent upon the location of the short-latency principal whisker. If a neuron's short-latency principal whisker was located in the dorsal part of the whisker pad, then the neuron was more likely to exhibit a dorsal-to-ventral spatiotemporal RF. If a neuron's short-latency principal whisker was in the more rostral part of the whisker pad, then it was more likely to exhibit a rostral-to-caudal spatiotemporal RF.

In summary, these results demonstrate that the RFs of both thalamic and cortical somatosensory neurons encompass a relatively large region of the facial whisker pad, integrating tactile information from many whiskers. Further, these results show that the properties of these RFs, in particular, the RF centers, are not static. Instead, the RFs of a large percentage of neurons in these areas demonstrate dynamic, time-dependent shifts in their spatial characteristics. In fact, data obtained in the visual, auditory and somatosensory systems indicate that RFs are better defined as spatiotemporal entities, since their spatial domain (or frequency tuning in the case of the auditory system) often varies as a function of post-stimulus time (Dinse et al., 1991; DeAngelis et al., 1993a,b; Ringach et al., 1997; Ghazanfar and Nicolelis, 1999). As described below, recent evidence suggests that the asynchronous convergence of multiple ascending (e.g., lemniscal and paralemniscal pathways), horizontal (e.g., corticocortical), and descending (e.g., corticothalamic pathways) afferents may account for the fact that neurons within both S1 cortex and VPM thalamus have highly dynamic RFs, whose spatial domain varies as a function of post-stimulus time (Nicolelis et al., 1993a; Nicolelis and Chapin, 1994; Ghazanfar and Nicolelis, 1999).

Network level properties of dynamic spatiotemporal receptive fields

Although the precise cellular and network level interactions that might give rise to these dynamic, spatiotemporal RFs are not presently clear, the data suggest that the asynchronous convergence of both feedforward and feedback projections onto VPM and S1 neurons may be critically involved (Nicolelis and Chapin, 1994; Nicolelis, 1997; Nicolelis et al., 1998). In this scenario, asynchronous inputs (i.e., arriving at different time epochs) from parallel, feedforward lemniscal and paralemniscal trigeminothalamic pathways converge at different locations on the dendritic tree and cell body of individual VPM neurons (Chiaia et al., 1991a,b). These pathways arise from different subdivisions of the trigeminal brainstem complex and are known to have different temporal lags, primarily due to the differences in conduction velocities of their axons (large and myelinated axons from the principal trigeminal nucleus, and small and unmyelinated axons from the spinal trigeminal complex). In principle, these differences in conduction velocity could account for at least some of the different latency components of the RFs of VPM neurons. Indeed, a similar time-lagged feedforward mechanism has been independently proposed for the genesis of spatiotemporal RFs in the lateral geniculate nucleus (Stevens and Gerstein, 1976; Cai et al., 1997) and in the primary visual cortex (DeAngelis et al., 1995).

In addition to these feedforward projections, corticofugal feedback projections would also be critically involved in the genesis of the spatiotemporal aspects of VPM RFs. The effects of corticofugal feedback on VPM RFs might include tonic inhibition, generated primarily by cortically driven excitation of γ-aminobutyric acid (GABA)ergic neurons located in the reticular nucleus of the thalamus and likely in the trigeminal brainstem complex (Jacquin et al., 1990; Lee et al., 1994). Further, direct corticothalamic glutamatergic projections may also account for a significant number of the long-latency, excitatory responses of VPM neurons which have a different long-latency RF center (or principal whisker) than the one observed at short-latency (Nicolelis and Chapin, 1994). In this scheme, the spatiotemporal structure of a single VPM neuron's RF would result from the temporally asynchronous convergence of parallel ascending trigeminothalamic projections and the dual excitatory and inhibitory effects of descending corticothalamic pathways.

Recently, we have begun to test these hypotheses, in particular, the role of corticothalamic projections in the genesis of VPM RFs. To address this issue, a microcannula was implanted in the S1 cortex, adjacent to microwire recording arrays, so that pharmacologically active compounds (such as muscimol, a $GABA_A$ agonist) could be infused in the vicinity of neurons located in a given brain region (cortical or subcortical) (Krupa et al., 1999). In this experimental paradigm (see Fig. 1), simultaneous recording of cortical neuronal activity was used to measure the effectiveness of muscimol (150 ng/150 nl in saline) infusion to block the activity of infragranular cortical neurons that project to the VPM thalamus. Throughout the duration of the cortical activity block (6–9 h under pentobarbital anesthesia), we were able to quantify the effects of removing the contribution of corticofugal pathways on the RFs of VPM neurons. The spatiotemporal RFs of VPM neurons were mapped prior to and during the S1 inactivation as described above. Overall, 70% of the VPM neurons exhibited changes in their RFs. In ~55% of the VPM neurons, cortical inactivation led to a significant reduction of the long-latency excitatory component of their RFs. This observation, which was consistent with our hypothesis, indicates that the reciprocal interactions with the S1 cortex account for a great portion of the dynamic organization of thalamic RFs in rats. Interestingly, in 38% of the VPM neurons, cortical inactivation led to unmasking of sensory responses to single whisker stimuli, suggesting that a source of tonic inhibitory influence to the VPM nucleus was removed. Since corticothalamic projections also reach the reticular nucleus of the thalamus, the main source of inhibitory feedback to the rat VPM, it is likely that interruption of the excitatory drive from the cortex to the reticular nucleus was responsible for these unmasked VPM responses. Finally, 26% of the VPM neurons exhibited modifications in the magnitude of their short-latency responses to single whisker inactivation following cortical inactivation. Significant shifts in sensory response latency were observed in 16% of the VPM neurons.

In summary, the results presented above demonstrate that somatosensory RFs are not static, time-invariant entities. Instead, these RFs are more accurately viewed as dynamic constructs comprised of spatiotemporally varying patterns of neuronal responses to peripheral stimulation. We have hypothesized that the spatiotemporally varying nature of RFs results from the asynchronous convergence of parallel ascending trigeminothalamic projections and the dual excitatory and inhibitory effects of descending corticothalamic pathways (Nicolelis and Chapin, 1994; Nicolelis, 1997; Nicolelis et al., 1998). In this scenario, an individual sensory neuron's RF is defined by the dynamic balance between these excitatory and inhibitory ascending and descending influences. This hypothesis predicts that perturbations in any one part of the system would result in an immediate rebalancing of these excitatory and inhibitory influences throughout the rest of the system. This rebalancing would be manifest by the emergence of shifts or modifications in the RF properties of sensory neurons located in different parts of the system. This hypothesis is supported by the results of cortical inactivation experiments which demonstrate that corticofugal feedback projections play a critical role in the definition of RF properties of VPM sensory neurons (Ergenzinger et al., 1998; Krupa et al., 1999; Parker and Dostrovsky, 1999). Further, the hypothesis described here suggests that the reorganization of RF properties seen immediately following a peripheral deafferentation is driven by a dynamic rebalancing of the ascending and descending projections that converge at all levels of the somatosensory system and define the organization of single neuronal RFs throughout the somatosensory system.

Sensory plasticity that immediately follows a peripheral deafferentation

Numerous studies have reported the occurrence of reorganization of RFs immediately following a peripheral sensory deafferentation. This sensory reorganization process that immediately follows a peripheral deafferentation is characterized by changes in both the spatial and temporal domains of the RFs of cortical, thalamic, and brainstem neurons. However, while these studies have consistently described the occurrence of this short-term plasticity, identification of the underlying cellular and network level mechanisms that might mediate this reorganization has remained elusive. In order to effectively quantify the changes that occur in RF properties immediately after a sensory loss, it is critical to record from the same sensory neurons before and immediately after the sensory deprivation so that any changes in RF properties can be precisely measured.

In an effort to obtain precise, quantifiable measures of how a peripheral deafferentation affects RFs immediately after the sensory loss, we have conducted a series of experiments in which simultaneous recordings from large ensembles of single neurons, located at multiple sites within the rat trigeminal somatosensory system, were obtained in the same rats both before and after the administration of a reversible peripheral nerve block (Nicolelis et al., 1993b; Faggin et al., 1997; Nicolelis, 1997; Katz et al., 1999; Krupa et al., 1999). This paradigm allows the accurate quantification of the RF reorganization that results from the reversible peripheral nerve block by comparing the changes in RF properties in the same neurons. In these experiments, arrays of microwire recording electrodes were implanted in multiple structures within the trigeminal somatosensory system including the spinal trigeminal nucleus, the VPM thalamus and the barrel region of the S1 cortex. Following surgical recovery, rats were lightly anesthetized and single unit activity was isolated on the microelectrodes. Up to 135 single units were isolated in a single rat. Once single unit activity was isolated, the RF properties of the recorded neurons were quantified as follows. A computer-controlled probe was used to mechanically stimulate individual facial whiskers (deflection amplitude 3–5°; stimulus duration 100 ms; stimulus frequency 1 Hz) one at a time, in random order. Up to 21 whiskers were stimulated in the control phase of these experiments to characterize the RFs of the brainstem, thalamic, and cortical neurons before the induction of a sensory deafferentation. In addition, the magnitude of sensory responses and latency distributions were also used to characterize the effects of the sensory deafferentation on the physiological properties of cortical and subcortical neuronal populations. Once this control phase was finished, a small amount of lidocaine (0.04 ml at 1% in saline) was injected in one location of the animal's face near the facial whisker pad. This lidocaine injection created a reversible block of normal trigeminal afferent fibers in the periphery. Immediately after this injection, the same set of individual whiskers was stimulated and the same measurements were obtained. The RF properties of each neuron were then compared before and after the nerve block using quantitative, statistical measures.

In the first of these studies, microelectrode recording arrays were implanted in the VPM thalamus and the activity of large populations of single VPM neurons was recorded prior to, during and after the lidocaine-induced peripheral nerve block (Nicolelis et al., 1993b). The results showed that both spatial and temporal components of VPM neuron RFs were modified immediately following the lidocaine-induced nerve block. This reorganization involved the immediate unmasking of new neuronal sensory responses, shifting of neuronal RFs away from the facial anesthetized zone, and significant alterations in VPM neuronal latencies. These results demonstrated that sensory reorganization occurs in thalamic sensory nuclei immediately after a peripheral sensory loss.

An important question that arose from this study was: since short-term plasticity is seen in thalamus immediately after a peripheral nerve block, how might this plasticity compare with reorganization seen in other regions of the trigeminal somatosensory system? For instance, is there a temporal progression of plasticity that appears first in brainstem trigeminal nuclei, then in thalamus and finally in S1 cortex? Alternatively, this order might be reversed. Also, how do the properties of reorganization at different levels in the system compare? The answers to these questions would be critically important to understanding the network level interactions that might mediate this sensory plasticity.

In order to address these questions, chronic and simultaneous multisite neural ensemble recordings were employed to compare the process of immediate sensory reorganization at three different levels of the rat trigeminal system: the pars interpolaris of the spinal subdivision of the trigeminal complex (SpV nucleus), the VPM nucleus of the thalamus, and the infragranular layers (primarily layer V) of the S1 cortex (Faggin et al., 1997).

Overall, data derived from twelve rats revealed that 65–70% of the recorded single neurons, distributed across the entire somatosensory system, exhibited signs of undergoing an immediate sensory reorganization following a lidocaine-induced reversible peripheral deafferentation. In these experiments, brainstem, thalamic, and cortical neurons consistently exhibited unmasking of novel sensory responses immediately after the peripheral block was initiated. Importantly, no clear sequence of establishment of these modifications was observed. In other words, even though one might expect that a gradient of changes moving from the brainstem toward the somatosensory cortex should be observed, we invariably noticed that the process of sensory reorganization unfolded concurrently at all processing levels of the rat trigeminal system.

In addition to their simultaneity, several measurements indicated that the immediate reorganization process was very similar at both cortical and thalamic levels. For example, no statistical difference was found between the number of neurons exhibiting novel sensory responses in the S1 cortex ($71.1 \pm 5.2\%$) and the VPM thalamus ($66.4 \pm 10.7\%$). Changes in neuronal sensory responses were also used to establish the size of both the region of the whisker pad that was anesthetized by the lidocaine injection and the region from which unmasked responses could be elicited after this peripheral block. Immediately after the lidocaine block, cortical and subcortical neurons rapidly lost their responsiveness to stimulation of the whiskers located in the 'anesthetized' zone and started to respond to neighboring whiskers located beyond their original RFs. The whiskers that generated unmasked sensory responses when stimulated defined the 'unmasking zone' of each subcortical and cortical somatotopic map. When the overall spatial extent (in number of whiskers) of the unmasked zones was quantitatively compared, again no statistical difference was found between the immediate reorganization process in the S1 cortex (9.1 ± 1.2 whiskers) and the VPM thalamus (6.1 ± 1.6 whiskers). Instead, the reorganization process was characterized by a great deal of overlap between the spatial distribution of unmasked responses in the S1 cortex and VPM thalamus, and to a lesser degree in the SpV nucleus. The only significant change detected by this analysis was that individual S1 cortical neurons exhibited larger expansions of their RFs (2.3 ± 0.12 whiskers) than VPM neurons (1.7 ± 0.13 whiskers) during the duration of the lidocaine block.

Two other analyses were employed to further compare parameters of cortical and thalamic unmasked responses: the magnitude and latencies of novel sensory responses. This analysis revealed that the magnitudes of unmasked sensory responses in the VPM thalamus (24.8 ± 2.2 spikes/s) were significantly higher than those of the S1 cortex (18.3 ± 0.8 spikes/s). Interestingly, the magnitudes of the cortical and thalamic unmasked responses were equivalent to the normal responses obtained when the surrounding regions (i.e., whiskers) of the original RFs of these neurons were stimulated during the control phase. These observations suggest that the immediate reorganization process either unmasked far surround areas of the original RFs or allowed subthreshold regions of these RFs to be expressed due to a lack of surround inhibition or an increase in excitatory drive to the neurons.

Our simultaneous thalamo-cortical recordings also indicated that profound changes in latency occurred concurrently across the somatosensory system immediately after the onset of a peripheral deafferentation. Shifts in thalamic and cortical latency distributions during the lidocaine block were very similar. The average minimal latency of cortical sensory responses was much longer (19.6 ± 0.3 ms) than normal (7–10 ms) and significantly longer than the equivalent unmasked thalamic responses (13.1 ± 0.6 ms), which were also much longer than normal (4–6 ms). These latency differences between thalamic and cortical unmasked sensory responses could not be explained by normal synaptic and conduction delays (around 2–3 ms) of the thalamocortical projection. A more detailed analysis of these latency distributions also revealed that cortical neurons exhibited a higher proportion of very long latency components in their unmasked responses when compared to thalamic neurons, suggesting that corticocortical interactions were also unmasked after the peripheral deafferentation.

Role of corticofugal feedback projections in sensory reorganization of subcortical structures

As mentioned above, data collected in our laboratory indicated that the excitatory drive provided by corticothalamic projections from the S1 cortex contributes significantly to the genesis of long-latency responses and spatiotemporal RFs of VPM neurons. Based on these findings, we decided to investigate the potential contribution of corticofugal projections to the process of immediate plastic reorganization in the VPM thalamus following a peripheral lidocaine block (Krupa et al., 1999). In these experiments, after a reversible cortical inactivation was produced by infusing muscimol in the rat S1 cortex, a small subcutaneous lidocaine block was created while the activity of populations of VPM neurons was recorded simultaneously. We observed that blocking neuronal activity in the infragranular layers of the rat primary somatosensory cortex leads to a 50% reduction in the percentage of thalamic neurons exhibiting immediate changes in their RFs following a subcutaneous injection of lidocaine (from $66 \pm 10.7\%$ to $29 \pm 3.2\%$; see Fig. 2). Inactivation of corticofugal projections from the S1 cortex also produces a statistically significant reduction in the size of neuronal RF unmasking observed in the VPM during the maintenance of the peripheral lidocaine blockade (from 1.7 ± 0.13 to 0.75 ± 0.11 whiskers). Finally, cortical activity block also leads to a small reduction

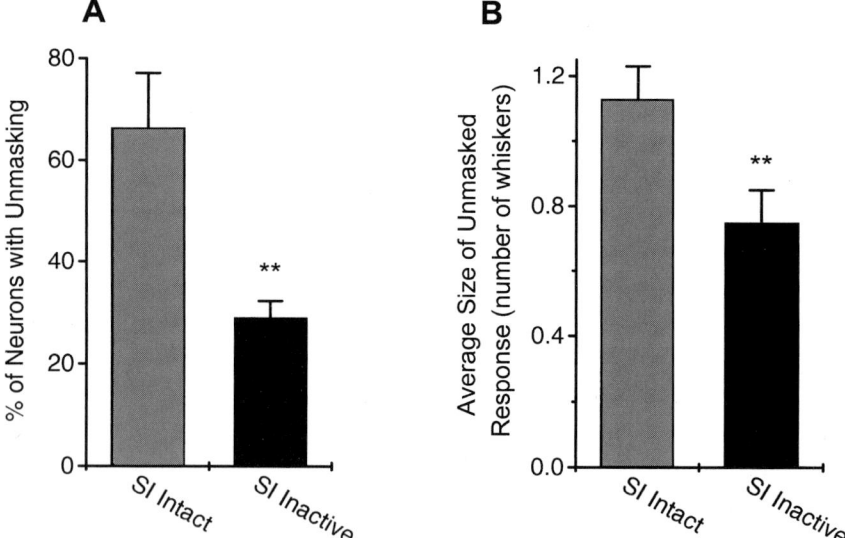

Fig. 2. (A) Percentage of VPM neurons (mean ± SEM) in the total recorded population showing unmasked responses to whisker stimulation following peripheral lidocaine deafferentation. Cortical inactivation resulted in a significant decrease in the percentage of VPM neurons showing unmasked responses following peripheral deafferentation (from $66 \pm 10.7\%$ to $29 \pm 3.2\%$; $t(7) = -4.5$, $P < 0.003$). S1 Intact = percentage of VPM neurons showing unmasked responses following facial lidocaine injections in which S1 cortex was not inactivated; S1 Inactive = percentage of VPM neurons showing unmasked responses following facial lidocaine injections while S1 cortex was inactivated with muscimol. (B) Average change in RF size (mean number of whiskers ± SEM) following facial lidocaine for each of the conditions described above. Cortical inactivation resulted in a significant decrease in the size of the unmasked RF following peripheral deafferentation (from 1.13 ± 0.11 to 0.75 ± 0.11 whiskers, $t(11) = 5.9$, $P < 0.005$). (Modified from Krupa et al., 1999.)

in the overall area of thalamic reorganization. The cortical inactivation additionally induced variations in the response latencies and magnitudes of VPM neurons during the reorganization process. In short, inactivation of descending corticofugal projections significantly reduced but did not eliminate the thalamic RF reorganization that immediately follows a lidocaine-induced peripheral nerve block.

Although these experiments alone cannot resolve the question of whether the reduction in thalamic reorganization was due to alterations in corticothalamic projections, or whether they reflected concurrent modifications in trigeminal brainstem nuclei due to alterations in corticobulbar projections, they provide the first evidence that corticofugal projections may contribute to the unfolding of the immediate reorganization process in the VPM thalamus. It is important to note that a sizable thalamic reorganization was still present during the blockade of S1 corticofugal projections. This suggests that parallel feedforward pathways are still capable of sustaining the reorganization process in the thalamus. One cannot eliminate the possibility that other corticofugal projections, which originate in distinct somatosensory cortical areas (such as S2, parietal ventral area, etc.) and converge on VPM neurons, could also contribute to the process of thalamic reorganization.

Conclusion

In conclusion, the results described above clearly indicate that the process of immediate sensory reorganization involves alterations in a variety of ascending and descending projections that converge on individual neurons and local circuitry at all levels of the somatosensory pathway. The occurrence of circuit-wide modifications could explain why the concurrent transformations observed at thalamic and cortical levels are so similar. Moreover, these data also underscore the relevance of taking into account the fact that subcortical plasticity may influence and be influenced by the process of cortical reorganization. To date, the debate concerning the degree to which subcortical structures, particularly different thalamic nuclei, influence the process of cortical reorganization that is triggered either by a peripheral deafferentation or by alterations in sensory experience is not settled. However, most laboratories that have investigated this issue have observed conclusive evidence for the occurrence of subcortical plasticity following a peripheral sensory loss.

More importantly, many of the main features of immediate plastic reorganization, such as unmasking of novel sensory responses, latency modifications, and expansion of RFs away from the deafferented zone are very similar to features of plasticity observed weeks or months after a peripheral deafferentation. This finding suggests that the sensory plasticity that immediately follows a peripheral sensory deprivation may provide the initial stimulus for triggering long-term modifications in the sensory pathway, such as changes in synaptic plasticity or sprouting of terminals, that might mediate the process of plastic reorganization seen weeks to months after the sensory loss. As described above, immediately after a peripheral deafferentation, there appears to be a rebalancing of both ascending and descending excitatory and inhibitory inputs at all levels of the trigeminal somatosensory system. These changes in the balance of excitation and inhibition lead to altered patterns of neural ensemble firing which can be seen as spatial and/or temporal changes in the RF properties of sensory neurons. These altered patterns of firing might induce, over time, changes in synaptic efficacy at one or more sites within the sensory system (Bear and Kirkwood, 1993; Kirkwood et al., 1996; Markram et al., 1997). Further, if altered patterns of neuronal activity persist over days or weeks, other levels of modifications may take place. For example, prolonged changes in neuronal activity could lead to the release of neurotrophins which are known to increase dendritic growth and arborization (McAllister et al., 1997) and to the establishment of new neural connections (Riddle et al., 1995). In the visual system, sprouting of this sort has been documented to occur in a matter of a few days (Antonini and Stryker, 1993). In addition, different circuits may be modified during the days and weeks that follow a peripheral deafferentation. For example, some authors have suggested that a selective functional strengthening or even sprouting of long-range horizontal corticocortical connections in the visual and somatosensory cortex could account for most of the cortical plasticity observed in adult animals (Darian-Smith and Gilbert, 1994; Florence et al., 1998). It is likely, however, that sprouting of many pathways, including ascending and descending projections, may take place during the long-term recovery from a peripheral deafferentation. Approaching the problem of understanding the system-wide changes that mediate both the immediate and long-term plasticity that follows sensory modifications, as well as those mechanisms that might mediate the transition from short- to long-term reorganization, is not a trivial matter and much more work in this area will be required to fully understand these processes.

References

Antonini, A. and Stryker, M.P. (1993) Rapid remodeling of axonal arbors in the visual cortex. *Science*, 260: 1819–1821.

Bear, M.F. and Kirkwood, A. (1993) Neocortical long-term potentiation. *Curr. Opin. Neurobiol.*, 3: 197–202.

Buonomano, D.V. and Merzenich, M.M. (1998) Cortical plasticity: from synapses to maps. *Annu. Rev. Neurosci.*, 21: 149–186.

Cai, D., DeAngelis, G.C. and Freeman, R.D. (1997) Spatiotemporal receptive field organization in the lateral geniculate nucleus of cats and kittens. *J. Neurophysiol.*, 78: 1045–1061.

Calford, M.B. and Tweedale, R. (1988) Immediate and chronic changes in responses of somatosensory cortex in adult flying-fox after digit amputation. *Nature*, 332: 446–448.

Chiaia, N.L., Rhoades, R.W., Bennett-Clarke, C.A., Fish, S.E. and Killackey, H.P. (1991a) Thalamic processing of vibrissal information in the rat, I. Afferent input to the medial ventral posterior and posterior nuclei. *J. Comp. Neurol.*, 314: 201–216.

Chiaia, N.L., Rhoades, R.W., Fish, S.E. and Killackey, H.P. (1991b) Thalamic processing of vibrissal information in the rat, II. Morphological and functional properties of medial ventral posterior nucleus and posterior nucleus neurons. *J. Comp. Neurol.*, 314: 217–236.

Chino, Y.M., Kaas, J.H., Smith, E.L.D., Langston, A.L. and Cheng, H. (1992) Rapid reorganization of cortical maps in adult cats following restricted deafferentation in retina. *Vis. Res.*, 32: 789–796.

Darian-Smith, C. and Gilbert, C.D. (1994) Axonal sprouting accompanies functional reorganization in adult cat striate cortex. *Nature*, 368: 737–740.

DeAngelis, G.C., Ohzawa, I. and Freeman, R.D. (1993a) Spatiotemporal organization of simple-cell receptive fields in the cat's striate cortex, I. General characteristics and postnatal development. *J. Neurophysiol.*, 69: 1091–1117.

DeAngelis, G.C., Ohzawa, I. and Freeman, R.D. (1993b) Spatiotemporal organization of simple-cell receptive fields in the cat's striate cortex, II. Linearity of temporal and spatial summation. *J. Neurophysiol.*, 69: 1118–1135.

DeAngelis, G.C., Ohzawa, I. and Freeman, R.D. (1995) Receptive-field dynamics in the central visual pathways. *Trends Neurosci.*, 18: 451–458.

Devor, M. and Wall, P.D. (1978) Reorganisation of spinal cord sensory map after peripheral nerve injury. *Nature*, 276: 75–76.

Dinse, H.R., Kruger, K., Mallot, H.A., Best, J., 1991. Temporal structure of cortical information processing: cortical architecture, oscillations, and non-separability of spatio-temporal receptive field organization. In: Kruger, J. (Ed.), Neuronal Cooperativity. Springer, Berlin, pp. 68–104.

Dostrovsky, J.O. (1999) Immediate and long-term plasticity in human somatosensory thalamus and its involvement in phantom limbs. *Pain Suppl.*, 6: S37–43.

Dostrovsky, J.O., Millar, J. and Wall, P.D. (1976) The immediate shift of afferent drive to dorsal column nucleus cells following deafferentation: a comparison of acute and chronic deafferentation in gracile nucleus and spinal cord. *Exp. Neurol.*, 52: 480–495.

Elbert, T., Flor, H., Birbaumer, N., Knecht, S., Hampson, S., Larbig, W. and Taub, E. (1994) Extensive reorganization of the somatosensory cortex in adult humans after nervous system injury. *NeuroReport*, 5: 2593–2597.

Ergenzinger, E.R., Glasier, M.M., Hahm, J.O. and Pons, T.P. (1998) Cortically induced thalamic plasticity in the primate somatosensory system [see comments]. *Nat. Neurosci.*, 1: 226–229.

Eysel, U.T. (1982) Functional reconnections without new axonal growth in a partially denervated visual relay nucleus. *Nature*, 299: 442–444.

Faggin, B.M., Nguyen, K.T. and Nicolelis, M.A. (1997) Immediate and simultaneous sensory reorganization at cortical and subcortical levels of the somatosensory system. *Proc. Natl. Acad. Sci. USA*, 94: 9428–9433.

Florence, S.L. and Kaas, J.H. (1995) Large-scale reorganization at multiple levels of the somatosensory pathway follows therapeutic amputation of the hand in monkeys. *J. Neurosci.*, 15: 8083–8095.

Florence, S.L., Taub, H.B. and Kaas, J.H. (1998) Large-scale sprouting of cortical connections after peripheral injury in adult macaque monkeys [see comments]. *Science*, 282: 1117–1121.

Garraghty, P.E. and Kaas, J.H. (1991a) Functional reorganization in adult monkey thalamus after peripheral nerve injury. *NeuroReport*, 2: 747–750.

Garraghty, P.E. and Kaas, J.H. (1991b) Large-scale functional reorganization in adult monkey cortex after peripheral nerve injury. *Proc. Natl. Acad. Sci. USA*, 88: 6976–6980.

Ghazanfar, A.A. and Nicolelis, M.A. (1999) Spatiotemporal properties of layer V neurons of the rat primary somatosensory cortex. *Cereb. Cortex*, 9: 348–361.

Ghazanfar, A.A., Nicolelis, M., 2000. The space–time continuum in mammalian sensory pathways. In: Miller, R. (Ed.), Time and the Brain. Harwood Academic Publishers, Sydney, in press.

Gilbert, C.D. and Wiesel, T.N. (1992) Receptive field dynamics in adult primary visual cortex. *Nature*, 356: 150–152.

Irvine, D.R. and Rajan, R. (1997) Injury-induced reorganization of frequency maps in adult auditory cortex: the role of unmasking of normally inhibited inputs. *Acta Otolaryngol. Suppl.*, 532: 39–45.

Jacquin, M.F., Wiegand, M.R. and Renehan, W.E. (1990) Structure–function relationships in the rat brain stem subnucleus interpolaris, VIII. Cortical inputs. *J. Neurophysiol.*, 64: 3–27.

Kaas, J.H. (1991) Plasticity of sensory and motor maps in adult mammals. *Annu. Rev. Neurosci.*, 14: 137–167.

Kaas, J.H., Krubitzer, L.A., Chino, Y.M., Langston, A.L., Polley, E.H. and Blair, N. (1990) Reorganization of retinotopic cortical maps in adult mammals after lesions of the retina. *Science*, 248: 229–231.

Kalaska, J. and Pomeranz, B. (1979) Chronic paw denervation causes an age-dependent appearance of novel responses from forearm in 'paw cortex' of kittens and adult cats. *J. Neurophysiol.*, 42: 618–633.

Katz, D.B., Simon, S.A., Moody, A. and Nicolelis, M.A. (1999) Simultaneous reorganization in thalamocortical ensembles evolves over several hours after perioral capsaicin injections. *J. Neurophysiol.*, 82: 963–977.

Kirkwood, A., Rioult, M.C. and Bear, M.F. (1996) Experience-dependent modification of synaptic plasticity in visual cortex [see comments]. *Nature*, 381: 526–528.

Krupa, D.J., Ghazanfar, A.A. and Nicolelis, M.A. (1999) Immediate thalamic sensory plasticity depends on corticothalamic feedback [see comments]. *Proc. Natl. Acad. Sci. USA*, 96: 8200–8205.

Lee, S.M., Friedberg, M.H. and Ebner, F.F. (1994) The role of GABA-mediated inhibition in the rat ventral posterior medial thalamus, I. Assessment of receptive field changes following thalamic reticular nucleus lesions. *J. Neurophysiol.*, 71: 1702–1715.

Markram, H., Lubke, J., Frotscher, M. and Sakmann, B. (1997) Regulation of synaptic efficacy by coincidence of postsynaptic APs and EPSPs [see comments]. *Science*, 275: 213–215.

McAllister, A.K., Katz, L.C. and Lo, D.C. (1997) Opposing roles for endogenous BDNF and NT-3 in regulating cortical dendritic growth. *Neuron*, 18: 767–778.

Merzenich, M.M., Kaas, J.H., Wall, J., Nelson, R.J., Sur, M. and Felleman, D. (1983a) Topographic reorganization of somatosensory cortical areas 3b and 1 in adult monkeys following restricted deafferentation. *Neuroscience*, 8: 33–55.

Merzenich, M.M., Kaas, J.H., Wall, J.T., Sur, M., Nelson, R.J. and Felleman, D.J. (1983b) Progression of change following median nerve section in the cortical representation of the hand in areas 3b and 1 in adult owl and squirrel monkeys. *Neuroscience*, 10: 639–665.

Merzenich, M.M., Nelson, R.J., Stryker, M.P., Cynader, M.S., Schoppmann, A. and Zook, J.M. (1984) Somatosensory cortical map changes following digit amputation in adult monkeys. *J. Comp. Neurol.*, 224: 591–605.

Metzler, J. and Marks, P.S. (1979) Functional changes in cat somatic sensory-motor cortex during short term reversible epidermal blocks. *Brain Res.*, 177: 379–383.

Moore, C.I. and Nelson, S.B. (1998) Spatio-temporal subthreshold receptive fields in the vibrissa representation of rat primary somatosensory cortex. *J. Neurophysiol.*, 80: 2882–2892.

Nicolelis, M.A. (1997) Dynamic and distributed somatosensory representations as the substrate for cortical and subcortical plasticity. *Semin. Neurosci.*, 9: 24–33.

Nicolelis, M.A. and Chapin, J.K. (1994) Spatiotemporal structure of somatosensory responses of many-neuron ensembles in the rat ventral posterior medial nucleus of the thalamus. *J. Neurosci.*, 14: 3511–3532.

Nicolelis, M.A., Lin, R.C., Woodward, D.J. and Chapin, J.K. (1993a) Dynamic and distributed properties of many-neuron ensembles in the ventral posterior medial thalamus of awake rats. *Proc. Natl. Acad. Sci. USA*, 90: 2212–2216.

Nicolelis, M.A., Lin, R.C., Woodward, D.J. and Chapin, J.K. (1993b) Induction of immediate spatiotemporal changes in thalamic networks by peripheral block of ascending cutaneous information. *Nature*, 361: 533–536.

Nicolelis, M.A., Katz, D. and Krupa, D.J. (1998) Potential circuit mechanisms underlying concurrent thalamic and cortical plasticity. *Rev. Neurosci.*, 9: 213–224.

Parker, J.L. and Dostrovsky, J.O. (1999) Cortical involvement in the induction, but not expression, of thalamic plasticity. *J. Neurosci.*, 19: 8623–8629.

Parker, J.L., Wood, M.L. and Dostrovsky, J.O. (1998) A focal zone of thalamic plasticity. *J. Neurosci.*, 18: 548–558.

Pettit, M.J. and Schwark, H.D. (1993) Receptive field reorganization in dorsal column nuclei during temporary denervation. *Science*, 262: 2054–2056.

Pons, T.P., Garraghty, P.E., Ommaya, A.K., Kaas, J.H., Taub, E. and Mishkin, M. (1991) Massive cortical reorganization after sensory deafferentation in adult macaques [see comments]. *Science*, 252: 1857–1860.

Rajan, R., Irvine, D.R., Wise, L.Z. and Hcil, P. (1993) Effect of unilateral partial cochlear lesions in adult cats on the representation of lesioned and unlesioned cochleas in primary auditory cortex. *J. Comp. Neurol.*, 338: 17–49.

Rasmusson, D.D. and Turnbull, B.G. (1983) Immediate effects of digit amputation on S-I cortex in the raccoon: unmasking of inhibitory fields. *Brain Res.*, 288: 368–370.

Riddle, D.R., Lo, D.C. and Katz, L.C. (1995) NT-4-mediated rescue of lateral geniculate neurons from effects of monocular deprivation. *Nature*, 378: 189–191.

Ringach, D.L., Hawken, M.J. and Shapley, R. (1997) Dynamics of orientation tuning in macaque primary visual cortex. *Nature*, 387: 281–284.

Robertson, D. and Irvine, D.R. (1989) Plasticity of frequency organization in auditory cortex of guinea pigs with partial unilateral deafness. *J. Comp. Neurol.*, 282: 456–471.

Stevens, J.K. and Gerstein, G.L. (1976) Interactions between cat lateral geniculate neurons. *J. Neurophysiol.*, 39: 239–249.

Wall, J.T. (1988) Variable organization in cortical maps of the skin as an indication of the lifelong adaptive capacities of circuits in the mammalian brain. *Trends Neurosci.*, 11: 549–557.

Wall, J.T. and Cusick, C.G. (1984) Cutaneous responsiveness in primary somatosensory (S-I) hindpaw cortex before and after partial hindpaw deafferentation in adult rats. *J. Neurosci.*, 4: 1499–1515.

Wall, P.D. and Egger, M.D. (1971) Formation of new connexions in adult rat brains after partial deafferentation. *Nature*, 232: 542–545.

Wilson, P. and Snow, P.J. (1987) Reorganization of the receptive fields of spinocervical tract neurons following denervation of a single digit in the cat. *J. Neurophysiol.*, 57: 803–818.

Yang, T.T., Gallen, C.C., Ramachandran, V.S., Cobb, S., Schwartz, B.J. and Bloom, F.E. (1994) Noninvasive detection of cerebral plasticity in adult human somatosensory cortex. *NeuroReport*, 5: 701–704.

CHAPTER 14

The reorganization of somatosensory and motor cortex after peripheral nerve or spinal cord injury in primates

Jon H. Kaas [*]

301 Wilson Hall, Department of Psychology, Vanderbilt University, 111 21st Avenue South, Nashville, TN 37240, USA

Introduction

Injuries to the spinal cord or peripheral nerves obviously impair an individual. The injury could involve, for example, a sensorimotor nerve that fails to regenerate completely and accurately, so that the flow of sensory information is sparse and disordered and motor control is disorganized and weak. More serious peripheral injuries may be followed by the therapeutic amputation of an arm or leg, resulting in a major loss of sensory inputs, muscle targets for motor control, and a useful part of the body. Spinal cord injuries vary in what fiber and neuron systems are damaged and they can deprive an individual in many ways. Damage to the dorsal or posterior half of the spinal cord severs ascending axons of peripheral nerve afferents. Such damage resembles a peripheral nerve injury, except that a motor component may be lacking. In addition, afferents entering the spinal cord also synapse on neurons in the dorsal horn of the spinal cord to form second-order sensory and reflex pathways. Thus, the sensory loss is not complete.

All such injuries are followed by immediate impairments that are related to the nature of the injury. There also may be some rapid recovery of lost function, possibly mediated by the recovery of partially damaged neurons, changes in synaptic strength in spinal cord circuits, or even by peripheral nerve or spinal cord regeneration. The immediate impairments and the peripherally mediated recoveries have been a concern and focus of research for those dealing with such injuries. In addition, there has been a growing awareness that such sensory and motor deprivations have additional impact on the organization and functions of more central parts of sensory and motor systems, the nuclei of the brainstem and thalamus, and especially the sensory and motor representations in postcentral and precentral cortex. Such nuclei and cortical centers have been shown to be extremely malleable in functional organization, even in the mature brain.

Changes in the activation pattern of a structure start a process of reorganization that involves altering the strengths of existing connections, and at least sometimes, the growth of neural connections. The changes in sensory and motor systems are probably often adaptive, especially when they are induced by sensory and motor experience during training and practice. With more extreme deprivation, unwanted outcomes may follow. Brain reorganization may be responsible for phantom sensations, including phantom pain, sensory mislocalizations and error, and problems of motor control. From a clinical point of view, an understanding of central changes that follow injuries to sensory and motor pathways could lead to better ways of promoting reorganizations that are useful and preventing those that have undesirable consequences.

---

[*] Corresponding author: Dr. Jon H. Kaas, 301 Wilson Hall, Department of Psychology, Vanderbilt University, 111 21st Avenue South, Nashville, TN 37240, USA. Fax: +1-615-343-4342; E-mail: Jon.Kaas@Vanderbilt.edu

This review concentrates on the major changes that occur in sensory and motor systems of primates after damage to peripheral nerves or sensory afferents in the spinal cord. Much of the relevant research is recent and from monkeys. Other important results have come from brain imaging studies in humans with injuries. Research related to the more investigated issue of how sensory systems respond to changes in training and sensory experience has been reviewed elsewhere (e.g., Salmon and Butters, 1995; Weinberger, 1995; Dykes, 1997; Ebner et al., 1997; Buonomano and Merzenich, 1998).

The somatosensory and motor systems of primates

Damage to sensory afferents or motor afferents potentially impact on all parts of an extensive sensorimotor megasystem in primates (see Kaas, 1993). For tactile discriminations, the relevant afferents signaling contact and finger and limb positions terminate in the ipsilateral dorsal column–trigeminal nuclear complex in the lower brainstem, where second-order neurons project to the contralateral thalamic relay nuclei, the ventroposterior nucleus for touch and the ventroposterior superior nucleus for position sense. These nuclei relay in turn to subdivisions of anterior parietal cortex, cytoarchitectonic areas 3a, 3b, 1 and 2, each of which contains a systematic map of body receptors. The ventroposterior (VP) nucleus relays activating inputs to area 3b, the homologue of primary somatosensory cortex, S1, in cats and rats (Kaas, 1983). Modulating projections go from VP to areas 1 and 2. The ventroposterior superior nucleus (VPS) sends activating inputs, largely those related to muscle spindle receptors, to areas 3a and 2. Area 3b projects to areas 1 and 2, and somewhat to area 3a, and these areas may depend on cortical connections with area 3b for responsiveness to tactile stimuli. Areas 3a and 2 are also responsive to the stimulation of muscle spindles and other deep receptors via direct thalamic inputs. Area 3a seems to provide important inputs to primary motor cortex, M1, and all four areas of anterior parietal cortex project to the second somatosensory area, S2, and the parietal ventral area, PV, in the lateral sulcus. S2 and PV appear to depend on these inputs for activation, and S2 and PV project to additional subdivisions of somatosensory cortex, as well as to motor cortex. Thus, a loss of peripheral nerve inputs in the peripheral nerve or dorsal spinal cord would deprive four areas of anterior parietal cortex of their major sources of activation, and this in turn would deprive S2, PV, and the other areas that may depend largely or completely on these fields. Peripheral nerves also activate spinal cord neurons that form the spinothalamic projections that terminate in the ventroposterior inferior nucleus (VPI). The projections of VPI to areas 3a, 3b, 1, 2, S2, and PV appear to modulate activities produced by inputs from VP and VPS.

Effects of peripheral nerve injury

Since the maps or representations of sensory receptors in somatosensory cortex are topographic, especially in area 3b, the loss of inputs from any given region of skin deactivates a portion of each representation, the size of which reflects the number of afferents that have been injured. The removal of a source of activation also negates the excitation of neurons providing 'laterally' spreading inhibition, so the complex system with excitatory and inhibitory neurons immediately adjusts. If only a few afferents and a limited region of skin are affected, neurons deprived of their major source of activation recover enough formerly subthreshold sources of activation so that slightly repositioned receptive fields appear (see Nicolelis, 1997). These changes reflect the rebalancing of a dynamically maintained system. However, slightly greater deafferentations produce larger regions of deactivated neurons, and only the outer fringe of deactivated neurons might have enough remaining disinhibited activating inputs to acquire new receptive fields. Instead, regions of unresponsive neurons might persist. The complete loss of afferents from a finger (Merzenich et al., 1984; Jain et al., 1998) or part of the glabrous hand in monkeys (Merzenich et al., 1983), for example, produces a zone of unresponsive cortex in area 3b and other fields. However, these neurons generally recover responsiveness to somatotopically adjacent intact inputs over a period of days to weeks. In the case of section of median nerve to part of the glabrous hand, the slowly emerging recovery is mediated in part by synaptic changes in the thalamus and cuneate nucleus of the brainstem (Garraghty and Kaas, 1991;

Xu and Wall, 1997), quite possibly through the process of local growth of active axon arbors and the local replacement of inactive with active synapses. If the deactivation is more extensive, involving all of the afferents from several fingers (Merzenich et al., 1984; Garraghty et al., 1994), the unresponsive zone in area 3b might persist for at least several months. Thus, with damage to sensory afferents we see (1) reorganizations in somatosensory cortex that are limited in size and involve immediate rebalancing of excitatory and inhibitory influences in a dynamic system so that some neurons immediately acquire new receptive fields, (2) more extensive but still moderate reorganizations that appear to involve the local replacement of inactive by active synapses over days to weeks of time, and (3) deprivations that exceed limits of these two mechanisms so that regions of cortex remain unresponsive to peripheral stimulation for months of time. These persisting deactivations can be overcome by the growth of longer, new connections, but only over many months of recovery (see below).

Effects of arm amputations on somatosensory cortex

Injuries sometimes occur in humans and monkeys where a limb is lost or damaged to the extent that a therapeutic amputation is needed. Years after such injuries, it has been possible to study the effects of such a major sensory loss on the somatosensory system in a few humans (Halligan et al., 1993; Lenz et al., 1994, 1998; Davis et al., 1998; Flor et al., 1998) and monkeys (Florence and Kaas, 1995; Florence et al., 1998; but also see Pons et al., 1991). In brief, recordings or functional imaging in cortex demonstrate a reactivation after the loss of an arm so that the extensive regions of area 3b and 1 that are normally responsive to the hand, wrist, forearm, and upper arm have become responsive instead to inputs from the stump of the arm, the face or both. The time course of this reorganization is not known from these cases, which have usually been studied years after injury. Nevertheless, it is known that reactivations include the deprived portions of the VP nucleus of the thalamus (Davis et al., 1998; Jones and Pons, 1998). In monkeys at least, the reactivation involves the growth of new connections from the part of the cuneate nucleus innervated by afferents from the stump to parts of the cuneate nucleus deprived of inputs by the amputation (Florence and Kaas, 1995) and from the trigeminal complex representing the face to the deprived parts of the cuneate nucleus (Jain et al., 2000). This new growth in the lower brainstem may be essential for the reactivation of the thalamus and cortex, although additional changes at these levels may be important as well. Most notably, the loss of an arm is also followed by a growth of horizontal connections in areas 3b and 1 of somatosensory cortex (Florence et al., 1998). Thus, we know from these and related results, that major losses of sensory inputs are followed by deactivations of large amounts of cortex that are only slightly recovered and reactivated by the mechanisms of rebalancing and local growth and synapse substitution. Large zones of deactivated cortex must persist for some time, but recordings years after injury do not tell us how long. Nevertheless, after years of recovery, the large zones of deprived cortex can be completely reactivated by inputs from the stump, face, or both.

Effects of dorsal column injury on somatosensory cortex

The complete section of the dorsal columns of ascending afferents in the spinal cord produces a profound loss of sensory information. If this section occurs at a high cervical level of the spinal cord, the loss can include afferents from all of the lower body and all but a few afferents from the anterior arm that enter above the injury. Otherwise, the ascending dorsal column inputs from the hand and the rest of the arm are lost. Afferents from the face enter above the damage and are preserved. Of course, afferents that enter the dorsal columns also have branches that terminate in the dorsal horn of the spinal cord, where they activate neurons that contribute to local reflex pathways and to the spinothalamic path. In addition, other second-order neurons from the lower limb might ascend ipsilaterally outside the dorsal columns. Largely because of these other pathways, the behavioral impact of damage restricted to the dorsal columns is not so severe and obvious (see for example, Wall, 1970). Reflexes important for locomotion remain, and only a few impairments in hand use and control are obvious. However, careful behavioral testing has revealed a number of sensory losses (see Vierck and Cooper, 1998).

When the effects on somatosensory cortex of sectioning the afferents in the dorsal columns of one side were studied in monkeys at various times after injury, it became possible to determine the time course of cortical reactivation after a major deafferentation (Jain et al., 1997, 1998). Immediately after a complete section of the dorsal columns of one side at a high cervical level in monkeys, areas 3a, 3b, and 1 of somatosensory cortex were completely deactivated throughout at least their hand and forearm portions. From recordings obtained at various times after the section, it was clear that the deactivation persisted over a period of months, during which the monkeys climbed about, and reached for food and used both hands, although the affected hand was used reluctantly and was less capable in retrieving or supporting objects. The hand cortex was completely unresponsive to tactile stimuli, and there was no evidence for reactivation by the intact spinothalamic system. Thus, considerable sensory and motor behavior was possible, and this behavior did not depend on the processing of somatosensory information in areas of the anterior parietal cortex, or on the areas such as S2 and PV that depend on inputs from anterior parietal cortex.

Incomplete lesions produced different results. Lesions that were intended to be complete often left a few inputs from a finger or part of the palm intact. The preservation of these few inputs was difficult to detect by reconstructing the lesion, but they were clearly revealed when somatosensory cortex was mapped in detail, and when harmless tracers were injected into skin of the hand to see if any tracer was transported by the site of the lesion to terminations in the cuneate nucleus of the lower brainstem. The difficulty in detecting such intact afferents in reconstructions of spinal cord lesions, and their dramatic impact on behavior, suggested that some afferents might have survived in some of the previous studies of dorsal column damage. Immediately after an incomplete section of the dorsal columns, and for several days thereafter, the surviving afferents activated only the portions of areas 3b and 1 that they normally activate. Over the next few weeks, however, these few afferents came to activate more and more of the hand representations in these fields, so that the hand regions soon become completely or nearly completely responsive to the preserved afferents. Hand use improved over the same time period, and it often appeared to approach that of normal monkeys. Presumably, having afferents from only part or parts of the hand proved to be very useful. The recovery could have been related to the expansion of the representations of these preserved inputs in cortex, or to other factors such as learning to use the normally innervated skin surfaces more effectively. Yet, the reactivation and reorganization of the hand cortex by a few intact afferents is a type of plasticity that may be functionally beneficial.

After longer recovery times, more extensive reactivations were observed. Six to eight months after complete unilateral dorsal column section, the hand cortex in areas 3a, 3b, and 1 was largely or completely reactivated by inputs from the lower face, largely the chin, and the little input from the upper anterior arm that typically entered above the section. The same neurons could be responsive to touch on both the arm and the face. Furthermore, six or more months after incomplete sections, neurons that were responsive to the few remaining inputs from the hand also became responsive to the face. Thus, hand cortex that had remained unresponsive to spinothalamic inputs for over half of a year finally became responsive to inputs from the face. Even if neurons had already, in the first few weeks, become responsive to any preserved inputs from the hand, they became additionally responsive to the face. Thus responsiveness to hand afferents did not protect them from subsequent, but slowly developing, plasticity.

Why did recovery take so long? After six months or more of recovery, monkeys demonstrated the growth of axons of face afferents from their normal terminations in the trigeminal complex to the cuneate nucleus. This new growth appears to be critical to the reactivation. However, it is uncertain why the new growth takes so long to appear, since the distances are short. Perhaps, deactivation of the cuneate nucleus must persist for some time before new growth is initiated by some molecular signal. New growth also occurs in somatosensory cortex, and this probably contributes to the process of reactivation. The new growth in cortex may also take a number of months to occur.

The functional consequences of the reactivation of hand cortex by face inputs are mild, but maladaptive. Basically, stimuli felt on the face may also be felt on the deprived hand. This conclusion stems from the misperceptions reported after arm amputations (see Davis et al., 1998; Lenz et al., 1998; Ramachandran and Hirstein, 1998), but misperceptions occur after deafferentations caused by spinal cord injuries as well (e.g., Lenz et al., 1994). The results are reminiscent of the early findings of Sperry (1943), where the crossing of peripheral nerves from one leg to the other in rats led to persisting sensory mislocalizations. Apparently, the somatosensory system does not easily adjust via feedback to correctly localizing inputs that have been translocated from quite different parts of the receptor sheet.

The effects of amputations on motor cortex

Arm amputations not only remove a large number of afferents, but they section the axons and remove the muscle targets for a large number of motor neurons of the spinal cortex. Long after such injuries, the organization of primary motor cortex has been studied by microstimulation in a few monkeys (Schieber and Deuel, 1997; Qi et al., 1999; Wu and Kaas, 1999). Primary motor cortex (M1) in monkeys has a large middle portion that is devoted to moving the shoulder, arm, forearm, wrist, and especially the fingers. More lateral parts of M1 represent the face, while more medial parts represent the trunk, hindlimb, and tail. The organization of motor cortex can be determined by placing microelectrodes into motor cortex and pulsing small levels of current until a movement of some part of the body is observed. The current level is reduced until the movement is just detectable. The site is then said to 'represent' the observed movement at that threshold level of current. By obtaining evoked movements at many electrode sites, the overall pattern of somatotopic organization of motor cortex can be determined. Stimulation of most sites throughout the forelimb sector of primary motor cortex, M1, evokes digit movements, but a scattering of other sites in the sector evoke shoulder, arm, or wrist movements. When the arm is amputated near the shoulder, digit and lower arm movements are, of course, no longer possible. Instead, at least after long recoveries, stimulation sites throughout this deprived cortex evoke stump or shoulder movements. These movements are evoked at normal thresholds for many more sites than in normal animals, and at slightly higher thresholds (more current) for other sites. Few or no sites in reorganized forelimb cortex evoke face or trunk movements. Thus, the major consequence of losing the motor targets in the arm is that cortical neurons formerly involved in evoking digit and lower arm movements become involved in stump and shoulder movements.

How motor cortex becomes more devoted to the upper arm and shoulder is uncertain. Injections of tracers in the spinal cord (Wu and Kaas, 1999) show that the corticospinal neurons devoted to hand movements remain in place, and they may grow to contact additional motor neurons in the spinal cord, especially those related to the upper arm. Additionally, motor neurons with severed axons may not die, but instead, acquire new muscle targets in the upper arm. Finally, horizontal connections may grow in cortex so that neurons in ineffective regions of hand cortex contact regions where neurons are controlling the shoulder. These and other possibilities remain to be investigated.

Motor cortex also appears to reorganize in humans with amputations (Cohen et al., 1991; also see Ojemann and Silbergeld, 1995). The reorganization of motor cortex may provide some limited benefit, in that the stump of the amputated limb may be under better motor control. Yet, the massive corticospinal projection from M1 and other fields that normally control fine digit movements are still in place, but they are at least underused. As others are exploring the possibility of the control of a robotic arm directly from the activities of neurons in cortex (Chapin, this volume; also see Baringa, 1999), it also seems possible that the outputs of preserved spinal cord motor neurons, with their massive cortical control, could be effectively used to guide mechanical arms.

Promoting or preventing plasticity

Brain mechanisms behind perceptual and skill learning, as well as other types of learning, are beginning to be understood, and this understanding may guide attempts to improve learning (Buonomano and Merzenich, 1998). Other types of activity and deprivation-dependent plasticity may also be subject to manipulation (e.g., Nudo et al., 1996), possibly with the aid of the timed release of neuromodulators (Bakin and Weinberger, 1996; Kilgard and Merzenich, 1998), but also by promoting or suppressing inhibitors and enhancers of neuronal growth (Schwab and Bartholdi, 1996). The considerable plasticity that exists in the motor and sensory systems of adult humans and other primates often serves us well in recoveries from brain, spinal cord, and peripheral nerve injuries, but some reorganizations, such as those leading to phantom pain (Flor et al., 1998) and focal dystonias (Bly et al., 1996), are highly undesirable. Future research on the mutability of the brain and the mechanisms that mediate and allow change can allow us to have more control over our brain organization, and promote desirable changes while preventing those we do not want.

References

Bakin, J.S. and Weinberger, N.M. (1996) Induction of a physiological memory in the cerebral cortex by stimulation of the nucleus basalis. *PNAS*, 43: 11219–11224.

Baringa, M. (1999) Turning thoughts into actions. *Science*, 286: 888–900.

Buonomano, D.V. and Merzenich, M.M. (1998) Cortical plasticity: from synapses to maps. *Annu. Rev. Neurosci.*, 21: 149–186.

Bly, N.N., Merzenich, M.M. and Jenkins, W.M. (1996) A primate genesis model of focal dystonia and repetitive strain injury, I. Learning-induced dedifferentiation of the representation of the hand in the primary somatosensory cortex in adult monkeys. *Neurology*, 47: 508–520.

Cohen, L.G., Bandinelli, S., Findley, T.W. and Hallett, M. (1991) Motor reorganization after upper limb amputation in man: a study with focal magnetic stimulation. *Brain*, 114: 615–627.

Davis, K.D., Kiss, Z.H., Luo, L., Tasker, R.R., Lozano, A.M. and Dostrovsky, J.O. (1998) Phantom sensations generated by thalamic microstimulation. *Nature*, 391: 385–387.

Dykes, R.W. (1997) Mechanisms controlling neuronal plasticity in somatosensory cortex. *Can. J. Physiol. Pharmacol.*, 75: 535–545.

Ebner, F.F., Rema, V., Sachdev, R. and Symons, F.J. (1997) Activity-dependent plasticity in adult somatic sensory cortex. *Semin. Neurosci.*, 9: 47–58.

Flor, H., Elbert, T., Muhlnickel, W., Pantev, C., Wienbruch, C. and Taub, E. (1998) Cortical reorganization and phantom phenomena in congenital and traumatic upper-extremity amputees. *Exp. Brain Res.*, 119: 205–212.

Florence, S.L. and Kaas, J.H. (1995) Large-scale reorganization at multiple levels of the somatosensory pathway follows therapeutic amputation of the hand in monkeys. *J. Neurosci.*, 15: 8083–8095.

Florence, S.L., Taub, H.B. and Kaas, J.H. (1998) Large-scale sprouting of cortical connections after peripheral injury in adult macaque monkeys. *Science*, 282: 1117–1121.

Garraghty, P.E. and Kaas, J.H. (1991) Functional reorganization in adult monkey thalamus after peripheral nerve injury. *NeuroReport*, 2: 747–750.

Garraghty, P.E., Hanes, D.P., Florence, S.L. and Kaas, J.H. (1994) Pattern of peripheral deafferentation predicts reorganizational limits in adult primate somatosensory cortex. *Somatosens. Motor Res.*, 11: 109–117.

Halligan, D.W., Marshall, W.C., Wade, D.T., Davey, J. and Marison, D. (1993) Thumb in cheek? Sensory organization and perceptual plasticity after limb amputation. *Neuroreport*, 4: 233–236.

Jain, N., Catania, K.C. and Kaas, J.H. (1997) Deactivation and reactivation of somatosensory cortex after dorsal spinal cord injury. *Nature*, 386: 495–498.

Jain, N., Catania, K.C. and Kaas, J.H. (1998) A histologically visible representation of the fingers and palm in primate area 3b and its immutability following long-term deafferentations. *Cereb. Cortex*, 8: 227–236.

Jain, N., Florence, S.L., Qi, H.K. and Kaas, J.H. (2000) Growth of new brainstem connections in adult monkeys with massive sensory loss. *PNAS*, 97: 5546–5550.

Jones, E.G. and Pons, T.P. (1998) Thalamic and brainstem contributions to large-scale plasticity of primate somatosensory cortex. *Science*, 282: 1121–1125.

Kaas, J.H. (1983) What, if anything is SI? Organization of first somatosensory area of cortex. *Physiol. Rev.*, 63: 206–231.

Kaas, J.H. (1993) The functional organization of somatosensory cortex in primates. *Ann. Anat.*, 175: 509–518.

Kilgard, M.P. and Merzenich, M.M. (1998) Cortical map reorganization enabled by nucleus basalis activity. *Science*, 279: 1714–1718.

Lenz, F.A., Gracely, R.H., Baker, F.H., Richardson, R.T. and Dougherty, P.M. (1998) Reorganization of sensory modalities evoked by microstimulation in region of the thalamic principal sensory nucleus in patients with pain due to nervous system injury. *J. Comp. Neurol.*, 399: 125–138.

Lenz, F.A., Kwan, H.C., Martin, R., Tasker, R.R., Richardson, R.T. and Dostrovsky, J.O. (1994) Characteristics of somatotopic organization and spontaneous neuronal activity in the region of the thalamic principal sensory nucleus in patients with spinal cord transection. *J. Neurophysiol.*, 72: 1570–1587.

Merzenich, M.M., Kaas, J.H., Wall, J.T., Nelson, R.J., Sur, M. and Felleman, D. (1983) Topographic reorganization of somatosensory cortical areas 3b and 1 in adult monkeys following restricted deafferentation. *Neuroscience*, 8: 33–55.

Merzenich, M.M., Nelson, R.J., Stryker, M.P., Cynader, M.S., Schoppmann, A. and Zook, J.M. (1984) Somatosensory cortical map changes following digit amputation in adult monkeys. *J. Comp. Neurol.*, 224: 591–605.

Nicolelis, M.A. (1997) Dynamic and distributed somatosensory representations as the substrate for cortical and subcortical plasticity. *Semin. Neurosci.*, 9: 24–33.

Nudo, R.J., Wise, B.M., SiFuentes, F. and Milliken, G.W. (1996) Neural substrates for the effects of rehabilitative training on motor recovery after ischemic infarct. *Science*, 272: 1791–1794.

Ojemann, J.G. and Silbergeld, D.L. (1995) Cortical stimulation mapping of phantom limb rolandic cortex; case report. *J. Neurosurg.*, 82: 641–644.

Pons, T.P., Garraghty, P.E., Ommaya, A.K., Kaas, J.H., Taub, E. and Mishkin, M. (1991) Massive cortical reorganization after sensory deafferentation in adult macaques. *Science*, 252: 1857–1860.

Qi, H.-X., Stepniewska, I. and Kaas, J.H. (1999) Reorganization of primary motor cortex in macaque amputees. *Soc. Neurosci. Abstr.*, 25: 385.

Ramachandran, V.S. and Hirstein, W. (1998) The perception of phantom limbs. The D.O. Hebb lecture. *Brain*, 121: 1603–1630.

Salmon, D.P. and Butters, N. (1995) Neurobiology of skill and habit learning. *Curr. Opin. Neurobiol.*, 5: 184–190.

Schieber, M.H. and Deuel, R.K. (1997) Primary motor cortex reorganization in a long-term monkey amputee. *Somatosens. Motor Res.*, 14: 157–167.

Schwab, M.E. and Bartholdi, D. (1996) Degeneration and regeneration of axons in the lesioned spinal cord. *Physiol. Rev.*, 76: 319–370.

Sperry, R.W. (1943) Functional results of crossing sensory nerves in rats. *J. Comp. Neurol.*, 78: 59–90.

Vierck Jr., C.J. and Cooper, B.Y. (1998) Cutaneous texture discrimination following transection of the dorsal spinal column in monkeys. *Somatosens. Motor Res.*, 15: 309–315.

Wall, P.D. (1970) The presence of ineffective synapses and the circumstances which unmask them. *Philos. Trans. R. Soc. London Ser. B*, 278: 361–372.

Weinberger, N.M. (1995) Dynamic regulation of receptive-fields and maps in the adult sensory cortex. *Annu. Rev. Neurosci.*, 18: 129–158.

Wu, C.W.H. and Kaas, J.H. (1999) The organization of motor cortex of squirrel monkeys with longstanding therapeutic amputations. *J. Neurosci.*, 19: 7679–7697.

Xu, J. and Wall, J.T. (1997) Rapid changes in brainstem maps of adult primates after peripheral injury. *Brain Res.*, 774: 211–215.

SECTION IV

Neurotrophins and activity-dependent plasticity

CHAPTER 15

Neurotrophins and activity-dependent plasticity

Hans Thoenen *

Max-Planck-Institute of Neurobiology, Department of Neurobiochemistry, Am Klopferspitz 18a, D-82152 Martinsried, Germany

Introduction

The modulation of activity-dependent neuronal plasticity is a relatively new facet in the ever-growing spectrum of biological actions of the nerve growth factor gene family called neurotrophins (reviewed by Bothwell, 1995; Lewin and Barde, 1996). The first suspicion for such an action evolved from the observation that the synthesis of neurotrophins (NTs), in particular brain-derived neurotrophic factor (BDNF) and nerve growth factor (NGF), is regulated very rapidly by neuronal activity (reviewed by Thoenen, 1991). These very effective regulatory mechanisms not only became apparent under extreme experimental conditions such as electroconvulsions or administration of kainic acid (Gall and Isackson, 1989; Zafra et al., 1990; Ballarín et al., 1991; Gall et al., 1991; Isackson et al., 1991), but also came into play through physiological stimuli such as visual input (Castrén et al., 1992; Rossi et al., 1999), whisker stimulation (Rocamora et al., 1996), osmotic stress (Castrén et al., 1995), learning (Falkenberg et al., 1992), or physical activity (Oliff et al., 1998). However, evidence that the modulatory role of activity-dependent neuronal plasticity is of physiological importance derived from the observation that the formation of ocular dominance columns can be strongly influenced by the local administration of not only NTs, particularly NGF and BDNF, but also the corresponding blocking agents, namely specific blocking antibodies or Trk receptor bodies (IgG antibodies, in which the variable domain has been replaced by the binding domain of the different Trk receptors) (Cabelli et al., 1995, 1997; Pizzorusso et al., 1999; reviewed by: Bonhoeffer, 1996; Cellerino and Maffei, 1996; Berardi and Maffei, 1999). Further evidence for a modulatory role of NTs in activity-dependent neuronal plasticity came from observations in a relatively simple, well analyzed in vitro system, namely hippocampal slices (reviewed by: Bliss and Collingridge, 1993; Collingridge and Bliss, 1995). Here, it was demonstrated that, in BDNF knockout mice, long-term potentiation (LTP) resulting from relatively gentle tetanic stimulation was markedly reduced, although not completely abolished (Korte et al., 1995; Patterson et al., 1996). Most remarkably, LTP was reduced in homo- and heterozygous BDNF knockout animals to the same extent, indicating that a critical minimal quantity of BDNF had to be available in order that it could fulfill its modulatory role. If the LTP was followed beyond a time period of 90–100 min after its initiation, the quantitative reduction became a qualitative one, in that LTP was completely abolished in hippocampal slices of both homo- and heterozygous knockout mice (Korte et al., 1996b). By contrast, in control animals, LTP was maintained at a stable level for up to 6 h, the longest period analyzed. In hippocampal slices of BDNF knockout animals, LTP could be reconstituted by the local re-expression of BDNF by adenoviral gene transfer (Korte et al., 1996a) or by bath administration of BDNF (Patterson et al., 1996). This indicated that the

---

* Corresponding author: Dr. Hans Thoenen, Max-Planck-Institute of Neurobiology, Department of Neurobiochemistry, Am Klopferspitz 18a, D-82152 Martinsried, Germany. Fax: +49-89-8578-3749.

observed changes in the formation and maintenance of LTP were due to the absence of BDNF, rather than to subtle, cumulative, developmental changes resulting from the absence or reduction of BDNF. Finally, the administration of BDNF to hippocampal slices and slices of the visual cortex had a similar effect as tetanic stimulation, i.e., it resulted in a long-term enhancement of synaptic transmission (Kang and Schuman, 1995; Figurov et al., 1996; Akaneya et al., 1997; Gottschalk et al., 1998).

In order to gain information on the cellular and molecular mechanisms that underlie the observed functional changes in vivo and in vitro, a great variety of suitable "reductionist" systems have been used to approach the different specific questions, ranging from synaptosomal preparations (Knipper et al., 1994a,b; Li et al., 1998; Sala et al., 1998) to dissociated neuronal cultures (Lessmann et al., 1994; Lessmann and Heumann, 1998; Meyerfranke et al., 1998; Berninger et al., 1999) and co-cultivation of spinal cord explants with skeletal muscle cells (Lohof et al., 1993). Additionally, neuronal cell lines have also been used, in particular the pheochromocytoma cell line PC12 (Heymach and Shooter, 1995; Krüttgen et al., 1998). These detailed analyses have demonstrated that the modulatory actions of NTs on synaptic transmission are mediated by both pre- and postsynaptic mechanisms. Presynaptically, NTs enhance activity-mediated neurotransmitter release (Lohof et al., 1993; Knipper et al., 1994a,b; Lessmann et al., 1994; Gottschalk et al., 1998; Lessmann and Heumann, 1998; Li et al., 1998). Postsynaptically, BDNF enhances transmission via N-methyl-D-aspartate (NMDA) receptors (Levine et al., 1995, 1998; Suen et al., 1997) and attenuates transmission via γ-aminobutyric acid (GABA)$_A$ receptors (Tanaka et al., 1997).

With the exception of Serge Marty's contribution, the following chapters in this section on neurotrophins and activity-dependent plasticity deal mainly with analysis of the exogenous administration of NTs. Serge Marty discusses differences in the regulation of neuropeptide Y (NPY), somatostatin, and parvalbumin levels in hippocampal neurons by neuronal activity and BDNF. BDNF mediates the activity-dependent regulation of NPY but not of somatostatin, whereas parvalbumin expression develops during activity blockade (Marty and Onténiente, 1999). McLean Bolton and colleagues describe the modulation of the function of non-NMDA receptors in cultures of individual hippocampal neurons on islands of glial cells, examining the changes in the function of α-amino-3-hydroxy-5-methyl-4-isoxazole-4-propionic acid (AMPA) receptors by the exogenous administration of BDNF (reviewed by McAllister et al., 1999).

Fredrick Seil and Rosemarie Drake-Baumann elaborate on studies of the promotion of activity-dependent inhibitory synaptogenesis in organotypic cerebellar cultures through the administration of BDNF and neurotrophin-4/5 (NT-4/5) (Seil, 1999). Bai Lu and Wolfram Gottschalk address details of functional changes in hippocampal slices in the context of LTP, in particular changes in presynaptic structures and functions in hippocampal slices resulting from the administration of exogenous BDNF (Figurov et al., 1996; Gottschalk et al., 1998). Finally, Arthur Konnerth and co-workers deal with a topic concerning the rapidity of the action of NTs in a different time scale, namely the activation of a tetrodotoxin (TTX)-resistant sodium channel within milliseconds (Kafitz et al., 1999). This promotes NTs to the category of the most potent excitatory agents yet identified. The short-term effects of NTs identified to date, in particular calcium influx and calcium mobilization from intraneuronal stores, occurred within a time frame of fractions of minutes, at best within several seconds, but never within milliseconds (reviewed by: Berninger and Poo, 1996; Canossa et al., 1997).

Site and extent of action of neurotrophins

Since, in this section, we are dealing almost exclusively with the effects of exogenous NTs added to the analytical systems used, I will also give a brief overview of the mechanisms that determine the site(s) and extent of the action of NTs under physiological conditions. The site of action of NTs, best analyzed for BDNF, is determined by the site of expression of the signal-transducing form of the TrkB receptor (reviewed by: Bothwell, 1995; Lewin and Barde, 1996; McAllister et al., 1999), be it pre- or postsynaptic. It has recently been demonstrated at the electron microscopic (EM) level that the localization of the expression of these receptors is both highly selective and restricted (Drake et al., 1999). This limited, highly selective expression of the full-length TrkB receptors is of particular importance for a model of more refined mechanisms of memory formation and retention (see below). A most important piece of information is still lacking, namely, whether there is an activity-dependent change in the sites and extent of the expression of TrkB receptors. At least at the mRNA level, the synthesis of TrkB receptors is regulated similarly to that of BDNF, although to a lesser extent, owing to functional changes mediated by physiological stimuli, such as the visual input (Castrén et al., 1992). Conversely, Meyerfranke et al. (1998) have demonstrated that, at least under specific culture conditions, depolarization and cyclic AMP elevation lead to a rapid recruitment of TrkB receptors to the plasma membrane. Moreover, it has yet to be demonstrated whether there is merely an increase in the TrkB receptor expression at the same sites as under conditions of basal activity, or whether there is also an increase in the sites of expression that, under resting conditions, are below the detection limit. This is clearly the case for BDNF, whereby an increase in neuronal activity results in not only an increased expression in those neurons that express relatively high levels under basal conditions, but also an increased number of neurons expressing BDNF as a consequence of increased neuronal activity (reviewed by Lindholm et al., 1994).

The second, very important mechanism determining the modulatory role of NTs, again best analyzed for BDNF, is the regulation of its synthesis, and the mechanism(s) and site(s) of secretion. In the analysis of the physiological functions of BDNF, it was demonstrated at a very early stage that its expression is regulated by neuronal activity (reviewed by Thoenen, 1991). Upregulation is mediated by glutamate and acetylcholine, and downregulation by GABA, at least in adult animals, in which GABA, in contrast to early developmental stages, results in a hyperpolarization of the neurons via the $GABA_A$ receptors (reviewed by: Lindholm et al., 1994; Marty et al., 1996). The upregulation of BDNF synthesis results from an enhanced calcium influx and is mediated via calmodulin (reviewed by Lindholm et al., 1994). More recently, it has been demonstrated that the calcium-regulated synthesis is mediated via cre-sequences in the promoter of the BDNF gene via the calcium-mediated phosphorylation of CREB (calcium/cAMP response element binding protein), a transcriptional activator acting in concert with other directly or indirectly acting cre-binding proteins (Shieh et al., 1998; Tao et al., 1998). The availability of BDNF acting via the TrkB signal-transducing receptors is determined by not only the regulation of synthesis, but also the mechanisms and sites of secretion of NTs. In previous experiments, it has been demonstrated that both NGF and BDNF are secreted according to the constitutive and regulated activity-dependent pathways (Blöchl and Thoenen, 1995, 1996; Goodman et al., 1996; Griesbeck et al., 1999). Moreover, regulated secretion is also mediated through the activation of Trk receptors (Canossa et al., 1997; Krüttgen et al., 1998) representing — in the physiological context — a potential positive feedback mechanism, in that the activity-mediated secretion, mainly via glutamate receptors, may be enhanced by a positive feedback via Trk receptors. For the analysis of the sites of secretion and the identification of the compartments from where NTs are secreted, it was essential that the activity-dependent secretion of NTs, analyzed for NGF and BDNF, was the same in both native, non-transduced hippocampal slices (Blöchl and Thoenen, 1995; Canossa et al., 1997; Griesbeck et al., 1999) and plasmid (Blöchl and Thoenen, 1995, 1996) or adenovirally transduced, dissociated hippocampal neurons (Canossa et al., 1997; Griesbeck et al., 1999). In contrast to all these observations supporting very similar or even identical sorting and secretion characteristics of NGF as compared to BDNF, Mowla et al. (1999) reported that there are distinct differences between the sorting and secretion characteristics of BDNF and NGF. Using a *Vaccinia* virus transduction system, they found that NGF was secreted exclusively according to the constitutive pathway, whereas BDNF was secreted exclusively according to the activity-dependent, regulated pathway. The potential reasons for these discrepant observations have been extensively discussed by Gärtner et al. (2000).

Activity-mediated secretory mechanisms

Adenovirally transduced hippocampal neurons have been used for the identification of the sites of synthesis, storage, and also secretion. Such studies proved impossible with non-transduced, cultivated neurons, since the levels of endogenous NTs are extremely low. As already stated above, the characteristics of activity-dependent secretion from native, non-transduced hippocampal slices proved to be the same as in plasmid or adenovirally transduced cultivated neurons. Moreover, the subcellular distribution in transduced, cultivated neurons remained the same, independent of the level of expression and the stage of differentiation (Gärtner, 2000; Gärtner et al., 2000). The activity-dependent secretion of both NGF and BDNF was analyzed for high potassium-mediated depolarization and by the administration of glutamate and also carbachol, a compound that shows a very similar spectrum of actions to that of acetylcholine, but is resistant to the very rapidly acting, powerful degradation by acetylcholinesterase (Blöchl and Thoenen, 1995). All the activity-mediated secretory mechanisms, including the Trk receptor-mediated NT secretion, showed very unusual characteristics. These regulated secretory mechanisms were not dependent on extracellular calcium, but instead proved to be strictly dependent on intact functional intracellular calcium stores that could be depleted by caffeine and/or thapsigargin (Blöchl and Thoenen, 1995; Canossa et al., 1997). With respect to the analysis of the underlying molecular mechanisms, the Trk receptor-mediated NT secretion proved most straightforward. It was shown that this secretion was initiated by the activation of phospholipase C (PLC)γ with a resulting, subsequent formation of inositol-1,4,5-triphosphate (IP_3), which, by interaction with the IP_3 receptor, then elicited the release of calcium from thapsigargin and caffeine-sensitive stores (Canossa et al., in prep.). These calcium stores are most likely identical to the smooth endoplasmic reticulum (ER) that extends throughout the entire neurite arborization, including axons and dendrites of fully differentiated neurons (reviewed by Berridge, 1998). The signal transduction pathway that is initiated by glutamate proved to be mediated via AMPA- and metabotropic glutamate receptors. Both AMPA- and metabotropic glutamate receptors activate G-proteins that may, in turn, activate PLC and the formation of IP_3. In any case, the AMPA- and metabotropic glutamate receptor-mediated secretion of NTs can be blocked by the high-affinity intracellular calcium chelator, BAPTA-AM, and the depletion of intracellular calcium stores by thapsigargin and/or caffeine (Blöchl and Thoenen, 1995; Canossa et al., in prep.).

Potential sites of neurotrophin secretion

In order to obtain information on the potential sites of secretion, we took advantage of the unique properties of neurotrophin-6 (NT-6) (Götz et al., 1994; Lai et al., 1998). NT-6 has an extra loop of 22 amino acids that render it, in contrast to the other members of the *NT* gene family, a heparin-binding molecule. Accordingly, NT-6 also binds tightly to the heparan sulfate proteoglycans, which are equally distributed all over the surface of cultivated hippocampal neurons (Gärtner et al., 2000). These unique properties of NT-6 provided the opportunity to use it as a tool for localizing the sites of activity-dependent secretion of NTs at the ultrastructural level. To this end, an adenoviral transduction construct was additionally provided with a C-terminal myc tag whose antigenic properties survive even a harsh fixation with glutaraldehyde. Glutaraldehyde fixation permitted an optimal preservation of membrane structures at the EM level and, in turn, the ultrastructural localization of the sites of activity-dependent NT-6myc secretion (Gärtner et al., 2000). However, as a prerequisite for such an approach, it was essential to demonstrate that NT-6myc showed subcellular distribution and secretion characteristics in hippocampal neurons that were indistinguishable from those of NGF and BDNF. All these NTs were localized identically in both dendrites and axons corresponding to the distribution in the smooth ER, which, as already mentioned above, forms a network throughout the entire dimensions of axons and dendrites, including dendritic spines and axonal nerve terminals (reviewed by Berridge, 1998). Additionally, it was important that NT-6 did not bind to any of the Trks expressed in dissociated hippocampal neurons. The very weak affinity to TrkA receptors, being about a thousand-fold lower than that of NGF, proved irrelevant, since no functional TrkA receptors are expressed in hippocampal neurons under the culture conditions used in our secretion experiments (Canossa et al., 1997; Gärtner et al., 2000). Since neurons are covered by heparan sulfate-carrying glycoproteins, the activity-mediated secretion of myc-tagged NT-6 is trapped at the neuronal surface immediately after release, resulting in the least possible diffusion artifacts after secretion. The surfaces of primary cultures of hippocampal neurons that were saturated with myc-tagged NT-6 revealed an equal distribution all over the neuronal surface (Gärtner et al., 2000). Importantly, the distribution showed no preferential binding to neurites. The label could be completely removed by washing the cells with heparin. Under the same experimental conditions, neither NGF nor BDNF was retained on the neuronal cell surface, and no detectable quantities of these two NTs, especially analyzed for NGF, could be removed from the cell surface by heparin (Gärtner et al., 2000). This is in distinct contrast to the substantial quantities of NT-6myc that could be released from the neuronal surface by the administration of heparin. A 3-min depolarization with 50 mM potassium, after cleaning the neuronal cell surface with heparin and thorough washing to remove residual traces of heparin, demonstrated a characteristic labeling of the neuronal surface. The myc-immunoreactivity was arranged in clusters, in contrast to the equal distribution of the saturating concentrations of NT-6myc. The quantitative evaluation of the distribution of the surface labeling demonstrated that the depolarization-mediated secretion of NT-6myc preferentially occurred on neurites. Compared with the very low basal levels, there was a sixteen-fold increase in neurites, partially identified unambiguously as dendrites or axons, whereas the increase on the surface of the cell bodies amounted to only a two-fold increase. This latter increase was consistent in three independent trials, but did not attain statistically significant levels (Gärtner et al., 2000). Nevertheless, the surface labeling after a 3-min depolarization was still much lower than that resulting from saturating concentrations of NT-6myc, which decorated the entire neuronal surface equally. Moreover, there was no preferential labeling at specialized structures such as synapses or dendritic spines.

The intracellular localization of NT-6myc and BDNFmyc by postembedding procedures was in agreement (Gärtner et al., 2000) with the light microscopic analysis of NGF (Blöchl and Thoenen, 1995, 1996) and BDNF localization (Griesbeck et al., 1999). NT-6myc showed a relatively dense, but not homogenous immunoreactivity in the perikaryon, including the Golgi, and a patchy labeling of all neuronal processes, as far as could be distinguished, in both dendrites and axons. Moreover, at the confocal light microscopic level, the immunoreactivity for all NTs was highly colocalized with luminal and membranous markers of the ER and Golgi apparatus (Blöchl and Thoenen, 1995, 1996; Griesbeck et al., 1999). The intracellular localization of NT-6myc and BDNFmyc at the EM level by postembedding procedures was in agreement with the light microscopic analysis, namely a preferential localization of gold immunoreactivity in ER-like structures, partially associated with microtubules, and partially in the vicinity of synaptic vesicles (Gärtner et al., 2000). In isolated neurons of hippocampal cell cultures, and also organotypic hippocampal cultures, large, dense core vesicles are very rare, and the few that were present, in axon terminals and extremely rarely in dendrites, were not even consistently labeled (Gärtner et al., 2000). Hence, large, dense core vesicles cannot be the compartment from which NTs are secreted in hippocampal neurons. This observation is of particular importance, since, in axon terminals of dorsal root ganglia neurons in the spinal cord, BDNF immunoreactivity was preferentially, although not exclusively, associated with large, dense core vesicles (Zhou and Rush, 1996; Michael et al., 1997). Accordingly, this compartment was considered to be the essential one from which the regulated secretion of BDNF occurs, as seems to be the case for a great number of neuropeptides (Hökfelt et al., 1980; Thureson-Klein and Klein, 1990). In recent, preliminary experiments, Annette Gärtner has also demonstrated that field stimulation of NT-6myc-transduced hippocampal neurons leads to a similar surface labeling as that resulting from depolarization by high potassium (Gärtner, 2000).

Activity-dependent neuronal plasticity and memory

In view of the very detailed analysis of the modulatory role of NTs, in particular BDNF, in activity-dependent neuronal plasticity, it has to be asked whether such intense efforts are in fact justified. Beyond the purpose of a most complete, comprehensive understanding of the whole spectrum of the physiological functions of NTs, their modulatory role in activity-dependent neuronal plasticity might be an attractive model for more refined mechanisms of memory formation and storage. To date, the mechanisms involved in short- and long-term modifications of synaptic transmission, linked to behavioral studies analyzing short- and long-term memory, have been rather general and ubiquitous ones, such as the influx of calcium via NMDA receptors, the activation and/or insertion of new AMPA receptors (silent synapses), the activation of Ca^{2+}/calmodulin-dependent protein kinases (CaM-kinases), in particular CaM-kinase II, and the activation of protein kinase A via cyclic AMP-dependent and -independent mechanisms (reviewed by: Mayford et al., 1996; Abel and Kandel, 1998). The highly selective modulatory role of NTs, in particular BDNF, could represent a model for a more refined formation and storage of memory if we bear in mind the almost unimaginable storage capacity of the human brain necessary to meet the requirements of remembering, e.g., thousands of words in different languages, hundreds of faces, and environmental cues to guarantee our orientation within the environment. Hence, more refined mechanisms must be postulated for the collection and retention of this extremely complex information over not only minutes, days, and weeks, but also many years. The highly selective, restricted modulatory role of activity-dependent neuronal plasticity by NTs, in particular BDNF, could be an attractive model for a more refined formation and retention of a huge number of engrams. However, I would like to emphasize that the importance of the function of BDNF should by no means be considered unique, and that BDNF should not be added to the already too-numerous molecules that have been tagged "memory molecules." The NT-Trk system modulating the activity-dependent neuronal plasticity is, indeed, one of the most complete that has been analyzed to date, but it is to be expected that a great number of other molecules, such as neuropeptides and cytokines, may fulfill very similar functions (reviewed by Jankowsky and Patterson, 1999). Even conventional neurotransmitters may have a preferentially modulatory role rather than a highly selective synapse-specific transmission function. This is, for instance, the case for the locus coeruleus and the raphe nucleus, which are formed by a relatively small number of neurons, projecting, however, to large areas of the entire brain. In these cases, the corresponding neurotransmitters, norepinephrine (locus coeruleus) and seretonin (raphe nucleus), have a modulatory role rather than a highly selective, synapse-specific transmission function.

Concluding remarks

In closing, I would like to address an additional aspect that is of interest to regeneration neuroscientists, one dealing predominantly with questions of the pathophysiology of degenerative disorders and corresponding therapeutic interventions. With respect to the use of native NTs and low-molecular derivatives, both peptidergic and non-peptidergic, for the treatment of degenerative disorders of the central nervous system, the information collected in the analysis of the modulatory role of NTs in activity-dependent neuronal plasticity is of considerable importance. It was, and is still believed that the main obstacle to be overcome in order to develop useful therapeutic tools is to bring these molecules across the blood-brain barrier. This is a very shortsighted, dangerous attitude, since an indiscriminate flooding of the whole brain with such molecules may result in serious side effects such as epileptic activity, hallucinations, and disruption of memory function (reviewed by Thoenen, 2000). In this context, it is worth mentioning that intraventricular infusion of NGF very impressively restored the structure and function of the basal forebrain cholinergic neurons in "impaired" aged rats, whereas the intraventricular infusion of NGF in young rats did not improve memory function, but instead impaired it. This indicates that an indiscriminate distribution of NTs or their derivatives throughout the brain, or even parts of the brain, by local administration may not only have the envisaged restorative effect on impaired function, but also interfere with normal brain function, for instance by the formation of "wrong" synapses, or imposing a wrong weight to the individual synapses. Indeed, after apparently encouraging results had been obtained from a first Alzheimer patient treated with NGF administered intraventricularly (Olson et al., 1992), a subsequent, more complete clinical study with several patients showed that the intraventricular infusion of NGF resulted in side effects that were so serious that the treatment had to be discontinued (Jonhagen et al., 1998).

References

Abel, T. and Kandel, E. (1998) Positive and negative regulatory mechanisms that mediate long-term memory storage. *Brain Res. Rev.*, 26: 360–378.

Akaneya, Y., Tsumoto, T., Kinoshita, S. and Hatanaka, H. (1997) Brain-derived neurotrophic factor enhances long-term potentiation in rat visual cortex. *J. Neurosci.*, 17: 6707–6716.

Ballarín, M., Ernfors, P., Lindefors, N. and Persson, H. (1991) Hippocampal damage and kainic acid injection induce a rapid increase in mRNA for BDNF and NGF in rat brain. *Exp. Neurol.*, 114: 35–43.

Berardi, N. and Maffei, L. (1999) From visual experience to visual function: roles of neurotrophins. *J. Neurobiol.*, 41: 119–126.

Berninger, B. and Poo, M.-m. (1996) Fast actions of neurotrophic factors. *Curr. Opin. Neurobiol.*, 6: 324–330.

Berninger, B., Schinder, A.F. and Poo, M.-m. (1999) Synaptic reliability correlates with reduced susceptibility to synaptic potentiation by brain-derived neurotrophic factor. *Learn. Mem.*, 6: 232–242.

Berridge, M.J. (1998) Neuronal calcium signaling. *Neuron*, 21: 13–26.

Bliss, T.V. and Collingridge, G.L. (1993) A synaptic model of memory: long-term potentiation in the hippocampus. *Nature*, 361: 31–39.

Blöchl, A. and Thoenen, H. (1995) Characterization of nerve growth factor (NGF) release from hippocampal neurons: evidence for a constitutive and an unconventional sodium-dependent regulated pathway. *Eur. J. Neurosci.*, 7: 1220–1228.

Blöchl, A. and Thoenen, H. (1996) Localization of cellular storage compartments and sites of constitutive and activity-dependent release of nerve growth factor (NGF) in primary cultures of hippocampal neurons. *Mol. Cell. Neurosci.*, 7: 173–190.

Bonhoeffer, T. (1996) Neurotrophins and activity-dependent development. *Curr. Opin. Neurobiol.*, 6: 119–126.

Bothwell, M. (1995) Functional interactions of neurotrophins and neurotrophin receptors. *Annu. Rev. Neurosci.*, 18: 223–253.

Cabelli, R.J., Hohn, A. and Shatz, C.J. (1995) Inhibition of ocular dominance column formation by infusion of NT-4/5 or BDNF. *Science*, 267: 1662–1666.

Cabelli, R.J., Shelton, D.L., Segal, R.A. and Shatz, C.J. (1997) Blockade of endogenous ligands of trkB inhibits formation of ocular dominance columns. *Neuron*, 19: 63–76.

Canossa, M., Griesbeck, O., Berninger, B., Campana, G., Kolbeck, R. and Thoenen, H. (1997) Neurotrophin release by neurotrophins: implications for activity-dependent neuronal plasticity. *Proc. Natl. Acad. Sci. USA*, 94: 13279–13286.

Castrén, E., Zafra, F., Thoenen, H. and Lindholm, D. (1992) Light regulates expression of brain-derived neurotrophic factor mRNA in rat visual cortex. *Proc. Natl. Acad. Sci. USA*, 89: 9444–9448.

Castrén, E., Thoenen, H. and Lindholm, D. (1995) Brain-derived neurotrophic factor messenger RNA is expressed in the septum, hypothalamus and in adrenergic brain stem nuclei of adult rat brain and is increased by osmotic stimulation in the paraventricular nucleus. *Neuroscience*, 64: 71–80.

Cellerino, A. and Maffei, L. (1996) The action of neurotrophins in the development and plasticity of the visual cortex. *Prog. Neurobiol.*, 49: 53–71.

Collingridge, G.L. and Bliss, T.V. (1995) Memories of NMDA receptors and LTP. *Trends Neurosci.*, 18: 54–56.

Drake, C.T., Milner, T.A. and Patterson, S.L. (1999) Ultrastructural localization of full-length trkB immunoreactivity in rat hippocampus suggests multiple roles in modulating activity-dependent synaptic plasticity. *J. Neurosci.*, 19: 8009–8026.

Falkenberg, T., Mohammed, A.K., Henriksson, B., Persson, H., Winblad, B. and Lindefors, N. (1992) Increased expression of brain-derived neurotrophic factor mRNA in rat hippocampus associated with improved spatial memory and enriched environment. *Neurosci. Lett.*, 138: 153–156.

Figurov, A., Pozzo-Miller, L.D., Olafsson, P., Wang, T. and Lu, B. (1996) Regulation of synaptic responses to high-frequency stimulation and LTP by neurotrophins in the hippocampus. *Nature*, 381: 706–709.

Gall, C.M. and Isackson, R.J. (1989) Limbic seizures increase neuronal production of messenger RNA for nerve growth factor. *Science*, 245: 758–761.

Gall, C., Murray, K. and Isackson, P.J. (1991) Kainic acid-induced seizures stimulate increased expression of nerve growth factor mRNA in rat hippocampus. *Mol. Brain Res.*, 9: 113–123.

Gärtner, A. (2000) Neurotrophine und aktivitätsabhängige synaptische Plastizität: Untersuchungen zum Ort und Mechanismus der regulierten Neurotrophin-Freisetzung in adenoviral transduzierten hippocampalen Primärkulturen. Doctoral thesis, University of Tübingen.

Gärtner, A., Shostak, Y., Hackel, N., Ethell, I.M. and Thoenen, H. (2000) Ultrastructural identification of storage compartments and localization of activity-dependent secretion of neurotrophin 6 in hippocampal neurons. *Mol. Cell. Neurosci.*, 15: 215–234.

Goodman, L.J., Valverde, J., Lim, F., Geschwind, M.D., Federoff, H.J., Geller, A.I. and Hefti, F. (1996) Regulated release and polarized localization of brain-derived neurotrophic factor in hippocampal neurons. *Mol. Cell. Neurosci.*, 7: 222–238.

Gottschalk, W., Pozzomiller, L.D., Figurov, A. and Lu, B. (1998) Presynaptic modulation of synaptic transmission and plasticity by brain-derived neurotrophic factor in the developing hippocampus. *J. Neurosci.*, 18: 6830–6839.

Götz, R., Köster, R., Winkler, C., Raulf, F., Lottspeich, F., Schartl, M. and Thoenen, H. (1994) Neurotrophin-6 is a new member of the nerve growth family. *Nature*, 272: 266–269.

Griesbeck, O., Canossa, M., Campana, G., Gärtner, A., Hoener, M.C., Nawa, H., Kolbeck, R. and Thoenen, H. (1999) Are there differences between the secretion characteristics of NGF and BDNF? Implications for the modulatory role of neurotrophins in activity-dependent neuronal plasticity. *Microsc. Res. Tech.*, 45: 262–275.

Heymach Jr., J.V. and Shooter, E.M. (1995) The biosynthesis of neurotrophin heterodimers by transfected mammalian cells. *J. Biol. Chem.*, 270: 12297–12304.

Hökfelt, T., Johansson, O., Ljungdahl, A., Lundberg, J.N. and

Schultzberg, M. (1980) Peptidergic neurons. *Nature*, 284: 515–520.

Jonhagen, M.E., Nordberg, A., Amberla, K., Backman, L., Ebendal, T., Meyerson, B., Olson, L., Seiger, A., Shigeta, M., Theodorsson, E., Viitanen, M., Winblad, B. and Wahlund, L.O. (1998) Intracerebroventricular infusion of nerve growth factor in three patients with Alzheimers-disease. *Dement. Geriatr. Cogn.*, 9: 246–257.

Isackson, P.J., Huntsman, M.M., Murray, K.D. and Gall, C.M. (1991) BDNF mRNA expression is increased in adult rat forebrain after limbic seizures: temporal pattern of induction distinct from NGF. *Neuron*, 6: 937–948.

Jankowsky, J.L. and Patterson, P.H. (1999) Cytokine and growth factor involvement in long-term potentiation. *Mol. Cell. Neurosci.*, 14: 273–286.

Kafitz, K.W., Rose, C.R., Thoenen, H. and Konnerth, A. (1999) Neurotrophin-evoked rapid excitation through TrkB receptors. *Nature*, 401: 918–921.

Kang, H. and Schuman, E.M. (1995) Long-lasting neurotrophin-induced enhancement of synaptic transmission in the adult hippocampus. *Science*, 267: 1658–1662.

Knipper, M., Da Penha Berzaghi, M., Blöchl, A., Breer, H., Thoenen, H. and Lindholm, D. (1994a) Positive feedback between acetylcholine and the neurotrophins nerve growth factor and brain-derived neurotrophic factor in the rat hippocampus. *Eur. J. Neurosci.*, 6: 668–671.

Knipper, M., Leung, L.S., Zhao, D. and Rylett, R.J. (1994b) Short-term modulation of glutamatergic synapses in adult rat hippocampus by NGF. *NeuroReport*, 5: 2433–2436.

Korte, M., Carroll, P., Wolf, E., Brem, G., Thoenen, H. and Bonhoeffer, T. (1995) Hippocampal long-term potentiation is impaired in mice lacking brain-derived neurotrophic factor. *Proc. Natl. Acad. Sci. USA*, 92: 8856–8860.

Korte, M., Griesbeck, O., Gravel, C., Carroll, P., Staiger, V., Thoenen, H. and Bonhoeffer, T. (1996a) Virus-mediated gene transfer into CA1 region restores LTP in BDNF mutant mice. *Proc. Natl. Acad. Sci. USA*, 93: 12547–12552.

Korte, M., Staiger, V., Griesbeck, O., Thoenen, H. and Bonhoeffer, T. (1996b) The involvement of brain-derived neurotrophic factor in hippocampal long-term potentiation revealed by gene targeting experiments. *J. Physiol. (Paris)*, 90: 157–164.

Krüttgen, A., Möller, J.C., Heymach, J.V. and Shooter, E.M. (1998) Neurotrophins induce release of neurotrophins by the regulated secretory pathway. *Proc. Natl. Acad. Sci. USA*, 95: 15867.

Lai, K.O., Fu, W.Y., Ip, F.C.F. and Ip, N.Y. (1998) Cloning and expression of a novel neurotrophin, NT-7, from carp. *Mol. Cell. Neurosci.*, 11: 64–76.

Lessmann, V. and Heumann, R. (1998) Modulation of unitary glutamatergic synapses by neurotrophin-4/5 or brain-derived neurotrophic factor in hippocampal microcultures: presynaptic enhancement depends on pre-established paired-pulse facilitation. *Neuroscience*, 86: 399–413.

Lessmann, V., Gottmann, K. and Heumann, R. (1994) BDNF and NT-4/5 enhance glutamatergic synaptic transmission in cultures of hippocampal neurons. *NeuroReport*, 6: 21–25.

Levine, E.S., Dreyfus, C.F., Black, I.B. and Plummer, M.R. (1995) Brain-derived neurotrophic factor rapidly enhances synaptic transmission in hippocampal neurons via postsynaptic tyrosine kinase receptors. *Proc. Natl. Acad. Sci. USA*, 92: 8074–8077.

Levine, E.S., Black, I.B. and Plummer, M.R. (1998) Neurotrophin modulation of hippocampal synaptic transmission. *Adv. Pharmacol.*, 42: 921–924.

Lewin, G.R. and Barde, Y.-A. (1996) Physiology of the neurotrophins. *Annu. Rev. Neurosci.*, 19: 289–317.

Li, Y.X., Xu, Y.F., Ju, D.S., Lester, H.A., Davidson, N. and Schuman, E.M. (1998) Expression of a dominant negative TrkB receptor, T1, reveals a requirement for presynaptic signaling in BDNF-induced synaptic potentiation in hippocampal neurons. *Proc. Natl. Acad. Sci. USA*, 95: 10884–10889.

Lindholm, D., Castrén, E., Berzaghi, M., Blöchl, A. and Thoenen, H. (1994) Activity-dependent and hormonal regulation of neurotrophin mRNA levels in the brain: implications on neuronal plasticity. *J. Neurobiol.*, 25: 1362–1373.

Lohof, A.M., Ip, N.Y. and Poo, M.-m. (1993) Potentiation of developing neuromuscular synapses by the neurotrophins NT-3 and BDNF. *Nature*, 363: 350–353.

Marty, S. and Onténiente, B. (1999) BDNF and NT-4 differentiate two pathways in the modulation of neuropeptide protein levels in postnatal hippocampal interneurons. *Eur. J. Neurosci.*, 11: 1647–1656.

Marty, S., Berzaghi, M.D. and Berninger, B. (1996) Neurotrophins and activity-dependent plasticity of cortical interneurons. *Trends Neurosci.*, 20: 198–202.

Mayford, M., Bach, M.E. and Kandel, E. (1996) CaMKII function in the nervous system explored from a genetic perspective. *Cold Spring Harb. Symp.*, 61: 219–224.

McAllister, A.K., Katz, L.C. and Lo, D.C. (1999) Neurotrophins and synaptic plasticity. *Annu. Rev. Neurosci.*, 22: 295–318.

Meyerfranke, A., Wilkinson, G.A., Kruttgen, A., Hu, M., Munro, E., Hanson, M.G., Reichardt, L.F. and Barres, B.A. (1998) Depolarization and cAMP elevation rapidly recruit trkB to the plasma membrane of CNS neurons. *Neuron*, 21: 681–693.

Michael, G.J., Averill, S., Nitkunan, A., Rattray, M., Bennett, D.L.H., Yan, Q. and Priestley, J.V. (1997) Nerve growth factor treatment increases brain-derived neurotrophic factor selectively in TrkA-expressing dorsal root ganglion cells and in their central terminations within the spinal cord. *J. Neurosci.*, 17: 8476–8490.

Mowla, S.J., Pareek, S., Farhadi, H.F., Petrecca, K., Fawcett, J.P., Seidah, N.G., Morris, S.J., Sossin, W.S. and Murphy, R.A. (1999) Differential sorting of nerve growth factor and brain-derived neurotrophic factor in hippocampal neurons. *J. Neurosci.*, 19: 2069–2080.

Oliff, H.S., Berchtold, N.C., Isackson, P. and Cotman, C.W. (1998) Exercise-induced regulation of brain-derived neurotrophic factor (BDNF) transcripts in the rat hippocampus. *Mol. Brain Res.*, 61: 147–153.

Olson, L., Nordberg, A., Von Holst, H., Bäckman, L., Ebendal, T., Alafuzoff, I., Amberla, K., Hartvig, P., Herlitz, A., Lilja, A., Lundqvist, H., Långström, B., Meyerson, B., Persson, A., Viitanen, M., Winblad, B. and Seiger, A. (1992) Nerve growth factor affects ^{11}C-nicotine binding, blood flow, EEG, and verbal episodic memory in an Alzheimer patient (Case Report). *J. Neural Transm.*, 4: 79–95.

Patterson, S.L., Abel, T., Deuel, T.A.S., Martin, K.C., Rose, J.C. and Kandel, E.R. (1996) Recombinant BDNF rescues deficits in basal synaptic transmission and hippocampal LTP in BDNF knockout mice. *Neuron*, 16: 1137–1145.

Pizzorusso, T., Berardi, N., Rossi, F.M., Viegi, A., Venstrom, K., Reichardt, L.F. and Maffei, L. (1999) TrkA activation in the rat visual cortex by antirat trkA IgG prevents the effect of monocular deprivation. *Eur. J. Neurosci.*, 11: 204–212.

Rocamora, N., Welker, E., Pascual, M. and Soriano, E. (1996) Upregulation of BDNF mRNA expression in the barrel cortex of adult mice after sensory stimulation. *J. Neurosci.*, 16: 4411–4419.

Rossi, F.M., Bozzi, Y., Pizzorusso, T. and Maffei, L. (1999) Monocular deprivation decreases brain-derived neurotrophic factor immunoreactivity in the rat visual cortex. *Neuroscience*, 90: 363–368.

Sala, R., Viegi, A., Rossi, F.M., Pizzorusso, T., Bonanno, G., Raiteri, M. and Maffei, L. (1998) Nerve growth factor and brain-derived neurotrophic factor increase neurotransmitter release in the rat visual cortex. *Eur. J. Neurosci.*, 10: 2185–2191.

Seil, F.J. (1999) BDNF and NT-4, but not NT-3, promote development of inhibitory synapses in the absence of neuronal activity. *Brain Res.*, 818: 561–564.

Shieh, P.B., Hu, S.C., Bobb, K., Timmusk, T. and Ghosh, A. (1998) Identification of a signaling pathway involved in calcium regulation of BDNF expression. *Neuron*, 20: 727–740.

Suen, P.C., Wu, K., Levine, E.S., Mount, H.T., Xu, J.L., Lin, S.Y. and Black, I.B. (1997) Brain-derived neurotrophic factor rapidly enhances phosphorylation of the postsynaptic N-methyl-D-aspartate receptor subunit 1. *Proc. Natl. Acad. Sci. USA*, 94: 8191–8195.

Tanaka, T., Saito, H. and Matsuki, N. (1997) Inhibition of $GABA_A$ synaptic responses by brain-derived neurotrophic factor (BDNF) in rat hippocampus. *J. Neurosci.*, 17: 2959–2966.

Tao, X., Finkbeiner, S., Arnold, D.B., Shaywitz, A.J. and Greenberg, M.E. (1998) Ca^{2+} influx regulates BDNF transcription by a CREB family transcription factor-dependent mechanism. *Neuron*, 20: 709–726.

Thoenen, H. (1991) The changing scene of neurotrophic factors. *Trends Neurosci.*, 14: 165–170.

Thoenen, H. (2000) Treatment of degenerative disorders of the nervous system: from helpless descriptive categorization to rational therapeutic approaches. In: Ingoglia, N.A. and Murray, M. (Eds.), *Regeneration in the Central Nervous System*. Marcel Dekker, in press.

Thureson-Klein, A. and Klein, R.L. (1990) Exocytosis from neuronal large dense-core vesicles. *Int. Rev. Cytol.*, 121: 67–126.

Zafra, F., Hengerer, B., Leibrock, J., Thoenen, H. and Lindholm, D. (1990) Activity dependent regulation of BDNF and NGF mRNAs in the rat hippocampus is mediated by non-NMDA glutamate receptors. *EMBO J.*, 9: 3545–3550.

Zhou, X.F. and Rush, R.A. (1996) Endogenous brain-derived neurotrophic factor is anterogradely transported in primary sensory neurons. *Neuroscience*, 74: 945–951.

CHAPTER 16

Differences in the regulation of neuropeptide Y, somatostatin and parvalbumin levels in hippocampal interneurons by neuronal activity and BDNF

Serge Marty [*]

INSERM Unité 421, Faculté de Médecine, 8 Rue du Général Sarrail, 94000 Créteil, France

Introduction

Inhibitory interneurons which use the neurotransmitter, γ-aminobutyric acid (GABA), play a crucial role in controlling excitatory transmission in the hippocampus (Miles and Wong, 1983). These interneurons are very diverse regarding their axonal projection, firing pattern and responses to neurotransmitters (Parra et al., 1998). Major subgroups can be recognized on the basis of their neurochemical characteristics, i.e., expression of specific calcium-binding proteins and neuropeptides. For instance, interneurons which express the calcium-binding protein, parvalbumin, innervate the cell body or axon initial segment of pyramidal neurons, while interneurons expressing the neuropeptide, somatostatin, innervate the most apical part of the dendrites of pyramidal neurons (Freund and Buzsáki, 1996). These different types of interneurons may differentially control sodium-dependent action potentials or calcium-dependent dendritic spikes (Miles et al., 1996).

Given the importance of interneurons in the control of excitatory transmission, their development is likely to be tightly regulated. Although hippocampal interneurons are generated prenatally, a considerable maturation of their morphological and neurochemical characteristics occurs during the postnatal period. Interneurons elaborate their dendritic trees during the first three postnatal weeks, in parallel with an increase in expression of the GABA synthesizing enzyme, glutamic acid decarboxylase (Seress and Ribak, 1988; Rozenberg et al., 1989; Lang and Frotscher, 1990). The expression of calcium-binding proteins and neuropeptides in interneurons also increases during the postnatal period (Bergmann et al., 1991; De Lecea et al., 1995; Jiang and Swann, 1997). For instance, the levels of neuropeptides such as neuropeptide Y (NPY), somatostatin or cortistatin reach a peak during the second or third postnatal week (Allen et al., 1984; Naus et al., 1988; De Lecea et al., 1997). However, the factors responsible for the maturation of these various aspects of the phenotype of interneurons remain largely unknown.

Neuronal activity is a likely candidate to pace the development of hippocampal interneurons, since it also exhibits a postnatal maturation in the hippocampus. During the first postnatal week, GABA depolarizes hippocampal neurons, before GABAergic neurotransmission becomes hyperpolarizing in parallel with the establishment of a non-N-methyl-D-aspartate (non-NMDA) glutamatergic transmission (Ben-Ari et al., 1989, 1997; Hosokawa et al., 1994; Durand et al., 1996). This

---

[*] Present address: INSERM U106, Hôpital de la Salpêtrière, Bâtiment de Pédiatrie, 47 Boulevard de l'Hôpital, 75651 Paris Cedex 13, France. Fax: +33-1-4570-9990; E-mail: marty@chups.jussieu.fr

switch is followed by transient hyperexcitability of hippocampal networks during the second and third postnatal weeks (Gómez-Di Cesare et al., 1997). Furthermore, modifications of neuronal activity in the adult hippocampus can change the neuropeptide content of interneurons. Seizure activity increases the expression of various neuropeptides in adult hippocampal interneurons (Gall et al., 1990; Schwarzer et al., 1996). Members of the neurotrophin gene family are other potential modulators of the maturation of interneurons (Marty et al., 1997). Particularly, Nawa and collaborators have observed strong effects of brain-derived neurotrophic factor (BDNF) on the levels of neuropeptides in interneurons, both in vitro and following in vivo injections (Nawa et al., 1993, 1994; Croll et al., 1994; Carnahan and Nawa, 1995). In agreement with these studies, immunoreactivity for NPY and various calcium-binding proteins was found to be decreased in hippocampal interneurons of BDNF-deficient mice (Jones et al., 1994). Since both the synthesis and the secretion of BDNF are regulated by neuronal activity (Thoenen, 1995), a causal relationship was proposed, whereby increased neuronal activity triggers an increase in BDNF availability, which in turn enhances neuropeptide protein levels (Carnahan and Nawa, 1995; Nawa et al., 1995).

In agreement with a role of neuronal activity in the maturation of the neurochemical characteristics of interneurons, studies summarized below indicate that depolarizing stimuli upregulate NPY expression in developing hippocampal interneurons. Moreover, BDNF is involved in the activity-dependent regulation of NPY expression, since depolarizing stimuli do not increase NPY expression in interneurons from BDNF knockout embryos. However, another neuropeptide, somatostatin, and a calcium-binding protein, parvalbumin, are not regulated following the same mechanism as NPY. Somatostatin expression is modulated by neuronal activity to the same extent as NPY, but BDNF does not mediate this activity-dependent regulation. Finally, parvalbumin expression develops despite activity blockade. Thus, factors responsible for the maturation of expression of the neurochemical characteristics of interneurons are very diverse.

BDNF mediates the activity-dependent modulation of NPY expression

The effects of depolarizing stimuli on the maturation of hippocampal interneurons, as well as the involvement of BDNF in these effects, were studied using dissociated cultures taken from E17 embryos (Berninger et al., 1995; Marty et al., 1996a). GABA, acting through $GABA_A$ receptors, depolarizes hippocampal neurons until the end of the first postnatal week, before GABAergic transmission switches to its classical hyperpolarizing effects (Ben-Ari et al., 1997). Neuronal cultures were therefore stimulated with the $GABA_A$ receptor agonist, muscimol.

Stimulation of neuronal cultures with muscimol after 5 days in vitro induced a transient Ca^{2+} influx, and an increased expression of BDNF mRNA in hippocampal neurons (Berninger et al., 1995). Treatment with the L-type voltage-gated Ca^{2+} channel blocker, nifedipine, suppressed the Ca^{2+} influx and also abolished the increase in BDNF mRNA levels, indicating that upregulation of BDNF mRNA was due to depolarization and calcium influx. During the first week in vitro, $GABA_A$ receptor stimulation with the agonist, muscimol, also promoted the morphological and neurochemical differentiation of hippocampal interneurons, as became apparent from a marked increase in size and NPY immunoreactivity (Marty et al., 1996a). The involvement of BDNF as a mediator of these effects of GABAergic stimulation on interneurons was suggested by the observation that exposure to exogenous BDNF mimicked the effects of GABAergic stimulation on interneurons. For the regulation of NPY, the involvement of BDNF was demonstrated by the fact that GABAergic stimulation failed to increase NPY immunoreactivity in cultures from BDNF knockout embryos. It was not due to an inability of neurons from BDNF knockout embryos to express NPY, because exposure to BDNF resulted in a similar upregulation of NPY in cultures from BDNF knockout and control embryos.

In contrast to younger cultures, in 3-weeks-old cultures, stimulation with the $GABA_A$ receptor blocker, bicuculline, induced a rise in intracellular calcium and in BDNF mRNA levels (Berninger et al., 1995). Thus, endogenously released GABA apparently suppressed BDNF mRNA at this later stage of development. Accordingly, treatment with the GABA$_A$ receptor agonist, muscimol, induced a decrease in cell size, dendritic arborization and NPY immunoreactivity of interneurons in these older cultures (Marty et al., 1996a).

Although hippocampal interneurons express the TrkB receptor for BDNF, they do not express BDNF, which is synthesized by their target cells, the pyramidal neurons (Rocamora et al., 1996; Schmidt-Kastner et al., 1996; Zachrisson et al., 1996). Thus, at early developmental stages, GABA released by interneurons may increase BDNF production and release by pyramidal neurons. BDNF may then act in a paracrine manner, modulating the levels of expression of NPY in interneurons as a function of the activity of pyramidal neurons. In more mature cultures, GABA represses BDNF production and release, leading to a reduced expression of NPY. Interestingly, recent studies using cultured cortical or hippocampal neurons indicate that chronic exposure to BDNF may also regulate the functional properties of cortical networks by increasing inhibitory synaptic transmission (Rutherford et al., 1997; Murphy et al., 1998; Vicario-Abejón et al., 1998).

In contrast to NPY, the activity-dependent modulation of somatostatin is not mediated by BDNF

The regulation of somatostatin by neuronal activity and BDNF was studied using organotypic cultures, and compared with the regulation of NPY under the same culture conditions (Marty and Onténiente, 1997, 1999). Explants were taken from 7-day-old rats and cultured for 2 weeks, a time during which maturation of synaptic transmission occurs with a time course resembling the in vivo situation (Muller et al., 1993). Glutamate, acting through non-NMDA receptors, is responsible for excitatory activity from the second postnatal week onward in the hippocampus, when GABAergic transmission exerts inhibitory effects (Ben-Ari et al., 1997). Explants were therefore cultured in the presence of either the GABA$_A$ receptor blocker, bicuculline, in order to block inhibitory transmission, or the non-NMDA glutamate receptor antagonist, 6,7-dinitroquinoxaline-2,3-dione (DNQX), in order to block excitatory transmission.

Explants cultured in the presence of bicuculline exhibited a strong increase in the number of NPY-immunoreactive neurons (Marty and Onténiente, 1999). NPY continued to be expressed at its organotypic location, with the labeled neurons being situated mainly in the stratum oriens. An opposite effect was observed following application of DNQX. The effects of these treatments were reversible upon removal of DNQX, indicating that they were due to modifications of expression of NPY rather than regulation of neuronal survival. The expression of the neuropeptide, somatostatin, was regulated by the interplay of excitatory and inhibitory activities to the same extent as NPY (Marty and Onténiente, 1997). Somatostatin also remained expressed in interneurons situated mostly in the stratum oriens, even though the number of somatostatin-immunoreactive neurons was strongly enhanced by bicuculline. Furthermore, double labeling studies demonstrated that somatostatin and the calcium-binding protein, calretinin, remained localized to different neurons in the stratum oriens, in agreement with the segregation of these markers in different types of interneurons in the adult hippocampus (Gulyás et al., 1996).

In order to study the involvement of BDNF in the activity-dependent modulation of expression of these neuropeptides, the duration of stimulation required to upregulate neuropeptides was initially examined. Bicuculline stimulation for the last 3 days in culture was sufficient to induce a strong upregulation of both NPY and somatostatin immunoreactivity (Marty and Onténiente, 1999). The involvement of BDNF as a mediator of the effects of bicuculline was then evaluated by studying the ability of BDNF treatment to mimic the effects of bicuculline. In agreement with the previously demonstrated involvement of BDNF in the activity-dependent modulation of NPY, application of BDNF for the last 3 days in culture also induced a strong upregulation of NPY. This upregulation occurred also when BDNF was applied to DNQX-treated explants. However, BDNF did not affect somatostatin immunoreactivity, indicating that this neurotrophic factor does not mediate the activity-dependent modulation of somatostatin levels.

These results indicate that the interplay of excitatory and inhibitory activity controls the levels of NPY and somatostatin in developing hippocampal interneurons. They are in agreement with a recent study demonstrating the importance of spontaneous

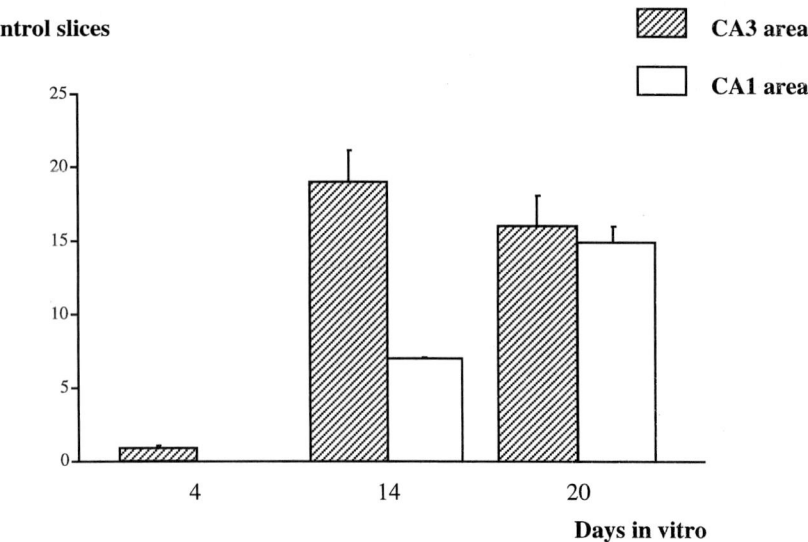

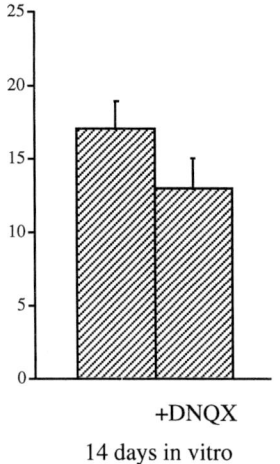

Fig. 1. Number of parvalbumin immunoreactive neurons in organotypic cultures explanted from 5-day-old rats. (A) Control explants. Note the increase in the number of immunolabeled cells in the CA3 area between 4 and 14 days. (B) Isolated CA3 areas cultured 14 days in the absence or presence of DNQX.

neuronal activity for the development of NPY expression in organotypic cultures of the visual cortex (Wirth et al., 1998). Furthermore, the absence of ectopic expression of these neuropeptides suggests that the regulation by neuronal activity is exerted over pools of interneurons which are committed to express particular neurochemical phenotypes. However, although both NPY and somatostatin are regulated by neuronal activity to a similar extent, BDNF appears to mediate only the effect of neuronal activity on the levels of NPY but not that on somatostatin levels. The effects of BDNF on NPY levels were observed in the presence DNQX, suggesting that they were independent of the actions of BDNF on synaptic transmission (Berninger and Poo, 1996). However, it was shown recently that BDNF induces a very rapid activation of a sodium ion conductance, leading to neuronal depolarization (Kafitz et al., 1999). Whether such depolarizing effects of BDNF are involved in the regulation of NPY levels remains to be studied.

Parvalbumin expression develops under blockade of excitatory transmission

Parvalbumin expression develops late in hippocampal interneurons, starting at postnatal day 7 (Bergmann et al., 1991; De Lecea et al., 1995). Factors responsible for this late development are at present unknown. Extracortical afferents or targets play a crucial role in the development of parvalbumin expression in the neocortex, because interneurons in organotypic cultures from the parietal cortex fail to develop parvalbumin expression when explanted a few days before the onset of expression in vivo (Vogt Weisenhorn et al., 1998). Afferent axons may exert at least part of their effects on parvalbumin expression by increasing neuronal activity in the neocortex.

This hypothesis is supported by the finding that tetrodotoxin injection in one eye of adult monkeys induces a reduction of parvalbumin immunoreactivity in the deprived area of the visual cortex (Carder et al., 1996). The involvement of extrahippocampal structures and neuronal activity in the development of parvalbumin expression in the hippocampus was therefore studied.

Explants were cultured according to the method of Stoppini et al. (1991) as described previously (Marty and Onténiente, 1997, 1999), except for the medium, which consisted of neurobasal medium (95 ml, Life Technologies 21103-031), B-27 serum-free supplement (2 ml, Life Technologies 17504-036), and 0.5 mM L-glutamine. Immunostaining was performed as described previously (Marty et al., 1996b), using an antibody raised against parvalbumin (1:1000, Sigma P-3171). The number of parvalbumin immunopositive cells was counted using a 10× objective, in a 1040 × 700 μm square positioned in the middle of the culture, over the CA1 field, or in the middle of the curved part of the Ammon's

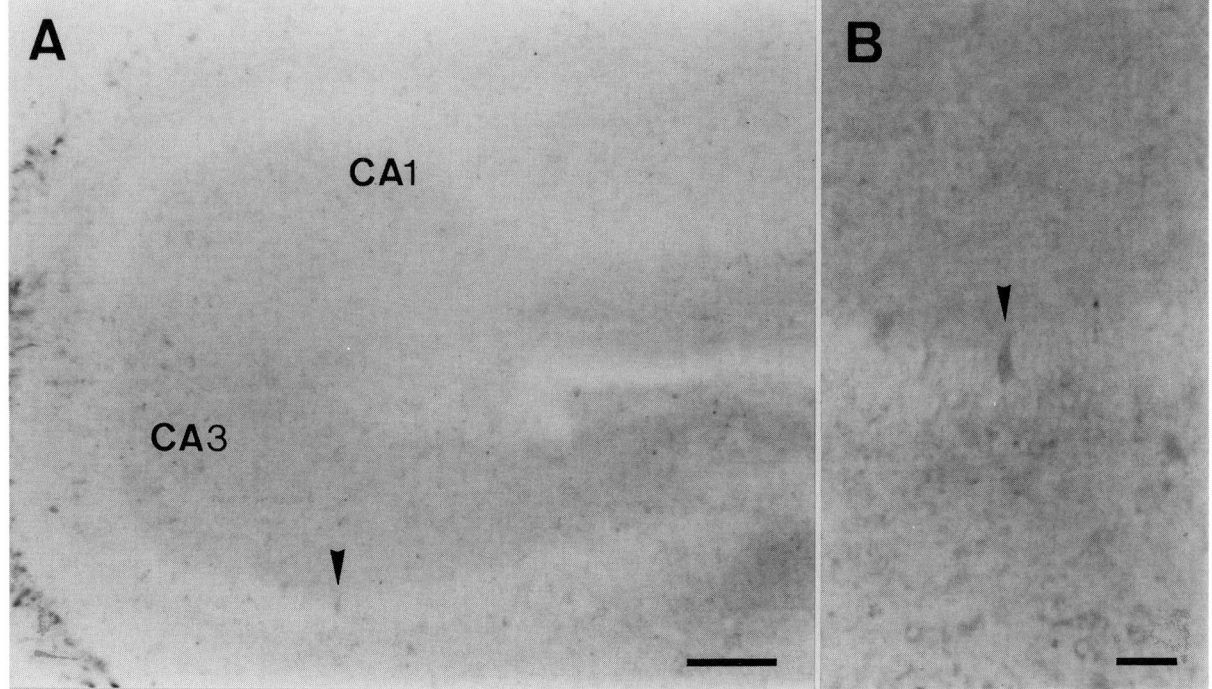

Fig. 2. Parvalbumin immunostaining in an explant derived from a 5-day-old rat and cultured for 4 days. (A) Note the absence of intensely stained neurons. The arrowhead points to one of the few labeled cells, which is reproduced at higher magnification in (B). Scale bar = 150 μm in (A), 50 μm in (B).

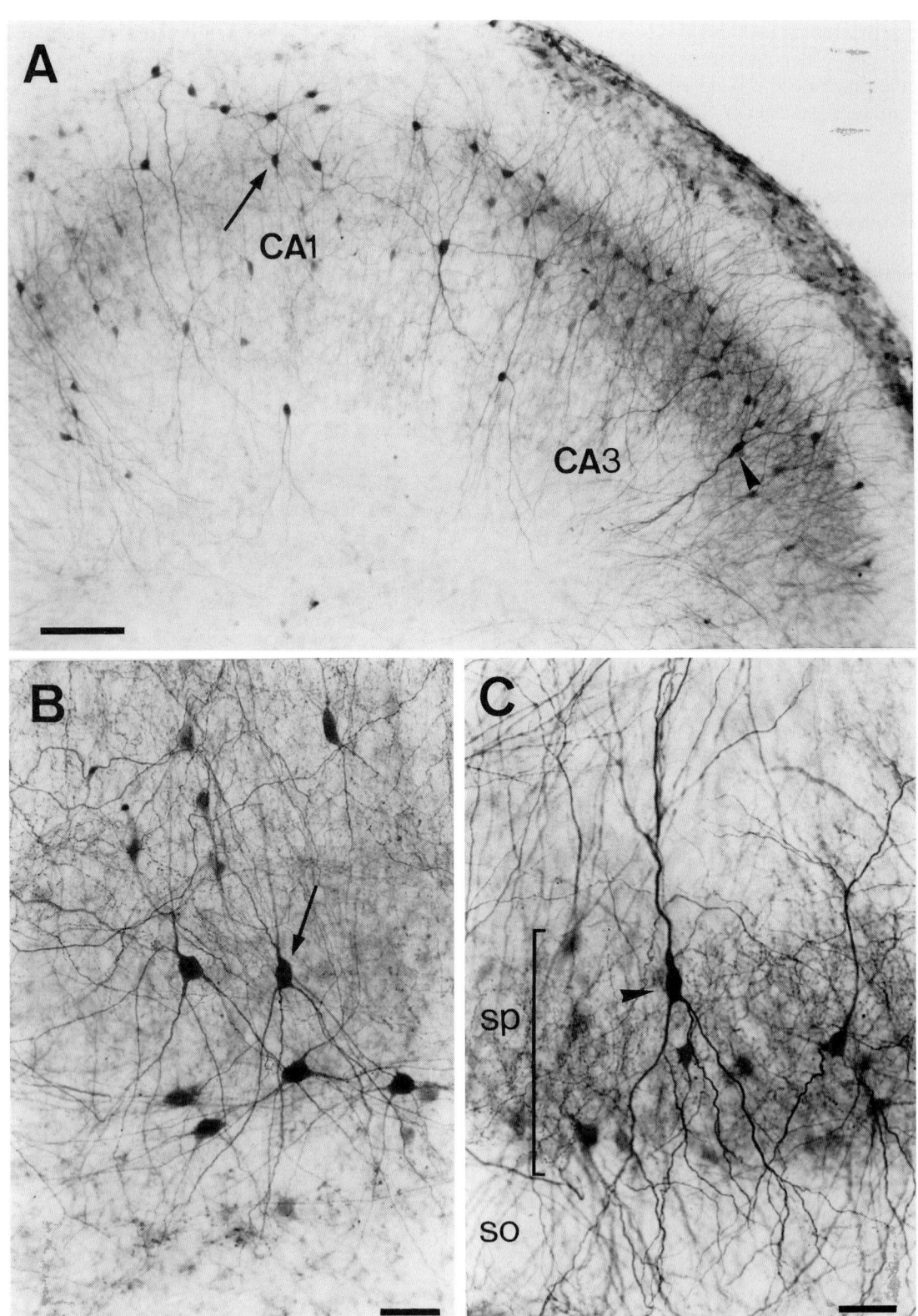

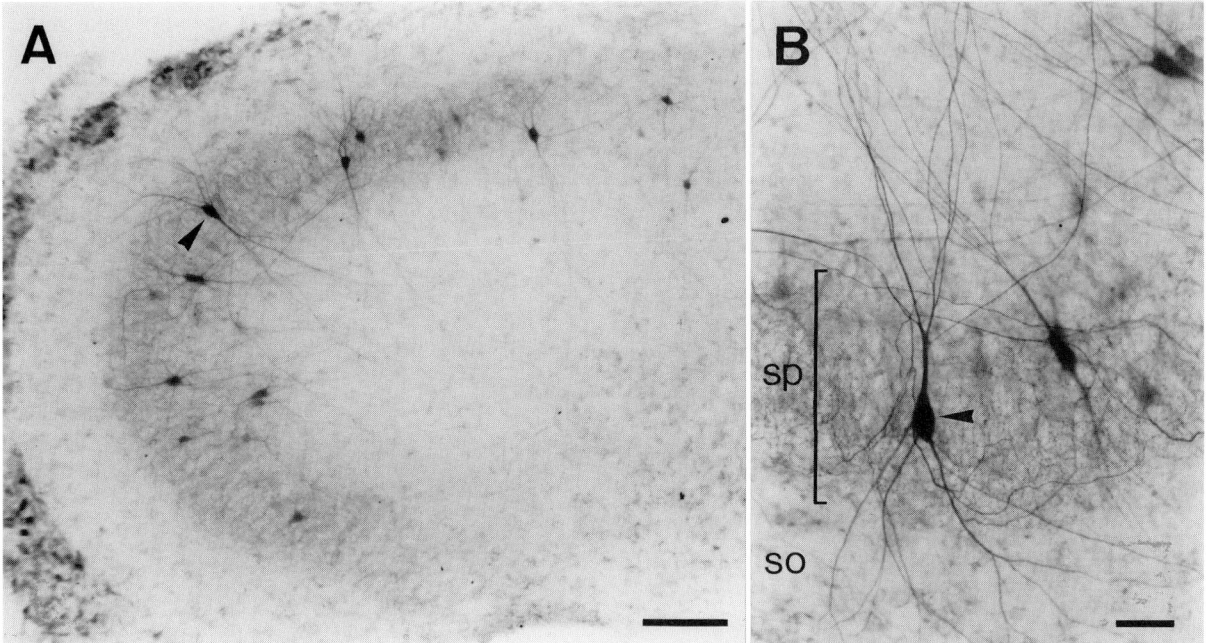

Fig. 4. Parvalbumin immunostaining in a CA3 isolated area explanted from a 5-day-old rat and cultured for 14 days in the presence of DNQX. The arrowhead in (A) points to a neuron which is reproduced at higher magnification in (B). Note the punctate labeling restricted to the pyramidal cell layer in the CA3 area; so = stratum oriens; sp = stratum pyramidale. The bracket delineates the stratum pyramidale. Scale bar = 150 μm in (A), 50 μm in (B).

horn, over the CA3 field. Only neurons exhibiting at least one labeled process were counted. Counts were performed in three explants per animal per time in culture. Mean values were first calculated for each animal. Mean values from three or four animals per time are shown in Fig. 1, together with the standard error to the mean.

The role of extrahippocampal afferents or targets in the development of parvalbumin expression in hippocampal interneurons was studied by following parvalbumin immunoreactivity in cultures explanted before the onset of in vivo expression. In explants taken from 5-day-old rats and cultured for 4 days, very few and weakly labeled parvalbumin-immunoreactive cells were observed, located in the CA3 area (Figs. 1A and 2). After 14 days in culture the number of immunoreactive neurons increased in the CA3 area, and immunoreactive neurons appeared in the CA1 area (Fig. 1A). Parvalbumin-immunoreactive neurons were more intensely stained in the CA3 area than in the CA1 area. In explants maintained 20 days in culture, the number and staining intensity of immunoreactive neurons in the CA1 area increased and reached a level equivalent to the one in the CA3 area (Figs. 1A and 3). The role of intrahippocampal afferents from the dentate gyrus and intrahippocampal targets in the CA1 area was studied by culturing explants of isolated CA3 area taken from 5-day-old rats. After 14 days in culture, the number and staining intensity of parvalbumin-immunoreactive neurons were similar in isolated CA3 area and in the CA3 area of explants containing all the hippocampal subfields (Fig. 1). A similar development of parvalbumin immunostaining

Fig. 3. Parvalbumin immunostaining in an explant derived from a 5-day-old rat and cultured for 20 days. (A) Note the intensely stained neurons in both the CA3 and the CA1 subfields. The arrow and arrowhead point to neurons which are reproduced at higher magnification in (B) and (C), respectively. Note the punctate labeling restricted to the pyramidal cell layer in the CA3 area; so = stratum oriens; sp = stratum pyramidale. The bracket delineates the stratum pyramidale. Scale bar = 150 μm in (A), 50 μm in (B) and (C).

was observed in isolated CA3 areas from 1-day-old rats maintained for 20 days in culture. Parvalbumin immunoreactivity therefore develops in hippocampal cultures explanted several days before the onset of parvalbumin expression. Furthermore, the developmental profile of parvalbumin expression in the explants is the same as in vivo expression in the CA3 area preceding expression in the CA1 subfield (Bergmann et al., 1991).

The role of ongoing neuronal activity in the development of parvalbumin expression was studied by comparing the development of parvalbumin immunoreactivity in control and DNQX-treated explants. After 14 days in culture, the number and staining intensity of parvalbumin-immunoreactive neurons was similar in DNQX-treated isolated CA3 areas and in control isolated CA3 areas (Figs. 1B and 4). Thus, blockade of neuronal activity did not prevent the development of parvalbumin expression.

These results indicate that parvalbumin expression in hippocampal interneurons develops in the absence of extrahippocampal afferents and targets, and under blockade of excitatory activity. Thus, despite the similarities of these neurons, parvalbumin expression does not develop in hippocampal interneurons following the same mechanisms as in neocortical interneurons. In the hippocampus, parvalbumin expression is either due to an intrinsic program of development of interneurons, or is triggered by local cellular interactions independent of neuronal activity.

Concluding remarks

The results described above indicate that the mechanisms responsible for the neurochemical maturation of hippocampal interneurons are very diverse. While parvalbumin expression develops despite blockade of excitatory transmission, the neuropeptide content of interneurons is very sensitive to the level of neuronal activity. Furthermore, although the neuropeptides NPY and somatostatin are regulated by neuronal activity to a similar extent, BDNF may mediate the effects of neuronal activity on the levels of NPY but not on somatostatin levels. From a functional point of view, NPY and somatostatin inhibit glutamatergic inputs on pyramidal neurons by a presynaptic action (McQuiston and Colmers, 1996; Boehm and Betz, 1997). It is not clear under which physiological conditions these neuropeptides are released, but they may be involved in preventing seizure activity in the adult brain (Vezzani et al., 1999). Particularly, NPY-deficient mice exhibit an increased susceptibility to seizures (Baraban et al., 1997). Regulation of the levels of neuropeptides by neuronal activity during postnatal maturation could thus set up appropriate neuroprotective mechanisms.

Acknowledgements

I thank Dr. Benedikt Berninger, Séverine Boillée, Dr. Isabelle Dusart, Dr. Marc Peschanski and Dr. Constantino Sotelo for comments on this paper.

References

Allen, J.M., McGregor, G.P., Woodhams, P.L., Polak, J.M. and Bloom, S.R. (1984) Ontogeny of a novel peptide, neuropeptide Y (NPY) in rat brain. *Brain Res.*, 303: 197–200.

Baraban, S.C., Hollopeter, G., Erickson, J.C., Schwartzkroin, P.A. and Palmiter, R.D. (1997) Knock-out mice reveal a critical antiepileptic role for neuropeptide Y. *J. Neurosci.*, 17: 8927–8936.

Ben-Ari, Y., Cherubini, E., Corradetti, R. and Gaiarsa, J.-L. (1989) Giant synaptic potentials in immature rat CA3 hippocampal neurones. *J. Physiol. (Lond.)*, 416: 303–325.

Ben-Ari, Y., Khazipov, R., Leinekugel, X., Caillard, O. and Gaiarsa, J.-L. (1997) $GABA_A$, NMDA and AMPA receptors: a developmentally regulated 'ménage à trois'. *Trends Neurosci.*, 20: 523–529.

Bergmann, I., Nitsch, R. and Frotscher, M. (1991) Area-specific morphological and neurochemical maturation of non-pyramidal neurons in the rat hippocampus as revealed by parvalbumin immunocytochemistry. *Anat. Embryol.*, 184: 403–409.

Berninger, B. and Poo, M.-m. (1996) Fast actions of neurotrophic factors. *Curr. Opin. Neurobiol.*, 6: 324–330.

Berninger, B., Marty, S., Zafra, F., Berzaghi, M., Thoenen, H. and Lindholm, D. (1995) GABAergic stimulation switches from enhancing to repressing BDNF expression in rat hippocampal neurons during maturation in vitro. *Development*, 121: 2327–2335.

Boehm, S. and Betz, H. (1997) Somatostatin inhibits excitatory transmission at rat hippocampal synapses via presynaptic receptors. *J. Neurosci.*, 17: 4066–4075.

Carder, R.K., Leclerc, S.S. and Hendry, S.H.C. (1996) Regulation of calcium-binding protein immunoreactivity in GABA neurons of macaque primary visual cortex. *Cereb. Cortex*, 6: 271–287.

Carnahan, J. and Nawa, H. (1995) Regulation of neuropeptide expression in the brain by neurotrophins. *Mol. Neurobiol.*, 10: 135–149.

Croll, S.D., Wiegand, S.J., Anderson, K.D., Lindsay, R.M. and Nawa, H. (1994) Regulation of neuropeptides in adult rat forebrain by the neurotrophins BDNF and NGF. *Eur. J. Neurosci.*, 6: 1343–1353.

De Lecea, L., Del Río, J.A. and Soriano, E. (1995) Developmental expression of parvalbumin mRNA in the cerebral cortex and hippocampus of the rat. *Mol. Brain Res.*, 32: 1–13.

De Lecea, L., Del Rio, J.A., Criado, J.R., Alcántara, S., Morales, M., Danielson, P.E., Henriksen, S.J., Soriano, E. and Sutcliffe, J.G. (1997) Cortistatin is expressed in a distinct subset of cortical interneurons. *J. Neurosci.*, 17: 5868–5880.

Durand, G.M., Kovalchuk, Y. and Konnerth, A. (1996) Long-term potentiation and functional synapse induction in developing hippocampus. *Nature*, 381: 71–75.

Freund, T.F. and Buzsáki, G. (1996) Interneurons of the hippocampus. *Hippocampus*, 6: 345–470.

Gall, C., Lauterborn, J., Isackson, P. and White, J. (1990) Seizures, neuropeptide regulation, and mRNA expression in the hippocampus. *Prog. Brain Res.*, 83: 371–390.

Gómez-Di Cesare, C.M., Smith, K.L., Rice, F.L. and Swann, J.W. (1997) Axonal remodeling during postnatal maturation of CA3 hippocampal pyramidal neurons. *J. Comp. Neurol.*, 384: 165–180.

Gulyás, A.I., Hájos, N. and Freund, T.F. (1996) Interneurons containing calretinin are specialized to control other interneurons in the rat hippocampus. *J. Neurosci.*, 16: 3397–3411.

Hosokawa, Y., Sciancalepore, M., Stratta, F., Martina, M. and Cherubini, E. (1994) Developmental changes in spontaneous GABA-A mediated synaptic events in rat hippocampal CA3 neurons. *Eur. J. Neurosci.*, 6: 805–813.

Jiang, M. and Swann, J.W. (1997) Expression of calretinin in diverse neuronal populations during development of rat hippocampus. *Neuroscience*, 81: 1137–1154.

Jones, K.R., Farinas, I., Backus, C. and Reichardt, L.F. (1994) Targeted disruption of the BDNF gene perturbs brain and sensory neuron development but not motor neuron development. *Cell*, 76: 989–999.

Kafitz, K.W., Rose, C.R., Thoenen, H. and Konnerth, A. (1999) Neurotrophin-evoked rapid excitation through TrkB receptors. *Nature*, 401: 918–921.

Lang, U. and Frotscher, M. (1990) Postnatal development of nonpyramidal neurons in the rat hippocampus (areas CA1 and CA3): a combined Golgi/electron microscope study. *Anat. Embryol.*, 181: 533–545.

Marty, S. and Onténiente, B. (1997) The expression pattern of somatostatin and calretinin by postnatal hippocampal interneurons is regulated by activity-dependent and -independent determinants. *Neuroscience*, 80: 79–88.

Marty, S. and Onténiente, B. (1999) BDNF and NT-4 differentiate two pathways in the modulation of neuropeptide protein levels in postnatal hippocampal interneurons. *Eur. J. Neurosci.*, 11: 1647–1656.

Marty, S., Berninger, B., Carroll, P. and Thoenen, H. (1996a) GABAergic stimulation regulates the phenotype of hippocampal interneurons through the regulation of brain-derived neurotrophic factor. *Neuron*, 16: 565–570.

Marty, S., Carroll, P., Cellerino, A., Castren, E., Staiger, V., Thoenen, H. and Lindholm, D. (1996b) Brain-derived neurotrophic factor promotes the differentiation of various hippocampal non-pyramidal neurons, including Cajal–Retzius cells, in organotypic slice cultures. *J. Neurosci.*, 16: 675–687.

Marty, S., Berzaghi, M. and Berninger, B. (1997) Neurotrophins and activity-dependent plasticity of cortical interneurons. *Trends Neurosci.*, 20: 198–202.

McQuiston, A.R. and Colmers, W.F. (1996) Neuropeptide Y_2 receptors inhibit the frequency of spontaneous but not miniature EPSCs in CA3 pyramidal cells of rat hippocampus. *J. Neurophysiol.*, 76: 3159–3168.

Miles, R. and Wong, R.K.S. (1983) Single neurones can initiate synchronized population discharge in the hippocampus. *Nature*, 306: 371–374.

Miles, R., Toth, K., Gulyás, A.I., Hajos, N. and Freund, T.F. (1996) Differences between somatic and dendritic inhibition in the hippocampus. *Neuron*, 16: 815–823.

Muller, D., Buchs, P.-A. and Stoppini, L. (1993) Time course of synaptic development in hippocampal organotypic cultures. *Dev. Brain Res.*, 71: 93–100.

Murphy, D.D., Cole, N.B. and Segal, M. (1998) Brain-derived neurotrophic factor mediates estradiol-induced dendritic spine formation in hippocampal neurons. *Proc. Natl. Acad. Sci. USA*, 95: 11412–11417.

Naus, C.C.G., Morrison, J.H. and Bloom, F.E. (1988) Development of somatostatin-containing neurons and fibers in the rat hippocampus. *Dev. Brain Res.*, 40: 113–121.

Nawa, H., Bessho, Y., Carnahan, J., Nakanishi, S. and Mizuno, K. (1993) Regulation of neuropeptide expression in cultured cerebral cortical neurons by brain-derived neurotrophic factor. *J. Neurochem.*, 60: 772–775.

Nawa, H., Pelleymounter, M.A. and Carnahan, J. (1994) Intraventricular administration of BDNF increases neuropeptide expression in newborn rat brain. *J. Neurosci.*, 14: 3751–3765.

Nawa, H., Carnahan, J. and Gall, C. (1995) BDNF protein measured by a novel enzyme immunoassay in normal brain and after seizure: partial disagreement with mRNA levels. *Eur. J. Neurosci.*, 7: 1527–1535.

Parra, P., Gulyás, A.I. and Miles, R. (1998) How many subtypes of inhibitory cells in the hippocampus?. *Neuron*, 20: 983–993.

Rocamora, N., Pascual, M., Acsàdy, L., De Lecea, L., Freund, T.F. and Soriano, E. (1996) Expression of NGF and NT-3 mRNAs in hippocampal interneurons innervated by the GABAergic septohippocampal pathway. *J. Neurosci.*, 16: 3991–4004.

Rozenberg, F., Robain, O., Jardin, L. and Ben-Ari, Y. (1989) Distribution of GABAergic neurons in late fetal and early postnatal rat hippocampus. *Dev. Brain Res.*, 50: 177–187.

Rutherford, L.C., DaWan, A., Lauer, H.M. and Turrigiano, G.G. (1997) Brain-derived neurotrophic factor mediates the activity-dependent regulation of inhibition in neocortical cultures. *J. Neurosci.*, 17: 4527–4535.

Schmidt-Kastner, R., Wetmore, C. and Olson, L. (1996) Comparative study of brain-derived neurotrophic factor messenger RNA and protein at the cellular level suggests multiple roles in hippocampus, striatum and cortex. *Neuroscience*, 74: 161–183.

Schwarzer, C., Sperk, G., Samanin, R., Rizzi, M., Gariboldi, M. and Vezzani, A. (1996) Neuropeptides-immunoreactivity and their mRNA expression in kindling: functional implications for limbic epileptogenesis. *Brain Res. Rev.*, 22: 27–50.

Seress, L. and Ribak, C.E. (1988) The development of GABAergic neurons in the rat hippocampal formation. An immunohistochemical study. *Dev. Brain Res.*, 44: 197–209.

Stoppini, L., Buchs, P.-A. and Muller, D. (1991) A simple method for organotypic cultures of nervous tissue. *J. Neurosci. Methods*, 37: 173–182.

Thoenen, H. (1995) Neurotrophins and neuronal plasticity. *Science*, 270: 593–598.

Vezzani, A., Sperk, G. and Colmers, W.F. (1999) Neuropeptide Y: emerging evidence for a functional role in seizure modulation. *Trends Neurosci.*, 22: 25–30.

Vicario-Abejón, C., Collin, C., McKay, R.D.G. and Segal, M. (1998) Neurotrophins induce formation of functional excitatory and inhibitory synapses between cultured hippocampal neurons. *J. Neurosci.*, 18: 7256–7271.

Vogt Weisenhorn, D.M., Celio, M.R. and Rickmann, M. (1998) The onset of parvalbumin-expression in interneurons of the rat parietal cortex depends upon extrinsic factor(s). *Eur. J. Neurosci.*, 10: 1027–1036.

Wirth, M.J., Obst, K. and Wahle, P. (1998) NT-4/5 and LIF, but not NT-3 and BDNF, promote NPY mRNA expression in cortical neurons in the absence of spontaneous bioelectrical activity. *Eur. J. Neurosci.*, 10: 1457–1464.

Zachrisson, O., Falkenberg, T. and Lindefors, N. (1996) Neuronal coexistence of trkB and glutamic acid decarboxylase$_{67}$ mRNAs in rat hippocampus. *Mol. Brain Res.*, 36: 169–173.

CHAPTER 17

Long-term regulation of excitatory and inhibitory synaptic transmission in hippocampal cultures by brain-derived neurotrophic factor

M. McLean Bolton *, Donald C. Lo and Nina T. Sherwood [†]

Department of Neurobiology, Box 3209, Duke University Medical Center, 101 Research Drive, Durham, NC 27710, USA

Introduction

One of the fundamental features of the central nervous system is the malleability of its synaptic connections. The activity-dependent modification of the strength and ultimately the existence of specific synapses is a principal way in which experiences and their consequences are etched into neural circuits. During development, refinement of synaptic connectivity is thought to involve the selective strengthening of synapses with correlated presynaptic and postsynaptic activity; similar processes are thought to occur in the adult as a basis for learning and memory. One mechanism by which appropriate connections may be strengthened is through the local action of synapse-modifying factors released in response to activity. Several lines of evidence suggest that brain-derived neurotrophic factor (BDNF), a member of the neurotrophin family of peptide growth factors, may be one of these elusive factors. In this review, we will outline some of the major arguments that have led to this idea, and summarize some recent experiments we have done to understand the biology of neurotrophins in regulating the strength of synaptic transmission in hippocampal neurons over long time scales.

We will also address the question of whether and how BDNF is involved in regulating the balance between excitatory and inhibitory synaptic drive. Despite ongoing activity-dependent modification of the strength of excitatory synapses throughout development and in the adult, some balance of excitation and inhibition must be maintained in order for the nervous system to function properly. One possible strategy for coupling excitation and inhibition is to have the same signals regulate both processes. There is increasing evidence that BDNF acts in this way: local changes in excitation result in fluctuations in BDNF expression levels, which in turn have been shown to regulate the strength of both excitatory and inhibitory transmission in several experimental settings. BDNF may thus effect selective strengthening of particular excitatory synapses while providing a concomitant component of homeostasis via the regulation of inhibition.

Why is BDNF an attractive molecule for regulating synaptic plasticity?

The expression patterns of BDNF and its receptor, TrkB, the characteristics of BDNF release, and the

---

* Corresponding author. M. McLean Bolton, Department of Neurobiology, Box 3209, Duke University Medical Center, 101 Research Drive, Durham, NC 27710, USA. Fax: +1-919-684-4431; E-mail: mcleanb@neuro.duke.edu

[†] Current address: Division of Biology 216-76, California Institute of Technology, Pasadena, CA 91125, USA.

signal transduction events initiated by TrkB activation strongly implicate these molecules in regulating long-term synaptic plasticity. BDNF and TrkB are expressed in regions of the brain that exhibit synaptic plasticity, such as the cortex and hippocampus (Chao, 1992; Barbacid, 1994). The developmental expression profiles of BDNF and TrkB are also consistent with a role in plasticity, in that periods of enhanced plasticity correlate with enhanced expression (reviewed in: Davies, 1994; McAllister et al., 1999).

In order to mediate activity-dependent synaptic plasticity, a candidate molecule must itself be regulated (e.g., produced, released, or activated) in response to neuronal activity. BDNF expression is dramatically increased by seizure activity in both the hippocampus and the cortex (Zafra et al., 1990; Ernfors et al., 1991; Isackson et al., 1991), and is elevated by electrical stimulation paradigms that induce long-term potentiation (LTP) in the hippocampus (Patterson et al., 1992; Castrén et al., 1993; Dragunow et al., 1993; for review see Castrén et al., 1998). More physiological signals such as sensory stimulation also increase BDNF and/or TrkB expression, as shown in the visual cortex in response to light exposure (Castrén et al., 1992; Bozzi et al., 1995) and in the somatosensory cortex following whisker stimulation (Rocamora et al., 1996b). Importantly, an increase in not only expression levels but also the release of neurotrophins has been observed in response to neuronal activity (Blöchl and Thoenen, 1995, 1996; Goodman et al., 1996; Canossa et al., 1997). Although its release mechanism remains unknown, BDNF has been localized to dense core vesicles in the dendrites of hippocampal neurons, suggesting the possibility of synapse-specific release (Fawcett et al., 1997; Smith et al., 1997; Moller et al., 1998), and thus a role in synapse-specific forms of developmental and adult plasticity.

The TrkB receptor signal transduction cascades initiated by BDNF binding result in the activation of transcription factors capable of inducing long-term changes in gene expression that would be expected to underlie persistent changes in neuronal function (reviewed in: Sheng and Greenberg, 1990; Segal and Greenberg, 1996; Friedman and Greene, 1999). In particular, the activation of the microtubule-associated protein (MAP) kinase cascade via the Src homology domain containing protein (SHC)/Ras pathway increases the transcription of several immediate early genes. In cultured dorsal root ganglion neurons, nerve growth factor (NGF) application to distal neurites leads to the rapid phosphorylation, within minutes, of TrkA and the transcription factor, calcium/cAMP response element binding protein (CREB). Subsequent expression of the immediate early gene product c-*fos*, however, is not observed until 2.5–5 h after NGF application to the neurites, and remains high at 7 h (Watson et al., 1999). Such slow time courses of c-*fos* induction indicate that the subsequent induction of delayed response genes modulating neuronal function can occur on even longer time scales.

While the characteristics of BDNF expression and release, as well as the expression patterns and signal transduction cascade activated by the TrkB receptor, are all indicative of a role for BDNF in synaptic plasticity, the challenge for the field has been to provide direct evidence that BDNF modulates synaptic strength. Such experiments have roughly fallen into three categories, which we will review in the following sections: (1) in vivo and transgenic manipulation of BDNF and TrkB receptor levels; (2) rapid effects of BDNF on synaptic transmission; (3) long-term regulation of synaptic transmission by BDNF.

Effects of BDNF and TrkB manipulation in vivo

In vivo studies manipulating BDNF and TrkB levels have begun to elucidate the role of BDNF in visual system development, LTP and spatial learning, and in the pathological plasticity associated with kindling-induced epileptogenesis. During visual system development, axons arising from the lateral geniculate nucleus (LGN) arrive in layer 4 of the visual cortex, and segregate into ocular dominance columns in which the neurons are driven primarily by either the right or left eye. The refinement of this pattern of connectivity is highly sensitive to visual experience and neuronal activity for a period of development known as the critical period. Blocking visual experience in one eye during the critical period permanently reduces the number of cortical neurons that eye is capable of activating.

The LGN afferents have been proposed to compete in an activity-dependent process for a limited amount of trophic factor produced by target neurons in the visual cortex. The initial evidence for an involvement of neurotrophins was shown for NGF: intraventricular infusion of NGF prevented the effects of monocular deprivation, while blocking endogenous NGF altered normal development (Domenici et al., 1991, 1994; Maffei et al., 1992; Berardi et al., 1993, 1994; Carmignoto et al., 1993). Neurotrophin-4/5 (NT-4/5), which also preferentially activates the TrkB receptor, was subsequently shown to prevent the shrinkage of LGN neurons due to monocular deprivation (Riddle et al., 1995). Furthermore, local perturbations of BDNF or NT-4/5 levels in visual cortex block the formation of ocular dominance columns in that region (Cabelli et al., 1995, 1997). Consistent with these experiments, transgenic mice overexpressing BDNF selectively in the early postnatal forebrain exhibit precocious maturation of the visual system both in terms of visual acuity and the closure of the critical period for ocular dominance plasticity (Huang et al., 1999).

Behavioral assays of learning in BDNF or TrkB knockouts have been limited by the early death of the homozygotes (Klein et al., 1993; Jones et al., 1994; Erickson et al., 1996). Although BDNF heterozygous mutant mice have been reported to have spatial learning deficits (Linnarsson et al., 1997), these findings have not been replicated in other mutant lines (Montkowski and Holsboer, 1997). The recent generation of conditional *trkB* knockout mice in which receptor expression is lost only in the postnatal forebrain has circumvented these limitations; these mice exhibit defects in spatial learning and CA1 hippocampal LTP (Minichiello et al., 1999). These results corroborate earlier findings in conventional BDNF knockout mice of diminished hippocampal LTP (Korte et al., 1995, 1996a,b; Patterson et al., 1996). Mechanistically, such reductions in LTP in these animals appear to arise from impairments in high-frequency firing and decreased synaptic vesicle docking and protein distribution (Pozzo-Miller et al., 1999). Conventional TrkB knockout mice have fewer synaptic vesicles at commissural/CA1 hippocampal synapses, as well as fewer axon collaterals and synaptic contacts (Martinez et al., 1998).

In vivo manipulation of BDNF levels also interferes with plasticity associated with epileptogenesis in the hippocampus. BDNF heterozygous knockout mice have an elevated kindling-induced seizure threshold (Kokaia et al., 1995), while intraventricular infusion of the Trk receptor body, TrkB-IgG to block endogenous BDNF, delays the onset of seizures induced by kindling (Binder et al., 1999). Other studies report, however, that BDNF infusion inhibits the development of kindled seizures (Larmet et al., 1995; Osehobo et al., 1999). Nonetheless, in both cases a role for BDNF in kindled seizures is implicated. In summary, the combined evidence from experiments manipulating BDNF function in vivo, either by direct infusion or by genetic manipulation, strongly implicate BDNF as a central mediator of experience-dependent synaptic plasticity.

Acute effects of BDNF on synaptic transmission

Several investigators have examined the effects of acute BDNF application on synaptic transmission directly. In the first such experiments, done in cultures of developing *Xenopus* neuromuscular synapses, BDNF and neurotrophin-3 (NT-3) were found to potentiate neuromuscular transmission within a few minutes of application (Lohof et al., 1993). Subsequent experiments have confirmed that the acute effects of BDNF are presynaptic in this system, inducing enhancement of neurotransmitter release (Wang et al., 1995; Stoop and Poo, 1996; Boulanger and Poo, 1999a,b). Rapid increases in excitatory synaptic transmission following BDNF application have also been demonstrated in dissociated hippocampal cultures (Lessmann et al., 1994; Levine et al., 1995, 1996, 1998; Lessmann and Heumann, 1998; Li et al., 1998; Song et al., 1998). Consistent with these observations, basal synaptic transmission at Schaffer collateral/CA1 synapses is acutely potentiated by BDNF in hippocampal slice preparations (Kang and Schuman, 1995a,b), although this effect has not been observed by all investigators (Figurov et al., 1996; Patterson et al., 1996; Frerking et al., 1998; Gottschalk et al., 1998). Similar potentiative effects of BDNF on excitatory transmission have been found in the hippocampal CA3 and dentate gyrus regions (Scharfman, 1997; Scharfman et al., 1999) and by some investigators in neocortex (Akaneya et al., 1996, 1997; Carmignoto et al., 1997; but see Huber et al., 1998; Kinoshita et al., 1999). BDNF has been demonstrated to interact with tetanus-induced LTP in hippocampus (Figurov et al., 1996; Gottschalk et al., 1998; Korte et al., 1998), and restores the ability to induce hippocampal LTP when acutely expressed or applied (Korte et al., 1996a,b; Patterson et al., 1996). Conversely, suppression of BDNF expression by the introduction of antisense oligonucleotides inhibits hippocampal LTP (Ma et al., 1998). Acutely applied BDNF also promotes the induction of LTP in neocortex (Akaneya et al., 1996, 1997; Huber et al., 1998; Kinoshita et al., 1999; Sermasi et al., 1999).

Long-term regulation of synaptic transmission by BDNF

Chronic BDNF treatment enhances excitatory postsynaptic currents in hippocampal neurons

Many alterations in synaptic connectivity associated with development and learning are long-lasting, and may ultimately be translated into structural modification of the synapse. The processes by which relatively short-term modifications in synaptic strength (occurring on a time scale of seconds to hours) are converted into long-term changes (days to years) are not well understood, nor is it even clear whether these are serial continua or distinct parallel processes. We have observed long-term synaptic strengthening by BDNF in both cultured autaptic excitatory hippocampal neurons (Sherwood and Lo, 1999) — single, isolated neurons that form synapses (autapses) onto themselves — as well as in conventional dissociated hippocampal cultures containing both excitatory and inhibitory neurons (Bolton et al., 2000). Each of these systems has its unique advantages: autaptic cultures enable the correlation of whole-cell and unitary current properties for a given cell, while conventional cultures enable the study of the circuit properties that arise from changes in single-cell characteristics. Both enable high-resolution electrophysiological recording and imaging at the level of individual cells.

Chronic BDNF treatment (7–14 days) of autaptic CA1 hippocampal neurons increased the amplitude of α-amino-3-hydroxy-5-methyl-4-isoxazole-4-propionic acid (AMPA) receptor-mediated synaptic currents evoked by action potential-induced transmitter release by 70% compared to untreated cultures or those in which endogenous BDNF was neutralized with TrkB-IgG (Fig. 1A,B; Shelton et al., 1995; McAllister et al., 1996, 1997). Importantly, these long-term effects were observed in the absence of BDNF during the recording period, and were distinct from the acute effects of BDNF on these cells (Fig. 1D).

The amplitude of excitatory postsynaptic currents (EPSCs) in the autaptic neuron reflects the spatiotemporal integration of all of the individual synaptic currents. The number of synapses, the amplitude of the unitary synaptic current, and the probability of vesicular release combined determine the amplitude of the synaptic response, but in addition the morphology of the neuron and membrane conductance properties contribute to the electrotonic filtering of the measured synaptic currents. No significant differences were observed with BDNF treatment in cell size or other passive membrane properties as measured by capacitance and input resistance, nor were there differences in the rise or decay kinetics of unitary synaptic currents (mEPSCs). Therefore, differential cable filtering of synaptic currents did not contribute significantly to the observed increase in evoked autaptic current amplitude. Whole-cell Na^+

Fig. 1. Long-term BDNF treatment enhances excitatory synaptic transmission in cultured autaptic neurons and is distinct from BDNF's short-term effects. (A) In this experiment, AMPA receptor-mediated EPSC (left) and mEPSC (right) traces were averaged for control (dashed lines) and BDNF-treated (solid lines) cells and superimposed. BDNF increased EPSCs and mEPSCs to similar extents. Scaling the averaged control traces to their respective BDNF traces (dotted lines) showed that the kinetics of activation and decay were indistinguishable between treatment groups. (B) EPSC amplitudes were 1.8-fold greater in BDNF-treated cells compared with controls; mEPSC amplitudes were increased to the same extent. Values normalized to control cells are shown (means ± SEM); asterisks denote statistical significance of $P < 0.05$ by ANOVA. (C) BDNF did not affect the average frequency of mEPSCs, measured continuously while cells were held at −70 mV. (D) In contrast to its chronic effects, acute BDNF exposure increased only the frequency of mEPSCs (right) and had no effect on the amplitude of either mEPSCs (center) or EPSCs (left). Each symbol represents an individual neuron recorded before and after BDNF application. Effects were observed within 10 min of BDNF addition, and were confirmed in recordings of populations of neurons briefly treated or untreated with BDNF. (Modified from Sherwood and Lo, 1999.)

and K+ currents were also equivalent in all conditions, supporting the idea that BDNF did not regulate intrinsic excitability, and thus the effectiveness of action potential propagation, in these neurons. Furthermore, BDNF treatment did not affect the number or size of synapses as determined by synapsin Ia,b immunostaining. Measurement of EPSCs and action potential-independent mEPSCs from the same

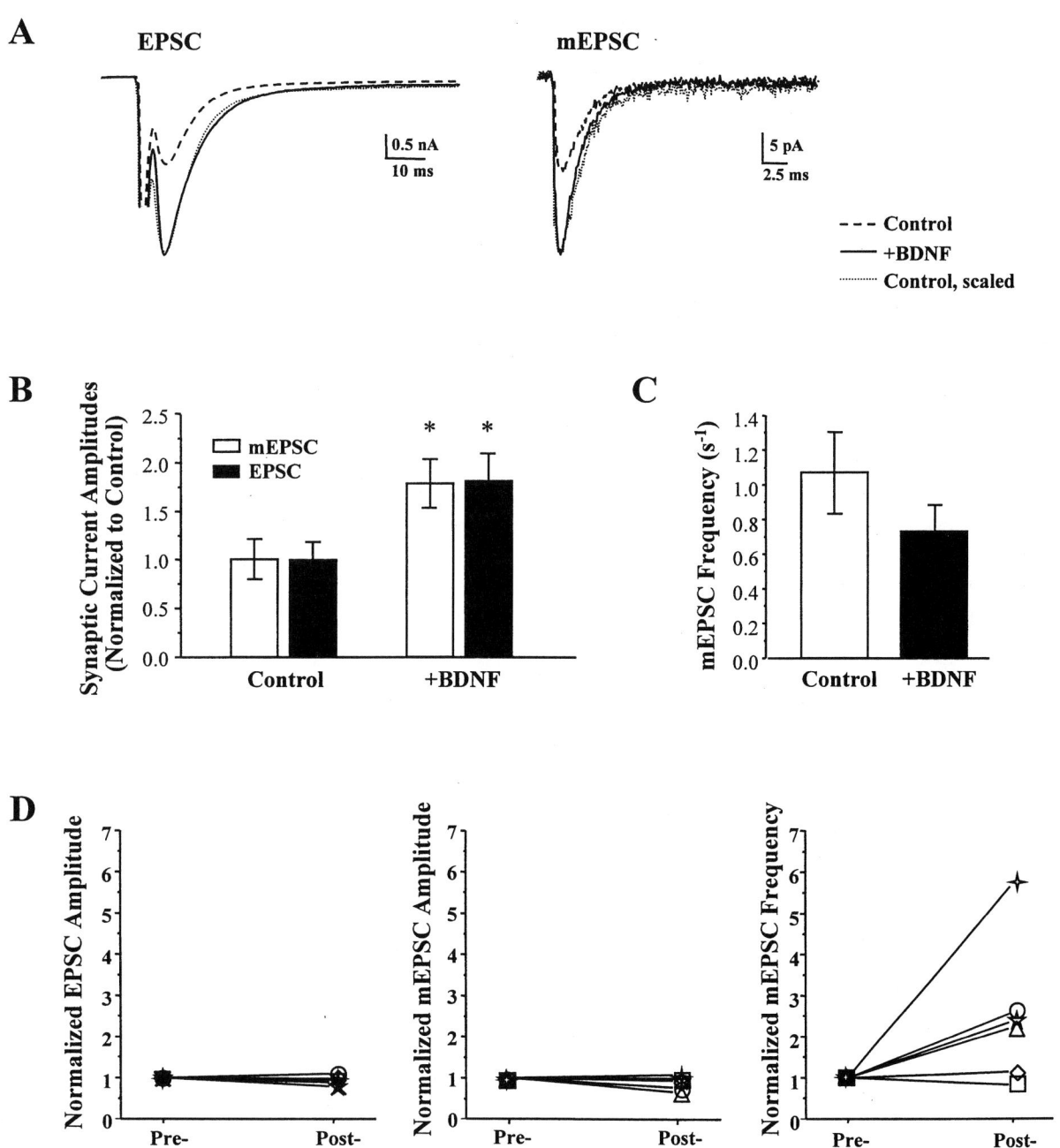

cells revealed instead that the mechanism underlying BDNF's effects on evoked synaptic transmission was an exclusive increase in the amplitude of mEPSCs (Fig. 1A,B). This increase in mEPSC amplitude occurred in the absence of an increase in mEPSC frequency (Fig. 1C), and completely accounted for the increase in EPSC amplitude. Thus, in these cultures BDNF enhanced the amplitude of evoked excitatory autaptic currents by increasing the amplitude of the current response to individual vesicles of glutamate.

We also observed an upregulation of mEPSC amplitude by chronic BDNF treatment in conventional mixed, dissociated cultures of hippocampal neurons (Bolton et al., 2000). In cultures treated for 4–7 days with BDNF compared to untreated or TrkB-IgG-treated cultures, BDNF increased the amplitude of AMPA receptor-mediated mEPSCs by ∼30% (Fig. 2A,B). That the magnitude of this increase was less than in the autaptic cultures may simply reflect differences in treatment duration or the developmental rate of the two systems, or it may be due to more complex interactions within a heterogeneous neuronal population. The increase in the amplitude of unitary synaptic transmission by BDNF in the conventional cultures was also accompanied by a significant increase in mEPSC frequency relative to TrkB-IgG-treated but not control cells (Fig. 2C).

BDNF has also been reported to increase the amplitude of spontaneous excitatory synaptic currents in dissociated cultures of early embryonic hippocampus, but at this very early developmental stage the predominant effect of BDNF is to increase the number of functional synaptic connections between neurons (Vicario-Abejon et al., 1998). In contrast, studies of long-term effects of BDNF in mixed, dissociated cultures of visual cortex suggest that BDNF regulates quantal amplitude of these pyramidal neurons in the opposite direction as observed for hippocampal neurons (Rutherford et al., 1998; Turrigiano et al., 1998; Turrigiano and Nelson, 1998; Turrigiano, 1999). Blocking endogenous BDNF in these cultures increases mEPSC amplitudes onto pyramidal neurons while exogenous BDNF has no effect. In addition, increases in mEPSC amplitude induced by activity blockade are prevented by the addition of BDNF. Interestingly, however, BDNF does increase the amplitude of mEPSCs onto bipolar interneurons.

Possible cellular mechanisms for the long-term regulation of excitatory transmission

Our finding that long-term treatment of excitatory hippocampal neurons with BDNF leads to an increase in the quantal size of AMPA receptor-mediated mEPSCs suggests a postsynaptic locus for these effects. Changes in mEPSC amplitude could be achieved by the modulation of receptor density by regulating AMPA receptor synthesis, insertion into or removal from the postsynaptic membrane, or degradation rate. Such regulation has been demonstrated in spinal neurons, where manipulating activity levels changes the half-life of AMPA receptors and is reflected in corresponding changes in quantal size (O'Brien et al., 1998). Similarly, rapid insertion and removal of AMPA receptors may underlie the potentiation and depression of synapses (Carroll et al., 1999; Lissin et al., 1999; Shi et al., 1999). BDNF has in fact been demonstrated to increase the expression of AMPA receptor subunits GluR1 and GluR2/3 in neocortical neurons (Narisawa-Saito et al., 1999a,b).

Changes in the composition or functional states of AMPA receptors are also possible mechanisms underlying BDNF-induced increases in excitatory responses. However, because the kinetics of the synaptic currents were not changed by BDNF treatment in our experiments, only certain mechanisms are plausible. One possibility, the post-translational modification of AMPA receptors, is supported by experiments in which Ca^{2+}/calmodulin-kinase II has been demonstrated to phosphorylate GluR1, resulting in an increase in receptor single-channel conductance (Tan et al., 1994; Barria et al., 1997a,b; Mammen et al., 1997; Derkach et al., 1999). The subunit composition of AMPA receptors also affects single-channel properties (reviewed in Hollmann and Heinemann, 1994). In addition to different homo- or heteromeric combinations of the four independently encoded subunits (GluR1–4), post-transcriptional modifications via alternative splicing or RNA editing provide further sources of variation in channel properties. Recombinant expression of AMPA receptors in human embryonic kidney (HEK)293 cells reveals, for example, that inclusion of the unedited form of GluR2 leads to increases in AMPA current amplitudes via an increase in channel conductance (Swanson et al., 1997).

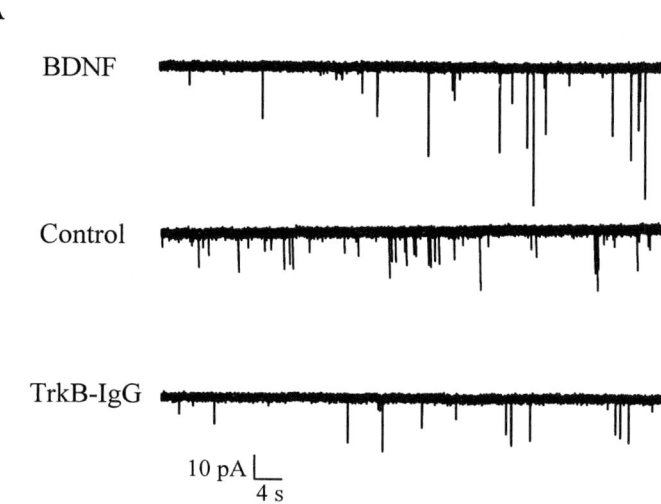
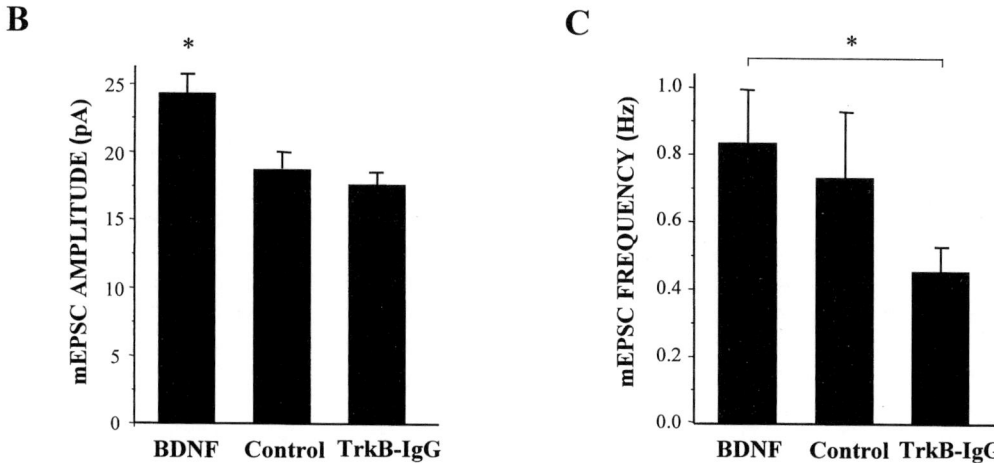

Fig. 2. BDNF increases AMPA receptor-mediated mEPSC amplitude in mixed hippocampal cultures. (A) Representative recordings of AMPA receptor-mediated mEPSCs on a compressed time base; traces from neurons in BDNF (top), control (middle), and TrkB-IgG (bottom) groups are shown. (B) BDNF increased mEPSC amplitude by ~30% compared to control cells and by ~40% compared to TrkB-IgG-treated neurons. (C) mEPSC frequency was elevated ~1.8-fold by BDNF compared to TrkB-IgG-treated cells but not compared to controls. (Adapted from Bolton et al., 2000.)

Formally, two presynaptic mechanisms could also underlie increases in mEPSC amplitude, provided that postsynaptic glutamate receptors are not typically saturated by the amount of glutamate released (Bekkers and Stevens, 1990; but see Clements et al., 1992; Tong and Jahr, 1994; Liu and Tsien, 1995; Forti et al., 1997). Under this assumption, both an increase in the amount of glutamate packaged per vesicle (reviewed in Reimer et al., 1998) or enhanced multivesicular release (Tong and Jahr, 1994) could also result in changes in mEPSC amplitude.

In conventional dissociated hippocampal cultures, BDNF increased mEPSC frequency almost twofold compared to TrkB-IgG-treated neurons (Bolton et al., 2000). Whether this increase in frequency arises from the same or a distinct molecular mechanism as the increase in mEPSC amplitude is unclear. One mechanism that could underlie increases in both amplitude and frequency is an upregulation of the number of postsynaptic AMPA receptors per synapse. At those synapses that contained AMPA receptors initially, increasing the number of AMPA receptors could increase mEPSC amplitude. At synapses that did not contain AMPA receptors initially, insertion of AMPA receptors would be observed as an increase in mEPSC frequency (Isaac et al., 1995; Liao et al., 1995; Durand et al., 1996; Wu et al., 1996).

However, increases in mEPSC frequency are most often associated with presynaptic mechanisms that enhance synaptic vesicle release probability. Interestingly, BDNF knockout mice have deficits in synaptic transmission with high-frequency repetitive stimulation due to a reduction in the number of docked vesicles in these animals (Pozzo-Miller et al., 1999). The reduction in mEPSC frequency observed with TrkB-IgG treatment in conventional cultures is consistent with the reduced number of docked vesicles in the BDNF knockout animals.

Although no changes in mEPSC frequency were observed in autaptic cultures, experiments investigating a paradigm of short-term synaptic plasticity, paired-pulse depression (PPD; Zucker, 1989), also support long-term presynaptic effects of BDNF (Sherwood and Lo, 1999). Autaptic neurons grown in BDNF were significantly less depressed in response to paired stimuli, again reminiscent of the effects observed by Pozzo-Miller et al. (1999) in BDNF knockout mice. PPD is attributed to the depletion of the pool of readily releasable vesicles, suggesting that BDNF acted presynaptically to enhance the size of this pool. Like its effects on mEPSC amplitude, BDNF-induced changes in PPD were not observed following acute BDNF treatment. In summary, these data provide evidence for both pre- and postsynaptic regulation of excitatory synaptic transmission by BDNF over long time scales.

Enhanced excitatory synaptic drive by BDNF leads to increased action potential firing rate

Using the conventional dissociated cultures, we investigated the consequences of BDNF-induced increases in glutamatergic synaptic transmission on excitatory neuronal activity within the circuit (Bolton et al., 2000). To evaluate the effect of BDNF on excitation directly, action potential firing rates in neurons with pyramidal morphology were measured using cell-attached patch clamp recording, with the contribution of inhibition to circuit activity eliminated during the recording period by blocking $GABA_A$ receptors with bicuculline. BDNF treatment increased the average spontaneous action potential firing rates of neurons approximately twofold (Fig. 3A,B). Furthermore, we found that the regulation of mEPSC amplitude by BDNF did not require concurrent action potential activity, and was therefore not a secondary consequence of increased activity levels induced by BDNF treatment. Blocking action potential activity with tetrodotoxin (TTX) for the entire duration of BDNF or TrkB-IgG treatment had no effect on the enhancement of mEPSC amplitude by BDNF (Fig. 3C).

Since the action potential firing properties of a network depend on the intrinsic excitability of its neuronal elements as well as the synaptic connectivity of the ensemble, we next investigated whether BDNF regulated intrinsic membrane excitability in these cultures. Accordingly, we measured several parameters of action potential activity that was generated by injecting a series of current pulses of increasing amplitude into current-clamped neurons; voltage responses to these current pulses were measured while all synaptic transmission was blocked pharmacologically. We found that BDNF did not change the firing rate of neurons at any of the current injection amplitudes examined. Additionally, BDNF affected neither action potential shape, as measured by action potential height and half-width, nor the voltage threshold at which the regenerative action potentials were first observed. Similarly, there were no differences between BDNF-treated and control or TrkB-IgG-treated neurons in resting membrane potential. A small increase in capacitance and a decrease in input resistance were detected, suggesting that BDNF may have had minor effects on the morphology of these neurons. However, it is unlikely that these differences contributed significantly to increasing spontaneous circuit activity since neither any aspect of intrinsic excitability measured nor the rise and decay kinetics of AMPA receptor-mediated synaptic currents were affected by BDNF. Thus, the predominant long-term effect of BDNF was to in-

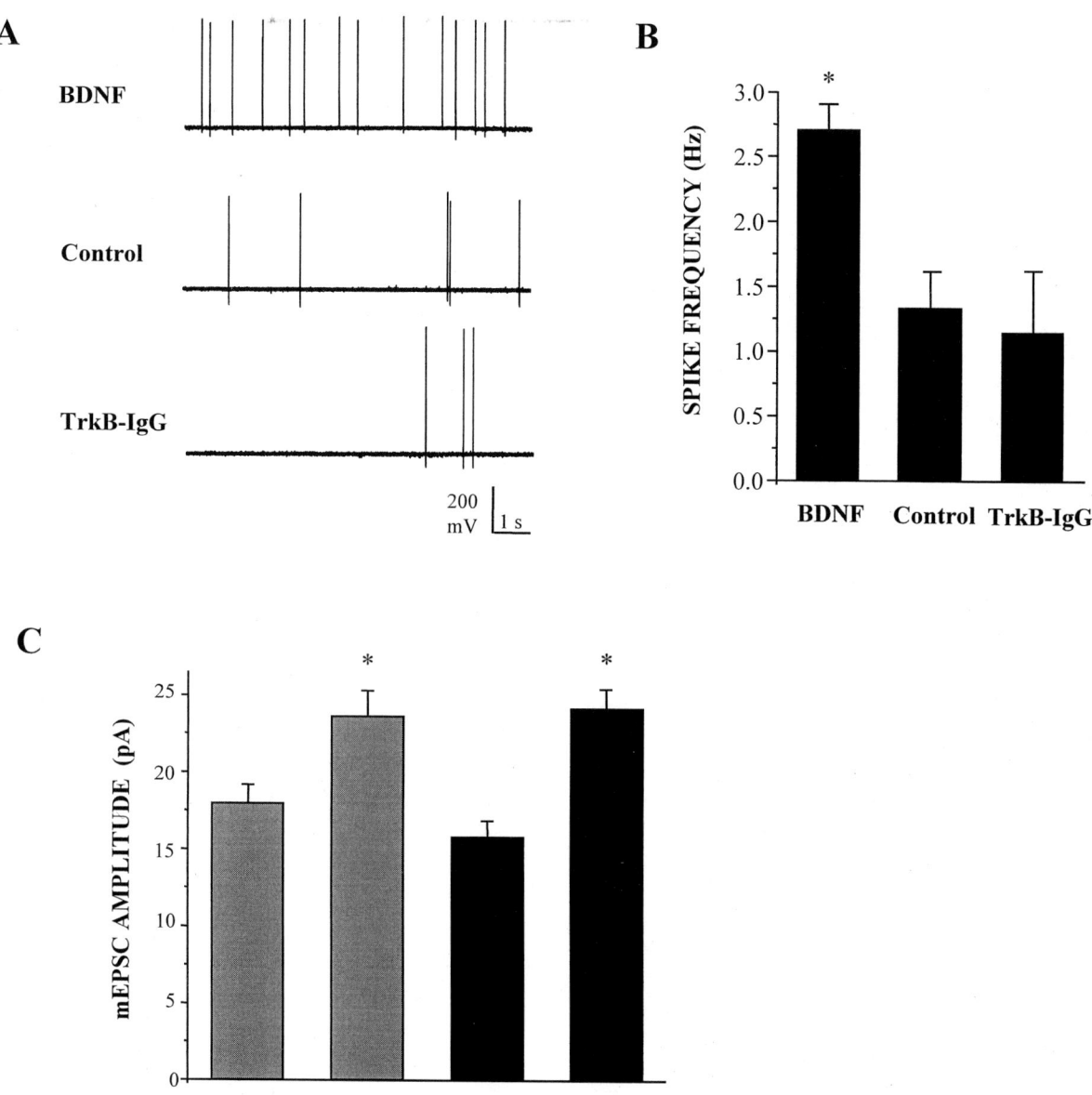

Fig. 3. Spontaneous action potential firing rates are increased in response to BDNF enhancement of mEPSC amplitudes. (A) Spontaneous action potential firing was increased in cultures treated with BDNF. Representative on-cell recordings measuring action potential frequency are shown at a compressed time base. (B) BDNF treatment increased the spontaneous firing rate of pyramidal neurons twofold compared to untreated or TrkB-IgG-treated controls. Firing rates of pyramidal neurons were measured during an acute blockade of inhibitory transmission by bicuculline to isolate the contribution of BDNF to excitatory synaptic drive. (C) BDNF regulation of mEPSC amplitude does not require activity, and is thus not a secondary effect of enhanced action potential firing. The addition of 5 μM TTX to block action potential activity for the entire duration of the BDNF treatment period did not block the enhancement of mEPSC amplitude by BDNF. (Adapted from Bolton et al., 2000.)

crease the action potential firing rates of excitatory hippocampal neurons via the enhancement of synaptic strength.

Regulation of inhibitory synaptic transmission in hippocampal cultures

The maturation of inhibitory circuitry is controlled in part by neuronal activity both in the cortex and hippocampus. There is increasing evidence that BDNF, released in an activity-dependent manner from the synaptic targets of interneurons (i.e., pyramidal neurons), promotes the development of inhibition. Interestingly, BDNF is not synthesized by γ-aminobutyric acid (GABA)ergic interneurons (Cellerino et al., 1996; Rocamora et al., 1996a; Schmidt-Kastner et al., 1996); instead, their source of BDNF is thought to be neighboring glutamatergic neurons (Nawa et al., 1995). BDNF therefore likely acts as a target-derived differentiation factor for inhibitory neurons. BDNF has been shown to regulate the morphology of interneurons, increasing the expression of the biosynthetic enzyme, glutamic acid decarboxylase (GAD), neurotransmitters GABA and neuropeptide Y, and the number of inhibitory synapses (Ip et al., 1993; Nawa et al., 1993, 1994; Croll et al., 1994; Marty et al., 1996; Marty, this volume; Seil and Drake-Baumann, this volume).

In our conventional culture experiments, we found that BDNF treatment increased the frequency of miniature inhibitory postsynaptic currents (mIPSCs) arising from GABAergic inputs onto neurons with pyramidal morphology (Bolton et al., 2000). During the recording period, TTX and 6-nitro-7-sulphamo-benzo[*f*]-quinoxaline-2,3-dione (NBQX) were added acutely to the extracellular solution to block action potentials and glutamatergic inputs, respectively. BDNF treatment increased the frequency of mIPSCs by almost twofold (Fig. 4A,C). In contrast to the increase in AMPA receptor-mediated mEPSC amplitudes described above, $GABA_A$ receptor-mediated mIPSC amplitudes were not affected by BDNF (Fig. 4B). As observed for excitatory synaptic currents, BDNF did not alter the kinetics of synaptic currents mediated by $GABA_A$ receptors.

Such frequency changes could have arisen from changes in either probability of transmitter release, numbers of inhibitory synaptic contacts, or both. We determined that the number of GABAergic terminals was not affected by BDNF but that the size of inhibitory terminals and intensity of GAD immunostaining was increased, suggesting that BDNF was likely to have enhanced the probability of transmitter release presynaptically. These findings indicated that while BDNF potentiated both excitatory and inhibitory synaptic transmission in these cultures, it did so through distinct physiological mechanisms. Furthermore, similar to previous reports of increases in GABAergic phenotypic differentiation but not inhibitory neuronal numbers after treatment with BDNF in vitro and in vivo (Ip et al., 1993; Nawa et al., 1993, 1994; Croll et al., 1994; Marty et al., 1996), we observed a 40% increase in the ratio of neurons that showed detectable anti-GAD staining after chronic BDNF treatment.

Consistent with these observations in the conventional cultures, measurement of inhibitory neurons in autaptic cultures revealed that long-term BDNF treatment induced a twofold increase in action-potential-driven IPSC amplitudes (Fig. 4D). Together, these observations continue to support a general role for BDNF in regulating inhibitory synaptic transmission and are consistent with BDNF acting as an activity-dependent, target-derived differentiation factor for GABAergic interneurons over long time scales.

Conclusion

A large body of work in the past decade has provided compelling, although sometimes contradictory, evidence for the role of BDNF in regulating neuronal plasticity in many different contexts. We have independently observed two predominant long-term effects of BDNF in both conventional and autaptic hippocampal cultures. Chronic BDNF treatment increases excitatory synaptic drive via an enhancement of unitary synaptic transmission at glutamatergic synapses, and also potentiates inhibitory synaptic transmission by increasing the probability of inhibitory neurotransmitter release.

In vivo, BDNF release is likely to be spatially precise and synapse-specific, rather than global as it is in these culture experiments. Depending on the spatial characteristics of BDNF release, and given the precision with which inhibitory and excitatory synapses

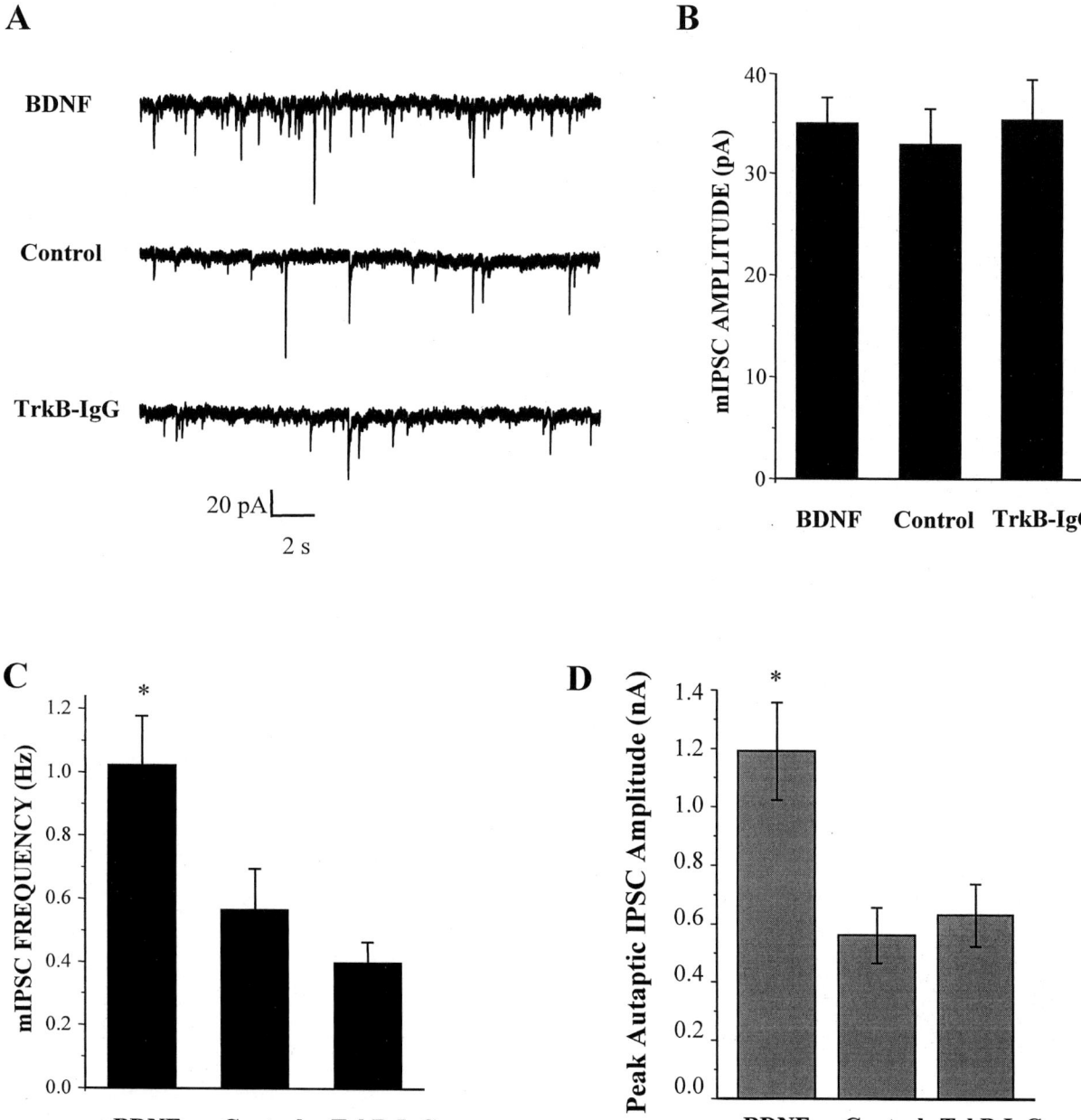

Fig. 4. BDNF increases GABA$_A$ receptor-mediated mIPSC frequency and the amplitude of evoked IPSCs. (A) Representative recordings from conventional hippocampal cultures of GABA$_A$ receptor-mediated mIPSCs on a compressed time base show the elevation of mIPSC frequency induced by BDNF treatment; traces from neurons in BDNF (top), control (middle), and TrkB-IgG (bottom) groups are shown. (B) In contrast to its effects on excitatory synapses, BDNF treatment had no effect on GABA$_A$ receptor-mediated mIPSC amplitudes. (C) BDNF treatment increased mIPSC frequency by ~1.8-fold compared to controls. (D) BDNF also increased evoked IPSC amplitudes recorded from autaptic inhibitory neurons by twofold ($n = 20$, 14, and 12 for BDNF, control, and TrkB-IgG-treated neurons, respectively), consistent with its effects on spontaneous inhibitory events in conventional cultures. ([A]–[C] from Bolton et al., 2000; [D], Sherwood and Lo, unpublished observations.)

are differentially formed onto distinct regions of target neurons (e.g., Craig et al., 1994), the effects of BDNF on excitatory and inhibitory synapses suggest precise mechanisms by which this molecule can simultaneously regulate long-term, synapse-specific changes in synaptic transmission while maintaining homeostasis of overall circuitry. Observations of BDNF's long-term effects in dissociated cultures of visual cortex neurons, however, have led to the alternative proposal that BDNF produced in response to neuronal activity globally and uniformly scales the strength of all excitatory inputs to a given neuron (Turrigiano, 1999). This enables the firing rate of a neuron to remain within the range where it remains sensitive to its individual synaptic inputs. This quantal scaling model implicitly requires that BDNF release levels are uniform over the entire dendritic arborization of the neuron, or that local elevations in BDNF concentration are translated into a nuclear response that alters all synapses equivalently. While no experiments have been published that address the global versus synapse-specific nature of BDNF's regulation of mEPSC amplitude, the answer to this question will have very different implications for the physiological role of BDNF. The differences observed between this system and in hippocampal neurons underscore the importance of examining the roles of such widely employed signaling molecules in multiple contexts, as the specifics of their downstream effects are likely to vary.

Acknowledgements

We thank Regeneron Pharmaceuticals for their generous provision of recombinant BDNF and TrkB-IgG. This work was supported by NIH National Research Service Awards (MH11519 to M.B. and 5F31MH11058 to N.S.) and by the Bryan Scholars Fund (N.S.).

References

Akaneya, Y., Tsumoto, T. and Hatanaka, H. (1996) Brain-derived neurotrophic factor blocks long-term depression in rat visual cortex. *J. Neurophysiol.*, 76: 4198–4201.

Akaneya, Y., Tsumoto, T., Kinoshita, S. and Hatanaka, H. (1997) Brain-derived neurotrophic factor enhances long-term potentiation in rat visual cortex. *J. Neurosci.*, 17: 6707–6716.

Barbacid, M. (1994) The Trk family of neurotrophin receptors. *J. Neurobiol.*, 25: 1386–1403.

Barria, A., Derkach, V. and Soderling, T. (1997a) Identification of the Ca^{2+}/calmodulin-dependent protein kinase II regulatory phosphorylation site in the alpha-amino-3-hydroxyl-5-methyl-4-isoxazole-propionate-type glutamate receptor. *J. Biol. Chem.*, 272: 32727–32730.

Barria, A., Muller, D., Derkach, V., Griffith, L.C. and Soderling, T.R. (1997b) Regulatory phosphorylation of AMPA-type glutamate receptors by CaM-KII during long-term potentiation. *Science*, 276: 2042–2045.

Bekkers, J.M. and Stevens, C.F. (1990) Presynaptic mechanism for long-term potentiation in the hippocampus. *Nature*, 346: 724–729.

Berardi, N., Domenici, L., Parisi, V., Pizzorusso, T., Cellerino, A. and Maffei, L. (1993) Monocular deprivation effects in the rat visual cortex and lateral geniculate nucleus are prevented by nerve growth factor (NGF), I. Visual cortex. *Proc. R. Soc. Lond. B. Biol. Sci.*, 251: 17–23.

Berardi, N., Cellerino, A., Domenici, L., Fagiolini, M., Pizzorusso, T., Cattaneo, A. and Maffei, L. (1994) Monoclonal antibodies to nerve growth factor affect the postnatal development of the visual system. *Proc. Natl. Acad. Sci. USA*, 91: 684–688.

Binder, D.K., Routbort, M.J., Ryan, T.E., Yancopoulos, G.D. and McNamara, J.O. (1999) Selective inhibition of kindling development by intraventricular administration of TrkB receptor body. *J. Neurosci.*, 19: 1424–1436.

Blöchl, A. and Thoenen, H. (1995) Characterization of nerve growth factor (NGF) release from hippocampal neurons: evidence for a constitutive and an unconventional sodium-dependent regulated pathway. *Eur. J. Neurosci.*, 7: 1220–1228.

Blöchl, A. and Thoenen, H. (1996) Localization of cellular storage compartments and sites of constitutive and activity-dependent release of nerve growth factor (NGF) in primary cultures of hippocampal neurons. *Mol. Cell. Neurosci.*, 7: 173–190.

Bolton, M.M., Pittman, A.J. and Lo, D.C. (2000) BDNF differentially regulates excitatory and inhibitory synaptic transmission in hippocampal cultures. *J. Neurosci.*, 20: 3221–3232.

Boulanger, L. and Poo, M. (1999a) Gating of BDNF-induced synaptic potentiation by cAMP. *Science*, 284: 1982–1984.

Boulanger, L. and Poo, M.-m. (1999b) Presynaptic depolarization facilitates neurotrophin-induced synaptic potentiation. *Nat. Neurosci.*, 2: 346–351.

Bozzi, Y., Pizzorusso, T., Cremisi, F., Rossi, F.M., Barsacchi, G. and Maffei, L. (1995) Monocular deprivation decreases the expression of messenger RNA for brain-derived neurotrophic factor in the rat visual cortex. *Neuroscience*, 69: 1133–1144.

Cabelli, R.J., Hohn, A. and Shatz, C.J. (1995) Inhibition of ocular dominance column formation by infusion of NT-4/5 or BDNF. *Science*, 267: 1662–1666.

Cabelli, R.J., Shelton, D.L., Segal, R.A. and Shatz, C.J. (1997) Blockade of endogenous ligands of trkB inhibits formation of ocular dominance columns. *Neuron*, 19: 63–76.

Canossa, M., Griesbeck, O., Berninger, B., Campana, G., Kolbeck, R. and Thoenen, H. (1997) Neurotrophin release by neurotrophins: implications for activity-dependent neuronal plasticity. *Proc. Natl. Acad. Sci. USA*, 94: 13279–13286.

Carmignoto, G., Canella, R., Candeo, P., Comelli, M.C. and Maffei, L. (1993) Effects of nerve growth factor on neuronal plasticity of the kitten visual cortex. *J. Physiol. (Lond.)*, 464: 343–360.

Carmignoto, G., Pizzorusso, T., Tia, S. and Vicini, S. (1997) Brain-derived neurotrophic factor and nerve growth factor potentiate excitatory synaptic transmission in the rat visual cortex. *J. Physiol. (Lond.)*, 498: 153–164.

Carroll, R.C., Lissin, D.V., von Zastrow, M., Nicoll, R.A. and Malenka, R.C. (1999) Rapid redistribution of glutamate receptors contributes to long-term depression in hippocampal cultures. *Nat. Neurosci.*, 2: 454–460.

Castrén, E., Zafra, F., Thoenen, H. and Lindholm, D. (1992) Light regulates expression of brain-derived neurotrophic factor mRNA in rat visual cortex. *Proc. Natl. Acad. Sci. USA*, 89: 9444–9448.

Castrén, E., Pitkanen, M., Sirvio, J., Parsadanian, A., Lindholm, D., Thoenen, H. and Riekkinen, P.J. (1993) The induction of LTP increases BDNF and NGF mRNA but decreases NT-3 mRNA in the dentate gyrus. *NeuroReport*, 4: 895–898.

Castrén, E., Berninger, B., Leingartner, A. and Lindholm, D. (1998) Regulation of brain-derived neurotrophic factor mRNA levels in hippocampus by neuronal activity. *Prog. Brain Res.*, 117: 57–64.

Cellerino, A., Maffei, L. and Domenici, L. (1996) The distribution of brain-derived neurotrophic factor and its receptor trkB in parvalbumin-containing neurons of the rat visual cortex. *Eur. J. Neurosci.*, 8: 1190–1197.

Chao, M.V. (1992) Neurotrophin receptors: a window into neuronal differentiation. *Neuron*, 9: 583–593.

Clements, J.D., Lester, R.A., Tong, G., Jahr, C.E. and Westbrook, G.L. (1992) The time course of glutamate in the synaptic cleft. *Science*, 258: 1498–1501.

Craig, A.M., Blackstone, C.D., Huganir, R.L. and Banker, G. (1994) Selective clustering of glutamate and gamma-aminobutyric acid receptors opposite terminals releasing the corresponding neurotransmitters. *Proc. Natl. Acad. Sci. USA*, 91: 12373–12377.

Croll, S.D., Wiegand, S.J., Anderson, K.D., Lindsay, R.M. and Nawa, H. (1994) Regulation of neuropeptides in adult rat forebrain by the neurotrophins BDNF and NGF. *Eur. J. Neurosci.*, 6: 1343–1353.

Davies, A.M. (1994) The role of neurotrophins in the developing nervous system. *J. Neurobiol.*, 25: 1334–1348.

Derkach, V., Barria, A. and Soderling, T.R. (1999) Ca^{2+}/calmodulin-kinase II enhances channel conductance of alpha-amino-3-hydroxy-5-methyl-4-isoxazolepropionate type glutamate receptors. *Proc. Natl. Acad. Sci. USA*, 96: 3269–3274.

Domenici, L., Berardi, N., Carmignoto, G., Vantini, G. and Maffei, L. (1991) Nerve growth factor prevents the amblyopic effects of monocular deprivation. *Proc. Natl. Acad. Sci. USA*, 88: 8811–8815.

Domenici, L., Cellerino, A., Berardi, N., Cattaneo, A. and Maffei, L. (1994) Antibodies to nerve growth factor (NGF) prolong the sensitive period for monocular deprivation in the rat. *NeuroReport*, 5: 2041–2044.

Dragunow, M., Beilharz, E., Mason, B., Lawlor, P., Abraham, W. and Gluckman, P. (1993) Brain-derived neurotrophic factor expression after long-term potentiation. *Neurosci. Lett.*, 160: 232–236.

Durand, G.M., Kovalchuk, Y. and Konnerth, A. (1996) Long-term potentiation and functional synapse induction in developing hippocampus. *Nature*, 381: 71–75.

Erickson, J.T., Conover, J.C., Borday, V., Champagnat, J., Barbacid, M., Yancopoulos, G. and Katz, D.M. (1996) Mice lacking brain-derived neurotrophic factor exhibit visceral sensory neuron losses distinct from mice lacking NT4 and display a severe developmental deficit in control of breathing. *J. Neurosci.*, 16: 5361–5371.

Ernfors, P., Bengzon, J., Kokaia, Z., Persson, H. and Lindvall, O. (1991) Increased levels of messenger RNAs for neurotrophic factors in the brain during kindling epileptogenesis. *Neuron*, 7: 165–176.

Fawcett, J.P., Aloyz, R., McLean, J.H., Pareek, S., Miller, F.D., McPherson, P.S. and Murphy, R.A. (1997) Detection of brain-derived neurotrophic factor in a vesicular fraction of brain synaptosomes. *J. Biol. Chem.*, 272: 8837–8840.

Figurov, A., Pozzo-Miller, L.D., Olafsson, P., Wang, T. and Lu, B. (1996) Regulation of synaptic responses to high-frequency stimulation and LTP by neurotrophins in the hippocampus. *Nature*, 381: 706–709.

Forti, L., Bossi, M., Bergamaschi, A., Villa, A. and Malgaroli, A. (1997) Loose-patch recordings of single quanta at individual hippocampal synapses. *Nature*, 388: 874–878.

Frerking, M., Malenka, R.C. and Nicoll, R.A. (1998) Brain-derived neurotrophic factor (BDNF) modulates inhibitory, but not excitatory, transmission in the CA1 region of the hippocampus. *J. Neurophysiol.*, 80: 3383–3386.

Friedman, W.J. and Greene, L.A. (1999) Neurotrophin signaling via Trks and p75. *Exp. Cell. Res.*, 253: 131–142.

Goodman, L.J., Valverde, J., Lim, F., Geschwind, M.D., Federoff, H.J., Geller, A.I. and Hefti, F. (1996) Regulated release and polarized localization of brain-derived neurotrophic factor in hippocampal neurons. *Mol. Cell. Neurosci.*, 7: 222–238.

Gottschalk, W., Pozzo-Miller, L.D., Figurov, A. and Lu, B. (1998) Presynaptic modulation of synaptic transmission and plasticity by brain-derived neurotrophic factor in the developing hippocampus. *J. Neurosci.*, 18: 6830–6839.

Hollmann, M. and Heinemann, S. (1994) Cloned glutamate receptors. *Annu. Rev. Neurosci.*, 17: 31–108.

Huang, Z.J., Kirkwood, A., Pizzorusso, T., Porciatti, V., Morales, B., Bear, M.F., Maffei, L. and Tonegawa, S. (1999) BDNF regulates the maturation of inhibition and the critical period of plasticity in mouse visual cortex. *Cell*, 98: 739–755.

Huber, K.M., Sawtell, N.B. and Bear, M.F. (1998) Brain-derived neurotrophic factor alters the synaptic modification threshold in visual cortex. *Neuropharmacology*, 37: 571–579.

Ip, N.Y., Li, Y., Yancopoulos, G.D. and Lindsay, R.M. (1993) Cultured hippocampal neurons show responses to BDNF, NT-3, and NT-4, but not NGF. *J. Neurosci.*, 13: 3394–3405.

Isaac, J.T., Nicoll, R.A. and Malenka, R.C. (1995) Evidence for silent synapses: implications for the expression of LTP. *Neuron*, 15: 427–434.

Isackson, P.J., Huntsman, M.M., Murray, K.D. and Gall, C.M. (1991) BDNF mRNA expression is increased in adult rat forebrain after limbic seizures: temporal patterns of induction distinct from NGF. *Neuron*, 6: 937–948.

Jones, K.R., Farinas, I., Backus, C. and Reichardt, L.F. (1994) Targeted disruption of the BDNF gene perturbs brain and sensory neuron development but not motor neuron development. *Cell*, 76: 989–999.

Kang, H. and Schuman, E.M. (1995a) Long-lasting neurotrophin-induced enhancement of synaptic transmission in the adult hippocampus. *Science*, 267: 1658–1662.

Kang, H.J. and Schuman, E.M. (1995b) Neurotrophin-induced modulation of synaptic transmission in the adult hippocampus. *J. Physiol. (Paris)*, 89: 11–22.

Kinoshita, S., Yasuda, H., Taniguchi, N., Katoh-Semba, R., Hatanaka, H. and Tsumoto, T. (1999) Brain-derived neurotrophic factor prevents low-frequency inputs from inducing long-term depression in the developing visual cortex. *J. Neurosci.*, 19: 2122–2130.

Klein, R., Smeyne, R.J., Wurst, W., Long, L.K., Auerbach, B.A., Joyner, A.L. and Barbacid, M. (1993) Targeted disruption of the trkB neurotrophin receptor gene results in nervous system lesions and neonatal death. *Cell*, 75: 113–122.

Kokaia, M., Ernfors, P., Kokaia, Z., Elmer, E., Jaenisch, R. and Lindvall, O. (1995) Suppressed epileptogenesis in BDNF mutant mice. *Exp. Neurol.*, 133: 215–224.

Korte, M., Carroll, P., Wolf, E., Brem, G., Thoenen, H. and Bonhoeffer, T. (1995) Hippocampal long-term potentiation is impaired in mice lacking brain-derived neurotrophic factor. *Proc. Natl. Acad. Sci. USA*, 92: 8856–8860.

Korte, M., Griesbeck, O., Gravel, C., Carroll, P., Staiger, V., Thoenen, H. and Bonhoeffer, T. (1996a) Virus-mediated gene transfer into hippocampal CA1 region restores long-term potentiation in brain-derived neurotrophic factor mutant mice. *Proc. Natl. Acad. Sci. USA*, 93: 12547–12552.

Korte, M., Staiger, V., Griesbeck, O., Thoenen, H. and Bonhoeffer, T. (1996b) The involvement of brain-derived neurotrophic factor in hippocampal long-term potentiation revealed by gene targeting experiments. *J. Physiol. (Paris)*, 90: 157–164.

Korte, M., Kang, H., Bonhoeffer, T. and Schuman, E. (1998) A role for BDNF in the late-phase of hippocampal long-term potentiation. *Neuropharmacology*, 37: 553–559.

Larmet, Y., Reibel, S., Carnahan, J., Nawa, H., Marescaux, C. and Depaulis, A. (1995) Protective effects of brain-derived neurotrophic factor on the development of hippocampal kindling in the rat. *NeuroReport*, 6: 1937–1941.

Lessmann, V. and Heumann, R. (1998) Modulation of unitary glutamatergic synapses by neurotrophin-4/5 or brain-derived neurotrophic factor in hippocampal microcultures: presynaptic enhancement depends on pre-established paired-pulse facilitation. *Neuroscience*, 86: 399–413.

Lessmann, V., Gottmann, K. and Heumann, R. (1994) BDNF and NT-4/5 enhance glutamatergic synaptic transmission in cultured hippocampal neurones. *NeuroReport*, 6: 21–25.

Levine, E.S., Dreyfus, C.F., Black, I.B. and Plummer, M.R. (1995) Brain-derived neurotrophic factor rapidly enhances synaptic transmission in hippocampal neurons via postsynaptic tyrosine kinase receptors. *Proc. Natl. Acad. Sci. USA*, 92: 8074–8077.

Levine, E.S., Dreyfus, C.F., Black, I.B. and Plummer, M.R. (1996) Selective role for trkB neurotrophin receptors in rapid modulation of hippocampal synaptic transmission. *Mol. Brain Res.*, 38: 300–303.

Levine, E.S., Crozier, R.A., Black, I.B. and Plummer, M.R. (1998) Brain-derived neurotrophic factor modulates hippocampal synaptic transmission by increasing N-methyl-D-aspartic acid receptor activity. *Proc. Natl. Acad. Sci. USA*, 95: 10235–10239.

Li, Y.X., Xu, Y., Ju, D., Lester, H.A., Davidson, N. and Schuman, E.M. (1998) Expression of a dominant negative TrkB receptor, T1, reveals a requirement for presynaptic signaling in BDNF-induced synaptic potentiation in cultured hippocampal neurons. *Proc. Natl. Acad. Sci. USA*, 95: 10884–10889.

Liao, D., Hessler, N.A. and Malinow, R. (1995) Activation of postsynaptically silent synapses during pairing-induced LTP in CA1 region of hippocampal slice. *Nature*, 375: 400–404.

Linnarsson, S., Björklund, A. and Ernfors, P. (1997) Learning deficit in BDNF mutant mice. *Eur. J. Neurosci.*, 9: 2581–2587.

Lissin, D.V., Carroll, R.C., Nicoll, R.A., Malenka, R.C. and von Zastrow, M. (1999) Rapid, activation-induced redistribution of ionotropic glutamate receptors in cultured hippocampal neurons. *J. Neurosci.*, 19: 1263–1272.

Liu, G. and Tsien, R.W. (1995) Properties of synaptic transmission at single hippocampal synaptic boutons. *Nature*, 375: 404–408.

Lohof, A.M., Ip, N.Y. and Poo, M.-m. (1993) Potentiation of developing neuromuscular synapses by the neurotrophins NT-3 and BDNF. *Nature*, 363: 350–353.

Ma, Y.L., Wang, H.L., Wu, H.C., Wei, C.L. and Lee, E.H. (1998) Brain-derived neurotrophic factor antisense oligonucleotide impairs memory retention and inhibits long-term potentiation in rats. *Neuroscience*, 82: 957–967.

Maffei, L., Berardi, N., Domenici, L., Parisi, V. and Pizzorusso, T. (1992) Nerve growth factor (NGF) prevents the shift in ocular dominance distribution of visual cortical neurons in monocularly deprived rats. *J. Neurosci.*, 12: 4651–4662.

Mammen, A.L., Kameyama, K., Roche, K.W. and Huganir, R.L. (1997) Phosphorylation of the alpha-amino-3-hydroxy-5-methylisoxazole-4-propionic acid receptor GluR1 subunit by calcium/calmodulin-dependent kinase II. *J. Biol. Chem.*, 272: 32528–32533.

Martinez, A., Alcantara, S., Borrell, V., Del Rio, J.A., Blasi, J., Otal, R., Campos, N., Boronat, A., Barbacid, M., Silos-Santiago, I. and Soriano, E. (1998) TrkB and TrkC signaling are required for maturation and synaptogenesis of hippocampal connections. *J. Neurosci.*, 18: 7336–7350.

Marty, S., Carroll, P., Cellerino, A., Castrén, E., Staiger, V., Thoenen, H. and Lindholm, D. (1996) Brain-derived neurotrophic factor promotes the differentiation of various hippocampal nonpyramidal neurons, including Cajal-Retzius cells, in organotypic slice cultures. *J. Neurosci.*, 16: 675–687.

McAllister, A.K., Katz, L.C. and Lo, D.C. (1996) Neurotrophin regulation of cortical dendritic growth requires activity. *Neuron*, 17: 1057–1064.

McAllister, A.K., Katz, L.C. and Lo, D.C. (1997) Opposing roles for endogenous BDNF and NT-3 in regulating cortical dendritic growth. *Neuron*, 18: 767–778.

McAllister, A.K., Katz, L.C. and Lo, D.C. (1999) Neurotrophins and synaptic plasticity. *Annu. Rev. Neurosci.*, 22: 295–318.

Minichiello, L., Korte, M., Wolfer, D., Kuhn, R., Unsicker, K., Cestari, V., Rossi-Arnaud, C., Lipp, H.P., Bonhoeffer, T. and Klein, R. (1999) Essential role for TrkB receptors in hippocampus-mediated learning. *Neuron*, 24: 401–414.

Moller, J.C., Kruttgen, A., Heymach Jr., J.V., Ghori, N. and Shooter, E.M. (1998) Subcellular localization of epitope-tagged neurotrophins in neuroendocrine cells. *J. Neurosci. Res.*, 51: 463–472.

Montkowski, A. and Holsboer, F. (1997) Intact spatial learning and memory in transgenic mice with reduced BDNF. *NeuroReport*, 8: 779–782.

Narisawa-Saito, M., Carnahan, J., Araki, K., Yamaguchi, T. and Nawa, H. (1999a) Brain-derived neurotrophic factor regulates the expression of AMPA receptor proteins in neocortical neurons. *Neuroscience*, 88: 1009–1014.

Narisawa-Saito, M., Silva, A.J., Yamaguchi, T., Hayashi, T., Yamamoto, T. and Nawa, H. (1999b) Growth factor-mediated Fyn signaling regulates alpha-amino-3-hydroxy-5-methyl-4-isoxazolepropionic acid (AMPA) receptor expression in rodent neocortical neurons. *Proc. Natl. Acad. Sci. USA*, 96: 2461–2466.

Nawa, H., Bessho, Y., Carnahan, J., Nakanishi, S. and Mizuno, K. (1993) Regulation of neuropeptide expression in cultured cerebral cortical neurons by brain-derived neurotrophic factor. *J. Neurochem.*, 60: 772–775.

Nawa, H., Pelleymounter, M.A. and Carnahan, J. (1994) Intraventricular administration of BDNF increases neuropeptide expression in newborn rat brain. *J. Neurosci.*, 14: 3751–3765.

Nawa, H., Carnahan, J. and Gall, C. (1995) BDNF protein measured by a novel enzyme immunoassay in normal brain and after seizure: partial disagreement with mRNA levels. *Eur. J. Neurosci.*, 7: 1527–1535.

O'Brien, R.J., Kamboj, S., Ehlers, M.D., Rosen, K.R., Fischbach, G.D. and Huganir, R.L. (1998) Activity-dependent modulation of synaptic AMPA receptor accumulation. *Neuron*, 21: 1067–1078.

Osehobo, P., Adams, B., Sazgar, M., Xu, Y., Racine, R.J. and Fahnestock, M. (1999) Brain-derived neurotrophic factor infusion delays amygdala and perforant path kindling without affecting paired-pulse measures of neuronal inhibition in adult rats. *Neuroscience*, 92: 1367–1375.

Patterson, S.L., Grover, L.M., Schwartzkroin, P.A. and Bothwell, M. (1992) Neurotrophin expression in rat hippocampal slices: a stimulus paradigm inducing LTP in CA1 evokes increases in BDNF and NT-3 mRNAs. *Neuron*, 9: 1081–1088.

Patterson, S.L., Abel, T., Deuel, T.A., Martin, K.C., Rose, J.C. and Kandel, E.R. (1996) Recombinant BDNF rescues deficits in basal synaptic transmission and hippocampal LTP in BDNF knockout mice. *Neuron*, 16: 1137–1145.

Pozzo-Miller, L.D., Gottschalk, W., Zhang, L., McDermott, K., Du, J., Gopalakrishnan, R., Oho, C., Sheng, Z.H. and Lu, B. (1999) Impairments in high-frequency transmission, synaptic vesicle docking, and synaptic protein distribution in the hippocampus of BDNF knockout mice. *J. Neurosci.*, 19: 4972–4983.

Reimer, R.J., Fon, E.A. and Edwards, R.H. (1998) Vesicular neurotransmitter transport and the presynaptic regulation of quantal size. *Curr. Opin. Neurobiol.*, 8: 405–412.

Riddle, D.R., Lo, D.C. and Katz, L.C. (1995) NT-4-mediated rescue of lateral geniculate neurons from effects of monocular deprivation. *Nature*, 378: 189–191.

Rocamora, N., Pascual, M., Acsady, L., De Lecea, L., Freund, T.F. and Soriano, E. (1996a) Expression of NGF and NT3 mRNAs in hippocampal interneurons innervated by the GABAergic septohippocampal pathway. *J. Neurosci.*, 16: 3991–4004.

Rocamora, N., Welker, E., Pascual, M. and Soriano, E. (1996b) Upregulation of BDNF mRNA expression in the barrel cortex of adult mice after sensory stimulation. *J. Neurosci.*, 16: 4411–4419.

Rutherford, L.C., Nelson, S.B. and Turrigiano, G.G. (1998) BDNF has opposite effects on the quantal amplitude of pyramidal neuron and interneuron excitatory synapses. *Neuron*, 21: 521–530.

Scharfman, H.E. (1997) Hyperexcitability in combined entorhinal/hippocampal slices of adult rat after exposure to brain-derived neurotrophic factor. *J. Neurophysiol.*, 78: 1082–1095.

Scharfman, H.E., Goodman, J.H. and Sollas, A.L. (1999) Actions of brain-derived neurotrophic factor in slices from rats with spontaneous seizures and mossy fiber sprouting in the dentate gyrus. *J. Neurosci.*, 19: 5619–5631.

Schmidt-Kastner, R., Wetmore, C. and Olson, L. (1996) Comparative study of brain-derived neurotrophic factor messenger RNA and protein at the cellular level suggests multiple roles in hippocampus, striatum and cortex. *Neuroscience*, 74: 161–183.

Segal, R.A. and Greenberg, M.E. (1996) Intracellular signaling pathways activated by neurotrophic factors. *Annu. Rev. Neurosci.*, 19: 463–489.

Sermasi, E., Tropea, D. and Domenici, L. (1999) A new form of synaptic plasticity is transiently expressed in the developing rat visual cortex: a modulatory role for visual experience and brain-derived neurotrophic factor. *Neuroscience*, 91: 163–173.

Shelton, D.L., Sutherland, J., Gripp, J., Camerato, T., Armanini, M.P., Phillips, H.S., Carroll, K., Spencer, S.D. and Levinson, A.D. (1995) Human trks: molecular cloning, tissue distribution, and expression of extracellular domain immunoadhesins. *J. Neurosci.*, 15: 477–491.

Sheng, M. and Greenberg, M.E. (1990) The regulation and function of c-*fos* and other immediate early genes in the nervous system. *Neuron*, 4: 477–485.

Sherwood, N.T. and Lo, D.C. (1999) Long-term enhancement of central synaptic transmission by chronic brain-derived neurotrophic factor treatment. *J. Neurosci.*, 19: 7025–7036.

Shi, S.H., Hayashi, Y., Petralia, R.S., Zaman, S.H., Wenthold, R.J., Svoboda, K. and Malinow, R. (1999) Rapid spine delivery and redistribution of AMPA receptors after synaptic NMDA receptor activation. *Science*, 284: 1811–1816.

Smith, M.A., Zhang, L.X., Lyons, W.E. and Mamounas, L.A. (1997) Anterograde transport of endogenous brain-derived neurotrophic factor in hippocampal mossy fibers. *NeuroReport*, 8: 1829–1834.

Song, D.K., Choe, B., Bae, J.H., Park, W.K., Han, I.S., Ho, W.K. and Earm, Y.E. (1998) Brain-derived neurotrophic factor rapidly potentiates synaptic transmission through NMDA, but suppresses it through non-NMDA receptors in rat hippocampal neuron. *Brain Res.*, 799: 176–179.

Stoop, R. and Poo, M.-m. (1996) Synaptic modulation by neurotrophic factors: differential and synergistic effects of brain-derived neurotrophic factor and ciliary neurotrophic factor. *J. Neurosci.*, 16: 3256–3264.

Swanson, G.T., Kamboj, S.K. and Cull-Candy, S.G. (1997) Single-channel properties of recombinant AMPA receptors depend on RNA editing, splice variation, and subunit composition. *J. Neurosci.*, 17: 58–69.

Tan, S.E., Wenthold, R.J. and Soderling, T.R. (1994) Phosphorylation of AMPA-type glutamate receptors by calcium/calmodulin-dependent protein kinase II and protein kinase C in cultured hippocampal neurons. *J. Neurosci.*, 14: 1123–1129.

Tong, G. and Jahr, C.E. (1994) Multivesicular release from excitatory synapses of cultured hippocampal neurons. *Neuron*, 12: 51–59.

Turrigiano, G.G. (1999) Homeostatic plasticity in neuronal networks: the more things change, the more they stay the same. *Trends Neurosci.*, 22: 221–227.

Turrigiano, G.G. and Nelson, S.B. (1998) Thinking globally, acting locally: AMPA receptor turnover and synaptic strength. *Neuron*, 21: 933–935.

Turrigiano, G.G., Leslie, K.R., Desai, N.S., Rutherford, L.C. and Nelson, S.B. (1998) Activity-dependent scaling of quantal amplitude in neocortical neurons. *Nature*, 391: 892–896.

Vicario-Abejon, C., Collin, C., McKay, R.D. and Segal, M. (1998) Neurotrophins induce formation of functional excitatory and inhibitory synapses between cultured hippocampal neurons. *J. Neurosci.*, 18: 7256–7271.

Wang, T., Xie, K. and Lu, B. (1995) Neurotrophins promote maturation of developing neuromuscular synapses. *J. Neurosci.*, 15: 4796–4805.

Watson, F.L., Heerssen, H.M., Moheban, D.B., Lin, M.Z., Sauvageot, C.M., Bhattacharyya, A., Pomeroy, S.L. and Segal, R.A. (1999) Rapid nuclear responses to target-derived neurotrophins require retrograde transport of ligand–receptor complex. *J. Neurosci.*, 19: 7889–7900.

Wu, G., Malinow, R. and Cline, H.T. (1996) Maturation of a central glutamatergic synapse. *Science*, 274: 972–976.

Zafra, F., Hengerer, B., Leibrock, J., Thoenen, H. and Lindholm, D. (1990) Activity dependent regulation of BDNF and NGF mRNAs in the rat hippocampus is mediated by non-NMDA glutamate receptors. *EMBO J.*, 9: 3545–3550.

Zucker, R.S. (1989) Short-term synaptic plasticity. *Annu. Rev. Neurosci.*, 12: 13–31.

CHAPTER 18

Neurotrophins and activity-dependent inhibitory synaptogenesis

Fredrick J. Seil * and Rosemarie Drake-Baumann

Office of Regeneration Research Programs and Neurology Research, VA Medical Center and Departments of Neurology and Cell and Developmental Biology, Oregon Health Sciences University, Portland, OR 97201, USA

Introduction

Neuronal activity plays an important role in development of the central nervous system (CNS), including the visual (Wiesel and Hubel, 1963; Harris, 1981; Shaw and Cynander, 1984; Reiter et al., 1986; Shatz and Stryker, 1988), auditory (Parks, 1979), olfactory (Meisami and Monsavi, 1981), and somatosensory (Woolsey and Wann, 1976) systems. There is evidence from both in vivo and in vitro studies to suggest that neuronal activity is essential for the full development and maintenance of inhibitory circuitry. Monocular deprivation of adult monkeys markedly reduced immunoreactivity of the inhibitory neurotransmitter, γ-aminobutyric acid (GABA), in visual cortex ocular dominance columns that had received input from the enucleated eye (Hendry and Jones, 1986). This effect was reproduced by intraocular injection of the sodium channel blocker, tetrodotoxin (TTX), to prevent retinal ganglion cell discharge (Hendry and Jones, 1988). A decrease in the number of GABA-reactive neurons in the visual cortex of dark-reared rats correlated with an increase in spontaneous cortical electrical activity (Benevento et al., 1995). Sensory deprivation by removal of whiskers in rats soon after birth led to a marked reduction of GABA-immunopositive synapses in layer IV of the barrel field cortex (Micheva and Beaulieu, 1995), correlating with an earlier report (Simons and Land, 1987) of increased spontaneous activity in the barrel field cortex of neonatally sensory-deprived animals.

A reduction of synapses was reported in dissociated cell and explant cultures of cerebral neocortex chronically exposed to activity-blocking agents (Janka and Jones, 1982; Van Huizen et al., 1985; Ruijter et al., 1991) and hyperactive discharges were evident after removal of the blocking agents in neocortical and hippocampal cultures (Ramakers et al., 1990; Baker and Ruijter, 1991; Furshpan, 1991). By contrast, continuous exposure of cerebral neocortex cultures to the anti-GABA agent, picrotoxin (PTX), resulted in accelerated synaptogenesis (Van Huizen et al., 1987) and a high incidence of very brief bursts of action potentials (Ramakers et al., 1991), suggesting that GABAergic inhibition was stronger in PTX exposed than in control cultures. These observations led Corner and Ramakers (1992) to conclude from their in vitro studies that spontaneous neuronal activity is critical for the development of adequate inhibitory transmission.

Neurotrophins appear to have essential roles in activity-dependent plastic changes in the CNS (reviewed in: Thoenen, 1995; Bonhoeffer, 1996; Marty et al., 1997; Shatz, 1997). Members of the neurotrophin family include nerve growth factor (NGF), brain-derived neurotrophic factor (BDNF),

---

* Corresponding author: Dr. Fredrick J. Seil, ORRP (P3-R&D-35), VA Medical Center, 3710 SW US Veterans Hospital Road, Portland, OR 97201, USA. Fax: +1-503-721-7906; E-mail: seilf@ohsu.edu

neurotrophin-3 (NT-3) and neurotrophin-4 (NT-4; also referred to as NT-4/5), and the Trk family of receptor tyrosine kinases with which they bind are, respectively, TrkA, TrkB, TrkC and TrkB (reviewed by: Lindsay, 1994; Bothwell, 1995; Lewin and Barde, 1996). Neurotrophins have also been reported to have functional effects at synapses (reviewed by: Schuman, 1999; Lu and Gottschalk, this volume). In a study with dissociated visual cortex cultures containing GABAergic interneurons and target pyramidal cells, Rutherford et al. (1997) found that blocking the cultures with TTX or exposing them continuously to the 'universal' Trk receptor inhibitor, K252a, reduced the percentage of GABA-immunopositive neurons without affecting neuronal survival, as determined by double labeling with the neuronal marker, microtubule-associated protein (MAP)2. Correlative electrophysiological studies revealed that GABA-mediated inhibition onto pyramidal neurons was decreased, and pyramidal cell discharge rates were increased. All of the effects of activity blockade were prevented by simultaneous exposure of the cultures to BDNF, but not to NGF or NT-3. The results of this study suggested to the authors that activity regulates cortical inhibition by regulation of BDNF.

Cerebellar culture models of activity-dependent inhibitory synaptogenesis

In the following sections of this chapter, we review our own studies of activity-dependent synaptogenesis in organotypic cerebellar cultures, based on ultrastructural observations and correlative electrophysiological studies, and of the effects of neurotrophins on the activity-dependent development of inhibitory synapses. The cultures were derived from newborn mice, at a time when there is very little synapse development in the cerebellar cortex, so that the synaptic development occurs in vitro, allowing an assessment of effects on synaptogenesis by manipulation of the culture system. As all major cortical neuronal types are included in these cultures, the expected synaptic interrelationships develop and reproduce the characteristic cerebellar cortical circuitry (reviewed by Seil, 1979, 1996). Missing from these isolated preparations are the extracerebellar afferents, the climbing and mossy fibers, and their synapses on Purkinje cell dendrites and granule cell dendrites, respectively, but the remaining elements are present.

Cultures continuously exposed to agents that increased neuronal activity

Our initial experiments in this series of studies consisted of exposing cerebellar cultures continuously to the anti-GABA agents, PTX or bicuculline (Seil et al., 1994). When mature cultures (e.g., 13 days in vitro [DIV] or older; Seil and Leiman, 1979; Herndon et al., 1981; Blank and Seil, 1982) were acutely exposed to either PTX or bicuculline, the rate of extracellularly recorded spontaneous cortical discharges was markedly increased. Cultures recorded from after 13–16 days of continuous exposure to anti-GABA agents since the time of explantation paradoxically had reduced rates of spontaneous cortical spike discharges following transfer to a physiological recording medium (buffered balanced salt solution) (Fig. 1). Quantitative assessment of cortical spikes (predominantly of Purkinje cell origin) showed that the discharge rates in PTX- or bicuculline-treated cultures were about half of the control rates. When cerebellar cultures exposed to anti-GABA agents were recorded from after only 3 DIV, spike discharges, although of low amplitude in immature cultures, were considerably more frequent than in control explants. By 12 DIV, however, the firing rate of PTX- and bicuculline-treated cultures was slow. With cortical stimulation, prolonged inhibition was evident in explants exposed to anti-GABA agents for 13 DIV or longer. Fig. 2A shows an inhibitory pause after a train of stimuli to the cortical surface of a control culture, while a long-lasting inhibition follows a similar train of stimuli to the cortex of a PTX-exposed explant in Fig. 2B. Cortical inhibition in response to electrical stimulation also developed earlier in cultures continuously exposed to anti-GABA agents, being consistently present from 8 DIV on, as opposed to 12 DIV for the consistent presence of inhibitory responses in untreated control explants. Thus cerebellar cultures continuously exposed to PTX or bicuculline had a slow rate of spontaneous cortical discharge, had prolonged inhibitory responses to cortical stimulation, and had an accelerated development of inhibitory responses.

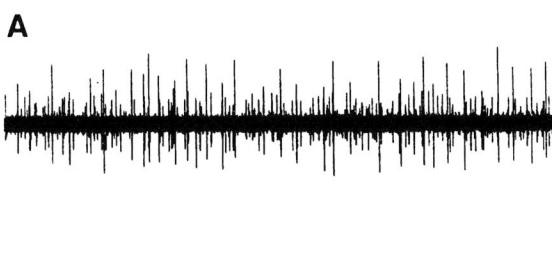

Fig. 1. Extracellularly recorded spontaneous cortical discharges in organotypic cerebellar cultures maintained in standard nutrient medium or continuously exposed since explantation to anti-GABA agents. For electrophysiological studies, the cultures were transferred to a buffered balanced salt solution and recorded at room temperature. (A) Control culture, 13 DIV. (B) Cerebellar culture, 13 DIV, continuously maintained in medium with 2×10^{-4} M PTX. The cortical discharge rate is appreciably reduced when compared with the control explant. (C) Cerebellar culture, 14 DIV, chronically exposed to 2×10^{-4} M bicuculline, with a resultant slow discharge rate similar to that of the PTX-treated explant. The time base marker equals 2 s.

Possible reasons for the electrophysiological changes became apparent on electron microscopic examination of cultures after 14–16 days of exposure to anti-GABA agents. The ratio of synapse profiles to Purkinje cell soma profiles was determined in sections containing nucleus, and only one section per Purkinje cell was counted. Twice as many axosomatic synapse profiles were evident on Purkinje cells in cultures treated with PTX or bicuculline than in untreated control explants (Fig. 3). The mean ratio of synapse to cell profiles was 2.2 for control

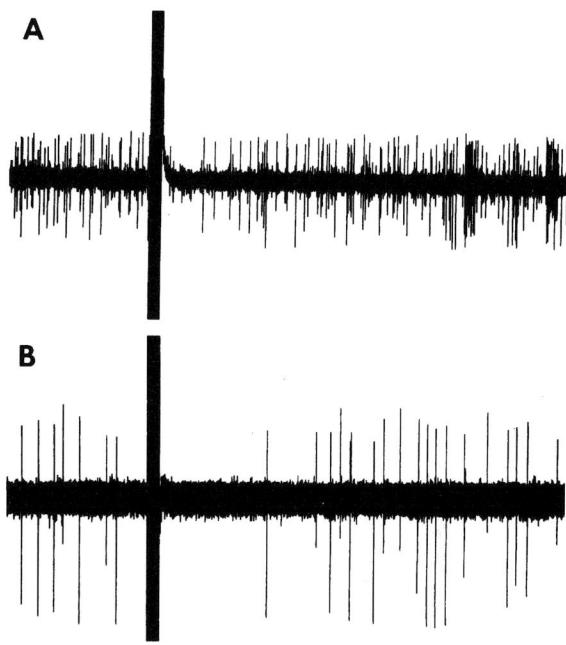

Fig. 2. Responses to repetitive cortical stimulation with 600 ms trains of pulses. (A) Control culture, 13 DIV. A stimulus train provokes a brief inhibition of spontaneous cortical activity. (B) Cerebellar culture, 13 DIV, continuously exposed to PTX since explantation. A prolonged inhibition of spontaneous cortical discharges follows the stimulus train. The time base marker equals 5 s. (From Seil et al., 1994.)

cultures, 4.8 for PTX-treated explants and 4.0 for cultures exposed to bicuculline. The axon terminals were identified by morphological criteria as basket cell terminals or Purkinje cell recurrent axon collateral terminals, the two types of synapses usually present on Purkinje cell somata, both of which are inhibitory (Palay and Chan-Palay, 1974). The ratio of synapses with basket cell terminals to recurrent axon collateral terminals was 1 : 1 in control cultures, but synapses with basket cell terminals outnumbered recurrent axon collateral synapses by 2 : 1 in explants continuously exposed to anti-GABA agents (Seil et al., 1994). This finding suggested that basket cell axons sprouted in order to increase the number of inhibitory terminals synapsing with Purkinje cell somata in response to exposure to anti-GABA agents that elevated neuronal discharge rates. The increased inhibition could have represented an exaggerated homeostatic response or a protective response to

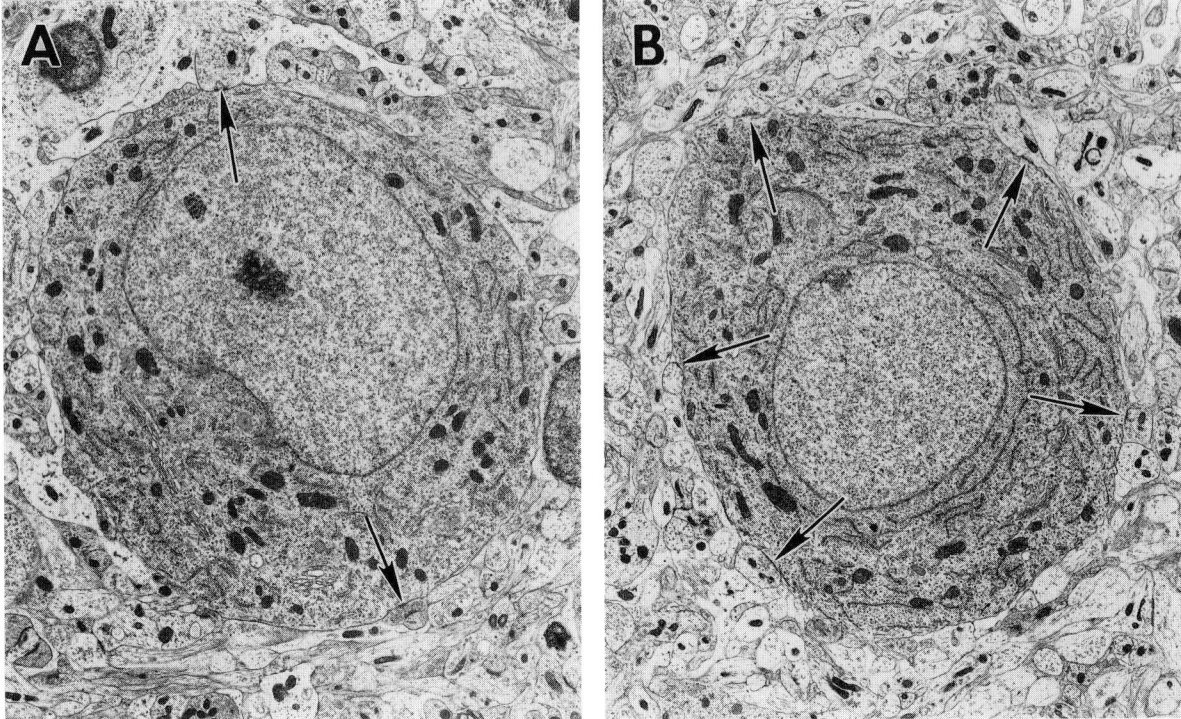

Fig. 3. Electron micrographs of Purkinje cell soma profiles from cerebellar cultures at 15 DIV. Arrows indicate inhibitory axosomatic synapse profiles. (A) Purkinje cell from a control culture. (B) Purkinje cell from a culture continuously exposed to PTX. The ratio of axosomatic synapse to soma profiles is increased. No other changes are evident. ×4000.

possible excitotoxic injury. In any case, a change in neuronal activity induced a structural change in the circuitry of a developing nervous system, which correlated with the observed functional changes.

Cerebellar cultures continuously exposed to activity-blocking agents

Our following study was of cerebellar cultures maintained in the absence of neuronal activity (Seil and Drake-Baumann, 1994). In order to silence Purkinje cells, both the somatic Na^+ spikes and the dendritic Ca^{2+} spikes must be blocked. For this purpose, both TTX and elevated levels of Mg^{2+} were incorporated into the nutrient medium at explantation and continuously thereafter. After 13–16 DIV, the cultures were transferred to a physiological medium for extracellular recording of cortical activity. When control cultures were transferred to the recording medium, they were immediately active (Fig. 4A). By contrast, cerebellar explants maintained in medium with incorporated TTX and elevated levels of Mg^{2+} were electrically silent for at least the first 10 min after transfer to the recording medium without activity-blocking agents (Fig. 4B). Then the cortical regions of the cultures began to slowly discharge (Fig. 4C), and the spikes increased in frequency until the preparations were in a state of hyperactivity 30–40 min after transfer (Fig. 4D), sometimes with bursts of rapid spikes (Fig. 4E), and the hyperactive discharges persisted for the duration of the recording sessions (up to 2 h). This activity was like that reported in earlier studies with culture preparations after release from activity blockade (Ramakers et al., 1990; Baker and Ruijter, 1991; Furshpan, 1991).

Ultrastructural examination of the activity-blocked cultures revealed a reduction of the ratio of Purkinje cell axosomatic synapse to cell profiles to half of the control value (Seil and Drake-Baumann, 1994) (Fig. 5A). The number of synapses with recurrent axon collateral terminals was the same as the number of synapses with basket cell

were counted. In the absence of climbing fibers in cerebellar cultures, axospinous cortical synapses are formed only with parallel fibers (granule cell axons), the remaining cortical excitatory elements. Axodendritic synapses consist of a mixture of excitatory synapses with parallel fiber terminals and inhibitory synapses with terminals from inhibitory interneuron axons and Purkinje cell axon collaterals. Axospinous synapses were as numerous in the cortical neuropil of activity-blocked cultures as in the cortex of untreated control explants, while half as many axodendritic synapses were present after activity blockade as were present in control cultures. The development of proportionally fewer axosomatic and axodendritic synapses than axospinous synapses in cultures exposed to TTX and elevated levels of Mg^{2+} is indicative of incomplete development of inhibitory circuitry. The stunted development of inhibition in activity-blocked cultures was consistent with the hyperactivity displayed by these preparations after release from activity blockade. Remarkably, the complement of excitatory synapses developed fully in the absence of neuronal activity.

Combining the results of the two studies, one with continuous exposure of cerebellar cultures to agents that increase neuronal activity and the other with similar exposure to activity-blocking agents, the concept proposed by Corner and Ramakers (1992) that neuronal activity is critical for the development of adequate inhibitory transmission is supported. Excess neuronal activity led to formation of an increased number of inhibitory synapses and a resultant slowing of spontaneous cortical discharges and prolongation of inhibitory responses in cerebellar explants, while the consequences of a lack of neuronal activity were reduced development of inhibitory synapses and sustained cortical hyperactivity following release of cultures from activity blockade. Neuronal activity appeared to have a significant effect on inhibitory synaptogenesis.

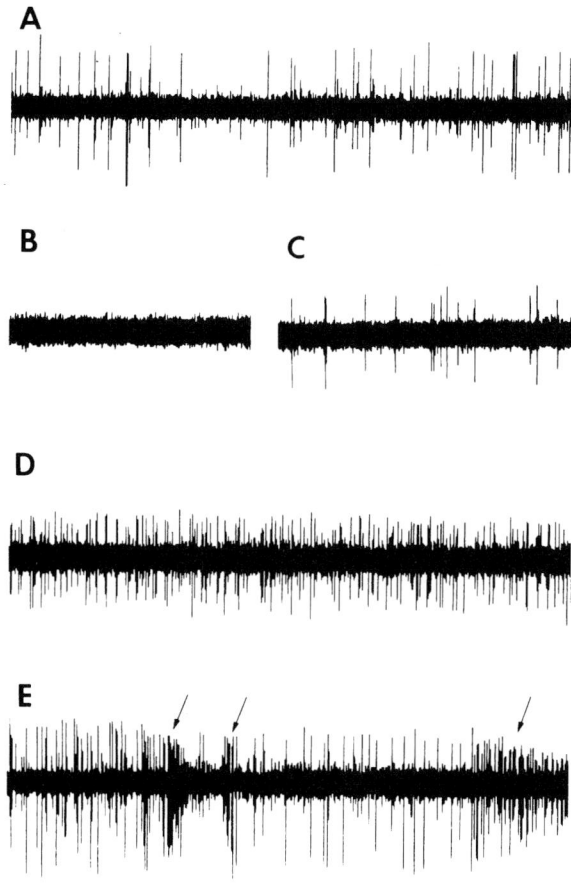

Fig. 4. Extracellularly recorded spontaneous cortical discharges in cerebellar cultures maintained in standard nutrient medium or in medium with incorporated activity-blocking agents, 11.1 mM Mg^{2+} plus 10^{-8} M TTX. (A) Control culture, 14 DIV. Cortical activity is immediately evident upon transfer of the culture to the recording medium. (B–D) Recording from an activity-blocked culture, 14 DIV. The culture is electrically silent immediately after transfer to the recording medium (B). Cortical discharges begin to appear 18 min later (C). By 40 min after transfer, the culture is in a state of sustained hyperactivity. (E) Recording from another activity-blocked explant, 15 DIV, after recovery from activity blockade. Bursts of rapid spikes are evident (arrows), which are characteristic of some activity-blocked cultures after recovery. The time base marker equals 5 s. (From Seil and Drake-Baumann, 1994.)

Effects of neurotrophins on activity-dependent inhibitory synaptogenesis

Neurotrophins appear to be synthesized and released in an activity-dependent manner (Zafra et al., 1991; Lindholm et al., 1994; Blöchl and Thoenen, 1995; Thoenen, 1995). In combination with

terminals, indicating that development of these inhibitory synapses was equally affected by blockade of neuronal activity. Further examination was extended to the cortical neuropil, where synapses on dendritic spines and smooth portions of dendrites

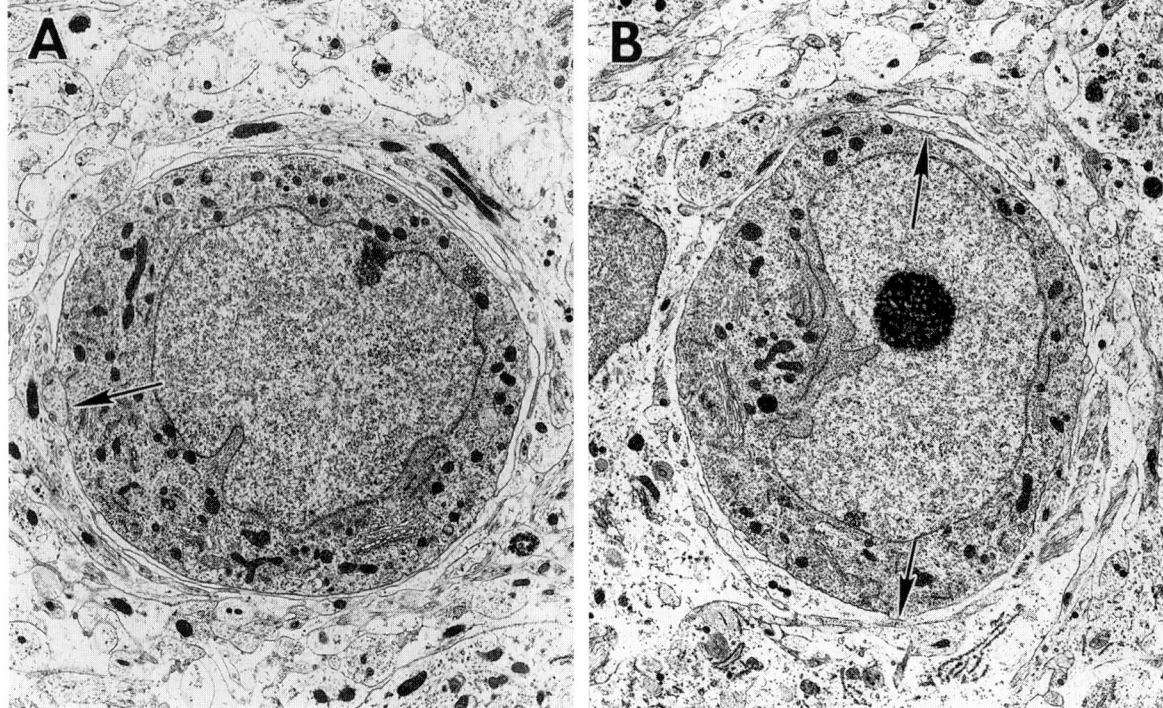

Fig. 5. Electron micrographs of Purkinje cell soma profiles from cerebellar cultures at 15 DIV. Arrows indicate inhibitory axosomatic synapse profiles. (A) Purkinje cell from a culture exposed to activity-blocking agents since explantation. The ratio of axosomatic synapse to soma profiles is reduced to half the control value. (B) Purkinje cell from a culture simultaneously exposed to activity-blocking agents and 25 ng BDNF/ml nutrient medium. The ratio of axosomatic synapse to soma profiles is similar to that of a Purkinje cell in a control culture (see Fig. 3A). ×4000.

their aforementioned functional effects at synapses (Schuman, 1999) and the prevention of some anti-inhibitory consequences of activity blockade by BDNF (Rutherford et al., 1997), these properties make them candidates for a role in inhibitory synaptogenesis. We did not consider that NGF was a likely candidate with regard to inhibitory synapses on Purkinje cells, however, as its high-affinity receptor, TrkA, has been reported to be only transiently expressed on Purkinje cells during development (Ernfors et al., 1992; Lärkfors et al., 1996). We therefore concentrated our efforts on the other three members of the neurotrophin family known to be present in mammalian systems, namely BDNF, NT-3 and NT-4, and determined their effects on the development of inhibitory Purkinje cell axosomatic synapses, initially by exogenous application to cultures during activity blockade and subsequently by investigating some of the functions of endogenous neurotrophins (Seil, 1999; Seil and Drake-Baumann, 2000).

Morphological effects of neurotrophins applied exogenously during activity blockade

Continuous application of the TrkB receptor ligands, BDNF or NT-4, to cerebellar cultures simultaneously with the activity-blocking agents, TTX and elevated levels of Mg^{2+}, resulted in development of the full complement of Purkinje cell axosomatic synapses (Fig. 5B). The mean ratio of synapse to cell profiles in the case of both untreated control cultures and activity-blocked cultures exposed to BDNF was 2.3, compared to 1.3 for untreated activity-blocked explants (Seil, 1999). Similarly, in the group of cultures treated with NT-4, the synapse to cell profile ratio was 2.3 for controls, 2.1 for NT-4-treated activity-blocked cultures and 1.2 for untreated activity-blocked explants. By contrast, no effect was evident from the exogenous application of the TrkC receptor ligand, NT-3, to activity-blocked cultures. The mean ratio of synapse to cell profiles in this group was

2.1 for controls, and 1.0 for both untreated activity-blocked cultures and NT-3-exposed activity-blocked explants. The synaptogenesis-promoting effect in the absence of neuronal activity appeared to be restricted to TrkB receptor ligands.

Correlative electrophysiological effects

All activity-blocked cultures were electrically silent for the first 10 min after transfer to the recording medium, whether or not they had been treated with neurotrophins, in contrast to the immediate cortical electrical activity of the control group (Seil and Drake-Baumann, 2000). The addition of any of the neurotrophins did not alter the effectiveness of the activity blockade by TTX and elevated levels of Mg^{2+}. After recovery from activity blockade, cultures treated with BDNF and NT-4 had spontaneous cortical discharge rates comparable to those of untreated control explants (Fig. 6), while recovered cultures exposed to NT-3 were hyperactive, with spontaneous cortical discharge rates like those of untreated activity-blocked cultures. The frequency of cortical spike discharges in untreated and NT-3 treated activity-blocked cultures after recovery was double that of control explants. The electrophysiological results correlated with the morphological findings, as BDNF and NT-4-treated activity-blocked cerebellar cultures had control numbers of inhibitory Purkinje cell axosomatic synapses and displayed control rates of spontaneous cortical activity, while untreated activity-blocked cultures and NT-3-treated activity-blocked explants had reduced numbers of inhibitory axosomatic synapses and exhibited hyperactive cortical discharges.

Role of endogenous neurotrophins in the presence of neuronal activity

The role of endogenous neurotrophins in the development of inhibitory synapses in the presence of control levels of neuronal activity was investigated in two ways. First, neurotrophins were added exogenously to otherwise untreated cerebellar cultures in the same dosage (25 ng/ml nutrient medium) as was used to promote synapse formation in the absence of neuronal activity, and inhibitory synaptogenesis was assessed. Second, otherwise untreated cultures

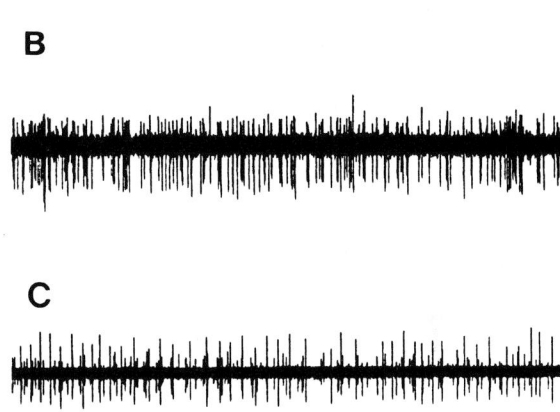

Fig. 6. Extracellularly recorded spontaneous cortical activity in control, activity-blocked, and BDNF-treated activity-blocked cultures after recovery from activity blockade. (A) Control culture, 14 DIV. (B) Hyperactive cortical discharges recorded at 14 DIV in a cerebellar culture 45 min after recovery from activity blockade. (C) Cortical activity recorded 40 min after recovery from activity blockade in a 15-DIV culture that had been treated with BDNF during activity blockade. The spontaneous cortical activity pattern is similar to that of untreated control cultures. The time base marker equals 5 s.

were continuously exposed to antibodies to TrkB receptor ligands, and Purkinje cell axosomatic synapse development was monitored after 15 DIV.

No differences from untreated controls in the mean ratio of Purkinje cell axosomatic synapse to cell profiles were evident with addition of any of the neurotrophins to the nutrient medium over the duration of the culture period (from 0 to 15 DIV). The added neurotrophins did not induce any additional inhibitory synaptogenesis under conditions of normal levels of spontaneous cortical activity. Possibilities as to why hyperinnervation of Purkinje cells with inhibitory terminals did not occur include (1) a feedback mechanism that may have inhibited endogenous neurotrophin release with the exogenous addition of neurotrophins, or (2) Purkinje cells discharging at control rates may have been insensitive to added neurotrophins. Marked changes were induced, however, by blocking TrkB receptor ligands with antibodies to BDNF and NT-4. Cultures continuously exposed to the combination of these antibodies since explantation developed only half of the control numbers of Purkinje cell axosomatic synapses. The effect was as complete as with blockade of neuronal activity for the same period of time. Antibody to BDNF alone only partially reduced inhibitory synapse formation, suggesting that both TrkB receptor ligands contributed to inhibitory synaptogenesis in cultures with control levels of neuronal activity.

Antibody inhibition of inhibitory hyperinnervation with increased neuronal activity

This same combination of antibodies was applied to cerebellar cultures simultaneously exposed to PTX to increase neuronal activity. As indicated earlier, Purkinje cell somata in cultures exposed to PTX alone became hyperinnervated with inhibitory terminals (Seil et al., 1994). Antibody blockade of BDNF and NT-4, presumably released in greater-than-normal quantities because of the increased neuronal activity, resulted in development of a complement of Purkinje cell axosomatic synapses more like that of control cultures (Seil and Drake-Baumann, 2000). The mean ratio of axosomatic synapse to cell profiles was 2.3 in this group of control cultures, 2.5 in explants exposed simultaneously to PTX and antibodies to BDNF and NT-4 from 0 to 15 DIV, and 3.5 in cultures treated with PTX alone for 15 DIV. Doubling the concentration of antibodies further reduced the development of Purkinje cell axosomatic synapses to approximately half of the control value, suggesting that the lower dose of antibodies did not bind all of the neurotrophin released by the PTX-induced increased neuronal activity. In either case, antibodies to BDNF and NT-4 mitigated the effect of increased neuronal activity on inhibitory synaptogenesis.

Discussion and conclusions

The results of these studies point strongly to a role for TrkB receptor ligands in activity-dependent inhibitory synaptogenesis. The specificity of TrkB receptor ligands for the promotion of inhibitory synapse development is reminiscent of the specificity of BDNF for preventing the reduction of GABA-immunopositive neurons and the increase in pyramidal cell discharges associated with activity blockade of visual cortex cultures (Rutherford et al., 1997) and the specificity of BDNF and NT-4 for influencing the development of ocular dominance columns in the visual cortex of cats (Cabelli et al., 1995). That each of the neurotrophins has a distinct biological role is further indicated by studies in which differences between TrkB receptor ligand actions have been defined (McAllister et al., 1995; Wirth et al., 1998).

The interplay of neuronal activity and neurotrophins is a subject of interest. Both activity blockade and antibody inhibition of Trk B receptor ligands achieve the same result, namely a reduction of inhibitory synaptogenesis. The effect of activity blockade can be overcome with the exogenous application of BDNF or NT-4. These findings are compatible with the concept that neurotrophins are released in an activity-dependent manner (Blöchl and Thoenen, 1995; Thoenen, 1995). In this scenario there is a background rate of neurotrophin release, which is reduced in the absence of neuronal activity, and which can be replaced with neurotrophins from an outside source to achieve the same effect as the endogenous release. It is this property that argues against the notion that activity blockade reduces the responsiveness of neurons to neurotrophins (McAllister et al., 1999). If responsiveness to neurotrophins were the critical factor, then the effects of activity blockade could not be overcome with addition of neurotrophins.

If, in the presence of neuronal activity, the released neurotrophins are neutralized by reaction with antibody, the effect is the same as if they had not been released. If one now presumes that greater quantities of neurotrophins are released with an increase in neuronal activity, such as might occur upon exposure to PTX, then application of antibodies to TrkB receptor ligands might bind the excess neurotrophins released and mitigate the neurotrophin effect. Thus our finding of an antibody dose-dependent reduction of the complement of Purkinje cell inhibitory axosomatic synapses in cultures simultaneously treated with PTX and antibodies to BDNF and NT-4 is also compatible with the concept of an activity-dependent release of neurotrophins.

The most interesting clue to the interdependence of neurotrophins and neuronal activity is provided by the lack of response of neurons discharging at control rates to exogenously added neurotrophins. An increase in the number of Purkinje cell axosomatic synapses, as happens with exposure to PTX, might have been expected with the addition of the same quantities of BDNF or NT-4 that promote inhibitory synaptogenesis in the absence of neuronal activity, but the Purkinje cells failed to respond to the presumed excess amounts of neurotrophins with increased synapse formation when discharging normally. Perhaps there is a change in the sensitivity of neurons to neurotrophins with different levels of neuronal activity, but it may be a more complex relationship than originally envisioned. Quantitative studies with varying levels of neurotrophins and neuronal activity might provide some insight into this relationship.

We have, in our studies, dealt with effects on numbers of synapses by changes in neuronal activity or levels of neurotrophins, and not on other aspects of synapse formation such as vesicle distribution, active zone length, etc. (e.g., see Lu and Gottschalk, this volume), all of which are relevant and subjects for future study. However, the resulting differences in synapse numbers were of such magnitudes (in the 50–100% range) that they could not fail to be of some significance, especially since they correlated with changes in spontaneous neuronal discharge rates. Thus cultures with increased numbers of inhibitory synapses discharged more slowly, and cultures with reduced numbers of inhibitory synapses were hyperactive. That inhibitory synapses were so profoundly affected by neuronal activity and neurotrophins in our culture system makes it tempting to think in terms of neurotrophin mediation of neuronal activity effects. However, synaptogenesis is a complex process, and we have only looked at changes in numbers of synapses. In a simpler optic axonal arborization system, Cohen-Cory (1999) has argued that BDNF modulates, but does not mediate, activity-dependent axon arborization because BDNF promotes axonal growth while neuronal activity is involved with stabilization of axonal branching. There are undoubtedly many opportunities for similar differences in mechanisms in the multiple steps involved in synaptogenesis.

Our studies have been focused on inhibitory synapses because of our finding that the full complement of excitatory cortical axospinous synapses developed in cerebellar cultures in the absence of neuronal activity, while inhibitory synapse formation was profoundly reduced (Seil and Drake-Baumann, 1994). Whereas excitation is the background state of the nervous system (Llinás, 1988), it appears that inhibitory circuitry does not develop fully unless there is something to inhibit. As we used the cerebellar culture model to study effects of neurotrophins, our observations were biased toward changes involving inhibitory synapses, so we hasten to say that we do not claim that neurotrophins affect only inhibitory synapse development. Vicario-Abejon et al. (1998) reported that BDNF promoted the formation of inhibitory synapses in developing hippocampal cultures, but BDNF and NT-3 also enhanced the development of excitatory glutamatergic synapses. Suffice it to say that there is more to be learned.

As a concluding statement, we would like to reiterate that the evidence for a role for TrkB receptor ligands in the promotion of activity-dependent inhibitory synaptogenesis is substantial. Further developments in this field could lead to possible therapeutic uses for neurotrophins to facilitate restoration of inhibitory circuitry after insults such as trauma or stroke, where recovery of an appropriate balance between excitation and inhibition may reduce the occurrence of such detrimental consequences as seizures. Much more definition is needed of the variety of effects of neurotrophins, both in acute and longer-term conditions, before further clinical applications of these factors are considered (see Thoenen, this volume), but their ultimate use as therapeutic tools to optimize functional recovery remains a hope.

Acknowledgements

We would like to dedicate this chapter to our late colleague, Arnold L. Leiman of the University of California, Berkeley. It was Dr. Leiman's inspiration to investigate the chronic effects of anti-GABA agents on cerebellar cultures, and it was with his collaborative efforts that the series of studies that we have reviewed was launched. We will miss his collaboration, his helpful discussions, and his friendship. The research was supported by the U.S. Department of

Veterans Affairs and National Institutes of Health Grant NS 17493.

References

Baker, R.E. and Ruijter, J.M. (1991) Chronic blockade of bioelectric activity in neonatal rat neocortex in vitro: physiological effects. *Int. J. Dev. Neurosci.*, 9: 321–329.

Benevento, L.A., Bakkum, B.W. and Cohen, R.S. (1995) Gamma-aminobutyric acid and somatostatin immunoreactivity in the visual cortex of normal and dark-reared rats. *Brain Res.*, 689: 172–182.

Blank, N.K. and Seil, F.J. (1982) Mature Purkinje cells in cerebellar tissue cultures. *J. Comp. Neurol.*, 208: 169–176.

Blöchl, A. and Thoenen, H. (1995) Characterization of nerve growth factor (NGF) release from hippocampal neurons: evidence for a constitutive and an unconventional sodium-dependent regulated pathway. *Eur. J. Neurosci.*, 7: 1220–1228.

Bonhoeffer, T. (1996) Neurotrophins and activity-dependent development of the neocortex. *Curr. Opin. Neurobiol.*, 6: 119–126.

Bothwell, M. (1995) Functional interactions of neurotrophins and neurotrophin receptors. In: Cowan, W.M., Shooter, E.M., Stevens, C.F. and Thompson, R.F. (Eds.), *Annual Review of Neuroscience, Vol. 18*. Annual Reviews, Palo Alto, CA, pp. 223–253.

Cabelli, R.J., Hohn, A. and Shatz, C.J. (1995) Inhibition of ocular dominance column formation by infusion of NT-4/5 or BDNF. *Science*, 267: 1662–1666.

Cohen-Cory, S. (1999) BDNF modulates, but does not mediate, activity-dependent branching and remodeling of optic axon arbors in vivo. *J. Neurosci.*, 19: 9996–10003.

Corner, M.A. and Ramakers, G.J.A. (1992) Spontaneous firing as an epigenetic factor in brain development — physiological consequences of chronic tetrodotoxin and picrotoxin exposure on cultured rat neocortex neurons. *Dev. Brain Res.*, 65: 57–64.

Ernfors, P., Merlio, J.-P. and Persson, H. (1992) Cells expressing mRNA for neurotrophins and their receptors during embryonic rat development. *Eur. J. Neurosci.*, 4: 1140–1158.

Furshpan, E.J. (1991) Seizure-like activity in cell culture. *Epilepsy Res.*, 10: 24–32.

Harris, W.A. (1981) Neural activity and development. *Annu. Rev. Physiol.*, 43: 698–710.

Hendry, S.A.C. and Jones, E.G. (1986) Reduction in number of immunostained GABAergic neurones in deprived-eye dominance columns of monkey area 17. *Nature*, 320: 750–753.

Hendry, S.A.C. and Jones, E.G. (1988) Activity-dependent regulation of GABA expression in the visual cortex of adult monkeys. *Neuron*, 1: 701–712.

Herndon, R.M., Seil, F.J. and Seidman, C. (1981) Synaptogenesis in mouse cerebellum: a comparative in vivo and tissue culture study. *Neuroscience*, 6: 2587–2598.

Janka, Z. and Jones, D.G. (1982) Junctions in rat neocortical explants cultured in TTX-, GABA-, and Mg^{2+}-environments. *Brain Res. Bull.*, 47: 289–292.

Lärkfors, L., Lindsay, R.M. and Alderson, R.F. (1996) Characterization of the responses of Purkinje cells to neurotrophin treatment. *J. Neurochem.*, 66: 1362–1373.

Lewin, G.R. and Barde, Y.-A. (1996) Physiology of the neurotrophins. In: Cowan, W.M., Shooter, E.M., Stevens, C.F. and Thompson, R.F. (Eds.), *Annual Review of Neuroscience, Vol. 19*. Annual Reviews, Palo Alto, Ca, pp. 289–317.

Lindholm, D., Castrén, E., Berzaghi, M., Blöchl, A. and Thoenen, H. (1994) Activity-dependent and hormonal regulation of neurotrophin mRNA levels in the brain — implications for neuronal plasticity. *J. Neurobiol.*, 25: 1362–1372.

Lindsay, R.M. (1994) Neurotrophins and receptors. In: Seil, F.J. (Ed.), *Neural Regeneration. Progress in Brain Research, Vol. 103*. Elsevier, Amsterdam, pp. 3–14.

Llinás, R.R. (1988) The intrinsic electrophysiological properties of mammalian neurons: insights into central nervous system function. *Science*, 242: 1654–1664.

Marty, S., Berzaghi, M.P. and Berninger, B. (1997) Neurotrophins and activity-dependent plasticity of cortical neurons. *Trends Neurosci.*, 20: 198–202.

McAllister, A.K., Lo, D.C. and Katz, L.C. (1995) Neurotrophins regulate dendritic growth in developing visual cortex. *Neuron*, 15: 791–803.

McAllister, A.K., Katz, L.C. and Lo, D.C. (1999) Neurotrophins and synaptic plasticity. In: Cowan, W.M., Shooter, E.M., Stevens, C.F. and Thompson, R.F. (Eds.), *Annual Review of Neuroscience, Vol. 22*. Annual Reviews, Palo Alto, CA, pp. 295-318.

Meisami, E. and Monsavi, R. (1981) Lasting effects of early olfactory deprivation on the growth, DNA, RNA and protein content, and Na-K-ATPase and AchE activity of the olfactory bulb. *Dev. Brain Res.*, 2: 217–229.

Micheva, K.D. and Beaulieu, C. (1995) An anatomical substrate for experience-dependent plasticity of the rat barrel field cortex. *Proc. Natl. Acad. Sci. USA*, 92: 11834–11838.

Palay, S. and Chan-Palay, V. (1974) *Cerebellar Cortex. Cytology and Organization*. Springer, New York.

Parks, T.N. (1979) Afferent influences on the development of the brainstem auditory nuclei of the chicken: otocyst ablation. *J. Comp. Neurol.*, 183: 665–678.

Ramakers, G.J.A., Corner, M.A. and Habets, A.M.M.C. (1990) Development in the absence of spontaneous bioelectric activity results in increased stereotype burst firing in cultures of dissociated cerebral cortex. *Exp. Brain Res.*, 79: 157–166.

Ramakers, G.J.A., Corner, M.A. and Habets, A.M.M.C. (1991) Abnormalities in the spontaneous firing patterns of cultured rat neocortical neurons after chronic exposure to picrotoxin during development in vitro. *Brain Res. Bull.*, 26: 429–432.

Reiter, H.O., Waitzman, D.M. and Stryker, M.P. (1986) Cortical activity blockade prevents ocular dominance plasticity in the kitten visual cortex. *Exp. Brain Res.*, 65: 182–188.

Ruijter, J.M., Baker, R.E., DeJong, B.M. and Romijn, H.J. (1991) Chronic blockade of bioelectric activity in neonatal rat cortex grown in vitro: morphological effects. *Int. J. Dev. Neurosci.*, 9: 331–338.

Rutherford, L.C., DeWan, A., Lauer, H.M. and Turrigiano, G.G. (1997) Brain-derived neurotrophic factor mediates the activity-dependent regulation of inhibition in neocortical cultures. *J. Neurosci.*, 17: 4527–4535.

Schuman, E. (1999) Neurotrophin regulation of synaptic activity. *Curr. Opin. Neurobiol.*, 9: 105–109.

Seil, F.J. (1979) Cerebellum in tissue culture. In: Schneider, D.M. (Ed.), *Reviews of Neuroscience, Vol. 4*. Raven Press, New York, pp. 105–177.

Seil, F.J. (1996) Neural plasticity in cerebellar cultures. *Prog. Neurobiol.*, 50: 533–556.

Seil, F.J. (1999) BDNF and NT-4, but not NT-3, promote development of inhibitory synapses in the absence of neuronal activity. *Brain Res.*, 818: 561–564.

Seil, F.J. and Drake-Baumann, R. (1994) Reduced cortical inhibitory synaptogenesis in organotypic cultures developing in the absence of neuronal activity. *J. Comp. Neurol.*, 342: 366–377.

Seil, F.J. and Drake-Baumann, R. (2000) TrkB receptor ligands promote activity-dependent inhibitory synaptogenesis. *J. Neurosci.*, 20: 5367–5373.

Seil, F.J. and Leiman, A.L. (1979) Development of spontaneous and evoked electrical activity of cerebellum in tissue culture. *Exp. Neurol.*, 64: 61–75.

Seil, F.J., Drake-Baumann, R., Leiman, A.L., Herndon, R.M. and Tiekotter, K.L. (1994) Morphological correlates of altered neuronal activity in organotypic cultures chronically exposed to anti-GABA agents. *Dev. Brain Res.*, 77: 123–132.

Shatz, C.J. (1997) Neurotrophins and visual system plasticity. In: Cowan, W.M., Jessell, T.M. and Zipursky, S.L. (Eds.), *Molecular and Cellular Approaches to Neural Development*. Oxford University Press, New York, pp. 509–524.

Shatz, C.J. and Stryker, M.P. (1988) Prenatal tetrodotoxin infusion blocks segregation of retinogeniculate afferents. *Science*, 242: 87–89.

Shaw, C. and Cynander, M. (1984) Disruption of cortical activity prevents ocular dominance changes in monocularly deprived kittens. *Nature*, 308: 731–734.

Simons, D.J. and Land, P.W. (1987) Early experience of tactile stimulation influences organization of somatic sensory cortex. *Nature*, 326: 694–697.

Thoenen, H. (1995) Neurotrophins and neuronal plasticity. *Science*, 270: 593–598.

Van Huizen, F., Romijn, H.J. and Habets, A.M.M.C. (1985) Synaptogenesis in rat cerebral cortex cultures is affected during chronic blockade of spontaneous bioelectric activity by tetrodotoxin. *Dev. Brain Res.*, 19: 67–80.

Van Huizen, F., Romijn, H.J., Van den Hooff, P. and Habets, A.M.M.C. (1987) Picrotoxin induced disinhibition of spontaneous bioelectric activity accelerates synaptogenesis in rat cerebral cortex cultures. *Exp. Neurol.*, 97: 280–288.

Vicario-Abejon, C., Collin, C., McKay, R.D.G. and Segal, M. (1998) Neurotrophins induce formation of functional excitatory and inhibitory synapses between cultured hippocampal neurons. *J. Neurosci.*, 18: 7256–7271.

Wiesel, T.N. and Hubel, D.H. (1963) Effects of visual deprivation on morphology and physiology of cells in the cat's lateral geniculate body. *J. Neurophysiol.*, 26: 978–993.

Wirth, M.J., Obst, K. and Wahle, P. (1998) NT-4/5 and LIF, but not NT-3 and BDNF, promote NPY mRNA expression in cortical neurons in the absence of spontaneous bioelectrical activity. *Eur. J. Neurosci.*, 10: 1457–1464.

Woolsey, T.A. and Wann, J.R. (1976) Areal changes in mouse cortical barrels following vibrissal damage at different postnatal ages. *J. Comp. Neurol.*, 170: 53–66.

Zafra, F., Castrén, H., Thoenen, H. and Lindholm, D. (1991) Interplay between glutamate and γ-aminobutyric acid transmitter systems in the physiological regulation of brain-derived neurotrophic factor and nerve growth factor synthesis in hippocampal neurons. *Proc. Natl. Acad. Sci. USA*, 88: 10037–10041.

CHAPTER 19

Modulation of hippocampal synaptic transmission and plasticity by neurotrophins

Bai Lu * and Wolfram Gottschalk

Unit on Synapse Development and Plasticity, Laboratory of Developmental Neurobiology, NICHD, NIH, Building 49, Room 5A38, 49 Convent Drive, MSC 4480, Bethesda, MD 20892-4480, USA

Introduction

The neurotrophin family of proteins include nerve growth factor (NGF), brain-derived neurotrophic factor (BDNF), neurotrophin-3 (NT-3), and neurotrophin-4/5 (NT-4/5) (Lewin and Barde, 1996). These proteins elicit their biological functions by interacting with their respective Trk receptor tyrosine kinases. NGF binds to TrkA, BDNF and NT-4/5 bind to TrkB, and NT-3 binds to TrkC (Chao, 1992; Barbacid, 1993). Substantial evidence now indicates that, in addition to their classic functional roles in neuronal survival and differentiation, neurotrophins also play an important role in synaptic transmission and plasticity. For example, two major modes of neurotrophin actions have been described at the neuromuscular synapse: acute enhancement of synaptic transmission (Lohof et al., 1993; Stoop and Poo, 1995, 1996; Wang and Poo, 1997; Xie et al., 1997), and long-term regulation of synapse maturation (Wang et al., 1995; Liou and Fu, 1997; Liou et al., 1997; Wang et al., 1998). In the visual cortex, neurotrophins may participate in the activity-dependent synaptic competition and the formation of ocular dominance columns (Domenici et al., 1991; Maffei et al., 1992; Cabelli et al., 1995, 1997; Riddle et al., 1995). Neurotrophins are also involved in complex and activity-dependent modulation of dendritic and axonal growth in the central nervous system (CNS) (Cohen-Cory and Fraser, 1995; McAllister et al., 1995, 1996, 1997; Cohen-Cory, 1999; Lom and Cohen-Cory, 1999). So far the most extensive studies regarding the role of neurotrophins in synaptic transmission and plasticity have been carried out in the hippocampus. These studies suggest that neurotrophins may serve as a new class of neuromodulators that mediate activity-dependent modifications of neuronal connectivity and synaptic efficacy. Thus, we need to take this new role of neurotrophins into account when considering neural regeneration and neurotrophin therapy. Further studies in this emerging field will not only provide valuable insights into the underlying mechanisms of synapse development and function, and how they are regulated, but also have important implications in the understanding and treatment of neurological diseases. This review will describe evidence for neurotrophic regulation of synaptic transmission and plasticity in the hippocampus. We will discuss in detail the physiological implications and the possible mechanism of neurotrophin-mediated modulation, and try to put complex and sometimes conflicting results into perspective. A more complete discussion of the role of neurotrophins in other systems can be found in a number of recent reviews (Lo, 1995; Thoenen, 1995;

---

* Corresponding author: Dr. Bai Lu, Unit on Synapse Development and Plasticity, Laboratory of Developmental Neurobiology, NICHD, NIH, Building 49, Room 5A38, 49 Convent Drive, MSC 4480, Bethesda, MD 20892-4480, USA. Fax: (301) 496-9939; E-mail: lub@codon.nih.gov

Berninger and Poo, 1996; Bonhoeffer, 1996; Lu and Figurov, 1997; McAllister et al., 1999).

Effect of NT-3 in hippocampal plasticity

The earliest evidence suggesting a link between neurotrophins and hippocampal LTP was the demonstration that both BDNF and NT-3 mRNA are upregulated by the high-frequency stimulation used to induce long-term potentiation (LTP) (Patterson et al., 1992; but see Castren et al., 1993). During development, NT-3 and TrkC are strongly expressed early on, but the expression of both becomes weaker, albeit still present, in the adult hippocampus (Ernfors et al., 1990; Maisonpierre et al., 1990; Lamballe et al., 1994; Martinez et al., 1998). Despite the upregulation of NT-3 by high-frequency stimulation, it has been suggested that endogenous NT-3 is not directly involved in hippocampal LTP. Anti-NT-3 antibodies have no effect on hippocampal CA1 LTP (Chen et al., 1999). Additionally, NT-3 is not required for synaptic transmission or LTP, as shown by studies using NT-3 conditional knockout mice in which the NT-3 gene has been deleted only in CNS neurons (Ma et al., 1999). However, these results should be interpreted with caution for two reasons. First, NT-3 may not affect CA1 LTP because its modulatory role may be 'region-specific' in the hippocampus. These studies looked at area CA1 synaptic plasticity. It has been found that high levels of NT-3 mRNA are present in dentate granule cells and in pyramidal neurons of CA2 and only the most medial part of the CA1 area (Ernfors et al., 1990; Maisonpierre et al., 1990). Therefore, while NT-3 may not be affecting synaptic plasticity in the CA1 region of the hippocampus, it may have a role in other regions of the hippocampus. This finding leads to the supposition that NT-3 may have a role in some forms of hippocampal synaptic plasticity in only specific regions of the hippocampus. Indeed, it was shown that at the lateral perforant path–granule cell synapse in the dentate gyrus of heterozygous (+/−) NT-3 knockout mice there was normal LTP, but diminished paired-pulse facilitation and post-tetanic potentiation (Kokaia et al., 1998). Second, the NT-3 conditional knockouts still have 30% NT-3 remaining in the brain, possibly derived from non-neuronal cells (Ma et al., 1999). This small amount of NT-3 could still exert its modulatory effect on CA1 synapses. A conditional knockout line with a complete deletion of the NT-3 gene in all cells in the CNS, such as the one recently reported by Bates et al. (1999), may be helpful in addressing this issue.

Role of BDNF in LTP and LTD

Much more consistent results have been obtained for the role of BDNF in hippocampal LTP. Studies from the CA1 area of the hippocampus have demonstrated that BDNF regulates the tetanus-induced LTP in both neonatal and adult rats (Figurov et al., 1996). The expression of BDNF and the TrkB receptor in the hippocampus increase gradually during hippocampal development (Maisonpierre et al., 1990; Friedman et al., 1991; Dugich-Djordjevic et al., 1993; Ringstedt et al., 1993). The level of BDNF in the hippocampus appears to be an important determinant for tetanus-induced LTP. In the neonate, when endogenous BDNF levels are low, application of exogenous BDNF facilitates LTP. In the adult, when the endogenous BDNF levels are high, inhibition of BDNF activity by the BDNF scavenger protein, TrkB-IgG, reduces the magnitude of LTP in the hippocampus. The effect of BDNF on LTP induction may be due to an enhanced ability of hippocampal synapses to follow the tetanic stimulation, such as theta burst stimulation (TBS) or a train of high-frequency stimulation (100 Hz, 1 s, HFS), used to induce LTP (Figurov et al., 1996; Gottschalk et al., 1998). LTP induced by the pairing of low-frequency stimulation with postsynaptic depolarization was not affected by BDNF, suggesting that BDNF does not modulate the LTP-triggering mechanisms (Figurov et al., 1996; but see Korte et al., 1995). BDNF may also be involved in the maintenance of LTP once it has been established; application of TrkB-IgG 30 min after LTP induction reverses the synaptic potentiation to baseline (Kang et al., 1997).

Experiments using BDNF knockout mice also demonstrate that BDNF modulates tetanus-induced LTP in the CA1 synapses. Two independent lines of BDNF knockout mice exhibit a severe impairment in hippocampal LTP (Korte et al., 1995; Patterson et al., 1996). Moreover, +/− mice showed the same degree of impairment as the −/− mice, suggesting that a critical level of BDNF in the hippocampus may be needed for LTP induction and/or maintenance. Introduction of BDNF back into the mutant slices, either by incubation with recombinant BDNF for a few hours (Patterson et al., 1996) or by virus-mediated BDNF gene transfer (Korte et al., 1996), rescue the deficits, suggesting that LTP impairment was due to mutation of the BDNF gene per se, rather than cumulative developmental abnormalities.

The role of BDNF in long-term depression (LTD) and LTP in layer II/III of the visual cortex has been examined in a number of laboratories. Slices treated with BDNF show little difference from control when the maximal level of LTP is induced by TBS, but exhibit significantly greater synaptic potentiation in response to a weak (20 Hz) tetanus (Huber et al., 1998). TBS-induced LTP is completely prevented after application of TrkB-IgG or K252a, a specific inhibitor for Trk receptor tyrosine kinases (Akaneya et al., 1997). LTD, induced by a train of low-frequency stimulation (1 Hz, 15 min), was inhibited by BDNF treatment (Huber et al., 1998; Kinoshita et al., 1999). Moreover, inhibition of BDNF activity, either by the Trk tyrosine kinase inhibitor K252a or by an anti-BDNF antibody, increases the magnitude of LTD (Akaneya et al., 1996; Kinoshita et al., 1999). These results suggest that BDNF acutely facilitates LTP and attenuates LTD in the visual cortex.

In marked contrast, long-term over-expression of BDNF in the visual cortex during development results in a very different effect on LTP. The maturation of the visual cortex occurs earlier (Huang et al., 1999) when the postnatal rise of BDNF levels in the cortex are genetically accelerated in certain transgenic mice. In these mice, over-expression of BDNF promotes the maturation of GABAergic inhibition and accelerates the age-dependent decline of cortical LTP, leading to an earlier termination of the critical period of ocular dominance plasticity in postnatal life. Thus, it is important to distinguish between acute and long-term modes of BDNF action, which may have dramatically different functional consequences.

Role of BDNF in a late phase of LTP and learning and memory

LTP can be divided into an early phase (E-LTP) that is short lasting (1–3 h) and independent of new protein synthesis, and a later phase (L-LTP) that requires activation of cAMP and new protein synthesis (Frey et al., 1988; Nguyen et al., 1994). Several lines of evidence suggest that BDNF may also be involved in hippocampal L-LTP. Tetanic stimulation used to induce L-LTP rapidly and selectively increases BDNF mRNA levels in the hippocampus, with little or no effects on other neurotrophins (Patterson et al., 1992; Castren et al., 1993; Dragunow et al., 1993; Kesslak et al., 1998; Morimoto et al., 1998). The delayed and sustained increase in BDNF transcription correlates well with the time course of L-LTP. TrkB receptor expression is also increased by LTP-inducing stimuli (Bramham et al., 1996; Dragunow et al., 1997). The transcription of BDNF appears to be mediated, in part, by CREB (calcium/cAMP response element binding protein), a cAMP-induced transcription factor required for L-LTP (Shieh et al., 1998; Tao et al., 1998). In the BDNF mutant mice, L-LTP can never be induced, although some animals exhibit E-LTP with reduced magnitudes (Korte et al., 1998). Moreover, application of the BDNF scavenger protein, TrkB-IgG, 30–70 min after induction of LTP reverses the already established E-LTP, and prevents the occurrence of L-LTP (Kang et al., 1997; Korte et al., 1998). Conceivably, tetanic stimulation could induce a Ca^{2+}- and CREB-mediated transcription and translation of BDNF, which in turn elicits structural and functional changes that underlie L-LTP in the hippocampal synapses. Taken together, these results suggest that enhanced BDNF production and secretion may contribute to the maintenance of L-LTP, which is dependent on both transcription and protein synthesis.

The finding that BDNF modulates hippocampal LTP implies a role in learning and memory. A reduction of BDNF levels by antisense oligonucleotides injected into adult hippocampus reduced the magnitude of LTP (Ma et al., 1998). Injection of the antisense before and during memory consolidation markedly impairs memory retention performance, while injection 6 h after training has no effect. Delivery of BDNF antibodies to the adult rat brain also impairs learning and memory (Mu et al., 1999). BDNF heterozygous knockout mice exhibit deficits in spatial learning, as reflected by poor performance in the Morris water maze test (Linnarsson et al., 1997; but see Montkowski and Holsboer, 1997).

Young adult +/− mice require twice the amount of training to reach normal performance ability, and aged +/− mice do not learn at all (Linnarsson et al., 1997). In a recent study, a conditionally knockout line was generated in which the deletion of the TrkB gene is restricted to the forebrain and hippocampus, and occurs only during postnatal development (Minichiello et al., 1999). The adult −/− mice, but not +/−, exhibit impairments in complex learning behavior when tested with the Morris water maze and eight-arm radial maze. However, simple passive avoidance learning is normal in both −/− and +/− mice. LTP at CA1 hippocampal synapses is substantially impaired in the −/− mice, but only partially reduced in the +/− mice. These results suggest that CA1 LTP may need to be reduced below a certain threshold before behavioral defects become apparent. Paradoxically, infusion of BDNF into the hippocampus of normal animals results in no change in the impairment in learning or memory (Fischer et al., 1994; Pelleymounter et al., 1996). Transgenic mice overexpressing BDNF by a β-actin promotor exhibit passive avoidance deficits along with difficulties in inducing LTP in the CA1 region of the hippocampus (Croll et al., 1999). Thus, although BDNF may modulate LTP and learning and memory in normal animals, too much BDNF may have adverse effects on these processes.

Effects of BDNF on basal synaptic transmission

An important and unresolved issue is whether neurotrophins acutely enhance basal, excitatory synaptic transmission. Several earlier studies using primary cultures of hippocampal neurons demonstrated that acute application of BDNF elicits a rapid enhancement of synaptic transmission and transmitter release (Knipper et al., 1994; Lessmann et al., 1994; Levine et al., 1995; Takei et al., 1997; Li et al., 1998a). More careful analysis, however, has revealed some complex effects of BDNF on cultured hippocampal neurons. Among the glutamatergic synapses, 30% are potentiated while 10% are inhibited by BDNF, and the remaining 60% show no response to the neurotrophin (Lessmann and Heumann, 1998). The intrinsic properties of the presynaptic neurons may determine whether and how they can respond to BDNF. BDNF preferentially enhances transmission in synapses that have a higher degree of paired-pulse facilitation (PPF) (Lessmann and Heumann, 1998). Moreover, the degree of BDNF-induced potentiation strongly correlates with the initial coefficient of variation (CV) of the amplitude of excitatory postsynaptic currents (EPSCs) (Berninger et al., 1999). Thus, BDNF preferentially potentiates relatively weak connections, but is less effective on stronger synapses. In cultured cortical neurons, NT-3 also potentiates neuronal excitability and synchronizes excitatory synaptic activity, though possibly indirectly by inhibiting GABAergic transmission (Kim et al., 1994).

It remains controversial whether BDNF enhances basal synaptic transmission at CA1 synapses in acute hippocampal slices. One group has shown that BDNF or NT-3 enhances basal synaptic transmission in the CA1 excitatory synapses within a time window of 5–10 min (Kang and Schuman, 1995, 1996; Kang et al., 1996). In contrast, other groups have found that BDNF does not affect basal excitatory transmission (Figurov et al., 1996; Patterson et al., 1996; Tanaka et al., 1997; Frerking et al., 1998; Gottschalk et al., 1998). In an effort to resolve this controversy, Kang et al. (1996) used different rates of perfusion to load BDNF into the recording chamber and found that BDNF elicits synaptic potentiation only when it is delivered to the slices with a high, but not low, perfusion rate. Because BDNF readily adheres to certain surfaces due to its quaternary structure, it is conceivable that slow perfusion rates might not elicit the potentiating effect of BDNF seen by Kang et al. (1996) within a short period of time. However, when using exactly the same experimental conditions, Frerking et al. (1998) reported that BDNF has no effect on basal, excitatory postsynaptic currents (EPSCs) for a long period of time, and elicits a small decrease in inhibitory postsynaptic currents (IPSCs). In other areas of the hippocampus, such as CA3 and dentate gyrus, it has been demonstrated that BDNF is able to potentiate excitatory synaptic transmission and neuronal excitability, but in a much slower time course (30–90 min) (Scharfman, 1997; Messaoudi et al., 1998; Scharfman et al., 1999). Thus, BDNF may modulate neuronal excitability by inhibiting GABAergic transmission, but whether or not it enhances basal, excitatory synaptic transmission in CA1 synapses is questionable.

Presynaptic modulation

A number of studies have demonstrated a presynaptic mechanism for BDNF-induced synaptic potentiation in cultured hippocampal neurons. The degree of BDNF-induced enhancement of excitatory synaptic transmission strongly correlates with the initial CV of EPSC amplitude, and is inversely correlated with the initial PPF and the frequency of miniature EPSCs (mEPSCs) (Lessmann and Heumann, 1998; Berninger et al., 1999). Biochemical experiments indicate that glutamate release is increased after application of BDNF to the cultured neurons, possibly through a non-exocytotic pathway (Takei et al., 1997, 1998). Targeting of a C-terminal truncated dominant negative TrkB receptor into presynaptic, but not postsynaptic neurons, prevents the BDNF effects on both evoked EPSCs and mEPSCs (Li et al., 1998a), while the amplitudes of glutamate-induced postsynaptic currents or mEPSCs are not affected (Li et al., 1998b). These experiments, however, did not address the relationship between the presynaptic effects and the modulation of LTP elicited by BDNF, nor did they completely rule out postsynaptic effects of BDNF. Indeed, potent modulation of postsynaptic glutamate receptors has been reported by a number of groups (see below).

An intrinsic problem of the studies using hippocampal cultures is that specific synaptic networks, such as the CA3–CA1 connections, are not preserved. Recent work using immuno-electron microscopy indicated that the TrkB proteins are localized both in the presynaptic terminals and postsynaptic dendrites and spines at the CA1 synapses. How does BDNF modulate the CA1 synapses in vivo? Experiments using hippocampal slices have elucidated a clear presynaptic action of BDNF in CA1 synapses, and have provided a potential mechanism for BDNF modulation of LTP (Gottschalk et al., 1998). First, BDNF only enhances synaptic transmission at CA1 synapses during high-frequency stimulation (HFS, >50 Hz), when synaptic fatigue is obvious, but has no effect on low-frequency (<20 Hz) transmission. Since HFS-induced fatigue is a known presynaptic phenomenon (Zucker, 1989; Larkman et al., 1991; Dobrunz and Stevens, 1997), BDNF must act presynaptically to attenuate the fatigue. Second, treatment with BDNF alters PPF, a simple and reliable measure of presynaptic properties with very few assumptions (Foster and McNaughton, 1991; Schultz et al., 1994; Dobrunz et al., 1997). This effect of BDNF is restricted to PPF with interpulse intervals shorter than 20 ms, suggesting again that the BDNF effect is frequency-dependent. Moreover, lowering the extracellular calcium concentration ($[Ca^{2+}]o$) is known to increase PPF and the synaptic responses to HFS. BDNF mimicked the effect of lowering $[Ca^{2+}]o$ in PPF and synaptic responses to HFS. Finally, the attenuation of synaptic fatigue by BDNF is not due to a reduction of the desensitization of postsynaptic non-NMDA receptors. It is conceivable that a better response to high-frequency tetanic stimulation elicited by BDNF will facilitate tetanus-induced LTP.

Modulation of high-frequency transmission and LTP

Abundant evidence suggests that BDNF facilitates LTP at hippocampal CA1 synapses by enhancing synaptic responses to high-frequency, tetanic stimulation. Treatment of neonatal slices with BDNF increases synaptic responses to HFS (Figurov et al., 1996; Gottschalk et al., 1998), while inhibition of endogenous BDNF activity by TrkB-IgG elicits a pronounced fatigue (Figurov et al., 1996). Similar modulation has been observed in the visual cortex (Huber et al., 1998). Studies using genetically modified mice further support the notion that BDNF acts presynaptically to modulate high-frequency synaptic transmission and LTP. Significant impairment of LTP correlates well with substantial synaptic fatigue in both $+/-$ and $-/-$ BDNF knockout mice (Pozzo-Miller et al., 1999), as well as in a trkB-loxP line in which the expression of TrkB is reduced by 70% (trkB hypomorphs, Xu et al., 2000). To exclude the postsynaptic involvement of the BDNF modulation, a trkB conditional knockout line (CA1-KO) was generated by crossing the trkB-loxP line with a transgenic line in which the expression of Cre recombinase is under the control of the CaMKII promotor (Xu et al., 2000). In this line of mice, the trkB gene is specifically deleted in CA1, but not in CA3 neurons. The CA1-KO exhibits the same responses to HFS and the same magnitude of LTP as the trkB hypomorphs, suggesting that the removal of TrkB in the postsynaptic CA1 pyramidal neurons does not affect the BDNF modulation at the CA1 synapses. Properties of postsynaptic AMPA and NMDA receptors are not affected by the TrkB reduction, indicating that BDNF does not modulate plasticity through TrkB in the postsynaptic CA1 neurons. Moreover, normal LTP is generated in the trkB hypomorphs by a depolarization–low-frequency stimulation pairing that puts minimal demands upon presynaptic terminal function. Finally, in the CA1-KO in which TrkB is exclusively eliminated in the CA1 neurons, application of TrkB-IgG further reduces the magnitude of LTP. Taken together, these results strongly suggest that modulation of LTP by BDNF is achieved by enhancing high-frequency transmission through activation of presynaptic, but not postsynaptic, TrkB receptors.

Rapid advances have also been made in understanding the signaling mechanisms underlying the synaptic actions of BDNF. Based on studies that were primarily using PC12 cells, three major signaling pathways for neurotrophins have been identified: the mitogen associated protein kinase (MAPK), phospholipase C-γ (PLC-γ), and phosphatidylinositol 3-kinase (PI3K) pathways (Kaplan and Stephens, 1994; Segal and Greenberg, 1996). We have examined the signaling mechanisms for BDNF-induced modulation of high-frequency transmission at the CA1 synapses in hippocampal slices (Gottschalk et al., 1999). Application of BDNF rapidly activated MAPK and PI3K, but not PLC-γ. Inhibition of MAPK and PI3K, but not PLC-γ, prevented the BDNF modulation of synaptic responses to HFS. NT-3 did not activate MAPK or PI3K, and had no effect on synaptic fatigue in the neonatal hippocampus. Neither forskolin, which activated MAPK but not PI3K, nor ciliary neurotrophic factor (CNTF), which activates PI3K but not MAPK, affected HFS-induced synaptic fatigue. Treatment of the slices with forskolin together with CNTF still had no effect on synaptic fatigue. Thus, activation of MAPK and PI3K is required but the two together are not sufficient to mediate the BDNF modulation of high-frequency transmission. Further investigation is needed to establish the signaling pathways that mediate BDNF modulation of LTP.

Modulation of synaptic vesicle docking and vesicle protein redistribution

The cellular and molecular mechanisms for BDNF modulation of high-frequency transmission are further analyzed using the BDNF knockout mice (Pozzo-Miller et al., 1999). Electron microscopic studies reveal a significant reduction in the number of vesicles docked at presynaptic active zones, but not the total vesicle number, active zone length, or terminal area, in the BDNF mutant mice. Quantitative analysis indicates that there are approximately 10.3 docked vesicles per active zone in $+/+$ CA1 synapses, but only 3–5 docked vesicles in the $+/-$ and $-/-$ synapses (Schikorski and Stevens, 1997; Pozzo-Miller et al., 1999). A selective reduction in the number of docked vesicles explains why the BDNF knockout mice exhibit more pronounced fatigue with no change in basal synaptic transmission. Moreover, hippocampal synaptosomes prepared from the mutant mice exhibit a marked decrease in the levels of synaptophysin as well as synaptobrevin (VAMP-2), a protein known to be involved in vesicle docking and fusion (Pozzo-Miller et al., 1999). Other synaptic proteins, including synaptotagmin, syntaxin-1 and SNAP-25, are unaffected. Treatment of the mutant slices with BDNF for a few hours reverses the electrophysiological and biochemical deficits at the hippocampal synapses. Thus, the synaptic defects in these mice do not simply reflect developmental consequences of BDNF gene knockout, but rather reflect an acute requirement for BDNF in high-frequency transmission. A BDNF-induced increase in synaptobrevin at nerve terminals may facilitate vesicle docking. Synapses with more docked vesicles in the presynaptic active zone will undoubtedly respond better to high-frequency, tetanic stimulation. An enhancement of synaptic responses to tetanus may contribute, at least in part, to BDNF modulation of LTP.

It is interesting to note that the reduction of synaptobrevin and synaptophysin in the BDNF mutant mice only occurs in particulate fractions from synaptosomes, but not in the SDS homogenates from the whole hippocampus (Pozzo-Miller et al., 1999). Moreover, BDNF still enhances high-frequency transmission (Gottschalk et al., 1999) and increases the levels of synaptophysin and synaptobrevin (N. Tartaglia and B. Lu, unpubl. observations) in the neonatal slices even in the presence of the protein synthesis inhibitor, anisomycin. These results suggest that the BDNF-induced increase in vesicle docking is probably mediated by redistribution of the synaptic proteins within the presynaptic terminals, rather than an increase in the absolute levels of these proteins. One possibility is that BDNF promotes the translocation of the synaptobrevin-containing vesicles from axons to the terminals. Alternatively, BDNF may facilitate the incorporation of synaptobrevin and synaptophysin into the vesicles. In addition, BDNF may also facilitate the formation of the synaptophysin–synaptobrevin complex, which appears to be associated with more mature synapses (Becher et al., 1999). Further investigations are required to elucidate the exact molecular mechanisms responsible for the BDNF effects on vesicle protein redistribution.

Postsynaptic modulatory effect

Several studies demonstrate that BDNF also modulates glutamate receptors postsynaptically. In cultured hippocampal neurons, rapid enhancement of synaptic transmission by BDNF has been attributed, at least in part, to an increase in postsynaptic modulation of glutamate receptors through activation of the TrkB receptor (Levine et al., 1995, 1996). BDNF increases the response to iontophoretically applied N-methyl-D-aspartate (NMDA), but not α-amino-3-hydroxy-5-methyl-4-isoxazole-4-propionic acid (AMPA) or acetylcholine, suggesting a selective enhancement of NMDA receptors (Levine et al., 1998; Song et al., 1998). Single-channel recordings demonstrate that BDNF increases the probability of opening of the NMDA channels in hippocampal neurons (Jarvis et al., 1997; Levine et al., 1998). Moreover, TrkB is localized to the postsynaptic density (Wu et al., 1996), and activation of the TrkB receptor leads to phosphorylation of the NMDA receptor subunits I and 2B (Lin et al., 1998; Suen et al., 1998).

While acute application of BDNF may enhance the activity of NMDA-type glutamate receptors in the hippocampus, long-term BDNF treatment of cultured cortical neurons has been shown to regulate synaptic currents mediated by AMPA-type glutamate receptors (Rutherford et al., 1998). Blockade of activity-dependent expression/secretion of BDNF by tetrodotoxin (TTX) for 2 days results in a marked increase in the amplitude of AMPA currents recorded from pyramidal–pyramidal cell synapses. BDNF prevents while TrkB-IgG mimics the TTX effect. In contrast, BDNF increases the quantal amplitude of pyramidal cell–interneuron synapses (Rutherford et al., 1998). Thus, BDNF differentially regulates the quantal amplitude of the AMPA-mediated synapses, depending on whether the postsynaptic cell is a pyramidal neuron or an interneuron.

Future perspectives

While substantial evidence supports the role of BDNF in hippocampal synaptic plasticity, a number of important issues remain to be elucidated. An interesting area that deserves more thorough investigation is the role of BDNF in the late, protein synthesis-dependent phase of LTP. Moreover, careful studies should be carried out to elucidate the underlying mechanisms as well as the physiological significance of the neurotrophic regulation of synaptic plasticity in vivo. For example, how does BDNF, or neurotrophins in general, achieve synapse-specific modulation? What is the relationship between neurotrophins and neuronal activity in various forms of synaptic plasticity? Further, the exact role of NT-3 and its relationship with BDNF and other neurotrophins in the development and plasticity of hippocampal neurons remains to be established. NT-3 and BDNF (or other neurotrophins) may act in concert or against each other to affect synaptic plasticity in the hippocampus. Activity-dependent regulation of neurotrophin secretion and neurotrophic responsiveness represent exciting areas for future research.

References

Akaneya, Y., Tsumoto, T. and Hatanaka, H. (1996) Brain-derived neurotrophic factor blocks long-term depression in rat visual cortex. *J. Neurophysiol.*, 76: 4198–4201.

Akaneya, Y., Tsumoto, T., Kinoshita, S. and Hatanaka, H. (1997) Brain-derived neurotrophic factor enhances long-term potentiation in rat visual cortex. *J. Neurosci.*, 17: 6707–6716.

Barbacid, M. (1993) Nerve growth factor: a tale of two receptors. *Oncogene*, 8: 2033–2042.

Bates, B., Rios, M., Trumpp, A., Chen, C., Fan, G., Bishop, J.M.

and Jaenisch, R. (1999) Neurotrophin-3 is required for proper cerebellar development. *Nat. Neurosci.*, 2: 115–117.

Becher, A., Drenckhahn, A., Pahner, I., Margittai, M., Jahn, R. and Ahnert-Hilger, G. (1999) The synaptophysin–synaptobrevin complex: a hallmark of synaptic vesicle maturation. *J. Neurosci.*, 19: 1922–1931.

Berninger, B. and Poo, M.-m. (1996) Fast actions of neurotrophic factors. *Curr. Opin. Neurobiol.*, 6: 324–330.

Berninger, B., Schinder, A.F. and Poo, M.-m. (1999) Synaptic reliability correlates with reduced susceptibility to synaptic potentiation by brain-derived neurotrophic factor. *Learn. Mem.*, 6: 232–242.

Bonhoeffer, T. (1996) Neurotrophins and activity-dependent development of the neocortex. *Curr. Opin. Neurobiol.*, 6: 119–126.

Bramham, C.R., Southard, T., Sarvey, J., Herkenham, M. and Brady, L.S. (1996) Unilateral LTP triggers bilateral increases in hippocampal neurotrophin and trk receptor mRNA expression in behaving rats: evidence for interhemispheric communication. *J. Comp. Neurol.*, 368: 371–382.

Cabelli, R.J., Horn, A. and Shatz, C.J. (1995) Inhibition of ocular dominance column formation by infusion of NT-4/5 or BDNF. *Science*, 267: 1662–1666.

Cabelli, R.J., Shelton, D.L., Segal, R.A. and Shatz, C.J. (1997) Blockade of endogenous ligands of trkB inhibits formation of ocular dominance columns. *Neuron*, 19: 63–76.

Castren, E., Pitkanen, M., Sirvio, J., Parsadanian, A., Lindholm, D., Thoenen, H. and Riekkinen, P.J. (1993) The induction of LTP increases BDNF and NGF mRNA but decreases NT-3 mRNA in the dentate gyrus. *NeuroReport*, 4: 895–898.

Chao, M.V. (1992) Neurotrophin receptors: a window into neuronal differentiation. *Neuron*, 9: 583–593.

Chen, G., Kolbeck, R., Barde, Y.A., Bonhoeffer, T. and Kossel, A. (1999) Relative contribution of endogenous neurotrophins in hippocampal long-term potentiation. *J. Neurosci.*, 19: 7983–7990.

Cohen-Cory, S. (1999) BDNF modulates, but does not mediate, activity-dependent branching and remodeling of optic axon arbors in vivo. *J. Neurosci.*, 19: 9996–10003.

Cohen-Cory, S. and Fraser, S.E. (1995) Effects of brain-derived neurotrophic factor on optic axon branching and remodelling in vivo. *Nature*, 378: 192–196.

Croll, S.D., Suri, C., Compton, D.L., Simmons, M.V., Yancopoulos, G.D., Lindsay, R.M., Wiegand, S.J., Rudge, J.S. and Scharfman, H.E. (1999) Brain-derived neurotrophic factor transgenic mice exhibit passive avoidance deficits, increased seizure severity and in vitro hyperexcitability in the hippocampus and entorhinal cortex. *Neuroscience*, 93: 1491–1506.

Dobrunz, L.E. and Stevens, C.F. (1997) Heterogeneity of release probability, facilitation and depletion at central synapses. *Neuron*, 18: 995–1018.

Dobrunz, L.E., Huang, E.P. and Stevens, C.F. (1997) Very short-term plasticity in hippocampal synapses. *Proc. Natl. Acad. Sci. USA*, 94: 14843–14847.

Domenici, L., Berardi, N., Carmignoto, G., Vantini, G. and Maffei, L. (1991) Nerve growth factor prevents the amblyopic effects of monocular deprivation. *Proc. Natl. Acad. Sci. USA*, 88: 8811–8815.

Dragunow, M., Beilharz, E., Mason, B., Lawlor, P., Abraham, W. and Gluckman, P. (1993) Brain-derived neurotrophic factor expression after long-term potentiation. *Neurosci. Lett.*, 160: 232–236.

Dragunow, M., Hughes, P., Mason-Parker, S.E., Lawlor, P. and Abraham, W.C. (1997) TrkB expression in dentate granule cells is associated with a late phase of long-term potentiation. *Mol. Brain Res.*, 46: 274–280.

Dugich-Djordjevic, M.M., Ohsawa, F. and Hefti, F. (1993) Transient elevation in catalytic trkB mRNA during postnatal development of the rat brain. *NeuroReport*, 4: 1091–1094.

Ernfors, P., Lee, K.-F. and Jaenisch, R. (1994) Mice lacking brain-derived neurotrophic factor develop with sensory deficits. *Nature*, 368: 147–150.

Figurov, A., Pozzo-Miller, L., Olafsson, P., Wang, T. and Lu, B. (1996) Regulation of synaptic responses to high-frequency stimulation and LTP by neurotrophins in the hippocampus. *Nature*, 381: 706–709.

Fischer, W., Sirevaag, A., Wiegand, S.J., Lindsay, R.M. and Björklund, A. (1994) Reversal of spatial memory impairments in aged rats by nerve growth factor and neurotrophins 3 and 4/5 but not by brain-derived neurotrophic factor. *Proc. Natl. Acad. Sci. USA*, 91: 8607–8611.

Foster, T.C. and McNaughton, B.L. (1991) Long-term synaptic enhancement in CA1 is due to increased quantal size, not quantal content. *Hippocampus*, 1: 79–91.

Frerking, M., Malenka, R.C. and Nicoll, R.A. (1998) Brain-derived neurotrophic factor (BDNF) modulates inhibitory, but not excitatory, transmission in the CA1 region of the hippocampus. *J. Neurophysiol.*, 80: 3383–3386.

Frey, U., Krug, M., Reymann, K.G. and Matthies, H. (1988) Anisomycin, an inhibitor of protein synthesis, blocks late phases of LTP phenomena in the hippocampal CA1 region in vitro. *Brain Res.*, 452: 57–65.

Friedman, W.J., Olson, L. and Persson, H. (1991) Cells that express brain-derived neurotrophic factor mRNA in the developing postnatal rat brain. *Eur. J. Neurosci.*, 3: 688–697.

Gottschalk, W., Pozzo-Miller, L.D., Figurov, A. and Lu, B. (1998) Presynaptic modulation of synaptic transmission and plasticity by brain-derived neurotrophic factor in the developing hippocampus. *J. Neurosci.*, 18: 6830–6839.

Gottschalk, W.A., Jiang, H., Tartaglia, N., Feng, L., Figurov, A. and Lu, B. (1999) Signaling mechanisms mediating BDNF modulation of synaptic plasticity in the hippocampus. *Learn. Mem.*, 6: 243–256.

Huang, Z.J., Kirkwood, A., Pizzorusso, T., Porciatti, V., Morales, B., Bear, M.F., Maffei, L. and Tonegawa, S. (1999) BDNF regulates the maturation of inhibition and the critical period of plasticity in mouse visual cortex. *Cell*, 98(6): 739–755.

Huber, K.M., Sawtell, N.B. and Bear, M.F. (1998) Brain-derived neurotrophic factor alters the synaptic modification threshold in visual cortex. *Neuropharmacology*, 37: 571–579.

Jarvis, C.R., Xiong, Z.G., Plant, J.R., Churchill, D., Lu, W.Y., MacVicar, B.A. and MacDonald, J.F. (1997) Neurotrophin modulation of NMDA receptors in cultured murine and isolated rat neurons. *J. Neurophysiol.*, 78: 2363–2371.

Kang, H. and Schuman, E.M. (1995) Long-lasting neurotrophin-induced enhancement of synaptic transmission in the adult hippocampus. *Science*, 267: 1658–1662.

Kang, H. and Schuman, E.M. (1996) A requirement for local protein synthesis in neurotrophin-induced hippocampal synaptic plasticity. *Science*, 273: 1402–1406.

Kang, H., Jia, L., Suh, K., Tang, L. and Schuman, E. (1996) Determinants of BDNF-induced hippocampal synaptic plasticity: role of the TrkB receptor and the kinetics of neurotrophin delivery. *Learn. Mem.*, 3: 188–196.

Kang, H., Welcher, A.A., Shelton, D. and Schuman, E.M. (1997) Neurotrophins and time: different roles for TrkB signaling in hippocampal long-term potentiation. *Neuron*, 19: 653–664.

Kaplan, D.R. and Stephens, R.M. (1994) Neurotrophin signal transduction by the Trk receptor. *J. Neurobiol.*, 25: 1404–1417.

Kesslak, J.P., So, V., Choi, J., Cotman, C.W. and Gomez-Pinilla, F. (1998) Learning upregulates brain-derived neurotrophic factor messenger ribonucleic acid: a mechanism to facilitate encoding and circuit maintenance?. *Behav. Neurosci.*, 112: 1012–1019.

Kim, H.G., Wang, T., Olafsson, P. and Lu, B. (1994) Neurotrophin 3 potentiates neuronal activity and inhibits γ-aminobutyrateric synaptic transmission in cortical neurons. *Proc. Natl. Acad. Sci. USA*, 91: 12341–12345.

Kinoshita, S., Yasuda, H., Taniguchi, N., Katoh-Semba, R., Hatanaka, H. and Tsumoto, T. (1999) Brain-derived neurotrophic factor prevents low-frequency inputs from inducing long-term depression in the developing visual cortex. *J. Neurosci.*, 19: 2122–2130.

Knipper, M., Da Penha Berzaghi, M., Blochl, A., Breer, H., Thoenen, H. and Lindholm, D. (1994) Positive feedback between acetylcholine and the neurotrophins nerve growth factor and brain-derived neurotrophic factor in the rat hippocampus. *Eur. J. Neurosci.*, 6: 668–671.

Kokaia, M., Asztely, F., Olofsdotter, K., Sindreu, C.B., Kullmann, D.M. and Lindvall, O. (1998) Endogenous neurotrophin-3 regulates short-term plasticity at lateral perforant path–granule cell synapses. *J. Neurosci.*, 18: 8730–8739.

Korte, M., Carroll, P., Wolf, E., Brem, G., Thoenen, H. and Bonhoeffer, T. (1995) Hippocampal long-term potentiation is impaired in mice lacking brain-derived neurotrophic factor. *Proc. Natl. Acad. Sci. USA*, 92: 8856–8860.

Korte, M., Griesbeck, O., Gravel, C., Carroll, P., Staiger, V., Thoenen, H. and Bonhoeffer, T. (1996) Virus-mediated gene transfer into hippocampal CA1 region restores long-term potentiation in brain-derived neurotrophic factor mutant mice. *Proc. Natl. Acad. Sci. USA*, 93: 12547–12552.

Korte, M., Kang, H., Bonhoeffer, T. and Schuman, E. (1998) A role for BDNF in the late-phase of hippocampal long-term potentiation. *Neuropharmacology*, 37: 553–559.

Lamballe, F., Smeyne, R.J. and Barbacid, M. (1994) Developmental expression of trkC, the neurotrophin-3 receptors, in the mammalian nervous system. *J. Neurosci.*, 14: 14–28.

Larkman, A., Stratford, K. and Jack, J. (1991) Quantal analysis of excitatory synaptic action and depression in hippocampal slices. *Nature*, 350: 344–347.

Lessmann, V. and Heumann, R. (1998) Modulation of unitary glutamatergic synapses by neurotrophin-4/5 or brain-derived neurotrophic factor in hippocampal microcultures: presynaptic enhancement depends on pre-established paired-pulse facilitation. *Neuroscience*, 86: 399–413.

Lessmann, V., Gottmann, K. and Heumann, R. (1994) BDNF and NT-4/5 enhance glutamatergic synaptic transmission in cultured hippocampal neurons. *NeuroReport*, 6: 21–25.

Levine, E.S., Dreyfus, C.F., Black, I.B. and Plummer, M.R. (1995) Brain-derived neurotrophic factor rapidly enhances synaptic transmission in hippocampal neurons via postsynaptic tyrosine kinase receptors. *Proc. Natl. Acad. Sci. USA*, 92: 8074–8077.

Levine, E.S., Dreyfus, C.F., Black, I.B. and Plummer, M.R. (1996) Selective role for trkB neurotrophin receptors in rapid modulation of hippocampal synaptic transmission. *Mol. Brain Res.*, 38: 300–303.

Levine, E.S., Crozier, R.A., Black, I.B. and Plummer, M.R. (1998) Brain-derived neurotrophic factor modulates hippocampal synaptic transmission by increasing N-methyl-D-aspartic acid receptor activity. *Proc. Natl. Acad. Sci. USA*, 95: 10235–10239.

Lewin, G.R. and Barde, Y.-A. (1996) Physiology of the neurotrophins. *Annu. Rev. Neurosci.*, 19: 289–317.

Li, Y.X., Xu, Y., Ju, D., Lester, H.A., Davidson, N. and Schuman, E.M. (1998a) Expression of a dominant negative TrkB receptor, T1, reveals a requirement for presynaptic signaling in BDNF-induced synaptic potentiation in cultured hippocampal neurons. *Proc. Natl. Acad. Sci. USA*, 95: 10884–10889.

Li, Y.X., Zhang, Y., Lester, H.A., Schuman, E.M. and Davidson, N. (1998b) Enhancement of neurotransmitter release induced by brain-derived neurotrophic factor in cultured hippocampal neurons. *J. Neurosci.*, 18: 10231–10240.

Lin, S.Y., Wu, K., Levine, E.S., Mount, H.T., Suen, P.C. and Black, I.B. (1998) BDNF acutely increases tyrosine phosphorylation of the NMDA receptor subunit 2B in cortical and hippocampal postsynaptic densities. *Mol. Brain Res.*, 55: 20–27.

Linnarsson, S., Björklund, A. and Ernfors, P. (1997) Learning deficit in BDNF mutant mice. *Eur. J. Neurosci.*, 9: 2581–2587.

Liou, J.C. and Fu, W.M. (1997) Regulation of quantal secretion from developing motoneurons by postsynaptic activity-dependent release of NT-3. *J. Neurosci.*, 17: 2459–2468.

Liou, J.C., Yang, R.S. and Fu, W.M. (1997) Regulation of quantal secretion by neurotrophic factors at developing motoneurons in *Xenopus* cell cultures. *J. Physiol. (Lond.)*, 503: 129–139.

Lo, D.C. (1995) Neurotrophic factors and synaptic plasticity. *Neuron*, 15: 979–981.

Lohof, A.M., Ip, N.Y. and Poo, M.-m. (1993) Potentiation of developing neuromuscular synapses by the neurotrophins NT-3 and BDNF. *Nature*, 363: 350–353.

Lom, B. and Cohen-Cory, S. (1999) Brain-derived neurotrophic factor differentially regulates retinal ganglion cell dendritic and axonal arborization in vivo. *J. Neurosci.*, 19: 9928–9938.

Lu, B. and Figurov, A. (1997) Role of neurotrophins in synapse development and plasticity. *Rev. Neurosci.*, 8: 1–12.

Ma, L., Reis, G., Parada, L.F. and Schuman, E.M. (1999) Neuronal NT-3 is not required for synaptic transmission or long-term potentiation in area CA1 of the adult rat hippocampus. *Learn. Mem.*, 6: 267–275.

Ma, Y.L., Wang, H.L., Wu, H.C., Wei, C.L. and Lee, E.H. (1998) Brain-derived neurotrophic factor antisense oligonucleotide impairs memory retention and inhibits long-term potentiation in rats. *Neuroscience*, 82: 957–967.

Maffei, L., Berardi, N., Domenici, L., Parisi, V. and Pizzorusso, T. (1992) Nerve growth factor (NGF) prevents the shift in ocular dominance distribution of visual cortical neurons in monocularly deprived rats. *J. Neurosci.*, 12: 4651–4662.

Maisonpierre, P.C., Belluscio, L., Friedman, B., Alderson, R.F., Wiegand, S.J., Furth, M.E., Lindsay, R.M. and Yancopoulos, G.D. (1990) NT-3, BDNF, and NGF in the developing rat nervous system: parallel as well as reciprocal patterns of expression. *Neuron*, 5: 501–509.

Martinez, A., Alcantara, S., Borrell, V., Del Rio, J.A., Blasi, J., Otal, R., Campos, N., Boronat, A., Barbacid, M., Silos-Santiago, I. and Soriano, E. (1998) TrkB and TrkC signaling are required for maturation and synaptogenesis of hippocampal connections. *J. Neurosci.*, 18: 7336–7350.

McAllister, A.K., Lo, D.C. and Katz, L.C. (1995) Neurotrophins regulate dendritic growth in developing visual cortex. *Neuron*, 15: 791–803.

McAllister, A.K., Katz, L.C. and Lo, D.C. (1996) Neurotrophin regulation of cortical dendritic growth requires activity. *Neuron*, 17: 1057–1064.

McAllister, A.K., Katz, L.C. and Lo, D.C. (1997) Opposing roles for endogenous BDNF and NT-3 in regulating cortical dendritic growth. *Neuron*, 18: 767–778.

McAllister, A.M., Katz, L.C. and Lo, D.C. (1999) Neurotrophins and synaptic plasticity. *Annu. Rev. Neurosci.*, 22: 295–318.

Messaoudi, E., Bardsen, K., Srebro, B. and Bramham, C.R. (1998) Acute intrahippocampal infusion of BDNF induces lasting potentiation of synaptic transmission in the rat dentate gyrus. *J. Neurophysiol.*, 79: 496–499.

Minichiello, L., Korte, M., Wolfer, D., Kuhn, R., Unsicker, K., Cestari, V., Rossi-Arnaud, C., Lipp, H.P., Bonhoeffer, T. and Klein, R. (1999) Essential role for TrkB receptors in hippocampus-mediated learning. *Neuron*, 24: 401–414.

Montkowski, A. and Holsboer, F. (1997) Intact spatial learning and memory in transgenic mice with reduced BDNF. *NeuroReport*, 8: 779–782.

Morimoto, K., Sato, K., Sato, S., Yamada, N. and Hayabara, T. (1998) Time-dependent changes in neurotrophic factor mRNA expression after kindling and long-term potentiation in rats. *Brain Res. Bull.*, 45: 599–605.

Mu, J.S., Li, W.P., Yao, Z.B. and Zhou, X.F. (1999) Deprivation of endogenous brain-derived neurotrophic factor results in impairment of spatial learning and memory in adult rats. *Brain Res.*, 835: 259–265.

Nguyen, P.T., Abel, T. and Kendal, E.R. (1994) Requirement of a critical period of transcription for induction of a later phase of LTP. *Science*, 265: 1104–1107.

Patterson, S., Grover, L.M., Schwartzkroin, P.A. and Bothwell, M. (1992) Neurotrophin expression in rat hippocampal slices: a stimulus paradigm inducing LTP in CA1 evokes increases in BDNF and NT-3 mRNAs. *Neuron*, 9: 1081–1088.

Patterson, S.L., Abel, T., Deuel, T.A., Martin, K.C., Rose, J.C. and Kandel, E.R. (1996) Recombinant BDNF rescues deficits in basal synaptic transmission and hippocampal LTP in BDNF knockout mice. *Neuron*, 16: 1137–1145.

Pelleymounter, M.A., Cullen, M.J., Baker, M.B., Gollub, M. and Wellman, C. (1996) The effects of intrahippocampal BDNF and NGF on spatial learning in aged Long Evans rats. *Mol. Chem. Neuropathol.*, 29: 211–226.

Pozzo-Miller, L., Gottschalk, W.A., Zhang, L., McDermott, K., Du, J., Gopalakrishnan, R., Oho, C., Shen, Z. and Lu, B. (1999) Impairments in high frequency transmission, synaptic vesicle docking and synaptic protein distribution in the hippocampus of BDNF knockout mice. *J. Neurosci.*, 19: 4972–4983.

Riddle, D.R., Lo, D.C. and Katz, L.C. (1995) NT-4-mediated rescue of lateral geniculate neurons from effects of monocular deprivation. *Nature*, 378: 189–191.

Ringstedt, T., Kagercrabtz, H. and Persson, H. (1993) Expression of members of the trk family in the developing postnatal rat brain. *Dev. Brain Res.*, 72: 119–131.

Rutherford, L.C., Nelson, S.B. and Turrigiano, G.G. (1998) BDNF has opposite effects on the quantal amplitude of pyramidal neuron and interneuron excitatory synapses. *Neuron*, 21: 521–530.

Scharfman, H.E. (1997) Hyperexcitability in combined entorhinal/hippocampal slices of adult rat after exposure to brain-derived neurotrophic factor. *J. Neurophysiol.*, 78: 1082–1095.

Scharfman, H.E., Goodman, J.H. and Sollas, A.L. (1999) Actions of brain-derived neurotrophic factor in slices from rats with spontaneous seizures and mossy fiber sprouting in the dentate gyrus. *J. Neurosci.*, 19: 5619–5631.

Schikorski, T. and Stevens, C.F. (1997) Quantitative ultrastructural analysis of hippocampal excitatory synapses. *J. Neurosci.*, 17: 5858–5867.

Schultz, P.E., Cook, E.P. and Johnston, D. (1994) Changes in paired-pulse facilitation suggest presynaptic involvement in long-term potentiation. *J. Neurosci.*, 14: 5325–5337.

Segal, R.A. and Greenberg, M.E. (1996) Intracellular signaling pathways activated by neurotrophic factors. *Annu. Rev. Neurosci.*, 19: 463–489.

Shieh, P.B., Hu, S.C., Bobb, K., Timmusk, T. and Ghosh, A. (1998) Identification of a signaling pathway involved in calcium regulation of BDNF expression. *Neuron*, 20: 727–740.

Song, D.K., Choe, B., Bae, J.H., Park, W.K., Han, I.S., Ho, W.K. and Earm, Y.E. (1998) Brain-derived neurotrophic factor rapidly potentiates synaptic transmission through NMDA, but suppresses it through non-NMDA receptors in rat hippocampal neurons. *Brain Res.*, 799: 176–179.

Stoop, R. and Poo, M.-m. (1995) Potentiation of transmitter release by ciliary neurotrophic factor requires somatic signaling. *Science*, 267: 695–699.

Stoop, R. and Poo, M.-m. (1996) Synaptic modulation by neurotrophic factors: differential and synergistic effects of brain-derived neurotrophic factor and ciliary neurotrophic factor. *J. Neurosci.*, 16: 3256–3264.

Suen, P.C., Wu, K., Xu, J.L., Lin, S.Y., Levine, E.S. and Black, I.B. (1998) NMDA receptor subunits in the postsynaptic density of rat brain: expression and phosphorylation by endogenous protein kinases. *Mol. Brain Res.*, 59: 215–228.

Takei, N., Sasaoka, K., Inoue, K., Takahashi, M., Endo, Y. and Hatanaka, H. (1997) Brain-derived neurotrophic factor increases the stimulation-evoked release of glutamate and the levels of exocytosis-associated proteins in cultured cortical neurons from embryonic rats. *J. Neurochem.*, 68: 370–375.

Takei, N., Numakawa, T., Kozaki, S., Sakai, N., Endo, Y., Takahashi, M. and Hatanaka, H. (1998) Brain-derived neurotrophic factor induces rapid and transient release of glutamate through the non-exocytotic pathway from cortical neurons. *J. Biol. Chem.*, 273: 27620–27624.

Tanaka, T., Saito, H. and Matsuki, N. (1997) Inhibition of GABAa synaptic responses by brain-derived neurotrophic factor (BDNF) in rat hippocampus. *J. Neurosci.*, 17: 2959–2966.

Tao, X., Finkbeiner, S., Arnold, D.B., Shaywitz, A.J. and Greenberg, M.E. (1998) Ca^{2+} influx regulates BDNF transcription by a CREB family transcription factor-dependent mechanism. *Neuron*, 20: 709–726.

Thoenen, H. (1995) Neurotrophins and neuronal plasticity. *Science*, 270: 593–596.

Wang, T., Xie, K.W. and Lu, B. (1995) Neurotrophins promote maturation of developing neuromuscular synapses. *J. Neurosci.*, 15: 4796–4805.

Wang, X., Berninger, B. and Poo, M.-m. (1998) Localized synaptic actions of neurotrophin-4. *J. Neurosci.*, 18: 4985–4992.

Wang, X.H. and Poo, M.-m. (1997) Potentiation of developing synapses by postsynaptic release of neurotrophin-4. *Neuron*, 19: 825–835.

Wu, K., Xu, J.L., Suen, P.C., Levine, E., Huang, Y.Y., Mount, H.T., Lin, S.Y. and Black, I.B. (1996) Functional trkB neurotrophin receptors are intrinsic components of the adult brain postsynaptic density. *Mol. Brain Res.*, 43: 286–290.

Xie, K., Wang, T., Olafsson, P., Mizuno, K. and Lu, B. (1997) Activity-dependent expression of NT-3 in muscle cells in culture: implication in the development of neuromuscular junctions. *J. Neurosci.*, 17: 2947–2958.

Xu, B., Gottschalk, W., Chow, A., Wilson, R., Schnell, E., Zang, K., Wang, D., Nicoll, R., Lu, B. and Reichardt, L.F. (2000) The role of BDNF receptors in mature hippocampus: modulation of long-term potentiation through a presynaptic mechanism involving TrkB. *J. Neurosci.*, in press.

Zucker, R.S. (1989) Short-term synaptic plasticity. *Annu. Rev. Neurosci.*, 12: 13–31.

CHAPTER 20

Neurotrophin-evoked rapid excitation of central neurons

Karl W. Kafitz, Christine R. Rose and Arthur Konnerth *

Institute for Physiology, Technical University of Munich, Biedersteinerstrasse 29, 80802 Munich, Germany

Introduction

Neurotrophins (NTs) represent a family of secreted proteins. These structurally related molecules include nerve growth factor (NGF), brain-derived neurotrophic factor (BDNF), neurotrophin-3 (NT-3) and neurotrophin-4/5 (NT-4/5). NTs bind to plasma membrane-delimited tyrosine kinase receptors (TrkA, B, or C) and/or to the p75-NGF receptor (Altar and DiStefano, 1998; Ibanez, 1998). The binding of NTs to their receptors triggers signaling cascades eventually resulting in altered gene expression and protein synthesis. NTs have usually been considered with regard to their involvement in the growth, survival and proliferation of developing neurons (Bothwell, 1995; Lewin and Barde, 1996; Ibanez, 1998).

While the actions mentioned above require a prolonged application of NTs in the range of hours or days, recent work demonstrated more rapid NT-evoked effects that are initiated within seconds to several minutes following the application of NTs and are probably too fast to be caused by changes in gene expression (Thoenen, 1995; Cellerino and Maffei, 1996; Schuman, 1999). These actions include morphological as well as physiological changes in neuronal properties (McAllister et al., 1999). Application of exogenous NTs leads to an increase in synaptic efficacy at the neuromuscular synapse (Lohof et al., 1993). Moreover, NTs play a pivotal role in the development of ocular dominance columns in the visual cortex (Cabelli et al., 1995; Cellerino and Maffei, 1996). In the rat hippocampus, BDNF leads to a long-lasting increase in postsynaptic field potentials (Kang and Schuman, 1995) and may play a crucial role in the initiation of long-term potentiation (Korte et al., 1996). On the other hand, increased synaptic activity results in an elevated expression of BDNF mRNA (Patterson et al., 1996).

The NT-induced modulation of synaptic function is likely be mediated by at least two different mechanisms. First, it may be due to the activation of presynaptic TrkB receptors which then causes an increase in transmitter release (Lessmann and Heumann, 1998). Alternatively, activation of postsynaptic TrkB receptors could lead to the initiation of intracellular signaling cascades that may mediate these effects (Levine et al., 1998; Li et al., 1998). In any case, it is likely that NT-evoked changes in the postsynaptic calcium concentration play a central role in the modulation of synaptic plasticity (Bonhoeffer, 1996).

Recently, our group could demonstrate a third, ultra-rapid mode of action of NTs on central neurons that happens in the range of milliseconds (Kafitz et al., 1999). It was found that BDNF and NT-4/5 depolarized neurons just as rapidly as the neurotransmitter, glutamate. The very rapid action suggests a direct interaction of the TrkB receptor and an as yet unidentified ion channel.

Transmitter-like excitation by neurotrophins

To study the ultra-rapid actions of NTs on central neurons, somatic whole-cell recordings were per-

---

* Corresponding author. Dr. Arthur Konnerth, Institut für Physiologie, Technische Universität München, Biedersteinerstrasse 29, 80802 Munich, Germany. Fax: +49-(0)89-4140-3377;
E-mail: konnerth@physiol.med.tu-muenchen.de

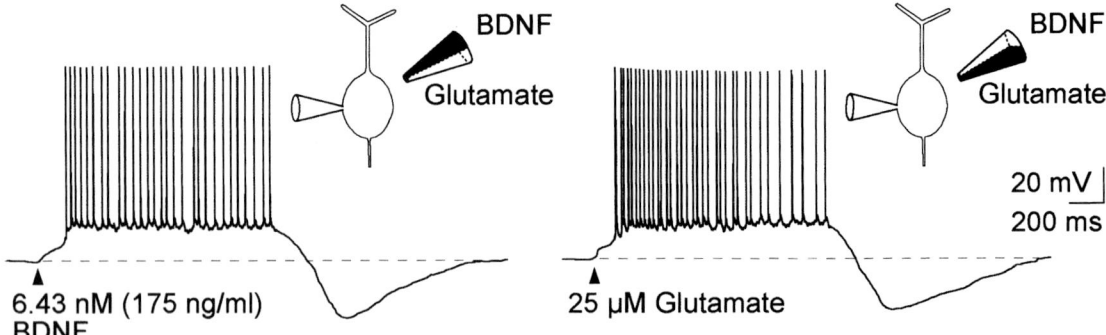

Fig. 1. Action potential firing produced by alternating BDNF (left) and glutamate application (right) in a hippocampal CA1 pyramidal neuron. Substances were focally applied (40 ms pulse duration) to the soma by a double-barrelled stimulation pipette. (This and all subsequent figures from Kafitz et al., 1999.)

formed from neuronal cell bodies of hippocampal, cortical, and cerebellar slices (250 μm) prepared from juvenile rats (postnatal days 9–14). Membrane or holding potentials were generally set to −60 mV. NTs were applied by a picospritzer coupled to glass capillaries that were placed at a distance of approximately 15 μm above the soma.

Fig. 1 shows an experiment in which 6.43 nM BDNF and 25 μM glutamate were alternately pressure-ejected from a double-barrelled pipette placed near the cell body of a CA1 pyramidal neuron in a hippocampal slice. BDNF, at a more than thousand-fold lower concentration than glutamate, had a depolarizing action that was similar to that of glutamate. A brief (40 ms) pulse-like application of BDNF evoked a rapid depolarization resulting in a train of action potentials (Fig. 1, left). When applied at intervals of 5 min or longer, these BDNF-mediated responses were remarkably reproducible for more than 100 min, without any noticeable rundown. At shorter application intervals, most likely due to desensitization, the responses were smaller or absent. Ejection of glutamate from the second barrel produced a virtually identical train of action potentials (Fig. 1, right). The latency of the BDNF-evoked depolarization, which includes the time needed for diffusion and the build-up of the effective concentration at the receptor site, was about 9 ms and similar to that of glutamate.

The lowest concentrations of BDNF that were found to cause depolarization ranged from 0.5 to 2 nM ($\cong$13 to 50 ng/ml) and are thus well within the range found for other BDNF-induced effects requiring activation of TrkB receptors (Ip et al.,

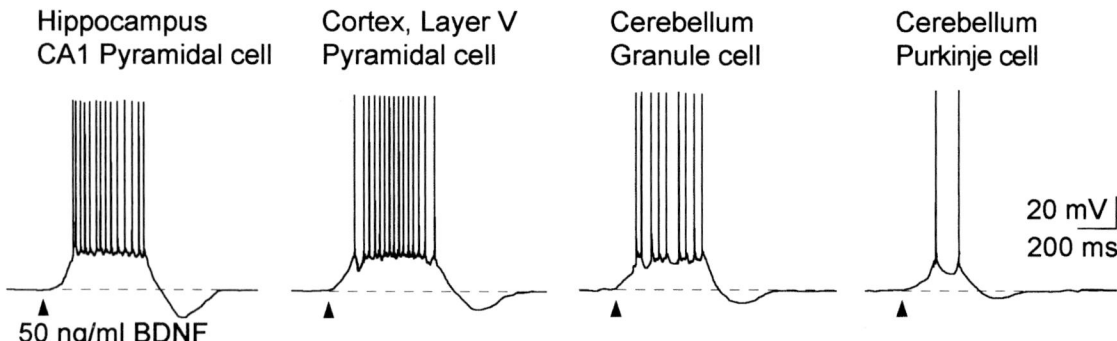

Fig. 2. BDNF evokes excitatory responses in different types of central neurons. Membrane potential was −60 mV; arrowheads indicate the time of the focal application (40 ms pulse duration).

1993; Lessmann et al., 1994, 1995; Figurov et al., 1996; Patterson et al., 1996; Canossa et al., 1997; Suen et al., 1997; Levine et al., 1998). However, it has to be borne in mind that these concentrations refer to those in the application pipettes, and that the concentrations at the site of action, namely at the TrkB receptors themselves, were most likely lower by a factor of 2 to 4 due to local dilution.

To test whether the fast depolarizing action of BDNF was also present in other brain regions, cortical and cerebellar slices were used and a 'standard' BDNF pulse (50 ng/ml, 40 ms duration) was applied to visually identified (Edwards et al., 1989) cortical-layer-V pyramidal cells, to cerebellar granule cells and to Purkinje cells (Fig. 2). While BDNF evoked depolarizing responses and/or action potential firing in every tested cell of each of these regions, there were nevertheless cell-type-specific differences in the magnitude of the response. In hippocampal and cortical pyramidal cells, rather similar BDNF-evoked responses were registered (15–16 action potentials). The action of BDNF appeared to be weaker in the cerebellum. In cerebellar granule cells, our standard stimulation pulse evoked 8–9 action potentials, while Purkinje cells responded with just 1–2 action potentials.

Depolarization requires TrkB receptors

In order to analyze whether the depolarizing action of BDNF resulted from an activation of the tyrosine kinase TrkB receptors, the protein kinase blocker, K-252a, was used. K-252a shows a relatively high preference for the blockade of transmembranous tyrosine kinase receptors of the trk gene family when applied at low concentrations (Knüsel and Hefti, 1992). Fig. 3 illustrates the results from a voltage-clamp experiment in a hippocampal CA1 pyramidal cell in which three subsequent applications of 50 ng/ml BDNF were performed. Application of K-252a for 10–11 min completely blocked BDNF-elicited inward currents. The block developed gradually and incubation with K-252a for less than 10 min resulted in an incomplete block. The BDNF-elicited current response returned to control values 10 min after stopping the K-252a administration. Another tyrosine kinase inhibitor, K-252b (200 nM), which has been reported to be less effective than K-252a (Knüsel and Hefti, 1992), blocked only 20% of the BDNF-evoked current.

To investigate whether other NTs exhibit the same excitatory properties as BDNF, the effect of NT-4/5, NT-3 and NGF on hippocampal CA1 pyramidal cells was compared to that of BDNF. Noteworthy, the most complete information on NT and Trk receptor expression and function is available for the hippocampus (Berninger et al., 1993; Ringstedt et al., 1993; Kang and Schuman, 1995; Patterson et al., 1996; Yan et al., 1997; Levine et al., 1998). To allow a direct comparison of all four NTs, a combination of two double-barreled application pipettes was used. As summarized in Fig. 4, NT-4/5, which, like BDNF, mediates its biological actions via TrkB receptors (Barbacid, 1995; Bothwell, 1995; Lewin and Barde, 1996), elicited inward currents, the amplitudes of which were indistinguishable from those evoked by comparable concentrations of BDNF. As with BDNF, K-252a completely blocked the

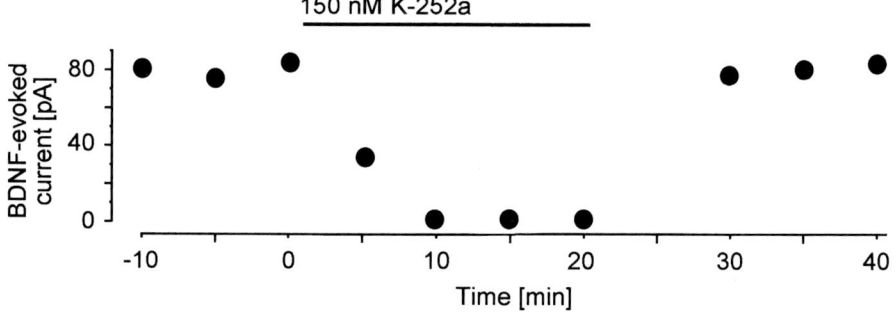

Fig. 3. BDNF (50 ng/ml)-mediated excitation of hippocampal CA1 pyramidal neurons requires TrkB receptors. The diagram shows the amplitudes of 10 consecutive BDNF-evoked currents before, during and after application of K-252a, a blocker of TrkB receptors.

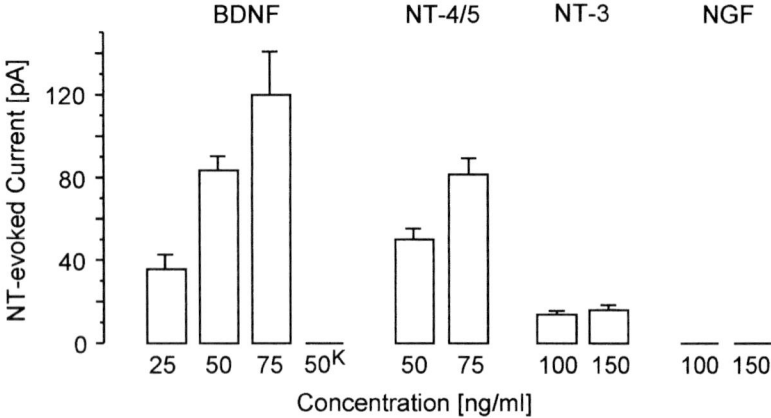

Fig. 4. Bar graph summarizing the current amplitudes (mean ± S.E.M) evoked by BDNF alone ($n = 12$ cells), BDNF in the presence of K-252a (50^K; $n = 8$), NT-4/5 ($n = 10$), NT-3 ($n = 13$), and NGF ($n = 12$) at different concentrations as indicated.

NT-4/5-evoked inward currents (not shown). In contrast, NT-3, which preferentially binds to TrkC receptors and only cross-reacts weakly with TrkB receptors (Barbacid, 1995; Bothwell, 1995; Lewin and Barde, 1996), elicited a much smaller inward current than BDNF or NT-4/5. Even at concentrations of 100–150 ng/ml, NT-3 produced barely detectable responses. Finally, NGF, which virtually exclusively binds to the TrkA receptors (Barbacid, 1995; Bothwell, 1995; Lewin and Barde, 1996) was applied. In agreement with the absence of TrkA receptors in hippocampal neurons (Merlio et al., 1992; Ip et al., 1993; Ringstedt et al., 1993; Kang and Schuman, 1995; Korte et al., 1996; Canossa et al., 1997; Yan et al., 1997), NGF (100, 150 ng/ml) produced no inward current in any of the CA1 pyramidal cells analyzed. Taken together, these results strongly suggest that the depolarizing actions of NTs in the hippocampus are mediated by TrkB receptors. The weak excitatory action of NT-3 may reflect a weak cross-activation of the TrkB receptor.

Ionic mechanisms of the fast neurotrophin action

In view of the BDNF-mediated increase in intracellular Ca^{2+} concentration reported earlier (Berninger et al., 1993; Berninger and Poo, 1996; Canossa et al., 1997), it seemed possible that the inward current was mediated by an influx of Ca^{2+} ions. However, Fig. 5a shows that the BDNF-induced current persisted without any detectable change in a Ca^{2+}-free extracellular solution, demonstrating that Ca^{2+} channels are not involved. Other possible candidates, based on earlier evidence of a NT-dependent regulation of synaptic excitation (Suen et al., 1997; Levine et al., 1998), were ionotropic glutamate receptor channels. However, a combination of receptor antagonists of these glutamate-gated ionotropic channels (10 μM 6-cyano-7-nitroquinoxaline-2,3-dione [CNQX] and 100 μM DL-2-amino-5-phosphovaleric acid [APV]) had no effect on the BDNF-induced inward current (Fig. 5b), indicating that glutamate receptor channels did not contribute to the NT-evoked inward current. It also seems unlikely that the NTs acted indirectly by mediating the release of glutamate (or some other transmitter) from neighboring

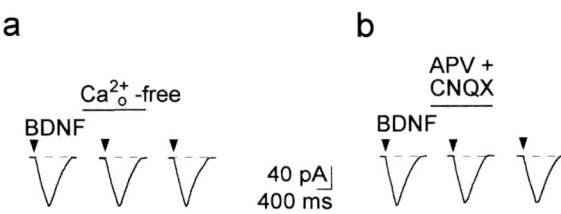

Fig. 5. The influence of the extracellular calcium concentration and glutamate receptor antagonists on the BDNF-evoked responses in CA1 pyramidal neurons. (a) Removal of extracellular Ca^{2+} (Ca_o^{2+}-free) did not alter the BDNF-induced inward current. (b) The BDNF-induced currents were not affected by the glutamate receptor antagonists, APV (100 μM) and CNQX (5 μM). In (a) and (b), BDNF (50 ng/ml) was applied for 100 ms (arrowheads).

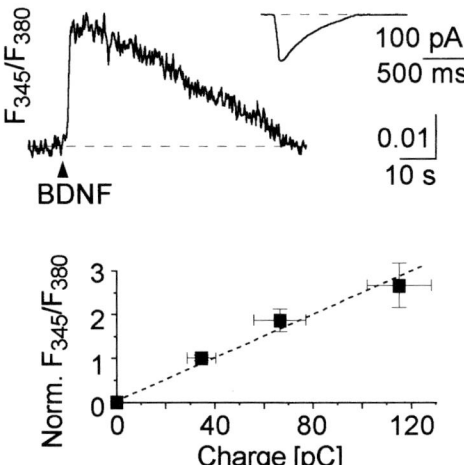

possibility, ratiometric Na^+ imaging was used in combination with whole-cell recordings (Rose and Ransom, 1997; Rose et al., 1999). Cells were loaded with the fluorescent Na^+ indicator dye, sodium-binding benzofuran-isophthalate (SBFI-TM, 1 mM, Molecular Probes), through the patch pipette. Fluorescence signals from the somata were obtained at excitation wavelengths of 345 nm (isosbestic point) and of 380 nm (Na^+-sensitive wavelength). Each BDNF-induced inward current was associated with a transient elevation of the intracellular Na^+ concentration (Fig. 6). Increasing the inward current by increasing the duration of the BDNF application pulse resulted in a directly proportional increase of both the inflowing charge and the amplitude of the Na^+-dependent fluorescence transient (Fig. 6, graph). These experiments were the first to support the involvement of a Na^+ channel. This was further supported by experiments in which the voltage dependence of the BDNF-evoked current was determined (Fig. 7). Thus, the measured reversal potential (60 mV) was remarkably similar to the calculated equilibrium potential for Na^+ (58 mV). Taken together, these experiments clearly indicate that the NT-evoked inward current was due to an influx of Na^+ ions.

Fig. 6. Changes in the intracellular Na^+ concentration evoked by stimulation with BDNF (125 ng/ml) in the soma of a hippocampal CA1 pyramidal cell. (Upper panel) Increase in the Na^+-dependent SBFI fluorescence and the corresponding inward current (inset). (Lower panel) Correlation between the BDNF-induced change in relative SBFI fluorescence (norm. F_{345}/F_{380}) and the evoked change in total charge. Data are expressed as means ± S.E.M. The first data point represents the application of vehicle alone ($n = 5$). The others were obtained by applying BDNF for 50, 100 and 200 ms, respectively.

cells, because tetrodotoxin, at a concentration (0.5 μM) at which it completely blocked action potential firing and hence presynaptic transmitter release, had no significant effect on the BDNF-induced inward current (not shown).

Thus, one of the few remaining alternatives for the rapid action of NTs in the millisecond range (Fig. 5a) was the activation of some Na^+ channel. To test this

Conclusions

The results presented here demonstrate a novel and unexpectedly rapid action of NTs. The NT-evoked depolarization has a latency of just a few milliseconds and is, therefore, orders of magnitude more

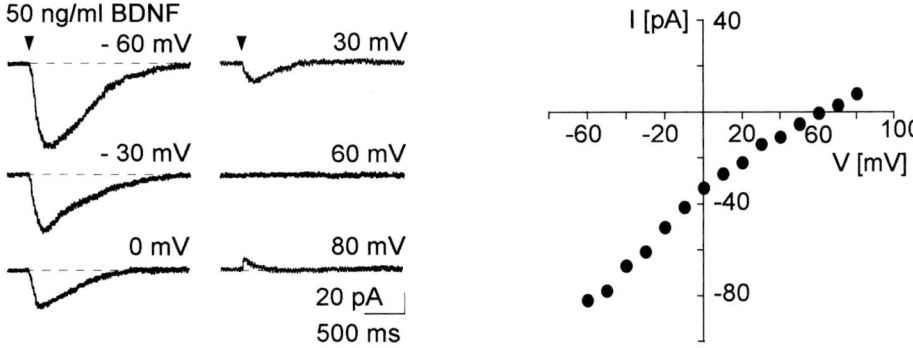

Fig. 7. Current–voltage relationship of the BDNF-evoked currents in CA1 pyramidal neurons. The left panel shows currents measured at 6 different holding potentials, as indicated. The right panel shows the I–V graph obtained from the same experiment. Note that the reversal potential of the BDNF-evoked currents was about 60 mV. Similar results were obtained in 4/4 experiments.

rapid than the previously reported fast effects of NTs (Berninger et al., 1993; Lohof et al., 1993; Berninger and Poo, 1996; Canossa et al., 1997; Suen et al., 1997; Levine et al., 1998). In fact, the onset of the BDNF-evoked response could not be distinguished from that of the excitatory neurotransmitter, glutamate. The extremely rapid onset of the response suggests a direct interaction of the transmembranous tyrosine kinase TrkB receptor and an as yet unidentified Na^+-permeable channel. The rapidity of the NT action combined with the observations that NTs were found in axon terminals (Blöchl and Thoenen, 1996; Zhou and Rush, 1996), that they activate postsynaptic dendrites (Canossa et al., 1997) and that they are secreted in an activity-dependent manner (Blöchl and Thoenen, 1996; Goodman et al., 1996), assigns to NTs functional properties similar to those of conventional excitatory neurotransmitters.

Acknowledgements

We thank R. Trautmann and E. Eilers for expert technical assistance. This study was supported by a fellowship from the DFG to K.W.K. and by grants from the DFG and HFSP to A.K.

References

Altar, C.A. and DiStefano, P.S. (1998) Neurotrophin trafficking by anterograde transport. *Trends Neurosci.*, 21: 433–437.

Barbacid, M. (1995) Neurotrophic factors and their receptors. *Curr. Opin. Cell Biol.*, 7: 148–155.

Berninger, B. and Poo, M. (1996) Fast actions of neurotrophic factors. *Curr. Opin. Neurobiol.*, 6: 324–330.

Berninger, B., Garcia, D.E., Inagaki, N., Hahnel, C. and Lindholm, D. (1993) BDNF and NT-3 induce intracellular Ca^{2+} elevation in hippocampal neurones. *NeuroReport*, 4: 1303–1306.

Blöchl, A. and Thoenen, H. (1996) Localization of cellular storage compartments and sites of constitutive and activity-dependent release of nerve growth factor (NGF) in primary cultures of hippocampal neurons. *Mol. Cell. Neurosci.*, 7: 173–190.

Bonhoeffer, T. (1996) Neurotrophins and activity-dependent development of the neocortex. *Curr. Opin. Neurobiol.*, 6: 119–226.

Bothwell, M. (1995) Functional interactions of neurotrophins and neurotrophin receptors. *Annu. Rev. Neurosci.*, 18: 223–253.

Cabelli, R.J., Hohn, A. and Shatz, C.J. (1995) Inhibition of ocular dominance column formation by infusion of NT-4/5 or BDNF. *Science*, 267: 1662-1666.

Canossa, M., Griesbeck, O., Berninger, B., Campana, G., Kolbeck, R. and Thoenen, H. (1997) Neurotrophin release by neurotrophins: implications for activity-dependent neuronal plasticity. *Proc. Natl. Acad. Sci. USA*, 94: 13279–13286.

Cellerino, A. and Maffei, L. (1996) The action of neurotrophins in the development and plasticity of the visual cortex. *Prog. Neurobiol.*, 49: 53–71.

Edwards, F.A., Konnerth, A., Sakmann, B. and Takahashi, T. (1989) A thin slice preparation for patch clamp recordings from neurones of the mammalian central nervous system. *Pflügers Arch.*, 414: 600–612.

Figurov, A., Pozzo-Miller, L.D., Olafsson, P., Wang, T. and Lu, B. (1996) Regulation of synaptic responses to high-frequency stimulation and LTP by neurotrophins in the hippocampus. *Nature*, 381: 706–709.

Goodman, L.J., Valverde, J., Lim, F., Geschwind, M.D., Fedoroff, H.J., Geller, A.I. and Hefti, F. (1996) Regulated release and polarized localization of brain-derived neurotrophic factor in hippocampal neurons. *Mol. Cell. Neurosci.*, 7: 222–238.

Ibanez, C.F. (1998) Emerging themes in structural biology of neurotrophic factors. *Trends Neurosci.*, 21: 438–444.

Ip, N.Y., Li, Y., Yancopoulos, G.D. and Lindsay, R.M. (1993) Cultured hippocampal neurons show responses to BDNF, NT-3, and NT-4, but not NGF. *J. Neurosci.*, 13: 3394–3405.

Kafitz, K.W., Rose, C.R., Thoenen, H. and Konnerth, A. (1999) Neurotrophin-evoked rapid excitation through TrkB receptors. *Nature*, 401: 918–921.

Kang, H. and Schuman, E.M. (1995) Long-lasting neurotrophin-induced enhancement of synaptic transmission in the adult hippocampus. *Science*, 267: 1658–1662.

Knüsel, B. and Hefti, F. (1992) K-252 compounds: modulators of neurotrophin signal transduction. *J. Neurochem.*, 59: 1987–1996.

Korte, M., Griesbeck, O., Gravel, C., Carroll, P., Staiger, V., Thoenen, H. and Bonhoeffer, T. (1996) Virus-mediated gene transfer into hippocampal CA1 region restores long-term potentiation in brain-derived neurotrophic factor mutant mice. *Proc. Natl. Acad. Sci. USA*, 93: 12547–12552.

Lessmann, V. and Heumann, R. (1998) Modulation of unitary glutamatergic synapses by neurotrophin-4/5 or brain-derived neurotrophic factor in hippocampal microcultures: presynaptic enhancement depends on pre-established paired-pulse facilitation. *Neuroscience*, 86: 399–413.

Lessmann, V., Gottmann, K. and Heumann, R. (1994) BDNF and NT-4/5 enhance glutamatergic synaptic transmission in cultured hippocampal neurones. *NeuroReport*, 6: 21–25.

Levine, E.S., Crozier, R.A., Black, I.B. and Plummer, M.R. (1998) Brain-derived neurotrophic factor modulates hippocampal synaptic transmission by increasing N-methyl-D-aspartic acid receptor activity. *Proc. Natl. Acad. Sci. USA*, 95: 10235–10239.

Lewin, G.R. and Barde, Y.-A. (1996) Physiology of the neurotrophins. *Annu. Rev. Neurosci.*, 19: 289–317.

Li, Y.X., Zhang, Y., Lester, H.A., Schuman, E.M. and Davidson, N. (1998) Enhancement of neurotransmitter release induced by brain-derived neurotrophic factor in cultured hippocampal neurons. *J. Neurosci.*, 98: 10231–10240.

Lohof, A.M., Ip, N.Y. and Poo, M.-m. (1993) Potentiation of

developing neuromuscular synapses by the neurotrophins NT-3 and BDNF. *Nature*, 363: 350–353.

McAllister, A.K., Katz, L.C. and Lo, D.C. (1999) Neurotrophins and synaptic plasticity. *Annu. Rev. Neurosci.*, 22: 295–318.

Merlio, J.P., Ernfors, P., Jaber, M. and Persson, H. (1992) Molecular cloning of rat trkC and distribution of cells expressing messenger RNAs for members of the trk family in the rat central nervous system. *Neuroscience*, 51: 513–532.

Patterson, S.L., Abel, T., Deuel, T.A., Malin, K.C., Rose, J.C. and Kandel, E.R. (1996) Recombinant BDNF rescues deficits in basal synaptic transmission and hippocampal LTP in BDNF knockout mice. *Neuron*, 16: 1137–1145.

Ringstedt, T., Lagercrantz, H. and Persson, H. (1993) Expression of members of the trk family in the developing postnatal rat brain. *Dev. Brain Res.*, 72: 119–131.

Rose, C.R. and Ransom, B.R. (1997) Regulation of intracellular sodium in cultured rat hippocampal neurones. *J. Physiol. (Lond.)*, 499: 573–587.

Rose, C.R., Kovalchuk, Y., Eilers, J. and Konnerth, A. (1999) Two-photon Na^+ imaging in spines and fine dendrites of central neurons. *Pflügers Arch.*, 439: 201–207.

Schuman, E.M. (1999) Neurotrophin regulation of synaptic transmission. *Curr. Opin. Neurobiol.*, 9: 105–109.

Suen, P.C., Wu, K., Levine, E.S., Mount, H.T., Xu, J.L., Lin, S.Y. and Black, I.B. (1997) Brain-derived neurotrophic factor rapidly enhances phosphorylation of the postsynaptic *N*-methyl-D-aspartate receptor subunit 1. *Proc. Natl. Acad. Sci. USA*, 94: 8191–8195.

Thoenen, H. (1995) Neurotrophins and neuronal plasticity. *Science*, 270: 593–598.

Yan, Q., Radeke, M.J., Matheson, C.R., Talvenheimo, J., Welcher, A.A. and Feinstein, S.C. (1997) Immunocytochemical localization of TrkB in the central nervous system of the adult rat. *J. Comp. Neurol.*, 378: 135–157. [Published erratum appears in (1997) *J. Comp. Neurol.*, 382: 546–547.]

Zhou, X.F. and Rush, R.A. (1996) Endogenous brain-derived neurotrophic factor is anterogradely transported in primary sensory neurons. *Neuroscience*, 74: 945–953.

SECTION V

Candidate cells for transplantation into the injured CNS

CHAPTER 21

Candidate cells for transplantation into the injured CNS

Itzhak Fischer [*]

Department of Neurobiology and Anatomy, MCP Hahnemann University, 3200 Henry Avenue, Philadelphia, PA 19129, USA

Introduction

The prospect of treating central nervous system (CNS) injury and restoring function is becoming increasingly brighter as earlier beliefs about the inability of the CNS to regenerate have been replaced by a better understanding of how to modify the environment to become permissive and how to provide therapeutic factors that promote regeneration. It is quite remarkable how rapidly and effectively recent developments in basic neuroscience, developmental biology, immunology, and molecular biology have been translated into useful experimental protocols for the study of CNS injury. This process has been facilitated by our ability to test hypotheses, developed in vitro, in experimental animal models of injury and degeneration. This section focuses on recent progress in identifying candidate cells for transplantation into the injured CNS, particularly in models of spinal cord injury. Candidate cells include autoimmune T-cells, olfactory ensheathing glia, neural stem cells, lineage-restricted precursors, stromal marrow cells, neurons derived from human embryonic teratocarcinoma cells and genetically modified fibroblasts. Transplantation is one of the principal therapeutic strategies that has captured the imagination of scientists working in repair of CNS injury. This approach has reinvented itself through incorporation of different cell types and new methods of gene transfer into these cells. The initial transplantation approach to spinal repair centered on the use of fetal grafts and peripheral nerves. Early studies showed that both of these intraspinal transplants provided a substrate for axon growth, but neither promoted long-distance regeneration in the adult. In the presence of neurotrophic factors, the limited regeneration was enhanced and some functional recovery has also been reported. There has been, however, a growing recognition of the need for alternative transplant sources, motivated not only by ethical issues, but also by practical limitations related to the availability and quality of the grafted tissue. Nevertheless, the pioneering studies with peripheral nerves and fetal grafts have set the stage for many of the cell transplantation strategies described in this chapter. For example, there is a natural progression and transition from fetal grafts to neural stem cells and neurons, and from peripheral nerve to Schwann cells and ensheathing glial cells. The growing choice of cells for transplantation presents both a challenge and an opportunity for development of future therapies, for it is now recognized that it will be necessary to use combined interventions to address the multiple needs of neuron rescue or replacement, robust regeneration, appropriate targeting and functional recovery.

What are the ideal cells for transplantation into the CNS?

The list shown in Table 1 includes specific properties that emphasize the potential clinical requirements concerning these cells. They should be easily

---

[*] Corresponding author: Dr. I. Fischer, Department of Neurobiology and Anatomy, MCP Hahnemann University, 2900 Queen Lane, Philadelphia, PA 19129, USA; Fax: +1-215-843-9082; E-mail: Itzhak.Fischer@drexel.edu

TABLE 1

Properties of candidate cells for transplantation into injured CNS

	Non-neural			Non-neuronal		Neuronal	
	modified fibroblasts	marrow stromal cells	macrophages	Schwann cells	olfactory ensheathing cells	immortalized neurons	neuronal precursors
Easily obtained	Yes	Yes	Yes	Yes	Yes	Yes	Yes
Autografts	Yes	Yes	Yes	Yes	No	No	No
In vitro expansion	Yes	Yes	Yes	Yes	?	Yes	Yes
Genetic modification	Yes	Yes	?	Yes	?	Yes	Yes
Neural phenotype	No	?	No	Yes	Yes	Yes[a]	Yes[a]

? = not known.
[a] Includes neuronal phenotype.

obtained, preferably as autologous cells from the host/patient. All of the cells described in this section are readily available, but only autoimmune T-cells, stromal marrow cells and fibroblasts can presently be obtained from patients. Another practical issue is the ability to culture, expand and store cells, not only to achieve the large number necessary for grafting, but also to guarantee a homogeneous population and high-level quality control standards. All the candidate cells, to various degrees, meet this requirement. Finally it is advantageous to be able to genetically modify cells and have them express therapeutic genes that will improve the graft environment and promote regeneration. Again, all of the candidate cells meet this requirement and, since they are proliferating cells, they can be modified by retrovirus vectors, which are relatively simple to prepare and use. At present only genetically modified fibroblasts and marrow stromal cells have been used for ex vivo gene therapy, while the potential of the other cells modified to produce bioactive molecules is yet to be explored.

What are the cells that are currently available for transplantation?

The cells discussed in this section are representative of the progress in identifying new sources for transplantation and demonstrating their beneficial effects in CNS injury. This is not intended to be a complete list, but rather a progress report on some of the current work to illustrate how diverse the field has become and how important it is to encourage interactions among different disciplines. The following is a short summary of the work presented in this section.

Autoimmune T-cells

The use of autoimmune T-cells for transplantation into injured CNS represents an increasing appreciation that not all forms of the immune response are detrimental and that some immune interventions can actually protect damaged tissue and promote repair. The beneficial effects of the immune system were initially demonstrated with the use of activated macrophages in spinal cord injury (Rapalino et al., 1998). More recently, autoimmune T-cells directed against myelin basic protein (MBP) were shown to protect injured retinal ganglion cells from secondary degeneration (Moalem et al., 1999). In experiments presented in this section, Michal Schwartz shows that the autoimmune T-cells, injected into rats with a contusive spinal cord injury at a thoracic level, increase the number of intact axons descending from the red nucleus, increase the mass of spared spinal cord tissue, and enhance locomotor activity. Schwartz hypothesizes that the autoimmune response normally triggered by the injury is weak and needs to be increased to provide therapeutic value, but at the same time the response needs to be carefully controlled to avoid induction of an autoimmune disease.

Olfactory ensheathing cells

The growth properties of olfactory axons stand as an exception to the inability of the adult CNS to regenerate. Both normal and injured olfactory axons can grow, find their normal target and form functional connections. To a large measure this unique behavior is dependent on the presence of olfactory ensheathing cells (OECs). These cells function not only as supporting cells that supply trophic factors, but also provide a permissive environment for axonal growth and can myelinate these axons. The development of methods to culture and characterize pure OECs, particularly from adult animals, has been the key to the success in the application of OECs in transplantation experiments (Ramón-Cueto and Avila, 1998). The beneficial effects of these cells in spinal cord injury are demonstrated by Almudena Ramón-Cueto in adult rats that were transected at the thoracic level and injected with OECs into the midline of both stumps. These studies show long-distance regeneration of supraspinal axons through the transection site, accompanied by migration of OECs in the same location, including the entire injury site where they intermingle with reactive glia. The ability of OECs to migrate appears to provide axons with a permissive environment to regenerate by counteracting the inhibitory signals of the adult CNS and possibly ensheathing the axons.

Neuronal progenitors

The discovery and characterization of multipotent neural stem cells in developing and adult CNS provides a very attractive candidate for cellular repair following CNS injury and degenerative disorders. Recent studies by Kalyani and Rao (1998) have demonstrated that during spinal cord development there is a sequential process of restriction of the multipotential stem cells that generates neuronal- and glial-restricted progenitors. The ability to isolate and expand these cells provides choices for transplantation by targeting specific types of cells to individual therapies ranging from replacement of an identified population of neurons in some disorders to the potential for repair of traumatic injury with multipotent stem cells. This is also an opportunity to study the properties of these cells in vivo, to examine the phenotypic potential of these cells under a variety of conditions and to re-examine the rules of neural cell developmental boundaries. Mahendra Rao discusses the different methods for preparation of multipotent stem cells, the properties of various lineage-restricted precursor cells and the differences among them. For example, multipotential neural stem cells can be prepared in the presence of fibroblast growth factor (FGF), epidermal growth factor (EGF) or both factors from a variety of CNS regions and at different developmental stages as well as in the adult. Glial precursors include glial-restricted tripotential precursor cells (Rao et al., 1999), as well as oligodendrocyte precursors (O-2A). Finally, it is important to note that the similarities in the properties of cells prepared from rodents and from human tissue hold great promise for future therapeutic applications of these fundamental studies.

Marrow stromal cells

Marrow stromal cells (MSCs) derived from the bone marrow are nonhematopoietic cells that can differentiate into a variety of mesenchymal cells. These cells are clinically attractive because they can be easily obtained from patients and genetically modified. The improved protocols for isolating and culturing MSCs made it possible to examine the properties of these cells following transplantation and to test their full potential for deafferentation in the CNS. Studies by Darwin Prockop and colleagues have shown that when human or murine MSCs are grafted into brain, they survive for several months with no evidence of inflammatory response and migrate in a pattern that resembles that of astrocytes. There is also evidence that these cells may differentiate into astrocytes and ongoing studies are focused on other neural phenotypes. Grafting of cells that were genetically modified to express two critical genes required for the production of L-DOPA demonstrate the potential of these cells for gene therapy application in treating degenerative CNS disorders. Grafting of human MSCs into injured spinal cord improves the permissivity of the lesion site for axonal growth, making these cells another promising candidate for delivery of therapeutic products in spinal cord repair.

Embryonic stem cells

Embryonic stem (ES) cells represent another exciting and theoretically unlimited source for transplantation. They have been isolated from different species, including humans, and are unique in their potential to generate all phenotypes (totipotential), including neural cells. Thus they can be isolated, expanded, maintained in culture and used to prepare ES cell lines and as a source for neural cells. Because of their totipotentiality, they may have less regional restriction and a better ability to become a universal precursor for different types of neural cells. Work presented from the laboratories of Virginia Lee and John Trojanowski suggests that a cell line derived from human embryonic carcinoma cells can be used for spinal cord injury. Neurons, prepared from the clonal NT2N cell line by treatment with retinoic acid, show properties of immature CNS neurons (Pleasure and Lee, 1993). When these neurons are implanted into the spinal cord of immunologically compromised mice, they integrate with the host, survive for months, differentiate into mature neurons and extend neurites. The behavior of the transplanted cells, however, is dependent on the area of grafting. In white matter, neurons extend long axon-like processes of over 2 cm, while in gray matter they have short processes that resemble dendrites, suggesting that the host exerts differential effects on these neurons, consistent with the properties of the local environment. Although these results do not directly demonstrate the therapeutic properties of human ES cells, they do suggest that cells that have properties of neuronal precursors and the potential to differentiate, integrate and replace lost function can be readily obtained and grafted.

Genetically modified fibroblasts

The grafting of genetically engineered cells into the injured CNS is an example of the ex vivo gene therapy approach, in which therapeutic genes are delivered via a transplant. One of the most successful applications of this strategy for spinal cord repair has been the grafting of genetically modified fibroblasts that express neurotrophic factors. The advantages of using fibroblasts include the ability to obtain the cells from the host, to genetically modify them with recombinant retroviral vectors, and then to expand and store large quantities for grafting. Work by Liu et al. (1999) demonstrates that modified fibroblasts can promote regeneration, rescue injured neurons and allow partial recovery of motor function.

Regeneration. Fibroblasts modified to produce BDNF were transplanted into a spinal cord injury model of a partial cervical hemisection at C3/4 that completely interrupted the rubrospinal tract coming from the brainstem. They fill the lesion cavity, form a continuous interface and promote axonal growth from dorsal roots and raphe–spinal axons. In addition, anterogradely labeled rubrospinal axons regenerate through and adjacent to the graft and extend distal to the transplant up to 3–4 cm, and terminate in appropriate laminae of the spinal gray matter. Retrograde labeling by fluorogold injections caudal to the BDNF-producing fibroblasts indicates that about 7% of rubrospinal neurons regenerate axons, representing a 7-fold increase relative to unmodified fibroblasts.

Rescue. After cervical hemisection, about 40% of neurons in the magnocellular division of the red nucleus undergo retrograde cell death or atrophy below the level of detection. Grafting of BDNF-producing fibroblasts rescues most of the injured red nucleus and many of the cells that are rescued retain their normal size and morphology.

Recovery of function. Rats that receive partial cervical hemisections show deficit of forelimb usage (the cylinder test) and of locomotion in a posture-challenging task (the rope-walking test). Rats receiving transplants of BDNF-producing fibroblasts show partial recovery of both fore- and hindlimb function, with the recovery abolished after a second lesion just rostral to the initial transplant site, indicating that some of the recovery is correlated with the presence of the graft. The challenge of improving the strategy of grafting genetically modified cells that express therapeutic genes is to develop methods to control the time and levels of transgene expression, and to test the efficacy of combination gene therapy.

References

Kalyani, A. and Rao, M.S. (1998) Cell lineage in the developing neural tube. *Biochem. Cell Biol.*, 76: 1051–1068.

Liu, Y., Kim, D., Himes, B.T., Chow, S.Y., Murray, M., Tessler, A. and Fischer, I. (1999) Transplants of fibroblasts genetically modified to express BDNF promote regeneration of adult rat rubrospinal axons. *J. Neurosci.*, 19: 4370–4387.

Moalem, G., Leibowitz-Amit, R., Yoles, E., Mor, F., Cohen, I.R. and Schwartz, M. (1999) Autoimmune T cells protect neurons from secondary degeneration after central nervous system axotomy. *Nat. Med.*, 5: 49–55.

Pleasure, S.J. and Lee, V.M.-Y. (1993) Ntera cells: a human cell line which displays characteristics expected of a human committed neuronal progenitor cell. *J. Neurosci. Res.*, 35: 585–602.

Ramón-Cueto, A. and Avila, J. (1998) Olfactory ensheathing glia: properties and function. *Br. Res. Bull.*, 46: 175–187.

Rao, M.S. (1999) Multipotent and restricted precursors in the central nervous system. *Anat. Rec.*, 257: 137–148.

Rapalino, O., Lazarov-Spiegler, O., Arganov, E., Velan, G.J., Yoles, E., Fraidakis, M., Solomon, A., Gepstein, R., Katz, A., Belkin, M., Hadani, M. and Schwartz, M. (1998) Implantation of stimulated homologous macrophages results in partial recovery of paraplegic rats. *Nat. Med.*, 4: 814–821.

CHAPTER 22

Autoimmune involvement in CNS trauma is beneficial if well controlled

Michal Schwartz *

Department of Neurobiology, The Weizmann Institute of Science, Rehovot 76100, Israel

Introduction

The central nervous system (CNS) has long been viewed as a site where any immune activity is detrimental. Moreover, if any immune activity is detected in the injured CNS, its effect was assumed to be negative. Yet, as is well known, the role of the immune system in general is to defend, to protect, and to repair. The question is: how can this apparent paradox be reconciled? Does it derive from a fundamental difference between the nervous system and the rest of the body with regard to their post-injury requirements for rescue and repair? Or is it an outcome of the immune system's ability to exert both beneficial and harmful effects, with the balance between them varying among different tissues? This article will summarize data suggesting that in cases of traumatic injury to the CNS, the role of the immune system is similar to its role in other tissues, but the unique nature of the CNS demands that immune involvement be more tightly controlled. Both innate and adaptive immune responses are needed for recovery after CNS axonal injury; macrophages are required for repair, and activated T-cells directed against CNS self-antigens are needed for protection. Immune cell therapy design should be based on timely intervention, with tight control of amounts and specificities so as to derive the maximal benefit with minimal risk.

In most tissues damaged by traumatic injury, the immune response has to do with the process of repair, a relatively simple task that does not require specificity to any particular pathogen and can therefore be mediated by the innate arm of the immune system, represented by the relatively non-specialized immune cells, the macrophages. These cells invade the injured tissue, remove dead cells and cell debris, and produce the factors needed to execute a myriad of processes that lead eventually to tissue repair (Clark, 1993a,b). This is the routine procedure when the insult is not pathogen-related and thus the primary need is for repair. The picture becomes more complicated, however, when pathogens are involved. The immune system must then provide not only repair and renewal of damaged tissues, but also defense against the infective organism and protection of the tissue from the progression of damage (Matzinger, 1994). In this case, the pathogen-induced damage evokes a 'stress' signal that recruits the acquired arm of the immune system, represented by the relatively more specialized T-cells.

It appears from our work that, at least in the CNS, even damage resulting from traumatic injury can benefit from the assistance of both arms of the immune system, the one involved in repair (macrophages) and the other concerned with defense and protection (T-cells). The T-cells mobilized in this case appear to be directed not against a specific

---

* Corresponding author: Dr. M. Schwartz, Department of Neurobiology, The Weizmann Institute of Science, Rehovot 76100, Israel; Fax: +972-8-934-4131; E-mail: bnschwar@wicc.weizmann.ac.il

pathogen, but against self-antigens expressed in the damaged CNS tissue (Schwartz et al., 1999). It thus appears, surprisingly, that what the CNS needs for recovery from trauma is not different from what is needed by other tissues, namely, the involvement of the immune system. Paradoxically, it also seems that the activity of the immune system in the injured CNS needs to be broader than in other injured tissues, while at the same time more stringently controlled. One possible reason for this, as discussed below, is the non-replicable character of the specialized nerve cells (neurons) in the CNS (Lotan and Schwartz, 1994; Cohen and Schwartz, 1999; Schwartz et al., 1999).

The role of macrophages in CNS regrowth and regeneration

Traumatic injury to CNS axons is followed by some immune activity, in the form of recruitment and activation of blood-borne monocytes and activation of resident microglia. However, this immune activity is limited to the site of injury in comparison with the massive spontaneous recruitment and activation of monocytes manifested by the process of Wallerian degeneration seen in injured peripheral nerves (Avellino et al., 1995). It is now clear that Wallerian degeneration, although apparently destructive in nature, is a prerequisite for subsequent regeneration (Brown et al., 1991; Lu and Richardson, 1991; Perry and Brown, 1992; Chen and Bisby, 1993; Perry et al., 1993; George and Griffin, 1994; La Fleur et al., 1996). The analogous process in non-nervous tissues is the clearing of dead cells and cell debris. In the nervous system, the process of tissue repair after axonal injury is not a matter of 'filling in' the space created by the damaged axon. This is because the damage is not intercellular but intracellular, involving axoplasmic disruption and therefore necessitating reconstruction of the axon all the way back to the target in the brain rather than mere bridging of the gap between the cut ends of the nerve. Thus, the apparently chaotic initial spread of damage to tissues beyond the site of the lesion is in fact a beneficial process — the price, so to speak, of preparing the tissue for repair. In the CNS, this spread of damage occurs more slowly than in the peripheral nervous system. The slow longitudinal degeneration in the CNS (i.e., along the injured axons themselves) may have both 'bad' and 'good' consequences. On the one hand, as proposed here and discussed below, it impedes the regeneration of damaged fibers. On the other hand, as a consequence of the above, it slows down the permeation of damage to viable neurons that escaped the primary injury.

The spread of degeneration (secondary degeneration) in the injured CNS has been shown to result from an extracellular imbalance in neuron-associated transmitters and other essential components, which become noxious when their physiological levels are exceeded owing to buffering impairment. Compounds such as glutamate and nitric oxide play a pivotal role in the maintenance and performance of the normal nervous system, but pose a threat when their levels exceed the system's buffering capacity. It is still a matter of debate whether immune involvement at an early post-traumatic stage has any negative effects on the delayed degeneration in the CNS (Popovich et al., 1999). The beneficial effect of methylprednisolone administration soon after spinal cord injury could be interpreted as an indication that local inflammation in the very early stages after trauma, even if mild, is bad for recovery (Blight, 1992; Constantini and Young, 1994). Somewhat later, however, when the process of degeneration becomes necessary for repair, the advantage of inflammation overrides the early disadvantage, but its mildness now makes it insufficiently effective (Lu and Richardson, 1991; Lazarov-Spiegler et al., 1996, 1998a,b; Prewitt et al., 1997; Lazar et al., 1999).

We have shown that after complete transection of the rat optic nerve or spinal cord, local administration of macrophages, following their activation by exposure to peripheral nerve tissue, promotes axonal regrowth (Lazarov-Spiegler et al., 1996; Rapalino et al., 1998). In the case of the spinal cord, treatment with the activated macrophages was effective even when applied as late as two weeks or one month after transection. Macrophage implantation resulted in partial recovery of function as well as some tissue restoration, indicated morphologically by fiber tracing, diffusion magnetic resonance imaging (MRI), and immunohistochemistry. All of the tests showed evidence of neural tissue bridging the gap between the cut ends of the nerve. How far the fibers grow, and what makes the macrophage-treated tissue amenable to regrowth and restoration of function, are still open questions. It seems clear, however, that the macrophages, as in all other injured tissues, prepare the CNS tissue for repair. Among the factors produced by the macrophages are cytokines and growth factors. In the context of CNS trauma, it is still a matter of controversy whether cytokines have a beneficial or a detrimental effect on the injured tissue, and there is evidence to support both of these possibilities (Hirschberg et al., 1994; Bethea et al., 1999). We suggest that viewing cytokine activity after brain and CNS trauma as 'good' or 'bad' is misleading, and we would do better to focus our attention on individual cytokines and other factors that may exert distinctive and possibly opposing effects, depending on the phase of recovery and on whether rescue or regrowth is predominantly required. Phase and requirements may be related, as treatment for protection has a narrower post-injury time window than treatment for growth. If properly regulated, local treatment of the damaged axons with a well controlled quantity of suitably activated macrophages, in a time window when the threat to the rescue of spared neuronal tissue is still low, might enable the damaged tissue to 'talk' with the macrophages in a way that allows them to self-regulate their activities according to need, while avoiding the undesirable effects of each factor. In cases of partial lesion of the nerve fibers rather than complete transection, it may be necessary to make suitable adjustments to the timing of macrophage application in order to exploit the optimal therapeutic window (Lazarov-Spiegler et al., 1998a).

The T-cell response to CNS damage

As discussed above, where damage to the tissue is of pathogenic origin, the specialized immune cells, the T-cells, are recruited to protect the tissue from the spread of damage. Studies have indicated that spinal cord injury triggers the activity of autoimmune T-cells. As these T-cells have traditionally been considered detrimental to CNS tissue, it was assumed that their effects on the damaged spinal cord are negative (Popovich et al., 1996). However, in a recent study showing that axonal injury in the CNS is followed by a transient, nonselective accumulation of T-cells at the site of injury, we made the astonishing discovery that of all the accumulated T-cells, only those which display immunity to a CNS self-antigen affect the damaged tissue in a manner that is not destructive, but protective (Moalem et al., 1999a, 2000a,b). Adoptive transfer of autoimmune T-cells directed against a cryptic epitope of the CNS antigen reduced the injury-induced spread of damage and thus resulted in a significantly improved outcome, manifested both functionally and morphologically, after optic nerve or spinal cord injury (Moalem et al., 1999a; Hauben et al., 2000). Thus, for example, a single injection of T-cells directed against myelin basic protein reduced the loss of fibers after partial lesion of the rat optic nerve or promoted recovery from partial injury (contusion) of the rat spinal cord. We found that after injury, the damaged nerve tissue became permissive to the accumulation of T-cells, regardless of their antigen specificity (Hirschberg et al., 1998; Moalem et al., 1999b), but only those T-cells directed against myelin-associated proteins had any effect on the damaged nerve, and that effect was beneficial, i.e., neuroprotective. T-cells of other specificities had no effect on the injured nerve. The neuroprotective effect could be achieved not only by passive immunization (transfer of T-cells) but also by active immunization, i.e., vaccination with relevant antigens.

This unexpected effect of the autoimmune T-cells challenges our understanding of autoimmunity in general and CNS autoimmunity in particular. We suspected, and recently proved, that the spontaneous autoimmune response triggered by the injury is potentially beneficial, but too weak to be effective, and therefore in need of therapeutic boosting. This would indicate that the autoimmune response triggered by injury to CNS tissue is analogous to the T-cell response triggered in other tissues by pathogen-associated damage. Just as the signal for this T-cell immune response is tissue stress, so the damage to the CNS is sufficiently threatening to justify the triggering of a T-cell response, even at the risk of its being directed against the self. Such a response would obviously need to be rigorously controlled to avoid incurring an autoimmune disease (see scheme in Fig. 1). In line with this view, it is tempting to suggest the existence of two potential stress signals in the CNS: (1) the threat of irreversible damage spread due to nerve-derived mechanisms of toxicity and the

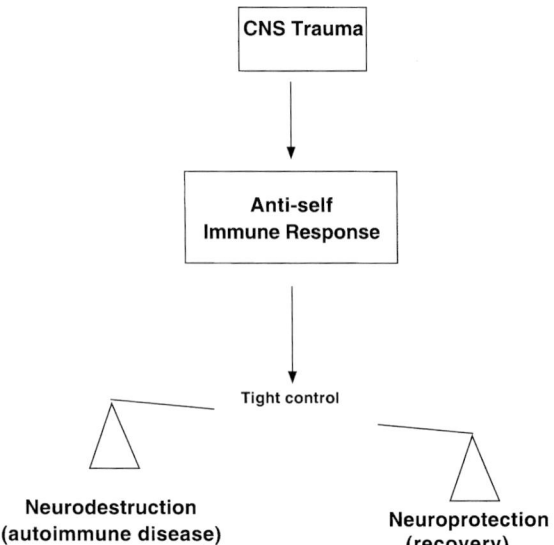

Fig. 1. Injury to CNS axons triggers an immune response against self that is neuroprotective rather than neurodestructive if well controlled, but is too weak to be effective and needs to be amplified.

irreplaceable character of the neural tissue, and (2) the threat of immune activity, including excessive autoimmune activity. We believe that the evolutionary development of limitation on immune activity in the CNS, in the form of immune privilege, came about because of the potentially harmful modulation of the intricate network of the healthy CNS as a result of any immune-related intervention. Thus, it seems that the mammalian brain developed physiological barriers to immune intervention, and that this development resulted in the acquiring of its status as an immune-privileged site. After injury, although the CNS may be in need of its protective immunity (in the form of autoimmunity), such autoimmunity is limited by the same mechanisms that limit immune activity in general in the CNS (Schwartz et al., 1999). Indeed, perhaps the existence of autoimmune T-cells to certain epitopes should be interpreted not as a failure of deletion of 'self', but rather as an indication that the individual needs those T-cells as part of its 'protective repertoire' (Cohen, 1992). If this is so, research efforts should be directed toward discovering how to regulate autoimmunity by awakening a protective response without incurring the risk of autoimmune disease.

References

Avellino, A.M., Hart, D., Dailey, A.T., MacKinnon, M., Ellegala, D. and Kliot, M. (1995) Differential macrophage responses in the peripheral and central nervous system during Wallerian degeneration of axons. *Exp. Neurol.*, 136: 183–198.

Bethea, J.R., Nagashima, H., Acosta, M.C., Briceno, C., Gomez, F., Marcillo, A.E., Loor, K., Green, J. and Dietrich, W.D. (1999) Systemically administered interleukin-10 reduces tumor necrosis factor-alpha production and significantly improves functional recovery following traumatic spinal cord injury in rats. *J. Neurotrauma*, 16: 851–863.

Blight, A.R. (1992) Spinal cord injury models: neurophysiology. *J. Neurotrauma*, 9: 147–149; discussion: 149–150.

Brown, M.C., Perry, V.H., Lunn, E.R., Gordon, S. and Heumann, R. (1991) Macrophage dependence of peripheral sensory nerve regeneration: possible involvement of nerve growth factor. *Neuron*, 6: 359–370.

Chen, S. and Bisby, M.A. (1993) Impaired motor axon regeneration in the C57BL/Ola mouse. *J. Comp. Neurol.*, 333: 449–454.

Clark, R.A. (1993a) Basics of cutaneous wound repair. *J. Dermatol. Surg. Oncol.*, 19: 693–706.

Clark, R.A. (1993b) Biology of dermal wound repair. *Dermatol. Clin.*, 11: 647–666.

Cohen, I.R. (1992) The cognitive paradigm and the immunological homunculus. *Immunol. Today*, 13: 490–494.

Cohen, I.R. and Schwartz, M. (1999) Autoimmune maintenance and neuroprotection of the central nervous system. *J. Neuroimmunol.*, 100: 111–114.

Constantini, S. and Young, W. (1994) The effects of methylprednisolone and the ganglioside GM1 on acute spinal cord injury in rats. *J. Neurosurg.*, 80: 97–111.

George, R. and Griffin, J.W. (1994) Delayed macrophage responses and myelin clearance during Wallerian degeneration in the central nervous system: the dorsal radiculotomy model. *Exp. Neurol.*, 129: 225–236.

Hauben, E., Nevo, U., Yoles, E., Moalem, G., Agranov, E., Mor, F., Akselrod, S., Neeman, M., Cohen, I.R. and Schwartz, M. (2000) Autoimmune T cells as potential neuroprotective therapy for spinal cord injury. *Lancet*, 355: 286–287.

Hirschberg, D.L., Yoles, E., Belkin, M. and Schwartz, M. (1994) Inflammation after axonal injury has conflicting consequences for recovery of function: rescue of spared axons is impaired but regeneration is supported. *J. Neuroimmunol.*, 50: 9–16.

Hirschberg, D.L., Moalem, G., He, J., Mor, F., Cohen, I.R. and Schwartz, M. (1998) Accumulation of passively transferred T cells independently of their antigen specificity following central nervous system trauma. *J. Neuroimmunol.*, 89: 88–96.

La Fleur, M., Underwood, J.L., Rappolee, D.A. and Werb, Z. (1996) Basement membrane and repair of injury to peripheral nerve: defining a potential role for macrophages, matrix metalloproteinases, and tissue inhibitor of metalloproteinases-I. *J. Exp. Med.*, 184: 2311–2326.

Lazar, D.A., Ellegala, D.B., Avellino, A.M., Dailey, A.T., Andrus, K. and Kliot, M. (1999) Modulation of macrophage and microglial responses to axonal injury in the peripheral and central nervous systems. *Neurosurgery*, 45: 593–600.

Lazarov-Spiegler, O., Solomon, A.S., Zeev Brann, A.B., Hirschberg, D.L., Lavie, V. and Schwartz, M. (1996) Transplantation of activated macrophages overcomes central nervous system regrowth failure. *FASEB J.*, 10: 1296–1302.

Lazarov-Spiegler, O., Rapalino, O., Agranov, E. and Schwartz, M. (1998a) Restricted inflammatory reaction in the CNS: a key impediment to axonal regeneration?. *Mol. Med. Today*, 4: 337–342.

Lazarov-Spiegler, O., Solomon, A.S. and Schwartz, M. (1998b) Peripheral nerve stimulated macrophages simulate a peripheral nerve-like regenerative response in rat transected optic nerve. *Glia*, 24: 329–337.

Lotan, M. and Schwartz, M. (1994) Cross talk between the immune system and the nervous system in response to injury: implications for regeneration. *FASEB J.*, 8: 1026–1033.

Lu, X. and Richardson, P.M. (1991) Inflammation near the nerve cell body enhances axonal regeneration. *J. Neurosci.*, 11: 972–978.

Matzinger, P. (1994) Tolerance, danger, and the extended family. *Annu. Rev. Immunol.*, 12: 991–1045.

Moalem, G., Leibowitz-Amit, R., Yoles, E., Mor, F., Cohen, I.R. and Schwartz, M. (1999a) Autoimmune T cells protect neurons from secondary degeneration after central nervous system axotomy. *Nat. Med.*, 5: 49–55.

Moalem, G., Monsonego, A., Shani, Y., Cohen, I.R. and Schwartz, M. (1999b) Differential T cell response in central and peripheral nerve injury: connection with immune privilege. *FASEB J.*, 13: 1207–1217.

Moalem, G., Leibowitz-Amit, R., Yoles, E., Muler-Gilor, S., Mor, F., Cohen, I.R. and Schwartz, M. (2000a) Autoimmune T cells retard the loss of function in injured rat optic nerves. *J. Neuroimmunol.*, in press.

Moalem, G., Gdalyahu, A., Shani, Y., Otten, V., Lazarovici, P., Cohen, I.R. and Schwartz, M. (2000b) Production of neurotrophins by activated T cells: Implications for neuroprotective autoimmunity. *J. Autoimmunity*, in press.

Perry, V.H. and Brown, M.C. (1992) Role of macrophages in peripheral nerve degeneration and repair. *Bioessays*, 14: 401–406.

Perry, V.H., Andersson, P.B. and Gordon, S. (1993) Macrophages and inflammation in the central nervous system. *Trends Neurosci.*, 16: 268–273.

Popovich, P.G., Stokes, B.T. and Whitacre, C.C. (1996) Concept of autoimmunity following spinal cord injury: possible roles for T lymphocytes in the traumatized central nervous system. *J. Neurosci. Res.*, 45: 349–363.

Popovich, P.G., Guan, Z., Wei, P., Huitinga, I., van Rooijen, N. and Stokes, B.T. (1999) Depletion of hematogenous macrophages promotes partial hindlimb recovery and neuroanatomical repair after experimental spinal cord injury. *Exp. Neurol.*, 158: 351–365.

Prewitt, C.M., Niesman, I.R., Kane, C.J. and Houlé, J.D. (1997) Activated macrophage/microglial cells can promote the regeneration of sensory axons into the injured spinal cord. *Exp. Neurol.*, 148: 433–443.

Rapalino, O., Lazarov-Spiegler, O., Agranov, E., Velan, G.J., Yoles, E., Fraidakis, M., Solomon, A.S., Gepstein, R., Katz, A., Belkin, M., Hadani, M. and Schwartz, M. (1998) Implantation of stimulated homologous macrophages results in partial recovery of paraplegic rats. *Nat. Med.*, 4: 814–821.

Schwartz, M., Moalem, G., Leibowitz-Amit, R. and Cohen, I.R. (1999) Innate and adaptive immune responses can be beneficial for CNS repair. *Trends Neurosci.*, 22: 295–299.

CHAPTER 23

Olfactory ensheathing glia transplantation into the injured spinal cord

Almudena Ramón-Cueto [*]

Neural Regeneration Group, Institute of Biomedicine, Spanish Council for Scientific Research (CSIC), Jaime Roig 11, 46010 Valencia, Spain

Introduction

The peripheral (PNS) and central nervous systems (CNS) of adult mammals do not respond equally to damage. Whereas injured axons are able to regenerate through the permissive cellular PNS environment (Ramón y Cajal, 1928; Fawcett and Keynes, 1990), the inhibitory nature of the CNS milieu prevents axons from growing (Ramón y Cajal, 1928; Schwab et al., 1993; Silver, 1994). This lack of axonal regeneration in the CNS causes a permanent disconnection of injured neurons with their targets, and thus an irreversible loss of all functions mediated by these neurons. The olfactory bulb constitutes an exception to this general rule. In this CNS structure, either normal or injured olfactory axons elongate and reestablish appropriate synaptic contacts with second-order neurons of the olfactory pathway (Graziadei and Monti Graziadei, 1980; Doucette et al., 1983). A noticeable difference between the olfactory bulb and other CNS regions non-permissive to axonal growth resides in the presence of olfactory ensheathing glia (OEG) in the olfactory bulb layers where uninjured and injured olfactory axons navigate. This unique glial type was first identified in the olfactory bulb by Golgi (1875) and Blanes (1898), and more recently studied by other authors (Barber and Lindsay, 1982; Doucette, 1984; reviewed in: Ramón-Cueto and Valverde, 1995; Ramón-Cueto and Avila, 1998). OEG exhibit features that could account for their axonal growth-promoting properties (reviewed in Ramón-Cueto and Avila, 1998). They produce neurotrophic factors and express molecules in their membranes known to be involved in cell adhesion and axonal elongation. Moreover, OEG ensheathe olfactory axons, preventing their exposure to inhibitory CNS molecules (Blanes, 1898; Barber and Lindsay, 1982; Doucette, 1984, 1990; Pasterkamp et al., 1998).

Spinal cord injury is a prevalent affliction that causes devastating functional consequences in patients. Injuries to the spinal cord (traumatic, ischemic, compressive) interrupt axonal tracts at the lesion site and, therefore, the sensory (ascending) and motor (descending) information carried by the damaged fibers. The functional consequences of this permanent disconnection between brain and spinal cord depend on the severity of the lesion and on the number of fibers affected. Over the past decades, several attempts have been made to find a repair strategy that circumvents and blocks the hostile CNS milieu, providing injured spinal cord axons with a supportive environment for their elongation (reviewed in: Olson, 1997; Bregman, 1998; Fawcett, 1998). Most of them have focused on the repair of incomplete spinal cord injuries, although some concentrate on

---

[*] Corresponding author: Dr. Almudena Ramón-Cueto, Neural Regeneration Group, Institute of Biomedicine, Spanish Council for Scientific Research (CSIC), Jaime Roig 11, 46010 Valencia, Spain. Fax: +34-963-690-800; E-mail: aramon@ibv.csic.es

healing complete lesions (Xu et al., 1995, 1997; Cheng et al., 1996; Menei et al., 1998; Ramón-Cueto et al., 1998, 2000; Rapalino et al., 1998). The axon growth-promoting properties of OEG have successfully been used to foster the regeneration of selectively injured tracts of the adult mammalian spinal cord (Ramón-Cueto and Nieto-Sampedro, 1994; Li et al., 1997; Navarro et al., 1999), and the repair of completely transected spinal cords (Ramón-Cueto et al., 1998, 2000). In the present article, we will review our results using transplants of pure OEG as tools to promote axonal regeneration into injured adult mammalian spinal cords and functional recovery of paraplegic rats. We will also discuss the results obtained by other authors in this field.

Pure OEG cultures for transplantation into the CNS

In all of our transplantation studies we have used pure OEG cultures obtained from adult rat olfactory bulbs (Ramón-Cueto and Nieto-Sampedro, 1994; Ramón-Cueto et al., 1998, 2000). Although these cells can also be cultured from embryonic and neonatal olfactory bulbs and epithelia (reviewed in Ramón-Cueto and Avila, 1998), adult OEG appear more suitable candidates for transplantation into the mature CNS for several reasons. OEG are originally located in an adult environment where they normally have an axonal growth-promoting role. Therefore, adult OEG might integrate and exert their effects better in the mature CNS after transplantation than young OEG. OEG are fully differentiated and committed in the adult olfactory bulb, and therefore they might be less susceptible to phenotypic changes after transplantation than immature OEG. In addition, adult OEG might open a possibility for autologous transplantation into the injured CNS in the future.

We have used adult Wistar (Ramón-Cueto and Nieto-Sampedro, 1992, 1994; Ramón-Cueto et al., 1993, 2000) and Fischer (Ramón-Cueto et al., 1998) male and female rats to prepare olfactory bulb primary cultures. The bulbs of these animals were dissected and the meninges completely removed to avoid fibroblast contamination of the cultures. Because OEG are located in the olfactory nerve and glomerular layers of this structure, we dissected, dissociated, and cultured only these two olfactory bulb layers. This will diminish the number of contaminating microglial cells from the cultures.

Primary cultures contain three different cell types identified by their immunocytochemical and ultrastructural properties: macrophages/microglia, endothelial cells, and OEG (Ramón-Cueto et al., 1993). If the final aim of the cultures is to use them for transplantation into the CNS, and to promote axonal elongation or survival, purification of OEG from the other contaminant cells seems necessary. Macrophages/microglia have opposing effects on axonal growth and neuronal survival that could interfere with those of OEG (Vaca and Wendt, 1992), including production of neurotoxic molecules that could affect neuron viability (Giulian, 1993). On the other hand, these cells are a source of neurotrophic factors (Elkabes et al., 1996), and after transplantation into lesioned spinal cords they enhance axonal growth (Rabchevsky and Streit, 1997) and promote partial functional recovery (Rapalino et al., 1998). We have developed a method to purify OEG from the other two cell types present in the primary cultures by immunoaffinity, using an antibody against p-75^{NFGR} (Ramón-Cueto and Nieto-Sampedro, 1994; Ramón-Cueto et al., 1998). OEG were purified from 7- to 10-days-old olfactory bulb cultures, and pure OEG were expanded in serum-containing medium, supplemented with forskolin and pituitary extract (Fig. 1). Recent studies by Pollock et al. (1999) indicated that neuregulin can also be used as a mitogen for OEG. We usually grow pure cultures (98% purity) for 15 days before transplantation, but they can be maintained in vitro for longer periods. However, after several passages of the cells and several months of survival, OEG may spontaneously immortalize (Sonigra et al., 1996). OEG were labeled with the nuclear fluorochrome, Hoechst 33342, and then injected into the CNS to stimulate the growth of injured axons.

OEG promotion of axonal regeneration into the adult mammalian spinal cord

Repair of selectively injured tracts of the spinal cord

The ability of OEG transplants to promote axonal regeneration in the adult rat CNS was first described in 1994 (Ramón-Cueto and Nieto-Sampedro, 1994).

In this pioneer study, sectioned dorsal root axons were able to regenerate through the non-permissive environment of the spinal cord after OEG transplantation (Fig. 2). Under normal conditions (with no treatments), injured dorsal root axons regenerate through the PNS milieu of the dorsal root but they are unable to overcome the inhibitory nature of the spinal cord, and cannot cross the CNS–PNS transitional zone (Ramón y Cajal, 1928; reviewed in Fraher, 1999). One thoracic dorsal root was transected in eight adult rats and the stump sutured to the spinal cord at its original place. A suspension of pure OEG was transplanted in the spinal cord at the root entry zone (Fig. 2A). Three weeks later dorsal root axons were observed reentering the spinal cord and regenerating through the CNS for several millimeters (Fig. 2B,C). Primary sensory afferents invaded the spinal cord laminae they innervate under normal conditions (laminae 1, 2, 3, 4, 5, and 10), but did not grow through the ventral horn and the dorsal columns. In a more recent study, Navarro et al. (1999) reported that OEG also promoted functional reconnection of regenerating sensory afferents. They performed multiple lumbar rhizotomies (from L3 to L6), close to their entrance into the spinal cord, and reapposed sectioned roots to their original place. OEG were purified using antibody-coated magnetic beads and transplanted at the dorsal root entry zone. Two months after transplantation, sensory fibers regenerated for several millimeters into the spinal cords and restored spinal reflex responses. Primary olfactory bulb cultures, containing macrophages/microglia, endothelial cells and OEG (Ramón-Cueto et al., 1993), have been used to repair other selectively injured tracts of the spinal cord (Li et al., 1997, 1998).

Repair of completely transected spinal cords

We have analyzed the repair properties of pure OEG in transected adult rat spinal cords because this type of spinal cord injury (traumatic) closely resembles those found in clinical situations. Traumatic injuries are the type of spinal cord lesion more frequently observed in human patients. Such injuries usually involve several tracts, cause a transection of spinal axons and create a gap at the injury site that regenerating axons have to cross. Moreover, after a contusive lesion, sometimes the neurosurgeon has to 'clean' the injured region and remove some pieces of bone and other debris by resecting the necrotic area. This causes an additional transection of part or the whole spinal cord depending on the severity of the lesion. In addition, we chose complete transection because we intended to model the most severe clinical scenario. Any technique repairing this type of lesion would have greater benefits in less severe injuries.

Two different experimental paradigms were used to test the efficacy of OEG as tools to repair completely sectioned spinal cords (Figs. 3 and 4; Ramón-Cueto et al., 1998, 2000). In one model, OEG were used to enhance the regenerative effect of Schwann cell-seeded guidance channels (Fig. 3A; Ramón-Cueto et al., 1998). One PAN/PVC tube filled with matrigel and Schwann cells was used to bridge both spinal cord stumps after removal of one thoracic segment (T9) (Xu et al., 1995, 1997). Schwann cells create a cable inside the tube that joins both cord stumps. Providing this repair strategy with no other treatments, only axons from some sensory and propriospinal neurons whose bodies are near the transection site regenerated into the cables. However, no extension of corticospinal or brainstem fibers was observed through the cables. OEG were transplanted into the midline of both spinal cord stumps (200,000 cells/stump), at 1 mm from the channel end, and animals allowed to survive for 6 weeks (Fig. 3A). After this period, OEG had migrated from the injection sites and invaded the entire injury region and the Schwann cell cable (Fig. 3B). Combining these two repair procedures, we observed that numerous regenerating axons invaded the Schwann cell and OEG-containing cables and crossed the injury site. Moreover, OEG transplantation elicited axonal growth from distant supraspinal (raphe) and propriospinal (from T10 to L3) neurons. These fibers crossed the transection site and regenerated for the longest distances analyzed (1.5 cm and 2.5 cm) within caudal and rostral spinal cord stumps, respectively (Fig. 3B). Strikingly, many regenerating fibers and essentially all of the serotonergic fibers did not invade Schwann cell-containing cables, and instead crossed the transection gap through the OEG-containing connective tissue surrounding the guidance channels (Fig. 3C,D). This indicates that for some axonal types, the environment created by OEG and fibroblasts is better than that created by Schwann cells, matrigel and OEG. It also suggests that OEG and fibroblasts provided a good substrate for the elongation of injured axons.

The center of a stab lesion is occupied by fi-

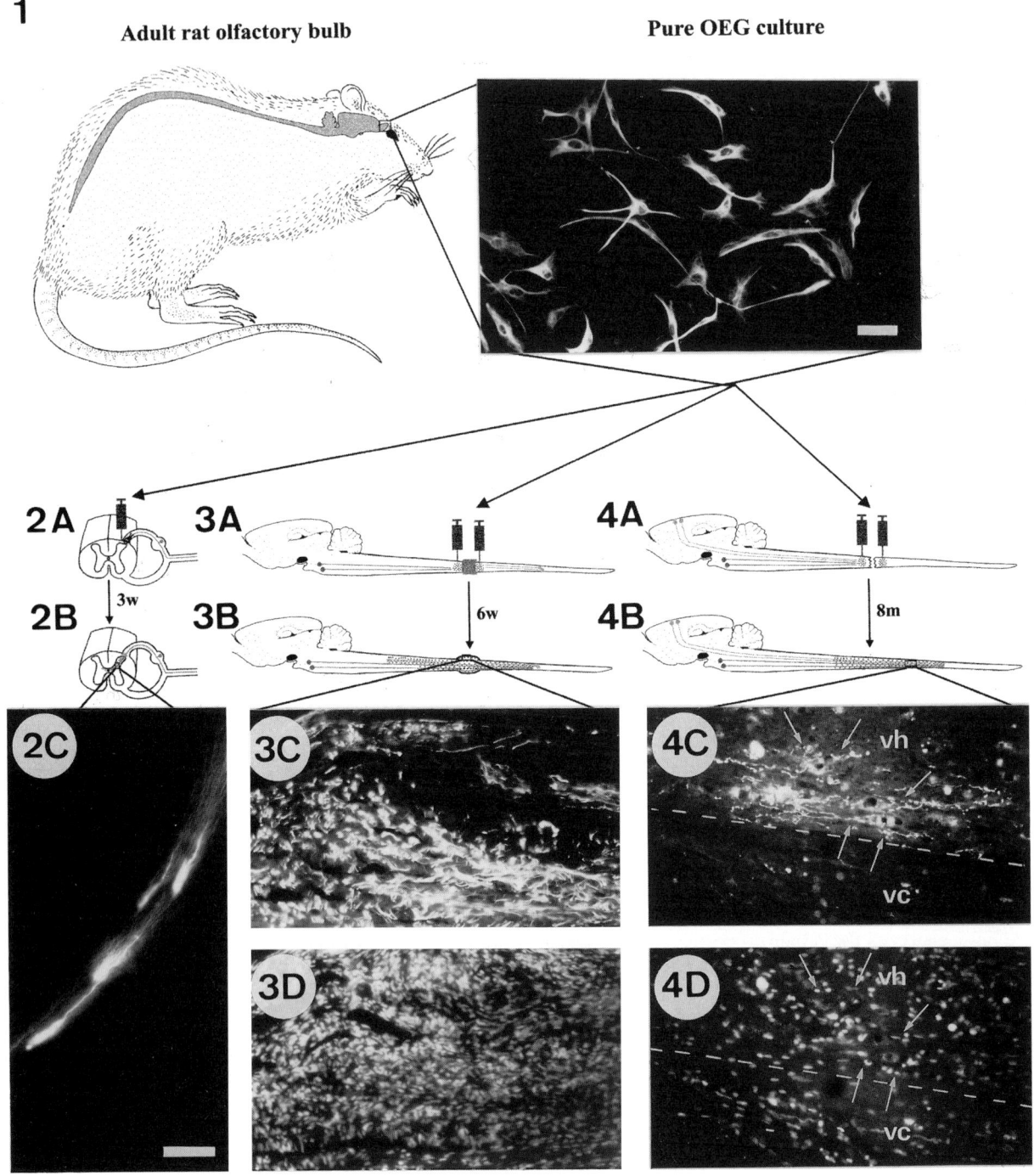

broblasts and reactive glia, and it is inhibitory to the growth of mature axons (Ramón y Cajal, 1928; Krikorian et al., 1981; Silver, 1994; Davies et al., 1999). OEG transplants enhanced the axonal growth-promoting properties of Schwann cells, and made possible the regeneration of axons through the gliotic and connective tissue of the injury region (see above, Ramón-Cueto et al., 1998). In a more recent study, we tested if OEG by themselves, with no additional treatments, could effectively promote functional recovery of paraplegic rats and axonal elongation in their completely transected cords (Fig. 4; Ramón-Cueto et al., 2000). Spinal cords of nine adult rats were sectioned at the thoracic level (T8) and suspensions of pure OEG were injected into the midline of both cord stumps, at 1 mm from their edge (200,000 cells/stump; Fig. 4A). From 3 to 7 months after surgery, recovery of voluntary hindlimb movement was assessed using a new climbing test with four different difficulty levels. This test allows a distinction to be made between local reflex activity and real voluntary movement. The rats' responses to light touch (contact placing) and joint bending (proprioceptive) stimuli were also evaluated. All OEG-transplanted paraplegic animals presented voluntary hindlimb movement, plantar placement of the paw, and body weight support, and also recovered light touch and proprioception. The regeneration of injured corticospinal, raphespinal (serotonergic), and coeruleospinal (noradrenergic) axons in the damaged cords was analyzed 8 months after surgery. In all OEG-transplanted rats, these injured axons crossed the lesioned area. They invaded the rostral glial fibrillary acidic protein (GFAP)-immunoreactive gliotic tissue, entered the GFAP-negative central core, and reached and crossed the caudal GFAP-positive glial scar (Fig. 4B). Sensory axons (calcitonin gene-related peptide-immunoreactive) elongated rostrally and crossed the GFAP-positive scar of the caudal stump, the GFAP-negative connective tissue, and the gliotic tissue of the rostral stump. OEG occupied the entire injury region, intermingled with reactive glia, and were observed in the same locations as regenerating axons (Fig. 4B). Corticospinal, raphespinal and coeruleospinal axons traversed the caudal glial scar and regenerated for long distances (up to 3 cm, L6) within the distal spinal cord stump (Fig. 4B,C). Strikingly, corticospinal axons avoided growing through the white matter, whereas serotonergic and noradrenergic fibers indistinctly elongated through white or gray matter. The latter two axonal types invaded those spinal cord regions they innervate under normal circumstances, indicating that regenerating axons were probably following specific cues. Moreover, these fibers were observed delineating some neurons in the ventral and dorsal horns,

---

Fig. 1. Pure OEG cultures (right) were obtained from adult rat olfactory bulbs (left) and transplanted into adult rat spinal cords. Right panel shows a photomicrograph of a pure OEG culture immunolabeled with antibody 4.11.C. (Heredia et al., 1998). Scale bar: 50 μm.

Fig. 2. (A) OEG injection into the spinal cord after rhizotomy of one thoracic dorsal root and anastomosis of the root to the spinal cord. (B) Three weeks later (3w), dorsal root axons (gray line) regenerated into the spinal cord and OEG (dots) migrated and surrounded them. Box corresponds to (C). (C) Photomicrograph of the region boxed in (B) showing DiI-labeled dorsal root axons regenerating through lamina IV and V of the dorsal horn ipsilateral to the lesion. Scale bar represents 70 μm in this and all subsequent photomicrographs.

Fig. 3. (A) OEG transplantation into the stumps of a completely transected spinal cord, at both ends of a Schwann cell-filled guidance channel. (B) Six weeks later (6w), injured axons (gray lines) crossed the transection site and regenerated into both spinal cord stumps. OEG (dots) migrated and axons and OEG were observed in the same spinal cord locations. Box represents (C) and (D). (C,D) Same spinal cord section showing neurofilament-immunoreactive fibers (C) and Hoechst-labeled OEG (D) invading the connective tissue at the injury site (region boxed in B).

Fig. 4. (A) OEG transplantation into both stumps of a spinal cord that was completely transected at T8 level. (B) Eight months after surgery (8m), corticospinal and brainstem injured axons (gray lines) regenerated through the transection site and into the caudal spinal cord stump. OEG (dots) migrated and were found in the same regions as axons. Box delimits the regions shown in the next two panels. (C) Spinal cord section showing serotonin-immunoreactive fibers regenerating through lamina VIII of the ventral horn (vh) near the ventral column white matter (vc), at 1 cm from the transection site. Notice some fibers delineating the bodies of some neurons (arrows). (D) Same field shown in (C) revealing Hoechst-labeled OEG. The labeled nuclei surround neuronal bodies and colocalize with the serotonergic fibers pointed to by arrows in (C) (compare arrows in C and D). Dashed lines in (C) and (D) delimit the ventral horn (vh) and column (vc).

indicating that some regenerating axons might have reached target neurons (Fig. 4C).

OEG migration after transplantation

In all the transplantation paradigms described above, OEG integrated successfully within the host parenchyma, and host spinal cords neither showed disruption of their cytoarchitecture nor any sign of inflammation. OEG migrated from the injection sites and were found in the same spinal cord locations as regenerating axons (Figs. 2B, 3B and 4B; compare Fig. 3C with 3D, and Fig. 4C with 4D). Therefore, it appears that OEG accompany growing axons through the inhibitory CNS environment and provide injured axons with the appropriate factors for their elongation. Migration of OEG in the spinal cord appeared to be greater after injection of pure OEG (Ramón-Cueto et al., 1998, 2000) than after transplantation of olfactory bulb cultures (Li et al., 1998). In the latter, the molecules produced by the other cell types present in the transplants (see above) might hinder the ability of OEG to freely move within the CNS. The migratory ability of OEG constitutes an advantage compared to other repair strategies in which grafts remain at the transplantation site or migrate just for short distances. A 'mobile graft', such as OEG, could provide axons with the appropriate molecular environment for their regeneration at different distances from the injury site, and thus help them to reach distant specific target fields.

After transplantation of primary olfactory bulb cultures, a suppression of scar formation by host astrocytes was reported (Li et al., 1998). In this paradigm, molecules released by the cells present in the transplants and migrating from the PNS (see above) might affect the response of host astrocytes to the injury. In our experimental models, GFAP labeling revealed that OEG did not prevent reactive astrogliosis at the injury site. However, it is still unknown if these cells could change the molecular composition of the gliotic tissue. Glial scars are a source of repulsive molecules such as tenascin, proteoglycans, ephrins and semaphorins, which prevent axons from growing through (Giulian, 1993; Silver, 1994; Pasterkamp et al., 1998; Davies et al., 1999; Miranda et al., 1999). Myelin from white matter tracts also constitutes a potent inhibitor to axonal elongation (Schwab et al., 1993). However, OEG migrated through glial scars and white matter and promoted the regeneration of injured axons through these repulsive chemical barriers. Although the specific mechanisms underlying OEG promotion of axonal regeneration remain to be determined, it is plausible that the molecules produced by OEG (reviewed in: Ramón-Cueto and Valverde, 1995; Ramón-Cueto and Avila, 1998) might counteract the effect of the chemorepellent molecules by shifting the balance between inhibitors and stimulators toward a growth-promoting effect. OEG might also ensheathe growing axons, preventing their exposure to these inhibitors (Pasterkamp et al., 1998).

Concluding remarks

There are now several indications showing that OEG provide a very promising tool to repair injured adult mammalian spinal cords. These cells promote functional and structural repair of selectively injured spinal tracts and completely transected adult rat spinal cords. OEG can be obtained from adult donors and this offers the future possibility of autologous transplantation. Moreover, OEG are originally located in the CNS, and integration into adult CNS and migration should be natural. Therefore, OEG open a new avenue in the search for an effective therapeutic procedure to treat spinal cord injury and CNS trauma. The next challenge is to extend our studies to primates, as this would determine to what extent our technique could also be applied to humans.

Acknowledgements

We are grateful to M. Bautista, J. Belio and J.A. Pérez for photographic work. This work was supported by A. García and M. Aguilar from 'Levantina del Calzado S.L.'.

References

Barber, P.C. and Lindsay, R.M. (1982) Schwann cells of the olfactory nerves contain glial fibrillary acidic protein and resemble astrocytes. *Neuroscience*, 7: 3077–3090.

Blanes, T. (1898) Sobre algunos puntos dudosos de la estructura del bulbo olfatorio. *Rev. Trim. Micrograf.*, 3: 99–127.

Bregman, B.S. (1998) Regeneration in the spinal cord. *Curr. Opin. Neurobiol.*, 8: 800–807.

Cheng, H., Cao, Y. and Olson, L. (1996) Spinal cord repair in adult paraplegic rats: partial restoration of hind limb function. *Science*, 273: 510–513.

Davies, S.J.A., Goucher, D.R., Doller, C. and Silver, J. (1999) Robust regeneration of adult sensory axons in degenerating white matter of the adult rat spinal cord. *J. Neurosci.*, 19: 5810–5822.

Doucette, J.R. (1984) The glial cells in the nerve fiber layer of the rat olfactory bulb. *Anat. Rec.*, 210: 385–391.

Doucette, R. (1990) Glial influences on axonal growth in the primary olfactory system. *Glia*, 3: 433–449.

Doucette, J.R., Kiernan, J.A. and Flumerfelt, B.A. (1983) The re-innervation of olfactory glomeruli following transection of primary olfactory axons in the central or peripheral nervous system. *J. Anat.*, 137: 1–19.

Elkabes, S., DiCicco-Bloom, E. and Black, I.B. (1996) Brain microglia/macrophages express neurotrophins that selectively regulate microglial proliferation and function. *J. Neurosci.*, 16: 2508–2521.

Fawcett, J.W. (1998) Spinal cord repair: from experimental models to human application. *Spinal Cord*, 36: 811–817.

Fawcett, J.W. and Keynes, R.J. (1990) Peripheral nerve regeneration. *Annu. Rev. Neurosci.*, 13: 43–60.

Fraher, J.P. (1999) The transitional zone and CNS regeneration. *J. Anat.*, 194: 161–182.

Giulian, D. (1993) Reactive glia as rivals in regulating neuronal survival. *Glia*, 7: 102–110.

Golgi, C. (1875) Sulla fina anatomia del bulbi olfatorii. *Ti Revista Sperimentale di Freniatria*, 1: 403–425.

Graziadei, P. and Monti Graziadei, G. (1980) Neurogenesis and neuron regeneration in the olfactory system of mammals, III. Deafferentation and reinnervation of the olfactory bulb following section of the fila olfactoria in rat. *J. Neurocytol.*, 9: 145–162.

Heredia, M., Gascuel, J., Ramón-Cueto, A., Santacana, M., Avila, J., Masson, C. and Valverde, F. (1998) Two novel monoclonal antibodies (1.9.E. and 4.11.C.) against olfactory bulb ensheathing glia. *Glia*, 24: 352–364.

Krikorian, J.C., Guth, L. and Donati, E.J. (1981) Origin of the connective tissue in the transected rat spinal cord. *Exp. Neurol.*, 72: 698–707.

Li, Y., Field, P.M. and Raisman, G. (1997) Repair of adult rat corticospinal tract by transplants of olfactory ensheathing cells. *Science*, 277: 2000–2002.

Li, Y., Field, P.M. and Raisman, G. (1998) Regeneration of adult rat corticospinal axons induced by transplanted olfactory ensheathing cells. *J. Neurosci.*, 18: 10514–10524.

Menei, P., Montero-Menei, C., Whittemore, S.R., Bunge, R.P. and Bunge, M.B. (1998) Schwann cells genetically modified to secrete human BDNF promote enhanced axonal regrowth across transected adult rat spinal cord. *Eur. J. Neurosci.*, 10: 607–621.

Miranda, J.D., White, L.A., Marcillo, A.E., Wilson, C.A., Jagrid, J. and Whittemore, S.R. (1999) Induction of Eph B3 after spinal cord injury. *Exp. Neurol.*, 156: 218–222.

Navarro, X., Valero, A., Gudino, G., Flores, J., Rodríguez, F.J., Verdú, E., Pascual, R., Cuadras, J. and Nieto-Sampedro, M. (1999) Ensheathing glia transplants promote dorsal root regeneration and spinal reflex restitution after multiple lumbar rhizotomy. *Ann. Neurol.*, 45: 207–215.

Olson, L. (1997) Regeneration in the adult central nervous system: experimental repair strategies. *Nat. Med.*, 3: 1329–1335.

Pasterkamp, R.J., De Winter, F., Holtmaat, A.J.G.D. and Verhaagen, J. (1998) Evidence for a role of the chemorepellent semaphorin III and its receptor neuropilin-1 in the regeneration of primary olfactory axons. *J. Neurosci.*, 18: 9962–9976.

Pollock, G.S., Franceschini, I.A., Graham, G., Marchionni, M. and Barnett, S.C. (1999) Neuregulin is a mitogen and survival factor for olfactory bulb ensheathing cells and an isoform is produced by astrocytes. *Eur. J. Neurosci.*, 11: 769–780.

Rabchevsky, A.G. and Streit, W.J. (1997) Grafting of cultured microglial cells into the lesioned spinal cord of adult rats enhances neurite outgrowth. *J. Neurosci. Res.*, 47: 34–48.

Ramón y Cajal, S. (1928) Degeneration and regeneration of the nervous system. In: DeFelipe, J. and Jones, E.G. (Eds.), *Cajal's Degeneration and Regeneration of the Nervous System.* Oxford University Press, New York, 1991.

Ramón-Cueto, A. and Avila, J. (1998) Olfactory ensheathing glia: properties and function. *Brain Res. Bull.*, 46: 175–187.

Ramón-Cueto, A. and Nieto-Sampedro, M. (1992) Glial cells from the adult rat olfactory bulb: immunocytochemical properties of pure cultures of ensheathing cells. *Neuroscience*, 47: 213–220.

Ramón-Cueto, A. and Nieto-Sampedro, M. (1994) Regeneration into the spinal cord of transected dorsal root axons is promoted by ensheathing glia transplants. *Exp. Neurol.*, 127: 232–244.

Ramón-Cueto, A. and Valverde, F. (1995) Olfactory bulb ensheathing glia: a unique cell type with axonal growth-promoting properties. *Glia*, 14: 163–173.

Ramón-Cueto, A., Pérez, J. and Nieto-Sampedro, M. (1993) In vitro enfolding of olfactory neurites by p75 NGF receptor positive ensheathing cells from adult rat olfactory bulb. *Eur. J. Neurosci.*, 5: 1172–1180.

Ramón-Cueto, A., Plant, G.W., Avila, J. and Bunge, M.B. (1998) Long-distance axonal regeneration in the transected adult rat spinal cord is promoted by olfactory ensheathing glia transplants. *J. Neurosci.*, 18: 3803–3815.

Ramón-Cueto, A., Cordero, M.I., Santos-Benito, F.F. and Avila, J. (2000) Functional recovery of paraplegic rats and motor axon regeneration in their spinal cords by olfactory ensheathing glia. *Neuron*, 25: 425–435.

Rapalino, O., Lazarov-Spiegler, O., Agranov, E., Velan, G.J., Yoles, E., Fraidakis, M., Solomon, A., Gepstein, R., Katz, A., Belkin, M., Hadani, M. and Schwartz, M. (1998) Implantation of stimulated homologous macrophages results in partial recovery of paraplegic rats. *Nat. Med.*, 4: 814–821.

Schwab, M.E., Kapfhammer, J.P. and Bandtlow, C.E. (1993) Inhibitors of neurite growth. *Annu. Rev. Neurosci.*, 16: 565–595.

Silver, J. (1994) Inhibitory molecules in development and regeneration. *J. Neurol.*, 242: 22–44.

Sonigra, R.J., Kandiah, S.S. and Wigley, C.B. (1996) Spontaneous immortalisation of ensheathing cells from adult rat olfactory nerve. *Glia*, 16: 247–256.

Vaca, K. and Wendt, E. (1992) Divergent effects of astroglia and microglia secretions on neuron growth and survival. *Exp. Neurol.*, 118: 62–72.

Xu, X.M., Guenard, V., Kleitman, N. and Bunge, M.B. (1995) Axonal regeneration into Schwann cell-seeded guidance channels grafted into transected adult rat spinal cord. *J. Comp. Neurol.*, 351: 145–160.

Xu, X.M., Chen, A., Guénard, V., Kleitman, N. and Bunge, M.B. (1997) Bridging Schwann cell transplants promote axonal regeneration from both the rostral and caudal stumps of transected adult rat spinal cord. *J. Neurocytol.*, 26: 1–16.

CHAPTER 24

Precursor cells for transplantation

Mahendra S. Rao [1,*] and Margot Mayer-Proschel [2]

[1] Department of Neurobiology and Anatomy and [2] Department of Oncological Sciences (HCI), University of Utah Medical School, 50 North Medical Drive, Salt Lake City, UT 84132, USA

Introduction

Acquisition of cell type-specific properties in the nervous system is a process of sequential restriction in developmental potential. Multiple classes of precursor cells have been identified and shown to be present in distinct spatial and temporal domains. Multipotent stem cells, neuronal precursors termed NRP (neuronal restricted precursor) cells, and multiple classes of glial precursors including GRP (glial restricted precursor) cells, O-2A (oligodendrocyte-type-2 astrocyte) precursor cells, APC (astrocyte precursor cells), etc., have been characterized. Antibodies to cell surface epitopes that distinguish among these cell types have been identified and used to purify specific populations of cells. Various precursor populations have been transplanted into neonatal and adult rat brains and their ability to integrate and differentiate into appropriate phenotypes has been analyzed. In this review the relative merits of different precursor populations for replacement and drug delivery therapy are discussed.

Cell therapy and spinal cord injury

The spinal cord constitutes less than 4% of the total central nervous system (CNS) volume. This 4%, however, includes all the major tracts carrying motor and sensory information between the body and the brain. Interruption of these connections after localized spinal cord trauma has functional consequences that far exceed any seen due to local neuronal/glial loss in other brain regions. Further, spinal cord injuries in humans and in other mammals are not followed by any significant regrowth of long axons, thus leading to long-term permanent deficits. The long-term consequences of these functional deficits are devastating economically, socially and psychologically for patients, their relatives and society. Recent advances in critical care of acute spinal injury and better long-term rehabilitation and chronic care have increased the life span of affected individuals, paradoxically increasing the number of patients who live with the consequence of a loss of voluntary motor control. This exacerbation of the burdens on patients and society has provided an important impetus for research into preventative, rehabilitative and reconstructive strategies.

Increased research effort has yielded considerable progress in understanding mechanisms that may limit cell injury, promote cell survival and activate regeneration. While an increased body of knowledge has led to insights concerning the mechanisms underlying damage, many questions remain to be answered to harness this knowledge and to design clinically effective replacement or regenerative strategies. In the past decade many investigators have focused on methods to coax regeneration by providing trophic support or by neutralizing inhibitory signals that restrict regrowth (reviewed in: Schwab,

---

* Corresponding author: Dr. Mahendra S. Rao, Department of Neurobiology and Anatomy, University of Utah Medical School, 50 North Medical Drive, Salt Lake City, UT 84132, USA. Fax: +1-801-581-4233; E-mail: Mahendra.Rao@hsc.utah.edu

TABLE 1

Potential roles of transplanted cells in spinal injury

Reduction of the inflammatory response/scar formation
Maintenance of synapses or reduction of axonal degeneration
Providing trophic support for surviving cells
Replacement of localized neuronal loss
Myelination of axons packing the cavity
Introduction of foreign genes
Reconstitution of an appropriate cellular environment
Mobilization of endogenous stem cell populations
Promoting revascularization

The potential roles of transplanted cells in therapeutic repair of spinal cord injuries are listed. The list was compiled from multiple reviews on the role of transplanted tissue in improving repair and regeneration.

1990; Schwab and Brösamle, 1997; Boonman and Isacson, 1999; Dunnett, 1999; Lu and Waite, 1999). Recently, several groups have attempted to provide a more permissive environment for regeneration by transplanting fetal tissue into the injured spinal cord (Reier et al., 1986; Katsuki et al., 1997; Miya et al., 1997; Akesson et al., 1998; Diener and Bregman, 1998). While a detailed discussion of fetal transplants is beyond the scope of this review, some results pertinent to this discussion of precursor cells are summarized in Table 1. In general it appears that embryonic tissue is more effective than adult tissue for promoting functional improvement (Reier et al., 1992; Theele et al., 1996). Spinal tissue appears to be superior to tissue from other regions (Asada et al., 1998). In addition, integration with host tissue is dependent on the degree of apposition of the graft with host, the time of transplantation, the development of revascularization, and the immune response (reviewed in: Stein, 1991; Reier et al., 1992). Fetal transplants seem to result in high rates of cell survival and functional integration with axonal regrowth across the damaged site, along with good evidence of reduction in glial scarring. Progress in using transplanted tissue, however, has been constrained by ethical considerations, the limited availability of appropriate tissue and the inability to maintain tissue fragments for prolonged periods in culture. The requirement that fresh tissue be readily available in close proximity to a transplant center has limited the number of studies examining such tissue grafts.

Dissociated cells offer the possibility of bypassing some of these problems. Recent research has identified multiple classes of precursor cells (reviewed in: Rao, 1999; Lee et al., 2000). These can be obtained from both adult and fetal tissue (reviewed in Kalyani and Rao, 1998). Cells can be maintained for prolonged periods and retain their abilities to differentiate and incorporate into the host environment. Dissociated cells also offer the potential of reduced damage and ease of transplantation when compared to tissue fragments, and in principle, offer greater control over the uniformity of cell number, cell type and factors that may be critical to successful repair. Indeed, early work using dissociated cells suggests that this is not simply a theoretical consideration but rather an important practical one (see below). The potential role of transplanted cells in promoting functional recovery is summarized in Table 1. Functions such as promoting revascularization or modulating the immune response are likely as critical as replacing cells lost to damage. It is clear, however, that the functionally diverse roles expected of cellular replacement cannot be fulfilled by the transplantation of any single universal donor cell type; rather, specialized cells or titered mixtures of cells may be important in replacement strategies.

Which cell types will work and how effective they will be is under active study, and the variety of cells that have been used is quite impressive. These include macrophages and activated T-cells, multiple classes of glial cells, neuronal precursors, multipotent CNS stem cells, pluripotent embryonic stem cells and their derivatives and differentiated postmitotic cells. Currently our operating assumption is that postmitotic cells do poorly in most transplant paradigms and that dividing immature cells that likely will respond to environmental signals to differentiate into region-specific phenotypes will be more valuable. As a result, multiple studies have focused on identifying dividing precursor cells. Indeed a large number of such cell types have been described and characterized. These cell types may be functionally classified as multipotent cells, neuronal precursors and glial precursors. In this review we discuss the relative advantages and disadvantages of different classes of cells, their potential therapeutic use as evident from transplant studies, and potential sources of cells for transplant therapy.

Multiple classes of precursor cells are available for cell replacement therapy

The choice of which cell type is most appropriate for transplantation in spinal cord injury depends on evaluating the functional benefit and the properties of the cells used for transplantation. For the purposes of this review we have classified precursor cells into three major categories based on the progeny of the cells any specific precursor population is able to generate in vivo (Figs. 1–3): (1) multipotent cells (neural stem cells [NSC]) that can differentiate into neurons, oligodendrocytes and astrocytes; (2) more restricted neuronal precursors that can generate multiple classes of neurons but cannot differentiate into astrocytes and oligodendrocytes; and (3) glial precursors that are restricted to differentiating into solely glial derivatives. Each of these categories can be further subdivided based on differences in culture conditions, methods of isolation, cell surface receptors expressed, response to cytokines, age of isolation and specific brain region of origin (reviewed in: Kalyani and Rao, 1998; Rao, 1999). Thus, several different multipotent precursor cells can be distinguished (Fig. 1). Cortical stem cells (Gensert and Goldman, 1996; Mehler and Gokhan, 1999) present in the adult can be distinguished from ventricular/subventricular stem cells based on antigen expression and differentiation bias. Likewise, spinal cord multipotent stem cells can be distinguished from cortical stem cells based on receptor expression and cytokine response (Kalyani et al., 1997; Shihabuddin et al., 1997). Examples of subclasses of neuronal and glial precursor cells are summarized in Figs. 2 and 3, respectively. We emphasize that a classification into neuronal or glial precursors should be made only if the precursor cell population under consideration fails to differentiate into neurons or glia under conditions in which multipotent or other precursors readily differentiate into these populations. Multiple conditions should be tested, and ideally, in vivo transplants should be used to confirm a limitation in differentiation potential.

Precursors like NSC, NRP, GRP or O-2A cells have all been shown to be capable of extensive self-renewal, and they can be maintained in culture for several passages, induced to express foreign genes, and, when transplanted, can differentiate into appropriate predicted phenotypes. Precursor cells therefore are likely candidates for cell replacement therapy, and the choice of an appropriate cell type depends on the therapeutic outcome required. The unique advantages and disadvantages of individual precursor populations are discussed in subsequent sections.

Multipotent stem cells

One approach to cellular replacement is to transplant multipotential cell populations (that can generate

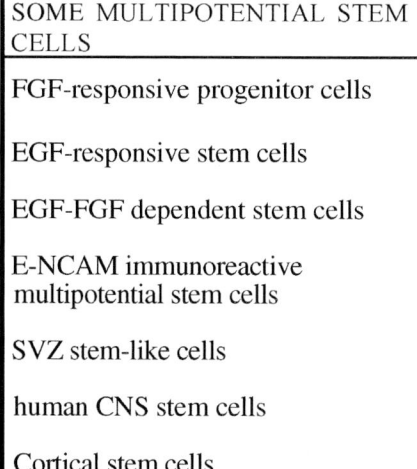
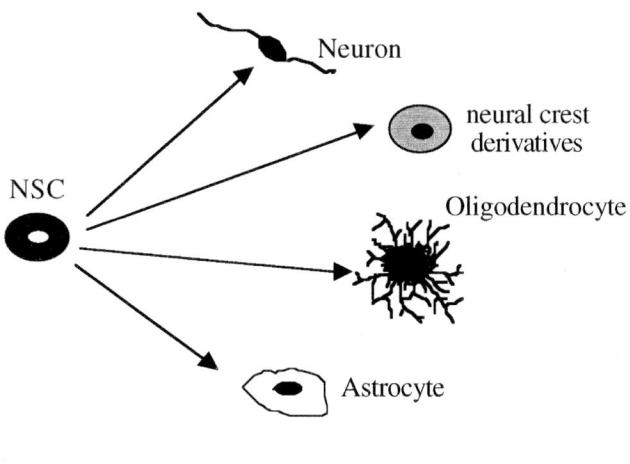

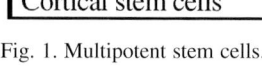
Fig. 1. Multipotent stem cells.

any required cell population) regardless of the specific need. The rationale behind such an approach is that while these cells are able to generate all differentiated cells types, restriction to the required phenotype will be regulated by the host environment. A recent study by Yandava et al. (1999) also demonstrated that multipotent cells seem to undergo a bias in their fate choice depending upon the 'need'. In these experiments, neural multipotential stem cells were transplanted into newborn dysmyelinated *shiverer* (*shi*) mice characterized by a 'global' lack of myelin. The authors suggest that a shift occurred toward oligodendrocyte differentiation and showed that a subgroup of the transplanted cells myelinated 40% of host neuronal processes. Modulating the environment in vitro has also been shown to bias the differentiation potential of stem cells (Davis and Temple, 1994; Snyder et al., 1995; Gritti et al., 1996; Johe et al., 1996; Burrows et al., 1997; Milward et al., 1997; Qian et al., 1997; Shen et al., 1998; Carpenter et al., 1999; Palmer et al., 1999; Vescovi and Snyder, 1999). With this view, several groups have worked on isolating multipotent stem cells (Fig. 1). Both fetal and adult stem cells have been isolated, have been maintained in culture for several years, and have been shown to retain their ability to differentiate even after prolonged passage in culture.

The potential advantages and disadvantages of using multipotent stem cells for transplantation are summarized in Table 2. Perhaps the single most important advantage of multipotent stem cells is the ability to generate large numbers of cells. NSC can be grown as floating spheres or adherent cells over long time periods. Cells continue to divide throughout this period, thus providing an economical way for generating the large cell numbers required for transplantation. Similar methods of isolation, generation and propagation of multipotent cells have been established in the human system (Chalmers-Redman et al., 1997; Carpenter et al., 1999; Vescovi et al., 1999; Yandava et al., 1999), suggesting a straightforward source of human cells. Cell availability also offers the opportunity to genetically manipulate cell populations. While it might be possible to genetically alter any type of cell, realistically multipotent cells represent the most likely candidate. The ability to immortalize stem cells offers the attractive possibility to generate clonal cell lines for transplantation, which, once established, could yield highly reproducible results (Snyder et al., 1992; McKay et al., 1993; Martínez-Serrano and Björklund, 1997). It should be noted, however, that the introduction of immortalizing genes might not only affect the cell life span but also alter the differentiation potential. For example, while primary O-2A progenitor cells seem to rarely give rise to astrocytes (Espinosa de los Monteros et al., 1993), the same cell population immortalized with an inducible SV40-large T antigen will yield a cell line that gives rise predominantly to astrocytes upon transplantation (Crang et al., 1991; Barnett et al., 1993a; Barnett and Crouch, 1995; Franklin et al., 1995, 1996; Franklin and Blakemore, 1997a; O'Leary and Blakemore, 1997). In addition to the introduction of immortalizing genes, genetic manipulation could also result in the engineering of cells that express different properties in order to enhance or control the outcome of transplantation. While the tools might be available to 'instruct' cells in a specific way, not enough is known about the signals that control cell division, differentiation, mi-

TABLE 2

Multipotent cells

Advantages	Disadvantages
Can be passaged and large numbers of cell readily obtained	Passaged cells lose their ability to differentiate into neurons
Primary non-passaged stem cells can be obtained in large numbers	Multiple neural types
Gene expression possible	Limited differentiation in vivo
Multiple neural phenotypes	Limited migration
	Possibility of heterotopias/tumors

Some of the advantages and disadvantages of using multipotent stem cells for transplantation are summarized. Note: The ability of stem cells to differentiate into multiple phenotypes can be both an advantage and a disadvantage, depending on the phenotypes of cells required for replacement.

gration or survival in various in vivo situations. One can reasonably expect that this information will become available and will further widen the therapeutic potential of NSC.

The ability of NSC to differentiate in vivo has been extensively documented (Snyder et al., 1992, 1995; Baetge, 1993; Cunningham and McKay, 1994; Hammang et al., 1997; Milward et al., 1997; Shetty and Turner, 1999; reviewed in Vescovi and Snyder, 1999) and functional improvement after transplantation has been demonstrated in many disease models including Parkinsonism (Svendsen et al., 1997; Studer et al., 1998), Huntington's disease (Brundin and Wictorin, 1993; Borlongan et al., 1999), cerebellar ataxia, stroke models (Snyder et al., 1995; Trojanowski et al., 1997) and various demyelination-associated disorders (reviewed in: Blakemore et al., 1995; Duncan, 1996).

There are, however, also potential shortcomings of using multipotent cells (Table 2). One major problem is the efficiency of differentiation and consequently the overall efficiency of transplantation. While many studies demonstrate 'proof-of-principle', surprisingly little is known about the quantitative aspect of cellular differentiation following transplantation. In some cases the cell type of interest represents less than 1% of the total number transplanted (Svendsen et al., 1996; Fricker et al., 1997; Lundberg et al., 1997; Vescovi et al., 1999) while others report a high degree of differentiation but a lack of migration (Friedrich and Lazzarini, 1993; O'Leary and Blakemore, 1997). Undifferentiated cells present in ectopic locations that do not respond to environmental cues could result in heterotopias or tumors (Table 2). The inability to migrate may restrict the area that can be reached by the transplant, thereby restricting functional recovery (Learish et al., 1999). A failure to respond appropriately to the environment can be anticipated based on our knowledge of the process of stem cell differentiation. Extensive studies in a variety of organisms (*Drosophila melanogaster*, zebrafish, chick) suggest that each step involved in stem cell to terminal differentiation requires a specific complex set of molecular cues (Jan and Jan, 1994). It has been well established that the embryonic environment provides different cues than an adult environment and allows for different developmental processes to take place.

Transplanted multipotential cells need to sense these appropriate cues before they become competent to differentiate into mature cell types. Indeed the degree of differentiation seen in transplants into the fetus and neonate is much more extensive than that seen in transplants into adults. The suggestion that the adult environment might lack appropriate cues is also supported by the observation that the stem cells present in the adult brain seem to be quiescent and unable to respond adequately to endogenous cues. This observation raises the concern that transplants of adult stem cells will perform unsatisfactorily in a transplantation paradigm. In order to design an optimal transplantation strategy, one has to take into account that transplantation will occur in a diseased brain; the host environment will most likely be compromised compared with that of a healthy organism, and the cues available to promote differentiation will be limited. If the necessary molecular cues that drive the terminal differentiation from a multipotent cell to a postmitotic cell are not available, the transition from a multipotent cell to the terminal differentiated cell will not occur. In such cases it might be more successful to transplant a cell population that is already committed to a specific lineage (see below).

The use of NSC as a reservoir of uncommitted cells that can be manipulated in vitro to yield more committed precursor cells may therefore represent an alternative and equally valid use of NSC. We and others have shown that more restricted precursors can be isolated from multipotent cells. For example, using cell surface markers, we have isolated GRP cells and NRP cells from the embryonic spinal cord (Mayer-Proschel et al., 1997; Rao and Mayer-Proschel, 1997). Similarly, Zhang et al. (1999) have used NSC, from which they were able to isolate and expand a cell population that predominately generated oligodendrocytes in vivo. Thus, it is possible that NSC can be grown in culture and their derivative selected just prior to transplantation. The final question of whether multipotent cells or more differentiated populations should be the 'cells of choice' for transplantation will only be answered by extensive quantitative and comparative studies using identical model systems.

Neuron restricted precursor cells

Historically, therapeutic neuronal replacement has not been a widely considered option in spinal cord injury. One important reason has been the observation that the adult CNS is inhibitory to axonal outgrowth and thus transplanted neurons are unlikely to be able to generate long axons or make appropriate connections. Recent studies have suggested, however, that this inhibitory environment can be overcome (Davies et al., 1997; Davies and Silver, 1998) and that regions distal to the injury site may not be as inhibitory as previously thought (Hatten et al., 1984; Lang et al., 1996). Further important advances have suggested that fetal neurons may not respond to extrinsic inhibitory influences (reviewed in Park et al., 1999). Such findings indicate that transplants of neurons for enhancing connectivity after spinal cord injury may be a viable option, and several recent studies have been undertaken to test this hypothesis.

Neuron restricted precursors or NRPs may have multiple uses in spinal cord injuries. NRPs may be a source for replacing neurons, and may provide an intermediate synaptic target to maintain projection neurons, or transplanted neurons may serve to bridge connections between axons upstream of the damaged site and targets present downstream. Integrated transplanted precursor cells may allow for the formation of novel connections that could subserve the function of lost long projection axons. Neurons may also provide trophic support for denervated axons as well as growth-promoting and regulatory molecules to regulate glial reaction and remyelination (Palmer et al., 1993; Martinez-Serrano et al., 1995; Tuszynski et al., 1996). The availability of NRP cells offers the potential of replacing solely neurons that are lost in a neurological disorder without the concomitant addition of potentially unwanted glial cells.

In spinal cord injury, in particular, NRPs may be better than multipotent stem cells as the environment may not provide appropriate signals for multipotent stem cells to differentiate. There is little evidence of ongoing neurogenesis in the spinal cord compared to that described in the olfactory bulb and hippocampus, and little data are provided to suggest that neurogenesis is stimulated after spinal cord injury. Thus it is likely that transplanting cells other than neuronal precursors will result in incomplete or limited neuronal differentiation. NRPs offer a further potential advantage over multipotent stem cells in their ability to migrate. In normal development, neuronal precursors differentiate from ventricular zone multipotent precursor cells and migrate considerable distances to appropriate sites before differentiation (Luskin, 1994). This is in contrast to multipotent cells, which are restricted to ventricular zones and may be actively inhibited from migration (reviewed in Alvarez-Buylla and Temple, 1998).

Although NRP cells represent a restricted progenitor cell population, their ability to self-renew extensively seems not to be compromised, thus providing a source for a large number of cells. For that reason, NRPs could also be used to pack the syrinx that may develop after traumatic spinal cord injury. Neurons generally do not express class-2 antigens and do not respond to inflammatory cytokines to upregulate HLA antigens. Thus neuronal replacement will likely be less immunogenic than transplanting a mixed population of cells. While NRP cells offer many potential advantages in vivo, more data are required before a convincing case for using these cells rather than multipotent stem cells in spinal cord injury can be made.

A potential disadvantage for using restricted cells could be their relative heterogeneity. Several important observations suggest that neuronal precursors isolated from different brain regions exhibit different biases in their fate choices (Fig. 2). For example, progenitors from the hippocampus, but not from the cerebellum or midbrain, produce hippocampal pyramidal neurons (Shetty and Turner, 1999), suggesting a bias in differentiation potential that is maintained in culture. Likewise, Li et al. (1998), analyzing mesencephalic neural precursors, showed that under appropriate conditions as many as 50% of the neurofilament-immunoreactive neurons appear dopaminergic, and this frequency is much higher than that obtained from any other precursor cell population. Similarly, Luskin and colleagues (Zigova et al., 1998) noted that transplanting subventricular zone (Svz[a]) neurons into the striatum resulted in predominantly γ-aminobutyric acid (GABA)ergic rather than dopaminergic differentiation, raising the possibility that the restriction in developmental potential cannot be reversed even in vivo.

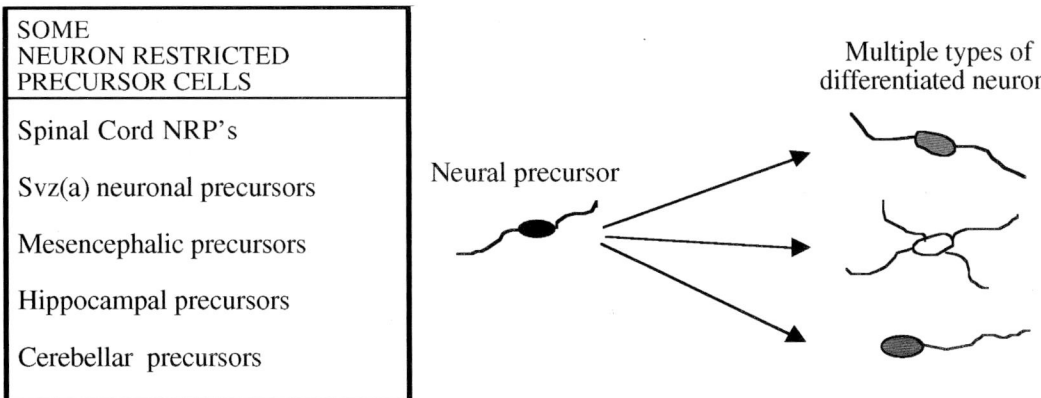

Fig. 2. Neuron restricted precursor cells.

Apart from a bias in differentiation potential, differences in migration ability have been noted as well. Alvarez-Byulla and colleagues (Wichterle et al., 1999) compared the migration ability of medial ganglion eminence (MGE) precursors and lateral ganglion eminence (LGE) precursors and noted that under identical transplant conditions, LGE neurons migrated less extensively than MGE neurons and appeared to be restricted in their ability to migrate to the cortex. A similar difference between Svz(a) precursor cell migration and spinal cord precursor migration was observed. Svz(a) precursors were restricted to migrating along the rostral migratory stream while spinal cord precursors migrated much more extensively but did not appear to migrate into the hippocampus (M. Luskin, pers. commun., 2000). Of importance is that the cells in these experiments were isolated from embryonic day (E)12–13 rodent embryos, a stage at which cells are still dividing and have neither migrated or received innervation. These data indicate that restriction in developmental potential has occurred early in development and, while neuronal precursors offer several advantages, it will be important to isolate precursors from appropriate regions for optimal results. In the case of spinal cord injuries, neuronal precursors isolated from the fetal spinal cord may be the best precursor population for neuronal replacement strategies.

Glial restricted precursors

Glial cells may play a role both in promoting growth and inhibiting regeneration (reviewed in: Olby and Blakemore, 1996; Gilmore and Sims, 1997; Lu and Waite, 1999). It has been suggested that myelin proteins and myelin-associated proteins produced by oligodendrocytes contain molecules inhibitory to axonal outgrowth (Cadelli et al., 1992; Bovolenta et al., 1993, 1997; Fernaud-Espinosa et al., 1998; Spillmann et al., 1998), and that reactive astrocytes are responsible for scar formation that forms an absolute barrier to axonal regrowth (Bovolenta et al., 1992; Houle, 1992; Sims et al., 1999). On the other hand, astrocytes may serve a critical role in promoting oligodendrocyte remyelination (Franklin et al., 1991; Hatten et al., 1991; Blakemore, 1992), regulating the blood–brain barrier and immune response, modulating local glutamate metabolism and providing trophic support for surviving neurons. Oligodendrocyte precursors likewise are critical for generating the myelin-producing oligodendrocytes to remyelinate surviving axons. Thus reconstitution of an appropriate glial environment is essential for promoting axonal regrowth and recovery. Both astrocytes and oligodendrocytes are necessary in maintaining appropriate axonal conduction. Recent studies have identified an impressive number of glial precursor cells able to generate mature glial phenotypes (see Fig. 3).

Among the studied progenitor cell populations, cells that exclusively produce astrocytes are relatively poorly characterized in transplant paradigms. The limited data available, however, clearly illustrate the importance of this class of cell. Blakemore and colleagues (Franklin et al., 1991) have shown that type-1 astrocytes facilitate the repair of de-

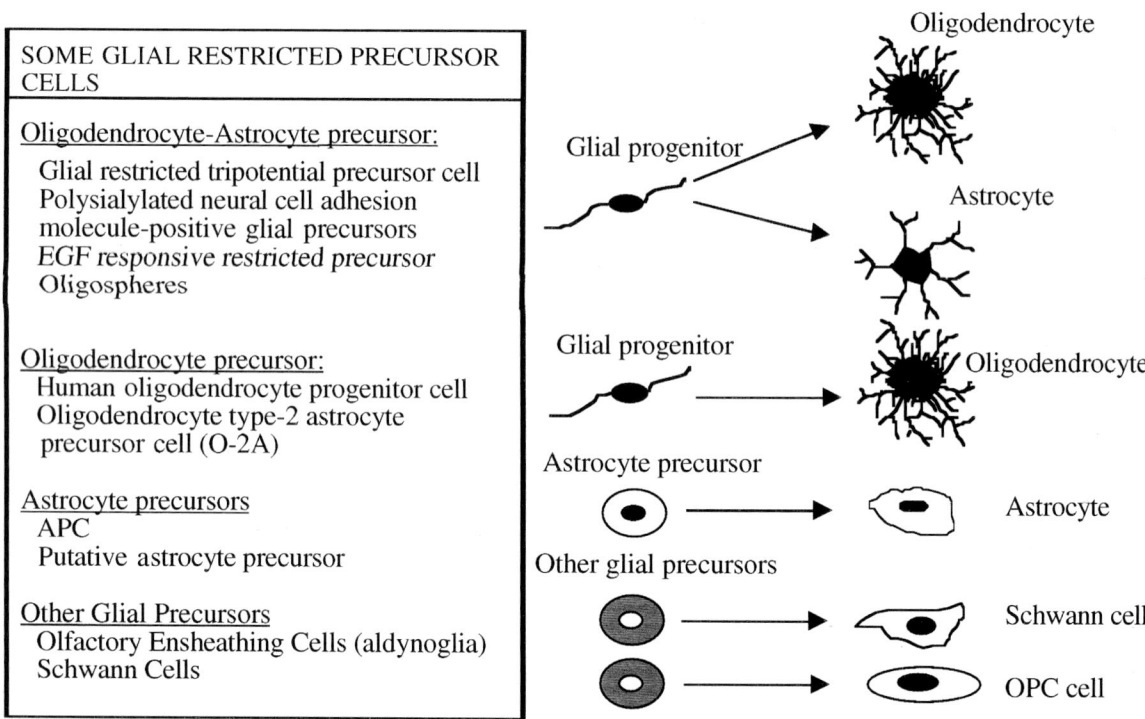

Fig. 3. Glial restricted precursor cells.

myelinating lesions by activating host oligodendrocytes/precursors. Wang et al. (1995) have shown that fetal astrocytes transplanted into a hemisected spinal cord will significantly reduce glial scar formation. Houle (1992) has shown that even in a chronic lesion with dense scar formation, a reduction in glial scar volume or a substantial reduction in new scar formation is seen with fetal transplants (presumably due to astrocytes). Smith et al. (1986) likewise have shown that in addition to a reduction in scar formation, type-1 astrocytes also migrate along blood vessels and may help reconstitute the blood–brain barrier and reduce the inflammatory response. Bernstein and others (Bernstein et al., 1993; Bernstein-Goral et al., 1997) have noted that astrocytes can provide trophic support for denervated neurons (for review see Houweling et al., 1998). In general, the data indicate that fetal astrocytes are better than adult astrocytes when used as cell replacement therapy. The availability of astrocyte specific precursor cells from fetal tissue in relatively large numbers will allow a better evaluation of the role of astrocytes in promoting spinal cord repair. Another important property of astrocytes is their ability to migrate extensively and to integrate seamlessly into the host parenchyma (Goldberg and Bernstein, 1987, 1988). Fetal astrocytes or their precursors will therefore be important candidates for drug and gene delivery and may be even better than oligodendrocyte precursors, which tend to be more difficult to infect/transfect (reviewed in Rao, 1999).

While astrocytes may be important in modulating endogenous myelinating precursors, in general it has been difficult to obtain sufficient remyelination without providing exogenous myelinating cells. Indeed an entire range of neuropathologies of the brain is defined by the lack or dysfunction of endogenous myelinating cell populations. This is equally important in spinal cord injuries, where even after successful nerve generation, the regenerating axons still need to be myelinated. Several candidate remyelinating cells have been used (Fig. 3). These include CNS myelinating cells such as O-2A and GRP cells, peripheral nervous system (PNS) myelinating Schwann cells and specialized aldynoglia (Gudiño-Cabrera and Nieto-Sampedro, 1999) such as olfactory ensheathing cells. Multiple transplant studies in which each of these cell types was transplanted into demyelinated spinal cord have shown moderate to extensive remyelination with measurable improvement in conduction velocity and frequency–response properties of the remyelinated axons (for review see Franklin and Barnett, 1997). In some cases values could be restored to near normal and the physiological improvement was accompanied with a reversal of functional deficits provided that axonal loss was minimal (Jeffery et al., 1999).

While each of these distinct glial precursor cell types can promote myelination, it is not clear which cell will ultimately be the best cell for such purposes. The efficiency of individual cell populations needs to be evaluated in side-by-side comparisons. In our opinion, providing additional precursors similar to resident precursors may in some cases be of greater benefit therapeutically. In other cases, providing a novel cell type which may be more resistant to ongoing damage or less responsive to the host immune system may be a better choice.

A better characterized glial precursor that has been used to remyelinate in vivo is the oligodendrocyte-type-2 astrocyte (O-2A) precursor cell (reviewed in: Crang et al., 1992; Blakemore et al., 1995; Franklin and Blakemore, 1995). Postnatally derived O-2A cells, when transplanted, are able to produce myelin in a number of transplantation paradigms in vivo, for example (Groves et al., 1993; Olby and Blakemore, 1996). In addition, Groves et al. (1993) showed that purified O-2A cells can be expanded in vitro in the presence of cooperating growth factors without losing their ability to extensively remyelinate axons after transplantation. One unresolved controversy in using O-2A cells in transplantation is the question of whether O-2A cells are able to generate astrocytes in vivo. Some investigators claim to be able to identify an astrocytic phenotype generated from the transplanted O-2A cells in certain transplantation paradigms (Marriott and Wilkin, 1993; Represa et al., 1993). It might well be the case that O-2A cells generate a specialized astrocytic phenotype that requires certain environmental conditions.

The uncertainties associated with the use of O-2A cells might be overcome by the use of an embryonic glial restricted cell type (GRP cell) that was recently identified in our laboratories and might be of particular interest for transplantation. GRP cells represent the earliest glial restricted precursors to date and may be the ancestor to all glia in the CNS (Rao and Mayer-Proschel, 1997; Rao et al., 1998). GRP cells not only give rise to differentiated glial phenotypes (oligodendrocytes and astrocytes), but can also generate other glial restricted precursor cells like the O-2A phenotype (Gregori et al., 2000). GRP cells might be a superior cell type for transplantation not only because of their embryonic phenotype that allows for extensive self-renewal, but also because these cells seem to both generate oligodendrocytes and astrocytic phenotypes in vivo (M. Mayer-Proschel, unpubl. observations). Likewise, oligospheres (Zhang et al., 1999), which can generate both oligodendrocytes and astrocytes and can remyelinate axons in a dysmyelinating mutant, may be a useful cell for spinal cord cell replacement therapy. As discussed above, it has been suggested that in order to promote extensive remyelination and axonal regrowth, both astrocytes and oligodendrocytes are likely to be required (Franklin and Blakemore, 1997b).

While the overall outcome of glial cell transplantation experiments show impressive and encouraging results, relatively little is known about the various factors that are responsible for the outcome of transplantation. It will be a challenge for the future to gain fundamental insight into the role of the host environment on glial cell differentiation. It is known for example, that transplantation of glial precursors into the normal brain results in failure of oligodendrocyte precursor survival and migration (Franklin et al., 1996; O'Leary and Blakemore, 1997). Cells derived from adult human brain also failed to remyelinate demyelinated rat axons in vitro (Targett et al., 1996). Using postnatally derived glial precursors as a graft population, although effectively remyelinating, may present another potential problem: postnatally derived O-2A cells seem to have a cell-intrinsic regulator of mitosis that brings them to the end of their normal mitotic capacity even in the presence of cytokines that promote self-renewal and inhibit differentiation (Bogler and Noble, 1994). If this intrinsic 'clock mechanism' is also functional in vivo, then cells expanded for weeks in vitro would only have a limited number of divisions left before terminal differentiation occurs. As one goal of transplantation will be to generate the largest possible volume of myelin with each transplanted cell, the importance of maintaining the capacity for self-renewal after transplantation is apparent. The relatively few studies that do compare young versus older cells in their capacity to remyelinate a lesion support the hypothesis that the relative age of a cell population might determine the success of a transplant. For example, Warrington et al. (1993) showed that postnatally derived O-2A cells produce more myelin over larger areas than later stages of the lineage (when cells express the surface antigen, O4) after transplantation into the myelin-deficient *shiverer* mouse. Archer et al. (1997) report a more dramatic remyelination in the canine system with fetal cells compared to postnatally derived cells.

An alternative for using progenitor cells that can generate either astrocytes or oligodendrocytes or both for transplantation is to use a cell that combines some aspects of both cell types. These include Schwann cells and olfactory ensheathing glia (OEG), and recent results suggest some unique advantages for using these two populations (for review see: Franklin and Barnett, 1997; Franklin and Blakemore, 1998). Autologous Schwann cells and OEG can be isolated, purified and amplified in tissue culture (Peden et al., 1990; Barnett et al., 1993b; Rutkowski et al., 1995). Transplantation with both cell types has produced extensive remyelination and functional improvement (for review see: Doucette, 1995; Xu et al., 1997). Indeed recent results using OEG by various investigators (Ramón-Cueto and Nieto-Sampedro, 1994; Li et al., 1997; Imaizumi et al., 1998; Pérez-Bouza et al., 1998; Ramón-Cueto and Avilla, 1998) have shown remyelination and axonal regeneration in various transplantation experiments. The ability of OEG to form myelin and support axon regeneration gives these cells the same range of properties that had been documented for Schwann cells.

Schwann cells exhibit a number of features that make them interesting candidates for transplantation strategies. The potential use of a patient's own peripheral nerve as a source of cells, a diminished autoimmune response to the graft (Hermanns et al., 1997), the ability of Schwann cells to migrate (Baron-Van Evercooren et al., 1996), the ability to support remyelination (Blakemore et al., 1987; Bunge, 1994; Baron-Van Evercooren et al., 1997; Franklin and Blakemore, 1998), and their role in axonal regeneration (Guénard et al., 1994; Guest et al., 1997), all contribute to the cells' potential as a superior graft population (for review see Franklin and Barnett, 1997). Taken together, it might be valuable to consider autologous Schwann cell transplantation alone or in combination with other cells to influence CNS regeneration. However, despite these successes, there are potential disadvantages to be considered. The complex interaction of astrocytes and Schwann cells that is required for successful remyelination places serious limitations on the potential usefulness of Schwann cells in therapeutic applications.

In light of all available data, it is premature to draw a conclusion whether OEG cells, Schwann cells or other glial progenitor populations are the most suitable and successful cells for use in transplantation approaches. It might become apparent that one population of glial precursor cells is very effective in a certain pathological paradigm but fails to achieve the desired effect in another experimental model. While available cell types need to be studied in comparable assay systems with a clear quantification strategy in mind, we believe that additional studies where specific combinations of cells are used will demonstrate further therapeutic improvement. Regardless of the type of cells used as a graft, mounting experimental data in the transplantation field suggest cautious optimism. Clear-cut evidence has accumulated to show that the normal environment is not as inhibitory as once supposed and that mechanisms are currently available to further reduce the inhibition to axonal outgrowth. Further specific cell types are available in sufficient number to modulate the blood–brain barrier, glial scar formation, provide trophic support and promote remyelination.

Controversy exists over the ability of glial cells to migrate in a host environment. Several studies have noted that glial cells migrate over extensive distances and, in addition, have the ability to migrate to specific target sites (Franklin et al., 1996; Young, 1996, 1997). Migration from one side of the brain to another or several centimeters from the site of injection has been reported (Tontsch et al., 1994; Lipsitz et al., 1995; Archer et al., 1997). More importantly, several laboratories (Gout and Dubois-Dalcq, 1993; Goldman et al., 1997) report the observation of directed migration either toward a transplanted growth factor or to the site of a lesion. Conversely, other reports claim that glial cells migrate only over a limited area (for review see: Franklin and Blakemore, 1997a,b) or are not able to migrate at all away from the transplantation site (O'Leary and Blakemore, 1997).

In any case, the ability of at least some precursor cells to migrate over long distances, coupled with the ability to express foreign genes in cultured cells prior to transplantation, allows therapeutic approaches previously considered difficult or impossible. For example, diffuse disorders such as Tay-Sachs disease (GM2 gangliosidosis) (Argov and Navon, 1984; Rosenberg and Iannaccone, 1995) can be theoretically cured by single or a few injections of genetically modified astrocytes. Likewise, it may become possible to target plaques of multiple sclerosis that are surgically inaccessible by taking advantage of the migration and homing ability of oligodendrocyte precursors. Trophic factors can be delivered using glial precursors that may co-migrate with regenerating axons to provide trophic support throughout the regenerative period. Studies testing the advantage of glial cells as delivery agents are in their infancy but the current results are grounds for optimism.

Stem and precursor cells for transplantation in humans

Much of the previous discussion has been based on data obtained in rodent models using rodent cells. However, for ultimate therapeutic use these results need to be validated with human cells or with cells from xenotransplant models such as pig or primate. Several laboratories have in recent years extended these studies to porcine, primate and human tissue (for review see Rao, 1999). These results are quite encouraging and the results from human tissue are summarized briefly. Multipotent human stem cells (HSC cells) have been identified by several groups (Chalmers-Redman et al., 1997; Carpenter et al., 1999; Vescovi et al., 1999). These cells share only some similarity with rodent cells. In particular the growth factor requirements appear distinct. For example, some groups have suggested that leukemia inhibitory factor (LIF) may be a required growth or trophic molecule (Carpenter et al., 1999). Of importance has been the observation that relatively large numbers of HSC cells can be harvested from donor cortex taken from middle-aged human samples (Goldman et al., 1997; Kukekov et al., 1999; Vescovi and Snyder, 1999). The number of cells obtained is quite large and several groups have shown that HSC cells harvested from adult tissue can be maintained for prolonged periods. HSC cells have been transplanted into animal models and have been shown to differentiate into multiple phenotypes, and in some cases have been shown to promote functional recovery (for review see Svendsen et al., 1997, 1999). Other groups have suggested that maintaining cells in culture reduces their differentiation potential or selects for cells that are biased toward developing into astrocytes (Quinn et al., 1999). Multipotent stem cells have also been identified from human fetal spinal cord (Moyer et al., 1997; Quinn et al., 1999), but to our knowledge not from adult human cord. This may reflect a bias in investigator interest or a difficulty in obtaining stem cells from adult spinal cord. Indeed, in general, spinal stem cells have been shown to be difficult to harvest even from adult rodents (Weiss et al., 1996).

More restricted precursors have also been identified. For example, we have used antibodies against polysialated neural cell adhesion molecule (NCAM) to harvest neuron restricted precursors from human fetal spinal cord tissue. These cells resemble their rodent counterparts and can be immortalized to obtain NRP cell lines (Li et al., 2000, and our unpublished results). Unlike rodent cells, human NRP cells appear to divide more slowly and we have been unable to maintain them for more than a few passages without immortalization (Li et al., 2000). Goldman and colleagues have used adenovirus constructs to drive cell type-specific expression of green fluorescent protein (GFP) using cell type-specific promoters. Using an alpha-1 tubulin promoter, the authors isolated dividing neuronal precursor populations and showed that such cells existed in multiple cortical regions (Wang et al., 1995).

Using cell surface markers that are unique to the glial lineage or using glia-specific promoters, several groups have isolated human glial progenitors (Scolding et al., 1999). We have used A2B5 (a ganglioside marker; Eisenbarth et al., 1979) to isolate a glial precursor population that can differentiate into astrocytes and oligodendrocytes, and Goldman and colleagues (Roy et al., 1999) have used 2′,3′-cyclic nucleoside phosphate (CNP) expression as a marker for a dividing oligodendrocyte precursor which differentiates predominantly into oligodendrocytes. To our knowledge, no astrocyte restricted precursor has been isolated as of yet but this likely reflects the nascent state of the human neural precursor field. The importance of these observations is that both multipotent and more restricted precursors co-exist in the human nervous system, and some of the same markers may be used to isolate these cells.

While both adult and fetal sources of stem cells have been identified, primary tissue nevertheless is often limited in availability and the degree of amplification achieved is still not sufficient for routine therapeutic use. Researchers therefore have attempted to mobilize endogenous stem cells and have evaluated other cells as potential substitutes or as sources of precursor cells for transplant therapy. Based on accumulating evidence that stem cells persist in the adult, another potential alternative to exogenous cell replacement has emerged. Rather than transplanting precursor cells, it may be possible to mobilize endogenous stem cells. Growth factor infusions have been shown to increase stem cell proliferation and exposure to complex environments leads to an increase in stem cell proliferation (Craig et al., 1996; Kuhn et al., 1997), clearly demonstrating our ability to manipulate endogenous stem cell proliferation. More recent evidence has shown that neural damage can also bias stem cell differentiation, raising the possibility of selective replenishment by endogenous stem cells (Parent et al., 1997; Pincus et al., 1998; Takagi et al., 1999). Mobilizing endogenous cells therefore offers the possibility of obviating the need for transplantation. While this is a promising development, we feel that many of these results are from mobilizing cells in regions where ongoing neurogenesis is in progress, and even in these regions appropriate cell numbers are not seen after damage and newborn cells are on occasion mislocalized. These observations suggest that appropriate migration cues no longer provide appropriate guidance mechanisms. Multipotent stem cells have been isolated from adult spinal cord (Weiss et al., 1996; Shihabuddin et al., 1997), but there is little evidence of spinal cord ependymal/ventricular zone cell division in adult rats and little evidence of increased proliferation after injury. Likewise, ongoing neurogenesis has not been observed in the spinal cord and does not appear to be induced even after injuries (however, see Kehl et al., 1997), suggesting that, at least in spinal cord injury, mobilizing multipotent or neuronal endogenous stem cells remains a relatively distant possibility. Glial progenitor proliferation has been shown to occur after injury (Zhang and Guth, 1997). However, this has not been sufficient to perform meaningful repair. It is not clear that further increasing glial proliferation will result in increased myelination, thus, the utility of mobilizing endogenous glial cells remains to be determined.

An alternative approach taken by investigators has been to identify autologous tissue that could substitute for required cells. Recently, for example, olfactory epithelium has been shown to contain multipotent precursors (Calof et al., 1998; Huard et al., 1998) that can be expanded in vitro and, when transplanted, can differentiate into neurons, astrocytes and oligodendrocytes. Since obtaining autologous olfactory epithelium is possible, olfactory tissue may serve as an alternative source of multipotent stem cells. Carotid body cells and adrenal chromaffin cells have been used as a source of catecholamine-secreting cells. However, these cells are likely to have a limited value in spinal cord injury. Olfactory ensheathing glia or aldynoglia and Schwann cells have been considered as alternative sources of myelinating glial precursors. As discussed above, these cells have been shown to be of therapeutic benefit in spinal cord injury models. Human Schwann cells have been harvested and have been maintained in culture for prolonged periods (Lopez and De Vries, 1999) and represent a viable autologous source of myelinating cells. Little, however, is known about human olfactory ensheathing cells, their characteristics in culture or their ability to myelinate.

Human ES cells as a source of neural precursors

A potentially limitless source of differentiated cells, embryonic stem (ES) cells, has become available. ES cells are totipotent cells that can generate all major phenotypes, including germ cells. ES cells have been identified in multiple species and their ability to differentiate into neurons and oligodendrocytes has been extensively characterized (reviewed in Rao, 1999). Thomson et al. (1998) have isolated at least two human ES cell lines and have shown that these cells can differentiate into CNS derivatives. Human ES cells have been maintained in continuous passage for at least a year, appear karyotypically normal, and retain their multipotential properties. W. Gerhart and colleagues (pers. commun., 2000) have generated primordial germ (PG) cell lines and shown that these can differentiate into neural and non-neural derivatives. ES cells and PG cells offer the possibility of either being transplanted directly or being differentiated in culture with selected cell populations harvested for transplantation. In the latter case, ES cells would serve as a bank or reservoir for cell replacement with appropriate populations being harvested as needed. The potential advantages and disadvantages of ES cells for transplant therapy are summarized in Table 3.

The idea that ES cells can be used as a reservoir of cells from which differentiated cells can be harvested as required has been validated in rodent models. Several groups have shown that cell surface markers (Mujtaba et al., 1999), manipulation of culture conditions (Okabe et al., 1996), or utilizing tissue-specific promoters (Li et al., 1998) can be used to isolate neural stem cells or more restricted precursors (Fig. 4). Recently, oligodendrocyte precursor cells isolated from ES cells have been shown to be functionally useful in a rat model of demyelination (Brüstle et al., 1998). Using cell type-specific markers, we have isolated neuronal and glial restricted precursors from mouse ES cells and have shown that these cells are antigenically and phenotypically identical to restricted precursors isolated from fetal tissue (Mujtaba et al., 1999). McKay and colleagues (Okabe et al., 1996) have manipulated culture conditions to isolate multipotent stem cells from ES cell cultures and Li and colleagues (Li et al., 1998) have used an elegant selection strategy to isolate neural precursor cells. Data on isolation of specific neural precursor types from human ES cell cultures are not available. However, preliminary results (M. Carpenter, pers. commun., 2000) suggest that markers utilized to isolate neuron and glial precursor populations from rodent ES cells are also expressed by differentiating human ES cell cultures. Thus it is likely that human ES cells could also serve a reservoir function.

One potential advantage of neurons, glia or precursor cells derived from ES cells is that they may not be regionally specified and thus may be more capable of site-specific integration. Further, since it is possible to obtain regionally specific phenotypes by manipulating ES cell culture conditions, it may be possible to obtain appropriately specified phenotypes. This raises the possibility that ES-cell-derived precursors may be preferred to fetal or adult precursor cells. McDonald and colleagues (1999) have presented suggestive evidence that ES-cell-derived cells may be useful in spinal cord injury. When embryoid bodies were transplanted in a spinal cord injury model, significant improvement and remyelination were seen. While the behavioral improvement was transient, the remyelination was clearly superior to that seen in other transplants. Suggestive evidence that neural cells derived from human ES cells will prove useful in spinal cord injury is provided by

TABLE 3

Embryonic stem cells

Advantages	Disadvantages
Can be passaged and large numbers of cell readily obtained	Generally passaged on a feeder layer that needs to be separated
Differentiated cells can be induced and selected cells transplanted immediately	Inappropriate phenotypes can develop
Stable gene expression is possible	Limited differentiation in vivo
Homologous recombination is possible	Limited migration
Multiple neural phenotypes can be obtained	Possibility of heterotopias/tumors
Cells are unlikely to be regionally specified	Limited knowledge of inducing molecules

Some of the advantages and disadvantages of using ES cells for transplants are summarized. It should be noted that recent results have shown that feeder-independent ES cell lines can be obtained and that data on human ES cells is quite limited. Additional information may force a reevaluation of the potential uses of human ES cells.

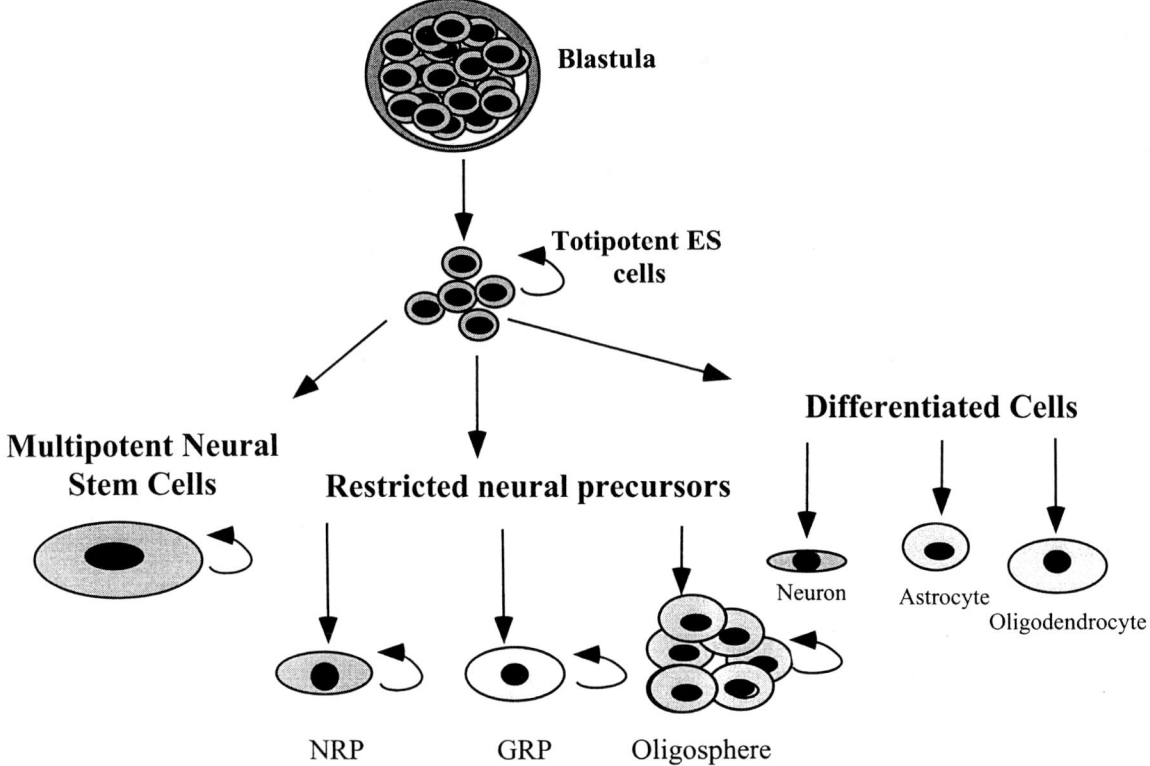

Fig. 4. ES cells as a source of neural precursors.

experiments using a neuron restricted precursor cell derived from a human embryonic carcinoma cell line. Trojanowski and colleagues (Trojanowski et al., 1997; Li et al., 2000) isolated a subclone (HNT-2) from a human embryonic carcinoma that showed exclusive differentiation into neurons. Neurons derived from this subclone have been used in rat models of spinal cord injury. Transplanted cells have been shown to differentiate into neurons that send long axonal projections which even project into the nerve roots. No tumors have been seen in any of the transplants done so far. Recent trials with transplantation into human lesions are in progress. These results suggest that it will be possible to obtain neuronal restricted precursors from human ES cell lines and that these are likely to be functionally useful.

An additional advantage of ES cells is the ability to be modified to express foreign genes. This has been done routinely in mouse ES cells and in principle should be possible in human cells also. Likewise, curing a genetic defect by homologous recombination in ES cells followed by transplanting differentiated cells that bear the normal gene is also feasible. It is important to emphasize that these techniques do not require cloning but simply manipulation in culture using well established techniques. Homologous recombination likely represents a unique advantage of ES cells over any other neural precursor population. Homologous recombination is possible in somatic cells but requires relatively large numbers of cells that are capable of long-term serial passage and several rounds of selection (Bunz et al., 1998). In general, genetic manipulation of somatic cells has proven to be relatively harder than that of ES cells (reviewed in Sedivy and Dutriaux, 1999). Few data on the ability to induce foreign gene expression or perform homologous recombination are available for human ES cells. Nevertheless, this represents a possibility that is well worth exploring.

Conclusion

In summary, current data suggest that transplants of cells may have a role in promoting recovery from spinal cord injury. Active management of the acute phase of injury followed by a combinatorial approach that likely will involve growth factors, transplants of selected cells, as well as attention to the immune system and revascularization, will be necessary to reduce loss of function. Cell therapy also is likely to be of value in the chronic phase of spinal cord injury both as a cell replacement therapy and as a mechanism of localized drug delivery. The ability to selectively replace defective populations of cells potentially allows an unprecedented degree of control of the repair process. The ability to obtain multiple classes of stem cells in relatively large numbers from fetal tissue, ES cells and from the adult provides us a possible basis of proceeding toward therapeutic intervention.

Acknowledgements

We thank Drs. Mark Noble, Tom Parks and Chris Proschel for their comments on this paper. We gratefully acknowledge the input of all members of our laboratory provided through discussions and constructive criticisms. This work was supported by the NIH, MDA March of Dimes and Acorda Therapeutics. MM-P was supported by the Multiple Sclerosis Society, the Keck Foundation and the Huntsman Cancer Center. MSR thanks Dr. S. Rao for her constant support.

References

Akesson, E., Kjaeldgaard, A. and Seiger, A. (1998) Human embryonic spinal cord grafts in adult rat spinal cord cavities: survival, growth, and interactions with the host. *Exp. Neurol.*, 149: 262–276.

Alvarez-Buylla, A. and Temple, S. (1998) Stem cells in the developing and adult nervous system. *J. Neurobiol.*, 36: 105–110.

Archer, D.R., Cuddon, P.A., Lipsitz, D. and Duncan, I.D. (1997) Myelination of the canine central nervous system by glial cell transplantation: a model for repair of human myelin disease. *Nat. Med.*, 3: 54–59.

Argov, Z. and Navon, R. (1984) Clinical and genetic variations in the syndrome of adult GM2 gangliosidosis resulting from hexosaminidase A deficiency. *Ann. Neurol.*, 16: 14–20.

Asada, Y., Kawaguchi, S., Hayashi, H. and Nakamura, T. (1998) Neural repair of the injured spinal cord by grafting: comparison between peripheral nerve segments and embryonic homologous structures as a conduit of CNS axons. *Neurosci. Res.*, 31: 241–249.

Baetge, E.E. (1993) Neural stem cells for CNS transplantation. *Ann. N.Y. Acad. Sci.*, 695: 285–291.

Barnett, S.C. and Crouch, D.H. (1995) The effect of oncogenes on the growth and differentiation of oligodendrocyte-type 2 astrocyte progenitor cells. *Cell Growth Differ.*, 6: 69–80.

Barnett, S.C., Franklin, R.J. and Blakemore, W.F. (1993a) In vitro and in vivo analysis of a rat bipotential O-2A progenitor cell line containing the temperature-sensitive mutant gene of the SV40 large T antigen. *Eur. J. Neurosci.*, 5: 1247–1260.

Barnett, S.C., Hutchins, A.M. and Noble, M. (1993b) Purification of olfactory nerve ensheathing cells from the olfactory bulb. *Dev. Biol.*, 155: 337–350.

Baron-Van Evercooren, A., Avellana-Adalid, V., Ben Younes-Chennoufi, A., Gansmuller, A., Nait-Oumesmar, B. and Vignais, L. (1996) Cell–cell interactions during the migration of myelin-forming cells transplanted in the demyelinated spinal cord. *Glia*, 16: 147–164.

Baron-Van Evercooren, A., Avellana-Adalid, V., Lachapelle, F. and Liblau, R. (1997) Schwann cell transplantation and myelin repair of the CNS. *Mult. Scler.*, 3: 157–161.

Bernstein, J.J., Willingham, L.A. and Goldberg, W.J. (1993) Migrated fetal astrocytes modulate nerve growth factor expression in host nucleus gracilis of the medulla after grafting in third cervical hindlimb dorsal columns of the spinal cord. *J. Neurosci. Res.*, 34: 394–400.

Bernstein-Goral, H., Diener, P.S. and Bregman, B.S. (1997) Regenerating and sprouting axons differ in their requirements for growth after injury. *Exp. Neurol.*, 148: 51–72.

Blakemore, W.F. (1992) Transplanted cultured type-1 astrocytes can be used to reconstitute the glia limitans of the CNS: the structure which prevents Schwann cells from myelinating CNS axons. *Neuropathol. Appl. Neurobiol.*, 18: 460–466.

Blakemore, W.F., Crang, A.J. and Patterson, R.C. (1987) Schwann cell remyelination of CNS axons following injection of cultures of CNS cells into areas of persistent demyelination. *Neurosci. Lett.*, 77: 20–24.

Blakemore, W.F., Olby, N.J. and Franklin, R.J. (1995) The use of transplanted glial cells to reconstruct glial environments in the CNS. *Brain Pathol.*, 5: 443–450.

Bogler, O. and Noble, M. (1994) Measurement of time in oligodendrocyte-type-2 astrocyte (O-2A) progenitors is a cellular process distinct from differentiation or division. *Dev. Biol.*, 162: 525–538.

Boonman, Z. and Isacson, O. (1999) Apoptosis in neuronal development and transplantation: role of caspases and trophic factors. *Exp. Neurol.*, 156: 1–15.

Borlongan, C.V., Sanberg, P.R. and Freeman, T.B. (1999) Neural transplantation for neurodegenerative disorders. *Lancet*, 353 (Suppl. 1): SI29–30.

Bovolenta, P., Wandosell, F. and Nieto-Sampedro, M. (1992) CNS glial scar tissue: a source of molecules which inhibit central neurite outgrowth. *Prog. Brain Res.*, 94: 367–379.

Bovolenta, P., Wandosell, F. and Nieto-Sampedro, M. (1993) Characterization of a neurite outgrowth inhibitor expressed after CNS injury. *Eur. J. Neurosci.*, 5: 454–465.

Bovolenta, P., Fernaud-Espinosa, I., Méndez-Otero, R. and Nieto-Sampedro, M. (1997) Neurite outgrowth inhibitor of gliotic brain tissue. Mode of action and cellular localization, studied with specific monoclonal antibodies. *Eur. J. Neurosci.*, 9: 977–989.

Brundin, P. and Wictorin, K. (1993) Neural transplantation in rat models of Parkinson's and Huntington's disease. *Sem. Neurosci.*, 5: 413–421.

Brüstle, O., Choudhary, K., Karram, K., Hüttner, A., Murray, K., Dubois-Dalcq, M. and McKay, R.D. (1998) Chimeric brains generated by intraventricular transplantation of fetal human brain cells into embryonic rats. *Nat. Biotechnol.*, 16: 1040–1044.

Bunge, R.P. (1994) The role of the Schwann cell in trophic support and regeneration. *J. Neurol.*, 242: S19–21.

Bunz, F., Dutriaux, A., Lengauer, C., Waldman, T., Zhou, S., Brown, J.P., Sedivy, J.M., Kinzler, K.W. and Vogelstein, B. (1998) Requirement for p53 and p21 to sustain G2 arrest after DNA damage. *Science*, 282: 1497–1501.

Burrows, R.C., Wancio, D., Levitt, P. and Lillien, L. (1997) Response diversity and the timing of progenitor cell maturation are regulated by developmental changes in EGFR expression in the cortex. *Neuron*, 19: 251–267.

Cadelli, D.S., Bandtlow, C.E. and Schwab, M.E. (1992) Oligodendrocyte- and myelin-associated inhibitors of neurite outgrowth: their involvement in the lack of CNS regeneration. *Exp. Neurol.*, 115: 189–192.

Calof, A.L., Mumm, J.S., Rim, P.C. and Shou, J. (1998) The neuronal stem cell of the olfactory epithelium. *J. Neurobiol.*, 36: 190–205.

Carpenter, M.K., Cui, X., Hu, Z.Y., Jackson, J., Sherman, S., Seiger, A. and Wahlberg, L.U. (1999) In vitro expansion of a multipotent population of human neural progenitor cells. *Exp. Neurol.*, 158: 265–278.

Chalmers-Redman, R.M.E., Priestley, T., Kemp, J.A. and Fine, A. (1997) In vitro propagation and inducible differentiation of multipotential progenitor cells from human fetal brain. *Neuroscience*, 76: 1121–1128.

Craig, C.G., Tropepe, V., Morshead, C.M., Reynolds, B.A., Weiss, S. and Van der Kooy, D. (1996) In vivo growth factor expansion of endogenous subependymal neural precursor cell populations in the adult mouse brain. *J. Neurosci.*, 16: 2649–2658.

Crang, A.J., Franklin, R.J., Blakemore, W.F., Trotter, J., Schachner, M., Barnett, S.C. and Noble, M. (1991) Transplantation of normal and genetically engineered glia into areas of demyelination. *Ann. N.Y. Acad. Sci.*, 633: 563–565.

Crang, A.J., Franklin, R.J., Blakemore, W.F., Noble, M., Barnett, S.C., Groves, A., Trotter, J. and Schachner, M. (1992) The differentiation of glial cell progenitor populations following transplantation into non-repairing central nervous system glial lesions in adult animals. *J. Neuroimmunol.*, 40: 243–253.

Cunningham, M. and McKay, R. (1994) Transplantation strategies for the analysis of brain development and repair. *J. Neurol.*, 242: S40–42.

Davies, S.J. and Silver, J. (1998) Adult axon regeneration in adult CNS white matter [letter]. *Trends Neurosci.*, 21: 515.

Davies, S.J., Fitch, M.T., Memberg, S.P., Hall, A.K., Raisman, G. and Silver, J. (1997) Regeneration of adult axons in white matter tracts of the central nervous system. *Nature*, 390: 680–683.

Davis, A.A. and Temple, S. (1994) A self-renewing multipotential stem cell in embryonic rat cerebral cortex. *Nature*, 362: 363–372.

Diener, P.S. and Bregman, B.S. (1998) Fetal spinal cord transplants support growth of supraspinal and segmental projections after cervical spinal cord hemisection in the neonatal rat. *J. Neurosci.*, 18: 779–793.

Doucette, R. (1995) Olfactory ensheathing cells: potential for glial cell transplantation into areas of CNS injury. *Histol. Histopathol.*, 10: 503–507.

Duncan, I.D. (1996) Glial cell transplantation and remyelination of the central nervous system. *Neuropathol. Appl. Neurobiol.*, 22: 87–100.

Dunnett, S.B. (1999) Repair of the damaged brain. The Alfred Meyer Memorial Lecture 1998. *Neuropathol. Appl. Neurobiol.*, 25: 351–362.

Eisenbarth, G.S., Walsh, F.S. and Nirenberg, M. (1979) Monoclonal antibody to plasma membrane antigen of neurons. *Proc. Natl. Acad. Sci. USA*, 76: 4913–4917.

Espinosa de los Monteros, A., Zhang, M. and De Vellis, J. (1993) O-2A progenitor cells transplanted into the neonatal rat brain develop into oligodendrocytes but not astrocytes. *Proc. Natl. Acad. Sci. USA*, 90: 50–54.

Fernaud-Espinosa, I., Nieto-Sampedro, M. and Bovolenta, P. (1998) A neurite outgrowth-inhibitory proteoglycan expressed during development is similar to that isolated from adult brain after isomorphic injury. *J. Neurobiol.*, 36: 16–29.

Franklin, R.J. and Barnett, S.C. (1997) Do olfactory glia have advantages over Schwann cells for CNS repair?. *J. Neurosci. Res.*, 50: 665–672.

Franklin, R.J. and Blakemore, W.F. (1995) Glial-cell transplantation and plasticity in the O-2A lineage — implications for CNS repair. *Trends Neurosci.*, 18: 151–156.

Franklin, R.J., Blakemore, W.F., 1997a. Transplanting oligodendrocyte progenitors into the adult CNS. J. Anat. 190 (Pt 1), 23–33.

Franklin, R.J. and Blakemore, W.F. (1997b) To what extent is oligodendrocyte progenitor migration a limiting factor in the remyelination of multiple sclerosis lesions?. *Mult. Scler.*, 3: 84–87.

Franklin, R.J. and Blakemore, W.F. (1998) Transplanting myelin-forming cells into the central nervous system: principles and practice. *Methods*, 16: 311–319.

Franklin, R.J., Crang, A.J. and Blakemore, W.F. (1991) Transplanted type-1 astrocytes facilitate repair of demyelinating lesions by host oligodendrocytes in adult rat spinal cord. *J. Neurocytol.*, 20: 420–430.

Franklin, R.J., Bayley, S.A., Milner, R., Ffrench-Constant, C. and Blakemore, W.F. (1995) Differentiation of the O-2A progenitor cell line CG-4 into oligodendrocytes and astrocytes following transplantation into glia-deficient areas of CNS white matter. *Glia*, 13: 39–44.

Franklin, R.J., Bayley, S.A. and Blakemore, W.F. (1996) Transplanted CG4 cells (an oligodendrocyte progenitor cell line) survive, migrate, and contribute to repair of areas of demyelination in X-irradiated and damaged spinal cord but not in normal spinal cord. *Exp. Neurol.*, 137: 263–276.

Fricker, R.A., Torres, E.M. and Dunnett, S.B. (1997) The effects of donor stage on the survival and function of embryonic striatal grafts in the adult rat brain, I. Morphological characteristics. *Neuroscience*, 79: 695–710.

Friedrich, V.L.J. and Lazzarini, R.A. (1993) Restricted migration of transplanted oligodendrocytes or their progenitors, revealed by transgenic marker M beta P. *J. Neural Transplant. Plast.*, 4: 139–146.

Gensert, J.M. and Goldman, J.E. (1996) In vivo characterization of endogenous proliferating cells in adult rat subcortical white matter. *Glia*, 17: 39–51.

Gilmore, S.A. and Sims, T.J. (1997) Glial–glial and glial–neuronal interfaces in radiation-induced, glia-depleted spinal cord. *J. Anat.*, 190: 5–21.

Goldberg, W.J. and Bernstein, J.J. (1987) Transplant-derived astrocytes migrate into host lumbar and cervical spinal cord after implantation of E14 fetal cerebral cortex into adult thoracic spinal cord. *J. Neurosci. Res.*, 17: 391–403.

Goldberg, W.J. and Bernstein, J.J. (1988) Migration of cultured fetal spinal cord astrocytes into adult host cervical cord and medulla following transplantation into thoracic spinal cord. *J. Neurosci. Res.*, 19: 34–42.

Goldman, J.E., Zerlin, M., Newman, S., Zhang, L. and Gensert, J. (1997) Fate determination and migration of progenitors in the postnatal mammalian CNS. *Dev. Neurosci.*, 19: 42–48.

Gout, O. and Dubois-Dalcq, M. (1993) Directed migration of transplanted glial cells toward a spinal cord demyelinating lesion. *Int. J. Dev. Neurosci.*, 11: 613–623.

Gregori, N., Bernard, S., Noble, M. and Mayer-Proschel, M. (2000) Tripotential glial-restricted precursor cells (GRP) can give rise to bipotential oligodendrocyte-type-2 astrocytes (O-2A) progenitor cells: evidence for a role of multiple layers of lineage restriction in oligodendrocyte development. *J. Neurosci.*, in press.

Gritti, A.G., Parati, E.A., Cova, L., Frolichsthal, P., Galli, R., Wanke, E., Faravelli, L., Morassutti, D.J., Roisen, F., Nickel, D. and Vescovi, L. (1996) Multipotential stem cells from the adult mouse brain proliferate and self-renew in response to basic fibroblast growth factor. *J. Neurosci.*, 16: 1091–1100.

Groves, A.K., Barnett, S.C., Franklin, R.J., Crang, A.J., Mayer, M., Blakemore, W.F. and Noble, M. (1993) Repair of demyelinated lesions by transplantation of purified O-2A progenitor cells. *Nature*, 362: 453–455.

Gudiño-Cabrera, G. and Nieto-Sampedro, M. (1999) Estrogen receptor immunoreactivity in Schwann-like brain macroglia. *J. Neurobiol.*, 40: 458–470.

Guénard, V., Aebischer, P. and Bunge, R.P. (1994) The astrocyte inhibition of peripheral nerve regeneration is reversed by Schwann cells. *Exp. Neurol.*, 126: 44–60.

Guest, J.D., Rao, A., Olson, L., Bunge, M.B. and Bunge, R.P. (1997) The ability of human Schwann cell grafts to promote regeneration in the transected nude rat spinal cord. *Exp. Neurol.*, 148: 502–522.

Hammang, J.P., Archer, D.R. and Duncan, I.D. (1997) Myelination following transplantation of EGF-responsive neural stem cells into a myelin-deficient environment. *Exp. Neurol.*, 147: 84–95.

Hatten, M.E., Mason, C.A., Liem, R.K., Edmondson, J.C., Bovolenta, P. and Shelanski, M.L. (1984) Neuron–astroglial interactions in vitro and their implications for repair of CNS injury. *CNS Trauma*, 1: 15–27.

Hatten, M.E., Liem, R.K., Shelanski, M.L. and Mason, C.A. (1991) Astroglia in CNS injury. *Glia*, 4: 233–243.

Hermanns, S., Wunderlich, G., Rosenbaum, C., Hanemann, C.O., Müller, H.W. and Stichel, C.C. (1997) Lack of immune responses to immediate or delayed implanted allogeneic and xenogeneic Schwann cell suspensions. *Glia*, 21: 299–314.

Houle, J. (1992) The structural integrity of glial scar tissue associated with a chronic spinal cord lesion can be altered by transplanted fetal spinal cord tissue. *J. Neurosci. Res.*, 31: 120–130.

Houweling, D.A., Bar, P.R., Gispen, W.H. and Joosten, E.A. (1998) Spinal cord injury: bridging the lesion and the role of neurotrophic factors in repair. *Prog. Brain Res.*, 117: 455–471.

Huard, J.M., Youngentob, S.L., Goldstein, B.J., Luskin, M.B. and Schwob, J.E. (1998) Adult olfactory epithelium contains multipotent progenitors that give rise to neurons and non-neural cells. *J. Comp. Neurol.*, 400: 469–486.

Imaizumi, T., Lankford, K.L., Waxman, S.G., Greer, C.A. and Kocsis, J.D. (1998) Transplanted olfactory ensheathing cells remyelinate and enhance axonal conduction in the demyelinated dorsal columns of the rat spinal cord. *J. Neurosci.*, 18: 6176–6185.

Jan, Y.N. and Jan, L.Y. (1994) Neuronal cell fate specification in *Drosophila*. *Curr. Opin. Neurobiol.*, 4: 8–13.

Jeffery, N.D., Crang, A.J., O'Leary, M.T., Hodge, S.J. and Blakemore, W.F. (1999) Behavioural consequences of oligodendrocyte progenitor cell transplantation into experimental demyelinating lesions in the rat spinal cord. *Eur. J. Neurosci.*, 11: 1508–1514.

Johe, K.K., Hazel, T.G., Muller, T., Dugich-Djordjevic, M.M. and McKay, R.D. (1996) Single factors direct the differentiation of stem cells from the fetal and adult central nervous system. *Genes Dev.*, 10: 3129–3140.

Kalyani, A.J. and Rao, M.S. (1998) Cell lineage in the developing neural tube. *Biochem. Cell Biol.*, 76: 1051–1068.

Kalyani, A., Hobson, K. and Rao, M.S. (1997) Neuroepithelial stem cells from the embryonic spinal cord: isolation, characterization and clonal analysis. *Dev. Biol.*, 187: 203–226.

Katsuki, M., Atsuta, Y. and Hirayama, T. (1997) Reinnervation of denervated muscle by transplantation of fetal spinal cord to transected sciatic nerve in the rat. *Brain Res.*, 771: 313–316.

Kehl, L.J., Fairbanks, C.A., Laughlin, T.M. and Wilcox, G.L. (1997) Neurogenesis in postnatal rat spinal cord: a study in primary culture. *Science*, 276: 586–589.

Kuhn, H.G., Winkler, J., Kempermann, G., Thal, L.J. and Gage,

F.H. (1997) Epidermal growth factor and fibroblast growth factor-2 have different effects on neural progenitors in the adult rat brain. *J. Neurosci.*, 17: 5820–5829.

Kukekov, V.G., Laywell, E.D., Suslov, O., Davies, K., Scheffler, B., Thomas, L.B., O'Brien, T.F., Kusakabe, M. and Steindler, D.A. (1999) Multipotent stem/progenitor cells with similar properties arise from two neurogenic regions of adult human brain. *Exp. Neurol.*, 156: 333–344.

Lang, D.M., Hille, M.G., Schwab, M.E. and Stuermer, C.A.O. (1996) Modulation of the inhibitory substrate properties of oligodendrocytes by platelet-derived growth factor. *J. Neurosci.*, 16: 5741–5748.

Learish, R.D., Brustle, O., Zhang, S.C. and Duncan, I.D. (1999) Intraventricular transplantation of oligodendrocyte progenitors into a fetal myelin mutant results in widespread formation of myelin. *Ann. Neurol.*, 46: 716–722.

Lee, J., Mayer-Proschel, M. and Rao, M. (2000) Gliogenesis in the central nervous system. *Glia*, 30: 105–121.

Li, M., Pevny, L., Lovell-Badge, R. and Smith, A. (1998) Generation of purified neural precursors from embryonic stem cells by lineage selection. *Curr. Biol.*, 8: 971–974.

Li, R., Thode, S., Zhou, J.Y., Richards, N., Pardinas, J., Rao, M.S. and Sah, D.W.Y. (2000) Motoneuron differentiation of immortalized human spinal cord cell lines. *J. Neurosci. Res.*, 59: 342–352.

Li, Y., Field, P.M. and Raisman, G. (1997) Repair of adult rat corticospinal tract by transplants of olfactory ensheathing cells [see Young (1997)]. *Science*, 277: 2000–2002.

Lipsitz, D., Archer, D.R. and Duncan, I.D. (1995) Acute dispersion of glial cells following transplantation into the myelin-deficient rat spinal cord. *Glia*, 14: 237–242.

Lopez, T.J. and De Vries, G.H. (1999) Isolation and serum-free culture of primary Schwann cells from human fetal peripheral nerve. *Exp. Neurol.*, 158: 1–8.

Lu, J. and Waite, P. (1999) Advances in spinal cord regeneration. *Spine*, 24: 926–930.

Lundberg, C., Martinez-Serrano, A., Cattaneo, E., McKay, R.D. and Björklund, A. (1997) Survival, integration, and differentiation of neural stem cell lines after transplantation to the adult rat striatum. *Exp. Neurol.*, 145: 342–360.

Luskin, M.B. (1994) Neuronal cell lineage in the vertebrate central nervous system. *FASEB J.*, 8: 722–730.

Marriott, D.R. and Wilkin, G.P. (1993) Substance P receptors on O-2A progenitor cells and type-2 astrocytes in vitro. *J. Neurochem.*, 61: 826–834.

Martínez-Serrano, A. and Björklund, A. (1997) Immortalized neural progenitor cells for CNS gene transfer and repair. *Trends Neurosci.*, 20: 530–538.

Martinez-Serrano, A., Lundberg, C., Horellou, P., Fischer, W., Bentlage, C., Campbell, K., McKay, R.D., Mallet, J. and Björklund, A. (1995) CNS-derived neural progenitor cells for gene transfer of nerve growth factor to the adult rat brain: complete rescue of axotomized cholinergic neurons after transplantation into the septum. *J. Neurosci.*, 15: 5668–5680.

Mayer-Proschel, M., Kalyani, A.J., Mujtaba, T. and Rao, M.S. (1997) Isolation of lineage-restricted neuronal precursors from multipotent neuroepithelial stem cells. *Neuron*, 19: 773–785.

McDonald, J.W., Liu, X.Z., Qu, Y., Liu, S., Mickey, S.K., Turetsky, D., Gottlieb, D.I. and Choi, D.W. (1999) Transplanted embryonic stem cells survive, differentiate and promote recovery in injured rat spinal cord. *Nat. Med.*, 5: 1410–1412.

McKay, R., Renfranz, P. and Cunningham, M. (1993) Immortalized stem cells from the central nervous system. *C. R. Acad. Sci. III*, 316: 1452–1457.

Mehler, M.F. and Gokhan, S. (1999) Postnatal cerebral cortical multipotent progenitors: regulatory mechanisms and potential role in the development of novel neural regenerative strategies. *Brain Pathol.*, 9: 515–526.

Milward, E.A., Lundberg, C.G., Ge, B., Lipsitz, D., Zhao, M. and Duncan, I.D. (1997) Isolation and transplantation of multipotential populations of epidermal growth factor-responsive, neural progenitor cells from the canine brain. *J. Neurosci. Res.*, 50: 862–871.

Miya, D., Giszter, S., Mori, F., Adipudi, V., Tessler, A. and Murray, M. (1997) Fetal transplants alter the development of function after spinal cord transection in newborn rats. *J. Neurosci.*, 17: 4856–4872.

Moyer, M.P., Johnson, R.A., Zompa, E.A., Cain, L., Morshed, T. and Hulsebosch, C.E. (1997) Culture, expansion, and transplantation of human fetal neural progenitor cells. *Transplant. Proc.*, 29: 2040–2041.

Mujtaba, T., Piper, D.R., Kalyani, A., Groves, A.K., Lucero, M.T. and Rao, M.S. (1999) Lineage-restricted neural precursors can be isolated from both the mouse neural tube and cultured ES cells. *Dev. Biol.*, 214: 113–127.

Okabe, S., Forsberg-Nilsson, K., Spiro, A.C., Segal, M. and McKay, R.D. (1996) Development of neuronal precursor cells and functional postmitotic neurons from embryonic stem cells in vitro. *Mech. Dev.*, 59: 89–102.

Olby, N.J. and Blakemore, W.F. (1996) Reconstruction of the glial environment of a photochemically induced lesion in the rat spinal cord by transplantation of mixed glial cells. *J. Neurocytol.*, 25: 481–498.

O'Leary, M.T. and Blakemore, W.F. (1997) Oligodendrocyte precursors survive poorly and do not migrate following transplantation into the normal adult central nervous system. *J. Neurosci. Res.*, 48: 159–167.

Palmer, M.R., Eriksdotter-Nilsson, M., Henschen, A., Ebendal, T. and Olson, L. (1993) Nerve growth factor-induced excitation of selected neurons in the brain which is blocked by a low-affinity receptor antibody. *Exp. Brain Res.*, 93: 226–230.

Palmer, T.D., Markakis, E.A., Willhoite, A.R., Safar, F. and Gage, F.H. (1999) Fibroblast growth factor-2 activates a latent neurogenic program in neural stem cells from diverse regions of the adult CNS. *J. Neurosci.*, 19: 8487–8497.

Parent, J.M., Yu, T.W., Leibowitz, R.T., Geschwind, D.H., Sloviter, R.S. and Lowenstein, D.H. (1997) Dentate granule cell neurogenesis is increased by seizures and contributes to aberrant network reorganization in the adult rat hippocampus. *J. Neurosci.*, 17: 3727–3738.

Park, K.I., Liu, S., Flax, J.D., Nissim, S., Stieg, P.E. and Snyder, E.Y. (1999) Transplantation of neural progenitor and stem cells: developmental insights may suggest new therapies for spinal cord and other CNS dysfunction. *J. Neurotrauma*, 16: 6756–6787.

Peden, K.W., Rutkowski, J.L., Gilbert, M. and Tennekoon, G.I. (1990) Production of Schwann cell lines using a regulated oncogene. *Ann. N.Y. Acad. Sci.*, 605: 286–293.

Pérez-Bouza, A., Wigley, C.B., Nacimiento, W., Noth, J. and Brook, G.A. (1998) Spontaneous orientation of transplanted olfactory glia influences axonal regeneration. *NeuroReport*, 9: 2971–2975.

Pincus, D.W., Keyoung, H.M., Harrison-Restelli, C., Goodman, R.R., Fraser, R.A., Edgar, M., Sakakibara, S., Okano, H., Nedergaard, M. and Goldman, S.A. (1998) Fibroblast growth factor-2/brain-derived neurotrophic factor-associated maturation of new neurons generated from adult human subependymal cells. *Ann. Neurol.*, 43: 576–585.

Qian, X., Davis, A.A., Goderie, S.K. and Temple, S. (1997) FGF2 concentration regulates the generation of neurons and glia from multipotent cortical stem cells. *Neuron*, 18: 81–93.

Quinn, S.M., Walters, W.M., Vescove, A.L. and Whittermore, S.R. (1999) Lineage restriction of neuroepithelial precursor cells from fetal human spinal cord. *J. Neurosci. Res.*, 57: 590–602.

Ramón-Cueto, A. and Avilla, J. (1998) Olfactory ensheathing glia: properties and function. *Brain Res. Bull.*, 46: 175–178.

Ramón-Cueto, A. and Nieto-Sampedro, M. (1994) Regeneration into the spinal cord of transected dorsal root axons is promoted by ensheathing glia transplants. *Exp. Neurol.*, 127: 232–244.

Rao, M.S. (1999) Multipotent and restricted precursors in the central nervous system. *Anat. Rec.*, 257: 137–148.

Rao, M. and Mayer-Proschel, M. (1997) Glial restricted precursors are derived from multipotent neuroepithelial stem cells. *Dev. Biol.*, 188: 48–63.

Rao, M., Noble, M. and Mayer-Proschel, M. (1998) A tripotential glial precursor cell is present in the developing spinal cord. *Proc. Natl. Acad. Sci. USA*, 95: 3996–4001.

Reier, P.J., Bregman, B.S. and Wujek, J.R. (1986) Intraspinal transplantation of embryonic spinal cord tissue in neonatal and adult rats. *J. Comp. Neurol.*, 247: 275–296.

Reier, P.J., Stokes, B.T., Thompson, F.J. and Anderson, D.K. (1992) Fetal cell grafts into resection and contusion/compression injuries of the rat and cat spinal cord. *Exp. Neurol.*, 115: 177–188.

Represa, A., Niquet, J., Charriaut Marlangue, C. and Ben Ari, Y. (1993) Reactive astrocytes in the kainic acid-damaged hippocampus have the phenotypic features of type-2 astrocytes. *J. Neurocytol.*, 22: 299–310.

Rosenberg, R.N. and Iannaccone, S.T. (1995) The prevention of neurogenetic disease. *Arch. Neurol.*, 52: 356–362.

Roy, N.S., Wang, S., Harrison-Restelli, C., Benraiss, A., Fraser, R.A.R., Graver, M., Braun, P.E. and Goldman, S.A. (1999) Identification, isolation and promoter-defined separation of mitotic oligodendrocyte progenitor cells from adult human subcortical white matter. *J. Neurosci.*, 15: 1–10.

Rutkowski, J.L., Kirk, C.J., Lerner, M.A. and Tennekoon, G.I. (1995) Purification and expansion of human Schwann cells in vitro [see comments]. *Nat. Med.*, 1: 80–83.

Schwab, M.E. (1990) Myelin-associated inhibitors of neurite growth and regeneration in the CNS. *Trends Neurosci.*, 13: 452–456.

Schwab, M.E. and Brösamle, C. (1997) Regeneration of lesioned corticospinal tract fibers in the adult rat spinal cord under experimental conditions. *Spinal Cord*, 35: 469–473.

Scolding, N.J., Rayner, P.J. and Compston, D.A. (1999) Identification of A2B5-positive putative oligodendrocyte progenitor cells and A2B5-positive astrocytes in adult human white matter. *Neuroscience*, 89: 1–4.

Sedivy, J.M. and Dutriaux, A. (1999) Gene targeting and somatic cell genetics — a rebirth or a coming of age?. *Trends Genet.*, 15: 88–90.

Shen, Q., Qian, X., Capela, A. and Temple, S. (1998) Stem cells in the embryonic cerebral cortex: their role in histogenesis and patterning. *J. Neurobiol.*, 36: 162–174.

Shetty, A.K. and Turner, D.A. (1999) Neurite outgrowth from progeny of epidermal growth factor-responsive hippocampal stem cells is significantly less robust than from fetal hippocampal cells following grafting onto organotypic hippocampal slice cultures: effect of brain-derived neurotrophic factor. *J. Neurobiol.*, 38: 391–413.

Shihabuddin, L.S., Ray, J. and Gage, F.H. (1997) FGF-2 is sufficient to isolate progenitors found in the adult mammalian spinal cord. *Exp. Neurol.*, 148: 577–586.

Sims, T.J., Durgun, M.B. and Gilmore, S.A. (1999) Transplantation of sciatic nerve segments into normal and glia-depleted spinal cords. *Exp. Brain Res.*, 125: 495–501.

Smith, G.M., Miller, R.H. and Silver, J. (1986) Changing role of forebrain astrocytes during development, regenerative failure, and induced regeneration upon transplantation. *J. Comp. Neurol.*, 251: 23–43.

Snyder, E.Y., Deitcher, D.L., Walsh, C., Arnold-Aldea, S., Hartwieg, E.A. and Cepko, C.L. (1992) Multipotent neural cell lines can engraft and participate in development of mouse cerebellum. *Cell*, 68: 33–51.

Snyder, E.Y., Taylor, R.M. and Wolfe, J.H. (1995) Neural progenitor cell engraftment corrects lysosomal storage throughout the MPS VII mouse brain. *Nature*, 374: 367–370.

Spillmann, A.A., Bandtlow, C.E., Lottspeich, F., Keller, F. and Schwab, M.E. (1998) Identification and characterization of a bovine neurite growth inhibitor (bNI-220). *J. Biol. Chem.*, 273: 19283–19293.

Stein, D.G. (1991) Fetal brain tissue grafting as therapy for brain dysfunctions: unanswered questions, unknown factors, and practical concerns. *J. Neurosurg. Anesthesiol.*, 3: 170–189.

Studer, L., Tabar, V. and McKay, R.D. (1998) Transplantation of expanded mesencephalic precursors leads to recovery in parkinsonian rats. *Nat. Neurosci.*, 1: 290–295.

Svendsen, C.N., Clarke, D.J., Rosser, A.E. and Dunnett, S.B. (1996) Survival and differentiation of rat and human epidermal growth factor-responsive precursor cells following grafting into the lesioned adult central nervous system. *Exp. Neurol.*, 137: 376–388.

Svendsen, C.N., Caldwell, M.A., Shen, J., Ter Borg, M.G., Rosser, A.E., Tyers, P., Karmiol, S. and Dunnett, S.B. (1997)

Long-term survival of human central nervous system progenitor cells transplanted into a rat model of Parkinson's disease. *Exp. Neurol.*, 148: 135–146.

Svendsen, C.N., Caldwell, M.A. and Ostenfeld, T. (1999) Human neural stem cells: isolation, expansion and transplantation. *Brain Pathol.*, 9: 499–513.

Takagi, Y., Nozaki, K., Takahashi, J., Yodoi, J., Ishikawa, M. and Hashimoto, N. (1999) Proliferation of neuronal precursor cells in the dentate gyrus is accelerated after transient forebrain ischemia in mice. *Brain Res.*, 831: 283–287.

Targett, M.P., Sussman, J., Scolding, N., O'Leary, M.T., Compston, D.A. and Blakemore, W.F. (1996) Failure to achieve remyelination of demyelinated rat axons following transplantation of glial cells obtained from the adult human brain. *Neuropathol. Appl. Neurobiol.*, 22: 199–206.

Theele, D.P., Schrimsher, G.W. and Reier, P.J. (1996) Comparison of the growth and fate of fetal spinal iso- and allografts in the adult rat injured spinal cord. *Exp. Neurol.*, 142: 128–143.

Thomson, J.A., Marshall, V.S. and Trojanowski, J.Q. (1998) Neural differentiation of rhesus embryonic stem cells. *Apmis*, 106: 149–156; discussion: 156–157.

Tontsch, U., Archer, D.R., Dubois-Dalcq, M. and Duncan, I.D. (1994) Transplantation of an oligodendrocyte cell line leading to extensive myelination. *Proc. Natl. Acad. Sci. USA*, 91: 11616–11620.

Trojanowski, J.Q., Kleppner, S.R., Hartley, R.S., Miyazono, M., Fraser, N.W., Kesari, S. and Lee, V.M.-Y. (1997) Transfectable and transplantable postmitotic human neurons: a potential platform for gene therapy of nervous system diseases. *Exp. Neurol.*, 144: 92–97.

Tuszynski, M.H., Roberts, J., Senut, M.C., HS, U., Gage, F.H., 1996. Gene therapy in the adult primate brain: intraparenchymal grafts of cells genetically modified to produce nerve growth factor prevent cholinergic neuronal degeneration. Gene Ther. 3, 305–314.

Vescovi, A.L. and Snyder, E.Y. (1999) Establishment and properties of neural stem cell clones: plasticity in vitro and in vivo. *Brain Pathol.*, 9: 569–598.

Vescovi, A.L., Parati, E.A., Gritti, A., Poulin, P., Ferrario, M., Wanke, E., Frölichsthal-Schoeller, P., Cova, L., Arcellana-Panlilio, M., Colombo, A. and Galli, R. (1999) Isolation and cloning of multipotential stem cells from the embryonic human CNS and establishment of transplantable human neural stem cell lines by epigenetic stimulation. *Exp. Neurol.*, 156: 71–83.

Wang, J.J., Chuah, M.I., Yew, D.T., Leung, P.C. and Tsang, D.S. (1995) Effects of astrocyte implantation into the hemisected adult rat spinal cord. *Neuroscience*, 65: 973–981.

Warrington, A.E., Barbarese, E. and Pfeiffer, S.E. (1993) Differential myelinogenic capacity of specific developmental stages of the oligodendrocyte lineage upon transplantation into hypomyelinating hosts. *J. Neurosci. Res.*, 34: 1–13.

Weiss, S., Dunne, C., Hewson, J., Wohl, C., Wheatley, M., Peterson, A.C. and Reynolds, B.A. (1996) Multipotent CNS stem cells are present in the adult mammalian spinal cord and ventricular neuroaxis. *J. Neurosci.*, 16: 7599–7609.

Wichterle, H., Garcia-Verdugo, J.M., Herrera, D.G. and Alvarez-Buylla, A. (1999) Young neurons from medial ganglionic eminence disperse in adult and embryonic brain. *Nat. Neurosci.*, 2: 461–466.

Xu, X.M., Chen, A., Guénard, V., Kleitman, N. and Bunge, M.B. (1997) Bridging Schwann cell transplants promote axonal regeneration from both the rostral and caudal stumps of transected adult rat spinal cord. *J. Neurocytol.*, 26: 1–16.

Yandava, B.D., Billinghurst, L.L. and Snyder, E.Y. (1999) Global cell replacement is feasible via neural stem cell transplantation: evidence from the dysmyelinated shiverer mouse brain. *Proc. Natl. Acad. Sci. USA*, 96: 7029–7034.

Young, W. (1996) Spinal cord regeneration [comment]. *Science*, 273: 451.

Young, W. (1997) Fear of hope [editorial; comment]. *Science*, 277: 1907.

Zhang, S.C., Ge, B. and Duncan, I.D. (1999) Adult brain retains the potential to generate oligodendroglial progenitors with extensive myelination capacity. *Proc. Natl. Acad. Sci. USA*, 96: 4089–4094.

Zhang, Z. and Guth, L. (1997) Experimental spinal cord injury: Wallerian degeneration in the dorsal column is followed by revascularization, glial proliferation, and nerve regeneration. *Exp. Neurol.*, 147: 159–171.

Zigova, T., Pencea, V., Betarbet, R., Wiegand, S.J., Alexander, C., Bakay, R.A. and Luskin, M.B. (1998) Neuronal progenitor cells of the neonatal subventricular zone differentiate and disperse following transplantation into the adult rat striatum. *Cell Transplant.*, 7: 137–156.

CHAPTER 25

Potential use of marrow stromal cells as therapeutic vectors for diseases of the central nervous system

Darwin J. Prockop *, S. Ausim Azizi, Donald G. Phinney, Gene C. Kopen and Emily J. Schwarz

Center for Gene Therapy, MCP Hahnemann University, 10118 New College Building, 245 North 15 Street, Philadelphia, PA 19102, USA

Introduction

In addition to precursors for hematopoietic cells, bone marrow contains stem-like cells for a variety of non-hematopoietic tissues. The cells are variously referred to as colony-forming fibroblasts, mesenchymal stem cells or marrow stromal cells (MSCs). Beginning over 20 years ago, Friedenstein et al. (1976, 1987) and many other investigators demonstrated that MSCs in culture can differentiate into osteoblasts, adipocytes, chondrocytes, fibroblasts and myoblasts (see Piersma et al., 1985; Owen and Friedenstein, 1988; Caplan, 1991; Prockop, 1997). Human MSCs are relatively easy to isolate from small aspirates of whole bone marrow by their adherence to tissue culture surfaces. Also, human MSCs are relatively easy to expand in culture by repeated passage. Some reports (Pittenger et al., 1999) suggest that human MSCs, in contrast to murine MSCs (Phinney et al., 1999), are homogeneous after one or two passages in culture. Other reports indicate that such cultures of human MSCs contain several phenotypically distinct cell types (Mets and Verdonk, 1981).

---

* Corresponding author: Dr. Darwin J. Prockop, Center for Gene Therapy, SL99, Tulane University Health Sciences Center, 1430 Tulane Avenue, New Orleans, LA 70112, USA. Fax: +1-504-988-7710; E-mail: dprocko@tulane.edu

Animal studies with systemic infusion of MSCs

In an initial series of experiments from our laboratory, we infused labeled MSCs systemically into isogenic mice with and without prior marrow ablation of the recipient mice in order to create a 'space' for engraftment of the cells (Pereira et al., 1995). After a delay of over one week, progeny of the cells appeared in a variety of tissues including bone, cartilage and lung. As the progeny of the MSCs appeared in the tissues, they appeared to take on the cellular phenotype of the tissue in that cells which appeared in bone expressed a marker gene for type I collagen, whereas they did not express the marker gene for type I collagen when they appeared in cartilage, a tissue in which type I collagen is not synthesized. Similar results reported from other laboratories demonstrated the appearance of progeny of infused MSCs in bone (Hou et al., 1999), lung, spleen and thymus (Keating et al., 1996). In addition, engraftment of donor MSCs in a repairing muscle was observed after either local injection or systemic injection (Ferrari et al., 1998). Also, engraftment into muscle of dystrophin-deficient mouse was seen after systemic infusion of rare marrow cells defined as a 'side population' or SP cells that may be precursors of MSCs (Goodell et al., 1997; Gussoni et al., 1999). In experiments in which male MSCs were infused into female recipient mice (Pereira et al., 1998), fluorescence in situ hybridization (FISH)

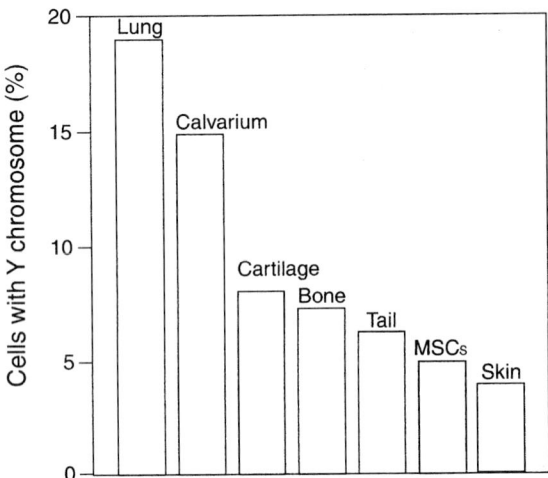

Fig. 1. FISH assay for Y chromosome to detect male cells in primary cultures of tissues from female mice that received systemic infusions of isogenic male MSCs. The recipient was a three-week female transgenic mouse that had a phenotype of osteogenesis imperfecta (OI) because it expressed a mutated gene for type I collagen. The mouse was irradiated with 700 cGy to enhance engraftment to cells. Because of the marrow ablation, the male MSCs (11×10^6) were infused intraperitoneally together with whole bone marrow (2×10^6) from an OI-transgenic mouse in order to provide hematopoietic stem cells. Primary cultures of various tissues were prepared 2.5 months after the infusion, and the cells assayed by FISH for presence of the Y chromosome. (From Pereira et al., 1998.)

analyses for the Y chromosome demonstrated the presence of progeny of MSCs in primary cultures obtained from a variety of tissues (Fig. 1).

Clinical trials

Two clinical trials have been initiated to use MSCs for the therapy of bone diseases. In one trial, allogeneic bone marrow transplantations were performed in children with severe osteogenesis imperfecta (Horwitz et al., 1999). The trial was based on the assumption that whole bone marrow may contain enough MSCs to improve the structure of bone by giving rise to wild-type osteoblasts in patients in whom the endogenous osteoblasts express a mutated gene for type I collagen. All three children who received marrow ablation followed by a bone marrow transplant from an HLA-compatible brother or sister showed objective signs of clinical improvement. However, donor cells accounted for only 1 to 2% of the osteoblasts in recipients and, therefore, there are several possible explanations for the results. A related clinical trial (Whyte et al., 1999) was carried out in an 8-month old child with infantile hypophosphatasia, a genetic deficiency of the alkaline phosphatase found in osteoblasts. The child was first treated with marrow ablation followed by a bone marrow transplant from an HLA-compatible sister. Subsequently, she was treated with a booster infusion of T-cell depleted MSCs from the same donor. The child showed temporary improvement from the bone marrow transplantation and then persistent improvement for nine months following the booster infusion of MSCs.

Animal studies with intracranial infusion of MSCs

To test further the potential of MSCs to engraft and differentiate, we infused either human or rat MSCs directly into the basal ganglia of adult rats (Azizi et al., 1998). Surprisingly, the infused cells did not aggregate and invoke immune responses as is seen with infusion of fibroblasts. Also, there was no evidence that the cells differentiated into mesenchymal cells. Instead, the cells integrated and migrated along known pathways in a manner that was similar to the integration and migration seen with infusion of paraventricular astrocytes that have many of the properties of embryonic neural stem cells (Fig. 2). The same pattern of integration and migration was seen with both rat MSCs and human MSCs infused into the brains of adult rats.

In another series of experiments, murine MSCs were infused into the periventricular region of the brains of newborn mice (Kopen et al., 1999). Again, the cells integrated and migrated in a manner similar to the integration and migration seen with infusions of embryonic neural stem cells. Some of the cells differentiated into astrocytes as demonstrated by the presence in the same cells of the pre-infusion marker, bromdeoxyuridine (BrdU), and immunoreactivity with an antibody to the astrocyte-specific protein, glial fibrillary acidic protein (GFAP). Other MSCs may have differentiated into neurons, since they appeared in large numbers in neuron-rich regions such as the internal granular layer of the cerebellum and the olfactory bulb.

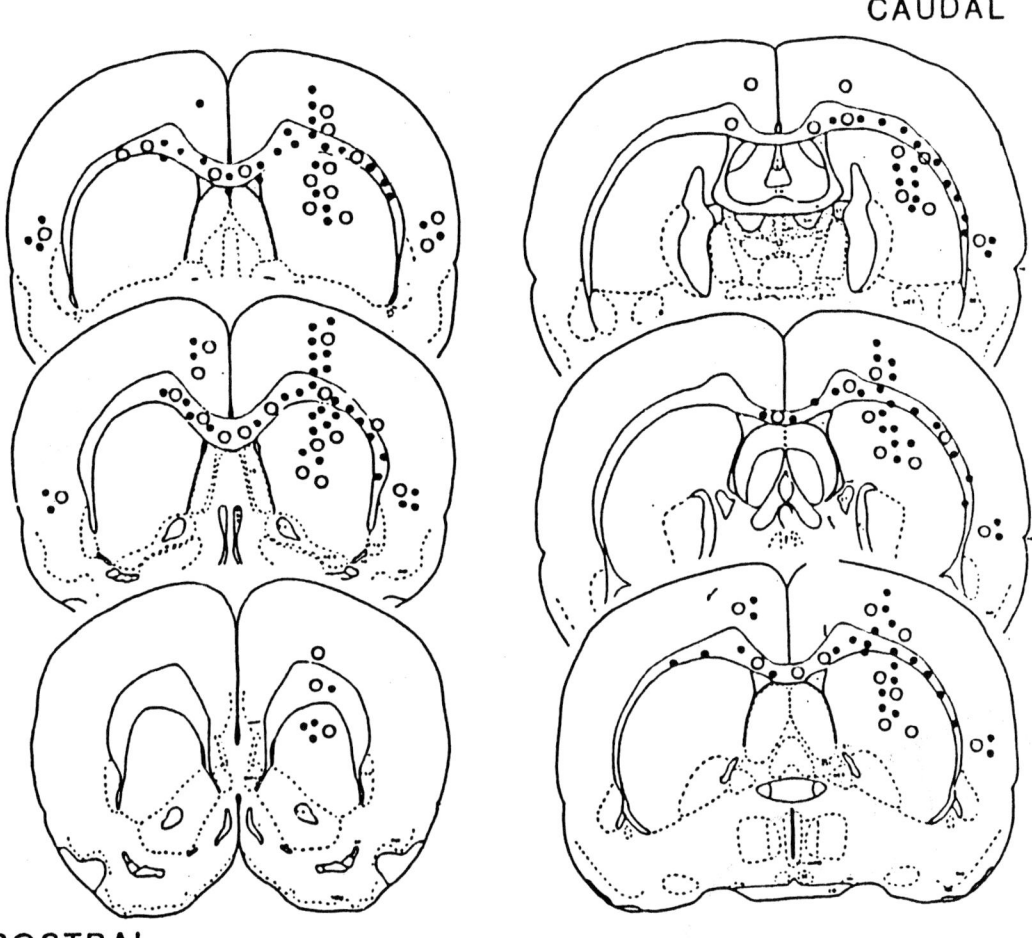

Fig. 2. Line drawings of rat forebrain to demonstrate the extent of integration and migration of human MSCs after implantation at the level of the bregma into the striatum of adult rats. The drawing is a composite from brains examined 4, 14, 30 and 72 days after cell infusion. Symbols: ●, clusters of human MSCs that were pre-labeled with the nuclear dye bis-benzimide; ○, clusters of infused paraventricular astrocytes that have many of the properties of neural stem cells; also pre-labeled with bis-benzimide. (From Azizi et al., 1998.)

In still another series of experiments, we tested the possibility of using MSCs as vectors for therapy of parkinsonism (Schwarz et al., 1999). We stably transfected rat MSCs with one retrovirus containing a gene for tyrosine hydroxylase and then with a second retrovirus containing a gene for guanosine 5′-triphosphate (GTP) cyclohydrase to provide the critical enzyme required to synthesize an essential cofactor for tyrosine hydroxylase. The transduced cells synthesized 3,4-dihydroxy-L-phenylalanine (L-DOPA) in vitro and maintained their multipotentiality to differentiate into osteoblasts and adipocytes in culture. The cells were then infused into the striatum of rats that had a phenotype of parkinsonism because they were lesioned with 6-hydroxydopamine (Fig. 3). Assays by microdialysis of the denervated striatum of the rats demonstrated that the transduced MSCs synthesized L-DOPA and metabolites of dopamine (Fig. 4). Also, there was a significant reduction in the behavioral phenotype of apomorphine-induced rotation when compared to controls (Fig. 5). The cells en-

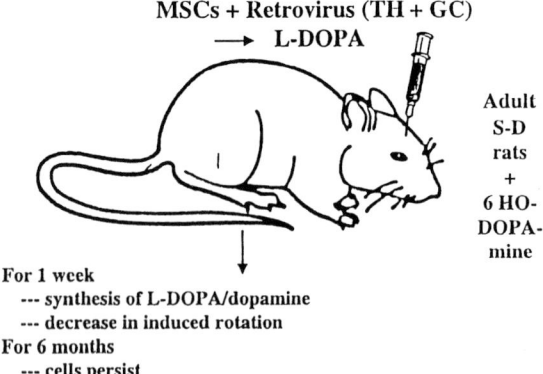

Fig. 3. Design of experiment in which MSCs were tested for the possible therapy of parkinsonism. The basal ganglia of the rats were ablated with 6-hydroxydopamine (6 HO-DOPA-mine). The genes for tyrosine hydroxylase (*TH*) and GTP cyclohydrase (*GC*) were introduced into rat MSCs with two retroviruses and were shown to synthesize and secrete L-DOPA in culture. (From Schwarz et al., 1999.)

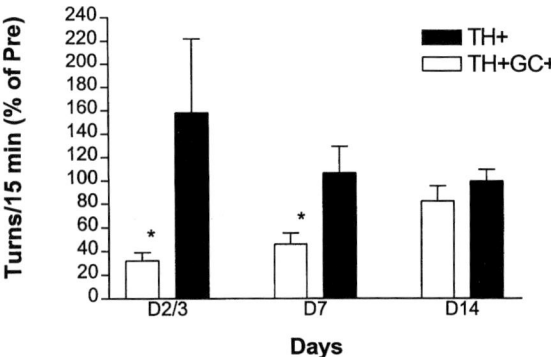

Fig. 4. Effects of genetically modified rMSCs on rotational behavior in 6-hydroxydopamine-lesioned rats. Apomorphine-induced contralateral rotation was measured before (Pre) and 2/3, 7 and 14 days after transplantation with either rMSCs expressing only the TH gene (*TH+*, that served as controls) or both the TH and GC gene (*TH+GC+*). The graph shows turns in a 15-min period as a percentage of that observed before transplantation. None of the compounds were detected in lesioned rats that were transplanted with rMSC-TH (control rMSCs). Data represent means ± SEM. Rotational behavior of rats given rMSC-THGC is significantly reduced on day 2/3 and day 7 ($*P < 0.02$) compared with Pre. (From Schwarz et al., 1999.)

grafted and survived for at least 87 days. However, expression of the transgene ceased after about 9 days, an observation consistent with reports from other laboratories in which similar retroviruses were used to express transgenes, and expression of the

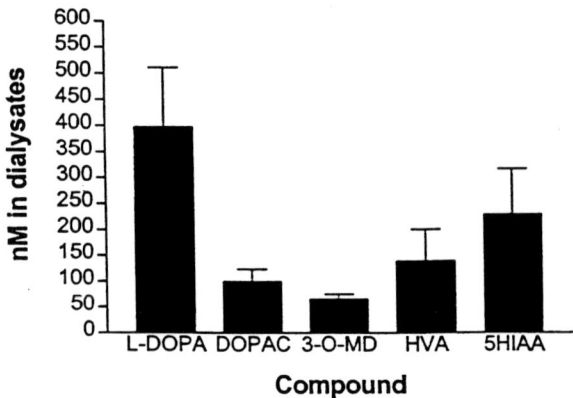

Fig. 5. Microdialysis in striatum of 6-hydroxydopamine-lesioned rats with transplants of rMSC-THGC. Dialysate samples were assayed by high-performance liquid chromatography (HPLC) at 30-min intervals for a total of 4 h. Compounds were measured in the lesioned striatum 3 days after transplantation. Data represent the concentration mean ± SEM ($n = 3$) in 30 μl dialysates. L-DOPA, 3,4-dihydroxy-L-phenylalanine; DOPAC, 3,4-dihydroxyphenylacetic acid; 3-0-MD, 3-*O*-methyl-DOPA; HVA, homovanillic acid; 5HIAA, 5-hydroxyindoleacetic acid. (From Schwarz et al., 1999.)

gene ceased after about one week (Bencsics et al., 1996). The mechanism of suppression of transgenes introduced into the brain with retroviruses is still not defined and is probably explained with either the presence of the long terminal repeats found in retroviruses, or the use of so-called ubiquitous promoters, such as the promoter of cytomegalic virus. Therefore, we are currently repeating these experiments with MSCs in which the same exogenous genes are introduced using non-viral techniques and promoters for housekeeping genes.

Based on these and related observations, we and others are exploring the possibility that MSCs can be used for cell and gene therapy of a variety of diseases of the musculoskeletal and central nervous systems.

Conclusion

MSCs have several features that make them attractive as vehicles for cell and gene therapy. They are relatively easy to isolate from bone marrow aspirates obtained under local anesthesia. The cells can be amplified in culture through many passages by their tight adherence to tissue culture plastic. In addition, it is relatively easy to gene-engineer the cells and obtain a large quantity of gene-engineered progenitor cells. For these reasons, the cells are being tested in many laboratories for therapy of a variety of diseases.

References

Azizi, S.A., Stokes, D.G., Augelli, B.J., DiGirolamo, C.M. and Prockop, D.J. (1998) Engraftment and migration of human bone marrow stromal cells implanted in the brains of albino rats — similarities to astrocyte grafts. *Proc. Natl. Acad. Sci. USA*, 95: 3908–3913.

Bencsics, C., Wachtel, S.R., Milstein, S., Hatakeyama, K., Becker, J.B. and Kang, U.J. (1996) Double transduction with GTP cyclohydrolase I and tyrosine hydroxylase is necessary for spontaneous synthesis of L-DOPA by primary fibroblasts. *J. Neurosci.*, 16: 4449–4456.

Caplan, A.I. (1991) Mesenchymal stem cells. *J. Orthoped. Res.*, 9: 641–650.

Ferrari, G., Cusella-De Angelis, G., Coletta, M., Paolucci, E., Stornaiuolo, A., Cossu, G. and Mavilio, F. (1998) Muscle regeneration by bone marrow-derived myogenic progenitors. *Science*, 279: 1528–1530.

Friedenstein, A.J., Gorskaja, U. and Kalugina, N.N. (1976) Fibroblast precursors in normal and irradiated mouse hematopoietic organs. *Exp. Hematol.*, 4: 267–274.

Friedenstein, A.J., Chailakhyan, R.K. and Gerasimov, U.V. (1987) Bone marrow osteogenic stem cells: in vitro cultivation and transplantation in diffusion chambers. *Cell Tiss. Kinet.*, 20: 263–272.

Goodell, M.A., Rosenzweit, M., Kim, H., Marks, D.F., DeMaria, M., Paradis, G., Grupp, S.A., Sieff, C.A., Mulligan, R.C. and Johnson, R.P. (1997) Dye efflux studies suggest that hematopoietic stem cells expressing low or undetectable levels of CD34 antigen exist in multiple species. *Nat. Med.*, 12: 1337–1345.

Gussoni, E., Soneoka, Y., Stickland, C.D., Buzney, E.A., Khan, M.K., Flint, A.F., Kunkel, L.M. and Mulligan, R.C. (1999) Dystrophin expression in the *mdx* mouse restored by stem cell transplantation. *Nature*, 401: 390–393.

Horwitz, E.M., Prockop, D.J., Fitzpatrick, L.A., Winston, W.K.K., Gordon, P.L., Neel, M., Sussman, M., Orchard, P., Marx, J.C., Pyeritz, R.E. and Brenner, M.K. (1999) Transplantability and therapeutic effects of bone marrow-derived mesenchymal cells in children with severe osteogenesis imperfecta. *Nat. Med.*, 5: 309–313.

Hou, Z., Nguyen, Q., Frenkel, B., Nilsson, S.K., Milne, M., Van Wijnen, A.J., Stein, J.L., Quesenberry, P., Lian, J.B. and Stein, G.S. (1999) Osteoblast-specific gene expression after transplantation of marrow cells: implications for skeletal gene therapy. *Proc. Natl. Acad. Sci. USA*, 96: 7294–7299.

Keating, A., Guinn, B., Larava, P. and Wang, X.-H. (1996) Human marrow stromal cells electrotransfected with human factor IX (FIX) cDNA engraft in Scid mouse marrow and transcribe human FIX. *Exp. Hematol.*, 24: 1056; Abs. 180.

Kopen, G.C., Prockop, D.J. and Phinney, D.J. (1999) Marrow stromal cells migrate throughout forebrain and cerebellum and they differentiate into astrocytes following injection into neonatal mouse brains. *Proc. Natl. Acad. Sci. USA*, 96: 10711–10716.

Mets, T. and Verdonk, G. (1981) In vitro aging of human bone marrow-derived stromal cells. *Mech. Ageing Dev.*, 16: 81–89.

Nilsson, S.K., Donnor, M.S., Weier, H.U., Frenkel, B., Lian, J.B., Stein, G.S. and Quesenberry, P.J. (1999) Cells capable of bone production engraft from whole bone marrow transplants in non-ablated mice. *J. Exp. Med.*, 189: 729–734.

Owen, M.E. and Friedenstein, A.J. (1988) Stromal stem cells: marrow-derived osteogenic precursors. In: *Cell and Molecular Biology of Vertebrate Hard Tissues*. Ciba Foundation Symp., Chichester, pp. 42–60.

Pereira, R.F., Halford, K.W., O'Hara, M.D., Leeper, D.B., Sokolov, B.P., Pollard, M.D., Bagasra, O. and Prockop, D.J. (1995) Cultured adherent cells from marrow can serve as long-lasting precursor cells for bone, cartilage and lung in irradiated mice. *Proc. Natl. Acad. Sci. USA*, 92: 4857–4861.

Pereira, R.F., O'Hara, M.D., Laptev, A.V., Halford, K.W., Pollard, M.D., Class, R., Simon, D., Livezey, K. and Prockop, D.J. (1998) Marrow stromal cells as a source of progenitor cells for non-hematopoietic tissues in transgenic mice with a phenotype of osteogenesis imperfecta. *Proc. Natl. Acad. Sci. USA*, 95: 1142–1147.

Phinney, D.G., Kopen, G., Isaacson, R.L. and Prockop, D.J. (1999) Plastic adherent stromal cells from the bone marrow of commonly used strains of inbred mice: variations in yield, growth and differentiation. *J. Cell. Biochem.*, 72: 570–585.

Piersma, A.H., Brockbank, K.G.M., Ploemacher, R.E., Van Vilet, E., Brakel-Van Peer, K.M.J. and Visser, P.J. (1985) Characterization of fibroblastic stromal cells from murine bone marrow. *Exp. Hematol.*, 13: 237–243.

Pittenger, M.F., Mackay, A.M., Beck, S.C., Jaiswal, R.K., Douglas, R., Mosca, J.E., Moorman, M.A., Simonetti, D.W., Craig, S. and Marshak, D.R. (1999) Multilineage potential of adult human mesenchymal stem cells. *Science*, 284: 143–147.

Prockop, D.J. (1997) Marrow stromal cells as stem cells for non-hematopoietic tissues. *Science*, 276: 71–74.

Schwarz, D.J., Alexander, G.M., Prockop, D.J. and Azizi, S.A. (1999) Multipotential marrow stromal cells transduced to produce L-DOPA: engraftment in a rat model of Parkinson's disease. *Hum. Gene Ther.*, 10: 2539–2549.

Whyte, M.P., Kurtzburg, J., McAlister, W.H., Coburn, S.P., Ryan, L.M., Miller, C.R., Gottesman, G.S. and Martin, P.L. (1999) Marrow cell transplantation for infantile hypophosphatasia. Abstracts of the 7th International Meeting on Osteogenesis Imperfecta, Montreal, August 1999.

CHAPTER 26

Neurobiology of human neurons (NT2N) grafted into mouse spinal cord: implications for improving therapy of spinal cord injury

Virginia M.-Y. Lee [1,*], Rebecca S. Hartley [1,2] and John Q. Trojanowski [1]

[1] *The Center for Neurodegenerative Disease Research, Department of Pathology and Laboratory Medicine, University of Pennsylvania School of Medicine, Hospital of the University of Pennsylvania, 3rd Floor Maloney Building, 3600 Spruce Street, Philadelphia, PA 19104-4283, USA*
[2] *Layton Bioscience, Inc., Sunnyvale, CA, USA*

Introduction

Current therapeutic interventions for severe spinal cord injury have limited efficacy, and this has prompted efforts to explore the use of transplanted neural cells to improve functional recovery in spinal cord injury patients (Tator, 1998; Kocsis, 1999; McDonald et al., 1999; Park et al., 1999; Whittemore, 1999). Although grafted neuron-like cells derived from embryonic progenitors integrate into the normal and injured central nervous system (CNS), where they can give rise to very long processes (> 1 cm) in white (Wictorin and Björklund, 1992; Wictorin et al., 1992; Davies et al., 1993, 1994) and gray matter (Björklund and Stenevi, 1977a,b; Björklund et al., 1980; Nornes et al., 1983; Foster et al., 1985; Gibbs et al., 1985; Zhou et al., 1985, 1989), these transplants also contain glia and other cell types.

However, since glial cells produce factors that influence neurons, the behavior of grafted neurons may be affected by host glial cells as well as by any exogenous glial cells that are present in the grafts (Lindsay, 1979; Lindsay and Raisman, 1984; Goldberg and Bernstein, 1987, 1988a,b; Emmett et al., 1991). Moreover, while immortalized and neoplastic stem cells can be induced to differentiate into neurons in vitro prior to implantation in the CNS, these progenitors almost always give rise to multiple different lineages, and cells belonging to other lineages will be co-grafted with the neurons generated from these stem cells (e.g., see Renfranz et al., 1991; Whittemore et al., 1991; Snyder et al., 1992; Onifer et al., 1993; Whittemore and White, 1993; Bain et al., 1995; Gage et al., 1995; Bruestle et al., 1998; Flax et al., 1998; Thomson et al., 1998a,b; and citations therein). Thus, it has been challenging to exploit many of the most common and well studied transplant model systems to gain insight into the effects of the host neural cells alone on grafted neurons.

For this reason, we studied spinal cord transplants of pure, postmitotic neurons (NT2N, hNT) generated from a clonal human embryonal carcinoma (Ntera-2, NT2) cell line in vitro by treating the NT2 cells with retinoic acid (RA) as well as previously de-

* Corresponding author: Dr. Virginia M.-Y. Lee, The Center for Neurodegenerative Disease Research, Department of Pathology and Laboratory Medicine, University of Pennsylvania School of Medicine, Hospital of the University of Pennsylvania, 3rd Floor Maloney Building, 3600 Spruce Street, Philadelphia, PA 19104-4283, USA; Fax: +1-215-349-5909; E-mail: vmylee@mail.med.vpenn.edu

scribed methods to obtain pure populations of NT2N neurons (Andrews et al., 1984; Lee and Andrews, 1986; Pleasure et al., 1992; Pleasure and Lee, 1993). Interestingly, NT2N neurons exhibit glutamatergic or GABAergic properties (Hartley et al., 1999), but they also are cholinergic (Zeller and Strauss, 1995), and some express tyrosine hydroxylase (Iacovitti and Stull, 1997; Miyazono et al., 1996). Further, transplanted NT2N neurons integrate into the host CNS of immune compromised mice, extend axon-like and dendrite-like processes, retain a neuronal phenotype for > 1 year, and do not form tumors or deleteriously affect the host (Kleppner et al., 1995; Mantione et al., 1995; Trojanowski et al., 1997). Indeed, RA-naïve NT2 cells grafted into the CNS of immune compromised mice can differentiate into postmitotic neurons and survive for > 1 year (Miyazono et al., 1995, 1996), but NT2N grafts are rejected in the CNS of immune competent rats (Trojanowski et al., 1993). Since grafted NT2N cells have been reported to ameliorate some of the impairments that result from ischemic stroke in experimental animals (Borlongan et al., 1998a,b,c), we review recent studies of the neurobiology of NT2N spinal cord grafts (Hartley et al., 1999), and the implications of these studies on research efforts to develop transplantation therapy to improve recovery of patients from spinal cord injury.

Design of studies to define the neurobiology of human neurons transplanted into nude mouse spinal cord

NT2N neurons were generated from the parent NT2 cells as described previously (Pleasure et al., 1992; Kleppner et al., 1995), and vehicle volume was adjusted to ∼10,000–40,000 cells/µl for injections of 1–2 µl at each spinal cord graft site. Homozygous athymic mice underwent transplantation using sterile pulled glass pipettes attached to a syringe pump. Briefly, NT2N neurons were implanted by introducing the pipette at an angle of 45–50° into the right side of the spinal cord at T13–L1 levels, and injecting NT2N neurons at sites throughout the dorsal–ventral extent of the spinal cord (Hartley et al., 1999). The mice were perfused (4% paraformaldehyde or 70% ethanol) under deep anesthesia (ketamine/xylazine), and the cords were embedded in paraffin (Trojanowski et al., 1989a,b). Near serial 6-µm thick sections were processed for light and confocal microscopic immunohistochemical studies as described elsewhere (Hickey et al., 1983; Lee et al., 1987; Schmidt et al., 1987, 1993; Trojanowski et al., 1989a,b; Tohyama et al., 1991; Miyazono et al., 1996). Antibodies similar to those used in previous studies of grafted NT2 cells and NT2N neurons were employed here (Trojanowski et al., 1993; Kleppner et al., 1995; Miyazono et al., 1995, 1996), including monoclonal antibodies (MAbs) that recognize human-specific epitopes in neuronal proteins (i.e., MOC-1, hSYN, HO14; see below) for unequivocal detection of the grafted human NT2N neurons (Lee et al., 1987; Tonder et al., 1988; Molenaar et al., 1991; Trojanowski et al., 1993; Kleppner et al., 1995; Miyazono et al., 1996).

We analyzed 86 mice with spinal cord grafts including 1 to 2 days old neonates and adults up to 85 days old, most of which were implanted at 24–25 days of age. Graft size was approximated using NIH Image software to outline the graft in every 20–30th section stained with the MOC-1 MAb specific for human neural cell adhesion molecules (NCAM). Cell size and shape were similarly assessed by measuring the largest and most clearly defined cell bodies in sections immunostained with MOC-1 or the HO14 MAb, which is specific for the human mid-size neurofilament (NF) subunit (NFM). Long, thin, non-tapering processes were classified as axons while thicker, tapered processes were judged to be dendrites. These findings were confirmed in selected sections immunostained with MAbs specific for axonal or dendritic proteins, i.e., HO14 to NFM or M13 to microtubule-associated protein 2 (MAP2). Maximal axonal outgrowth was measured in every 20–30th section through grafts immunostained for human NCAM or human NFM, and these sections also were used to monitor the percentage of axons extending the longest distances in fifteen cases with survival times of approximately 1, 6 and 15 months. To do this, we counted the number of human axons at ∼400 mm from the graft ('proximal to graft') and the number of human fibers in the area containing the longest measured axon ('distal to graft'). The number of fibers found at the farthest locations divided by those found near the graft represents the percentage of graft-derived fibers that extend the longest distances.

Salient neurobiological properties of human NT2N neuron grafts in spinal cord of immunocompromised nude mice

The human-specific MOC-1 MAb enabled accurate localization of the grafted human NT2N neurons. Although the volume of the grafts ranged from 0.036×10^{-3} mm^3 to 0.254 mm^3, these volumes did not correlate with post-implantation survival times or host age at implantation, and there was no evidence of graft rejection or tumorigenicity. Notably, cultured NT2N cells express protein markers that are characteristic of immature human neurons in situ in spinal cord and other regions of the developing human CNS including NCAM, glutamic acid decarboxylase (GAD), NF subunits, synaptophysin, etc., but they do not express protein markers of fully mature human neurons in situ such as all six adult CNS tau isoforms and the most highly phosphorylated (P^{3+}) NFM and high molecular weight NF (NFH) subunits (Pleasure et al., 1992; Pleasure and Lee, 1993). A marker protein profile similar to that detected in NT2N cells in vitro also was seen in the NT2N neurons after short survival times in vivo. With increasing post-implantation survival times, the grafted NT2N cells began to express P^{3+} NFH and adult tau isoforms typical of fully mature human CNS neurons, and by seven months post-implantation, the expression of these and other neuronal proteins by the grafted NT2N neurons resembled that seen in neurons of the fully mature human spinal cord or other regions of the adult human CNS (Tohyama et al., 1991; Yachnis et al., 1993; Arnold and Trojanowski, 1996). Notably, the onset of the expression of synaptophysin in vivo correlates closely with synapse formation and synaptic activity (Wiedenmann and Franke, 1985; Knaus et al., 1986; Leclerc et al., 1989; Bergmann et al., 1991; Fletcher et al., 1991; Tohyama et al., 1991), and we observed that synaptophysin and GAD immunoreactivity also increased with longer survival times in the grafts, but none of the grafted NT2N neurons expressed tyrosine hydroxylase. Neither the graft site nor the age of the host at implantation affected the phenotype or morphology of the implanted NT2N neurons, and there was no evidence that the NT2N neurons migrated from the site of the initial implants.

NT2N spinal cord grafts showed outgrowth of abundant processes into host gray and white matter, and this appeared to be unrelated to the age of the host at implantation. Notably, the outgrowth, trajectory and other features of these processes varied depending on the location of the processes, and the properties of these processes appeared to recapitulate those of the host axons and dendrites in the same spinal cord region. For example, axon-like NT2N processes extended into white matter and they followed a trajectory similar to the adjacent host white matter axons. In contrast, dendrite-like processes of grafted NT2N neurons in gray matter regions of the host spinal cord were randomly arrayed near the graft. Moreover, most NT2N axonal processes that could be traced over long distances were found to travel within large host spinal cord white matter tracts, especially when grafts of the parent NT2N neurons were located at the gray/white matter interface. However, some graft-derived NT2N neurites also extended along axonal bundles in the gray matter and they frequently crossed the midline of spinal cord gray matter where host fibers normally decussate.

Immediately (<2 weeks) after implantation, grafted NT2N cells had a uniform size and shape within the needle track, but by 2 weeks after implantation, the needle track was no longer evident and the appearance of the NT2N neuronal perikarya was more variable. However, the cell soma size distributions of the NT2N neurons remained uniform beyond this time point regardless of the density or position of the NT2N cells in the graft, the CNS implantation site, or the age of the host at the time of implantation. Further, the average size of the grafted NT2N neurons did not exceed that of most of the host motor neurons, or that of the host dorsal spinal cord neurons.

Some of the processes of the NT2N neurons extended for distances of >2 cm in the host spinal cord white matter. However, NT2N neurons implanted at neonatal, juvenile or adult ages all showed similar axonal outgrowth. Long-distance outgrowth was assessed by measuring the longest unequivocally immunoreactive process in each section. Single graft fibers grew along host fibers, similar to axons of embryonic human or murine cells implanted into rat spinal cords (Wictorin and Björklund, 1992; Li and Raisman, 1993). When outgrowth was assessed with MOC-1, distal immunoreactivity decreased at post-implantation times of >6 weeks. When assessed with HO14, however, distal immunoreactivity did not decrease over the same time. The rate of maximal outgrowth seen with the HO14 MAb was ∼1.4 mm/week for the first 6 weeks, followed by a rate of at least 1 mm/month thereafter. Indeed, some NT2N neurites grew for distances of >2 cm, but could not be followed further because they continued beyond the spinal cord segments analyzed here. The extent of neurite outgrowth was similar in regions rostral and caudal to the graft site as well as in both the dorsal and ventral white matter tracts. Finally, the number of processes identified at locations 'distal to graft' divided by the number of processes identified at sites 'proximal to graft' overall for survival times of ∼1, 6 and 15 months was estimated to be 82%, suggesting that many NT2N neuronal processes attain very long trajectories, while data from each of these survival intervals suggested that NT2N neurons extend processes at a rate of ∼1.4 mm/week in the first 6 weeks after implantation and at ∼1 mm/month thereafter.

Myelination was examined using confocal double-label immunofluorescence with an anti-myelin-associated glycoprotein (MAG) antibody (Pedraza et al., 1990) and HO14 (to identify NT2N axons), and these studies showed that at least some NT2N axons were ensheathed by MAG positive oligodendrocytes. The close proximity of myelinated host fibers to NT2N processes in host white matter tracts precluded definitive assessment of the NT2N fibers within the tracts. The trajectory of long NT2N processes varied within white matter tracts. Axons were observed to continue within the white matter tracts, cross into spinal roots, or turn back into gray matter. Many dendrite-like processes were found immediately adjacent to the grafts, while large numbers of axon-like processes traveled along host white matter tracts or bundles. In three of the 15-month post-implantation cases and in one of the 7-month cases, rare NT2N axonal processes were also found within spinal nerves. Other processes were observed to follow large white matter tracts, but then turn abruptly, exit the tract, and then reenter gray matter, and many of these processes appeared to terminate within the gray matter. It was not clear whether these processes elaborated terminal arborizations similar to host axons since NFM cannot be detected in axon terminal arborizations. However, a human-specific anti-synaptophysin antibody (hSYN) showed punctate immunoreactivity near HO14 immunoreactive processes in host gray matter. In graft areas, confocal double-label immunofluorescence studies with hSYN and HO14 showed robust punctate hSYN immunoreactivity within the grafts and in the surrounding host. The hSYN immunoreactivity in the host around the graft may indicate graft–host connections, or connections between graft processes. Confocal double-label immunofluorescence studies in host areas devoid of grafted cell bodies or dendritic processes also showed punctate hSYN immunoreactivity near HO14 immunoreactive processes which is suggestive of graft–host synaptic connections.

Therapeutic implications of human NT2N neuron grafts for spinal cord injury patients

The studies reviewed here demonstrate that human NT2N neurons implanted into the spinal cord of nude mice: (1) integrate and survive for >15 months in gray and white matter sites with the phenotypic features of mature human CNS neurons; (2) exhibit a neuronal phenotype that is similar at different graft sites regardless of the host age at implantation; (3) extend dendrite-like and axon-like processes; (4) give rise to axons that are differentially influenced by cues intrinsic to the gray and white matter regions into which these processes extend, and establish a trajectory that appears to be 'molded' by the axonal projection patterns in a given spinal cord region. Additionally, we identified some NT2N axons that were ensheathed by host oligodendrocytes, and others that grew through white matter tracts into spinal nerves. Finally, the morphological appearance of some NT2N processes expressing synaptophysin suggests that these processes may form synaptic contacts with host cells.

While diverse CNS regions may differentially influence neurotransmitter expression by neurons that arise from transplanted RA-naïve NT2 cells (Miyazono et al., 1996), and the properties of NT2N processes differ in spinal cord white versus gray matter, other phenotypic features of these grafted neurons did not vary in distinct spinal cord locations. This is consistent with the behavior of some neural precursor cells, although, unlike the NT2 cells, neural precursors isolated from phenotypically different embryonic brain regions can have the capacity to generate multiple lineages of differentiated CNS cells (Björklund and Stenevi, 1977a; Tonder et al., 1988, 1990; Wictorin et al., 1990, 1992; Fujii, 1994; Stromberg et al., 1992; Wictorin and Björklund, 1992; Davies et al., 1993, 1994; Li and Raisman, 1993; Barbe, 1996; McDonald et al., 1999). Notably, the long-distance outgrowth of NT2N processes described here is similar to that of murine embryonic hippocampal cells implanted into spinal cord white matter (Li and Raisman, 1993), and by human embryonic spinal cord cells implanted into lesioned rat spinal cord (Wictorin and Björklund, 1992). It is not clear if all neurons are capable of generating lengthy axons, or if these processes arise from a subset of grafted neurons (Wictorin and Björklund, 1992). However, at least to some extent, the outgrowth of neuronal processes (including those derived from NT2N neurons) also is regulated by the host since the NT2N processes appeared to be 'molded' by the host CNS architecture in the studies reviewed here.

Processes that extend from regenerating endogenous neurons and implanted cells can be seen within spinal nerves that exit into the host peripheral nervous system (PNS) (Clowry and Vrbova, 1992; Sieradzan and Vrbova, 1993; Wu et al., 1994; Houle et al., 1996; Nogradi and Vrbova, 1996a,b; Carlstedt, 1997), but the outgrowth of processes from neurons grafted into the CNS has not been reported to extend through an intact CNS–PNS transition zone. However, neurons in human embryonic dorsal root ganglia (DRGs) implanted at the site of excised rat DRGs were shown to generate processes that grew from the PNS into the spinal cord (Kozlova et al., 1994, 1995). Early embryonic neurites have also been shown to cross mature dorsal root entry zones (Golding et al., 1996). Nonetheless, it remains to be determined what mechanisms enable the axons of implanted NT2N neurons to extend long distances within the spinal cord and cross into the PNS.

Based on data from the studies (Hartley et al., 1999) reviewed here, the NT2N neurons appear to be an attractive model system in which to pursue the development of novel strategies for investigating the neurobiology of transplanted human neurons as well as the discovery of new transplant therapies for patients with spinal cord injuries. Although the therapeutic potential of neural cell transplants for treating spinal cord injury patients remains to be explored in well designed and rigorously controlled studies of experimental animal models, encouraging support for the use of transplanted neural cells as a therapeutic intervention to augment recovery from spinal cord injury is beginning to emerge (Kocsis, 1999; McDonald et al., 1999). In addition, the potential use of transplanted NT2N neurons as a therapeutic intervention for other CNS diseases is beginning to be investigated now in studies conducted on experimental animal models (Borlongan et al., 1998a,b,c; Muir et al., 1999; Philips et al., 1999). Moreover, emerging preliminary data from a Phase I clinical trial of NT2N neurons implanted adjacent to the stroke cavity of a resolved infarct in a small number of patients suggest that these grafts do not have deleterious effects (Kondziolka et al., 1999). However, further assessment of these patients is needed to determine if these grafts have any beneficial therapeutic effects on the residual motor impairments in these stroke patients (Kondziolka et al., 1999). While the mechanisms underlying recovery from ischemic stroke in experimental animals that receive human NT2N neuron grafts are not clear at this time, the beneficial effects of these grafts may result from as yet poorly understood 'neurotrophic' properties of the grafted NT2N neurons (Borlongan et al., 1998a,b,c). These uncertainties notwithstanding, it is plausible that the functional capacity of some remaining, but effete endogenous neurons surviving in the region of the stroke cavity might be augmented by sustained interactions between the host neurons and the grafted NT2N neurons. While the neuropathological sequelae of injuries due to cerebrovascular accidents in the brain compared to those due to trauma in the spinal cord are different, as are the mechanisms of repair and the extent of recovery in each of these regions of the CNS, it appears that some of the delayed functional improvement in spinal cord injury patients results from regenerative mechanisms that might be augmented or enhanced by transplanted neurons (Tator, 1998). For example, the robust outgrowth of lengthy processes from grafted NT2N neurons could be exploited therapeutically if these processes provide a 'scaffold' to guide/promote the successful regeneration of transected or otherwise damaged host axons in spinal cord injury patients. Thus, by mechanisms similar to those proposed for stroke (Borlongan et al., 1998a,b,c), or by other as yet unknown mechanisms, NT2N neurons implanted into the damaged region of the spinal cord may augment functional recovery in patients who are impaired as a consequence of spinal cord injury.

Summary and conclusion

Emerging data suggest that current strategies for the treatment of spinal cord injury might be improved or augmented by spinal cord grafts of neural cells, and it is possible that grafted neurons might have therapeutic potential. Thus, here we have summarized recent studies of the neurobiology of clonal human (NT2N) neurons grafted into spinal cord of immunodeficient athymic nude mice. Postmitotic human NT2N neurons derived in vitro from an embryonal carcinoma cell line (NT2) were transplanted into spinal cord of neonatal, adolescent and adult nude mice where they became integrated into the host gray and white matter, did not migrate from the graft site, and survived for >15 months after implantation. The neuronal phenotype of the grafted NT2N cells was similar in gray and white matter regardless of host age at implantation, and some of the processes extended by the transplanted NT2N neurons became ensheathed by oligodendrocytes. However, there were consistent differences between NT2N processes traversing white versus gray matter. Most notably, NT2N processes with a trajectory in white matter extended over much longer distances (some for >2 cm) than those confined to gray matter. Thus, NT2N neurons grafted into spinal cord of nude mice integrated into gray as well as white matter, where they exhibited and maintained the morphological and molecular phenotype of mature neurons for >15 months after implantation. Also, the processes extended by grafted NT2N neurons differentially responded to cues restricted to gray versus white matter. Further insight into the neurobiology of grafted human NT2N neurons in the normal and injured spinal cord of experimental animals may lead to novel and more effective strategies for the treatment of spinal cord injury.

Acknowledgements

The studies reviewed here were supported in part by grants from the National Institute on Aging, National Institute of Neurological Disorders and Stroke and the National Cancer Institute of the National Institutes of Health. Dr. Lee is the John H. Ware, 3rd chair of Alzheimer's Disease Research. Drs. Lee and Trojanowski are founding scientists of and consult for Layton Bioscience, Inc., the commercial producer of NT2N (hNT) neurons.

References

Andrews, P.W., Damjanov, I., Simon, D., Banting, G.S., Carlin, C., Dracopoli, N.C. and Fogh, J. (1984) Pluripotent embryonal carcinoma clones derived from the human teratocarcinoma cell line N-tera-2. *Lab. Invest.*, 50: 147–162.

Arnold, S.E. and Trojanowski, J.Q. (1996) Human fetal hippocampal development, II. The neuronal cytoskeleton. *J. Comp. Neurol.*, 367: 293–307.

Bain, G., Kitchens, D., Yao, M., Huettner, J.E. and Gottlieb, D.I. (1995) Embryonic stem cells express neuronal properties in vitro. *Dev. Biol.*, 168: 342–357.

Barbe, M.F. (1996) Tempting fate and commitment in the developing forebrain. *Neuron*, 16: 1–4.

Bergmann, M., Lahr, G., Mayerhofer, A. and Gratzl, M. (1991) Expression of synaptophysin during the prenatal development of the rat spinal cord: correlation with basic differentiation processes of neurons. *Neuroscience*, 42: 569–582.

Björklund, A. and Stenevi, U. (1977a) Experimental reinnervation of the rat hippocampus by grafted sympathetic ganglia, I. Axonal regeneration along the hippocampal fimbria. *Brain Res.*, 138: 259–270.

Björklund, A. and Stenevi, U. (1977b) Reformation of the severed septohippocampal cholinergic pathway in the adult rat by transplanted septal neurons. *Cell Tissue Res.*, 185: 289–302.

Björklund, A., Dunnett, S.B., Stenevi, U., Lewis, M.E. and Iversen, S.D. (1980) Reinnervation of the denervated striatum by substantia nigra transplants: functional consequences as revealed by pharmacological and sensorimotor testing. *Brain Res.*, 199: 307–333.

Borlongan, C.V., Shimizu, T., Trojanowski, J.Q., Watanabe, S., Lee, V.M.-Y., Tajima, Y., Cahaill, D., Freeman, T.B., Nishino, H., Sanberg, P.R., 1998a. Animal models of cerebral ischemia: neurodegeneration and cell transplantation. In: Freeman, T.B., Widner, H. (Eds.), Cell Transplantation for Neurological Diseases: Toward Reconstruction of the Human Central Nervous System. Humana Press, Totowa, NJ, pp. 211–230.

Borlongan, C.V., Tajima, Y., Trojanowski, J.Q., Lee, V.M.-Y. and Sanberg, P.R. (1998b) Transplantation of cryopreserved human embryonal carcinoma-derived neurons (NT2N cells) promotes functional recovery in ischemic rats. *Exp. Neurol.*, 149: 310–321.

Borlongan, C.V., Tajima, Y., Trojanowski, J.Q., Lee, V.M.-Y. and Sanberg, P.R. (1998c) Cerebral ischemia and CNS transplantation: differential effects of grafted fetal rat striatal cells and human neurons derived from a clonal cell line. *NeuroReport*, 9: 3703–3709.

Bruestle, O., Choudhary, K., Karram, K., Huettner, A., Murray, K., Dubois-Dalcq, M. and McKay, R.D.G. (1998) Chimeric brains generated by intraventricular transplantation of fetal human brain cells into embryonic rats. *Nat. Biotechnol.*, 16: 1040–1044.

Carlstedt, T. (1997) Nerve fibre regeneration across the peripheral–central transitional zone. *J. Anat.*, 190: 51–56.

Clowry, G.J. and Vrbova, G. (1992) Observations on the development of transplanted embryonic ventral horn neurones grafted into adult rat spinal cord and connected to skeletal muscle implants via a peripheral nerve. *Exp. Brain Res.*, 91: 249–258.

Davies, S.J., Field, P.M. and Raisman, G. (1993) Long fibre growth by axons of embryonic mouse hippocampal neurons microtransplanted into the adult rat fimbria. *Eur. J. Neurosci.*, 5: 95–106.

Davies, S.J., Field, P.M. and Raisman, G. (1994) Long interfascicular axon growth from embryonic neurons transplanted into adult myelinated tracts. *J. Neurosci.*, 14: 1596–1612.

Emmett, C.J., Lawrence, J.M., Raisman, G. and Seeley, P.J. (1991) Cultured epithelioid astrocytes migrate after transplantation into the adult rat brain. *J. Comp. Neurol.*, 311: 330–341.

Flax, J.D., Aurora, S., Yang, C., Simonin, C., Wills, A.M., Billinghurst, L.L., Jendoubi, M., Sidman, R.L., Wolfe, J.H., Kim, S.U. and Snyder, E.Y. (1998) Engraftable human neural stem cells respond to developmental cues, replace neurons, and express foreign genes. *Nat. Biotechnol.*, 16: 1033–1039.

Fletcher, T.L., Cameron, P., De Camilli, P. and Banker, G. (1991) The distribution of synapsin I and synaptophysin in hippocampal neurons developing in culture. *J. Neurosci.*, 11: 1117–1126.

Foster, G.A., Schultzberg, M., Gage, F.H., Björklund, A., Hökfelt, T., Nornes, H., Cuello, A.C., Verhofstad, A.A. and Visser, T.J. (1985) Transmitter expression and morphological development of embryonic medullary and mesencephalic raphe neurones after transplantation to the adult rat central nervous system, I. Grafts to the spinal cord. *Exp. Brain Res.*, 60: 427–444.

Fujii, M. (1994) Transplant-to-host neuron migration and neurite projection from homotopically transplanted olfactory bulb as demonstrated by mouse allelic Thy-1 form. *Exp. Neurol.*, 128: 97–102.

Gage, F.H., Coates, P.W., Palmer, T.D., Kuhn, H.G., Fisher, L.J., Suhonen, J.O., Peterson, D.A., Suhr, S.T. and Ray, J. (1995) Survival and differentiation of adult neuronal progenitor cells transplanted to the adult brain. *Proc. Natl. Acad. Sci. USA*, 92: 11879–11883.

Gibbs, R.B., Harris, E.W. and Cotman, C.W. (1985) Replacement of damaged cortical projections by homotypic transplants of entorhinal cortex. *J. Comp. Neurol.*, 237: 47–64.

Goldberg, W.J. and Bernstein, J.J. (1987) Transplant-derived astrocytes migrate into host lumbar and cervical spinal cord after implantation of E14 fetal cerebral cortex into adult thoracic spinal cord. *J. Neurosci. Res.*, 17: 391–403.

Goldberg, W.J. and Bernstein, J.J. (1988a) Fetal cortical astrocytes migrate from cortical homografts throughout the host brain and over the glia limitans. *J. Neurosci. Res.*, 20: 38–45.

Goldberg, W.J. and Bernstein, J.J. (1988b) Migration of cultured fetal spinal cord astrocytes into adult host cervical cord and medulla following transplantation into thoracic spinal cord. *J. Neurosci. Res.*, 19: 34–42.

Golding, J.P., Shewan, D., Berry, M. and Cohen, J. (1996) An in vitro model of the rat dorsal root entry zone reveals developmental changes in the extent of sensory axon growth into the spinal cord. *Mol. Cell. Neurosci.*, 7: 191–203.

Hartley, R.S., Trojanowski, J.Q. and Lee, V.M.-Y. (1999) Differential effects of spinal cord gray and white matter on process outgrowth from grafted human NTERA2 neurons (NT2N, hNT). *J. Comp. Neurol.*, 415: 404–418.

Hickey, W.F., Lee, V.M.-Y., Trojanowski, J.Q., McMillan, L.J., McKearn, T.J., Gonatas, J. and Gonatas, N.K. (1983) Immunohistochemical application of monoclonal antibodies against myelin basic protein and neurofilament triple protein subunits: advantages over antisera and technical limitations. *J. Histochem. Cytochem.*, 31: 1126–1135.

Houle, J.D., Skinner, R.D., Garcia-Rill, E. and Turner, K.L. (1996) Synaptic evoked potentials from regenerating dorsal root axons within fetal spinal cord tissue transplants. *Exp. Neurol.*, 139: 278–290.

Iacovitti, L. and Stull, N.D. (1997) Expression of tyrosine hydroxylase in newly differentiated neurons from a human cell line (hNT). *NeuroReport*, 8: 1471–1474.

Kleppner, S.R., Robinson, K.A., Trojanowski, J.Q. and Lee, V.M.-Y. (1995) Transplanted human neurons derived from a teratocarcinoma cell line (NTera-2) mature, integrate, and survive for over 1 year in the nude mouse brain. *J. Comp. Neurol.*, 357: 618–632.

Knaus, P., Betz, H. and Rehm, H. (1986) Expression of synaptophysin during postnatal development of the mouse brain. *J. Neurochem.*, 47: 1302–1304.

Kocsis, J.D. (1999) Restoration of function by glial cell transplantation into demyelinated spinal cord. *J. Neurotrauma*, 16: 695–703.

Kondziolka, D., Wechsler, L., Meltzer, C., Thulborn, K., Jannetta, P., Slagel, C., Goldstein, S., Rakela, J. and Elder, S. (1999) Phase I safety and effectiveness trial of the cerebral transplantation of LBS neurons in patients with substantial fixed motor deficit following cerebral infarction. *J. Neurosurg.*, 90: 435A.

Kozlova, E.N., Stromberg, I., Bygdeman, M. and Aldskogius, H. (1994) Peripherally grafted human foetal dorsal root ganglion cells extend axons into the spinal cord of adult host rats by circumventing dorsal root entry zone astrocytes. *NeuroReport*, 5: 2389–2392.

Kozlova, E.N., Rosario, C.M., Stromberg, I., Bygdeman, M.

and Aldskogius, H. (1995) Peripherally grafted human foetal dorsal root ganglion cells extend axons into the spinal cord of adult host rats by circumventing dorsal root entry zone astrocytes. *NeuroReport*, 6: 269–272.

Leclerc, N., Beesley, P.W., Brown, I., Colonnier, M., Gurd, J.W., Paladino, T. and Hawkes, R. (1989) Synaptophysin expression during synaptogenesis in the rat cerebellar cortex. *J. Comp. Neurol.*, 280: 197–212.

Lee, V.M.-Y. and Andrews, P.W. (1986) Differentiation of NTERA-2 clonal human embryonal carcinoma cells into neurons involves the induction of all three neurofilament proteins. *J. Neurosci.*, 6: 514–521.

Lee, V.M.-Y., Carden, M.J., Schlaepfer, W.W. and Trojanowski, J.Q. (1987) Monoclonal antibodies distinguish several differentially phosphorylated states of the two largest rat neurofilament subunits (NF-H and NF-M) and demonstrate their existence in the normal nervous system of adult rats. *J. Neurosci.*, 7: 3474–3488.

Li, Y. and Raisman, G. (1993) Long axon growth from embryonic neurons transplanted into myelinated tracts of the adult rat spinal cord. *Brain Res.*, 629: 115–127.

Lindsay, R.M. (1979) Adult rat brain astrocytes support survival of both NGF-dependent and NGF-insensitive neurones. *Nature*, 282: 80–82.

Lindsay, R.M. and Raisman, G. (1984) An autoradiographic study of neuronal development, vascularization and glial cell migration from hippocampal transplants labeled in intermediate explant culture. *Neuroscience*, 12: 513–530.

Mantione, J.R., Kleppner, S.R., Miyazono, M., Wertkin, A.M., Lee, V.M.-Y. and Trojanowski, J.Q. (1995) Human neurons that constitutively secrete Aβ do not induce Alzheimer's disease pathology following transplantation and long-term survival in the rodent brain. *Brain Res.*, 671: 333–337.

McDonald, J.W., Liu, X.-Z., Qu, Y., Liu, S., Mickey, S.K., Turetsky, D., Gottlieb, D.I. and Choi, D.W. (1999) Transplanted embryonic stem cells survive, differentiate and promote recovery in injured rat spinal cord. *Nat. Biotechnol.*, 5: 1410–1412.

Miyazono, M., Lee, V.M.-Y. and Trojanowski, J.Q. (1995) Proliferation, cell death and neuronal differentiation in transplanted human embryonal carcinoma (NTera2) cells depend on the graft site in nude and SCID mice. *Lab. Invest.*, 73: 273–283.

Miyazono, M., Nowell, P.C., Finan, J.L., Lee, V.M.-Y. and Trojanowski, J.Q. (1996) Long-term integration and neuronal differentiation of human embryonal carcinoma cells (NTera-2) transplanted into the caudoputamen of nude mice. *J. Comp. Neurol.*, 376: 603–613.

Molenaar, W.M., De Leij, L. and Trojanowski, J.Q. (1991) Neuroectodermal tumors of the peripheral and the central nervous system share neuroendocrine N-CAM-related antigens with small cell lung carcinomas. *Acta Neuropathol.*, 83: 46–54.

Muir, J.K., Raghupathi, R., Saatman, K.E., Wilson, C.S., Lee, V.M.-Y., Trojanowski, J.Q., Philips, M.F. and McIntosh, T.K. (1999) Terminally differentiated human neurons survive and integrate following transplantation into the traumatically injured rat brain. *J. Neurotrauma*, 16: 403–414.

Nogradi, A. and Vrbova, G. (1996a) Improved motor function of denervated rat hindlimb muscles induced by embryonic spinal cord grafts. *Eur. J. Neurosci.*, 8: 2198–2203.

Nogradi, A. and Vrbova, G. (1996b) Reinnervation of denervated hindlimb muscles by axons of grafted motoneurons via the reimplanted L4 ventral root. *Neurobiology*, 4: 231–232.

Nornes, H., Björklund, A. and Stenevi, U. (1983) Reinnervation of the denervated adult spinal cord of rats by intraspinal transplants of embryonic brain stem neurons. *Cell Tissue Res.*, 230: 15–35.

Onifer, S.M., Whittemore, S.R. and Holets, V.R. (1993) Variable morphological differentiation of a raphe-derived neuronal cell line following transplantation into the adult rat CNS. *Exp. Neurol.*, 122: 130–142.

Park, K.I., Liu, S., Flax, J.D., Nissim, S., Stieg, P.E. and Snyder, E.Y. (1999) Transplantation of neural progenitor and stem cells: developmental insights may suggest new therapies for spinal cord and other CNS dysfunction. *J. Neurotrauma*, 16: 675–687.

Pedraza, L., Owens, G.C., Green, L.A. and Salzer, J.L. (1990) The myelin-associated glycoproteins: membrane disposition, evidence of a novel disulfide linkage between immunoglobulin-like domains, and posttranslational palmitylation. *J. Cell. Biol.*, 111: 2651–2661.

Philips, M.F., Muir, J.K., Saatman, K.E., Raghupathi, R., Lee, V.M.-Y., Trojanowski, J.Q. and McIntosh, T.K. (1999) Survival and integration of transplanted postmitotic human neurons (hNT) following experimental brain injury in immunocompetent rats. *J. Neurosurg.*, 90: 116–124.

Pleasure, S.J. and Lee, V.M.-Y. (1993) NTera-2 cells: a human cell line which displays characteristics expected of a human committed neuronal progenitor cell. *J. Neurosci. Res.*, 35: 585–602.

Pleasure, S.J., Page, C. and Lee, V.M.-Y. (1992) Pure, postmitotic, polarized human neurons derived from NTera 2 cells provide a system for expressing exogenous proteins in terminally differentiated neurons. *J. Neurosci.*, 12: 1802–1815.

Renfranz, P.J., Cunningham, M.G. and McKay, R.D. (1991) Region-specific differentiation of the hippocampal stem cell line HiB5 upon implantation into the developing mammalian brain. *Cell*, 66: 713–729.

Schmidt, M.L., Carden, M.J., Lee, V.M.-Y. and Trojanowski, J.Q. (1987) Phosphate dependent and independent neurofilament epitopes in the axonal swellings of patients with motor neuron disease and controls. *Lab. Invest.*, 56: 282–294.

Schmidt, M.L., Murray, J.M. and Trojanowski, J.Q. (1993) Continuity of neuropil threads with tangle-bearing and tangle-free neurons in Alzheimer disease cortex. A confocal laser scanning microscopy study. *Mol. Chem. Neuropathol.*, 18: 299–312.

Sieradzan, K. and Vrbova, G. (1993) The ability of developing spinal neurons to reinnervate a muscle through a peripheral nerve conduit is enhanced by cografted embryonic spinal cord. *Exp. Neurol.*, 122: 232–243.

Snyder, E.Y., Deitcher, D.L., Walsh, C., Arnold-Aldea, S., Hartwieg, E.A. and Cepko, C.L. (1992) Multipotent neural cell lines can engraft and participate in development of mouse cerebellum. *Cell*, 68: 33–51.

Stromberg, I., Bygdeman, M. and Almqvist, P. (1992) Target-specific outgrowth from human mesencephalic tissue grafted to cortex or ventricle of immunosuppressed rats. *J. Comp. Neurol.*, 315: 445–456.

Tator, C.H. (1998) Biology of neurological recovery and functional restoration after spinal cord injury. *Neurosurgery*, 42: 696–707.

Thomson, J.A., Itskovitz-Eldor, J., Shapiro, S.S., Waknitz, M.A., Swiergiel, J.J., Marshall, V.S. and Jones, J.M. (1998a) Embryonic stem cell lines derived from human blastocysts. *Science*, 282: 1145–1147.

Thomson, J.A., Marshall, V.S. and Trojanowski, J.Q. (1998b) Neural differentiation of rhesus embryonic stem cells. *APMIS*, 106: 149–157.

Tohyama, T., Lee, V.M.-Y., Rorke, L.B. and Trojanowski, J.Q. (1991) Molecular milestones that signal axonal maturation and the commitment of human spinal cord precursor cells to the neuronal or glial phenotype in development. *J. Comp. Neurol.*, 310: 285–299.

Tonder, N., Sorensen, J.C., Bakkum, E., Danielsen, E. and Zimmer, J. (1988) Hippocampal neurons grafted to newborn rats establish efferent commissural connections. *Exp. Brain Res.*, 72: 577–583.

Tonder, N., Sorensen, T. and Zimmer, J. (1990) Grafting of fetal CA3 neurons to excitotoxic, axon-sparing lesions of the hippocampal CA3 area in adult rats. *Prog. Brain Res.*, 83: 391–409.

Trojanowski, J.Q., Schuck, T., Schmidt, M.L. and Lee, V.M.-Y. (1989a) Distribution of phosphate-independent MAP2 epitopes revealed with monoclonal antibodies in microwave-denatured human nervous system tissues. *J. Neurosci. Methods*, 29: 171–180.

Trojanowski, J.Q., Schuck, T., Schmidt, M.L. and Lee, V.M.-Y. (1989b) Distribution of tau proteins in the normal human central and peripheral nervous system. *J. Histochem. Cytochem.*, 37: 209–215.

Trojanowski, J.Q., Mantione, J.R., Lee, J.H., Seid, D.P., You, T., Inge, L.J. and Lee, V.M.-Y. (1993) Neurons derived from a human teratocarcinoma cell line establish molecular and structural polarity following transplantation into the rodent brain. *Exp. Neurol.*, 122: 283–294.

Trojanowski, J.Q., Kleppner, S.R., Hartley, R.S., Miyazono, M., Fraser, N.W., Kesari, S. and Lee, V.M.-Y. (1997) Transfectable and transplantable post-mitotic human neurons: a potential 'platform' for gene therapy of nervous system diseases. *Exp. Neurol.*, 144: 92–97.

Whittemore, S.R. (1999) Neuronal replacement strategies for spinal cord injury. *J. Neurotrauma*, 16: 667–673.

Whittemore, S.R. and White, L.A. (1993) Target regulation of neuronal differentiation in a temperature-sensitive cell line derived from medullary raphe. *Brain Res.*, 615: 27–40.

Whittemore, S.R., Holets, V.R., Keane, R.W., Levy, D.J. and McKay, R.D. (1991) Transplantation of a temperature-sensitive, nerve growth factor-secreting, neuroblastoma cell line into adult rats with fimbria-fornix lesions rescues cholinergic septal neurons. *J. Neurosci. Res.*, 28: 156–170.

Wictorin, K. and Björklund, A. (1992) Axon outgrowth from grafts of human embryonic spinal cord in the lesioned adult rat spinal cord. *NeuroReport*, 3: 1045–1048.

Wictorin, K., Brundin, P., Gustavii, B., Lindvall, O. and Björklund, A. (1990) Reformation of long axon pathways in adult rat central nervous system by human forebrain neuroblasts. *Nature*, 347: 556–558.

Wictorin, K., Brundin, P., Sauer, H., Lindvall, O. and Björklund, A. (1992) Long distance directed axonal growth from human dopaminergic mesencephalic neuroblasts implanted along the nigrostriatal pathway in 6-hydroxydopamine lesioned adult rats. *J. Comp. Neurol.*, 323: 475–494.

Wiedenmann, B. and Franke, W.W. (1985) Identification and localization of synaptophysin, an integral membrane glycoprotein of M_r 38,000 characteristic of presynaptic vesicles. *Cell*, 41: 1017–1028.

Wu, W., Han, K., Li, L. and Schinco, F.P. (1994) Implantation of PNS graft inhibits the induction of neuronal nitric oxide synthase and enhances the survival of spinal motoneurons following root avulsion. *Exp. Neurol.*, 129: 335–339.

Yachnis, A.T., Rorke, L.B., Lee, V.M.-Y. and Trojanowski, J.Q. (1993) Expression of neuronal and glial polypeptides during histogenesis of the human cerebellar cortex including observations on the dentate nucleus. *J. Comp. Neurol.*, 334: 356–369.

Zeller, M. and Strauss, W.L. (1995) Retinoic acid induces cholinergic differentiation of NTera 2 human embryonal carcinoma cells. *Int. J. Dev. Neurosci.*, 13: 437–445.

Zhou, C.F., Raisman, G. and Morris, R.J. (1985) Specific patterns of fibre outgrowth from transplants to host mice hippocampi, shown immunohistochemically by the use of allelic forms of Thy-1. *Neuroscience*, 16: 819–833.

Zhou, C.F., Li, Y. and Raisman, G. (1989) Embryonic entorhinal transplants project selectively to the deafferented entorhinal zone of adult mouse hippocampi, as demonstrated by the use of Thy-1 allelic immunohistochemistry. Effect of timing of transplantation in relation to deafferentation. *Neuroscience*, 32: 349–362.

CHAPTER 27

Grafting of genetically modified fibroblasts into the injured spinal cord

Yi Liu, Marion Murray, Alan Tessler and Itzhak Fischer [*]

Department of Neurobiology and Anatomy, MCP Hahnemann University, 3200 Henry Avenue, Philadelphia, PA 19129, USA

Introduction

Functional deficits following mammalian spinal cord injury (SCI) occur because some injured neurons die and surviving neurons fail to regenerate damaged axons (reviewed in: Tetzlaff et al., 1994; Tessler et al., 1997). Understanding the mechanisms responsible for survival and regeneration will identify appropriate strategies for repair of spinal injury and therapeutic agents that can improve recovery of function. For example, axotomy is thought to deprive the injured neurons of target-derived trophic factors, particularly members of the neurotrophin family (Levi-Montalcini, 1987; DiStefano et al., 1992; Korsching, 1993; Bothwell, 1995; Mocchetti and Wrathall, 1995). Indeed, previous studies have shown that direct application of neurotrophic factors to injured axons can prevent retrograde neuron death and atrophy of surviving axotomized neurons (Hofer and Barde, 1988; Sendtner et al., 1992; Yan et al., 1992; Hefti et al., 1993; Arenas and Persson, 1994; Diener and Bregman, 1994; Bregman et al., 1998; Shibayama et al., 1998) and also promote axon regeneration (Xu et al., 1995; Oudega and Hagg, 1996; Bregman et al., 1997). Development of clinically applicable methods to provide neurotrophic factors to injured central nervous system (CNS) has therefore become an active field of inquiry. Transplantation techniques have long been employed as a therapeutic strategy for experimental models of spinal cord injury (reviewed in: Goldberger et al., 1993; Tessler and Murray, 1996; Tessler et al., 1997). Transplants of fetal CNS tissue are effective in preventing neuron death and promoting regeneration (Bregman and Reier, 1986; Bernstein-Goral and Bregman, 1993; Himes et al., 1994; Bregman et al., 1997, 1998; Miya et al., 1997; Mori et al., 1997; Diener and Bregman, 1998), probably because they provide neurotrophic factors sufficient to support the axotomized neurons and may function as a bridge or relay across the injury site (Goldberger et al., 1993; Tessler and Murray, 1996; Tessler et al., 1997). However, there are serious practical and ethical problems associated with the use of fetal tissue and regeneration of adult axons into fetal tissue is extremely limited and requires supplementing fetal grafts with neurotrophic factors to improve regeneration (Bregman et al., 1997, 1998).

Grafting genetically modified cells that produce therapeutic factors is an alternative strategy in treatment of spinal cord injury with several advantages over the use of fetal tissue. With this strategy ex vivo gene therapy cells are modified using vectors encoding therapeutic genes and then grafted into the injury site. This biological method of delivery of neurotrophin products can enhance both neuronal survival and regeneration (Rosenberg et al., 1988; Kawaja et al., 1992; Himes et al., 1995; Grill et al., 1997; Tessler et al., 1997; Menei et al., 1998; Liu

---

[*] Corresponding author: Dr. Itzhak Fischer, Department of Neurobiology and Anatomy, MCP Hahnemann University, 2900 Queen Lane, Philadelphia, PA 19129, USA. Fax: +1-215-843-9082; E-mail: Itzhak.Fischer@drexel.edu

et al., 1999). For example, genetically engineered fibroblasts that expressed nerve growth factor (NGF) promoted axon regeneration in brain (Rosenberg et al., 1988; Kawaja et al., 1992) and intraspinal grafts of neurotrophin-3 (NT-3) expressing fibroblasts promoted corticospinal tract (CST) regeneration (Grill et al., 1997). The regenerating CST axons, however, failed to grow into the graft or host white matter and the length of the growing axons was very limited (Grill et al., 1997).

We have used a similar ex vivo gene therapy strategy to study survival and regeneration by axotomized neurons in the red nucleus (RN) and then examined the effect of these grafts on recovery of motor function. Neurons located in the caudal magnocellular portion of the RN provide the major component of the rubrospinal tract (RST) (Tracey, 1995). A further advantage is that more than 99% of RST axons cross in the ventral tegmental decussation and descend in the superficial dorsolateral funiculus (Tracey, 1995). A lesion of the right dorsolateral funiculus axotomizes almost the entire left magnocellular neuron population but leaves the contralateral nucleus largely intact to serve as an internal control. About 45% of the RN neurons die following this cervical lesion, the surviving RN neurons show a mean decrease in soma size (atrophy) of 40% and they do not regenerate their axons (Mori et al., 1997). Rubrospinal neurons express the full-length TrkB receptor, which accounts for their regenerative response to the application of the neurotrophins, brain-derived neurotrophic factor (BDNF) and neurotrophin-4/5 (NT-4/5) (Xu et al., 1995; Kobayashi et al., 1997; Ye and Houle, 1997). BDNF and NT-4/5 are thus candidates for introduction by ex vivo gene therapy into the injured spinal cord of adult rats. The RN model thus enables us to identify effects of interventions on neuronal survival and regeneration. In addition, the rubrospinal tract is part of the extrapyramidal motor system and injury can be expected to produce motor deficits, while successful interventions will show an amelioration of the deficits.

We review here our studies (Liu et al., 1999, 2000) indicating that intraspinal grafts of BDNF-producing fibroblasts (1) reduced RN neuron loss, (2) minimized the atrophy of surviving RN neurons, (3) promoted axonal growth, including regeneration of rubrospinal axons, and (4) resulted in recovery of forelimb and hindlimb function.

Procedures

All procedures were approved by the institutional animal welfare committee and conformed to the NIH guidelines for the care and use of laboratory animals. Groups of adult female Sprague–Dawley rats received a right partial cervical hemisection that destroyed the rubrospinal tract and a transplant of either BDNF-producing fibroblasts (Fb/BDNF), primary fibroblasts (Fb) or gelfoam was introduced into the lesion cavity. Sham-operated animals were used as additional controls. All animals were immunosuppressed with cyclosporin A, beginning 3 days before surgery and continuing for the duration of the experiment, and were treated with methylprednisolone, injected intravenously at the time of surgery. Rats were tested weekly for recovery of control of forelimb (cylinder) and locomotor (rope test) function. Animals were scored in the cylinder test according to the frequency of use of one or both forepaws in spontaneous exploration. Locomotion across a rope was scored according to frequency of slips or falls from the rope during a crossing. A subgroup of rats then received a second lesion rostral to the graft at the cervical (C)2 dorsolateral quadrant to eliminate descending axons stimulated to grow by the graft and were tested weekly for another 2 months before sacrifice and anatomical analysis.

Genetically engineered fibroblasts

Primary skin fibroblasts were modified using a retroviral construct BDNF.IRES.GEO that contained the coding region of human BDNF, the internal ribosomal entry site (IRES) and a GEO gene which is a fusion gene of the *lacZ* (encoding *Escherichia coli* β-galactosidase) and neomycin resistance (neo) genes. The long terminal repeat (LTR) promoter transcribes a multicistronic mRNA and the IRES directs the translation of the BDNF and GEO genes. Transgene expression by the engineered fibroblasts was monitored by X-gal histological staining for β-galactosidase and immunocytochemical analysis for BDNF showing a very high level of co-expression. To measure the production of BDNF and to demonstrate its biological activity, homogenates of Fb/BDNF and media conditioned by them were analyzed using Western blotting, slot blot, immunocytochemistry and bioassay with embryonic chicken dorsal root ganglia (DRG) explants. We calculated that the Fb/BDNF cells secreted BDNF at a rate of 10–15 ng/10^6 cells/24 h.

Axonal regeneration

For retrograde tracing, animals received another laminectomy 3–4 segments caudal to the initial injury site and 1 μl of 2% fluorogold (FG) was pressure injected into each side of the spinal cord. Animals were euthanized 3 days later. For anterograde labeling, biotinylated dextran amine (BDA) was injected into the RN 14 days prior to sacrifice. Animals were allowed to survive 1 or 2 months after surgery (Fig. 1).

Cell counting and cell size analysis

Neurons in the magnocellular portion of the RN were counted from cresyl violet-stained sections using unbiased stereological counting methods (West, 1993). FG-labeled RN neurons were counted similarly. The cross-sectional area of FG-labeled or cresyl violet-stained RN neurons was measured and neurons were classified according to cross-sectional area as being <100 μm^2 (very small), 100–300 μm^2 (small), 300–500 μm^2 (medium), 500–1000 μm^2 (large) and >1000 μm^2 (giant).

Effects of transplants

Lesion site and graft survival

Spinal cord tissue from control lesion and transplant groups was analyzed for size of lesion, transplant survival, host–graft apposition and scar tissue formation. We verified that all lesions destroyed the lateral funiculus, including the entire rubrospinal pathway, and also invaded the ventral funiculus and part of the gray matter. Gelfoam implants were reabsorbed by 1 month, whereas transplants of unmodified or genetically modified fibroblasts survived, completely filling the lesion cavity in the host spinal cord. We found that the expression of β-gal in the genetically modified fibroblasts was high a few weeks after the grafting, but markedly reduced at 2 months. There was otherwise no obvious morphological difference between the two types of fibroblast grafts. Both consisted of densely packed cells with the morphological characteristics of fibroblasts that were apposed to the host tissue without interruption by cysts or formation of a large scar. The reduced immune response was demonstrated by immunocytochemical staining for astrocytes, microglia and macrophages (using glial fibrillary acidic protein [GFAP], OX-42 and ED-1 antibodies), which also showed that these cells were present on the border of the transplants, but not within the transplant tissue.

Effects of transplants on RN neuron survival

We compared the number of neurons remaining in the injured RN to the number on the intact side in cresyl violet-stained preparations. In control animals, the left and right RN contained roughly equal numbers of neurons (Table 1). One and two months after spinal cord hemisection, cell counts in animals receiving gelfoam implants showed a 45% cell loss in the injured RN (Table 1). In animals receiving Fb/BDNF transplants about 85% of the cell survived. Fb/BDNF transplants therefore reduced RN neuron loss from 45% to 15%, providing a 60–70% rescue of RN neurons that would otherwise have died (Table 1). In contrast, animals receiving transplants of unmodified fibroblasts showed a 45% cell loss that was not significantly different from that seen after

TABLE 1

Comparison of the number and size of RN neurons

Group	Ratio of cell numbers	Ratio of mean cell size	Ratio of mean cell size (FG-labeled)
Normal	1.00	1.00	1.00
Fb/BDNF	0.85	0.81	1.07
Fb	0.53	0.55	1.04
Hx	0.57	0.59	1.04

The results are expressed as the ratio between the lesion (left) and the control red nucleus (right).
Fb/BDNF = transplant of BDNF-producing fibroblasts; Fb = transplant of unmodified fibroblasts; Hx = axotomized animals without transplant.

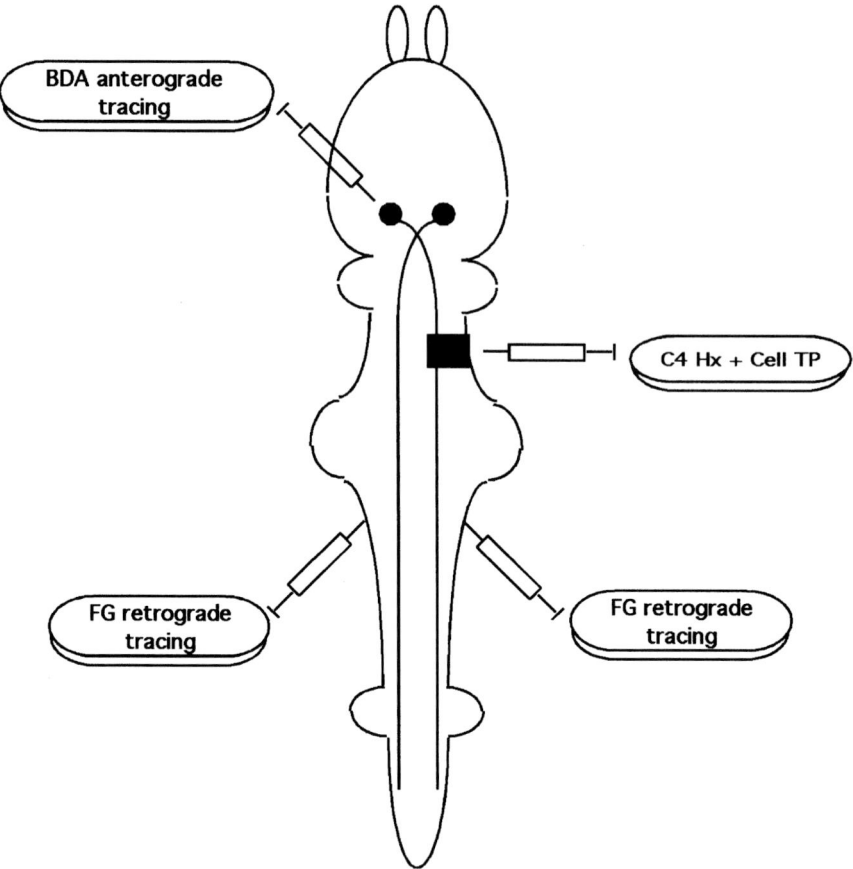

Fig. 1. Schematic diagram of the experiment.

gelfoam implants. Thus, unmodified fibroblasts offered no protection against axotomy-induced neuron death in the RN (Table 1) but Fb/BDNF transplants rescued axotomized RN neurons that were destined to die.

Effects of transplants on RN neuron atrophy

In control animals, the left and right RN contained equal numbers of small, medium, and large neurons. Less than 5% of RN neurons were giant cells (>1000 μm^2) and no neurons were smaller than 100 μm^2. The mean soma size was about 430 μm^2, with a left/right ratio of 1.0 (Table 1). The cross-sectional areas of RN neurons in control, lesion and transplant animals were compared to determine whether transplants prevented atrophy following axotomy. After 1 and 2 months survivals, the mean soma area in animals receiving gelfoam implants was reduced to about 220 μm^2 and left/right ratio was reduced to 0.58, indicating >40% decrease in mean cell size of the surviving RN neurons (Table 1). The size distribution of the neurons was dramatically shifted toward the small size, with 80% of the neurons in the small size range. Neurons smaller than 100 μm^2, which are not present in the normal RN, comprised about 10% of the total surviving RN population. Giant neurons were absent in the injured RN, and the number of medium to large neurons was also significantly reduced. In animals receiving unmodified Fb transplants, the cell size distribution was very similar to that of the gelfoam implants, indicating that unmodified fibroblasts offered no protection against axotomy-induced neuron atrophy. In contrast, in recipients of Fb/BDNF transplants, the mean soma area was about 330 μm^2 and the left/right ratio was about 0.81, indicating <20% decrease in mean soma size. Fb/BDNF transplants prevented >55% of the atrophy in surviving RN neurons (Table 1). Their effect on reducing neuron atrophy was also evident in the soma size distribution. The general distribution was restored almost to normal except for minor changes that included the absence of giant neurons, more cells in the small size category and a small fraction of very small cells. Thus Fb/BDNF grafts ameliorated atrophy of axotomized RN neurons.

Effects of transplants on rubrospinal regeneration

Anterograde tracing with BDA. To identify axons regenerating from axotomized RN neurons, we injected the anterograde tracer BDA into the magnocellular portion of the lesioned RN to label RST axons (Fig. 1). Numerous BDA-labeled RST axons regenerated into, through and around the BDNF/Fb transplants and axon branches could be seen to enter laminae V–VII in the gray matter below the level of the lesion/transplant. BDA-labeled axons regenerated as far caudally as the upper thoracic spinal cord, 4–5 segments caudal to the transplant, although their number was small compared to normal. In general, the location of the regenerated RST axons differed only slightly from normal. The maximum length of RST regeneration and the number and distribution of regenerating axons differed among animals. There was no consistent difference in the most caudal spinal cord level in which BDA-labeled axons were detected between animals surviving 1 and 2 months, suggesting that regeneration was largely completed within the first month. In animals receiving unmodified Fb or gelfoam grafts, few or no BDA-labeled axons were present either in the transplants or at the host–graft interface and none were detected caudal to the transplant. In animals receiving Fb/BDNF transplants, RST axons rostral to the transplants also showed a considerable amount of sprouting, whereas sprouting was minimal in animals with gelfoam or unmodified fibroblast transplants.

Retrograde tracing with FG. FG injected caudal to the lesion site was also used to identify neurons whose axons had regenerated (Fig. 1). The number of RN neurons retrogradely labeled was counted and compared in gelfoam, unmodified fibroblast and Fb/BDNF groups. About 7% of RN neurons were retrogradely labeled in Fb/BDNF recipients (Liu et al., 1999). In contrast <1% (the percentage of uncrossed projections) RN neurons were retrogradely labeled in unmodified Fb or gelfoam recipients. Our results thus indicated that there was no sparing of rubrospinal axons after implants of gelfoam or unmodified fibroblasts but significant regeneration after Fb/BDNF transplants.

We then measured the cross-sectional area of FG-labeled RN neurons. The mean soma size measurement of FG-labeled neurons in intact RN of operated and normal animals did not differ significantly from that of the FG-labeled cells in the injured RN in operated animals (Table 1). The ratio of sizes of FG-labeled neurons between injured RN and intact RN approximated 1.0 in all animal groups studied (Table 1), and soma size distribution showed a similar distribution pattern in each group. These results indicated that RN neurons that had been retrogradely labeled by FG maintained their soma size throughout the survival time, regardless of the treatment group to which they belonged. The <1% cells that were labeled after gelfoam or unmodified fibroblast grafts represented the uncrossed portion, and therefore cells that had escaped injury. In animals receiving Fb/BDNF transplants, however, about 7% RN neurons labeled by FG had been axotomized by the spinal cord hemisection and had therefore regenerated their severed axons. That these 7% RN neurons also maintained their soma size indicates a strong relationship between protection from atrophy and the ability to regenerate. The maintenance of normal soma size in the population of neurons that regenerated thus contributes to, although it does not entirely account for, the overall effect of BDNF on prevention of atrophy of axotomized RN neurons.

Behavior analysis. Animals were tested in a variety of tasks. BBB scores (Basso et al., 1995) of 19–20 (out of 21) indicated good recovery of overground locomotion in animals of all groups. Locomotion on a narrow (2″) beam also indicated good recovery of a challenging locomotor task (data not shown). We also studied recovery in two tasks, the cylinder test and the rope-walking test, in which a deficit was seen following the cervical injury and in which differential recovery occurred. When placed in a clear cylinder, normal rats spontaneously rear and explore the wall of a cylinder using a single forepaw alone (50%), or both forepaws together (50%). Hemisection at the upper cervical level produced asymmetry in forelimb use; hemisected rats or rats with non-modified transplants rarely used the forelimb ipsilateral to the injury and did not show recovery during the 8-week observation period. Animals with Fb/BDNF transplants used the injured forelimb more frequently, resulting in more symmetrical limb use, than did animals in the gelfoam or unmodified Fb groups. Improvement in use of the affected forelimb was seen in animals 1 week after transplantation and reached a plateau by week 4.

A second test was locomotion along a rope. This test challenges foot placement, balance, and posture to a greater extent than locomotion along a narrow beam. Rats quickly learn to cross the rope but animals with gelfoam or unmodified fibroblast grafts commit numerous errors; in contrast, animals with Fb/BDNF transplants cross the rope with far fewer errors.

To study whether functional recovery was mediated by axonal growth stimulated by the transplant, animals were subjected to a second lesion, which removed the right dorsolateral quadrant at C2, just rostral to the transplant. Rats were then retested on the cylinder and rope tests for another 2 months. The second lesion almost completely abolished the recovered function in Fb/BDNF animals but had little effect on the Fb or gelfoam recipients.

Discussion

Because of its well-defined anatomy, the RN–RST system has long been a focus of studies of CNS neuron survival and regeneration. The present report employed an ex vivo gene therapy paradigm to study RN neuron survival, RST regeneration and recovery of motor function. The results demonstrated that axotomy at the cervical level caused about 45% loss of rubrospinal neurons, presumably by retrograde cell death. Significant cell atrophy occurred among the surviving RN neurons, which on average lost >40% of their mean soma size. BDNF-producing fibroblasts transplanted into the lesion cavity prevented 60–70% of the retrograde neuron loss and >55% of the neuron atrophy. Furthermore, the Fb/BDNF transplants also promoted axon regeneration in about 7% of RN neurons and completely prevented the atrophy of these neurons. In contrast, grafts of unmodified fibroblasts failed to prevent RN neuron death and atrophy; they also failed to promote RST axon regeneration. Finally, recovery of two behavioral tests was improved in animals receiving BDNF-Fb and this recovery was abolished after a second lesion eliminated the effects of the graft.

Transplants of BDNF-producing fibroblasts partially prevented RN neuron loss and atrophy following cervical axotomy

Fetal spinal cord transplants completely prevented RN neuron loss after a thoracic injury in newborn rats (Bregman and Reier, 1986) and partially prevented RN neuron death following a cervical hemisection in adult rats, but they failed to protect the surviving RN neurons from atrophy (Mori et al., 1997). Fetal tissue transplants are thought to rescue axotomized RN neurons by acting as a surrogate source of neurotrophins, especially BDNF (Mori et al., 1997; Tessler et al., 1997). Virtually all RN neurons express full-length TrkB (Kobayashi et al., 1997), the specific receptor for BDNF, and BDNF applied either at the spinal cord injury site or near RN neuron cell bodies offers protection against death and atrophy (Diener and Bregman, 1994; Kobayashi et al., 1997). We observed that Fb/BDNF transplants rescued most (60–70%) RN neurons that would have died, prevented much (55%) of the soma atrophy among the surviving cells and promoted axon regeneration by 7% of RN neurons. In contrast, Mori et al. (1997) found that fetal transplants alone produced only a 50% rescue from cell death and no prevention of atrophy and no RST axon regeneration. The smaller quantities of neurotrophic factors provided by fetal tissue may be sufficient for neuron survival but inadequate to maintain perikaryal size or to promote regeneration. We found that Fb/BDNF transplants expressed high levels of the transgene at 1 week and continued expression at reduced levels for at least 2 months. We suggest that the engineered cells sustained the survival of axotomized RN neurons, prevented their atrophy and promoted axon regeneration by continuously secreting bioactive BDNF for a prolonged period of time. Grafting cells genetically modified to express a therapeutic gene would therefore appear to have an advantage over fetal tissue transplantation because the engineered cells can provide the needed factors to the injured neurons in a more abundant, homogeneous and sustained fashion.

Fb/BDNF transplants into adult rats did not rescue all axotomized RN cells or prevent all soma atrophy. In contrast, fetal tissue transplants completely prevented RN cell death in newborn rats subjected to a mid-thoracic hemisection (Bregman and Reier, 1986). The cervical lesion axotomizes more RST axons than the thoracic lesion and RN neuron survival may have more stringent requirements in adults than in neonates. Nevertheless, BDNF, the only therapeutic factor produced by the Fb/BDNF transplants, may not be sufficient to prevent all RN neuron death. NT-3 may also be a survival factor for RN neurons (Diener and Bregman, 1994; Kobayashi et al., 1997). In addition, RN neurons may be a more heterogeneous population than some other CNS structures, such as the Clarke's nucleus (CN), in their requirement for neurotrophic factors for survival and their expression of neurotrophin receptors. Fetal CNS tissues expressing high levels of NT-3 mRNA and cells genetically modified to produce NT-3 completely prevented axotomy-induced CN neuron death but not atrophy (Himes et al., 1994, 1995; Tessler et al., 1997).

Neither fetal transplants nor administration of BDNF or NT-3 completely prevented RN neuron atrophy following a mid-thoracic hemisection in adult rats, but a combination of transplants with either factor permitted complete maintenance of neuron size (Bregman et al., 1998). A combination of factors may therefore be necessary for maintaining perikaryal size of RN neurons. The infusion of BDNF or NT-4/5 directly into the vicinity of RN neuron cell bodies also completely prevented atrophy induced by a cervical lesion similar to that used in the present study (Kobayashi et al., 1997). Therefore it is possible that even though all RN neurons express full-length TrkB on their cell bodies (Kobayashi et al., 1997), only a portion of them have the receptor on their axons. Alternatively, axon receptors may be less accessible to the factor because of the presence of truncated TrkB receptors on reactive astrocytes at the lesion site (Frisen et al., 1992, 1993). We suggest that transplants of cells genetically modified to secrete multiple neurotrophic factors will offer more complete protection against axotomy-induced RN neuron death and atrophy.

It is, of course, difficult to determine conclusively whether a neuronal population has undergone retrograde cell death or severe atrophy. Some of the apparent loss of axotomized RN neurons may be the result of atrophy, which has caused them to shrink below our limits of detection. In our experimental groups that received grafts of gelfoam alone or unmodified fibroblasts, however, we recognized both significant cell loss and shrinkage of many of the surviving neurons. We can therefore recognize two populations of axotomized RN neurons: those that die (or show profound shrinkage) and those that survive in an atrophic state. This second population could retain the ability to respond to delayed administration of BDNF by an increase in soma size. Such an effect has been described for axotomized adult DRG neurons (Verge et al., 1989).

Transplants of BDNF-producing fibroblasts prevented soma atrophy in RN neurons that regenerated

Fb/BDNF transplants promoted axon regeneration in 7% of RN neurons following cervical axotomy, as shown by retrograde FG labeling. We found no atrophy among FG-labeled cells. Several mechanisms may account for this finding. First, these cells may represent a population of RN neurons that respond to BDNF treatment by maintaining their soma size and by regenerating axons at least several segments caudal to the transplant. Other studies have demonstrated that exogenous BDNF can promote axon regeneration in a small population of RN neurons (Xu et al., 1995; Kobayashi et al., 1997; Ye and Houle, 1997; Menei et al., 1998). The relationship between maintenance of soma size and axon regeneration was also suggested by studies using a combination of fetal tissue transplants and administration of neurotrophins (Bregman et al., 1997, 1998). Based on the results from fetal tissue transplants, Mori et al. (1997) have proposed that larger quantities of neurotrophic factors may be necessary to maintain soma size of neurons than to rescue them. We now propose that the ability of neurotrophic factors to promote axon regeneration may be still more restricted than their ability to prevent cell death and atrophy, because we find that Fb/BDNF transplants rescue 60–70% of RN neurons and prevent 55% of cell atrophy among the survivors, but promote axon regeneration in only 7% of RN neurons. An alternative explanation for the finding that regenerated RN neurons maintained soma size is that these cells were able to reinnervate their original or new targets and thereby obtained trophic support from the targets, which may be more comprehensive and appropriate than the support from the Fb/BDNF transplants. Such innervation may also provide activity-induced survival. It is also possible that the regenerating RN neurons represent a population of cells that do not die or atrophy after axotomy. It is difficult, however, to be certain whether these mechanisms are responsible together or alone for the normal soma size of regenerated RN neurons.

Elements permissive for RST regeneration

Several features of the grafts and their interaction with the host are likely to have contributed to RST regeneration. First, the excellent graft survival and tissue apposition provided the regenerating RST axons with a continuous terrain for growth. Almost all of the transplants entirely filled the lesion cavity, were intimately apposed to the host tissue and thus formed an interface with the host that was not interrupted by cysts or scars. The absence of scar formation may reflect the mild host immune reaction due to immunosuppression by cyclosporin A and the neuronal protective effects of methylprednisolone and BDNF (Novikova et al., 1996). The second crucial feature was the relatively permissive environment provided by the fibroblast grafts (Tuszynski et al., 1994, 1996; Nakahara et al., 1996). The most important feature, however, was the local delivery of BDNF. Fb/BDNF transplants were homogeneously and robustly stained by X-gal histochemistry at 1 week and many cells remained X-gal-positive for at least 2 months, suggesting that the engineered cell transplants acted as an abundant and sustained source of BDNF. Previous reports have demonstrated the effectiveness of BDNF in promoting RST regeneration. For example, as many as 200 neurons regenerated into a peripheral nerve graft when BDNF was administered adjacent to RN perikarya (Kobayashi et al., 1997), and RN neurons failed to regenerate into Schwann cell grafts unless BDNF was administered or the Schwann cells were engineered to express BDNF (Xu et al., 1995; Menei et al., 1998). Exogenous BDNF also dramatically increased the number of chronically injured RN neurons that regenerated into a peripheral nerve graft (Ye and Houle, 1997). At least part of the mechanism by which BDNF promotes regeneration is a direct effect on gene expression by RN neurons. RN neurons respond to BDNF by upregulation of two regeneration-associated genes, GAP-43 and Tα-1 tubulin, which supports a regenerative response (Tetzlaff et al., 1994; Kobayashi et al., 1997).

Although BDNF may also have exerted a tropic influence, the greater length of regeneration through white matter that we observed for RST axons requires additional comment. We found that the regenerated axons occupied the general location of the RST in normal white matter and terminated in regions of gray matter that are the normal targets of these axons. This observation is consistent with the notion that the regenerating axons responded to cues similar to those that operate during development. The regenerating axons appear to have been able to grow in response to BDNF and these or other cues despite the well known inhibitory influence of CNS myelin. The numerous in vivo studies of myelin-associated neurite growth inhibitors have primarily focused on CST axons (Savio and Schwab, 1990; Schnell and Schwab, 1990, 1993; Schnell et al., 1994; Z'Graggen et al., 1998). Whether they exert similar effects on RST axons has not been examined. It is possible that RST axons are less susceptible to inhibition by myelin or, as suggested for other developing and adult axons that grow through adult white matter (Wictorin and Björklund, 1992; Wictorin et al., 1992; Li and Raisman, 1993; Davies et al., 1994, 1997; Li et al., 1997, 1998; Oudega and Hagg, 1999), that regenerating RST axons lack or downregulate the relevant receptors. The RST regeneration that we observed shares several characteristics with the growth through white matter reported for these other axons, including rapid growth rate (up to 1–2 mm/day; Davies et al., 1994), growth of at least several centimeters (Wictorin et al., 1992), and the ability to find targets (Wictorin et al., 1992; Davies et al., 1994, 1997). We speculate that, in response to BDNF and the favorable substrate provided by the transplants, the intrinsically robust regenerative capacity of RN neurons allowed their axons to grow past the lesion site and through host white matter in response to cues that remained after injury and relatively unaffected by the inhibitory influence of CNS myelin. Future experiments will determine whether additional systems of supraspinal axons are similarly responsive to neurotrophic factors supplied by transplants.

Transplants promote recovery of function

Rats with Fb/BDNF showed significant recovery of motor function in two tests, while rats receiving gelfoam on unmodified fibroblast grafts showed no recovery. The re-lesion experiments that abolished the recovery in Fb/BDNF recipients indicated that the presence of the graft was responsible for the recovery. Our anatomical studies concentrated on the RN–RST system but sprouting, regeneration and rescue of other neuronal systems may well have contributed to the recovery that we observed.

Conclusion

We have demonstrated that transplants of fibroblasts, genetically modified by a retroviral vector encoding BDNF, into a partial cervical spinal cord hemisection (1) partially prevented axotomy-induced neuron loss and atrophy in the RN, (2) promoted rubrospinal axon regeneration for long distances, and (3) induced recovery of forelimb and hindlimb function, which was abolished following a re-lesion (Liu et al., 1999). The enhanced survival of the remaining RN neurons that did not regenerate axons caudal to the transplants may also have contributed to functional recovery by sprouting and local segmental mechanisms, because we found considerable rubrospinal axon sprouting rostral to a transplant in animals receiving BDNF-producing fibroblasts. In contrast, unmodified fibroblasts or gelfoam alone failed to offer protection against neuron loss and atrophy or to promote axon regeneration and recovery of function. Ex vivo gene therapy may therefore prove to be an effective treatment of human SCI by both promoting axon regeneration and rescuing neurons from axotomy-induced cell death.

References

Arenas, E. and Persson, H. (1994) Neurotrophin-3 prevents the death of adult central noradrenergic neurons in vivo. *Nature*, 367: 368–371.

Basso, D.M., Beattie, M.S. and Bresnahan, J.C. (1995) A new sensitive locomotor rating scale for locomotor recovery after spinal cord contusion using the NYU weight-drop device versus transection. *J. Neurotrauma*, 12: 1–21.

Bernstein-Goral, H. and Bregman, B.S. (1993) Spinal cord transplants support the regeneration of axotomized neurons after spinal cord lesions at birth: a quantitative double-labeling study. *Exp. Neurol.*, 123: 118–132.

Bothwell, M. (1995) Functional interactions of neurotrophins and neurotrophin receptors. *Annu. Rev. Neurosci.*, 18: 223–253.

Bregman, B.S. and Reier, P.J. (1986) Neural tissue transplants rescue axotomized rubrospinal cells from retrograde death. *J. Comp. Neurol.*, 244: 86–95.

Bregman, B.S., McAtee, M., Dai, H.N. and Kuhn, P.L. (1997) Neurotrophic factors increase axonal growth after spinal cord injury and transplantation in the adult rat. *Exp. Neurol.*, 148: 475–494.

Bregman, B.S., Broude, E., McAtee, M. and Kelley, M.S. (1998) Transplants and neurotrophic factors prevent atrophy of mature CNS neurons after spinal cord injury. *Exp. Neurol.*, 149: 13–27.

Davies, S.J., Field, P.M. and Raisman, G. (1994) Long interfascicular axon growth from embryonic neurons transplanted into adult myelinated tracts. *J. Neurosci.*, 14: 1596–1612.

Davies, S.J., Fitch, M.T., Memberg, S.P., Hall, A.K., Raisman, G. and Silver, J. (1997) Regeneration of adult axons in white matter tracts of the central nervous system. *Nature*, 390: 680–683.

Diener, P.S. and Bregman, B.S. (1994) Neurotrophic factors prevent the death of CNS neurons after spinal cord lesions in newborn rats. *NeuroReport*, 5: 1913–1917.

Diener, P.S. and Bregman, B.S. (1998) Fetal spinal cord transplants support growth of supraspinal and segmental projections after cervical spinal cord hemisection in the neonatal rat. *J. Neurosci.*, 18: 779–793.

DiStefano, P.S., Friedman, B., Radziejewski, C., Alexander, C., Boland, P., Schick, C.M., Lindsay, R.M. and Wiegand, S.J. (1992) The neurotrophins BDNF, NT-3, and NGF display distinct patterns of retrograde axonal transport in peripheral and central neurons. *Neuron*, 8: 983–993.

Frisen, J., Verge, V.M.K., Cullheim, S., Persson, H., Fried, K., Middlemas, D.S., Hunter, T., Hokfelt, T. and Risling, M. (1992) Increased levels of trkB mRNA and trkB protein-like immunoreactivity in the injured rat and cat spinal cord. *Proc. Natl. Acad. Sci. USA*, 89: 11282–11286.

Frisen, J., Verge, V.M.K., Fried, K., Risling, M., Persson, H., Trotter, J., Hokfelt, T. and Lindholm, D. (1993) Characterization of glial trkB receptors: differential response to injury in the central and peripheral nervous systems. *Proc. Natl. Acad. Sci. USA*, 90: 4971–4975.

Goldberger, M.E., Murray, M. and Tessler, A. (1993) Sprouting and regeneration in the spinal cord: their roles in recovery of function after spinal injury. In: Gorio, A. (Ed.), *Neuroregeneration*. Raven Press, New York, pp. 241–264.

Grill, R., Murai, K., Blesch, A., Gage, F.H. and Tuszynski, M.H. (1997) Cellular delivery of neurotrophin-3 promotes corticospinal axonal growth and partial functional recovery after spinal cord injury. *J. Neurosci.*, 17: 5560–5572.

Hefti, F., Knusel, B. and Lapchak, P.A. (1993) Protective effects of nerve growth factor and brain-derived neurotrophic factor on basal forebrain cholinergic neurons in adult rats with partial fimbrial transections. *Prog. Brain Res.*, 98: 257–263.

Himes, B.T., Goldberger, M.E. and Tessler, A. (1994) Grafts of fetal central nervous system tissue rescue axotomized Clarke's nucleus neurons in adult and neonatal operates. *J. Comp. Neurol.*, 339: 117–131.

Himes, B.T., Solowska-Baird, J., Boyne, L., Snyder, E.Y., Tessler, A. and Fischer, I. (1995) Grafting of genetically modified cells that produce neurotrophins in order to rescue axotomized neurons in rat spinal cord. *Soc. Neurosci. Abstr.*, 21: 537.

Hofer, M.M. and Barde, Y.-A. (1988) Brain-derived neurotrophic factor prevents neuronal death in vivo. *Nature*, 331: 261–262.

Kawaja, M.D., Rosenberg, M.B., Yoshida, K. and Gage, F.H. (1992) Somatic gene transfer of nerve growth factor promotes the survival of axotomized septal neurons and the regeneration of their axons in adult rats. *J. Neurosci.*, 12: 2849–2864.

Kobayashi, N.R., Fan, D.P., Giehl, K.M., Bedard, A.M., Wiegand, S.J. and Tetzlaff, W. (1997) BDNF and NT-4/5 prevent atrophy of rat rubrospinal neurons after cervical axotomy, stimulate GAP-43 and Tα1-tubulin mRNA expression, and promote axonal regeneration. *J. Neurosci.*, 17: 9583–9595.

Korsching, S. (1993) The neurotrophic factor concept: a reexamination. *J. Neurosci.*, 13: 2739–2748.

Levi-Montalcini, R. (1987) The nerve growth factor: thirty-five years later. *Biosci. Rep.*, 7: 681–699.

Li, Y. and Raisman, G. (1993) Long axon growth from embryonic neurons transplanted into myelinated tracts of the adult rat spinal cord. *Brain Res.*, 629: 115–127.

Li, Y., Field, P.M. and Raisman, G. (1997) Repair of adult rat corticospinal tract by transplants of olfactory ensheathing cells. *Science*, 277: 2000–2002.

Li, Y., Field, P.M. and Raisman, G. (1998) Regeneration of adult rat corticospinal axons induced by transplanted olfactory ensheathing cells. *J. Neurosci.*, 18: 10514–10524.

Liu, Y., Kim, D., Himes, B.T., Chow, S.Y., Murray, M., Tessler, A. and Fischer, I. (1999) Transplants of fibroblasts genetically modified to express BDNF promote regeneration of adult rat rubrospinal axons. *J. Neurosci.*, 19: 4370–4387.

Liu, Y., Himes, B.T., Murray, M., Tessler, A., Fischer, I., 2000. Grafts of BDNF-producing fibroblasts that promote regeneration of axotomized rubrospinal neurons also rescue most neurons form retrograde death and prevent their atrophy. J. Comp. Neurol., submitted.

Menei, P., Monteromenei, C., Whittemore, S.R., Bunge, R.P. and Bunge, M.B. (1998) Schwann cells genetically modified to secrete human BDNF promote enhanced axonal regrowth across transected adult rat spinal cord. *Eur. J. Neurosci.*, 10: 607–621.

Miya, D., Giszter, S., Mori, F., Adipudi, V., Tessler, A. and Murray, M. (1997) Fetal transplants alter the development of function after spinal cord transection in newborn rats. *J. Neurosci.*, 17: 4856–4872.

Mocchetti, I. and Wrathall, J.R. (1995) Neurotrophic factors in central nervous system trauma. *J. Neurotrauma*, 12: 853–870.

Mori, F., Himes, B.T., Kowada, M., Murray, M. and Tessler, A. (1997) Fetal spinal cord transplants rescue some axotomized rubrospinal neurons from retrograde cell death in adult rats. *Exp. Neurol.*, 143: 45–60.

Nakahara, Y., Gage, F.H. and Tuszynski, M.H. (1996) Grafts of fibroblasts genetically modified to secrete NGF, BDNF, NT-3, or basic FGF elicit differential responses in the adult spinal cord. *Cell Transplant.*, 5: 191–204.

Novikova, L., Novikov, L. and Kellerth, J.O. (1996) Brain-derived neurotrophic factor reduces necrotic zone and supports neuronal survival after spinal cord hemisection in adult rats. *Neurosci. Lett.*, 220: 203–206.

Oudega, M. and Hagg, T. (1996) Nerve growth factor promotes regeneration of sensory axons into adult rat spinal cord. *Exp. Neurol.*, 140: 218–229.

Oudega, M. and Hagg, T. (1999) Neurotrophins promote regeneration of sensory axons in the adult rat spinal cord. *Brain Res.*, 818: 431–438.

Rosenberg, M.B., Friedmann, T., Robertson, R.C., Tuszynski, M., Wolff, J.A., Breakefield, X.O. and Gage, F.H. (1988) Grafting genetically modified cells to the damaged brain: restorative effects of NGF expression. *Science*, 242: 1575–1578.

Savio, T. and Schwab, M.E. (1990) Lesioned corticospinal tract axons regenerate in myelin-free rat spinal cord. *Proc. Natl. Acad. Sci. USA*, 87: 4130–4133.

Schnell, L. and Schwab, M.E. (1990) Axonal regeneration in the rat spinal cord produced by an antibody against myelin-associated neurite growth inhibitors. *Nature*, 343: 269–272.

Schnell, L. and Schwab, M.E. (1993) Sprouting and regeneration of lesioned corticospinal tract fibers in the adult rat spinal cord. *Eur. J. Neurosci.*, 5: 1156–1171.

Schnell, L., Schneider, R., Kolbeck, R., Barde, Y.A. and Schwab, M.E. (1994) Neurotrophin-3 enhances sprouting of corticospinal tract during development and after adult spinal cord lesion. *Nature*, 367: 170–173.

Sendtner, M., Holtmann, B., Kolbeck, R., Thoenen, H. and Barde, Y.-A. (1992) Brain-derived neurotrophic factor prevents the death of motoneurons in newborn rats after nerve section. *Nature*, 360: 757–759.

Shibayama, M., Hattori, S., Himes, B.T., Murray, M. and Tessler, A. (1998) Neurotrophin-3 prevents death of axotomized Clarke's nucleus neurons in adult rat. *J. Comp. Neurol.*, 390: 102–11.

Tessler, A. and Murray, M. (1996) Neural transplantation: spinal cord. In: Chick, W.L. and Lanza, R.P. (Eds.), *Yearbook of Cell and Tissue Transplantation*. Kluwer, The Hague, pp. 175–182.

Tessler, A., Fischer, I., Giszter, S., Himes, B.T., Miya, D., Mori, F. and Murray, M. (1997) Embryonic spinal cord transplants enhance locomotor performance in spinalized newborn rats. In: Seil, F.J. (Ed.), *Neuronal Regeneration, Reorganization and Repair. Advances in Neurology, Vol. 72.* Lippincott–Raven Press, Philadelphia, pp. 291–303.

Tetzlaff, W., Kobayashi, N.R., Giehl, K.M., Tsui, B.J., Cassar, S.L. and Bedard, A.M. (1994) Response of rubrospinal and corticospinal neurons to injury and neurotrophins. *Prog. Brain Res.*, 103: 271–286.

Tracey, D.J. (1995) Ascending and descending pathways in the spinal cord. In: Paxinos, G. (Ed.), *The Rat Nervous System*, 2nd ed. Academic Press, San Diego, CA, pp. 67–80.

Tuszynski, M.H., Peterson, D.A., Ray, J., Baird, A., Nakahara, Y. and Gage, F.H. (1994) Fibroblasts genetically modified to produce nerve growth factor induce robust neuritic ingrowth after grafting to the spinal cord. *Exp. Neurol.*, 126: 1–14.

Tuszynski, M.H., Gabriel, K., Gage, F.H., Suhr, S., Meyer, S. and Rosetti, A. (1996) Nerve growth factor delivery by gene transfer induces differential outgrowth of sensory, motor, and noradrenergic neurites after adult spinal cord injury. *Exp. Neurol.*, 137: 157–173.

Verge, V.M.K., Riopelle, R.J. and Richardson, P.M. (1989) Nerve growth factor receptors on normal and injured sensory neurons. *J. Neurosci.*, 9: 914–922.

West, M.J. (1993) New stereological methods for counting neurons. *Neurobiol. Aging*, 14: 275–285.

Wictorin, K. and Björklund, A. (1992) Axon outgrowth from grafts of human embryonic spinal cord in the lesioned adult rat spinal cord. *NeuroReport*, 3: 1045–1048.

Wictorin, K., Brundin, P., Sauer, H., Lindvall, O. and Björklund, A. (1992) Long distance directed axonal growth from human dopaminergic mesencephalic neuroblasts implanted along the nigrostriatal pathway in 6-hydroxydopamine lesioned adult rats. *J. Comp. Neurol.*, 323: 475–494.

Xu, X.M., Guenard, V., Kleitman, N., Aebischer, P. and Bunge, M.B. (1995) A combination of BDNF and NT-3 promotes supraspinal axonal regeneration into Schwann cell grafts in adult rat thoracic spinal cord. *Exp. Neurol.*, 134: 261–272.

Yan, Q., Elliott, J. and Snider, W.D. (1992) Brain-derived neurotrophic factor rescues spinal motor neurons from axotomy-induced cell death. *Nature*, 360: 753–755.

Ye, J.H. and Houle, J.D. (1997) Treatment of the chronically injured spinal cord with neurotrophic factors can promote axonal regeneration from supraspinal neurons. *Exp. Neurol.*, 143: 70–81.

Z'Graggen, W.J., Metz, G.A.S., Kartje, G.L., Thallmair, M. and Schwab, M.E. (1998) Functional recovery and enhanced corticofugal plasticity after unilateral pyramidal tract lesion and blockade of myelin-associated neurite growth inhibitors in adult rats. *J. Neurosci.*, 18: 4744–4754.

SECTION VI

New directions in regeneration research

CHAPTER 28

Delivery of therapeutic molecules into the CNS

Marina E. Emborg and Jeffrey H. Kordower [*]

Department of Neurological Sciences and Research Center for Brain Repair, Rush University, 2242 West Harrison Street, Chicago, IL 60612, USA

Introduction

If a molecule is therapeutic for a particular neurodegenerative disease, its utility will ultimately be determined by, among other things, its ability to be effectively delivered to the critical neuronal populations. Experimentally, effective delivery to the brain has been achieved by systemic administration of substances that are able to cross the blood–brain barrier (BBB) or by disrupting barriers to reach specific vulnerable neuronal populations. Each one of these methods has advantages and disadvantages according to the problem to be solved. A successful therapy will be one that matches the properties of the therapeutic molecule and its delivery method with the characteristics of the pathology to be treated and the therapeutic goal.

In this chapter we will review traditional and novel systems of delivery to the brain. Because treatment for Parkinson's disease (PD) has been revolutionized through the last decades, on several occasions we will refer to PD as an example for the use of new technologies.

Systemic administration

Systemic administration of therapeutic molecules can be achieved via a number of routes including by mouth, intranasally, or by injection. Delivery in these ways is simple, relatively economic and non-invasive. However, the lack of specific administration often prevents optimal dosing as neuronal systems in both the central nervous system (CNS) and peripheral nervous system (PNS) unrelated to the disease in question will also be influenced by systemic drug administration. Further, therapeutic proteins are often ineffective after systemic administration, due to the poor transport of large molecules across the BBB. The capillary endothelial cells of the cerebral vasculature form the BBB and are closely apposed to one another in connections called 'tight junctions'. These tight junctions limit many hydrophilic molecules and larger substances from passing between the endothelial cells. Recently, several methods have been devised to circumvent the BBB.

Altering the aperture of the BBB

Tight junctions of the BBB can be temporarily opened by osmotic shock, with leukotrines, with glycopeptides or with bradykinin agonists such as RMP-7. Nevertheless the opening in a normal brain allows only the passage of molecules in the range of approximately 1000 Da. This clearly limits the systemic delivery of a number of classes of compounds, including trophic factors in chronic degenerative diseases. In addition, the opening of the BBB achieved using these methods is transient and subject to strong tachyphylaxis; thus even with continuous infusion of bradykinin or RMP-7 into the carotid artery (an invasive procedure), maximal duration of opening is less

---

[*] Corresponding author: Dr. Jeffrey H. Kordower, Department of Neurological Sciences, Rush University, 2242 West Harrison Street, Chicago, IL 60612, USA. Fax: +1-312-633-1564; E-mail: jkordowe@rush.edu

than 30 to 60 min (Inamura et al., 1994; Sanovich et al., 1995; Bartus et al., 1996a,b).

Crossing the BBB

Molecules with trophic activity are being tested for the treatment of many neurodegenerative diseases to protect and regenerate vulnerable neuronal systems. But trophic factors in general are large molecules that can not cross the BBB. Only small lipophilic substances can enter the brain, diffusing through the endothelial cells that comprise the cerebral capillaries. New compounds have been synthesized that show some promise, such as immunophilins or gangliosides. Immunophilins are fat-soluble peptides that can readily cross the BBB to induce its central effects. Gangliosides are complex acidic glycosphingolipids that are an important component of mammalian plasma and intracellular membranes and are also able to cross the BBB. Although studies in non-human primate models suggest that the success of immunophilins and gangliosides could be limited to early stages of PD and can be affected by individual variations (Emborg and Colombo, 1994; Emborg et al., 2000), these drugs are being evaluated in clinical trials targeting parkinsonian patients (Schneider et al., 1998). More information is necessary to know how much the peripheral and central circulation of the drug could affect other systems, as well as the bioavailability of the drug in the brain. Molecules attempting to cross the BBB can be metabolized in peripheral tissues and/or at the cerebral micro- and macrovessels (Riachi and Harik, 1992). For example, oral administration of 3,4-dihydroxy-L-phenylalanine (L-DOPA) is widely used for dopamine (DA) replacement treatment in PD. Although DA is the catecholamine that is needed, DA cannot cross the BBB. However, L-DOPA, the immediate metabolic precursor of DA, does penetrate the brain, where it is decarboxylated to DA (Nutt and Fellman, 1984). Because decarboxylation can also occur in peripheral tissues, high doses of L-DOPA are required to reach therapeutic levels in the brain. Oral L-DOPA loses initially 70% of the dose during the absorption at the gastrointestinal tract. Although lower doses of the drug are required with intravenous injections, peripheral metabolism affects the amount that reaches the brain. Administration of L-DOPA in combination with a dopa-decarboxylase inhibitor facilitates penetrance of L-DOPA into the brain by preventing peripheral breakdown. Thus a higher amount of L-DOPA will reach the brain and a lower dose of L-DOPA will be required. L-DOPA crosses biological membranes by a saturable, sodium-independent, facilitated transport mechanism for the aromatic and branched-chain amino acids (Oldendorf, 1971; Daniel et al., 1976). The importance of this shared system is that the capacity of the transport process is limited. Therefore, the competition between substrates for the carrier system determines the flux of each competing substrate across the membrane, which is important in the passage of L-DOPA, as well as other therapeutic molecules, across the intestinal endothelium and the BBB.

The route of systemic delivery affects the absorption of the compound

Searching for a direct but non-invasive delivery method to the CNS, Sakane et al. (1999) studied the extent of transnasal drug delivery to the brain through the cerebrospinal fluid (CSF) in the rat, using [^{3}H]5-fluorouracil (5FU) as a model drug. It was confirmed first that the concentration of 5FU in the CSF was significantly higher following nasal administration compared with intravenous injection, indicating direct transport of 5FU from the nasal cavity to the CSF. Consequently, a significant amount of 5FU was transported from the nasal cavity to the brain through the CSF, and thus the delivery of the hydrophilic drug to the brain is augmented by nasal drug application. Although the ability of a compound to cross into the CNS using this delivery route is affected by the molecular characteristics of the molecule, thus limiting the therapeutic agents that can be used in this fashion, transnasal delivery of specific substances can prove to be a valid alternative.

Lipophilic transporters

Lipophilic vesicles called liposomes have been used as carriers to deliver hydrophilic molecules to the brain (Gregoriadis et al., 1985). Because lipophilic molecules diffuse through the lipid phase of the BBB endothelial cells, carriers with a lipophilic membrane can be used to package hydrophilic molecules. An alternative method is to shield the hydrophilic portions of molecules with lipophilic molecules (Tsuzuki et al., 1991; Alyautdin et al., 1997). These methods present limitations because passage through the BBB is dependent not only on lipophilicity, but also restricted by size. Large lipophilic containers cannot diffuse through the capillary cells limiting the molecules that can be delivered in this way.

Active transport systems

Active transport through the capillary cells allows some hydrophilic molecules or large lipophilic substances to enter the brain from the bloodstream (Broadwell et al., 1996). Facilitated transport of therapeutic molecules across the BBB can be achieved by linking the protein to a vehicle molecule that is actively transported across the BBB. Insulin (Duffy and Pardridge, 1987) and insulin-like growth factors (Duffy et al., 1988) have been used as vehicles to transport drugs across the BBB. Some proteins that have been positively charged, such as albumin (Kumagai et al., 1989; Kang and Pardridge, 1994) or histone (Pardridge et al., 1989), have also been tested as delivery vehicles (Triguero et al., 1989). The iron-transferrin transport system has been used successfully by directing a specific monoclonal antibody against the transferrin receptor, OX-26. Some OX receptors are present in peripheral organs, but most of them are present at the brain endothelial cells (Jeffries et al., 1984). The OX-26 antibody has been used as a vehicle to deliver drugs such as methotrexate (Friden et al., 1991) as well as trophic factors such as nerve growth factor (NGF) (Granholm et al., 1994; Kordower et al., 1994), brain-derived neurotrophic factor (BDNF) (Pardridge et al., 1994) and glial cell line-derived neurotrophic factor (GDNF) (Albeck et al., 1997) across the BBB of the rat. OX-26 can also be conjugated with liposomes to carry larger amounts of the same substance (Huwyler et al., 1996). The conjugates are transported by transcytosis through the brain endothelial cells (Broadwell et al., 1996) to influence specific neuronal populations.

Clinical importance of systemic delivery

Clinically, systemic delivery of drugs is still the delivery method of choice. A compound that is able to cross the BBB in general has the following advantages: (1) easy administration and follow-up; (2) decreased risk of infections or complications related to the administration procedure, because of the non-invasive characteristic of the method; and (3) dose can be adjusted according to response and drug levels in blood. However, as we have discussed previously, there are severe limitations: (1) high doses are needed because of peripheral metabolism of the compound; (2) side effects can be observed because of the metabolism of the compound by unintended PNS or CNS structures; and (3) adverse effects can occur due to unregulated availability of the drug at the target area.

Novel delivery methods to the CNS

The direct release of biologically active substances into the brain can sidestep the problems and limitations associated with systemic delivery. Compounds that do not easily penetrate the BBB, that are metabolized by peripheral tissue, or that have side effects related to global brain effects, can be regionally released to gain access to their target area. These methods have produced positive results in animal models of disease and some are already in clinical use.

Intraventricular administration of compounds

Direct intraventricular (ICV) injections have been accepted as a suitable method for clinical trials. However, substances infused into the ventricular system will disperse throughout the brain, predominantly affecting the brain parenchyma adjacent to the ventricular system, often limiting the dose available and producing unacceptable side effects. There have been significant failures in the clinical delivery of trophic factors using the intraventricular route. In this regard, most trophic factors such as BDNF, NGF, and GDNF diffuse poorly within brain parenchyma following ICV infusion (Lapchak, 1993; Lapchak et al., 1996, 1997; Kordower et al., 1999a). NGF ICV delivery for treatment of Alzheimer's disease failed to stop the progression of the disease and its administration was complicated by side effects such as a pain syndrome and pial hyperplasia (Olson, 1993; Olson et al., 1994; Eriksdotter-Jonhagen et al., 1998).

Because of the striking success in animal models of PD, ICV GDNF was tested clinically in patients with PD (Kordower et al., 1999a). The outcome proved unsuccessful with side effects that included nausea, loss of appetite and abnormal social behavior. The postmortem analysis of GDNF expression revealed that GDNF was undetectable within the striatum or substantia nigra. The data indicate that short-term monthly injections of ICV GDNF at low doses are not associated with clinical improvement or enhancement of nigrostriatal circuitry. These findings support the concept that site-specific delivery is critical to the success of therapies with trophic factors.

Diffusion of compounds in the brain

The limited extracellular space of the brain, and the physical barrier engendered by the presence of neuronal and glial structures limit the diffusion of a substance through brain parenchyma. In comparison, the parallel extracellular spaces that are found in white matter tracts can allow diffusion of substances. Limited diffusion is essential to achieve localized delivery into small regions (e.g., subthalamic nucleus), but it can complicate therapies aimed at large structures such as the striatum, where multiple site delivery may be required. Nevertheless, compounds may be distributed further if they are taken up by neurons and retrogradely or anterogradely transported from the injection site. High- and low-affinity uptake sites and catabolizing enzymes affect the diffusion gradients and, consequently, diffusion distance. Sendelbeck and Urquhart (1985) studied the spatial distributions of different drugs (lipid soluble and insoluble) after continuous microperfusion of the diencephalon of rabbits. The basic pattern of distribution was the same for each drug: the tissue concentration of the perfusate was maximal at the cannula tip, and declined sharply with radial distance from the tip. However, at any given distance, concentrations of the ionized, lipid-insoluble drugs were one to two orders of magnitude higher than those derived from a lipid-soluble drug. The results demonstrated that intracerebrally microperfused drugs may have quantitatively different spatial distributions related to their physicochemical characteristics and/or their binding and metabolism in brain tissue. Consequently, spatial distribution of a compound will not solely depend on where in the brain a substance will be delivered, but on the unique characteristics of the agent under study.

Direct injection of compounds into the CNS

Target-specific delivery can be achieved by direct injection of the desired compound into the area of interest using stereotaxic guidance. This technique is broadly used in basic research for single and multiple injections. When several injections are required over time, a guide cannula system attached to the skull can secure a portal of entry into the brain (Carvey et al., 1994). Continuous chronic delivery can be obtained by connecting a cannula to a delivery pump placed in a subcutaneous pocket. Osmotic pumps can constantly infuse during 1 to 28 days, pumping at a rate of 0.5 to 10.0 μl/min, depending on the selected device (Alzet). This delivery method avoids extensive manipulation of the animals and, in case of complications, can be interrupted by disconnecting the pump. A more sophisticated device has been developed for human therapy. In this case, a cannula is attached to an implantable battery-powered programmable pump (Medtronic Synchromed Programmable pump) that is also placed in a subcutaneous pocket. The external part of the system is a small computer that is used to non-invasively program and alter the characteristics of the implanted pump by telemetry. The refilling of the pump is done through the skin following a sterile technique with a Hubber needle that fits the pump port. Although the smart design of this pump facilitates the dosing of chronic treatments, complications related to the pump position and connected tubing are possible. This type of chronic infusion system has been limited for CNS human use to continuous epidural/intrathecal delivery of analgesics. In general, the invasive nature of intraparenchymal injections has limited their use in humans to few clinical studies and to acute administration (e.g., Penn et al., 1998).

Implantation of microspheres

Alternative systems that can achieve local sustained release of therapeutic molecules are biodegradable preparations in which an agent is incorporated in a biocompatible polymeric matrix. Such a matrix can be directly implanted in the target site. Several polymer devices that allow local delivery of chemotherapeutic agents have been successfully tested in animal models of neurogliomas. Macroscopic non-biodegradable devices and carmustine-loaded biodegradable polymer disks are already in clinical trial (Oda et al., 1982; Kubo et al., 1986; Brem et al., 1991). Menei et al. (1996) developed poly(lactic acid-co-glycolic acid) microspheres that are biocompatible with the brain, and that are totally biodegraded within 2 months. Poly-(DL-lactide)co-glycolide was also used as the matrix material to deliver dopamine to dopamine-depleted striatum in a rat model of PD (McRae-Degueurce et al., 1988). Implantation of this delivery system resulted in sustained release of DA into the basal ganglia over a 6-week period with minimal reaction in the surrounding brain tissue. Using a similar principle, NGF delivery was also tested in PD rat models to increase survival of implants of adrenergic cells (Stromberg et al., 1985).

Cell transplantation

The surgical implantation of cells that produce and release therapeutic molecules has also been used as a way to deliver compounds to the brain. PD has provided the ideal system for testing transplantation techniques. A basic principle of neural transplantation for PD is to provide DA from a graft in a stable fashion directly into the striatum, where the intrinsic dopaminergic system has been destroyed by degeneration. Different sources of dopaminergic cells have been investigated. Chromaffin cells of the adrenal medulla (Freed et al., 1981, 1990; Björklund and Stenevi, 1985), and lately, glomus or type I cells of the carotid body (Espejo et al., 1998; Luquin et al., 1999), have been tested in rodent and non-human primate models of PD. While somewhat successful in rodent studies, the outcome of clinical trials employing adrenal grafts has been poor (Backlund et al., 1985; Madrazo et al., 1987), and this approach has largely been abandoned.

Fetal substantia nigra was first used as donor tissue in experimental models of PD (Björklund and Stenevi, 1979). It has been proven in rodents, non-human primates and human subjects that grafts of fetal nigral neurons consistently survive, produce DA, form synaptic connections and ameliorate behavioral deficits due to lesions of the nigrostriatal pathway (see Björklund and Stenevi, 1985; Björklund et al., 1987; Sladek and Gash, 1988; Yurek and Sladek, 1990; Zigmond et al., 1990; Freed et al., 1992; Kordower et al., 1995a; Nakao et al., 1995; Schwarting and Huston, 1996). The greatest limitation is the availability of donors and the ethical and moral implications of the use of human fetuses. Xenotransplantation of bovine (Zawada et al., 1998) or porcine (Barker et al., 1999; Friedrich, 1999; Habeck, 2000) dopaminergic embryonic neurons could overcome the difficulties of obtaining human embryonic tissue, but their clinical use is tempered by the need of life-long immunosuppression, the possibility of intraspecies viral infections, and ethical issues.

Neural stem cells have also been shown to survive for long periods of time in vivo and to integrate well within the host tissue without being tumorigenic (Gage et al., 1995; Brustle and McKay, 1996). Although the source of these cells is a developing embryonic brain, its advantage compared with fetal tissue is that stem cells are able to multiply and become a new source of cells (Svendsen, 1997). Because they are pluripotential, researchers are investigating the possibility that microenvironmental signals in the target area of the brain will guide the development of the transplanted cells. In addition, neural stem cells can be genetically modified to obtain a relatively homogeneous expression of a desired gene product (Kordower et al., 1997).

The success observed with fetal tissue transplantation in PD patients has inspired the use of grafts in other pathologies such as Huntington's disease (HD), Alzheimer's disease (AD) and spinal cord injury (for review see Seventh International Meeting on Neural Transplantation and Repair, 2000). Epidermal growth factor (EGF) responsive stem cells derived from transgenic mice in which the glial fibrillary acidic protein (GFAP) promoter directs the expression of hNGF has been successfully used in a rodent model of HD to prevent the striatal lesion induced by a quinolinic acid injection (Kordower et al., 1997). Similar studies in animal models of AD are also being evaluated, targeting the cholinergic cell population. Future analysis will reveal the possibilities of cellular methods of delivery for clinical application.

Implantation of encapsulated cells

Encapsulation methods evolved from the need to ensure (1) a limited cell growth, (2) a decreased immunogenic response by limiting the exposure of foreign particles to the system, and (3) the security of allowing the retrieval of the capsule in case of adverse effects. The cells are embedded in a polymer-based matrix that allows for the bidirectional diffusion of nutrients into the capsule and the therapeutic compound out to the brain. However, the pores of the capsule are too small to allow immune cells to enter the capsule and attack the genetically modified cells. Thus this technology allows for allo- or xenografting without immunosuppression.

Encapsulated cells such as PC12 have been used as biological pumps to deliver DA (Kordower et al., 1995b), while encapsulated genetically engineered baby hamster kidney (BHK) cells have been used to deliver NGF (Date et al., 1996; Kordower et al., 1998, 1999b) neurotrophin-3 (NT-3), neurotrophin-4 (NT-4) (Emerich et al., 1998a) or ciliary neurotrophic factor (CNTF) (Emerich et al., 1998b). The advantage of these systems is that gene transfer traditionally has proven to be easier in replicating cells such as BHK.

Gene therapy

The development of molecular genetic techniques to deliver therapeutic molecules for the treatment of any disease, genetic or non-genetic, is proving to overcome the delivery limitations to the CNS. Gene transfer can be applied to cells inside the brain (in vivo gene therapy) or to cells in vitro that will be grafted at a later time (ex vivo gene therapy). For neurodegenerative diseases, gene therapy can be used to deliver neurotransmitters or enzymes associated with the synthesis of neurotransmitters, trophic factors to support the viability of a cell, or antiapoptotic genes that can prevent cell death (for review see: Ebendal et al., 1994; Wim et al., 1998).

There are two categories of vectors: non-viral and viral (Robbins and Ghivizzani, 1998). Non-viral methods range from direct injection of DNA to mixing DNA with polylysine or cationic lipids that allow the gene to cross the cell membrane. Most of these approaches have poor delivery efficiency and transient expression of the transgene, but have the clear advantage of avoiding immunogenic reactions. If more efficient non-viral transfer methods are developed, they may provide a valuable alternative to viral vectors.

Viral vectors are becoming increasingly important tools to explore the feasibility of gene therapy to treat diseases of the nervous system (Verma and Somia, 1997). This gene transfer technology is based on the use of a virus as a gene delivery vehicle. Optimally, vectors to be used for clinical applications should have (1) high titer, (2) low particle ratio or absence of helper virus, (3) sustained levels of transgene expression, (4) regulatable expression, (5) minimal immune response, and (6) specific cell targeting (for review see Choi-Lundberg and Bohn, 1998). Presently, four viral vectors are more commonly used for gene transfer in the nervous system: herpes simplex virus, adenovirus, adeno-associated virus and lentivirus.

A number of gene therapy approaches have been tested clinically, all without success. However, many of these have not received detailed scrutiny in non-human primate models of disease, a step which could be highly predictive of clinical findings. Davidson et al. (1994) initially demonstrated short-term (7 day) gene transfer using adenovirus in Rhesus monkeys. Bohn et al. (1999) attempted in vivo gene delivery to African green monkeys. Successful transduction was highly variable and accompanied by significant immunogenicity and cytotoxicity using adenoviral vectors. Bankiewicz et al. (1998) have successfully achieved β-galactosidase (βGal) and tyrosine hydroxylase expression in Rhesus monkeys for up to 3 months after injection. We have recently shown that lentiviral delivery of βGal and GDNF genes can be successfully achieved in the nigrostriatal system of Rhesus monkeys (Kordower et al., 1999c). Robust lentiviral transfection of striatal and nigral cells was observed up to 3 months after the vector injections. These results suggest that adeno-associated and lentiviral vectors hold promise for gene transfer approaches in neurological diseases because (1) they infect quiescent cells such as neurons, (2) in vivo gene expression in the CNS is essentially restricted to neurons, (3) infected cells are not recognized by the host immune system, (4) infected areas do not show significant toxicity, (5) they allow long-term, sustained expression of the transgene, and (6) they can incorporate a transgene of at least 9 kb (Verma and Somia, 1997; Bankiewicz et al., 1998; Kordower et al., 1999b). Currently, vector systems encoding information to externally regulate the expression of the desired protein are being evaluated to add flexibility to gene delivery therapies (Robbins and Ghivizzani, 1998).

A final comment

New systems and compounds are being developed following the advances in our understanding of the pathophysiology of disease. Systematic analyses of the advances provide us with the tools to decide between delivery systems. But in the end, a successful method of delivery will be the one best suited to solve the problem on hand with the least number of side effects or complications.

References

Albeck, D.S., Hoffer, B.J., Quissell, D., Sanders, L.A., Zerbe, G. and Granholm, A.C. (1997) A non-invasive transport system for GDNF across the blood–brain barrier. *NeuroReport*, 8: 2293–2298.

Alyautdin, R.N., Petrov, V.E., Langer, K., Berthold, A., Kharkevich, D.A. and Kreuter, J. (1997) Delivery of loperamide across the blood–brain-barrier with polysorb 80-coated polybutylcyanoacrylate nanoparticles. *Pharm. Res.*, 14: 325–328.

Backlund, E.O., Neal, J.H., Waters, C.H., Appley, A.J., Boyd, S.D., Couldwell, W.T., Wheelock, V.H. and Weiner, L.P. (1985) Transplantation of adrenal medullary tissue to striatum in parkinsonism; first clinical trials. *J. Neurosurg.*, 62: 169–173.

Bankiewicz, K.S., Bringas, J.R., McLaughlin, W.W., Pivirotto, P., Hundal, R., Emborg, M.E. and Nagy, D. (1998) Application of gene therapy for Parkinson's disease: nonhuman primate experience. *Adv. Pharmacol.*, 42: 801–806.

Barker, R.A., Ratcliffe, E., Richards, A. and Dunnett, S.B. (1999) Fetal porcine dopaminergic cell survival in vitro and its relationship to embryonic age. *Cell Transplant.*, 8: 593–599.

Bartus, R.T., Elliot, P., Dean, R.L., Hayward, N.J., Nagle, T.L., Huff, M.R., Snodgras, P.A. and Blunt, D.G. (1996a) Controlled modulation of BBB permeability using the bradykinin agonist, RMP-7. *Exp. Neurol.*, 142: 14–28.

Bartus, R.T., Elliott, P., Hayward, N., Dean, R. and Fisher, S. (1996b) Permeability of the BBB by the B2 agonist RMP-7: evidence for a sensitive, autoregulated, receptor-mediated system. *Immunopharmacology*, 33: 270–278.

Björklund, A. and Stenevi, U. (1979) Reconstruction of the nigrostriatal dopamine pathway by intracerebral nigral transplants. *Brain Res.*, 177: 555–560.

Björklund, A. and Stenevi, U. (1985) Intracerebral neural grafting. In: Björklund, A. and Stenevi, U. (Eds.), *Neural Grafting in the Mammalian CNS*. Elsevier, Amsterdam, pp. 3–14.

Björklund, A., Lindvall, O., Isacson, O., Brundin, P., Wictorin, K., Strecker, R.E., Clarke, D.J. and Dunnett, S.B. (1987) Mechanisms of action of intracerebral neural implants: studies on nigral and striatal grafts to the lesioned striatum. *Trends Neurosci.*, 10: 509–516.

Bohn, M.C., Choi-Lundberg, D.L., Davidson, B.L., Leranth, C., Kozlowski, D.A., Smith, J.C., O'Banion, M.K. and Redmond Jr., D.E. (1999) Adenovirus-mediated transgene expression in nonhuman primate brain. *Hum. Gene Ther.*, 10: 1175–1184.

Brem, H., Mahaley Jr., M.S., Vick, N.A., Black, K.L., Schold Jr., S.C., Burger, P.C., Friedman, A.H., Ciric, I.S., Eller, T.W., Cozzens, J.W. and Kenealy, J.N. (1991) Interstitial chemotherapy with drug polymer implants for the treatment of recurrent gliomas. *J. Neurosurg.*, 74: 44144–44146.

Broadwell, R.D., Baker-Cairns, B.J., Friden, P.M., Oliver, C. and Villegas, J.C. (1996) Transcytosis of protein through the mammalian cerebral epithelium and endothelium, III. Receptor-mediated transcytosis through the blood–brain barrier of blood borne transferrin and antibody against the transferrin receptor. *Exp. Neurol.*, 142: 47–65.

Brustle, O. and McKay, R.D. (1996) Neuronal progenitors as tools for cell replacement in the nervous system. *Curr. Opin. Neurobiol.*, 6: 688–695.

Carvey, P.M., Maag, T.J. and Lin, D. (1994) Injection of biologically active substances into the brain. In: Flanagan, T.R., Emerich, D.F. and Winn, S.R. (Eds.), *Providing Pharmacological Access to the Brain: Alternate Approaches. Methods in Neurosciences, Vol. 21*, Academic Press, San Diego, CA, pp. 214–233.

Choi-Lundberg, D.L., Bohn and M.C., 1998. Applications of gene therapy to neurological diseases and injuries. In: Quesenberry, P.J., Stein, G.S., Forget, B. and Weissman, S. (Eds.), *Stem Cell Biology and Gene Therapy*. Wiley, New York, pp. 503–553.

Daniel, P.M., Moorhouse, S.R. and Pratt, O.E. (1976) Do changes in blood levels of other aromatic amino acids influence levodopa therapy?. *Lancet*, 1: 95.

Date, I., Ohmoto, T., Imaoka, T., Ono, T., Hammang, J.P., Francis, J. and Greco, C. (1996) Cografting with polymer-encapsulated human nerve growth factor-secreting cells and chromaffin cell survival and behavioral recovery in hemiparkinsonian rats. *J. Neurosurg.*, 84: 1006–1012.

Davidson, B.L., Doran, S.E., Shewach, D.S., Latta, J.M., Hartman, J.W. and Roessler, B.J. (1994) Expression of *Escherichia coli* beta-galactosidase and rat HPRT in the CNS of *Macaca mulatta* following adenoviral mediated gene transfer. *Exp. Neurol.*, 125: 258–267.

Duffy, K.R. and Pardridge, W.M. (1987) Blood–brain barrier

transcytosis of insulin in developing rabbits. *Brain Res.*, 420: 32–38.

Duffy, K.R., Pardridge, W.M. and Rosenfeld, R.G. (1988) Human blood–brain barrier insulin-like growth factor receptor. *Metabolism*, 37: 136–140.

Ebendal, T., Lonnerberg, P., Pei, G., Kylberg, A., Kullander, K., Persson, H. and Olson, L. (1994) Engineering cells to secrete growth factors. *J. Neurol.*, 242: S5–S7.

Emborg, M.E. and Colombo, J.A. (1994) Long-term MPTP-treated monkeys are resistant to GM1 systemic therapy. *Mol. Chem. Neuropathol.*, 21: 75–82.

Emborg, M.E., Shin, P., Roitberg, B., Sramek, J.G., Chu, Y., Stebbing, G., Hamilton, J.S., Suzdak, P.D., Steiner, J.P. and Kordower, J.H. (2000) Systemic administration of the immunophilin ligand GPI 1046 in MPTP treated monkeys. In revision.

Emerich, D.F., Bruhn, S., Chu, Y. and Kordower, J.H. (1998a) Cellular delivery of CNTF but not NT-4/5 prevents degeneration of striatal neurons in a rodent model of Huntington's disease. *Cell Transplant.*, 7: 213–225.

Emerich, D.F., Winn, S.R., Hantraye, P.M., Peschanski, M., Chen, E.-Y., Chu, Y., McDermott, P., Baetge, E.E. and Kordower, J.H. (1998b) Protective effect of encapsulated cells producing neurotrophic factor CNTF in a monkey model of Huntington's disease. *Nature*, 386: 395–399.

Eriksdotter-Jonhagen, M., Nordberg, A., Amberla, K., Backman, L., Ebendal, T., Meyerson, B., Olson, L., Seiger, A., Shigeta, M., Theodorsson, E., Viitanen, M., Winblad, B. and Wahlund, L.O. (1998) Intracerebroventricular infusion of nerve growth factor in three patients with Alzheimer's disease. *Dement. Geriatr. Cogn. Disord.*, 9: 246–257.

Espejo, E.F., Montoro, R.J., Armengol, J.A. and Lopez-Barneo, J. (1998) Cellular and functional recovery of Parkinsonian rats after intrastriatal transplantation of carotid body cell aggregates. *Neuron*, 20: 197–206.

Freed, C.R., Breeze, R.E., Rosenberg, N.L., Schneck, S.A., Kriek, E., Qi, J.X., Lone, T., Zhang, Y.B., Snyder, J.A. and Wells, T.H. (1992) Survival of implanted fetal dopamine cells and neurologic improvement 12 to 46 months after transplantation in Parkinson's disease. *N. Engl. J. Med.*, 327: 1549–1555.

Freed, W.J., Morihisa, J.M., Spoor, E., Hoffer, B.J., Olson, L., Seiger, A. and Wyatt, R.J. (1981) Transplanted adrenal chromaffin cells in rat brain reduce lesion-induced rotational behaviour. *Nature*, 292: 351–352.

Freed, W.J., Poltorack, M. and Becker, J.B. (1990) Intracerebral adrenal medulla grafts: a review. *Exp. Neurol.*, 110: 139–166.

Friden, P.M., Walus, L.R., Musso, G.F., Taylor, M.A., Malfroy, B. and Starzyk, R.M. (1991) Anti-transferrin receptor antibody and antibody–drug conjugates cross the blood–brain barrier. *Proc. Natl. Acad. Sci. USA*, 88: 4771–4775.

Friedrich, M.J. (1999) Fetal pig neural cells for Parkinson disease. *JAMA*, 282: 2198–2199.

Gage, F.H., Ray, J. and Fisher, L.J. (1995) Isolation, characterization, and use of stem cells from the CNS. *Annu. Rev. Neurosci.*, 18: 159–192.

Granholm, A.C., Backman, C., Bloom, F., Ebendal, T., Gerhardt, G.A., Hoffer, B., Mackerlova, L., Olson, L., Soderstrom, S., Walus, L.R. and Friden, P.M. (1994) NGF and anti-transferrin receptor antibody conjugate: short and long-term effects on survival of cholinergic neurons in intraocular septal transplants. *J. Pharmacol. Exp. Ther.*, 268: 448–459.

Gregoriadis, G., Senior, J., Wolff, B. and Kirby, M.C. (1985) Targeting of liposomes to accessible cells in vivo. *Ann. N.Y. Acad. Sci.*, 446: 319–340.

Habeck, M. (2000) Xenotransplanted neurons show potential to treat spinal injuries and Parkinson's disease. *Mol. Med. Today*, 6: 46–47.

Huwyler, J., Wu, D. and Pardridge, W.M. (1996) Brain drug delivery of small molecules using immunoliposomes. *Proc. Natl. Acad. Sci. USA*, 93: 14164–14169.

Inamura, T., Nomura, T., Bartus, R.T. and Black, K.L. (1994) Intracarotid infusion of RMP-7, a bradykinin analog: a method for selective drug delivery to brain tumors. *J. Neurosurg.*, 81: 752–758.

Jeffries, W.A., Brandon, M.R., Hunt, S.V., Williams, A.F., Gatter, K.C. and Mason, D.Y. (1984) Transferrin receptor on endothelium of brain capillaries. *Nature*, 312: 162–163.

Kang, Y.S. and Pardridge, W.M. (1994) Brain delivery of biotin bound to a conjugate of neutral avidin and cationized human albumin. *Pharmacol. Res.*, 11: 1257–1264.

Kordower, J.H., Charles, V., Bayer, R., Bartus, R.T., Putney, S., Walus, L.R. and Friden, P.M. (1994) Intravenous administration of a transferrin receptor antibody–nerve growth factor conjugate prevents degeneration of cholinergic striatal neurons in a model of Huntington's disease. *Proc. Natl. Acad. Sci. USA*, 91: 9077–9088.

Kordower, J.H., Freeman, T.B., Snow, B.J., Vingerhoets, F., Mufson, E.J., Sanberg, P.R., Hauser, R.A., Smith, D.A., Nauert, G.M., Perl, D.P. and Olanow, C.W. (1995a) Neuropathological evidence of graft survival and striatal reinnervation after the transplantation of fetal mesencephalic tissue in a patient with Parkinson's disease. *N. Engl. J. Med.*, 332: 1118–1124.

Kordower, J.H., Liu, Y.-T., Winn, S. and Emerich, D.F. (1995b) Encapsulated PC12 cell transplants into hemiparkinsonian monkeys: a behavioral, neuroanatomical and neurochemical analysis. *Cell Transplant.*, 4: 155–171.

Kordower, J.H., Chen, E.-Y., Winkler, C., Fricker, R., Charles, V., Messing, A., Mufson, E., Woing, S.C., Rosenstein, J.M., Björklund, A., Emerich, D., Hammang, J. and Carpenter, M.K. (1997) Grafts of EGF-responsive neural stem cells derived from GFAP-hNGF transgenic mice: trophic and tropic effects in a rodent model of Huntington's disease. *J. Comp. Neurol.*, 387: 96–113.

Kordower, J.H., Chen, E.-Y., Mufson, E.J., Winn, S.R. and Emerich, D.F. (1998) Intrastriatal implants of polymer encapsulated cells genetically modified to secrete human nerve growth factor: trophic effects upon cholinergic and noncholinergic striatal neurons. *Neuroscience*, 72: 63–77.

Kordower, J.H., Palfi, S., Chen, E.-Y., Ma, S., Sendera, T., Cochran, E.J., Mufson, E.J., Penn, R., Goetz, C. and Comella, C.D. (1999a) Clinico-pathological findings following intraventricular GDNF treatment in a patient with Parkinson's disease. *Ann. Neurol.*, 46: 419–424.

Kordower, J.H., Isacson, O. and Emerich, D.F. (1999b) Cellular delivery of trophic factors for the treatment of Huntington's disease: is neuroprotection possible?. *Exp. Neurol.*, 159: 4–20.

Kordower, J.H., Bloch, J., Chu, Y.P., Palfi, S., Roitberg, B., Emborg, M.E., Hantraye, P., Deglon, N. and Aebischer, P. (1999c) Lentiviral gene transfer to the non-human primate nigrostriatal system. *Exp. Neurol.*, 160: 1–16.

Kubo, O., Himuro, H., Inoue, N., Tajika, Y., Tajika, T., Tohyama, T., Sakairi, M., Yoshida, M., Kaetsu, I. and Kitamura, K. (1986) Treatment of malignant brain tumors with slowly releasing anticancer drug–polymer composites. *No Shinkei Geka*, 14: 1189–1195.

Kumagai, A.K., Eisenberg, J.B. and Pardridge, W.M. (1989) Absorptive-mediated endocytosis of cationized albumin and a beta-endorphin-cationized albumin chimeric peptide by isolated brain capillaries. Model system of blood–brain barrier transport. *J. Biol. Chem.*, 262: 15214–15219.

Lapchak, P.A. (1993) Nerve growth factor pharmacology: application to the treatment of cholinergic neurodegeneration in Alzheimer's disease. *Exp. Neurol.*, 124: 16–20.

Lapchak, P.A., Jiao, S., Miller, P.J., Williams, L.R., Cummins, V., Inouye, G., Matheson, C.R. and Yan, Q. (1996) Pharmacological characterization of glial cell line-derived neurotrophic factor (GDNF): implications for GDNF as a therapeutic molecule for treating neurodegenerative diseases. *Cell Tissue Res.*, 286: 179–189.

Lapchak, P.A., Jiao, S.S., Collins, F. and Miller, P.J. (1997) Glial cell line-derived neurotrophic factor: distribution and pharmacology in the rat following a bolus intraventricular injection. *Brain Res.*, 747: 92–102.

Luquin, M.R., Montoro, R.J., Guillen, J., Saldise, L., Insausti, R., Del Rio, J. and Lopez-Barneo, J. (1999) . *Neuron*, 22: 743–750.

Madrazo, I., Drucker-Colin, R., Diaz, V., Martinez-Mata, J., Torres, C. and Becerril, J.J. (1987) Open microsurgical autograft of adrenal medulla to the right caudate nucleus in two patients with intractable Parkinson's disease. *N. Engl. J. Med.*, 316: 831–834.

McRae-Degueurce, A., Hjorth, S., Dillon, D.L., Mason, D.W. and Tice, T.R. (1988) Implantable microencapsulated dopamine (DA): a new approach for slow-release DA delivery into brain tissue. *Neurosci. Lett.*, 92: 303–309.

Menei, P., Boisdron-Celle, M., Croue, A., Guy, G. and Benoit, J.P. (1996) Effect of stereotactic implantation of biodegradable 5-fluorouracil-loaded microspheres in healthy and C6 glioma-bearing rats. *Neurosurgery*, 39: 117–123.

Nakao, N., Itakura, T., Uematsu, Y. and Komai, N. (1995) Transplantation of cultured sympathetic ganglionic neurons into parkinsonian rat brain: survival and function of graft. *Acta Neurochir. (Wien)*, 133: 61–67.

Nutt, J.G. and Fellman, J.H. (1984) Pharmacokinetics of levodopa. *Clin. Neuropharmacol.*, 7: 35–49.

Oda, Y., Uchida, Y., Murata, T., Mori, K., Tokuriki, Y., Handa, H., Kobayashi, A., Hashi, K. and Kieler, J. (1982) Treatment of brain tumors with anticancer pellet — experimental and clinical study. *No Shinkei Geka*, 10: 375–381.

Oldendorf, W.H. (1971) Brain uptake of radiolabeled amino acids, amines and hexoses after arterial injection. *Am. J. Physiol.*, 221: 1629–1639.

Olson, L. (1993) NGF and the treatment of Alzheimer's disease. *Exp. Neurol.*, 124: 5–15.

Olson, L., Backman, L., Ebendal, T., Eriksdotter-Jonhagen, M., Hoffer, B., Humpel, C., Freedman, R., Giacobini, M., Meyerson, B. and Nordberg, A. (1994) Role of growth factors in degeneration and regeneration in the central nervous system: clinical experiences with NGF in Parkinson's and Alzheimer's diseases. *J. Neurol.*, 242: S12–S15.

Pardridge, W.M., Triguero, D. and Buciak, J. (1989) Transport of histone through the blood–brain barrier. *Pharmacol. Exp. Ther.*, 251: 821–826.

Pardridge, W.M., Kang, Y.S. and Buciak, J.L. (1994) Transport of human recombinant brain-derived neurotrophic factor (BDNF) through the rat blood–brain barrier in vivo using vector-mediated peptide drug delivery. *Pharmacol. Res.*, 11: 738–746.

Penn, R.D., Kroin, J.S., Reinkesmeyer, A. and Corcos, D. (1998) Injection of GABA-agonist into globus pallidus in patient with Parkinson's disease. *Lancet*, 351: 340–341.

Riachi, N.J. and Harik, S.I. (1992) Monoamine oxidases of the brains and livers of macaque and cercopithecus monkeys. *Exp. Neurol.*, 115: 212–217.

Robbins, P.D. and Ghivizzani, S.C. (1998) Viral vectors for gene therapy. *Pharmacol. Ther.*, 80: 35–47.

Sakane, T., Yamashita, S., Yata, N. and Sezaki, H. (1999) Transnasal delivery of 5-fluorouracil to the brain in the rat. *J. Drug Target*, 7: 233–240.

Sanovich, E., Bartus, R.T., Friden, P.M., Dean, R.L., Le, H.Q. and Brightman, M.W. (1995) Pathway across blood–brain barrier opened by the bradykinin agonist, RMP-7. *Brain Res.*, 705: 125–135.

Schneider, J.S., Roeltgen, D.P., Mancall, E.L., Chapas-Crilly, J., Rothblat, D.S. and Tatarian, G.T. (1998) Parkinson's disease: improved function with GM1 ganglioside treatment in a randomized placebo-controlled study. *Neurology*, 50: 1630–1636.

Schwarting, R.K.W. and Huston, J.P. (1996) The unilateral 6-hydroxydopamine lesion model in behavioral brain research. Analysis of functional deficits, recovery and treatments. *Prog. Neurobiol.*, 50: 275–331.

Sendelbeck, S.L. and Urquhart, J. (1985) Spatial distribution of dopamine, methotrexate, and antipyrine during continuous intracerebral microperfusion. *Brain Res.*, 328: 251–258.

Seventh International Meeting on Neural Transplantation and Repair (2000). *Exp. Neurol.*, 161: 397–431.

Sladek Jr., J.R. and Gash, D.M. (1988) Nerve cell grafting in Parkinson's disease. *J. Neurosurg.*, 68: 337–351.

Stromberg, I., Herrera-Marschitz, M., Ungerstedt, U., Ebendal, T. and Olson, L. (1985) Chronic implants of chromaffin tissue into the dopamine-denervated striatum. Effects of NGF on graft survival, fiber growth and rotational behavior. *Exp. Brain Res.*, 60: 335–349.

Svendsen, C.N. (1997) Neural stem cells for brain repair. *Alzheimer's Res.*, 3: 131–135.

Triguero, D.J., Buciak, B., Yang, J. and Pardridge, W.M. (1989) Blood–brain-barrier transport of cationized immunoglobulin

G. Enhanced delivery compared to native protein. *Proc. Natl. Acad. Sci. USA*, 86: 4761–4765.

Tsuzuki, N., Hama, T., Hibi, T., Konishi, R., Futaki, S. and Kitagawa, K. (1991) Adamantane as a brain-directed drug carrier for poorly absorbed drug: anti-nociceptive effects of [D-Ala2] leu-enkephalin derivatives conjugated with 1-adamantane moiety. *Biochem. Pharmacol.*, 41: R5–R8.

Verma, I.M. and Somia, N. (1997) Gene therapy — promises, problems and prospects. *Nature*, 389: 239–242.

Wim, T.J., Hermens, M.C. and Verhaagen, J. (1998) Viral vectors, tools for gene transfer in the nervous system. *Prog. Neurobiol.*, 55: 399–432.

Yurek, D.M. and Sladek Jr., J.R. (1990) Dopamine cell replacement: Parkinson's disease. *Annu. Rev. Neurosci.*, 13: 415–440.

Zawada, W.M., Cibelli, J.B., Choi, P.K., Clarkson, E.D., Golueke, P.J., Witta, S.E., Bell, K.P., Kane, J., Ponce de Leon, F.A., Jerry, D.J., Robl, J.M., Freed, C.R. and Stice, S.L. (1998) Somatic cell cloned transgenic bovine neurons for transplantation in parkinsonian rats. *Nat. Med.*, 4: 569–574.

Zigmond, M.J., Abercrombie, E.D., Berger, T.W., Grace, A.A. and Stricker, E.M. (1990) Compensations after lesions of central dopaminergic neurons: some clinical and basic implications. *Trends Neurosci.*, 13: 290–296.

CHAPTER 29

Neurotrophin small-molecule mimetics

Youmei Xie and Frank M. Longo *

Department of Neurology, VA Medical Center and University of California San Francisco, 4150 Clement Street, San Francisco, CA 94121, USA

Introduction

Neurotrophins have highly potent biological effects preventing neuronal death and promoting neurite outgrowth; however, like most proteins they do not have optimal pharmacological properties. Factors limiting their clinical application include stability, nervous system penetration and their wide array of local and systemic biological activities. An important approach for addressing these limitations is the development of synthetic, small-molecule neurotrophin mimetics with optimal profiles of stability, tissue penetration and targeted biological actions. Neurotrophin mimetic strategies include the following: development of (1) agents that induce endogenous synthesis and secretion of neurotrophins; (2) agents that act directly at neurotrophin receptors as agonists, partial agonists or antagonists; and (3) agents that augment neurotrophin-induced signal transduction. Compounds targeted to specific neurotrophin receptors have the potential to mimic the entire range of functions or a subset of functions of a given neurotrophin. For example, prevention of neuronal death in the absence of stimulating neurite outgrowth might constitute a desired activity profile in certain applications. The identification of specific neurotrophin protein domains likely to modulate receptor interaction has guided synthesis of neurotrophin small-molecule peptidomimetics corresponding to individual domains and functioning via selected receptors to trigger neurotrophin-like signal transduction. Synthesis of domain-specific mimetics has also provided a key proof-of-concept that it may be possible to design neurotrophin antagonists that would inhibit neurotrophin actions in the contexts of neurotrophin-induced cell death, aberrant sprouting, etc. Current neurotrophin small-molecule studies provide a basis and proof-of-concept to guide programs of rational drug design and large-scale screening for compounds with medicinal properties and targeted neurotrophin activities.

Potential roles of neurotrophins in spinal cord injury

Neurotrophins are proteins secreted by neuronal targets or neurons themselves that regulate a broad spectrum of neuronal function and response to injury. The fundamental nature of these processes and their relevance to neurodegenerative disease, neural injury and neural regeneration has encouraged pharmacologic development and therapeutic application of neurotrophins (Ibanez, 1995; Hefti, 1997; McInnes and Sykes, 1997; Connor and Dragunow, 1998; Skaper and Walsh, 1998; Hughes and O'Leary, 1999). In the context of spinal cord injury, there are multiple fundamental mechanisms by which neurotrophins might contribute to functional recovery. Neurotrophins are likely to regulate cell atrophy and death, neurite outgrowth, remyelination, collateral sprouting, synaptogenesis, maintenance of orig-

---

* Corresponding author: Dr. Frank M. Longo, VAMC/UCSF V-127, 4150 Clement Street, San Francisco, CA 94121, USA. Fax: +1-415-750-2273; E-mail: LFM@itsa.UCSF.edu

inal and regenerated fiber networks, neurotransmitter and other biochemical differentiation and training-induced plasticity. Other processes regulated by neurotrophins that are particularly relevant to spinal cord injury and recovery include inflammation and the pain-related mechanisms of afferent and sympathetic fiber sprouting and neurotransmitter function. In recent years, a number of groups have applied neurotrophins to animal models of spinal cord injury via catheter infusion or secretion by transplanted cells. While these methods of application have a number of disadvantages for routine clinical use, they have been instrumental in demonstrating that neurotrophins can contribute to cell survival and to regeneration of ascending and descending fibers (Diener and Bregman, 1994; Tetzlaff et al., 1994; Grill et al., 1997; Kobayashi et al., 1997; Shibayama et al., 1998; Stichel and Muller, 1998; Liu et al., 1999; Oudega and Hagg, 1999).

Neurotrophins and their receptors

The development of neurotrophin small-molecule mimetics is dependent upon the known mechanisms of action of the neurotrophins. Neurotrophins and their receptors have been the subject of a number of recent reviews (Greene and Kaplan, 1995; Segal and Greenberg, 1996; Kaplan and Miller, 1997; Bredesen et al., 1998; Chao et al., 1998; Frade and Barde, 1998; Friedman and Greene, 1999; Yuen and Mobley, 1999). The mammalian neurotrophin family consists of nerve growth factor (NGF), brain-derived neurotrophic factor (BDNF), neurotrophin-3 (NT-3) and neurotrophin-4/5 (NT-4/5). Neurotrophin monomeric proteins contain approximately 120 amino acids each and form non-covalent homodimers that constitute the bioactive form. Neurotrophins bind with high affinity (kD $\approx 10^{-11}$ M) to Trk tyrosine kinase receptors and with lower affinity (kD $\approx 10^{-9}$ M) to the p75 receptor (p75^{NTR}), a member of the tumor necrosis factor (TNF) receptor superfamily. Binding to Trk receptors by neurotrophin family members is relatively selective: NGF with TrkA; BDNF and NT-4/5 with TrkB and NT-3 primarily with TrkC. Interaction of neurotrophins with Trk receptors induces both conformational changes and dimerization which lead to autophosphorylation of specific Trk tyrosine residues and internalization of ligand–receptor complexes. Signaling components activated by neurotrophin/Trk interactions include: members of the Ras/Raf-extracellular receptor-activated kinase (ERK) pathway; phospholipase Cγ-1 and phosphatidylinositol-3 (PI-3) kinase. Ras/ERK and phospholipase Cγ-1 signaling are thought to contribute to neuronal differentiation and PI-3 kinase signaling has been proposed to primarily modulate neuronal survival; however, these distinctions remain to be further established.

Mechanisms of p75^{NTR} function have also been recently reviewed (Barker, 1998; Bredesen et al., 1998; Chao et al., 1998; Frade and Barde, 1998). Depending on the cellular context including cell type, developmental stage and ratio of p75^{NTR} to Trk receptors, neurotrophin interaction with p75^{NTR} can either prevent or promote cell death. The interaction of neurotrophins with p75^{NTR} can modulate neurotrophin binding and affinity to Trk receptors and can also positively or negatively regulate signaling via Trk receptors (Barker and Shooter, 1994; Maliartchouk and Saragovi, 1997; Ross et al., 1998). In cells not expressing Trk receptors, it has been established that neurotrophin binding to p75^{NTR} can directly regulate cell survival via modulation of p75^{NTR} signaling independently of Trk receptors (Casaccia-Bonnefil et al., 1996; Frade et al., 1996). In cells expressing Trk, it is also likely that ligand interaction with p75^{NTR} can modulate intracellular signaling independently of direct effects on Trk signaling (Rabizadeh et al., 1993; Barrett and Bartlett, 1994; Taglialatela et al., 1996; Bredesen et al., 1998). As described below, these potential Trk-independent modes of p75^{NTR} signaling are especially relevant to elucidating the mechanisms of action of neurotrophin small-molecule mimetics acting in a p75^{NTR}-dependent, Trk-independent manner. Signaling intermediates potentially regulated by neurotrophin–p75^{NTR} interactions include ceramide and the AKT and JNK kinases.

In considering the mechanism of action of small-molecule mimetics acting via neurotrophin receptors, it is important to point out that inhibition and mutational analyses of the known signaling components regulated by NGF-induced Trk and p75^{NTR} signaling indicate that additional, currently unidentified, components also play a role in neurotrophin signaling. The establishment of more complete neurotrophin signaling intermediate networks will facilitate analysis of signaling induced by neurotrophin small-molecule mimetics.

Domains of NGF interacting with its receptors

A neurotrophin agonist or antagonist would be expected to contain structural determinants of a neurotrophin's active sites that interact with its receptors. The overall structure of NGF is reviewed in Fig. 1. In the case of NGF several approaches have been used to deduce which protein domains interact with NGF receptors (Bradshaw et al., 1994; McDonald and Chao, 1995). A peptide mapping strategy in which synthetic peptides with sequences corresponding to specific NGF regions were tested for their ability to inhibit NGF activity indicated that residues 29–35 formed a key active site (Longo et al., 1990). Subse-

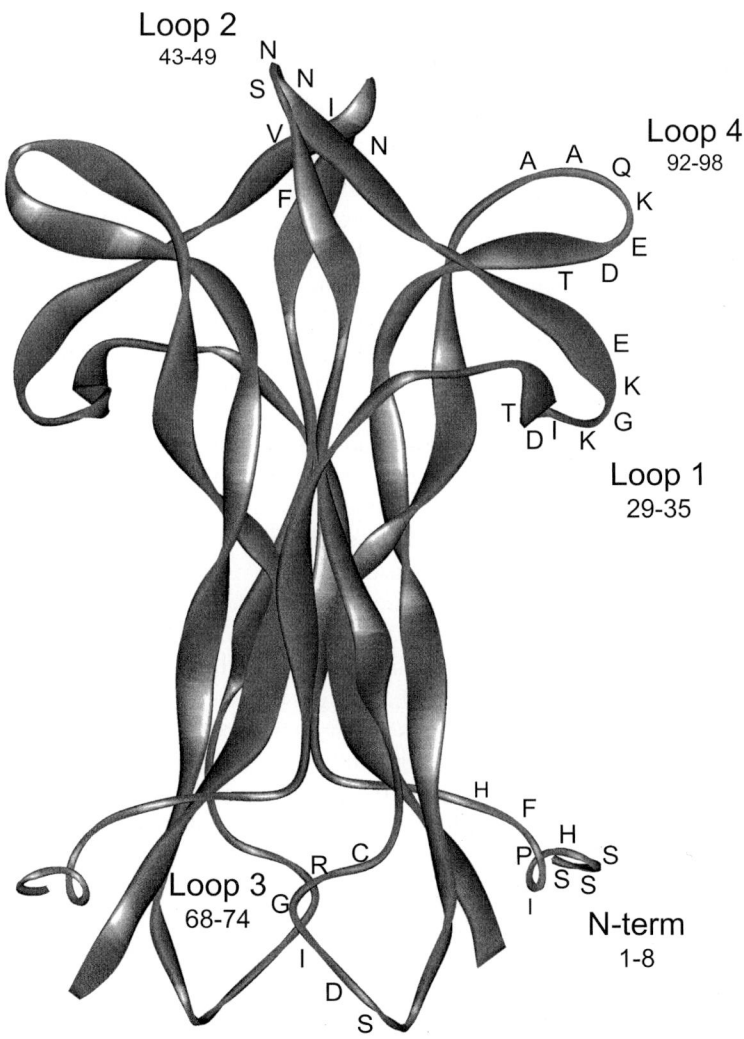

Fig. 1. Structure of the nerve growth factor homodimer. This illustration is derived from coordinates obtained from the crystal structural of the human NGF dimer in complex with recombinant TrkA (Wiesmann et al., 1999). Coordinates are available in the Protein Data Bank. Amino acids are shown using single-letter nomenclature and residue number is based on numbering commonly used for mouse NGF (McDonald and Chao, 1995). The N-terminal along with loops 2 and 4 mediate TrkA binding. Loop 1, along with a small number of residues in other regions, mediate p75[NTR] binding and loop 3 is important for NGF homodimer formation.

quent NGF crystallography and molecular modeling studies revealed that NGF contained three surface hydrophilic β-hairpin loops (loops 1, 2 and 4) that were likely candidates for receptor interaction sites (McDonald et al., 1991; Holland et al., 1994). Loop 1 consists of residues 29–35 and subsequent studies have confirmed that region 29–35 synthetic peptides inhibit NGF activity and NGF–p75^{NTR} receptor binding (LeSauteur et al., 1995; Van der Zee et al., 1996). Recombinant substitution studies indicate that residues Lys^{32} and Lys^{34} are likely to interact with p75^{NTR} receptors (Ibanez et al., 1992). Molecular modeling studies of NGF and p75^{NTR} interaction also point to loop 1 residues as critical receptor interacting sites (Shamovsky et al., 1999). The NGF sites interacting with TrkA have also been derived via recombinant protein, chemical modification and NGF–TrkA co-crystal approaches. TrkA binding sites primarily consist of residues in NGF loop 2 (residues 40–49), loop 4 (residues 91–97) and the N-terminus (residues 1–8) (McDonald and Chao, 1995; Woo and Neet, 1996; Kullander et al., 1997; Woo et al., 1998; Wiesmann et al., 1999).

Rationale for developing small-molecule mimetics of neurotrophins

Factors limiting therapeutic applications of the neurotrophin proteins in the settings of acute and long-term spinal cord injury include restricted penetration of the CNS and the poor medicinal properties characteristic of most proteins (Poduslo and Curran, 1996; Frey et al., 1997; Saltzman et al., 1999). In the clinical areas of thrombocytopenia, granulocytopenia and diabetes, limitations in the therapeutic applications of thrombopoietin (TPO), granulocyte-colony-stimulating factor (G-CSF) and insulin proteins have spurred programs that have successfully identified synthetic small molecules inducing dimerization and activation of their respective receptors (Cwirla et al., 1997; Tian et al., 1998; Zhang et al., 1999). The development of small-molecule mimetics with favorable chemical properties that function as agonists or antagonists that mimic or inhibit neurotrophin functions in the appropriate biological context will likewise be critical in the therapeutic application of neurotrophin effects. Moreover, in settings or post-injury stages in which neurotrophins might contribute to cell death (Casaccia-Bonnefil et al., 1996; Frade et al., 1996), inflammation (Levi-Montalcini et al., 1996), pain-related mechanisms of sympathetic (Nauta et al., 1999) or afferent (Krenz et al., 1999) fiber sprouting or upregulation of pain-mediating neurotransmitters (Malcangio et al., 1997), neurotrophin antagonists, rather than agonists, may be particularly relevant.

Another potential therapeutic advance derived from the development of small-molecule mimetics stems from the principle that differences in ligand–receptor interactions between different ligands for a given receptor can result in changes in the overall pattern of receptor signal transduction. Differences in pattern, duration or intensity of post-receptor signaling can differentially influence the biological endpoints including survival versus mitogenesis versus neurite outgrowth (Qiu and Green, 1992; Greene and Kaplan, 1995; Marshall, 1995). For example, comparison of signaling profiles and promotion of survival versus neurite outgrowth between NGF and NGF in association with an anti-NGF monoclonal antibody demonstrates differential signaling and activity patterns (Saragovi et al., 1998). These findings raise the possibility that alterations in TrkA ligand interactions lead to the formation of differentially oriented TrkA dimers resulting in distinct signaling profiles. Mutation of the Shc site on TrkB receptors causes a decreased survival response for NT-4 with a relatively spared response for BDNF (Minichiello et al., 1998). This study also suggests that NT-4 and BDNF might promote differentially oriented TrkB dimers that trigger different signaling profiles. These findings point to the possibility that differences in the orientation or conformational changes of p75^{NTR} or Trk receptor multimerization induced by small-molecule mimetics versus those induced by native ligands might create opportunities for stimulating selected patterns of biological effects. Creation of single-domain neurotrophin mimetics will also constitute a powerful approach for linking specific neurotrophin domains with specific patterns of intracellular signal transduction.

The concept of differential receptor activation by native ligands versus small-molecule mimetics can also be applied across dual-receptor systems. Neurotrophin mimetics preferentially acting via p75^{NTR} versus Trk receptors might be used to promote differential neurotrophin signaling and activities. This possibility is supported by the finding that at similar levels of TrkA activation, NT-3 promotes neuronal survival two- to threefold less well than NGF and that NT-3 binds to $p75^{NTR}$ with greater affinity compared to NGF (Belliveau et al., 1997). As discussed below, studies in our laboratory suggest that NGF mimetics acting preferentially via $p75^{NTR}$ or preferentially via TrkA receptors demonstrate differential effects on survival versus neurite outgrowth.

Neurotrophin mimetics: definition of terms

The term 'mimetic' is widely used to describe a number of strategies for promoting the effects of neurotrophin proteins. A pharmacological and precise application of the term 'mimetic' would describe compounds that share structural features of a given neurotrophin protein that allow them to interact with neurotrophin receptors via the same receptor site used by neurotrophins. Members of this group of mimetics might function as either receptor antagonists or agonists. Another group of potential 'mimetic' compounds consists of agents that may or may not share structural features of neurotrophins that nevertheless act at neurotrophin receptors in a noncompetitive (i.e., at receptor sites distinct from neurotrophin-interacting sites) manner to up- or downregulate receptor activity. Other strategies for 'mimicking' the effects of neurotrophins can include increasing levels of endogenous neurotrophins or augmenting the effects of endogenous or administered neurotrophins. Each of these mimetics strategies is summarized in Table 1 and discussed below.

The terms 'agonist', 'antagonist' and 'partial agonist' are additional pharmacological terms describing fundamental ligand–receptor interaction mod-

TABLE 1

Categories of neurotrophin small-molecule mimetics

Category	Compound	Reference
Neurotrophin antagonists	NGF loop-1 peptidomimetics	Longo et al., 1990
		LeSauteur et al., 1995
		Van der Zee et al., 1996
	NGF loop-4 peptidomimetics	LeSauteur et al., 1995
	BDNF loop-2 peptidomimetics	O'Leary and Hughes, 1998
Neurotrophin agonists (compete with NGF)	NGF loop-1 peptidomimetics	Longo et al., 1997
	NGF loop-4 peptidomimetics	Xie and Longo, 1997
		Xie et al., 2000
Neurotrophin mimetics interacting with neurotrophin receptors (additive to NGF)	NGF β-turn peptidomimetic	Maliartchouk et al., 2000
Other compounds interacting with neurotrophin receptors	K252 derivatives	Knusel and Hefti, 1992
		Angeles et al., 1998
	ALE-0540	Owolabi et al., 1999
Compounds interacting with neurotrophins	PD 90780	Spiegel et al., 1995
	kynurenic acid derivatives	Jaen et al., 1995
	NGF loop-3 mimetic	Rashid et al., 1995
	zinc	Ross et al., 1997
Compounds augmenting neurotrophin effects	AIT-082	Middlemiss et al., 1995
	SR57746A	Pradines et al., 1995
Compounds inducing neurotrophin synthesis/secretion	propentofylline	Shinoda et al., 1990
	catechol derivatives	Takeuchi et al., 1990
		Carswell et al., 1992
	sesquiterpenes	Kawagishi et al., 1997
	idebenone	Nitta et al., 1994
	purine transport inhibitors	Yamada et al., 1997
	AIT-082	Glasky et al., 1997
	scabronines A and G	Obara et al., 1999
	vit. D3 derivative CB1093	Riaz et al., 1999

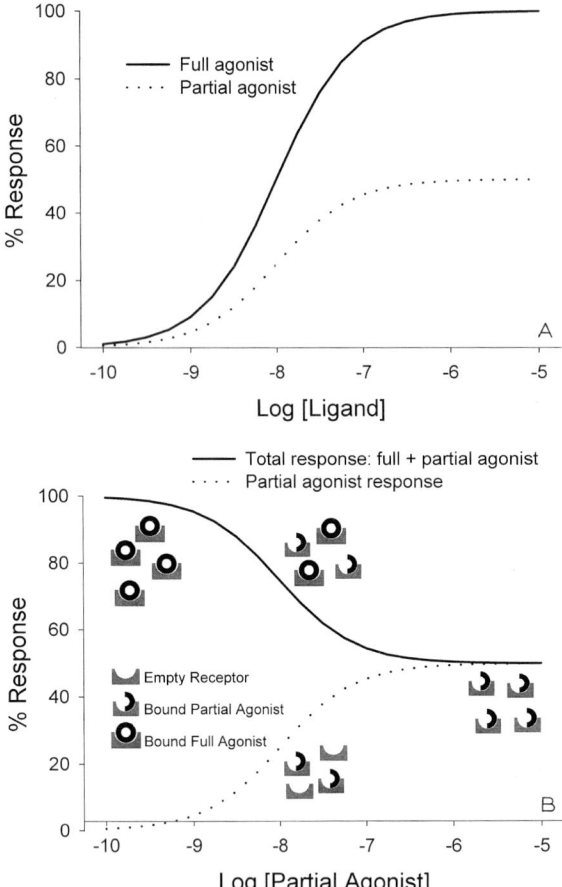

Fig. 2. Profiles of full and partial agonists. (A) A partial agonist functions at a maximum efficacy that is less than the maximum efficacy of the full agonist or native ligand. (B) In the presence of a fixed concentration of a full agonist or native ligand, increasing concentrations of a partial agonist leads to partial inhibition. Thus a partial agonist also functions as a partial antagonist in the presence of the native ligand.

ner similar to that of the native ligand and thereby demonstrates competitive binding but demonstrates no ligand-like activity. In the presence of the native ligand, a partial agonist would be expected to compete with native ligand binding and result in a decrease in efficacy (Fig. 2B). In this context, the partial agonist functions as a 'partial antagonist'.

Peptidomimetics: an important intermediate-stage tool in drug development

Traditional approaches based on random screening of chemical compound libraries for receptor binding activity have been relatively unproductive in identifying small-molecule growth factor agonists. The discovery of synthetic peptides mimicking specific domains of native protein ligands (peptidomimetics) that function as antagonists or agonists is emerging as a powerful intermediate-stage strategy in small-molecule drug design (Longo and Mobley, 1996; Cunningham and Wells, 1997; Kieber-Emmons et al., 1997; McInnes and Sykes, 1997; Hughes and O'Leary, 1999). Small synthetic peptides have been identified that mimic the ability of protein growth factor ligands to induce dimerization, signal transduction and biological activity characteristic of the native ligands. Examples include peptides activating the receptors for thrombopoietin (TPO) and erythropoietin (EPO). For both TPO and EPO peptidomimetics, the ability of the mimetics to activate receptors is highly dependent on the dimeric state (Cwirla et al., 1997; Wrighton et al., 1997). The contribution of mimetic dimerization for achieving agonist activity was also found during development of NGF mimetics (described below). Structure studies of co-crystals formed with EPO dimeric peptidomimetics and the EPO receptor extracellular domain are contributing to defining the small-molecule structural features required for EPO receptor binding (Livnah et al., 1999; Middleton et al., 1999). These EPO peptidomimetics have served as starting points for the design of non-peptide EPO small-molecule compounds.

Peptidomimetic-based development of non-peptide small-molecule agonists mimicking protein ligands has been described (Freidinger, 1999). Peptidomimetics of the RGD domain of fibrinogen that mediate the binding of fibrinogen to the platelet gly-

els (Bourne, 1998) that are critical for addressing and understanding neurotrophin mimetic strategies. These models are illustrated in Fig. 2. A small-molecule mimetic functioning as an agonist interacts with the receptor in a manner similar to that of the native ligand such that the agonist and native ligand, if present concomitantly, would compete for binding sites. A 'full agonist' has 100% of the efficacy of the native ligand. A 'partial agonist' also demonstrates competitive binding but has less than 100% efficacy (Fig. 2A). A small-molecule functioning as an 'antagonist' also interacts with the receptor in a man-

coprotein IIb/IIIa complex were used to guide the creation of a non-peptide small-molecule mimetic (Nicholson et al., 1991). Replacement of the RGD residues with guanido-octanoate and phenethylamine moieties as part of a rational design strategy led to synthesis of the SC-47643 mimetic. SC-47643 has been shown to inhibit platelet aggregation in preclinical studies (Nicholson et al., 1991). Using similar strategies, it is possible that the neurotrophin peptidomimetics described below will lead to non-peptide medicinal compounds functioning as neurotrophin antagonists or agonists in vivo.

There are other approaches by which peptidomimetics of protein ligands can contribute to the identification of non-peptide mimetics. The proof-of-concept, provided by peptidomimetic agonists corresponding to specific protein domains and functioning as agonists, can encourage and guide chemical library or phage-display screening programs. For example, the finding that a dimerized structure is required would suggest a focus on bivalent compounds. Structural coordinates of conformationally constrained peptidomimetics, particularly those in a receptor-bound state, can be used to screen structural databases of small-molecule compounds. In addition, the signal transduction patterns and biological activity profiles stimulated by peptidomimetics can provide a more efficient framework for evaluating 'hits' derived from small-molecule screens.

Neurotrophin antagonists

As part of a peptide mapping strategy to identify NGF-active sites, linear, monomeric peptides corresponding to specific NGF regions were tested for their ability to block NGF action (Longo et al., 1990). Peptides corresponding to NGF residues 29–35 were found to inhibit NGF, but not other survival-promoting factors, in a sequence-dependent manner. Structure–activity peptide studies narrowed down the key residues required for inhibitory activity to Lys^{32}, Gly^{33}, Lys^{34}, Glu^{35}. Although functioning at low potencies, these peptidomimetics constituted the first NGF antagonists. Subsequent NGF crystallization studies revealed that residues Lys^{32}, Gly^{33}, Lys^{34}, Glu^{35} constitute the core of NGF loop 1 (McDonald et al., 1991). Cyclization of linear peptides is a standard approach for applying conformational constraint and inducing structure more closely matching those of β-loops present in native proteins (Hruby et al., 1990). As predicted by this conformational constraint strategy, subsequent studies of monomeric cyclized loop-1 NGF peptides have led to the development of peptidomimetics with more potent NGF-inhibiting activity compared to that of linear peptides (LeSauteur et al., 1995; Van der Zee et al., 1996). These monomeric cyclized loop-1 peptides demonstrate no NGF-like activity and thus constitute NGF antagonists rather than full or partial agonists.

Cyclized monomeric peptidomimetics corresponding to NGF loop 4 were also found to inhibit NGF bioactivity and [^{125}I]NGF binding (LeSauteur et al., 1995). NMR studies of these loop-4 peptides demonstrate a striking similarity between calculated peptidomimetic structures and the structure of the corresponding loop derived from NGF X-ray studies (Beglova et al., 1998). Administration of the loop-4 cyclized monomer NGF antagonist (C92–96) via cannulas implanted within rat cortex led to an inhibition of NGF-induced upregulation of choline acetyltransferase (ChAT) activity and a decrease in the size of cholinergic boutons (Debeir et al., 1999). This study demonstrates that NGF small-molecule antagonists can be used to block the function of NGF in vivo. It will be of interest to determine whether this or more potent NGF antagonists can block NGF-induced cell death, NGF-regulated inflammatory process or pain-mediating afferent or sympathetic fiber sprouting in the context of neural injury.

Peptidomimetic antagonists corresponding to BDNF loop 2 have also been characterized (O'Leary and Hughes, 1998). Cyclized monomers were identified that inhibit the bioactivity of BDNF while having no effect on NGF. Inhibitory activity is dependent on cyclization and varies with systematic substitution of individual residues with alanine. These BDNF mimetics do not show survival-promoting activity in the absence of BDNF. Thus, similar to the NGF loop-1 and loop-4 cyclized monomeric peptidomimetics described above, these BDNF loop-2 mimetics constitute BDNF antagonists rather than full or partial agonists. Based on molecular modeling studies, these investigators suggest that the structural homology between BDNF loop 2 and these loop-2 peptidomimetics will contribute to the design of dimeric peptidomimetics that may act as BDNF agonists.

Neurotrophin loop-1 agonists

Synthesis of a NGF loop-1 peptidomimetic in a cyclized dimeric rather than monomeric form led to the discovery that a loop-1 dimeric peptidomimetic exhibits the NGF activity of preventing neuronal death of cultured dorsal root ganglion (DRG) sensory neurons (Longo et al., 1997). This loop-1 peptidomimetic was termed 'P7' and constitutes the first NGF small molecule corresponding to a specific domain of NGF and demonstrating agonist activity. The maximum survival-promoting effect of P7 (efficacy) is well under that of NGF, suggesting the profile of a partial agonist. As expected of a NGF partial agonist, addition of P7 to a fixed dose of NGF demonstrates a dose-dependent partial inhibition of NGF activity.

Analogs of P7 containing changes in amino acid sequence, or synthesized in the monomeric or linear forms, have no agonist activity. This dimeric requirement for P7 activity resembles that found for TPO and EPO peptidomimetics. As might be expected of a NGF loop-1 mimetic, agonist activity of P7 is dependent on $p75^{NTR}$ but not Trk receptor function. Activity is lost with the addition of antibody directed against the $p75^{NTR}$ or in assays conducted with $p75^{NTR}$ −/− versus +/+ neurons. Activity of NGF, but not P7, can be inhibited by the addition of the Trk inhibitor, K252a. Given the role of $p75^{NTR}$ in regulating neuronal survival, it is of particular interest to note that P7 primarily supports neuronal survival and relatively minimal neurite outgrowth. Further structure–function studies, signal transduction profiling and structural comparisons of NGF–$p75^{NTR}$ and P7–$p75^{NTR}$ complexes will be required to further elucidate receptor modulation mechanisms of P7. Completed and ongoing structure–function studies of P7 will also guide screening strategies of phage display libraries and available chemical directories in the search for more potent and stable NGF mimetics acting via the $p75^{NTR}$ receptor.

Neurotrophin loop-4 agonists

Using a similar approach for that developed for loop-1 dimeric peptidomimetics, we tested the hypothesis that similarly designed loop-4 dimeric peptides might function as NGF agonists via TrkA receptors. A dimeric peptidomimetic termed 'P92' was found to promote survival and neurite outgrowth of cultured DRG sensory neurons (Xie and Longo, 1997; Xie et al., 2000). As in the case of the P7 peptidomimetic, the maximum survival-promoting activity of P92 is well under that of NGF, suggesting a partial agonist profile. Eight lines of evidence support the hypothesis that P92 acts via TrkA receptors. (1) P92 corresponds to NGF loop 4, one of the two primary loop domains interacting with TrkA. Deviations from loop-4 sequence eliminated activity. (2) P92 activity is highly structure-dependent. (3) As expected of a NGF partial agonist, P92 partially inhibits NGF bioactivity in a structure-dependent manner. Moreover, P92 inhibits NGF over the same concentration range that a cyclized peptide containing NGF residues 92–96 was shown to block [^{125}I]NGF binding (LeSauteur et al., 1995). (4) The biological profile elicited by P92 of both preventing neuronal death and stimulating neurite outgrowth parallels that triggered by NGF interaction with TrkA. (5) The absolute requirement of dimerization for P92 activity is consistent with models in which NGF induces receptor dimerization. (6) P92 activity is blocked by the TrkA inhibitors, K252a and AG879. While inhibitors entirely specific for TrkA are not available, the largely non-overlapping tyrosine kinase target profiles of these Trk inhibitors suggests that P92 activity is TrkA-dependent. (7) P92 stimulates ERK activation over a time course that parallels that of NGF. As found with NGF, P92-induced ERK activation was blocked by K252a. (8) The neurite outgrowth activity, but not the survival-promoting activity, of both NGF and P92 was blocked by the ERK inhibitor, U0126. This effect of ERK inhibition is consistent with the role of ERK in mediating ligand–receptor-induced neurite outgrowth (Perron and Bixby, 1999).

The partial agonist nature of P92 activity is of particular interest. In general, partial agonism can result from either 'incorrect' fit between the partial agonist and receptor or 'correct' but incomplete receptor binding (Pliska, 1999). NGF interaction with TrkA induces both dimerization and conformational changes of TrkA (Woo et al., 1998); therefore, absence of full agonist activity by a NGF small-molecule mimetic could result from a lack of either of these binding properties. Parallel NGF/TrkA and P92/TrkA co-crystallization and physical–chemical studies will play an important role in elucidating binding mechanisms and the iterative process of designing P92 derivatives more closely mimicking NGF–TrkA binding. The NGF two-receptor system points to other potential mechanisms contributing to the partial agonist profile of P92. Several studies suggest that unbound p75 negatively regulates and bound p75 positively regulates TrkA function (Barker and Shooter, 1994; Maliartchouk and Saragovi, 1997; Ross et al., 1998). Thus, the absence of p75 ligand interaction by a TrkA agonist might result in reduced TrkA signaling. A similar mechanism was suggested by the finding that TrkA-directed antibodies achieve only partial NGF responses (Clary et al., 1994; Maliartchouk and Saragovi, 1997).

Another important feature of the partial agonist profile is the well established precedent that the partial agonist nature of some compounds makes possible their application in vivo as partial antagonists while maintaining low levels of receptor stimulation. For example, drugs such as clonidine, oxymetazoline and tamoxafin are partial agonists with efficacies well under that of their full agonist counterparts (Lipworth and Grove, 1997; Kenakin, 1999; Norris et al., 1999). In pathological states such as neural regeneration, NGF agonists might be of benefit. However, given the important functions of endogenous NGF, it is possible that downregulation of the contribution of NGF to states of chronic pain or inflammation (Levi-Montalcini et al., 1996; Levine, 1998; Shu and Mendell, 1999) will require NGF partial agonists rather than pure antagonists. The proof-of-concept that NGF loop-4 small-molecule mimetics can function as partial agonists points to important novel and physiologically relevant signal transduction and biological activity endpoints that should be included in the screening and design of NGF small-molecule mimetics.

The distinct profiles in receptor dependency ($p75^{NTR}$ versus Trk) and biological profiles between the NGF loop-1 P7 mimetic (primarily survival-promoting) and the loop-4 P92 mimetic (survival- and neurite-promoting) constitute a first reduction-to-practice of the theoretical possibility that neurotrophin small molecules mimicking different domains might act via differential mechanisms to trigger differential responses. This distinction in biological activity profile is consistent with studies demonstrating that TrkA regulates both survival and neurite outgrowth while the $p75^{NTR}$ primarily regulates cell death. These small-molecule differential effects indicate that it may be possible to design small-molecule NGF mimetics acting via partly non-overlapping signaling networks to promote (or prevent) different sets of biological endpoints.

In another study, a focused β-turn peptidomimetic library was synthesized and screened for the property of binding to TrkA (Maliartchouk et al., 2000). The mimetic D3 was found to bind to cells expressing TrkA but not to non-TrkA-bearing cells. As found with P92, D3 promoted survival of DRG neurons with an efficacy less than that of NGF. In contrast to P92, D3 did not inhibit NGF as would be expected if D3 were acting as an NGF partial agonist but instead demonstrated an additive effect. This additive profile along with the finding that D3 promoted TrkA : TrkA homodimer formation points to the possibility that D3 acts at a site on the TrkA receptor that is distinct from that of the NGF-interacting site to stabilize the signaling conformation of preformed TrkA homodimers.

Other small molecules modulating Trk or $p75^{NTR}$ signaling

A category of small molecules acting via Trk and non-Trk signaling mechanisms to modulate neurotrophic function consists of the indolocarbazole alkaloid, K252a, and its derivatives. K252a has been shown to inhibit Trk autophosphorylation via competition with ATP at the Trk kinase domain (Angeles et al., 1998). This Trk inhibitory activity of K252a makes it a useful tool for blocking NGF-induced Trk activation. Interestingly, K252a, in the absence of NGF, promotes survival of dorsal root ganglion sensory neurons over a concentration range (100–800 nM) slightly higher than concentrations (100–200 nM) typically used to inhibit Trk activation (Borasio, 1990). The closely related alkaloid, K252b, at concentrations lower than those inhibiting TrkA, has been shown to potentiate the stimulatory effect of NT-3 on cholinergic cell differentiation and TrkA phosphorylation in PC12 cells (Knusel and Hefti, 1992; Knusel et al., 1992). Whether the survival-promoting target of K252a and K252b is TrkA or another signaling component is unknown. CEP-

1347 (KT7515), a 3,9-bis[ethyl(thio)methyl]K252a derivative, has been found to rescue motoneurons from apoptotic death in vitro, presumably via its ability to block activation of death-inducing JNK1 activation rather than via activation of a MAP kinase pathway (Maroney et al., 1998). CEP-1347 has also been found to support the survival of embryonic sensory, sympathetic and parasympathetic neurons (Borasio et al., 1998). The direct intracellular target(s) of CEP-1347 mediating these effects remains to be identified.

Screening of non-peptide small molecules for the property of inhibiting binding of NGF to PC12 cells led to the identification of the quinoline derivative, ALE-0540 (Owolabi et al., 1999). ALE-0540 inhibits binding of NGF to TrkA and p75^{NTR} as well as NGF-induced TrkA signal transduction. Intraperitoneal or spinal intrathecal administration of ALE-0540 to rats subjected to nerve ligation injury led to a dose-dependent decrease in tactile allodynia. While preliminary experiments demonstrated that ALE-0540 did not bind to NGF itself, additional studies will be required to determine whether it functions as a competitive or noncompetitive antagonist of NGF binding.

Another strategy for modulating NGF interaction with its receptors consists of administering compounds that interact directly with NGF to alter its structure and thereby modulate its receptor binding. Ross et al. (1997) found that zinc binds to NGF to change its conformation, thus rendering it unable to bind to p75^{NTR} and TrkA or to activate signal transduction. In another study, a depsibicyclic peptide corresponding to two noncontiguous elements of the NGF loop-3 domain was found to prevent NGF binding to both p75^{NTR} and TrkA and to block NGF-mediated neurite outgrowth in vivo (Rashid et al., 1995). This peptide is thought to inhibit NGF binding by altering NGF dimer confirmation, a process consistent with a loop-3 role in NGF homodimer formation. Screening of the Parke–Davis chemical bank for compounds that block binding of NGF to p75^{NTR} led to the identification of PD 90780 (Spiegel et al., 1995). PD 90780 appears to function by binding to NGF itself rather than via p75^{NTR} binding. Preliminary molecular docking studies suggest that PD 90780 binds to NGF loop 1, the domain primarily interacting with p75^{NTR}. Additional studies at Parke–Davis have identified kynurenic acid derivatives with structural similarities to PD 90780 that also inhibit NGF binding to p75^{NTR} (Jaen et al., 1995).

Small molecules potentiating neurotrophin effects

Studies demonstrating that purine nucleosides and nucleotides are released extracellularly after CNS injury, and that guanosine and GTP stimulate neurite outgrowth of hippocampal neurons co-cultured with astrocytes and enhance NGF-mediated neurite outgrowth of PC12 cells, raised the possibility that purine derivatives might be used to promote neurotrophic effects (Gysbers and Rathbone, 1992; Middlemiss et al., 1995). A synthetic purine derivative (hypoxanthine derivative AIT-082) has been shown to enhance NGF-induced neurite outgrowth of PC12 cells (Middlemiss et al., 1995). The receptors and signal transduction mechanisms modulated by AIT-082 remain unknown. The findings that oral or intraperitoneal administration of AIT-082 leads to improved memory in normal and aged memory-deficient mice (Glasky et al., 1994) have led to current ongoing trials of AIT-082 for Alzheimer's disease. The non-peptide compound, SR57746A, has been shown to potentiate the neurite-promoting activity of NGF on PC12 cells but has no similar activity in the absence of NGF (Pradines et al., 1995). The SR57746A target mediating this activity remains to be identified.

Small molecules inducing neurotrophin production

Application of small molecules in vivo and to cultured cells has been reported to induce the production of endogenous neurotrophins. In some cases these small molecules are described as neurotrophin 'mimetics'. The use of the term 'mimetic' in this context implies a degree of mechanistic specificity that in some cases may not exist. Purine transport inhibitors were reported to increase NGF production by astrocytes and improve cognitive performance of impaired rats (Yamada et al., 1997). Administration of purine derivative AIT-082 in memory-impaired animals has been associated with increased production of multiple neurotrophic factors (Glasky et al., 1997). Oral administration of idebenone to basal forebrain-lesioned rats induces increased NGF levels and improves learning and memory function (Nitta et al., 1994). Oral administration of the vitamin D3 derivative, CB1093, was shown to upregulate NGF levels in streptozoticin-diabetic rats (Riaz et al., 1999). Addition of the xanthine derivative, propentofylline, to cultured astroglia cells was found to cause a tenfold increase in NGF levels in the conditioned medium, presumably via stimulation of synthesis and secretion of NGF by astroglial cells (Shinoda et al., 1990). Other compounds found to stimulate synthesis and secretion of NGF by cultured cells include the sesquiterpenes, dictyophorines A and B (Kawagishi et al., 1997), the catechol derivatives, 1,4-benzoquinone (Takeuchi et al., 1990) and 4-methylcatechol (Carswell et al., 1992), and the recently identified diterpenoids, scabronines A and G (Obara et al., 1999).

One key advantage of this type of small-molecule approach is that many of these lead compounds have favorable physical–chemical properties, including stability and CNS penetration. The disadvantages of the NGF secretagogue approach is that in most cases the direct targets, as well as the nature and specificity of the mechanisms by which a given small molecule stimulates growth factor production and secretion, remain unknown. Moreover, it is possible that the synthesis and secretion of many other categories of factors are also affected. In in vivo studies, the cause–effect relationships between increased neurotrophic factor levels and changes in learning and memory function will have to be established.

Conclusions

Small-molecule mimetics of neurotrophins are likely to play a key role in modulating the biological effects of neurotrophins in the context of spinal cord injury therapeutics. Depending on the targeted biological process, agents that either augment or block neurotrophin function will be required. Neurotrophin mimetic strategies include development of compounds that promote neurotrophin synthesis and secretion, compounds that function as agonists or antagonists at neurotrophin receptors, compounds that interact with neurotrophins to alter receptor interactions, and compounds that modulate neurotrophin signaling to augment or inhibit neurotrophin activity.

A first generation of NGF peptidomimetics functioning as antagonists and agonists via $p75^{NTR}$ and TrkA receptors has already been developed. The impressive recent advances in small-molecule development for other proteins, including insulin, TPO and EPO, that function by inducing receptor dimerization and conformational change, suggest that the development of more potent NGF small-molecule mimetics with favorable medicinal properties will be possible.

Acknowledgements

Studies conducted in our laboratory were supported by the Alzheimer's Association, the John Douglas French Foundation, the Veterans Administration and the National Institute of Aging. We thank Drs. William Mobley and Gregory Ross for helpful discussions.

References

Angeles, T.S., Yang, S.X., Steffler, C. and Dionne, C.A. (1998) Kinetics of trkA tyrosine kinase activity and inhibition by K252a. *Arch. Biochem. Biophys.*, 349: 267–274.

Barker, P.A. (1998) $p75^{NTR}$: a study in contrasts. *Cell Death Differ.*, 5: 346–356.

Barker, P.A. and Shooter, E.M. (1994) Disruption of NGF binding to the low affinity neurotrophin receptor p75LNTR reduces NGF binding to TrkA on PC12 cells. *Neuron*, 13: 203–215.

Barrett, G.L. and Bartlett, P.F. (1994) The p75 nerve growth factor receptor mediates survival or death depending on the stage of sensory neuron development. *Proc. Natl. Acad. Sci. USA*, 91: 6501–6505.

Beglova, N., LeSauteur, L., Ekiel, I., Saragovi, H.U. and Gehring, K. (1998) Solution structure and internal motion of a bioactive peptide derived from nerve growth factor. *J. Biol. Chem.*, 273: 23652–23658.

Belliveau, D.J., Krivko, I., Kohn, J., Lachance, C., Pozniak, C., Rusakov, D., Kaplan, D. and Miller, F.D. (1997) NGF and neurotrophin-3 both activate TrkA on sympathetic neurons but differentially regulate survival and neuritogenesis. *J. Cell Biol.*, 136: 375–388.

Borasio, G.D. (1990) Differential effects of the protein kinase inhibitor K252a on the in vitro survival of chick embryonic neurons. *Neurosci. Lett.*, 108: 207–212.

Borasio, G.D., Horstmann, S., Anneser, J.M., Neff, N.T. and Glicksman, M.A. (1998) CEP-1347/KT7515, a JNK pathway inhibitor, supports the in vitro survival of chick embryonic neurons. *NeuroReport*, 9: 1435–1439.

Bourne, H.R., 1998. Drug receptors and pharmacodynamics. In: Katzung, B.G. (Ed.), Basic and Clinical Pharmacology. 7th ed., Ch. 2, Appleton and Lange, Stamford, CN, pp. 9–33.

Bradshaw, R.A., Murray-Rust, J., Ibanez, C.F., McDonald, N.Q.,

Lapatto, R. and Blundell, T.L. (1994) Nerve growth factor: structure/function relationships. *Protein Sci.*, 3: 1901–1913.

Bredesen, D.E., Ye, X., Tasinato, A., Sperandio, S., Wang, J.J.L., Assa-Munt, N. and Rabizadeh, S. (1998) p75^{NTR} and the concept of cellular dependence: seeing how the other half die. *Cell Death Differ.*, 5: 365–371.

Carswell, S., Hoffman, E.K., Clopton-Hartpence, K., Wilcox, H.M. and Lewis, M.E. (1992) Induction of NGF by isoproterenol, 4-methylcatechol and serum occurs by three distinct mechanisms. *Mol. Brain Res.*, 15: 145–150.

Casaccia-Bonnefil, P., Carter, B.D., Dobrowsky, R.T. and Chao, M.V. (1996) Death of oligodendrocytes mediated by the interaction of nerve growth factor with its receptor p75. *Nature*, 383: 716–719.

Chao, M., Casaccia-Bonnefil, P., Carter, B., Chittka, A., Kong, H. and Yoon, S.O. (1998) Neurotrophin receptors: mediators of life and death. *Brain Res. Rev.*, 26: 295–301.

Clary, D.O., Weskamp, G., Austin, L.R. and Reichardt, L.F. (1994) TrkA cross-linking mimics neuronal responses to nerve growth factor. *Mol. Biol. Cell*, 5: 549–563.

Connor, B. and Dragunow, M. (1998) The role of neuronal growth factors in neurodegenerative disorders of the human brain. *Brain Res. Rev.*, 27: 1–39.

Cunningham, B.C. and Wells, J.A. (1997) Minimized proteins. *Curr. Opin. Struct. Biol.*, 7: 457–462.

Cwirla, S.E., Balasubramanian, P., Duffin, D.J., Wagstrom, C.R., Gates, C.M., Singer, S.C., Davis, A.M., Tansik, R.L., Mattheakis, L.C., Boytos, C.M., Schatz, P.J., Baccanari, D.P., Wrighton, N.C., Barrett, R.W. and Dower, W.J. (1997) Peptide agonist of the thrombopoietin receptor as potent as the natural cytokine. *Science*, 276: 1696–1699.

Debeir, T., Saragovi, H.U. and Cuello, A.C. (1999) A nerve growth factor mimetic TrkA antagonist causes withdrawal of cortical cholinergic boutons in the adult rat. *Proc. Natl. Acad. Sci. USA*, 96: 4067–4072.

Diener, P.S. and Bregman, B.S. (1994) Neurotrophic factors prevent the death of CNS neurons after spinal cord lesions in newborn rats. *NeuroReport*, 5: 1913–1917.

Frade, J.M. and Barde, Y.-A. (1998) Nerve growth factor: two receptors, multiple functions. *BioEssays*, 20: 137–145.

Frade, J.M., Rodriguez-Tebar, A. and Barde, Y.-A. (1996) Introduction of cell death by endogenous nerve growth factor through its p75 receptor. *Nature*, 383: 166–168.

Freidinger, R.M. (1999) Nonpeptidic ligands for peptide and protein receptors. *Curr. Opin. Chem. Biol.*, 3: 395–406.

Frey, W.H., Liu, J., Chen, X., Thorne, R.G., Fawcett, J.R., Ala, T.A. and Rahman, Y.E. (1997) Delivery of ^{125}I-NGF to the brain via the olfactory route. *Drug Delivery*, 4: 8792.

Friedman, W.J. and Greene, L.A. (1999) Neurotrophin signaling via Trks and p75. *Exp. Cell Res.*, 253: 131–142.

Glasky, A.J., Melchior, C.L., Pirzadeh, B., Heydari, N. and Ritzmann, R.F. (1994) Effect of AIT-082, a purine analog on working memory in normal and aged mice. *Pharm. Biochem. Behav.*, 47: 325–329.

Glasky, A.J., Glasky, M.S., Ritzmann, R.F. and Rathbone, M.P. (1997) AIT-082, a novel purine derivative with neuroregenerative properties. *Exp. Opin. Invest. Drugs*, 6: 1413–1417.

Greene, L.A. and Kaplan, D.R. (1995) Early events in neurotrophin signalling via Trk and p75 receptors. *Curr. Opin. Neurobiol.*, 5: 579–587.

Grill, R., Murai, K., Blesch, A., Gage, F.H. and Tuszynski, M.H. (1997) Cellular delivery of neurotrophin-3 promotes corticospinal axonal growth and partial recovery after spinal cord injury. *J. Neurosci.*, 17: 5560–5572.

Gysbers, J.W. and Rathbone, M.P. (1992) Guanosine enhances NGF-stimulated neurite outgrowth in PC12 cells. *NeuroReport*, 3: 997–1000.

Hefti, F. (1997) Pharmacology of neurotrophic factors. *Annu. Rev. Pharmacol. Toxicol.*, 37: 239–267.

Holland, D.R., Cousens, L.S., Meng, W. and Mathews, B.W. (1994) Nerve growth factor in different crystal forms displays structural flexibility and reveals zinc binding sites. *J. Mol. Biol.*, 239: 385–400.

Hruby, V.J., Al-Obeidi, F. and Kazmierski, W. (1990) Emerging approaches in the molecular design of receptor-selective peptide ligands: conformational, topographical and dynamic considerations. *Biochem. J.*, 268: 249–262.

Hughes, R.A. and O'Leary, P.D. (1999) Exploiting neurotrophic factors for the treatment of neurodegenerative conditions: an Australian perspective. *Drug Dev. Res.*, 46: 268–276.

Ibanez, C.F. (1995) Neurotrophic factors: from structure–function studies to designing effective therapeutics. *Trends Biotechnol.*, 13: 217–227.

Ibanez, C.F., Ebendal, T., Barbany, G., Murray-Rust, J., Blundell, T.L. and Persson, H. (1992) Disruption of the low affinity receptor-binding site in NGF allows neuronal survival and differentiation by binding to the trk gene product. *Cell*, 69: 329–341.

Jaen, J.C., Laborde, E., Bucsh, R.A., Caprathe, B.W., Sorenson, R.J., Fergus, J., Spiegel, K., Marks, J., Dickerson, M.R. and Davis, R.E. (1995) Kynurenic acid derivatives inhibit the binding of nerve growth factor (NGF) to the low-affinity p75 NGF receptor. *J. Med. Chem.*, 38: 4439–4445.

Kaplan, D.R. and Miller, F.D. (1997) Signal transduction by the neurotrophin receptors. *Curr. Opin. Cell Biol.*, 9: 213–221.

Kawagishi, H., Ishiyama, D., Mori, H., Sakamoto, H., Ishiguro, Y., Furukawa, S. and Li, J. (1997) Dictyophorines A and B, two stimulators of NGF-synthesis from the mushroom *Dictyophora indusiata*. *Phytochemistry*, 45: 1203–1205.

Kenakin, T. (1999) Efficacy in drug receptor theory: outdated concept or under-valued tool?. *Trends Pharm.*, 20: 400–405.

Kieber-Emmons, T., Murali, R. and Greene, M.I. (1997) Therapeutic peptide and peptidomimetics. *Curr. Opin. Biotechnol.*, 8: 435–441.

Kobayashi, N.R., Fan, D.P., Giehl, K.W., Bedard, A.M., Wiegand, S.J. and Tetzlaff, W. (1997) BDNF and NT-4/5 prevent atrophy of rat rubrospinal neurons after cervical axotomy, stimulate GAP-43 and Tα1-tubulin mRNA expression and promote axonal regeneration. *J. Neurosci.*, 17: 9583–9595.

Knusel, B. and Hefti, F. (1992) K-252a compounds: modulators of neurotrophin signal transduction. *J. Neurochem.*, 59: 1987–1996.

Knusel, B., Kaplan, D.R., Winslow, J.W., Rosenthal, A., Burton, L.E., Beck, K.D., Rabin, S., Nikolics, K. and Hefti, F. (1992)

K-252b selectively potentiates cellular actions and trk tyrosine phosphorylation mediated by neurotrophin-3. *J. Neurochem.*, 59: 715–722.

Krenz, N.R., Meakin, S.O., Krassioukov, A.V. and Weaver, L.C. (1999) Neutralizing intraspinal nerve growth factor blocks autonomic dysreflexia caused by spinal cord injury. *J. Neurosci.*, 19: 7405–7414.

Kullander, K., Kaplan, D. and Ebendal, T. (1997) Two restricted sites on the surface of the nerve growth factor molecule independently determine TrkA receptor binding activation. *J. Biol. Chem.*, 272: 9300–9307.

LeSauteur, L., Wei, L., Gibbs, B.F., Saragovi, H.U., 1995. Small peptide mimetics of nerve growth factor bind TrkA receptors and affect biological responses. [Published erratum appears in J. Biol. Chem. 270, 1249, 1995.] *J. Biol. Chem.* 270, 6564–6569.

Levi-Montalcini, R., Skaper, S.D., Dal Toso, R., Petrelli, L. and Leon, A. (1996) Nerve growth factor: from neurotrophin to neurokine. *Trends Neurosci.*, 19: 514–520.

Levine, J.D. (1998) New directions in pain research: molecules to maladies. *Neuron*, 20: 649–654.

Lipworth, B.J. and Grove, A. (1997) Evaluation of partial beta-adrenoceptor agonist activity. *Br. J. Clin. Pharmacol.*, 43: 9–14.

Liu, Y., Kim, D., Himes, B.T., Chow, S.Y., Schallert, T., Murray, M., Tessler, A. and Fischer, I. (1999) Transplants of fibroblasts genetically modified to express BDNF promote regeneration of adult rat rubrospinal axons and recovery of forelimb function. *J. Neurosci.*, 19: 4370–4387.

Livnah, O., Stura, E.A., Middleton, S.A., Johnson, D.L., Jolliffe, L.K. and Wilson, I.A. (1999) Crystallographic evidence for preformed dimers of erythropoietin receptor before ligand activation. *Science*, 283: 987–990.

Longo, F.M. and Mobley, W.C. (1996) Minimized hormones grow in stature. *Nat. Biotechnol.*, 14: 1092.

Longo, F.M., Vu, T.K. and Mobley, W.C. (1990) The in vitro biological effect of nerve growth factor is inhibited by synthetic peptides. *Cell Regul.*, 1: 189–195.

Longo, F.M., Manthorpe, M., Xie, Y. and Varon, S. (1997) Synthetic NGF peptide derivatives prevent death via a p75 receptor-dependent mechanism. *J. Neurosci. Res.*, 48: 1–17.

Malcangio, M., Garrett, N.E., Cruwys, S.C. and Tomlinson, D.R. (1997) Nerve growth factor and neurotrophin-3 induced changes in nociceptive threshold and the release of substance P from the rat isolated spinal cord. *J. Neurosci.*, 17: 8459–8467.

Maliartchouk, S. and Saragovi, H.U. (1997) Optimal nerve growth factor trophic signals mediated by synergy of TrkA and p75 receptor-specific ligands. *J. Neurosci.*, 17: 6031–6037.

Maliartchouk, S., Feng, Y., Ivanisevic, L., Debeir, T., Cuello, A.C., Burgess, K. and Saragovi, H.U. (2000) A designed peptidomimetic agonistic ligand of TrkA nerve growth factor receptors. *Mol. Pharmacol.*, 57: 385–391.

Maroney, A.C., Glicksman, M.A., Basma, A.N., Walton, K.M., Knight, E., Murphy, C.A., Bartlett, B.A., Finn, J.P., Angeles, T., Matsuda, Y., Neff, N.T. and Dionne, C.A. (1998) Motoneuron apoptosis is blocked by CEP-1347 (KT 7515), a novel inhibitor of the JNK signaling pathway. *J. Neurosci.*, 18: 104–111.

Marshall, C.J. (1995) Specificity of receptor tyrosine kinase signaling: transient versus sustained extracellular signal-related kinase activation. *Cell*, 80: 179–185.

McDonald, N.Q. and Chao, M.V. (1995) Structural determinants of neurotrophic action. *J. Biol. Chem.*, 270: 19669–19672.

McDonald, N.Q., Lapatto, R., Murray-Rust, J., Gunning, J., Wlodawer, A. and Blundell, T.L. (1991) New protein fold revealed by a 2.3-A resolution crystal structure of nerve growth factor. *Nature*, 354: 411–414.

McInnes, C. and Sykes, B.D. (1997) Growth factor receptors: structure, mechanism, and drug discovery. *Biopolymers*, 43: 339–366.

Middlemiss, P.J., Glasky, A.J., Rathbone, M.P., Werstuik, E., Hindley, S. and Gysbers, J. (1995) AIT-082, a unique purine derivative, enhances nerve growth factor mediated neurite outgrowth from PC12 cells. *Neurosci. Lett.*, 199: 131–134.

Middleton, S.A., Barbone, F.P., Johnson, D.L., Thurmond, R.L., You, Y., McMahon, F.J., Jin, R., Livnah, O., Tullai, J., Farrell, F.X., Goldsmith, M.A., Wilson, I.A. and Jolliffe, L.K. (1999) Shared and unique determinants of the erythropoietin (EPO) receptor are important for binding EPO and EPO mimetic peptide. *J. Biol. Chem.*, 274: 14163–14169.

Minichiello, L., Casagranda, F., Soler-Tatche, R., Stucky, C.L., Postigo, A., Lewin, G.R., Davies, A.M. and Klein, R. (1998) Point mutation in trkB causes loss of NT4-dependent neurons without major effects on diverse BDNF responses. *Neuron*, 21: 335–345.

Nauta, H.J., Wehman, J.C., Koliatsos, V.E., Terrell, M.A. and Chung, K. (1999) Intraventricular infusion of nerve growth factor as the cause of sympathetic fiber sprouting in sensory ganglia. *J. Neurosurg.*, 91: 447–453.

Nicholson, N.S., Panzer-Knodle, S.G., Salyers, A.K., Taite, B.B., King, L.W., Miyano, M., Gorczynski, R.J., Williams, M.H., Zupec, M.E., Tjoeng, F.S., Adams, S.P. and Feigen, L.P. (1991) In vitro and in vivo effects of a peptide mimetic (SC-47643) of RGD as an antiplatelet and antithrombotic agent. *Thromb. Res.*, 62: 567–578.

Nitta, A., Murakami, Y., Furukawa, Y., Kawatsura, W., Hayashi, K., Yamada, K., Hasegawa, T. and Nabeshima, T. (1994) Oral administration of idebenone induces nerve growth factor in the brain and improves learning and memory in basal forebrain-lesioned rats. *Naunyn-Schmiedebergs Arch. Pharmacol.*, 349: 401–407.

Norris, J.D., Paige, L.A., Christensen, D.J., Chang, C.-Y., Huacani, M.R., Fan, D., Hamilton, P.T., Fowlkes, D.M. and McDonnell, D.P. (1999) Peptide antagonists of the human estrogen receptor. *Science*, 285: 744–746.

Obara, Y., Nakahata, N., Kita, T., Takaya, Y., Kobayashi, H., Hosoi, S., Kiuchi, F., Ohta, T., Oshima, Y. and Ohizumi, Y. (1999) Stimulation of neurotrophic factor secretion from 1321N1 human astrocytoma cells by novel diterpenoids, scabronines A and G. *Eur. J. Pharmacol.*, 370: 79–84.

O'Leary, P.D. and Hughes, R.A. (1998) Structure–activity relationships of conformationally constrained peptide analogues of loop 2 of brain-derived neurotrophic factor. *J. Neurochem.*, 70: 1712–1721.

Oudega, M. and Hagg, T. (1999) Neurotrophins promote regeneration of sensory axons in the adult rat spinal cord. *Brain Res.*, 818: 431–438.

Owolabi, J.B., Rizkalla, G., Tehim, A., Ross, G.M., Riopelle, R.J., Kamboj, R., Ossipov, M., Bian, D., Wegert, S., Porreca, F. and Lee, D.K.H. (1999) Characterization of antiallodynic actions of ALE-0540, a novel nerve growth factor receptor antagonist, in the rat. *J. Pharmacol. Exp. Ther.*, 289: 1271–1276.

Perron, J.C. and Bixby, J.L. (1999) Distinct neurite outgrowth signaling pathways converge on ERK activation. *Mol. Cell. Neurosci.*, 13: 362–378.

Pliska, V. (1999) Partial agonism: mechanisms based on ligand–receptor interactions and on stimulus–response coupling. *J. Recept. Signal Transduct. Res.*, 19: 597–629.

Poduslo, J.F. and Curran, G.L. (1996) Permeability at the blood–brain and blood–nerve barriers of the neurotrophic factors: NGF, CNTF, NT-3, BDNF. *Mol. Brain Res.*, 36: 280–286.

Pradines, A., Magazin, M., Schilitz, P., Le Fur, G., Caput, D. and Ferrara, P. (1995) Evidence for nerve growth factor-potentiating activities of the non-peptidic compound SR57746A in PC12 cells. *J. Neurochem.*, 64: 1954–1964.

Qiu, M.-S. and Green, S.H. (1992) PC12 cell neuronal differentiation is associated with prolonged p21ras activity and consequent prolonged ERK activity. *Neuron*, 9: 705–717.

Rabizadeh, R., Oh, J., Zhong, L.-t., Yang, J., Bitler, C.M., Butcher, L.L. and Bredesen, D.E. (1993) Induction of apoptosis by the low-affinity NGF receptor. *Science*, 261: 345–348.

Rashid, K., Van der Zee, C.E.E.M., Ross, G.M., Chapman, C.A., Stanisz, J., Riopelle, R.J., Racine, R.J. and Fahnestock, M. (1995) A nerve growth factor peptide retards seizure development and inhibits neural sprouting in a rat model of epilepsy. *Proc. Natl. Acad. Sci. USA*, 92: 9495–9499.

Riaz, S., Malcangio, M., Miller, M. and Tomlinson, D.R. (1999) A vitamin D3 derivative (CB1093) induces nerve growth factor and prevents neurotrophic deficits in streptozotocin-diabetic rats. *Diabetologia*, 42: 1308–1313.

Ross, G.M., Shamovsky, I.L., Lawrance, G., Solc, M., Dostaler, S.M., Jimmo, S.L., Weaver, D.F. and Riopelle, R.J. (1997) Zinc alters conformation and inhibits biological activities of nerve growth factor and related neurotrophins. *Nat. Med.*, 3: 872–878.

Ross, G.M., Shamovsky, I.L., Lawrance, G., Solc, M., Dostaler, S.M., Weaver, D.F. and Riopelle, R.J. (1998) Reciprocal modulation of TrkA and p75[NTR] affinity states is mediated by direct receptor interactions. *Eur. J. Neurosci.*, 10: 890–898.

Saltzman, W.M., Mak, M.W., Mahoney, M.J., Duenas, E.T. and Cleland, J.L. (1999) Intracranial delivery of recombinant nerve growth factor: release kinetics and protein distribution for three delivery systems. *Pharm. Res.*, 16: 232–240.

Saragovi, H.U., Zheng, W., Maliartchouk, S., DiGugliemo, G.M., Mawal, Y.R., Kamen, A., Woo, S.B., Cuello, A.C., Debeir, T. and Neet, K.E. (1998) A TrkA-selective, fast internalizing nerve growth factor–antibody complex induces trophic but not neuritogenic signals. *J. Biol. Chem.*, 273: 34933–34940.

Segal, R.A. and Greenberg, M.E. (1996) Intracellular signaling pathways activated by neurotrophic factors. *Annu. Rev. Neurosci.*, 19: 463–489.

Shamovsky, I.L., Ross, G.M., Riopelle, R.J. and Weaver, D.F. (1999) The interaction of neurotrophins with the p75[NTR] common neurotrophin receptor: a comprehensive molecular modeling study. *Protein Sci.*, 8: 1–12.

Shibayama, M., Hattori, S., Himes, B.T., Murray, M. and Tessler, A. (1998) Neurotrophin-3 prevents death of axotomized Clarkes nucleus neurons in adult rat. *J. Comp. Neurol.*, 390: 102–111.

Shinoda, I., Furukawa, Y. and Furukawa, S. (1990) Stimulation of nerve growth factor synthesis/secretion by propentofylline in cultured mouse astroglial cells. *Biochem. Pharmacol.*, 39: 1813–1816.

Shu, X.-Q., Mendell, L.M., 1999. Neurotrophins and hyperalgesia. Proc. Natl. Acad. Sci. USA 96, 7693–7696.

Skaper, S.D. and Walsh, F.S. (1998) Neurotrophic molecules: strategies for designing effective therapeutic molecules in neurodegeneration. *Mol. Cell. Neurosci.*, 12: 179–193.

Spiegel, K., Agrafiotis, D., Caprathe, B., Davis, R.E., Dickerson, M.R., Fergus, J.H., Hepburn, T.W., Marks, J.S., Van Dorf, M., Wieland, D.M. and Jaen, J.C. (1995) PD 90780, nonpeptide inhibitor of nerve growth factor's binding to the p75 NGF receptor. *Biochem. Biophys. Res. Commun.*, 217: 488–494.

Stichel, C.C. and Muller, H.W. (1998) Experimental strategies to promote axonal regeneration after traumatic central nervous system injury. *Prog. Neurobiol.*, 56: 119–148.

Takeuchi, R., Murase, K., Furukawa, Y., Furukawa, S. and Hayashi, K. (1990) Stimulation of nerve growth factor synthesis/secretion by 1,4-benxoquinone and its derivatives in cultured mouse astroglial cells. *FEBS Lett.*, 261: 63–66.

Taglialatela, G., Hibbert, C.J., Hutton, L.A., Werrbach-Perez, K. and Perez-Polo, J.R. (1996) Suppression of pl40trkA does not abolish nerve growth factor-mediated rescue of serum-free PC12 cells. *J. Neurochem.*, 66: 1826–1835.

Tetzlaff, W., Kobayashi, N.R., Giehl, K.M., Tsui, B.J., Cassar, S.L. and Bedard, A.M. (1994) Response of rubrospinal and corticospinal neurons to injury and neurotrophins. *Prog. Brain Res.*, 103: 271–286.

Tian, S.-S., Lamb, P., King, A.G., Miller, S.G., Kessler, L., Luengo, J.I., Johnson, R.K., Gleason, J.G., Pelus, L.M., Dillon, S.B. and Rosen, J. (1998) A small, nonpeptidyl mimic of granulocyte-colony-stimulating factor. *Science*, 281: 257–259.

Van der Zee, C.E.E.M., Ross, G.M., Riopelle, R.J. and Hagg, T. (1996) Survival of cholinergic forebrain neurons in developing p75NFGR-deficient mice. *Science*, 274: 1729–1732.

Wiesmann, C., Ultsch, M.H., Bass, S.H. and De Vos, A.M. (1999) Crystal structure of nerve growth factor in complex with the ligand-binding domain of the TrkA receptor. *Nature*, 401: 184–188.

Woo, S.B. and Neet, K.E. (1996) Characterization of histidine residues essential for receptor binding and activity of nerve growth factor. *J. Biol. Chem.*, 271: 24433–24441.

Woo, S.B., Whalen, C. and Neet, K.E. (1998) Characterization of the recombinant extracellular domain of the neurotrophin receptor TrkA and its interaction with nerve growth factor (NGF). *Protein Sci.*, 7: 1006–1016.

Wrighton, N.C., Balasubramian, P., Barbone, F.P., Kashyap, A.K., Farrel, F.X., Jolliffe, L.K., Barrett, R.W. and Dower, W.J. (1997) Increased potency of an erythropoietin peptide mimetic through covalent dimerization. *Nat. Biotechnol.*, 15: 1261–1265.

Xie, Y.M. and Longo, F.M. (1997) Neurotrophic activity of synthetic peptide derivatives corresponding to NGF loop region 92–98. *Soc. Neurosci. Abstr.*, 23: 1701.

Xie, Y., Tisi, M.A., Yeo, T.T. and Longo, F.M. (2000) NGF loop 4 dimeric mimetics activate ERK and AKT and promote NGF-like neurotrophic effects. *J. Biol. Chem.*, in press.

Yamada, K., Nitta, A., Hasegawa, T., Fuji, K., Hiramatsu, M., Kameyama, T., Furukawa, Y., Hayashi, K. and Nabeshima, T. (1997) Orally active NGF synthesis stimulators: potential therapeutic agents in Alzheimer's disease. *Behav. Brain Res.*, 83: 117–122.

Yuen, E.C. and Mobley, W.C. (1999) Early BDNF, NT-3, and NT-4 signaling events. *Exp. Neurol.*, 159: 296–308.

Zhang, B., Salituro, G., Szalkowski, D., Li, A., Zhang, Y., Royo, I., Vilella, D., Diez, M.T., Pelaez, F., Ruby, C., Kendall, R.L., Mao, X., Griffin, P., Calaycay, J., Zierath, J.R., Heck, J.V., Smith, R.G. and Moller, D.E. (1999) Discovery of a small molecule insulin mimetic with antidiabetic activity in mice. *Science*, 284: 974–977.

CHAPTER 30

Tissue engineering strategies for nervous system repair

Patrick A. Tresco [*]

W.M. Keck Center for Tissue Engineering, Department of Bioengineering, University of Utah, 20 South 2030 East, Biopolymers Research Building, Room 108D, Salt Lake City, UT 84112, USA

Introduction

Tissue engineering is a rapidly developing cross-disciplinary field that integrates discoveries from cell biology, the medical sciences and various engineering fields to produce technology for replacing or correcting poorly functioning parts in humans or in animals. Tissue engineered products apply to virtually every tissue and organ system, as well as many of the major medical problems that afflict mankind. Specially designed materials and more complex devices are combined with living cells to restore function in diseased, damaged or genetically defective tissues. The technology is rapidly developing, having taken advantage of dramatic recent advances in cell and developmental biology, immunology, genetic engineering, and various fields of engineering.

At present, several approaches are being developed that involve the implantation of a biomaterial that either contains living cells or, following implantation, is colonized by host cells. For our purposes, we define 'biomaterial' as any material, natural or synthetic, that is not rejected by the organism following implantation into a body cavity or tissue. The biomaterial may consist of extracellular matrix, or may be made of any of a number of plastic or synthetic materials, or may consist of a combination of the two. Irrespective of the type of biomaterial, the objective is biological acceptance or achieving symbiosis at the interface of implant and host tissue. Basically, two types of implant have been developed that differ mainly with respect to how they interact with host tissue at the site of implantation.

At one end of the spectrum are the cell encapsulation systems, somewhat autonomous implants in which the transplanted cells are physically isolated from host tissue, extracellular matrix (ECM) and larger solutes by a selectively permeable barrier. These types of implants have been generally made of non-degradable materials and are dependent upon host physiological mechanisms for controlling pH, metabolic waste removal, electrolytes, and nutrient availability within the encapsulation chamber. Concentration gradients provide the force for the transfer of solutes across the encapsulation membrane. In general, encapsulation membranes offer greater resistance to the diffusive transport of molecules as they increase in size. Available information suggests that such barriers can prevent immune cells and immune related soluble complexes from killing the transplanted cells in the absence of pharmacological immune suppression. In this regard, cell encapsulation technology provides a means to expand the potential pool of donor cells and tissues, while avoiding the toxic secondary effects of immune suppression. Typically, such implants are used as focal, sustained delivery systems or to replace endocrine or enzymatic function.

---

[*] Corresponding author: Dr. Patrick A. Tresco, W.M. Keck Center for Tissue Engineering, Department of Bioengineering, University of Utah, 20 South 2030 East, Biopolymers Research Building, Room 108D, Salt Lake City, UT 84112, USA. Fax: +1-801-585-5151; E-mail: patrick.tresco@m.cc.utah.edu

At the other end of the spectrum are the biodegradable scaffolds, three-dimensional, highly porous implants that generally consist of a meshwork of bonded filaments or may exist as an open-celled foam-like material that functions as a temporary substrate that may or may not contain living cells. In general, such implants become integrated with host tissue at the site of implantation, eventually sharing a host-derived blood supply as well as host-derived immune and mesenchymal cells. In such cases, the transplanted cells, or the host cells that subsequently colonize the implant, secrete ECM and gradually remodel or replace the slowly degrading original material. The term 'scaffold' has been used to describe such temporary substrates that are eventually fully reabsorbed by the host so that no permanent foreign body is associated with long-term treatment. The optimal result is a biologically transparent interface that becomes indistinguishable from the surrounding tissue. Such implants have been designed to replace tissues that perform mechanical functions such as ligament, cartilage, bone, and skin or tubular tissue such as a blood vessel. In such cases, the eventual development of normal tissue architecture is paramount if normal tissue function is to be achieved.

Discoveries and new technology developed in this emerging field have been directed toward the treatment of a number of disorders including cardiovascular disease, liver disease, diabetes, burn wounds, skin ulcers, joint injuries and bone fractures. Although most of the advances are still a long way from the clinic, recent studies suggest the possibility of restoring a diverse number of tissues including bone (Vacanti et al., 1993; Reddi, 1994; Puelacher et al., 1996), breast (Cao et al., 1998b), cartilage (Puelacher et al., 1994; Cao et al., 1997, 1998a; Britt and Park, 1998; Freed et al., 1998; Kaab et al., 1998; Riesle et al., 1998; Sims et al., 1998), heart valves (Zund et al., 1997; Shinoka et al., 1998), intestine (Kaihara et al., 1999; Kim et al., 1999), kidney (Humes, 1996), liver (Mooney et al., 1997; Pollok et al., 1998), pancreas (Zekorn et al., 1996), muscle (Vandenburgh et al., 1996, 1998; Chromiak et al., 1998; Mulder et al., 1998; Okano and Matsuda, 1998), skin (Triglia et al., 1991; Hansbrough et al., 1992; Naughton et al., 1997; Trent and Kirsner, 1998), trachea (Sakata et al., 1994), urinary bladder (Yoo and Atala, 1997; Oberpenning et al., 1999), and vascular tissue (L'Heureux et al., 1998; Shinoka et al., 1998). The reader may find it helpful to read any of a number of papers that have addressed the issues and developments related to this emerging discipline (Cima et al., 1991; Aebischer et al., 1993; Cima and Langer, 1993; Langer and Vacati, 1993; Bellamkonda and Aebischer, 1994; Ezzell, 1995; Galletti et al., 1995; Hubbell, 1995; Lanza and Chick, 1995, 1997a; Colton, 1996; Lanza et al., 1996; Woerly et al., 1996; Maysinger and Morinville, 1997; Hudson et al., 1999; Lysaght and Aebischer, 1999). Similar strategies are being developed to restore or replace function throughout the nervous system.

In this chapter, we examine some of the ways in which these strategies are being applied to restore nervous system function. Where possible we attempt to limit our discussion to issues related to central nervous system (CNS) repair. We first cover the use of encapsulated cell implants, and then focus our attention on technologies that attempt to guide or target the growth of regenerating axons to appropriate targets. Due to the expanding volume of literature, we have chosen to illustrate certain key issues or developments through a limited number of selected examples.

Cell encapsulation

Encapsulated cell technology initially developed as an alternative to implanting purely synthetic delivery systems and as a potential solution to the donor supply problem associated with conventional cell transplantation approaches. Early studies focused on reversing the behavioral deficits associated with various animal models of neurodegenerative disease (Emerich et al., 1992). Later, in response to advances in genetic engineering, the technology evolved as a means of delivering growth factors (Maysinger and Morinville, 1997). The major advantage of the method is the ability to achieve relatively high, sustained local concentrations of a biotherapeutic at a target site, while limiting the systemic dose so as to decrease the likelihood of peripheral side effects or systemic toxicity. As a CNS-related therapeutic technology, its development has largely focused on implantable medical devices that enable sustained delivery of small molecules directly to deep brain sites, the spinal cord or to adjacent tissues.

It appears that the approach can be traced to experiments conducted in the early 1980s that involved the intracranial implantation of rat, sheep or human pituitary cells into hypophysectomized (hypox) rats (Hymer et al., 1981). The method as originally described involved simply infusing a cell suspension into preformed, 'off the shelf' semi-permeable plastic tubes that were subsequently crimped on the ends and then heat-sealed shut. Implantation of cell-loaded capsules into the third ventricle was followed by host growth for three weeks before weight gains eventually plateaued. Histological analysis of retrieved capsules revealed the presence of viable cells of somatotrophic and corticotrophic phenotype. In hindsight, the most significant result was the observation that such devices created an 'immunologically privileged site' deep within the brain, most likely due to the fact that the membrane hindered the diffusion of molecules the size of immunoglobulins and larger into the intracapsular space. Growth of hypox rats receiving capsules containing xenogeneic sheep cells or pieces of human postmortem pituitary gland supported the concept of physical 'immunoprotection'. Furthermore, rats implanted with encapsulated human prolactin (hPRL) secreting adenoma cells displayed detectable quantities of circulating hPRL 100 days after implantation without immune suppression.

Interestingly, little additional work was performed in this general area until it was picked up again in the early 1990s by Aebischer and colleagues, who developed the technology as a therapeutic alternative to implantable, purely synthetic, sustained-delivery systems, and as a solution to the donor supply problem associated with conventional cell transplantation therapies (Aebischer et al., 1988; Jaeger et al., 1990). The therapeutic potential of cell encapsulation was supported by several early studies that examined capsule material biocompatibility, immune protection, sustained release of neurotransmitters, and preclinical efficacy in several models of neurodegenerative disease (Jaeger et al., 1990; Emerich et al., 1992; Tresco et al., 1992a; Lindner and Emerich, 1998).

The first report on biocompatibility examined the brain tissue reaction to polymer capsules implanted in rats for up to 54 weeks. It was shown that such materials were well tolerated following implantation into cortical parenchyma, being associated with no recognizable neurological or behavioral lesion (Winn et al., 1989). In general, the host response was consistent with that reported following a mechanical stab wound. It involved minimal necrosis at the brain–polymer interface that was characterized by a glial fibrillary acidic protein (GFAP)-positive astrocytic reaction that varied in thickness along the capsule. Endogenous neuronal cell bodies were frequently preserved in the compartment immediately adjacent to the implant, and normal synapses at the transmission electron microscopy (TEM) level were observed within a few microns of the brain–biomaterial interface.

In other studies, the Aebischer group established that encapsulated intracranial, xeno-transplants could survive throughout a 12-week study period in the absence of host immune suppression, and showed that such implants containing catecholaminergic cell types continued to display appropriate neurotransmitter-releasing phenotype following long-term encapsulation (Aebischer et al., 1991a,b,c; Jaeger et al., 1992). When maintained in vitro over equivalent time intervals, encapsulated cells exhibited good survival, proliferated, and spontaneously released neurotransmitter for periods up to six months.

Additional early work revealed that intrastriatal implantation of encapsulated catecholamine-releasing cells could reduce the degree of apomorphine-induced rotations in 6-hydroxydopamine (6-OHDA) lesioned rats (Winn et al., 1991; Tresco et al., 1992b), which was sustained for as long as six months after capsule implantation (Tresco et al., 1992b). Microdialysis studies revealed the presence of dopamine (DA) near cell-containing capsules, which was undetectable in lesioned animals, and near empty control capsules. Histological analysis for time points up to 24 weeks in vivo demonstrated that encapsulated PC12 cells consistently survived and expressed tyrosine hydroxylase. These findings were extended by the demonstration that encapsulated cells secreting catecholamines promoted symptomatic recovery in aged rodents (Emerich et al., 1993, 1994a), and hemiparkinsonian primates (Aebischer et al., 1994; Kordower et al., 1995). A subsequent study in unilaterally lesioned rats showed that intrastriatal implants were effective in reducing apomorphine-induced rotations, but placement in the lateral ventricle was not accompanied by a similar reduction, indicating perhaps the shortcoming of such a diffusion-based system (Emerich et al., 1996a). In all, these studies demonstrated that the sustained release of dopamine from the implant was sufficient to exert a long-term functional influence upon 6-OHDA unilaterally lesioned rats, and suggested that capsules containing secretory cells, in general, may be an effective method of long-term neurotransmitter delivery in the CNS.

The therapeutic potential of encapsulated cells was strengthened by a study that showed that delivery of nerve growth factor (NGF) to the lateral ventricles of fimbria-fornix-lesioned rats prevented reduction in choline acetyltransferase (ChAT) expression in medial forebrain cholinergic neurons (Hoffman et al., 1993). Encapsulated rat fibroblasts genetically modified to produce NGF were implanted in the lateral ventricle near the lesion cavity of adult rats that had previously received unilateral aspirative fimbria-fornix lesions. As in earlier studies, it was shown that rats with cell-containing capsules showed no undue reaction to the implants. The cells remained viable, were confined to the capsule space, and released sufficient NGF to prevent the lesion-induced loss of septal ChAT expression, whereas animals receiving unmodified control fibroblasts had no effect on endogenous ChAT immunoreactivity. This study was the first to demonstrate that encapsulated genetically engineered cells could be used to provide a means for delivering neurotrophic factors to the damaged CNS. The therapeutic potential was further strengthened by a number of subsequent studies that confirmed that encapsulated genetically engineered cell lines could also deliver NGF to the rat and primate CNS with functional effects observed over significant time frames (Emerich et al., 1994b; Kordower et al., 1994; Winn et al., 1994, 1996).

Of particular interest is one study that examined the long-term therapeutic potential of encapsulated cells for growth factor delivery using genetically modified xenotransplants in the absence of host immunesuppression (Winn et al., 1996). The gene encoding human nerve growth factor (hNGF) was introduced in a dihydrofolate reductase-based pNUT expression vector system into a baby hamster kidney (BHK) cell line. The BHK-hNGF and mock-transfected cells were placed into capsules and transplanted into the lateral ventricles of young adult rats. Remarkably, following removal from the lateral ventricles after 13.5 months, secretory levels of hNGF and long-term gene expression were maintained and comparable to preimplantation levels. The study reported that biologically active hNGF was released by the encapsulated cells at 3.6 ± 0.8 ng/device per 24 h prior to implantation and following device retrieval at 2.2 ± 0.4 ng/device per 24 h. Furthermore, genomic DNAs (hNGF transgene), as determined by polymerase chain reaction (PCR) analyses, revealed that the transgene copy number from the BHK-hNGF cells recovered after 13.5 months was equivalent to preimplantation levels. Moreover, no deleterious effects from the released hNGF were detectable by analysis of body weight, mortality rate, motor/ambulatory function, hyperalgesia, or cognitive function as assessed with the Morris water maze. Animals receiving BHK-hNGF implants exhibited a marked hypertrophy of cholinergic neurons within the striatum and nucleus basalis, but not the medial septum ipsilateral to the capsule. In addition, a robust sprouting response of cholinergic fibers was observed within the frontal cortex and lateral septum proximal to the implant. The study demonstrated that encapsulated xenogeneic cells could be used to provide long-term delivery of hNGF, and from a technological standpoint argued that encapsulated cells may be useful for delivering other neurotrophic factors within the CNS for prolonged time periods.

The broad applicability of the approach is suggested by the number of studies reporting functional effects using cells secreting other neuroactive agents and employing other relevant animal models (Langer and Vacati, 1993; Bellamkonda and Aebischer, 1994; Lysaght et al., 1994; Ezzell, 1995; Hubbell, 1995; Lanza and Chick, 1995; Lanza et al., 1996; Woerly et al., 1996; Maysinger and Morinville, 1997; Lindner and Emerich, 1998; Lysaght and Aebischer, 1999). The available literature suggests that such implants have the potential for treating a number of conditions including age-related degeneration (Emerich et al., 1993, 1994a, 1996b), Alzheimer's disease (Hoffman et al., 1993; Emerich et al., 1994b, 1997; Winn et al., 1994, 1996; Schinstine et al., 1995; Kordower et al., 1996, 1997; Lindner et al., 1996), amyotrophic lateral sclerosis (Aebischer and Kato, 1995; Sagot et al., 1995; Aebischer et al., 1996a,b; Tan et al., 1996), chronic pain (Sagen et al., 1993; Buchser et al., 1996; Decosterd et al., 1998), neuroprotection (Deglon et al., 1996; Vejsada et al., 1998), Huntington's disease (Hammang et al., 1995; Emerich et al., 1996c, 1998), improving survival of fetal transplants (Sautter et al., 1998), and Parkinson's disease (Jaeger et al., 1990; Aebischer et al., 1991a; Winn et al., 1991; Tresco et al., 1992a,b; Tresco, 1994; Kordower et al., 1995; Lindner et al., 1995, 1997; Subramanian et al., 1997). To date, such implants have been used to deliver several putative therapeutic agents to CNS tissue including ciliary neurotrophic factor (CNTF) (Aebischer and Kato, 1995; Aebischer et al., 1996a,b; Emerich et al., 1997), DA and other catecholamines (Jaeger et al., 1990; Aebischer et al., 1991a; Winn et al., 1991; Tresco et al., 1992a,b; Sagen et al., 1993; Tresco, 1994; Kordower et al., 1995; Lindner et al., 1995, 1997; Buchser et al., 1996; Subramanian et al., 1997; Decosterd et al., 1998), endorphins and enkephalins (Sagen et al., 1993; Buchser et al., 1996; Decosterd et al., 1998), glial cell-derived neurotrophic factor (GDNF) (Lindner et al., 1995; Emerich et al., 1996a), and NGF (Hoffman et al., 1993; Emerich et al., 1994b, 1997; Winn et al., 1994, 1996; Schinstine et al., 1995; Kordower et al., 1996, 1997; Lindner et al., 1996). In addition, the approach also has been used as a basic science tool to establish the mechanism by which transplanted embryonic suprachiasmatic nucleus tissue restores host circadian locomotory function (Lehman et al., 1995; Silver et al., 1996). Although most of the aforementioned studies have been carried out in pre-clinical models, clinical trials using xenogeneic implants for the treatment of chronic pain have been conducted (Buchser et al., 1996).

Many polymeric materials have been evaluated as encapsulation membranes including poly(sulfone) (Sun et al., 1980), AN69 (Kessler et al., 1992; Honiger et al., 1994; Serguera et al., 1999), poly(urethane) (Zondervan et al., 1992), poly(vinyl alcohol) (Segawa et al., 1987), poly(amide) (Catapano et al., 1990), and a variety of hydrogels (Lim and Sun, 1980; Winn and Tresco, 1994). However, for nervous system applications, plastic tubular membranes made of poly(acrylonitrile co-vinyl chloride) (PAN-PVC) have been used almost exclusively. Most of the studies cited above have used a simple device that consists of a preformed hollow fiber membrane in which cells are infused into the lumen and the ends are subsequently sealed in a manner similar to the original Hymer study (Hymer et al., 1981). For brain implants, capsule size has varied from 5 to 7 mm in length and approximately 0.5 to 1.0 mm in outer diameter. Devices developed for delivering cells to the subdural space of the spinal cord have been longer. In general, cells have been introduced into such devices as a cell suspension in growth media or suspended in a variety of immobilization matrices that are used to keep the cells optimally separated in the encapsulation chamber (Winn and Tresco, 1994; Zielinski and Aebischer, 1994). The tubes are sealed by crimping the ends, and either thermowelding the material by resistance heating or by sealing with an ultraviolet (UV) curable adhesive. Capsules have been placed in brain parenchyma, within the ventricles, or in the subarachnoid space adjacent to the spinal cord. Issues related to mass production, safety, and long-term clinical utility have fueled the development of more sophisticated devices.

For certain treatments such as neurodegenerative disease, cell encapsulation devices may need to function for years or longer. In other cases, such as transient delivery of growth factors, the therapy may only be needed for a short time frame. Therefore devices have been designed to enable retrieval at various time intervals. Designs have evolved to reinforce the fragile cell encapsulation membrane and to impart column strength to the device to handle better the physical demands of the retrieval process. One design for delivering cells near the spinal cord employs a two-part device where the cell containing section is attached to an adjacent tethering section that can be grasped by the surgeon and pulled to remove the implant (Ezzell, 1995; Buchser et al., 1996; Lysaght and Aebischer, 1999). Included in such designs is an axially, oriented strain relief element that provides tensile reinforcement for the device and shields the fragile membrane from excessive loads encountered during the retrieval process.

Feasibility studies or technology?

Despite the encouraging results, fundamental questions remain unanswered regarding cell encapsulation technology. For instance, although several studies report long-term survival of encapsulated cells from several months to one year following implantation into the CNS of several species, little information is available that describes how these devices actually work. No study has examined how device performance is related to the number of surviving cells within the device. In fact, few studies provide any information as to the number of viable cells and their arrangement within such devices as a function of implantation site or as a function of time. This raises a number of questions. How is cell homeostasis achieved? How far will agents diffuse into host parenchyma? Do the transport properties of the membrane change with time? When cells die within the device do large molecules that are released inside the capsule raise the osmotic pressure? Does volume regulation become a problem as necrosis increases over time? Is membrane transport influenced by host matrix deposition or by the reactive layer of astrocytes that has been reported to envelop such implants? In short, is biofouling a problem? Can the host response to such implants be modified? Can other materials be used? What capsule transport characteristics are needed for immune protection in the ventricles, parenchyma or in subdural space? Can stronger devices be produced from such fragile plastic membranes? What are the effects of retrieving such implants from brain tissue should such a situation be warranted? Is hemorrhage an issue? What kind of scarring response occurs following retrieval of such a device and are there ways of minimizing the damage? In short, while the available information supports the idea that encapsulation devices have therapeutic potential for focal sustained delivery, and clearly supports the use of such devices in basic science, fundamental issues remain to be answered before this technique is reduced to widely used medical technology.

Bridging technologies

Tissue engineering also focuses on the development of substrates that enhance and direct the neural regeneration process. At present the treatment of nerve injury is limited to microsurgical realignment of the damaged nerve fascicles or repair using an autologous nerve graft to bridge a damaged nerve segment. In cases were the defect is extensive, autografts are the only alternative. Given the limited availability of donor grafts and the morbidity associated with removing a viable nerve, nerve autografts still serve as the clinical 'gold standard'. However, despite continued optimization of these approaches, the combination of a shortage of suitable donor grafts and suboptimal functional recovery has motivated the search for suitable alternative bridging biomaterials.

At one end of the spectrum are nerve guidance channels, plastic hollow conduits that are used like an autograft to bridge the damaged nerve segment. The original and simplest devices were developed from commercially available or 'off the shelf' plastic tubing that was slipped over each end of the injured nerve segment and sutured or glued in place. Such devices physically bridge the defect and allow the natural wound healing response to proceed in a sheltered microenvironment. Various iterations of such materials have been investigated and are described.

The other general area involves the engineering of materials that are explicitly designed to participate or directly enhance the regeneration process. This new generation of 'biointeractive materials' is being designed and developed as scaffolding based upon biological principles derived from the study of the development and organization of the nervous system. Selected examples of this emerging area of technology are also described.

Guidance channels

The first guidance devices that were systematically developed employed hollow tubular conduits to physically guide or funnel sprouting axonal processes toward distal targets. In addition to restricting the migration of cells and sprouting processes within the confines of the tube, such systems created a space for increasing local growth factor concentration, and acted as a barrier to restrict or to minimize the infiltration of provisional matrix, fibroblasts, and other host inhibitory or cell scar-forming entities that would otherwise inhibit the normal regenerative process. Guidance channels, also called 'entubulation or coupling devices', were initially developed to repair peripheral nerves. To this end a wide variety of commercially available synthetic and naturally occurring biomaterials have been examined including ethylene vinyl acetate (Aebischer et al., 1989), polyethylene (Madison et al., 1988), polyvinylchloride (Scaravilli, 1984a,b), polyvinylidenefluoride (Aebischer et al.,

1987a,b), polytetraflouroethylene (PTFE) (Valentini et al., 1989), silicone (Longo et al., 1983; Williams et al., 1983; Kerns et al., 1991), polyacrylonitrile-vinyl chloride (Uzman and Villegas, 1983), expanded PTFE (Young et al., 1984), collagen (Archibald et al., 1991), as well as some tubular living tissues such as intestine (Wang et al., 1999), and vein (Chiu, 1999). In addition, several studies indicate that synthetic, resorbable materials such as polyglycolic acid (Molander et al., 1983) and polylactic acid (Nyilas et al., 1983) can be used with favorable results.

As the therapeutic potential of growth factors emerged, second-generation devices were developed that incorporated the release of specific growth factors either directly from the walls of the guidance channel (Aebischer et al., 1989; Hadlock et al., 1999) or from an external pumping device (Santos et al., 1995). To further augment the regenerative capacity of such devices, 'off the shelf' presumptive growth-promoting matrices were included into the lumen of the guidance channel including agarose (Labrador et al., 1995), collagen (Madison et al., 1988), fibrin (Williams et al., 1987), and Matrigel (Valentini et al., 1987). Enhanced regeneration in the peripheral nervous system sparked efforts to evaluate the technology for clinical potential in spinal cord repair (Guenard et al., 1992; Xu et al., 1995a,b, 1996, 1997, 1999; Ramón-Cueto et al., 1998).

As an alternative to the acellular approaches outlined above, glial cells have been incorporated into the lumen of guidance channels to promote axonal regeneration in the peripheral and central nervous systems (Guenard et al., 1992; Keeley et al., 1993; Bunge, 1994; Levi et al., 1994; Xu et al., 1995a,b, 1996, 1997, 1999). Investigators argued that such an approach would increase the therapeutic potential of such devices. In particular, a case was made for the use of Schwann cells where several advantages have been cited, including the fact that they can be isolated from the adult, expanded in vitro, genetically modified, and can serve as both a substrate for regeneration and as a source of trophic factors. One of the first studies in this area used Schwann cells derived from adult nerves that were seeded in permselective guidance channels to promote rat sciatic nerve regeneration across an 8-mm gap (Guenard et al., 1992). Guidance channels containing Schwann cells in Matrigel were compared to sciatic nerve autografts, empty channels, or channels filled with Matrigel alone. One day after seeding the channels with the Schwann cell–Matrigel suspension, a central cable of Schwann cells was observed oriented along the axis within the lumen of the hollow fiber. It was envisioned that the combination of Schwann cells would serve as a scaffold or contact guidance mechanism to direct regeneration along the long axis of the channel. The ability of channels containing syngeneic Schwann cells to foster regeneration was shown to be dependent on the Schwann cell seeding density and histocompatibility of the donor cells. More importantly, under appropriate seeding conditions, regenerating nerve cables in Schwann cell-containing tubes consistently yielded larger, more organotypic fascicle architecture than control channels. The study was the first to demonstrate that cultured adult syngeneic Schwann cells seeded in permselective synthetic guidance channels supported extensive nerve regeneration.

The Bunge group examined the usefulness of Schwann cells enclosed within semipermeable polyacrylonitrile/polyvinylcholoride (PAN/PVC) guidance channels to bridge the transected rat spinal cord following the removal of T9–11 segments (Bunge, 1994; Xu et al., 1995b). In this whole cord entubulation model, rat spinal cords were transected at the T8 cord level and the next caudal segment was removed. Each cut stump was inserted 1 mm into either end of the semipermeable plastic channel. One month later, a bridge between the severed stumps was observed as determined by the gross and histological appearance, and the ingrowth of propriospinal axons from both stumps. The consistent observation with this model has been significantly more myelinated axonal fibers compared to controls, associated with an increased supply of regenerated host vasculature. However, none of the regrowth penetrated the distal host interface.

To enhance regeneration in the transected cord using the entubulation approach, growth factors have been delivered into the channel lumen (Xu et al., 1995a). Brain-derived neurotrophic factor (BDNF) and neurotrophin-3 (NT-3) were delivered simultaneously into the channel using an Alzet minipump at a rate of 12 μg/day and compared to phosphate-buffered saline as the vehicle control. They reported nearly twice as many myelinated nerve fibers with the Schwann cell–neurotrophin combination compared to controls. In addition, it was reported that some nerve fibers were immunoreactive for serotonin, suggesting the possibility that raphe-derived axons regenerated. More significantly, in the Schwann cell–neurotrophin treated group, retrograde labeling methods revealed neurons in ten brain stem nuclei or 67% of the lateral and spinal vestibular nuclei of the brain stem. The investigators found 15 times more labeled brain stem neurons in the neurotrophin treated group than in the control group that received Schwann cells and vehicle alone. The results clearly demonstrated that the combination of growth factor infusion, BDNF and NT-3, into the guidance channel enhanced the regeneration of specific populations of brain stem neurons over that observed with Schwann cell grafts alone.

As an alternative to entubulating the entire transected spinal cord, smaller 'mini-channels' also seeded with Schwann cells have been examined to bridge a right spinal cord hemisection at the eighth thoracic segment (Xu et al., 1999). Another difference from whole cord entubulation was the fact that this group explicitly closed the dura to restore cerebral spinal fluid circulation. As observed in the larger entubulation model, the group reported that a tissue cable containing grafted Schwann cells established a bridge between the two stumps of the hemicord one month after transplantation. Approximately 10,000 myelinated and unmyelinated axons (1:9) per cable were found at the midpoint of the channel. In addition to axons of peripheral nervous system origin, axons from as many as nineteen brain stem regions also grew caudally into the graft. More significantly, using anterograde axonal labeling, the group showed that some portion of the regenerating axons in the Schwann cell grafts were able to penetrate through the distal graft–host interface to reenter the host tissue. This population entered the grey matter where terminal bouton-like structures were observed. Controls displayed only limited regrowth and no penetration at the distal host interface. These encouraging findings support the notion that Schwann cells are strong promotors of axonal regeneration and that engineering smaller guidance channels may overcome some of the inhibition encountered with axonal reentry, synaptic reconnection and functional recovery following spinal cord injury.

Biointeractive materials for nerve regeneration

Although many types of commercially available or 'off the shelf' materials have been investigated as bridging substrates, none has outperformed the autologous nerve graft, either in the peripheral nervous system or in the damaged spinal cord. As a consequence, researchers have focused on developing 'designer materials' to enhance nerve regeneration. One area involves the development of materials with interfacial properties designed to mimic favorable biological activity at the site of repair.

The composition and organization of the ECM is known to play a major role in the development and patterning of nervous system tissue. Specifically, the direction or orientation of ECM fibers has long been recognized as providing critical topographic information that guides morphogenesis. Structural proteins and multiadhesive proteins such as laminin, fibronectin and tenascin, as well as various proteoglycans, have been identified that provide adhesive and repulsive cues that help delineate boundaries of cell migration and long tract development. Accordingly, a number of groups have exploited this information to drive the development of three-dimensional substrates for use in nerve repair.

Several groups are developing hydrogel systems that are specifically engineered to mimic certain functions of the ECM microenvironment (Bellamkonda et al., 1995a,b; Borkenhagen, 1998; Yu et al., 1999). Several studies have appeared that employ a similar strategy in which oligopeptides derived from ECM components such as laminin are covalently linked to the hydrogel using the coupling reagents. In vitro studies using dorsal root ganglion and PC12 cells demonstrated neurite extension over underivatized controls. Hydrogels modified with laminin motifs containing the amino acid sequence CDPGYIGSR and PEPMIX enhanced neurite outgrowth from dorsal root ganglia while GRGDSP- and IKVAV-derivatized gels inhibited neurite extension. In another series of studies, agarose hydrogel scaffolds were engineered to stimulate and guide neuronal processes through the covalent incorporation of the intact laminin protein. Laminin-modified agarose gels significantly enhanced neurite extension from embryonic day 9 (E9) chick dorsal root ganglia and PC12 cells. After incubation of dorsal root ganglia or PC12 cells with YIGSR peptide or integrin beta1 antibody, respectively, the neurite outgrowth promoting effects in laminin-modified agarose gels were significantly decreased or abolished.

Substrate-induced cytoarchitectural responses or contact guidance plays an important role in tissue development, including establishing precursor migratory patterns and guiding axonal trajectories in the developing nervous system. Probably the best example of the phenomenon is the effect of ECM on the migration of neural crest, which migrates along the dorsal aspect of the developing neural tube by following aligned matrix fibrils associated with adjacent somites (Newgreen, 1989). While the underlying cause of the fibril alignment is speculative, tensional or cell to matrix forces may be involved. The importance of substrate topography is underscored by recent studies in which white matter tracts provide directional guidance to neurite outgrowth in cultured and transplanted neurons (Davies et al., 1997; Pettigrew and Crutcher, 1999). In these studies, neurite outgrowth was restricted along the long axis of the white matter tracts.

Substrate-induced cytoarchitectural responses can be engineered through the use of externally applied magnetic fields that align matrix fibrils for directing neurite outgrowth (Ceballos et al., 1999; Dubey et al., 1999). In these studies, high-strength magnetic fields were used to induce the alignment of collagen fibrils from collagen solutions placed in tubular molds undergoing fibrillogenesis. Neurite outgrowth from dorsal root ganglia explants placed at the end of aligned rod-shaped collagen gels were substantially greater than that observed in unaligned control gels and increased with an increase in magnetic field strength, as did the collagen gel rod birefringence, an indicator of collagen fibril alignment along the rod axis. Bioresorbable collagen nerve guides filled with magnetically aligned type I collagen gel exhibited significantly greater nerve regeneration across 4- or 6-mm surgical gaps of mouse sciatic nerve compared to unaligned gels analyzed at 30 and 60 days.

Conclusions

As indicated above, a significant number of studies have established the feasibility of using encapsulated cells for delivering small molecules throughout the CNS of a variety of species of different developmental ages. Preclinical studies support the therapeutic potential of such devices for periods from six months to one year. With further technological development, the use of encapsulated cells may prove useful especially in cases where cell-based therapeutics are required for a limited period of time in a continuous manner such as over several weeks or months, as may be the case for certain focal lesions of the brain or spinal cord.

The technology of designer substrates that specifically target nerve regeneration is at its earliest stages of development. Probably the most significant shortcoming of the technology at present is the fact that most of the approaches have focused on the use of materials and ligands that nonselectively bind a variety of cell types. That is, while many materials support nerve cell attachment and permit neurite outgrowth, to the best of our knowledge no one has demonstrated that materials can be made that are selective for neurons alone. In fact most of the ligands that have been coupled to substrates are derived from ECM components that interact with a large number of integrin receptors. Since these receptors are expressed by a number of different cell types at the lesion site, including astrocytes, fibroblasts, macrophages and cells derived from the meninges, these materials permit and encourage traffic jams at the material–host interface. In fact it is not clear that ligand coating alone is the most desirable route to explore. It may be more desirable to pre-seed materials with cellular coatings. Among the attractive candidates are appropriate glial or accessory cell phenotypes, precursors or perhaps genetically engineered cells with regulated expression of trophic factors and/or cell adhesion molecules. Lastly, an important but as yet unexplored area is the development of technology that facilitates regenerating neurons to get off of such substrates so that they can integrate with distal targets and restore functionality.

The progress made by combining biomaterials engineering with the study of the nervous system during the last decade has sparked new strategies for repairing the nervous system. Our increasing knowledge of biology coupled with advances in biomaterials engineering will provide new technological platforms that will enable the next generation of novel basic research tools and medical devices. As we go forward, it is important to remind ourselves of how little we actually know. Most of the interventions that are being pioneered today will undoubtedly come to be seen as incredibly naïve in the not too distant future. Nonetheless, we are making significant progress in our search for more effective ways of restoring function in the nervous system.

References

Aebischer, P. and Kato, A.C. (1995) Treatment of amyotrophic lateral sclerosis using a gene therapy approach. *Eur. Neurol.*, 35: 65–68.

Aebischer, P., Valentini, R.F., Dario, P., Domenici, C. and Galletti, P.M. (1987a) Piezoelectric guidance channels enhance regeneration in the mouse sciatic nerve after axotomy. *Brain Res.*, 436: 165–168.

Aebischer, P., Valentini, R.F., Dario, P., Domenici, C., Guenard, V., Winn, S.R. and Galletti, P.M. (1987b) Piezoelectric nerve guidance channels enhance peripheral nerve regeneration. *ASAIO Trans.*, 33: 456–458.

Aebischer, P., Winn, S.R. and Galletti, P.M. (1988) Transplantation of neural tissue in polymer capsules. *Brain Res.*, 448: 364–368.

Aebischer, P., Salessiotis, A.N. and Winn, S.R. (1989) Basic fibroblast growth factor released from synthetic guidance channels facilitates peripheral nerve regeneration across long nerve gaps. *J. Neurosci. Res.*, 23: 282–289.

Aebischer, P., Tresco, P.A., Winn, S.R., Greene, L.A. and Jaeger, C.B. (1991a) Long-term cross-species brain transplantation of a polymer-encapsulated dopamine-secreting cell line. *Exp. Neurol.*, 111: 269–275.

Aebischer, P., Wahlberg, L., Tresco, P.A. and Winn, S.R. (1991b) Macroencapsulation of dopamine-secreting cells by coextrusion with an organic polymer solution. *Biomaterials*, 12: 50–56.

Aebischer, P., Winn, S.R., Tresco, P.A., Jaeger, C.B. and Greene, L.A. (1991c) Transplantation of polymer encapsulated neurotransmitter secreting cells: effect of the encapsulation technique. *J. Biomech. Eng.*, 113: 178–183.

Aebischer, P., Goddard, M. and Tresco, P.A. (1993) Cell encapsulation for the nervous system. In: Goosen, M.F.A. (Ed.), *Fundamentals of Animal Cell Encapsulation and Immobilization*. Academic Press, San Diego, CA, pp. 197–223.

Aebischer, P., Goddard, M., Signore, A.P. and Timpson, R.L. (1994) Functional recovery in hemiparkinsonian primates transplanted with polymer-encapsulated PC12 cells. *Exp. Neurol.*, 126: 151–158.

Aebischer, P., Pochon, N.A., Heyd, B., Deglon, N., Joseph, J.M., Zurn, A.D., Baetge, E.E., Hammang, J.P., Goddard, M., Lysaght, M., Kaplan, F., Kato, A.C., Schluep, M., Hirt, L., Regli, F., Porchet, F. and De Tribolet, N. (1996a) Gene therapy for amyotrophic lateral sclerosis (ALS) using a polymer encapsulated xenogenic cell line engineered to secrete hCNTF. *Hum. Gene Ther.*, 7: 851–860.

Aebischer, P., Schluep, M., Deglon, N., Joseph, J.M., Hirt, L., Heyd, B., Goddard, M., Hammang, J.P., Zurn, A.D., Kato, A.C., Regli, F. and Baetge, E.E. (1996b) Intrathecal delivery of CNTF using encapsulated genetically modified xenogeneic cells in amyotrophic lateral sclerosis patients. *Nat. Med.*, 2: 696–699; erratum in *Nat. Med.*, 2: 1041.

Archibald, S.J., Krarup, C., Shefner, J., Li, S.T. and Madison, R.D. (1991) A collagen-based nerve guide conduit for peripheral nerve repair: an electrophysiological study of nerve regeneration in rodents and nonhuman primates. *J. Comp. Neurol.*, 306: 685–696.

Bellamkonda, R. and Aebischer, P. (1994) Review: tissue engineering in the nervous system. *Biotechnol. Bioeng.*, 43: 543–554.

Bellamkonda, R., Ranieri, J.P. and Aebischer, P. (1995a) Laminin oligopeptide derivatized agarose gels allow three-dimensional neurite extension in vitro. *J. Neurosci. Res.*, 41: 501–509.

Bellamkonda, R., Ranieri, J.P., Bouche, N. and Aebischer, P. (1995b) Hydrogel-based three-dimensional matrix for neural cells. *J. Biomed. Mater. Res.*, 29: 663–671.

Britt, J.C. and Park, S.S. (1998) Autogenous tissue-engineered cartilage: evaluation as an implant material. *Arch. Otolar. Head Neck Surg.*, 124: 671–677.

Buchser, E., Goddard, M., Heyd, B., Joseph, J.M., Favre, J., De Tribolet, N., Lysaght, M., Aebischer, P. (1996) Immunoisolated xenogenic chromaffin cell therapy for chronic pain. Initial clinical experience. *Anesthesiology*, 85: 1005–1012; discussion: pp. 29A–30A; erratum in *Anesthesiology*, 86 (1997) 509.

Bunge, M.B. (1994) Transplantation of purified populations of Schwann cells into lesioned adult rat spinal cord. *J. Neurol.*, 242: S36–39.

Cao, Y., Vacanti, J.P., Paige, K.T., Upton, J., Vacanti, C.A., 1997. Transplantation of chondrocytes utilizing a polymer-cell construct to produce tissue-engineered cartilage in the shape of a human ear (see comments). *Plast. Reconstr. Surg.* 100, 297–302; discussion pp. 303–304.

Cao, Y., Rodriguez, A., Vacanti, M., Ibarra, C., Arevalo, C. and Vacanti, C.A. (1998a) Comparative study of the use of poly(glycolic acid), calcium alginate and pluronics in the engineering of autologous porcine cartilage. *J. Biomater. Sci. Polym. Ed.*, 9: 475–487.

Cao, Y.L., Lach, E., Kim, T.H., Rodriguez, A., Arevalo, C.A. and Vacanti, C.A. (1998b) Tissue-engineered nipple reconstruction. *Plast. Reconstr. Surg.*, 102: 2293–2298.

Catapano, G., Iorio, G., Driolio, E., Lombardi, C.P., Crucitti, F., Dogleitto, G.B. and Bellantone, M. (1990) Theoretical and experimental analysis of a hybrid bioartificial membrane pancreas. A distributed parameter model taking into account starling fluxes. *J. Membrane Sci.*, 52: 351–378.

Ceballos, D., Navarro, X., Dubey, N., Wendelschafer-Crabb, G., Kennedy, W.R. and Tranquillo, R.T. (1999) Magnetically aligned collagen gel filling a collagen nerve guide improves peripheral nerve regeneration. *Exp. Neurol.*, 158: 290–300.

Chiu, D.T. (1999) Autogenous venous nerve conduits. A review. *Hand Clin.*, 15: 667–671.

Chromiak, J.A., Shansky, J., Perrone, C. and Vandenburgh, H.H.

(1998) Bioreactor perfusion system for the long-term maintenance of tissue-engineered skeletal muscle organoids. *In Vitro Cell. Dev. Biol. Anim.*, 34: 694–703.

Cima, L.G. and Langer, R. (1993) Engineering human tissue. *Chem. Eng. Progr.*, 6: 46–54.

Cima, L.G., Vacanti, J.P., Vacanti, C., Ingber, D., Mooney, D. and Langer, R. (1991) Tissue engineering by cell transplantation using degradable polymer substrates. *J. Biomech. Eng.*, 113: 143–151.

Colton, C.K. (1996) Engineering challenges in cell-encapsulation technology. *Trends Biotechnol.*, 14: 158–162.

Davies, S.J., Fitch, M.T., Memberg, S.P., Hall, A.K., Raisman, G. and Silver, J. (1997) Regeneration of adult axons in white matter tracts of the central nervous system. *Nature*, 390: 680–683.

Decosterd, I., Buchser, E., Gilliard, N., Saydoff, J., Zurn, A.D. and Aebischer, P. (1998) Intrathecal implants of bovine chromaffin cells alleviate mechanical allodynia in a rat model of neuropathic pain. *Pain*, 76: 159–166.

Deglon, N., Heyd, B., Tan, S.A., Joseph, J.M., Zurn, A.D. and Aebischer, P. (1996) Central nervous system delivery of recombinant ciliary neurotrophic factor by polymer encapsulated differentiated C2C12 myoblasts. *Hum. Gene Ther.*, 7: 2135–2146.

Dubey, N., Letourneau, P.C. and Tranquillo, R.T. (1999) Guided neurite elongation and Schwann cell invasion into magnetically aligned collagen in simulated peripheral nerve regeneration. *Exp. Neurol.*, 158: 338–350.

Emerich, D.F., Winn, S.R., Christenson, L., Palmatier, M.A., Gentile, F.T. and Sanberg, P.R. (1992) A novel approach to neural transplantation in Parkinson's disease: use of polymer-encapsulated cell therapy. *Neurosci. Biobehav. Rev.*, 16: 437–447.

Emerich, D.F., McDermott, P.E., Krueger, P.M., Frydel, B., Sanberg, P.R. and Winn, S.R. (1993) Polymer-encapsulated PC12 cells promote recovery of motor function in aged rats. *Exp. Neurol.*, 122: 37–47.

Emerich, D.F., McDermott, P.E., Krueger, P.M. and Winn, S.R. (1994a) Intrastriatal implants of polymer-encapsulated PC12 cells: effects on motor function in aged rats. *Prog. Neuropsychopharmacol. Biol. Psychiatry*, 18: 935–946.

Emerich, D.F., Winn, S.R., Harper, J., Hammang, J.P., Baetge, E.E. and Kordower, J.H. (1994b) Implants of polymer-encapsulated human NGF-secreting cells in the nonhuman primate: rescue and sprouting of degenerating cholinergic basal forebrain neurons. *J. Comp. Neurol.*, 349: 148–164.

Emerich, D.F., Winn, S.R. and Lindner, M.D. (1996a) Continued presence of intrastriatal but not intraventricular polymer-encapsulated PC12 cells is required for alleviation of behavioral deficits in Parkinsonian rodents. *Cell Transplant.*, 5: 589–596.

Emerich, D.F., Plone, M., Francis, J., Frydel, B.R., Winn, S.R. and Lindner, M.D. (1996b) Alleviation of behavioral deficits in aged rodents following implantation of encapsulated GDNF-producing fibroblasts. *Brain Res.*, 736: 99–110.

Emerich, D.F., Lindner, M.D., Winn, S.R., Chen, E.Y., Frydel, B.R. and Kordower, J.H. (1996c) Implants of encapsulated human CNTF-producing fibroblasts prevent behavioral deficits and striatal degeneration in a rodent model of Huntington's disease. *J. Neurosci.*, 16: 5168–5181.

Emerich, D.F., Cain, C.K., Greco, C., Saydoff, J.A., Hu, Z.Y., Liu, H. and Lindner, M.D. (1997) Cellular delivery of human CNTF prevents motor and cognitive dysfunction in a rodent model of Huntington's disease. *Cell Transplant.*, 6: 249–266.

Emerich, D.F., Bruhn, S., Chu, Y. and Kordower, J.H. (1998) Cellular delivery of CNTF but not NT-4/5 prevents degeneration of striatal neurons in a rodent model of Huntington's disease. *Cell Transplant.*, 7: 213–225.

Ezzell, C. (1995) Tissue engineering and the human body shop: encapsulated-cell transplants enter the clinic. *J. NIH Res.*, 7: 47–51.

Freed, L.E., Hollander, A.P., Martin, I., Barry, J.R., Langer, R. and Vunjak-Novakovic, G. (1998) Chondrogenesis in a cell–polymer–bioreactor system. *Exp. Cell Res.*, 240: 58–65.

Galletti, P.M., Aebischer, P. and Lysaght, M.J. (1995) The dawn of biotechnology in artificial organs. *ASAIO J.*, 41: 49–57.

Guenard, V., Kleitman, N., Morrissey, T.K., Bunge, R.P. and Aebischer, P. (1992) Syngeneic Schwann cells derived from adult nerves seeded in semipermeable guidance channels enhance peripheral nerve regeneration. *J. Neurosci.*, 12: 3310–3320.

Hadlock, T., Sundback, C., Koka, R., Hunter, D., Cheney, M. and Vacanti, J. (1999) A novel, biodegradable polymer conduit delivers neurotrophins and promotes nerve regeneration. *Laryngoscope*, 109: 1412–1416.

Hammang, J.P., Emerich, D.F., Winn, S.R., Lee, A., Lindner, M.D., Gentile, F.T., Doherty, E.J., Kordower, J.H. and Baetge, E.E. (1995) Delivery of neurotrophic factors to the CNS using encapsulated cells: developing treatments for neurodegenerative diseases. *Cell Transplant.*, 4: S27–28.

Hansbrough, J.F., Cooper, M.L., Cohen, R., Spielvogel, R., Greenleaf, G., Bartel, R.L. and Naughton, G. (1992) Evaluation of a biodegradable matrix containing cultured human fibroblasts as a dermal replacement beneath meshed skin grafts on athymic mice. *Surgery*, 111: 438–446.

Hoffman, D., Breakefield, X.O., Short, M.P. and Aebischer, P. (1993) Transplantation of a polymer-encapsulated cell line genetically engineered to release NGF. *Exp. Neurol.*, 122: 100–106.

Honiger, J., Darquy, S., Reach, G., Muscat, E., Thomas, M. and Collier, C. (1994) Preliminary report on cell encapsulation in a hydrogel made of a biocompatible material, AN69, for the development of a bioartificial pancreas. *Int. J. Artif. Organs*, 17: 46–52.

Hubbell, J.A. (1995) Biomaterials in tissue engineering. *Biotechnology*, 13: 565–576.

Hudson, T.W., Evans, G.R. and Schmidt, C.E. (1999) Engineering strategies for peripheral nerve repair. *Clin. Plast. Surg.*, 26: 617–628.

Humes, H.D. (1996) Tissue engineering of a bioartificial kidney: a universal donor organ. *Transplant. Proc.*, 28: 2032–2035.

Hymer, W.C., Wilbur, D.L., Page, R., Hibbard, E., Kelsey, R.C. and Hatfield, J.M. (1981) Pituitary hollow fiber units in vivo and in vitro. *Neuroendocrinology*, 32: 339–349.

Jaeger, C.B., Greene, L.A., Tresco, P.A., Winn, S.R. and Aebischer, P. (1990) Polymer encapsulated dopaminergic cell lines as 'alternative neural grafts'. *Prog. Brain Res.*, 82: 41–46.

Jaeger, C.B., Aebischer, P., Tresco, P.A., Winn, S.R. and Greene, L.A. (1992) Growth of tumour cell lines in polymer capsules: ultrastructure of encapsulated PC12 cells. *J. Neurocytol.*, 21: 469–480.

Kaab, M.J., Ito, K., Clark, J.M. and Notzli, H.P. (1998) Deformation of articular cartilage collagen structure under static and cyclic loading. *J. Orthop. Res.*, 16: 743–751.

Kaihara, S., Kim, S.S., Benvenuto, M., Choi, R., Kim, B.S., Mooney, D., Tanaka, K. and Vacanti, J.P. (1999) Successful anastomosis between tissue-engineered intestine and native small bowel. *Transplantation*, 67: 227–233.

Keeley, R., Atagi, T., Sabelman, E., Padilla, J., Kadlcik, S., Keeley, A., Nguyen, K. and Rosen, J. (1993) Peripheral nerve regeneration across 14-mm gaps: a comparison of autograft and entubulation repair methods in the rat. *J. Reconstr. Microsurg.*, 9: 349–358; discussion: pp. 359–360.

Kerns, J.M., Fakhouri, A.J., Weinrib, H.P. and Freeman, J.A. (1991) Electrical stimulation of nerve regeneration in the rat: the early effects evaluated by a vibrating probe and electron microscopy. *Neuroscience*, 40: 93–107.

Kessler, L., Aprahamian, M., Keipes, M., Damge, C., Pinget, M. and Poinsot, D. (1992) Diffusion properties of an artificial membrane used for Langerhans islets encapsulation: an in vitro test. *Biomaterials*, 13: 44–49.

Kim, S.S., Kaihara, S., Benvenuto, M., Choi, R.S., Kim, B.S., Mooney, D.J., Taylor, G.A. and Vacanti, J.P. (1999) Regenerative signals for tissue-engineered small intestine. *Transplant. Proc.*, 31: 657–660.

Kordower, J.H., Winn, S.R., Liu, Y.T., Mufson, E.J., Sladek Jr., J.R., Hammang, J.P., Baetge, E.E. and Emerich, D.F. (1994) The aged monkey basal forebrain: rescue and sprouting of axotomized basal forebrain neurons after grafts of encapsulated cells secreting human nerve growth factor. *Proc. Natl. Acad. Sci. USA*, 91: 10898–10902.

Kordower, J.H., Liu, Y.T., Winn, S. and Emerich, D.F. (1995) Encapsulated PC12 cell transplants into hemiparkinsonian monkeys: a behavioral, neuroanatomical, and neurochemical analysis. *Cell Transplant.*, 4: 155–171.

Kordower, J.H., Chen, E.Y., Mufson, E.J., Winn, S.R. and Emerich, D.F. (1996) Intrastriatal implants of polymer encapsulated cells genetically modified to secrete human nerve growth factor: trophic effects upon cholinergic and noncholinergic striatal neurons. *Neuroscience*, 72: 63–77.

Kordower, J.H., Mufson, E.J., Fox, N., Martel, L. and Emerich, D.F. (1997) Cellular delivery of NGF does not alter the expression of beta-amyloid immunoreactivity in young or aged nonhuman primates. *Exp. Neurol.*, 145: 586–591.

Labrador, R.O., Buti, M. and Navarro, X. (1995) Peripheral nerve repair: role of agarose matrix density on functional recovery. *NeuroReport*, 6: 2022–2026.

Langer, R. and Vacati, J.P. (1993) Tissue engineering. *Science*, 260: 920–926.

Lanza, R.P. and Chick, W.L. (1995) Encapsulated cell therapy. *Sci. Am.*, 7/8: 16–25.

Lanza, R.P. and Chick, W.L. (1997a) Transplantation of encapsulated cells and tissues. *Surgery*, 121: 1–9.

Lanza, R.P. and Chick, W.L. (1997b) Immunoisolation: at a turning point. *Immunol. Today*, 18: 135–139.

Lanza, R.P., Hayes, J.L. and Chick, W.L. (1996) Encapsulated cell technology. *Nat. Biotechnol.*, 14: 1107–1111.

Lehman, M.N., LeSauter, J., Kim, C., Berriman, S.J., Tresco, P.A. and Silver, R. (1995) How do fetal grafts of the suprachiasmatic nucleus communicate with the host brain?. *Cell Transplant.*, 4: 75–81.

Levi, A.D., Guenard, V., Aebischer, P. and Bunge, R.P. (1994) The functional characteristics of Schwann cells cultured from human peripheral nerve after transplantation into a gap within the rat sciatic nerve. *J. Neurosci.*, 14: 1309–1319.

L'Heureux, N., Paquet, S., Labbe, R., Germain, L. and Auger, F.A. (1998) A completely biological tissue-engineered human blood vessel [see comments]. *FASEB J.*, 12: 47–56.

Lim, F. and Sun, A.M. (1980) Microencapsulated islets as bioartificial endocrine pancreas. *Science*, 210: 908–910.

Lindner, M.D. and Emerich, D.F. (1998) Therapeutic potential of a polymer-encapsulated L-DOPA and dopamine-producing cell line in rodent and primate models of Parkinson's disease. *Cell Transplant.*, 7: 165–174.

Lindner, M.D., Winn, S.R., Baetge, E.E., Hammang, J.P., Gentile, F.T., Doherty, E., McDermott, P.E., Frydel, B., Ullman, M.D., Schallert, T. and Emerich, D.F. (1995) Implantation of encapsulated catecholamine and GDNF-producing cells in rats with unilateral dopamine depletions and parkinsonian symptoms. *Exp. Neurol.*, 132: 62–76.

Lindner, M.D., Kearns, C.E., Winn, S.R., Frydel, B. and Emerich, D.F. (1996) Effects of intraventricular encapsulated hNGF-secreting fibroblasts in aged rats. *Cell Transplant.*, 5: 205–223.

Lindner, M.D., Plone, M.A., Mullins, T.D., Winn, S.R., Chandonait, S.E., Stott, J.A., Blaney, T.J., Sherman, S.S. and Emerich, D.F. (1997) Somatic delivery of catecholamines in the striatum attenuate parkinsonian symptoms and widen the therapeutic window of oral sinemet in rats. *Exp. Neurol.*, 145: 130–140.

Longo, F.M., Skaper, S.D., Manthorpe, M., Williams, L.R., Lundborg, G. and Varon, S. (1983) Temporal changes of neuronotrophic activities accumulating in vivo within nerve regeneration chambers. *Exp. Neurol.*, 81: 756–769.

Lysaght, M.J. and Aebischer, P. (1999) Encapsulated cells as therapy. *Sci. Am.*, 280: 76–82.

Lysaght, M.J., Frydel, B., Gentile, F., Emerich, D. and Winn, S. (1994) Recent progress in immunoisolated cell therapy. *J. Cell. Biochem.*, 56: 196–203.

Madison, R.D., Da Silva, C.F. and Dikkes, P. (1988) Entubulation repair with protein additives increases the maximum nerve gap distance successfully bridged with tubular prostheses. *Brain Res.*, 447: 325–334.

Maysinger, D. and Morinville, A. (1997) Drug delivery to the nervous system. *Trends Biotechnol.*, 15: 410–418.

Molander, H., Engkvist, O., Hagglund, J., Olsson, Y. and Torebjork, E. (1983) Nerve repair using a polyglactin tube and nerve graft: an experimental study in the rabbit. *Biomaterials*, 4: 276–280.

Mooney, D.J., Sano, K., Kaufmann, P.M., Majahod, K., Schloo, B., Vacanti, J.P. and Langer, R. (1997) Long-term engraftment of hepatocytes transplanted on biodegradable polymer sponges. *J. Biomed. Mater. Res.*, 37: 413–420.

Mulder, M.M., Hitchcock, R.W. and Tresco, P.A. (1998) Skeletal myogenesis on elastomeric substrates: implications for tissue engineering. *J. Biomater. Sci. Polym. Ed.*, 9: 731–748.

Naughton, G., Mansbridge, J. and Gentzkow, G. (1997) A metabolically active human dermal replacement for the treatment of diabetic foot ulcers. *Artif. Organs*, 21: 1203–1210.

Newgreen, D.F. (1989) Physical influences on neural crest cell migration in avian embryos: contact guidance and spatial restriction. *Dev. Biol.*, 131: 136–148.

Nyilas, E., Chiu, T.H., Sidman, R.L., Henry, E.W., Brushart, T.M., Dikkes, P. and Madison, R. (1983) Peripheral nerve repair with bioresorbable prosthesis. *Trans. Am. Soc. Artif. Intern. Organs*, 29: 307–313.

Oberpenning, F., Meng, J., Yoo, J.J. and Atala, A. (1999) De novo reconstitution of a functional mammalian urinary bladder by tissue engineering [see comments]. *Nat. Biotechnol.*, 17: 149–155.

Okano, T. and Matsuda, T. (1998) Tissue engineered skeletal muscle: preparation of highly dense, highly oriented hybrid muscular tissues. *Cell Transplant*, 7: 71–82.

Pettigrew, D.B. and Crutcher, K.A. (1999) White matter of the CNS supports or inhibits neurite outgrowth in vitro depending on geometry. *J. Neurosci.*, 19: 8358–8366.

Pollok, J.M., Kluth, D., Cusick, R.A., Lee, H., Utsunomiya, H., Ma, P.X., Langer, R., Broelsch, C.E. and Vacanti, J.P. (1998) Formation of spheroidal aggregates of hepatocytes on biodegradable polymers under continuous-flow bioreactor conditions. *Eur. J. Pediatr. Surg.*, 8: 195–199.

Puelacher, W.C., Mooney, D., Langer, R., Upton, J., Vacanti, J.P. and Vacanti, C.A. (1994) Design of nasoseptal cartilage replacements synthesized from biodegradable polymers and chondrocytes. *Biomaterials*, 15: 774–778.

Puelacher, W.C., Vacanti, J.P., Ferraro, N.F., Schloo, B. and Vacanti, C.A. (1996) Femoral shaft reconstruction using tissue-engineered growth of bone. *Int. J. Oral Maxillofac. Surg.*, 25: 223–228.

Ramón-Cueto, A., Plant, G.W., Avila, J. and Bunge, M.B. (1998) Long-distance axonal regeneration in the transected adult rat spinal cord is promoted by olfactory ensheathing glia transplants. *J. Neurosci.*, 18: 3803–3815.

Reddi, A.H. (1994) Symbiosis of biotechnology and biomaterials: applications in tissue engineering of bone and cartilage [see comments]. *J. Cell. Biochem.*, 56: 192–195.

Riesle, J., Hollander, A.P., Langer, R., Freed, L.E. and Vunjak-Novakovic, G. (1998) Collagen in tissue-engineered cartilage: types, structure, and crosslinks. *J. Cell. Biochem.*, 71: 313–327.

Sagen, J., Wang, H., Tresco, P.A. and Aebischer, P. (1993) Transplants of immunologically isolated xenogeneic chromaffin cells provide a long-term source of pain-reducing neuroactive substances. *J. Neurosci.*, 13: 2415–2423.

Sagot, Y., Tan, S.A., Baetge, E., Schmalbruch, H., Kato, A.C. and Aebischer, P. (1995) Polymer encapsulated cell lines genetically engineered to release ciliary neurotrophic factor can slow down progressive motor neuronopathy in the mouse. *Eur. J. Neurosci.*, 7: 1313–1322.

Sakata, J., Vacanti, C.A., Schloo, B., Healy, G.B., Langer, R. and Vacanti, J.P. (1994) Tracheal composites tissue engineered from chondrocytes, tracheal epithelial cells, and synthetic degradable scaffolding. *Transplant. Proc.*, 26: 3309–3310.

Santos, F.X., Bilbao, G., Rodrigo, J., Fernandez, J., Martinez, D., Mayoral, E. and Rodriguez, J. (1995) Experimental model for local administration of nerve growth factor in microsurgical nerve reconnections. *Microsurgery*, 16: 71–76.

Sautter, J., Tseng, J.L., Braguglia, D., Aebischer, P., Spenger, C., Seiler, R.W., Widmer, H.R. and Zurn, A.D. (1998) Implants of polymer-encapsulated genetically modified cells releasing glial cell line-derived neurotrophic factor improve survival, growth, and function of fetal dopaminergic grafts. *Exp. Neurol.*, 149: 230–236.

Scaravilli, F. (1984a) The influence of distal environment on peripheral nerve regeneration across a gap. *J. Neurocytol.*, 13: 1027–1041.

Scaravilli, F. (1984b) Regeneration of the perineurium across a surgically induced gap in a nerve encased in a plastic tube. *J. Anat.*, 139: 411–424.

Schinstine, M., Fiore, D.M., Winn, S.R. and Emerich, D.F. (1995) Polymer-encapsulated Schwannoma cells expressing human nerve growth factor promote the survival of cholinergic neurons after a fimbria-fornix transection. *Cell Transplant.*, 4: 93–102.

Segawa, M., Kakano, H., Nakagawa, K., Kanahiro, H., Nakajima, Y. and Shiratori, T. (1987) Effect of hybrid artificial pancreas on glucose regulation in diabetic dogs. *Transplant. Proc.*, 19: 985.

Serguera, C., Bohl, D., Rolland, E., Prevost, P. and Heard, J.M. (1999) Control of erythropoietin secretion by doxycycline or mifepristone in mice bearing polymer-encapsulated engineered cells. *Hum. Gene Ther.*, 10: 375–383.

Shinoka, T., Shum-Tim, D., Ma, P.X., Tanel, R.E., Isogai, N., Langer, R., Vacanti, J.P. and Mayer Jr., J.E. (1998) Creation of viable pulmonary artery autografts through tissue engineering. *J. Thorac. Cardiovasc. Surg.*, 115: 536–545; discussion: pp. 545–546.

Silver, R., LeSauter, J., Tresco, P.A. and Lehman, M.N. (1996) A diffusible coupling signal from the transplanted suprachiasmatic nucleus controlling circadian locomotor rhythms. *Nature*, 382: 810–813.

Sims, C.D., Butler, P.E., Cao, Y.L., Casanova, R., Randolph, M.A., Black, A., Vacanti, C.A. and Yaremchuk, M.J. (1998) Tissue engineered neocartilage using plasma derived polymer substrates and chondrocytes. *Plast. Reconstr. Surg.*, 101: 1580–1585.

Subramanian, T., Emerich, D.F., Bakay, R.A., Hoffman, J.M., Goodman, M.M., Shoup, T.M., Miller, G.W., Levey, A.I., Hubert, G.W., Batchelor, S., Winn, S.R., Saydoff, J.A. and Watts, R.L. (1997) Polymer-encapsulated PC-12 cells demonstrate high-affinity uptake of dopamine in vitro and 18F-Dopa uptake and metabolism after intracerebral implantation in nonhuman primates. *Cell Transplant.*, 6: 469–477.

Sun, A.M., Parasious, W., Macmorine, H., Sefton, M. and Stone, R. (1980) An artificial endocrine pancreas containing cultured islets of Langerhans. *Artif. Organs*, 4: 275.

Tan, S.A., Deglon, N., Zurn, A.D., Baetge, E.E., Bamber, B., Kato, A.C. and Aebischer, P. (1996) Rescue of motoneurons from axotomy-induced cell death by polymer encapsulated cells genetically engineered to release CNTF. *Cell Transplant.*, 5: 577–587.

Trent, J.F. and Kirsner, R.S. (1998) Tissue engineered skin: Apligraf, a bi-layered living skin equivalent. *Int. J. Clin. Pract.*, 52: 408–413.

Tresco, P.A. (1994) Encapsulated cells for sustained neurotransmitter delivery to the central nervous system. *J. Controlled Release*, 28: 253–258.

Tresco, P.A., Winn, S.R. and Aebischer, P. (1992a) Polymer encapsulated neurotransmitter secreting cells. Potential treatment for Parkinson's disease. *ASAIO J.*, 38: 17–23.

Tresco, P.A., Winn, S.R., Tan, S., Jaeger, C.B., Greene, L.A. and Aebischer, P. (1992b) Polymer-encapsulated PC12 cells: long-term survival and associated reduction in lesion-induced rotational behavior. *Cell Transplant.*, 1: 255–264.

Triglia, D., Braa, S.S., Yonan, C. and Naughton, G.K. (1991) In vitro toxicity of various classes of test agents using the neutral red assay on a human three-dimensional physiologic skin model. *In Vitro Cell. Dev. Biol.*, 27A: 239–244.

Uzman, B.G. and Villegas, G.M. (1983) Mouse sciatic nerve regeneration through semipermeable tubes: a quantitative model. *J. Neurosci. Res.*, 9: 325–338.

Vacanti, C.A., Kim, W., Upton, J., Vacanti, M.P., Mooney, D., Schloo, B. and Vacanti, J.P. (1993) Tissue-engineered growth of bone and cartilage. *Transplant. Proc.*, 25: 1019–1021.

Valentini, R.F., Aebischer, P., Winn, S.R. and Galletti, P.M. (1987) Collagen- and laminin-containing gels impede peripheral nerve regeneration through semipermeable nerve guidance channels. *Exp. Neurol.*, 98: 350–356.

Valentini, R.F., Sabatini, A.M., Dario, P. and Aebischer, P. (1989) Polymer electret guidance channels enhance peripheral nerve regeneration in mice. *Brain Res.*, 480: 300–304.

Vandenburgh, H., Del Tatto, M., Shansky, J., Lemaire, J., Chang, A., Payumo, F., Lee, P., Goodyear, A. and Raven, L. (1996) Tissue-engineered skeletal muscle organoids for reversible gene therapy. *Hum. Gene Ther.*, 7: 2195–2200.

Vandenburgh, H., Del Tatto, M., Shansky, J., Goldstein, L., Russell, K., Genes, N., Chromiak, J. and Yamada, S. (1998) Attenuation of skeletal muscle wasting with recombinant human growth hormone secreted from a tissue-engineered bioartificial muscle. *Hum. Gene Ther.*, 9: 2555–2564.

Vejsada, R., Tseng, J.L., Lindsay, R.M., Acheson, A., Aebischer, P. and Kato, A.C. (1998) Synergistic but transient rescue effects of BDNF and GDNF on axotomized neonatal motoneurons. *Neuroscience*, 84: 129–139.

Wang, K.K., Cetrulo Jr., C.L. and Seckel, B.R. (1999) Tubulation repair of peripheral nerves in the rat using an inside-out intestine sleeve. *J. Reconstr. Microsurg.*, 15: 547–554.

Williams, L.R., Longo, F.M., Powell, H.C., Lundborg, G. and Varon, S. (1983) Spatial–temporal progress of peripheral nerve regeneration within a silicone chamber: parameters for a bioassay. *J. Comp. Neurol.*, 218: 460–470.

Williams, L.R., Danielsen, N., Muller, H. and Varon, S. (1987) Exogenous matrix precursors promote functional nerve regeneration across a 15-mm gap within a silicone chamber in the rat. *J. Comp. Neurol.*, 264: 284–290.

Winn, S.R. and Tresco, P.A. (1994) Hydrogel applications for encapsulated cellular transplants. In: Flanagan, T.R., Emerich, D.F. and Winn, S.R. (Eds.), *Providing Pharmacological Access to the Brain: Alternate Approaches*. Academic Press, San Diego, CA, pp. 387–402.

Winn, S.R., Aebischer, P. and Galletti, P.M. (1989) Brain tissue reaction to permselective polymer capsules. *J. Biomed. Mater. Res.*, 23: 31–44.

Winn, S.R., Tresco, P.A., Zielinski, B., Greene, L.A., Jaeger, C.B. and Aebischer, P. (1991) Behavioral recovery following intrastriatal implantation of microencapsulated PC12 cells. *Exp. Neurol.*, 113: 322–329.

Winn, S.R., Hammang, J.P., Emerich, D.F., Lee, A., Palmiter, R.D. and Baetge, E.E. (1994) Polymer-encapsulated cells genetically modified to secrete human nerve growth factor promote the survival of axotomized septal cholinergic neurons. *Proc. Natl. Acad. Sci. USA*, 91: 2324–2328.

Winn, S.R., Lindner, M.D., Lee, A., Haggett, G., Francis, J.M. and Emerich, D.F. (1996) Polymer-encapsulated genetically modified cells continue to secrete human nerve growth factor for over one year in rat ventricles: behavioral and anatomical consequences. *Exp. Neurol.*, 140: 126–138.

Woerly, S., Plant, G.W. and Harvey, A.R. (1996) Neural tissue engineering: from polymer to biohybrid organs. *Biomaterials*, 17: 301–310.

Xu, L.H., Owens, L.V., Sturge, G.C., Yang, X., Liu, E.T., Craven, R.J. and Cance, W.G. (1996) Attenuation of the expression of the focal adhesion kinase induces apoptosis in tumor cells. *Cell Growth Differ.*, 7: 413–418.

Xu, X.M., Guenard, V., Kleitman, N., Aebischer, P. and Bunge, M.B. (1995a) A combination of BDNF and NT-3 promotes supraspinal axonal regeneration into Schwann cell grafts in adult rat thoracic spinal cord. *Exp. Neurol.*, 134: 261–272.

Xu, X.M., Guenard, V., Kleitman, N. and Bunge, M.B. (1995b) Axonal regeneration into Schwann cell-seeded guidance channels grafted into transected adult rat spinal cord. *J. Comp. Neurol.*, 351: 145–160.

Xu, X.M., Chen, A., Guenard, V., Kleitman, N. and Bunge, M.B. (1997) Bridging Schwann cell transplants promote axonal regeneration from both the rostral and caudal stumps of transected adult rat spinal cord. *J. Neurocytol.*, 26: 1–16.

Xu, X.M., Zhang, S.X., Li, H., Aebischer, P. and Bunge, M.B. (1999) Regrowth of axons into the distal spinal cord through a Schwann-cell-seeded mini-channel implanted into hemisected adult rat spinal cord. *Eur. J. Neurosci.*, 11: 1723–1740.

Yoo, J.J. and Atala, A. (1997) A novel gene delivery system using urothelial tissue engineered neo-organs. *J. Urol.*, 158: 1066–1070.

Young, B.L., Begovac, P., Stuart, D.G. and Goslow Jr., G.E.

(1984) An effective sleeving technique in nerve repair. *J. Neurosci. Methods*, 10: 51–58.

Yu, X., Dillon, G.P. and Bellamkonda, R.B. (1999) A laminin and nerve growth factor-laden three-dimensional scaffold for enhanced neurite extension. *Tissue Eng.*, 5: 291–304.

Zekorn, T.D., Horcher, A., Mellert, J., Siebers, U., Altug, T., Emre, A., Hahn, H.J. and Federlin, K. (1996) Biocompatibility and immunology in the encapsulation of islets of Langerhans (bioartificial pancreas). *Int. J. Artif. Organs*, 19: 251–257.

Zielinski, B.A. and Aebischer, P. (1994) Chitosan as a matrix for mammalian cell encapsulation. *Biomaterials*, 15: 1049–1056.

Zondervan, G.J., Hoppen, H.J., Pennings, A.J., Fritschy, W., Wolters, G. and Van Schilfgaarde, R. (1992) Design of a polyurethane membrane for the encapsulation of islets of Langerhans. *Biomaterials*, 13: 136–143.

Zund, G., Breuer, C.K., Shinoka, T., Ma, P.X., Langer, R., Mayer, J.E. and Vacanti, J.P. (1997) The in vitro construction of a tissue engineered bioprosthetic heart valve. *Eur. J. Cardiothorac. Surg.*, 11: 493–497.

CHAPTER 31

In vivo neuroprotection of injured CNS neurons by a single injection of a DNA plasmid encoding the *Bcl-2* gene

Raul A. Saavedra [1,*], Marion Murray [1], Sonsoles de Lacalle [2] and Alan Tessler [1,3]

[1] *Department of Neurobiology and Anatomy, MCP Hahnemann University, 2900 Queen Lane, Philadelphia, PA 19129, USA*
[2] *Department of Biology and Microbiology, California State University, Los Angeles, CA, USA*
[3] *V.A. Medical Center, Philadelphia, PA, USA*

Introduction

Spinal cord injury is a devastating event for which there is no effective treatment or cure. Axons that descend to the spinal cord from neurons located in the brain and axons that ascend from the spinal cord to higher centers of the nervous system are severed or otherwise damaged. Many of these neurons are lost by retrograde death, an apoptotic process, and those that survive are not able to regenerate their axons to their normal targets. Additional cells are lost at the injury site as the lesion expands during the days and weeks following injury (bystander death). The goals of therapeutic approaches to spinal cord injury should include rescue of neurons otherwise destined to die, which could increase the population of neurons available for axon regeneration and permit the reestablishment of functional cell to cell interactions with suitable targets. Previous studies have shown that supplying trophic factors prevented loss of neurons that express the cell surface receptors for those factors. This chapter describes recent experiments designed to rescue injured neurons in animal models of central nervous system (CNS) mechanical and chemical injuries by delivering a DNA plasmid that codes for the antiapoptotic B-cell lymphoma 2 (*Bcl-2*) gene. The product of this gene acts downstream of trophic factors and prevents activation of the late stages of apoptosis. This method, therefore, could be an effective means to rescue different populations of neurons.

Experimental therapies for spinal cord injury

Possible therapies for spinal cord injury include transplants of embryonic spinal cord. These transplants have been shown to improve locomotor recovery after several different types of experimental spinal cord injury in adult animals and to enhance the development of locomotion in rats and cats that have had their spinal cords injured at birth (Howland et al., 1995; Miya et al., 1997; Mori et al., 1997; Tessler et al., 1997). The beneficial effects of these transplants include the rescue of neurons that would otherwise be lost. These effects are likely to depend on the supply of neurotrophic factors by the transplanted tissue, since it has been demonstrated that the delivery of neurotrophins enhances survival of injured neurons in the brain and spinal cord (Xu et al., 1995; Ye and Houle, 1997; Shibayama et al., 1998). Embryonic CNS transplants, however, are presently not a practical means for treating human spinal cord injury because they are legally, politically

---

*Corresponding author: Dr. Raul A. Saavedra, National Institute of Neurological Diseases and Stroke, National Institute of Health, Neuroscience Center, Suite 3208, 6001 Executive Boulevard, Bethesda, MD 20892, USA. Fax: +1 301-402-0182; E-mail: raul_saavedra@nih.gov

and ethically controversial, and sufficient quantities of fresh healthy tissue are unlikely to be available when required. Moreover, systemic delivery and targeting of proteins that promote survival and regeneration to the spinal cord are difficult because these macromolecules do not normally cross the blood–brain barrier, and direct administration of therapeutic agents to the spinal cord by indwelling pump can cause additional trauma at the catheter site.

The advent of gene therapy technologies has raised the real possibility of applying the powerful techniques of molecular biology and genetics to human patients with CNS injuries. Well characterized animal models of CNS injury allow the rapid identification and assessment of the therapeutic potential of numerous genes, as well as the precise time and mode of application, doses, and duration of treatment. There are at least two general strategies for gene therapy. The first approach, ex vivo gene therapy, consists in genetically modifying cells to express therapeutic genes and then transplanting these cells into the injured CNS (Liu et al., 1999). The second approach, in vivo gene therapy, is to provide the therapeutic gene directly by administration of a genetically engineered virus (Liu et al., 1997) or of DNA molecules (Saavedra, 1999; Takahashi et al., 1999). Although all these strategies are promising approaches to the treatment of spinal cord injury, DNA plasmids offer the additional advantages of simplicity of design and construction, low cost, safety to the patient and health care provider, capacity to carry large therapeutic genes, and capacity to transduce postmitotic, differentiated neurons. Moreover, DNA molecules can be easily complexed to cationic lipids or polypeptides to neutralize the negative charge of DNA and enhance their internalization by cells of the nervous system. Intraspinal injections of DNA plasmids are also less likely than viruses to generate an adverse host immunologic response and can be efficiently combined with other clinical interventions such as systemic administration of methylprednisolone, a pharmacologic agent currently used in the early hours after CNS injury.

A gene with potential therapeutic value

The antiapoptotic *Bcl-2* gene is a potential therapeutic agent, because it is evolutionarily conserved and its expression can prevent death of cells exposed to many deleterious stimuli, including growth factor deprivation, actinomycin D, exposure to glucocorticoids, ultraviolet radiation and staurosporin (Adams and Cory, 1998; Green and Reed, 1998). Furthermore, overexpression of the *Bcl-2* gene product decreases apoptotic death of neurons which occurs during normal mouse development (Martinou et al., 1994). Bcl-2 and the related protein, Bcl-XL, are thought to preserve both the morphological and physiological integrity of cellular organelles (Polyak et al., 1997; Green, 1998; Green and Reed, 1998; Pettmann and Henderson, 1998). In mitochondria, Bcl-2 hinders the release of cytochrome c and apoptosis-inducing factor (AIF) (Green and Reed, 1998) from the mitochondrial intermembrane space to the cytosol. Cytochrome c is an essential coactivator of caspase proteases that participate as effectors of cell death, and AIF is a flavoprotein that plays a role in nuclear DNA fragmentation (Susin et al., 1999). Bcl-2 and Bcl-XL suppress the release to the cytosol of sequestered matrix Ca^{2+} induced by uncouplers of respiration, and Bcl-2 can prevent changes of the mitochondrial permeability transition pore complex, a large conductance channel involved in the maintenance of the electrochemical potential generated across the inner mitochondrial membrane (Zamzami et al., 1997). Bcl-2 also protects cells from reactive oxygen species and lipid peroxidation.

Red nucleus neurons are protected from axotomy in a strain of mice that overexpresses the *Bcl-2* gene from human

We have tested the hypothesis that overexpression of the *Bcl-2* gene prevents retrograde loss of axotomized red nucleus (RN) neurons in a strain of transgenic mice which constitutively overexpress this gene in neurons of the CNS (Martinou et al., 1994; Zhou et al., 1999) (Fig. 1). Groups of adult wild-type and Bcl-2 overexpressing mice were unilaterally hemisected on the left side of the spinal cord at cervical segment 4/5 (C4/5), and allowed to survive for 1, 2 or 3 months after the lesion. This lesion affects the RN neurons of the contralateral (right) side and leaves the RN neurons of the ipsilateral side unaffected. Wild-type and *Bcl-2* mice that were not operated served as intact controls. In lesioned

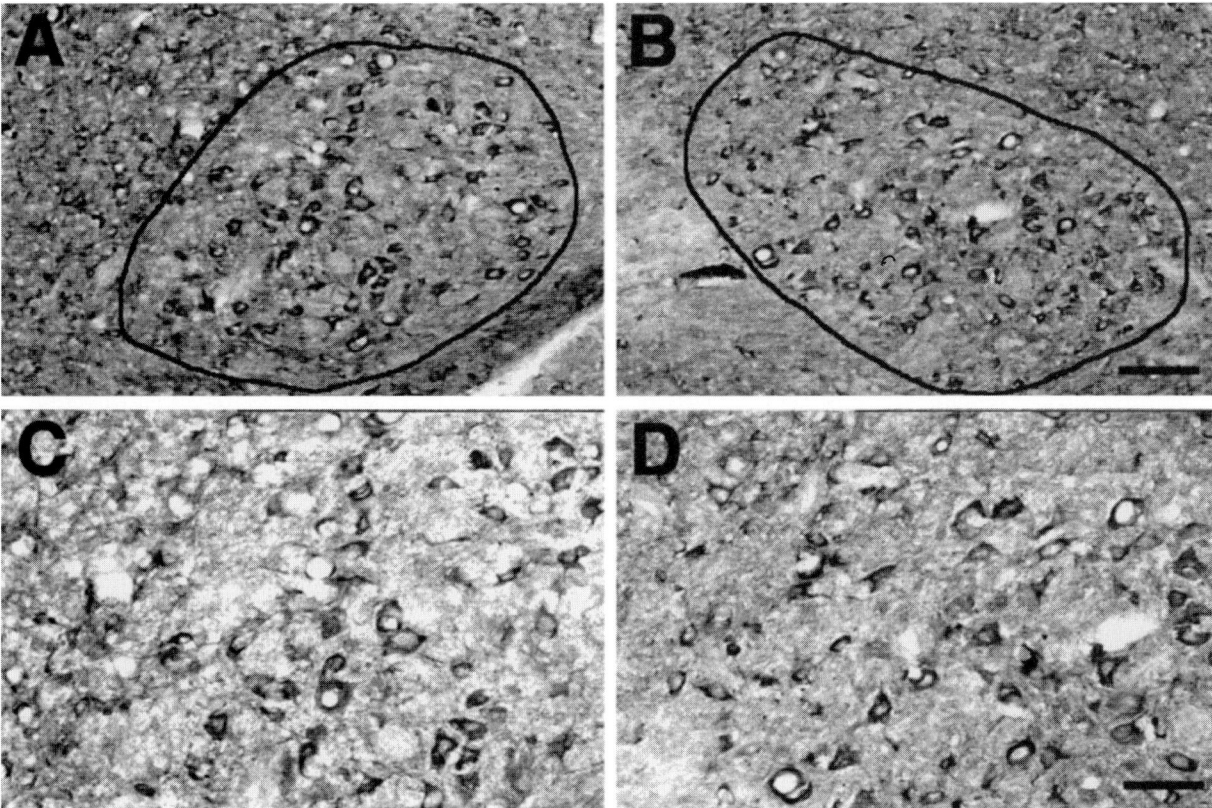

Fig. 1. Overexpression of the human Bcl-2 gene product in RN neurons in the midbrain of an adult transgenic mouse. Tissue sections were reacted with an anti Bcl-2 antibody and the reaction product was visualized with a Vetastain kit (Vector Laboratories, Burlingame, CA). A and B show the left and right RN at a low magnification. Bar = 200 μm. The RN is encircled by a continuous line. C and D show a higher magnification of the RN tissue presented in A and B. Bar = 100 μm.

wild-type mice at 1, 2 or 3 months after subtotal hemisection, 45% of total RN neurons were lost (could no longer be detected) on the lesioned side, whereas in lesioned Bcl-2 overexpressing mice only 25% of total RN neurons were lost on the lesioned side (Table 1). Thus constitutive overexpression of the Bcl-2 gene product provided partial protection from axotomy-induced apoptosis. We distinguished three classes of RN neurons in mice according to their soma areas. The number of large (>200 μm^2) RN neurons on the lesioned side decreased to 35% in wild-type mice and to 23% in *Bcl-2* mice at 3 and 2 months after surgery, respectively. These results indicate that the constitutive overexpression of the *Bcl-2* gene in transgenic mice prevents loss of axotomized RN neurons but does not prevent their atrophy. These results are consistent with the observations of others that axotomized retinal ganglion cells (Bonfanti et al., 1996), axotomized neonatal facial motor neurons (Coulpier et al., 1996) and trophic support-deprived olivary neurons (Zanjani et al., 1998) from Bcl-2 overexpressing mice also exhibit increased survival compared to wild-type mice, but the rescued neurons survive in an atrophic state.

A single intraspinal injection of a DNA plasmid that codes for the Bcl-2 gene product prevents loss and atrophy of axotomized Clarke's nucleus neurons

The neuroprotective effect of Bcl-2 in transgenic mice led us to test the hypothesis that an intraspinal injection of this gene can alter the genetic program of injured CNS neurons and prevent their loss. We used in vivo gene therapy to transduce the product of the *Bcl-2* gene to prevent loss of axotomized

TABLE 1

The ratio of RN neurons in lesioned (right) over intact (left) side in adult wild-type and Bcl-2 overexpressing mice [1]

Experimental group	<100 μm²	100–200 μm²	>200 μm²	Total
Wild-type:				
Normal (n = 7)	0.79 ± 0.11	1.00 ± 0.06	1.32 ± 0.12	1.01 ± 0.01
Lesioned 1 month (n = 7)	2.20 ± 0.44 *	0.98 ± 0.14	0.39 ± 0.06 *	0.55 ± 0.01 *
Lesioned 2 months (n = 7)	1.79 ± 0.25 *	1.21 ± 0.20	0.47 ± 0.16 *	0.54 ± 0.01 *
Lesioned 3 months (n = 7)	2.16 ± 0.45 *	1.20 ± 0.25	0.46 ± 0.15 *	0.56 ± 0.01 *
Bcl-2:				
Normal (n = 7)	1.25 ± 0.25	0.92 ± 0.08	1.27 ± 0.37	1.00 ± 0.01
Lesioned 1 month (n = 7)	2.02 ± 0.09 *	0.96 ± 0.09	0.64 ± 0.23 *	0.75 ± 0.01 *
Lesioned 2 months (n = 7)	2.14 ± 0.20 *	1.03 ± 0.09	0.29 ± 0.08 *	0.75 ± 0.01 *

[1] Values are expressed as the average ratio of number of RN neurons on the lesioned side over the control side ± standard error.
* Indicates statistically significant difference from Normal (Control), as determined by one-way ANOVA followed by t-test analyses ($p < 0.05$) (Zhou et al., 1999).

Clarke's nucleus (CN) neurons in rat (Takahashi et al., 1999). Groups of adult rats were hemisected on the right side of the spinal cord at thoracic segment 8 (T8), and allowed to survive for 2 months after the lesion. This lesion affects CN neurons of the ipsilateral (right) side and leaves CN neurons of the contralateral side unaffected. In a group of lesioned rats, 30% of total CN neurons were lost in the ipsilateral (right) lumbar segment 1 (L1) (Table 2). We distinguished three classes of CN neurons in rat according to their soma areas. About 85% of large (>400 μm²) CN neurons, which are less than 10% of all CN neurons at L1 and whose axons project to the cerebellum, atrophied and/or were lost. In another group of lesioned rats, we injected a DNA plasmid that encodes the human *Bcl-2* gene and the bacterial reporter gene, *LacZ* (pBcl-2/LacZ) (Lawrence et al., 1996) complexed with the mix of cationic lipids, Lipofectamine (GIBCO-BRL, Gaithersburg, MD), into the right side of segment T8 just caudal to the hemisection site. Two months following T8 hemisection and pBcl-2/LacZ DNA injection, we observed no statistically significant loss of CN neurons at L1 (compared to normal rats) and, therefore, a significant rescue of axotomized CN neurons. The population of large CN neurons at L1 was decreased by 39% (compared to an 85% decrease in hemisected alone), indicating a partial, but robust, protection from atrophy. In contrast, a DNA plasmid that codes for the *LacZ* reporter gene, but not *Bcl-2* (pSV-β-gal; Table 2), did not prevent CN neuron loss or atrophy.

To determine the time course of expression of the reporter gene, *LacZ*, in CN neurons at L1, we injected another group of rats with DNA plasmid pBcl-2/LacZ complexed to Lipofectamine into the right side of segment T8 of normal spinal cord or just caudal to the hemisection site. The expression of the reporter gene was transient, since its product was detected in the perikarya of ipsilateral CN neurons for up to 7 days, but not at 14 days.

TABLE 2

The ratio of CN neurons in lesioned (right) over intact (left) side [1]

Experimental group	<200 μm²	200–400 μm²	>200 μm²	Total
Normal (n = 6)	0.98 ± 0.04	0.96 ± 0.06	1.02 ± 0.05	0.98 ± 0.03
Hemisected (n = 6)	0.77 ± 0.03 *	0.60 ± 0.06 *	0.15 ± 0.02 *	0.68 ± 0.03 *
Hemisected + pBcl-2/LacZ (n = 6)	0.95 ± 0.06	0.76 ± 0.04	0.63 ± 0.06	0.87 ± 0.03
Hemisected + pSV-β-gal (n = 4)	0.75 ± 0.03 *	0.51 ± 0.09 *	0.19 ± 0.02 *	0.63 ± 0.03 *

[1] Values are expressed as the average ratio of number of RN neurons on the lesioned side over the control side ± standard error.
* Indicates statistically significant difference from Normal (Control), as determined by one-way ANOVA followed by Duncan New Multiple Range and Student–Newman–Keuls post-hoc analyses ($p < 0.05$) (Takahashi et al., 1999).

An intraspinal injection of the Bcl-2 DNA plasmid also prevents loss and atrophy of axotomized red nucleus neurons

We tested the hypothesis that an intraspinal injection of the DNA plasmid, pBcl-2/LacZ, would also prevent loss and atrophy of axotomized RN neurons in rat (Shibata et al., 2000). Groups of adult rats were subtotally hemisected on the right side of the spinal cord at C3/4, and allowed to survive for 2 months after surgery. This lesion affects the RN neurons of the contralateral (left) side and leaves the RN neurons of the ipsilateral side unaffected. Two months after subtotal hemisection, 50% of the total RN neurons in the magnocellular portion of the nucleus were lost on the lesioned side (Table 3). We distinguished three classes of RN neurons in rat according to their soma areas. More than 96% of large RN neurons (>500 μm^2), which are about 29% of all RN neurons in the studied region, were lost, indicating extensive atrophy and/or death of these neurons. Another group of lesioned rats was injected at the time of injury with DNA plasmid pBcl-2/LacZ complexed to Lipofectamine into the right side of C3 just rostral to the subtotal hemisection site. In these rats, only 16% of total RN neurons in the magnocellular region contralateral to the lesion were lost, indicating a significant rescue of these neurons. Interestingly, only 76% of large RN neurons were lost, indicating a partial, but significant, protection from atrophy. In contrast, a control DNA plasmid (pBcl-2Δ/LacZ) that contains a deletion mutation that renders the human *Bcl-2* gene nonfunctional but encodes a wild-type *LacZ* reporter gene (Table 3), and another plasmid (pSV-β-gal) that contains the *LacZ* gene but not the *Bcl-2* gene, did not prevent RN neuron loss or atrophy (data not shown).

To determine the time course of expression of the reporter gene, *LacZ*, in RN neurons, we injected another group of rats with DNA plasmid pBcl-2/LacZ complexed to Lipofectamine into the right side of the normal spinal cord or just rostrally to the subtotal hemisection site. The expression pattern of the reporter gene in RN neurons differed from that observed in CN neurons (Takahashi et al., 1999), since its product was detected in the perikarya of left and right side RN neurons at 3, 7 and 14 days after DNA injection. An analysis of the number of labeled RN neurons at 7 days after the injection indicated that at least 52% of these cells exhibited β-galactosidase activity in the perikaryon (Fig. 2).

Bcl-2 protects neurons of the diagonal band of Broca after an immunotoxic lesion

We tested the hypothesis that a single injection of DNA plasmid pBcl-2/LacZ also protected a distinct population of neurons that has been injured by a chemical insult. We selected the horizontal limb of the diagonal band of Broca (HDB), located in the rostral-medial region of the basal forebrain, because it contains a well defined population of cholinergic neurons. HDB neurons project ipsilaterally to the entorhinal cortex and the olfactory bulb. These cholinergic neurons can be lesioned by a single injection of the toxin, Saporin, conjugated with an antibody against p75, a low-affinity neurotrophin receptor (192 IgG-Saporin). This immunotoxin selectively affects HDB neurons because they specifically internalize the conjugate by means of the p75 receptor expressed on their cell surfaces. A group of adult rats was stereotaxically injected with 192 IgG-Saporin into the right HDB and survived for 2 weeks.

TABLE 3

The ratio of RN neurons on lesioned (left) over non-lesioned (right) side in adult rat [1]

Experimental group	<300 μm^2	300–500 μm^2	>500 μm^2	Total
Intact control (n = 6)	1.07 ± 0.12	1.02 ± 0.08	1.19 ± 0.23	1.04 ± 0.02
Hemisected (n = 6)	1.72 ± 0.24	0.29 ± 0.02 [*,+]	0.04 ± 0.02 [*]	0.52 ± 0.02 [*,+]
Hemisected + pBcl-2/LacZ (n = 6)	1.55 ± 0.11	0.89 ± 0.05	0.28 ± 0.03 [*]	0.87 ± 0.02 [*]
Hemisected + pBcl-2 Δ/LacZ (n = 6)	1.56 ± 0.14	0.23 ± 0.05 [*,+]	0.07 ± 0.03 [*]	0.52 ± 0.02 [*,+]

[1] Values are expressed as the average ratio of number of RN neurons on the lesioned side over the control side ± standard error.
[*] Indicates statistically significant difference from Intact control, as determined by ANOVA followed by Dunnett's test ($p < 0.01$);
[+] indicates statistically significant difference from Hemisected plus pBcl-2/LacZ (Shibata et al., 2000).

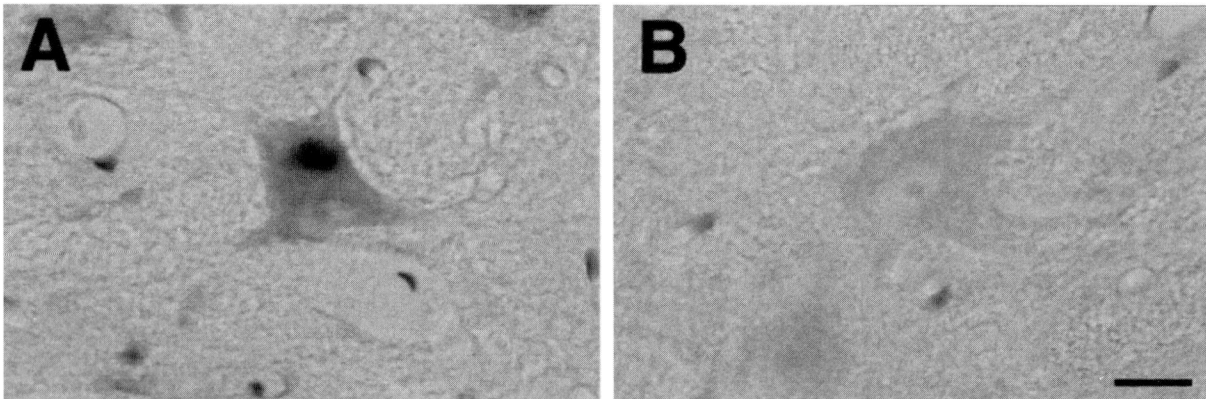

Fig. 2. β-Galactosidase expression in RN neurons from adult rat. A shows a RN neuron from a rat subtotally hemisected at C3/C4 and injected just rostral to the lesion with DNA plasmid pBcl-2/LacZ. B shows a RN neuron from a control (not injected) rat. Sections were counterstained with Cresyl Violet for 2 min. Bar = 25 μm.

TABLE 4

Percentages (lesioned side versus control side) of choline acetyltransferase (ChAT)-positive neurons 2 weeks after an immunotoxic lesion produced by an injection of 192 IgG-Saporin into HDB Nucleus in adult rat

Experimental group	Percentages of ChAT-positive neurons
Intact side ($n = 4$)	100.0
192-Saporin + lipofectamine ($n = 2$)	14.1
192-Saporin + Bcl-2 DNA ($n = 2$)	52.6

This lesion affects the right HDB and leaves the contralateral HDB as an intact control. Two of the rats injected with 192 IgG-Saporin also received an injection of DNA plasmid pBcl-2/LacZ complexed with Lipofectamine into the right HDB. About 85% of choline acetyltransferase-positive neurons were lost in the right HDB of the rats which received the 192 IgG-Saporin injection, whereas only about 47% of these neurons were lost in rats that received an injection of the immunotoxin followed by an injection of DNA plasmid pBcl-2/LacZ (Table 4). These results suggest that a single injection of a plasmid coding for *Bcl-2* gene can rescue a population of chemically injured neurons in the rat brain.

Discussion

The administration of trophic factors can rescue axotomized neurons which express receptors for those factors. Different neuronal populations, however, depend on distinct trophic factors for their survival. For example, CN, RN and HDB neurons depend on neurotrophin-3 (NT-3) (Himes et al., 1994; Shibayama et al., 1998), brain-derived neurotrophic factor (BDNF) (Liu et al., 1999) and nerve growth factor (NGF) (Lucidi-Phillipi et al., 1996), respectively. A more efficient approach to rescue different populations of axotomized neurons is to supply genes that code for products which play central metabolic roles downstream of trophic factor receptors. Antiapoptotic gene products such as Bcl-2 may be more suitable than trophic factors to rescue a broader range of injured neurons. Our observations indicate that constitutive overexpression of the antiapoptotic gene, *Bcl-2*, from human in adult transgenic mice prevents loss, but not atrophy, of axotomized RN neurons. These results suggest that an active *Bcl-2* gene in adult CNS neurons of wild-type mammals, which normally do not express the endogenous *Bcl-2* gene, could rescue diverse populations of axotomized neurons.

The cytoarchitecture of the CNS facilitates the targeted delivery of exogenous genetic information into distinct neuron populations. We have developed a simple and safe protocol that consists of injecting a DNA plasmid encoding the human *Bcl-2* gene into the CNS. The DNA molecules can be taken up by the soma or processes of cells at or near the injection site and transported to their nuclei, where the therapeutic gene is expressed. The therapeutic gene can protect both distantly located neurons whose axons cross the site of injection, as well as cells located at or near the injection site. Expression of the therapeutic gene by local cells may have additional neuroprotective effects that ameliorate the hostile environment generated by the injury. By using this method, we observed that a single injection of the Bcl-2 DNA plasmid protected axotomized CN, RN and HDB neurons in adult rats. A significant number of large CN and RN neurons from lesioned rats injected with the Bcl-2 DNA plasmid also remained. This observation indicates that the injection of Bcl-2 DNA plasmid can protect CN and RN neurons from atrophy. The reason for the different effect on atrophy observed in Bcl-2 overexpressing mice and in Bcl-2 DNA injected rats is not known. It could relate, however, to distinct interactions of the Bcl-2 gene product with other gene products in these two species, to different levels of transgene expression, to differential post-translational modifications, or to a compensatory response to the potential adverse effect of the constitutive overexpression of Bcl-2.

Expression of the transgene in rats appeared to be transient, but the neuroprotective effect on CN and RN neurons was long-lasting (at least two months). Bcl-2 prevents apoptosis by acting as a molecular switch which regulates a central metabolic pathway that integrates extracellular signals, mitochondrial physiology and caspase/protease activities (Cenni et al., 1996; Zamzami et al., 1997; Pettmann and Henderson, 1998). Therefore, activation of this switch within the first few days, before the apoptotic process becomes irreversible, may be sufficient to preserve different types of axotomized neurons from the metabolic crisis generated by injury. Our observations also suggest that an injection of the *Bcl-2* gene can rescue brain neurons injured by an immunotoxin. The fact that several neuron types injured by mechanical or chemical means can be rescued by an injection of the *Bcl-2* gene raises the possibility of developing safe and effective gene therapies for CNS injuries in human patients.

Future prospects

Injuries to the CNS constitute a most serious medical problem. For example, more than 250,000 people are currently affected by spinal cord injury in the USA alone. These conditions impose on the patients, their families and society enormous human suffering and economic burden. Despite the progress made in the management of spinal cord injury over the past 20 years, no effective cure or treatment has yet been developed. The frustratingly slow progress in this area emphasizes the need to develop novel and safe approaches toward treatment and cure. Recent advancements in gene therapy methodologies offer a unique opportunity to develop new and effective treatments for spinal cord injury. Our observations indicate that one can manipulate CNS neurons in vivo to prevent the loss and atrophy which occur following axotomy by using a single intraspinal injection of a DNA plasmid coding for an antiapoptotic gene. A rational step forward includes the application of this simple technique to introduce genes that promote axon regeneration in survivor neurons. Several potential therapeutic genes can be used for this purpose, including neurotrophins such as BDNF (Liu et al., 1999), and axon growth and guidance factors (Tessier-Levigne and Goodman, 1996). The establishment of cell to cell interactions between regenerating axons and their proper targets is, however, essential for effective recovery of function, because misguided regeneration of axotomized axons may not lead to functional recovery and can be potentially detrimental to the patient (Calancie et al., 1996). Therefore, a future challenge to the designing of gene therapy strategies for CNS injuries also includes the selective manipulation of internal and external cues to guide regenerating axons to their proper targets.

Acknowledgements

We are most grateful to Ms. Theresa Connors and Dr. Masato Shibata for assistance with the experiments and preparation of the manuscript. This work was supported by grants from the National Institutes of Health (NS24707), Eastern Paralyzed Veterans Association, International Spinal Research Trust, International Research for Paraplegia, Department of Veterans Affairs, and National Multiple Sclerosis Society (PP0604).

References

Adams, J.M. and Cory, S. (1998) The Bcl-2 protein family: arbiters of cell survival. *Science*, 281: 1322–1326.

Bonfanti, L., Strettoi, E., Chierzi, S., Cenni, M.C., Liu, X.-H., Martinou, J.-C., Maffei, L. and Rabacchi, S.A. (1996) Protection of retinal ganglion cells from natural and axotomy-induced cell death in neonatal transgenic mice overexpressing bcl-2. *J. Neurosci.*, 16: 4186–4194.

Calancie, B., Lutton, S. and Broton, J.G. (1996) Central nervous system plasticity after spinal cord injury in man: interlimb reflexes and the influence of cutaneous stimulation. *Electroencephalogr. Clin. Neurophysiol.*, 101: 304–315.

Cenni, M.C., Bonfanti, L., Martinou, J.C., Ratto, G.M., Strettoi, E. and Maffei, L. (1996) Long-term survival of retinal ganglion cells following optic nerve section in adult bcl2 transgenic mice. *Eur. J. Neurosci.*, 8: 1735–1745.

Coulpier, M., Junier, M.-P., Peschanski, M. and Dreyfus, P.A. (1996) Bcl-2 sensitivity differentiates two pathways for motoneuronal death in the wobbler mutant mouse. *J. Neurosci.*, 16: 5897–5904.

Green, D.R. (1998) Apoptotic pathways: the road to ruin. *Cell*, 94: 695–698.

Green, D.R. and Reed, J.C. (1998) Mitochondria and apoptosis. *Science*, 281: 1309–1312.

Himes, B.T., Goldberger, M.E. and Tessler, A. (1994) Grafts of fetal central nervous system tissue rescue axotomized Clarke's nucleus neurons in adult and neonatal operates. *J. Comp. Neurol.*, 339: 117–131.

Howland, D.R., Bregman, B.S., Tessler, A. and Goldberger, M.E. (1995) Development of locomotor behavior in the spinal kitten. *Exp. Neurol.*, 135: 108–122.

Lawrence, M.S., Ho, D.H., Sun, G.H., Steinberg, G.K. and Sapolsky, R.M. (1996) Overexpression of bcl-2 with herpes simplex virus vectors protects CNS neurons against neurological insults in vitro and in vivo. *J. Neurosci*, 16: 486–496.

Liu, Y., Himes, B.T., Moul, J., Huang, W., Chow, S.Y., Tessler, A. and Fischer, I. (1997) Application of recombinant adenovirus for in vivo gene delivery to spinal cord. *Brain Res.*, 768: 19–29.

Liu, Y., Kim, D., Himes, B.T., Chow, S., Murray, M., Tessler, A. and Fischer, I. (1999) Transplants of fibroblasts genetically modified to express BDNF promote regeneration of adult rat rubrospinal axons. *J. Neurosci.*, 19: 4370–4387.

Lucidi-Phillipi, C.A., Clary, D.O., Reichardt, L.F. and Gage, F.H. (1996) TrkA activation is sufficient to rescue axotomized cholinergic neurons. *Neuron*, 16: 653–663.

Martinou, J.-C., Dubois-Dauphin, M., Staple, J.K., Rodriquez, I., Frankowski, H., Missotten, M., Albertini, P., Talabot, D., Catsicas, S., Pietra, C. and Huarte, J. (1994) Overexpression of Bcl-2 in transgenic mice protects neurons from naturally occurring cell death and experimental ischemia. *Neuron*, 13: 1017–1030.

Miya, D., Giszter, S., Mori, F., Adipudi, V., Tessler, A. and Murray, M. (1997) Fetal transplants alter the development of function after spinal cord transection in newborn rats. *J. Neurosci.*, 17: 4856–4872.

Mori, F., Himes, B.T., Kowada, M., Murray, M. and Tessler, A. (1997) Fetal spinal cord transplants rescue some axotomized rubrospinal neurons from retrograde cell death in adult rats. *Exp. Neurol.*, 143: 45–60.

Pettmann, B. and Henderson, C.E. (1998) Neuronal cell death. *Neuron*, 20: 633–647.

Polyak, K., Xia, Y., Zweier, J.L., Kinzler, K.W. and Vogelstein, B. (1997) A model for p53-induced apoptosis. *Nature*, 389: 300–305.

Saavedra, R.A. (1999) Can we cure spinal cord injury using a single injection of DNA?. *BioMedicina*, 2: 377–380.

Shibata, M., Murray, M., Tessler, A., Ljubetic, C. and Saavedra R.A. (2000) Single injections of a DNA plasmid that contains the human bcl-2 gene prevent loss and atrophy of distinct neuronal populations after spinal cord injury in adult rats. Submitted.

Shibayama, M., Hattori, S., Himes, B.T., Murray, M. and Tessler, A. (1998) Neurotrophin-3 prevents death of axotomized Clarke's nucleus neurons in adult rats. *J. Comp. Neurol.*, 390: 102–111.

Susin, S.A., Lorenzo, H.K., Zamzami, N., Marzo, I., Snow, B.E., Brothers, G.M., Mangion, J., Jacotot, E., Costantini, P., Loeffler, M. and Larochette, N. (1999) Molecular characterization of mitochondrial apoptosis-inducing factor. *Nature*, 397: 441–446.

Takahashi, K., Shwarz, E., Ljubetic, C., Murray, M., Tessler, A. and Saavedra, R.A. (1999) A DNA plasmid that codes for human bcl-2 gene preserves axotomized Clarke's Nucleus neurons and reduces atrophy after spinal cord hemisection in adult rats. *J. Comp. Neurol.*, 404: 159–171.

Tessier-Levigne, M. and Goodman, C. (1996) The molecular biology of axon guidance. *Science*, 247: 1123–1133.

Tessler, A., Fisher, I., Giszter, S., Himes, B.T., Miya, D., Mori, F. and Murray, M. (1997) Embryonic spinal cord transplants enhance locomotor performance in spinalized newborn rats. In: Seil, F.J. (Ed.), *Neuronal Regeneration, Reorganization and Repair. Advances in Neurology, Vol. 72.* Lippincott–Raven, Philadelphia, PA, pp. 291–303.

Ye, J.H. and Houle, J.D. (1997) Treatment of the chronically injured spinal cord with neurotrophic factors can promote axonal regeneration from supraspinal neurons. *Exp. Neurol.*, 143: 70–81.

Xu, X.M., Guenard, V., Kleitman, N., Aebischer, P. and Bunge, M. (1995) A combination of BDNF and NT-3 promotes supraspinal axonal regeneration into Schwann cell grafts in adult rat thoracic spinal cord. *Exp. Neurol.*, 134: 261–272.

Zamzami, N., Hirsch, T., Dallaporta, B., Petit, P.X. and Kroemer, G. (1997) Mitochondrial implication in accidental and programmed cell death: apoptosis and necrosis. *J. Bioenerg. Biomembr.*, 29: 185–193.

Zanjani, H.S., Vogel, M.W., Martinou, J.C., Delhaye-Bouchaud, N. and Mariani, J. (1998) Postnatal expression of Hu-bcl-2 gene in Lurcher mutant mice fails to rescue Purkinje cells but protects inferior olivary neurons from target-related cell death. *J. Neurosci.*, 18: 319–327.

Zhou, L., Connors, T., Chen, D., Murray, M., Tessler, A., Kambin, P. and Saavedra, R.A. (1999) Red Nucleus neurons of Bcl-2 overexpressing mice are protected from cell death induced by axotomy. *NeuroReport*, 10: 3417–3421.

Subject Index

Subjects listed are discussed in the chapters that start on pages referenced here

activity blockade, 219
activity-dependent neurotrophin secretion, 183
Ammon's horn, 193
AMPA receptors, 203
animal models, 43
anterograde tracing, 309
apoptosis, 9, 365
astrocyte, 273
astrocytes, 293
autoimmune T-cells, 253
autoimmunity, 43, 259
axon regeneration, 309

Bcl-2, 365
biomaterials, 349
brain-derived neurotrophic factor (BDNF), 183, 203, 231, 243, 309
bridging materials, 349

calcitonin gene-related peptide (CGRP), 9
calcium-binding proteins, 193
cell encapsulation, 349
cell transplantation, 309
cerebellar cultures, 219
chemokines, 33
Clarke's nucleus, 365
clinical neurophysiology, 71
clonal human neurons, 299
CNS injury, 259
cortical mapping, 135
cytokine, 3, 33, 43

demyelination, 3, 9
dendritic secretion, 183
depolarization, 243
development, 193, 219
diagonal band of Broca, 365
directional tuning, 135
domains, 23
DNA plasmids, 365
dorsal column, 173

ensemble, 161
extracellular matrix, 3

feedback, 161
functional electrical stimulation, 115
functional magnetic resonance imaging, 99
functional neuroimaging, 99
functional recovery, 265

GABA, 9
$GABA_A$ receptors, 203
GABAergic neurons, 193
gene delivery, 323, 365
gene therapy, 293, 365
glia, 273
glial cell line-derived neurotrophic factor (GDNF), 323
gray matter, 299

hippocampal neurons, 203
hippocampus, 231
human, 273

immune response, 33
inflammation, 3, 43
inhibitory transmission, 203
integrins, 23
interleukin-10, 33
intraparenchymal, 323
intraventricular, 323
ion channel, 243

laminin-1, 23
Laufband (treadmill), 89
locomotion, 89, 99
locomotor networks, 99
locomotor programs, 89
long-term potentiation (LTP), 183, 231
LTD, 231
lymphocyte, 43

macrophage, 3, 43, 259
marrow stem cells, 293
marrow stromal cells, 253
matrix metalloproteinases, 33
memory formation and retention, 183
microglia, 43
mimetic, 333
modified fibroblasts, 253
monoclonal antibodies, 299
motor control, 135
motor cortex, 115, 135
motor function, 71
motor learning, 61, 99
motor neuron, 173
mouse, 193
movement vectors, 135
multi-neuron recording, 115
muscle, 61

Na^+ current, 243
nerve chamber, 121
nerve cuff, 121
nerve growth factor (NGF), 183, 333
nerve recording, 121
nerve regeneration, 121
nerve stimulation, 121
nerve tubulization, 121
neural precursors, 293
neural repair, 265
neural stem cells, 253
neuroimmunology, 259
neuron, 273
neuron growth, 173
neuron loss, 365
neuronal activity, 219
neuronal ensembles, 115
neuronal growth, 23
neuronal precursors, 253
neuronal processes, 299
neuronal transplant therapy, 299
neuroprostheses, 115
neuroprotection, 33, 259, 365
neurorobotics, 115
neurotrophins (NT), 193, 203, 231, 309, 333
non-human primate, 323
NT-3, 243
NT-4/5, 243
NT-6, 183

olfactory ensheathing glia, 253
olfactory glia, 265
oligodendrocyte, 9
oligodendrocyte progenitors, 3
Onuf's nucleus, 9
organotypic culture, 193
outcome measures, 71

P75, 333
paralysis, 115
paraplegia, 89
Parkinson's disease, 323
parkinsonism, 293
peptidomimetic, 333
peripheral nerve, 61
plasticity, 61, 99, 161, 173
population vector, 135
precursor cells, 273
prosthetic device, 135
protein fragments, 23
Purkinje cells, 219

rat, 193
receptive field, 161
receptors, 23
recovery, 71
recovery of function, 309
red nucleus, 365
reflexes, 71
regeneration, 259, 265, 333, 349
rehabilitation, 61
reorganization, 161
repair, 273, 349
representation, 173
retrograde tracing, 309
retrovirus, 293, 309

sacral parasympathetic nucleus, 9
Schwann cells, 253
secondary injury, 9
somatosensory, 161
spinal contusion, 3
spinal cord, 61, 259, 273
spinal cord injury, 71, 89, 99, 265, 365
spinal cord trauma, 43, 299
stem cells, 273
stepping, 61
substrates, 349

sustained delivery, 349
synapses, 219
synaptic plasticity, 231
synaptic transmission, 203, 231
synaptic vesicle docking, 231

tenascin-C, 23
tetraplegia, 89
tetrodotoxin, 219
thalamus, 173
tissue engineering, 349
transgenes, 293, 365

transplantation, 265, 273
Trk, 333
Trk receptors, 219
TrkB, 243
trophic factors, 323
tumor necrosis factor-α, 33

ventral lateral thalamus, 115
voluntary contractions, 71

walking, 61
white matter, 299